CW00894089

Handbook of Electroluminescent Materials

Series in Optics and Optoelectronics

Series Editors: **R G W Brown**, University of Nottingham, UK
E R Pike, Kings College, London, UK

Other titles in the series

Applications of Silicon–Germanium Heterostructure Devices
C K Maiti and G A Armstrong

Optical Fibre Devices
J-P Goure and I Verrier

Laser-Induced Damage of Optical Materials
R M Wood

Optical Applications of Liquid Crystals
L Vicari (ed)

Stimulated Brillouin Scattering
M Damzen, V I Vlad, V Babin and A Mocofanescu

Handbook of Moiré Measurement
C A Walker (ed)

Forthcoming titles in the series

High Speed Photonic Devices
N Dagli (ed)

Diode Lasers
D Sands

High Aperture Focusing of Electromagnetic Waves and Applications in Optical Microscopy
C J R Sheppard and P Torok

Transparent Conductive Coatings
C I Bright

Photonic Crystals
M Charlton and G Parker (eds)

Other titles of interest

Thin-Film Optical Filters (third edition)
H Angus Macleod

Series on Optics and Optoelectronics

Handbook of Electroluminescent Materials

Edited by

D R Vij

Department of Physics, Kurukshetra University, Kurukshetra, India

Institute of Physics Publishing
Bristol and Philadelphia

British Library Cataloguing-in-Publication Data

A catalogue record for this book is available from the British Library.

ISBN 0 7503 0923 7

Library of Congress Cataloging-in-Publication Data are available

Commissioning Editor: Tom Spicer
Production Editor: Simon Laurenson
Production Control: Sarah Plenty and Leah Fielding
Cover Design: Victoria Le Billon
Marketing: Nicola Newey and Verity Cooke

Published by Institute of Physics Publishing, wholly owned by The Institute of Physics, London

Institute of Physics Publishing, Dirac House, Temple Back, Bristol BS1 6BE, UK

US Office: Institute of Physics Publishing, The Public Ledger Building, Suite 929, 150 South Independence Mall West, Philadelphia, PA 19106, USA

Typeset by Academic + Technical, Bristol
Printed in the UK by MPG Books Ltd, Bodmin, Cornwall

Contents

Preface

Fundamental and applied research into luminescence phenomena is advancing on a broad front, involving advanced experimental and theoretical methods, with increasing interactions with other branches of solid state physics and molecular science. In particular, electroluminescence (EL), i.e. emitting of electromagnetic radiations (visible or near visible) due to the application of an electric field (a.c. or d.c.), has found ever-increasing commercial applications over the past two decades, especially, because of the EL materials being used in light emitting diodes (LEDs), diode arrays, illumination engineering and display devices such as flat information display devices (TV display), memory devices, solid state indicators (e.g. digital clocks, meter read-outs), high resolution x-ray screens, storage CRTs, devices for controlled geometry of the light emitting area, multi-element scales for data recording on photosensitive materials, photographic recording of information (as a binary code), injection lasers etc. This reference level handbook has been planned to review the present status and trends in applied research on EL materials, and to emphasize their tremendous applications of commercial interest.

A brief introduction to EL phenomena is given in the first chapter of the book. Various EL materials of commercial and scientific importance are then described in separate chapters. Each chapter is self-contained, with a brief description of the historical background, preparation methods, physicochemical structure, EL and related characteristics (theoretical models and mechanisms) of the concerned material(s), along with their device fabrication, applications and future trends.

It is hoped that the book will be useful to a wide circle of students, researchers, teachers and professionals studying and working in the areas of electroluminescence; especially physics, materials science, solid state semiconductor electronics, thin-film electrooptic devices, optoelectronic devices (lasers), display and illumination engineering.

This handbook is a valuable collection by a diverse group of outstanding and experienced researchers, and has an international flavour in as much as the contributors are from the UK, United States, Germany, Japan, India, Poland and Russia.

It is my pleasant duty to express my thanks to the contributors, who made this handbook possible. I am indebted to all those authors and publishers who freely granted permissions to reproduce their copyrighted works. The patient encouragement of the staff of the Institute of Physics Publishing Ltd is also acknowledged with gratitude. Last but not the least, I wish to express my deep appreciation to my wife, Meenakshi, and my daughters Surabhi and Monika, for their indispensable support, encouragement and patience throughout the time-consuming task of writing and editing this handbook.

D R Vij
Kurukshetra
Spring 2004

Acknowledgments

The editor and contributing authors of this book are indebted to the publishers of scientific journals and books, and various authors for freely granting permissions to reproduce their copyrighted works. The editor is especially thankful to the following publishers without whose permissions this book would not have been a reality.

American Institute of Physics, American Physical Society, American Scientific Publishers, The Electrochemical Society, IEEE, Elsevier, Oxford University Press, Society for Information Display, Nikkei Electronics, Japan, CRC Press, Springer-Verlag, Wykeham Publications, London, Elsevier Science, USA, Plenum Press, World Scientific Publishing Company, Singapore, Materials Research Society, Wissenschaft und Technik, Germany, Elsevier Science, B.V., Akademie-Verlag Germany, Chapman & Hall, London, Institute für Physik/Kristallographite, Humboldt University, Germany, Clarendon Press, UK, John Wiley & Sons, USA, Macmillan Magazines, UK, ESPRIT, Sweden, NISCOM and Indian National Science Academy, India.

Chapter 1

Electroluminescence: an introduction

***D Haranath*[1], *Virendra Shanker*[1] *and D R Vij*[2]**
[1]National Physical Laboratory, New Delhi, India
[2]Department of Physics, Kurukshetra University, India

1.1 Introduction

Unlike earlier definitions, electroluminescence (EL) nowadays is considered as non-thermal (excluding incandescence, which is light generation due to resistance or Joule heating) generation of light from a material when a high electric field is applied to it. The material can be inorganic, semiconductor or organic. EL, in fact, is the conversion of electrical energy into optical energy by an electronic relaxation process.

The EL in inorganic materials is classified into two groups: high-field EL and injection EL. The high-field EL is further divided into two types: powder phosphor EL and thin-film EL. The categories of EL with respect to typical device applications are summarized in figure 1.1.

High-field EL consists of excitation of luminescence centres by majority charge carriers accelerated under the action of strong electric (a.c. or d.c.) fields $\sim 10^6$ V/cm. This type of EL mechanism relies on inter- and intra-quantum shell transitions at luminescent centres/ions. High-energy electrons raise the luminescent centres to the excited quantum states via impact ionization and/or impact excitation. The excited centres must eventually relax to ground state emitting photons, known as the radiative relaxation process. The electron excitation and radiative relaxation are atomic transitions localized at the luminescent centre. The active layer can consist of a doped semiconductor of II–VI compounds and, in addition, either a powder (embedded in a matrix) or an organic or inorganic thin film.

Injection EL is where the light emission is generated by the direct injection and subsequent recombination of electron–hole pairs at a p–n homo- or hetero-junction. This is the light generation mechanism in light emitting diodes (LEDs) and semiconductor lasers. The material need not

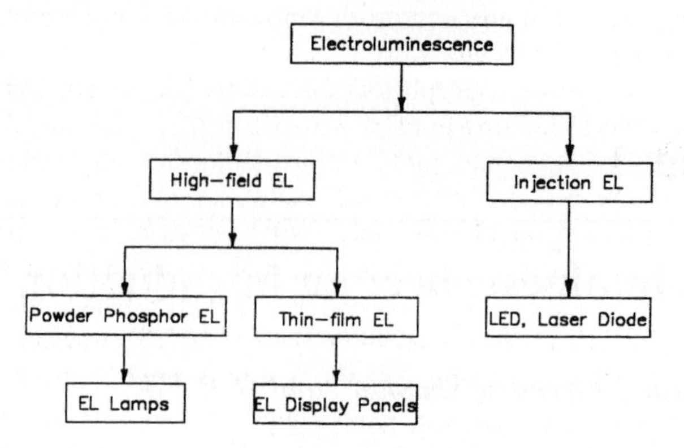

Figure 1.1. Representation of various categories of EL with respect to typical device applications.

be in direct contact with the electrodes and no net current passes through the materials. In injection EL, light is emitted upon recombination of minority and majority carriers across the band gap of crystals. The III–V compounds are of outstanding value for injection luminescence because of their large, direct band gaps and high luminescence efficiency throughout the visible region of the spectrum.

Two more phenomena, i.e., photoelectroluminescence (PEL) [1–4] and electrophotoluminescence (EPL) [1, 5], are also used nowadays. PEL is used whenever the excitation mechanism of the electric field is influenced by additional irradiation, and EPL is the term employed when the luminescence by irradiation (with UV, x-rays, cathode rays etc.) is controlled by electric fields, leading to an enhancement or to a quenching effect of the luminescence emission. This is generally termed as Gudden–Pohl effect [5]. The main difference between EPL and EL is that EPL is an emptying of traps by the applied field whereas EL is the field excitation of luminescence centres.

The beginning of EL is attributed to Destriau [6] who reported that upon application of a high field to ZnS:Cu powder phosphor suspended in liquid dielectric (oil) emitted light. Henish [7], Ivey [8] and Pankove [9] have given a complete bibliography of first period of EL research and a comprehensive description of the state-of-the-art in their books.

EL phosphors are dilute solid solutions. For example, in the case of zinc sulfide (ZnS), the oldest, best understood and most widely used phosphor host in high-field EL devices, the ZnS component is the solvent known as the host. Solute components of ZnS phosphors constitute activators (transition metal and rare-earth ions) known as luminescent centres. The electrical and electro-optical characteristics of EL phosphor devices are determined primarily by the host, whereas colour or emission wavelength, generated from impact excitation and/or impact ionization of luminescent centres or activators by high-energy electrons, is determined by the activator(s).

Many early powder phosphor devices were d.c. driven, and the efforts focused for general lighting applications. EL in thin film was described as early as 1960 by Vlasenko and Popkov [10]. Though Vecht *et al* [11] demonstrated the first d.c. powder EL panel, the majority of development efforts were focused on thin-film EL (TFEL) devices. Russ and Kennedy [12] were the first to propose a structure with phosphor layer sandwiched between two insulating layers for TFEL displays which was a breakthrough in the TFEL display technology. Kahng [13], however, employed an a.c. driven device using Lumocen (luminescence from molecular centres). Inoguchi *et al* [14], who brought about a turning point in the studies on EL, were able to establish TFEL cells based on ZnS:Mn phosphors exhibiting high efficiency and a reasonable lifetime for the flat panels.

While powder EL is used mainly for lighting applications, thin-film EL is for flat panel display (FPD) technology. The FPD group includes other technologies such as liquid crystal displays (LCD), plasma displays (PD) and field emission displays (FED). Though liquid crystal displays dominate the FPD sector ($\sim$85%), thin-film EL displays offer several advantages such as wide viewing angle ($>160°$), high contrast in high ambient illumination conditions, broad operating-temperature range ($-60\,°C$ to $+100\,°C$), fast response times, and the capability of very high resolution and legibility. They are rugged and insensitive to shock and vibration. Alternating current driven TFEL (ACTFEL) displays have become the most reliable, longest running ($>$50 000 h) devices on the market [10], making them ideal for military, avionics, medical and industrial applications.

In the structure of the EL cell developed by Inoguchi *et al* [14], the insulators serve as buffers since they completely encapsulate (cover) the active layer of EL phosphor, shielding the device from breakdown under high local current densities, with their ensured reliable operation, and hence these structures are promising candidates as flat video-screens/panels, TV imaging systems [15] etc. Barrow *et al* [16] have displayed a polychrome (green, yellow, red) display based on a filter system. They used a stacked structure with the red and green phosphors on one substrate, and blue-emitting cerium-doped calcium thiogallate phosphor on a separate substrate. Due to the reliability of the structure, fair reproducibility of the device characteristics and improved thin film deposition, the physical aspects of high-field EL can be investigated in a precise manner.

Initially, flat panels with ZnS:Mn as the active layer were commercialized. Full-colour display panels could not be made only with ZnS or ZnSe as hosts because they emit light in the yellow region. That is why many more hosts, such as alkaline earth chalcogenides (AECs) like SrS, SrSe and CaS were also tried for full-colour display. Though green- and yellow-emitting phosphors were investigated in order to get full-colour displays, the problem of getting phosphors emitting blue have not been solved completely.

This chapter deals with the fundamental processes (physical and electrical as well as optical) involved in the phosphor materials exhibiting EL. Since the book deals with the EL and related properties of various types of materials, such as inorganics, organics, semiconductors, liquid crystals etc. their detailed EL and related characteristics will be discussed in their respective chapters.

1.2 Fundamental physical processes

There are four fundamental processes in EL behaviour which may lead to light emission. These processes are: (i) injection of charge carriers into the phosphor layer, (ii) acceleration of charge carriers to optical energies, (iii) excitation of luminescence centres by energetic electrons (impact excitation), and (iv) radiative relaxation of luminescent centres.

The mechanism of EL behaviour of materials, though with a slight difference, is almost the same. Therefore, much of the physics of ZnS-type devices can be applied to devices based on other types of materials such as AECs used for full-colour display devices.

1.2.1 Atomic transitions

The EL properties of luminescent centres (ions) are determined by their quantum mechanical properties through intershell (change of principal quantum number n) and intrashell (without change in n) electronic transitions following selection rules. There are three properties that are of primary importance: radiative kinetics, the cross-section for impact excitation, and radiative transition energy (colour).

The electrical energy supplied to the phosphor layer raises the luminescent centre to its excited state. This centre eventually relaxes to the ground state by dissipating the absorbed energy. The relaxation process may emit: (i) a photon (radiative relaxation), (ii) a phonon to the lattice (non-radiative dissipation of absorbed energy to the lattice), or (iii) transfer energy to another luminescent centre. Transitions allowed by selection rules possess short decay times (about a nanosecond), whereas forbidden transitions are of long decay times (about a millisecond or longer). The internal quantum efficiency, defined as the ratio of the radiative relaxation to the total relaxation, decreases with the rise of temperature and is termed 'thermal quenching' [17]. Details of fundamentals of atomic transitions, cross-section for impact excitation, radiative transition energy etc. are described in details in section 2.2 of Chapter 2.

1.3 High-field EL

1.3.1 Powder phosphor EL

The EL phenomenon was first observed by Destriau [6] in 1936. The first EL cell fabricated by Destriau was very delicate to handle. The ZnS powder phosphor was suspended in a liquid dielectric (castor oil) and the transparent electrode necessary to see the emission was a sheet of mica covered with salt water. A strong electric field was applied across the electrodes to speculate the EL process. The EL cell originally used by Destriau is shown in figure 1.2. Today, this type of EL is classified as powder phosphor EL.

After the preparation of powder EL phosphor it is essential to determine the amount of dopants/impurities incorporated in the host lattice. Table 1.1 shows synthesis parameters and chemical analysis of some of the EL powder phosphors developed under various preparative conditions [18].

No efforts were made to develop practical devices using this phenomenon until transparent electrically conductive films made of SnO_2 were developed. With the use of these transparent conductive films, a.c. powder EL cells were made possible for commercial applications as solid-state, low-power, uniform area light sources. For the first time, EL devices were developed and marketed by Sylvania under the trade name 'Panelite'. This step by Sylvania had triggered the research and development of EL devices and panels all over the world. However, the display applications using this type of EL suffered from low luminance output and significant luminance degradation during the operational times of not more than 20 days.

Meanwhile, in the early 1980s, there was a huge demand to develop flat panel displays instead of cathode-ray tubes (CRTs) for alphanumeric and graphic information displays. Development of flat panel devices resulted in the investigation of thin-film EL (TFEL) devices, liquid crystal displays (LCDs), plasma display panels (PDPs) and light emitting diode (LED) arrays. In 1974, a.c.-driven, high-field, thin-film EL devices with high luminance output and lifetimes as long as 10 000 h were developed. It has

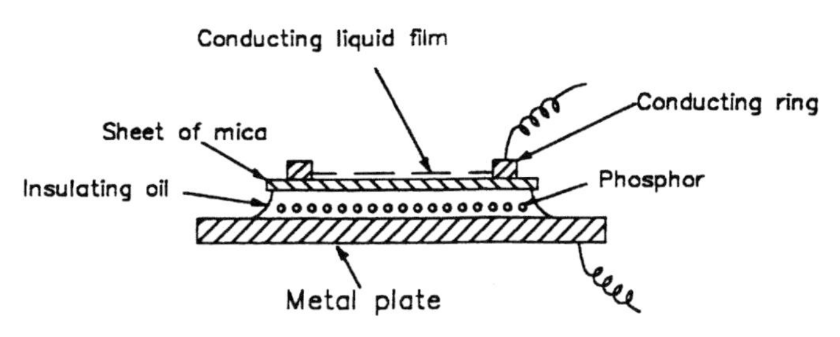

Figure 1.2. Electroluminescent cell originally developed by Destriau [6].

Table 1.1. Typical parameters of electroluminescent ZnS:Cu phosphor prepared by various routes and their chemical analysis [18].

Sample No.	Sample name	Carrier gas	Chemicals added to ZnS	Impurities detected in the product by chemical analysis
1	EL-22	H_2S	Cu: 800 ppm Al: 500 ppm	Cu: 446 ± 10 ppm Al: 203 ± 2 ppm
2	EL-50	$H_2S + HBr$	Cu: 800 ppm	Cu: 421 ± 10 ppm Br: 350 ± 10 ppm
3	EL-63	CS_2	Cu: 800 ppm NH_4Br: 5 wt% KBr: 5 wt% NaCl: 2 wt%	Cu: 416 ± 10 ppm Br: 290 ± 10 ppm Cl: 98 ± 10 ppm
4	EL-69	N_2	Cu: 1500 ppm NH_4Br: 5 wt% KBr: 10 wt% S: 5 wt%, NH_2CSNH_2: 10 wt%	Cu: 581 ± 10 ppm Br: 350 ± 10 ppm
5	EL-75	N_2	Cu: 1500 ppm Al: 500 ppm NH_4Br: 5 wt% KBr: 10 wt% S: 5 wt%, NH_2CSNH_2: 10 wt%	Cu: 788 ± 10 ppm Al: 252 ± 2 ppm Br: 780 ± 10 ppm

also been demonstrated that these EL panels can be used for information displays with high information contents. Nowadays a.c. thin-film EL (ACTFEL) display panels are commercially available.

The research efforts on making efficient a.c. and d.c. powder EL phosphors are continuing. Alternating current powder EL panels are generally used in back lighting of liquid crystal displays, whereas d.c. powder EL panels find maximum usage in displays with multiplex capability for matrix addressing. For further improvement of characteristics of powder EL devices, a continuous and active investigation on the preparation techniques and structure of EL devices is required.

1.3.2 Alternating current powder EL

The structure of typical a.c.-driven powder EL device is shown in figure 1.3. The EL active layer consists of crystalline powder phosphor with particle sizes of 10–50 μm dispersed in a plastic dielectric, and a condenser-like system is made by sandwiching between two electrodes. It is required that one electrode be transparent to see the emission. The typical thickness of

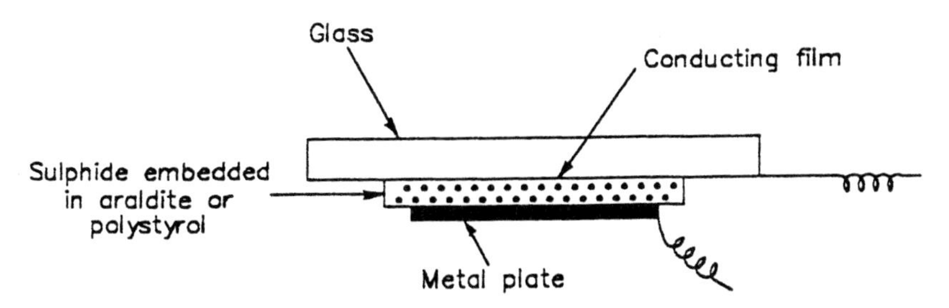

Figure 1.3. Electroluminescent cell made using conducting glass as a transparent electrode [6].

the EL active layer must be 50–100 μm. The colour of the emission depends on the activators doped to the host lattice. The most common EL phosphor used is the green-emitting ZnS:Cu,Cl (or Al) in which Cu acts as activator. The thickness of the dielectric layer with phosphor should also be about 100 μm. The dielectric used in the EL cell fabrication must have a relatively high dielectric constant. It can be an epoxy resin, polystyrene, cyanoethylcellulose or even a ceramic.

Usually, about 200 V is applied to the EL cell, which gives an average field of $2 \times 10^5\,\mathrm{V\,cm^{-1}}$. However, the field inside the film on the ZnS crystal grain will depend on the dielectric material used. If the voltage V is applied to the EL cell of thickness t then the mean field $E_m = V/t$. For spherical grains of ZnS of dielectric constant ε_{r1} in a dielectric of dielectric constant ε_{r2}, the field E_{ZnS} on the grains is given by

$$E_{ZnS} = E_m \left[\frac{3\varepsilon_{r2}}{2\varepsilon_{r2} + \varepsilon_{r1} - f(\varepsilon_{r1} - \varepsilon_{r2})} \right] \tag{1.1}$$

where f is the fraction of the total volume occupied by the ZnS grains.

1.3.2.1 EL characteristics and mechanism

EL is observed when an a.c. voltage of about 50–200 V corresponding to an electric field of the order of $10^5\,\mathrm{V\,cm^{-1}}$ is applied across the electrodes. The brightness–voltage (B–V) characteristics of some EL devices prepared under different preparative conditions are shown in figure 1.4. The observed dependence of the brightness (B) on the applied voltage (V) is expressed by the relation:

$$B = B_0 \exp(-b/V^{0.5}) \tag{1.2}$$

where B_0 and b are the constants. The values of these constants depend on the particle size of the phosphor, the concentration of the EL powder in the dielectric, the dielectric constant of the embedding medium and the device thickness [19].

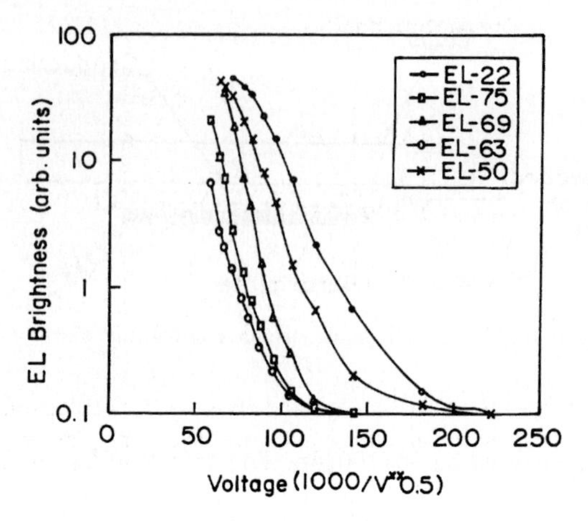

Figure 1.4. Variation of EL brightness with applied voltage for a typical EL device [18].

It has been established empirically that one of the important parameters affecting the EL characteristics is particle size, which in turn decides the EL efficiency and lifetimes of the EL device. The efficiency increases in proportion to $1/d^{0.5}$, where d is the particle size. This leads to the nonlinearity of the B–V curve. However, the operational lifetime decreases in proportion to d.

EL characteristics of Cu-activated ZnS phosphors can be understood if one considers the microscopic nature of the EL phenomenon. The best experimental observations and theoretical interpretation of EL were presented by Fischer [20–22] in the early 1960s and summarized by Ono [23]. Careful examination of ZnS:Cu,Cl particles was done under an optical microscope [21]. It was observed that the shape of the light-emitting region within a single EL particle took the form of double lines with the shapes similar to twinkling tails of a comet, as shown in figure 1.5.

Fischer found that each half of the double comet lines lit up alternately whenever the nearest electrode turned positive. More than 20 such comet line

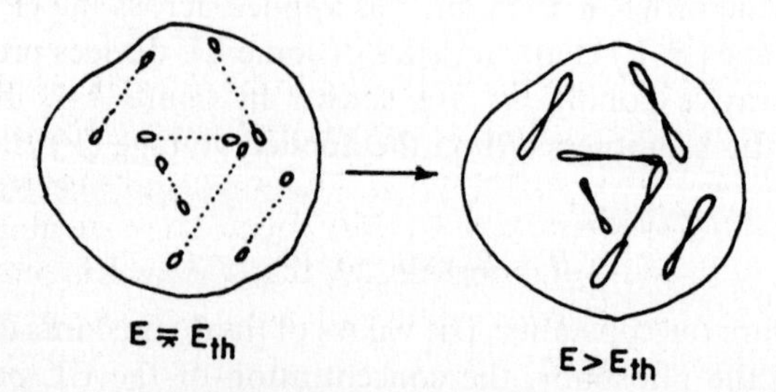

Figure 1.5. Typical view of EL from ZnS:Cu,Cl particles. Double comet lines at threshold voltage and above threshold voltage are illustrated [20].

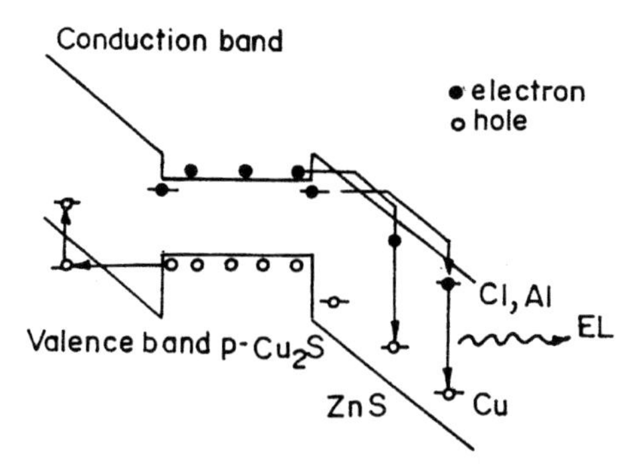

Figure 1.6. Schematic band model of simultaneous injections of electrons and holes from opposite ends of a Cu_2S needle into the ZnS:Cu,Cl surrounding lattice [21].

pairs are observed in one particle. The comet lines follow definite crystal orientations so that network aligned with 60° angles are often formed. Those lines making acute angles to the applied field are observed to be the brightest.

Based on these observed facts, Fischer [20, 22] proposed the following model for the EL mechanism. Zinc sulfide EL powders are typically prepared at high temperatures (>1000 °C) where the hexagonal wurtzite phase dominates. When the powders are cooled, there is a phase transition to the cubic zincblende structure. Copper, which is being used as an activator in the EL phenomenon, preferentially precipitates on defects formed in the hexagonal-to-cubic transformation with a concentration exceeding their solubility limit in ZnS. The Cu forms thin embedded Cu_2S conducting needles in the ZnS crystal lattice. Cu_2S is known to be a p-type semiconductor and ZnS an n-type, thus creating a junction similar to the Schottky barrier [24]. The B–V relation can be explained thoroughly using this bipolar tunnel emission (field emission) model [20–22, 24], which states that the applied field is concentrated at the tips of these Cu_2S needles. When voltages are applied just above the threshold point, simultaneous tunnelling of electrons and holes from both ends of Cu_2S needles occurs through the barrier into the ZnS lattice [22]. Emitted electrons and holes are illustrated in figure 1.6.

At higher voltages, even the isolated phosphor particles along with the bound ones give rise to emission, which may be due to the depletion of the Schottky barrier and direct excitation of luminescent centres by the collision of highly accelerated electrons [25–28]. Moreover, experimentally it has been established by Antonov-Romanovsky [29] that for an EL cell, the relation $\exp(-b/V^{0.5})$ is valid at lower voltages and the relation $\exp(-b/V)$ is applicable at higher voltages.

In addition, the dependence of EL brightness on frequency shows a mixed behaviour and deviations from linear relationship at lower and

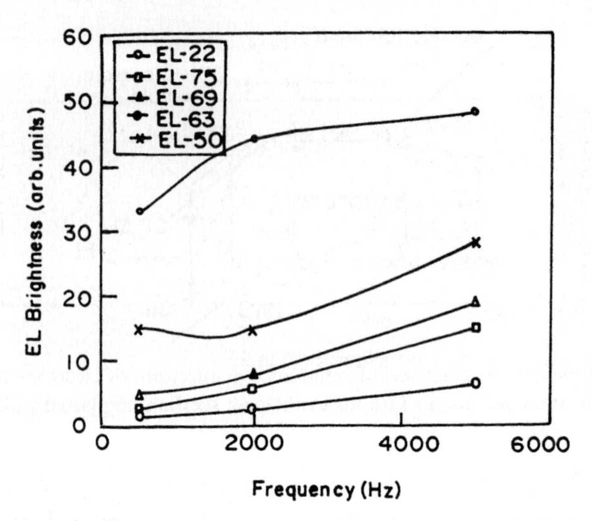

Figure 1.7. Frequency variation of brightness of different EL samples [18].

higher frequencies (see figure 1.7). It has been assumed that for each cycle of applied frequency the donor ionization, electron acceleration and ionization of luminescent centres begin afresh [30]. Hence, within wide limits one can say that EL brightness is directly proportional to the applied frequency. However, at higher frequencies saturation effects occur especially for the case of ZnS:Cu phosphors. This is because if the frequency is too high the electrons liberated during any half-cycle will not get recombined before the following half-cycle occurs. Hence, the recombination of freed electrons with empty luminescent centres gives rise to saturation effects. At lower frequencies (<50 Hz), certainly there arise systematic deviations from linear relationship, which may be due to the polarization effects [31].

But Tanaka [32] and Rennie and Sweet [33] came out with different explanations regarding brightness–frequency (*B–F*) measurements. According to them, the extent of crystallinity of the phosphor during the preparation was the major factor which governs the frequency dependence of EL brightness.

1.3.2.2 EL emission spectra

The emission spectra of a.c. powder EL devices are shown in figure 1.8. Emission colours can be controlled by incorporating different activators in the host lattices [34]. When a ZnS lattice is doped with Cu (activator) and Al or halide (co-activator), donor–acceptor pairs are formed. The EL is caused by the radiative recombination of electron–hole pairs at donor–acceptor pair sites. The introduction of Cu and Al in ZnS (i.e. ZnS:Cu,Al) produces a green ($\sim$550 nm) emission colour. The combination of Cu and

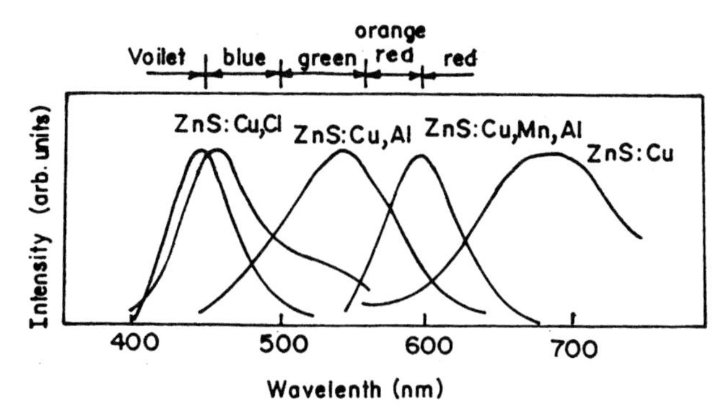

Figure 1.8. Alternating current EL spectra of various kinds of ZnS powder phosphors [24].

Cl in ZnS (i.e. ZnS:Cu,Cl) gives blue (~460 nm) and green emission bands, their relative intensity depending on the relative amount of Cu to Cl. ZnS:Cu,I phosphor shows a blue emission. ZnS:Cu phosphor in which no co-activator is incorporated shows a red emission. Incorporation of Mn^{2+} ions into a ZnS:Cu,Cl phosphor system shows a yellow emission (~580 nm) due to Mn^{2+}. The emission colour can be controlled by mixing phosphors emitting different colours [see Chapter 3 of ref. 23].

1.3.2.3 Lifetime determination of an a.c. powder EL cell

Operational lifetime is an important aspect of a.c. powder EL devices. The durability of an EL cell depends on the driving conditions (such as voltage, frequency and brightness levels) and on environmental conditions, especially on temperature and humidity. A lifetime, defined as the time required for the brightness to decrease to half its initial value, of 2000 h can be achieved only by careful protection of the device from the harmful effects of moisture. The decay of brightness with time is generally expressed as

$$B/B_0 = (1 + \alpha t)^{-1} \tag{1.3}$$

where α is nearly proportional to the driving frequency.

Recently, the lifetimes of EL cells have been improved tremendously. Using power driving techniques that maintain constant power output, the lifetime of EL device producing 200 cd m^{-2} has been extended up to >3000 h. These lifetimes meet the requirements for applications involving back lighting of liquid crystal displays (LCDs).

1.3.3 Direct current powder EL

The structure of a typical d.c. powder EL device is shown in figure 1.9.

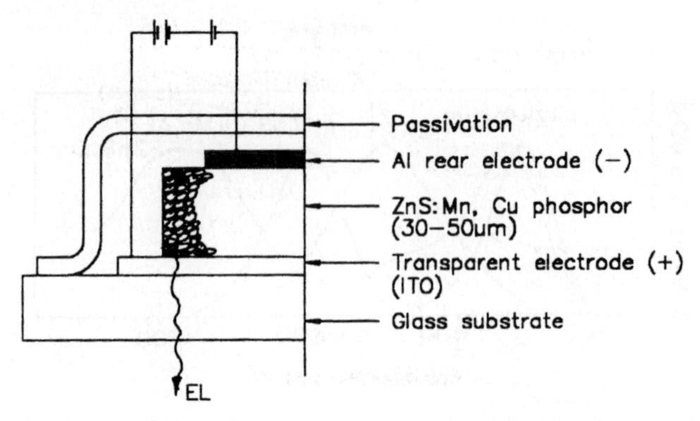

Figure 1.9. Typical structure of a d.c. powder phosphor EL device [11].

Vecht *et al* [11] reported pioneering work on d.c. powder EL. In the case of an a.c. powder EL device, phosphor particles of 5–30 μm are generally used, whereas in the case of d.c. powder EL device, phosphor powder of 0.5–1 μm are used in the device fabrication. The most studied d.c. powder EL phosphor is ZnS:Mn. A conductive layer of Cu_2S is chemically formed on the ZnS:Mn particles by immersing the powder in a hot $CuSO_4$ solution. This treatment is often referred as 'copper coating'. Thus Cu-coated ZnS:Mn phosphor powder is deposited on a conducting glass (ITO) with a small amount of binder mixed creating an EL active layer of about 50 μm. Aluminium metal is evaporated on to the phosphor layer, which acts as the rear electrode. The transparent ITO and Al rear electrodes are positive and negative respectively. But the devices thus fabricated are highly conducting and do not produce EL. To complete the fabrication process, a treatment called 'forming' is necessary [11, 35]. In this process a d.c. voltage is applied to the device in the forward bias condition. A large current flows through the device and heats up the phosphor layer, allowing Cu^+ ions to migrate towards the negatively charged Al electrode. This produces a narrow Cu-free region adjacent to the ITO anode. Since the Cu-free region is less conductive, the greater part of the applied voltage is concentrated at this place, and a high electric field of the order of $10^6\,V\,cm^{-1}$ develops across this region. When the thickness of the region becomes 1–2 μm, the area begins to emit light. However, the formed region keeps growing slowly with time. As a result, the applied field decreases gradually and the process becomes self-limiting. An energy band model [35] corresponding to the structure of a d.c. powder EL device is shown in figure 1.10.

Electrons are injected from the Cu_2S layer by tunnelling into the Cu-free ZnS:Mn layer. These electrons are accelerated by the high-field and excite the Mn^{2+} centres by impact. Thus, the excitation mechanism of Mn^{2+} centres in d.c. powder EL is essentially the same as that of a.c. thin-film EL.

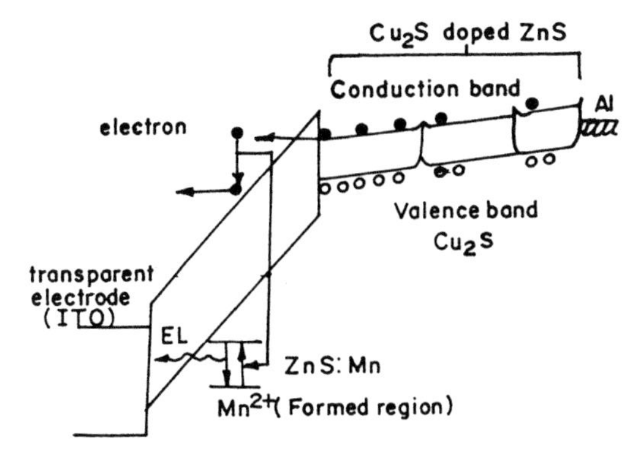

Figure 1.10. Energy band model of a d.c. powder EL device [35].

1.3.3.1 EL characteristics

The dependence of brightness (B) of a d.c. powder EL device on applied d.c. voltage (V) is expressed by the relation

$$B = B_0 \exp(-b/V^{0.5}). \tag{1.4}$$

Direct current powder EL devices are suitable for use in display panels with X–Y matrix electrodes as these devices have a higher discrimination ratio, thus allowing simple matrix addressing. The discrimination ratio is defined as the ratio of the brightness obtained at an applied voltage V to that obtained at $V/3$.

In the case of EL display panels with N row electrodes, the EL cell in each row is addressed in a fraction of the field time. This results in a low duty ratio of $(1/N)$ to drive EL cells. In pulsed excitation, d.c. powder EL devices can be operated beyond the forming voltage without producing further forming. This brings about a better discrimination ratio and higher brightness levels than those expected based on B–V curves for d.c. drive [35].

1.3.3.2 Direct current powder colour EL

There can be a variety of EL colours using different host materials with different activators. There exist many EL devices based on ZnS and alkaline earth sulphides (SrS, CaS etc.) doped with rare-earth activators that have been reported and demonstrated in these materials [36, 37]. For example, ZnS doped with Tm^{3+}, (Er^{3+} or Tb^{3+}), Nd^{3+} and Sm^{3+} gives blue, green, orange and red, respectively. CaS:Ce^{3+},Cl based devices produce green EL with brightness about one-third of the standard value of the ZnS:Mn yellow EL device. CaS:Eu^{2+},Cl and SrS:Ce^{3+},Cl devices show red and blue-green EL, respectively. To further improve the EL characteristics of phosphors for industrial display applications, an understanding of basic concepts of luminescent materials is needed.

1.3.3.3 Lifetime determination of d.c. powder EL cells

The basic reason for the brightness degradation is not clear yet. In general, pulsed operation and low brightness level operation lead to longer lifetimes. Devices operated at about 85 cd m^{-2} can retain more than half of their initial brightness for more than 10 000 h. This long lifetime is good enough for many practical applications.

1.4 Injection EL

In 1952, Haynes and Briggs [38] first reported the infrared EL from a forward-biased p–n junction in Ge and Si diodes. The injection of charge carriers through a p–n junction (or through rectifying contact on a crystal) is termed an 'injection EL'. These diodes are usually called light-emitting diodes (LEDs) and have been widely used since late 1960s. Semiconductors with a wide bandgap show this type of EL. In 1962, semiconductor lasers were made using GaAs diodes, which operate by stimulating an injection mechanism. The LEDs commercially available after the 1960s are green-emitting GaP:N and red-emitting GaP:Zn,O. It is well known that GaP is a semiconductor with an indirect bandgap; and the N and (Zn,O) centres in GaP are iso-electronic traps that provide efficient recombination for electrons and holes to produce good EL. In early 1990s, very bright LEDs used for outdoor displays were developed using III–V compounds that have a direct bandgap; for example, GaAlAs (red), GaAsP (orange), InGaAlP (green) etc. In 1994, another compound (GaInN) with a direct bandgap was developed that gavc very bright blue and green LEDs. Thus, LEDs covering the entire visible range with high brightness levels are now commercially available.

1.4.1 EL mechanism

The mechanism of light generation in injection EL and high-field EL are quite different from each other. There are two methods with which injection of minority carriers can occur.

In the first method, the injection occurs from an electrode (rectifying contact) into the crystal. The electrons are accelerated and excite the luminescence centres similar to the case of high-field EL. Only the source of electrons is different. This type of injection EL can be observed in single crystals of ZnS or in thin films that have contact with electrodes.

In the second method, light emission occurs at a p–n junction. At thermal equilibrium, a depletion layer is formed and a diffusion potential V_d across the junction is produced. When the p–n junction is forward-biased, the diffusion potential decreases to $(V_d - V)$ and the electrons are

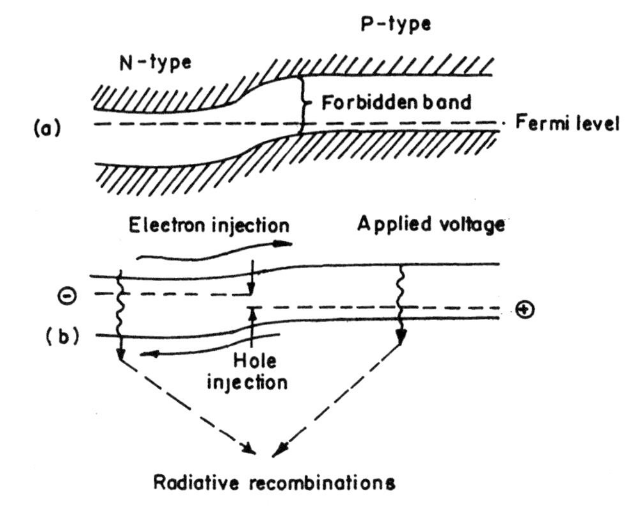

Figure 1.11. Energy band diagram for a p–n junction: (a) without applied field; (b) field applied in forward direction. Injection of minority carriers is inferred from radiative recombination [39].

injected from the n-region into the p-region while holes are injected from the p-region into the n-region; i.e. minority carrier injection takes place. Subsequently, the minority carriers diffuse and recombine with majority carriers directly or through trapping at various kinds of recombination centres, producing injection EL. An energy band diagram [39] for both the methods is shown in figure 1.11.

The following relation gives the total diffusion current on a p–n junction:

$$J = J_{\rm h} + J_{\rm e} + J_{\rm s}[\exp(qV/nkT) - 1], \qquad J_{\rm s} = q\left[\frac{D_{\rm h}p_{\rm ec}}{L_{\rm h}} + \frac{D_{\rm e}n_{\rm hc}}{L_{\rm e}}\right] \tag{1.5}$$

where $D_{\rm h}$ and $D_{\rm e}$ are diffusion coefficients for holes and electrons, $p_{\rm ec}$ and $n_{\rm hc}$ are the concentrations of holes and electrons respectively as minority carriers at thermal equilibrium, and $L_{\rm h}$ and $L_{\rm e}$ are diffusion lengths given by $(D\tau)^{0.5}$, where τ is the lifetime of the minority carriers.

1.5 Thin-film EL

TFEL devices are classified into two types: a.c. and d.c. type EL devices. Because of low efficiency and reliability related problems, the DCTFEL devices are not much in use. The ACTFEL devices are also divided into single-insulating and double-insulating layer types, as shown in figure 1.12.

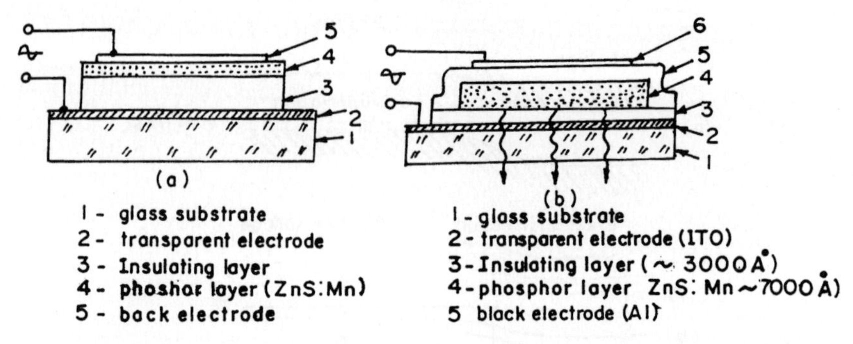

Figure 1.12. Schematic structure of the thin-film EL devices: (a) a single-insulating layer type EL devices; (b) a double-insulating layer type EL device [40].

Kahng *et al* [13] and Chase *et al* [40] made devices of the former type (DCTFEL) with a ZnS lattice doped with various rare-earth ions but they failed because of poor reliability. On the other hand, the latter type (ACTFEL) structure has been the subject of recent attention.

1.5.1 Direct current thin-film EL

Vlasenko and Popkov [10] studied DCTFEL in ZnS:Mn phosphors and explored physics of EL exhibited by this material. Direct current driven TFEL devices have a simple metal–insulator–metal (MIM) structure, and provide the clear picture of electrical conduction in ZnS films. The current–voltage (I–V) characteristics of d.c. MIM devices have been observed to be superlinear in the high field region, whereas at low field the linear behaviour is exhibited. The passage of large current densities by an insulator is due to field-dependent carrier density and quantum mechanical tunnelling.

Impurities and dopants in the hosts have been found to give n-type conductivity. This one-carrier injection current due to injected electrons is a space charge current which varies with applied voltage at a rate proportional to the passage of charge [4, 41]. This space charge results in a distorted or non-uniform EL, as mentioned earlier.

The distribution of the defects being non-uniform over the thickness of the phosphor layer, the electric field profile will depend on the bias polarity. Hence, the nonlinearity of I–V characteristics is not only due to difference in barrier height but also related to the features of the bulk as well.

1.5.2 Alternating current thin-film EL

Inoguchi *et al* [14, 42] reported that a triple-layered, thin-film EL (TFEL) devices, consisting of a ZnS:Mn^{2+} electroluminescent layer sandwiched by a pair of insulating layers, showed high brightness levels and an unusually longer life, which were the unsolved problems during the first era of EL development. Further, Mito *et al* [15] showed that these EL panels could be used as a TV imaging system. Since then the research and development of TFEL got much attention and triggered world-wide investigations because of the high probability of these devices finding practical applications in information displays. Uede *et al* [43] and Takeda *et al* [44] reported the successful production of practical EL display units and practical application technologies of $ZnS:Mn^{2+}$ EL display panels. With the use of these technologies [45], production of six-inch diagonal matrix EL panels started in Japan during 1983.

Since then, the ACTFEL panels with triple-layered structure have been continuously under investigation for improvement of colour and contrast, fabrication techniques for new phosphor materials and insulating materials. At present one can observe both multicolour and full-colour TFEL units in the market. In the following sections, emphasis is given to describing the ACTFEL devices in detail.

1.5.2.1 Structure of ACTFEL device

A typical structure of ACTFEL device is shown in figure 1.12(b), where a thin-film phosphor layer is sandwiched between thin-film insulating layers. These thin films are subsequently deposited on an ITO-coated substrate, and an Al rear electrode is deposited on the second insulating layer. When an a.c. voltage of about 200 V is applied between the ITO and Al electrodes, then EL emission is observed.

The host materials that are widely used for TFEL device fabrication are ZnS, ZnSe, SrS, CaS etc. Apart from Mn^{2+}, the rare-earth ions suitable to act as luminescent centres are Tb^{3+}, Sm^{3+}, Pr^{3+}, Eu^{2+} and Ce^{3+}.

1.5.2.2 Insulating materials

The insulating layers play an important role in the stability of the double-insulating layered TFEL devices and also control the EL characteristics. The materials that can be used for insulation should have critical properties such as high breakdown electric field strength, high dielectric constant and good adhesion. There are two types of dielectric insulating material that are mainly used: (1) amorphous oxides and nitrides [44], such as Al_2O_3, SiO_2, TiO_2, Ta_2O_5 and Si_3N_4, and (2) ferroelectric materials [46–48], such as $BaTiO_3$, $SrTiO_3$ and $PbTiO_3$. Figure 1.13 shows the relative dielectric constant versus breakdown electric field characteristics of typical dielectric materials used for the double-insulating TFEL devices [48]. In practical devices, a complex film such as Al_2O_3/TiO_2 and a composite film such as Si_3N_4/SiO_2 are used to obtain the best dielectric properties.

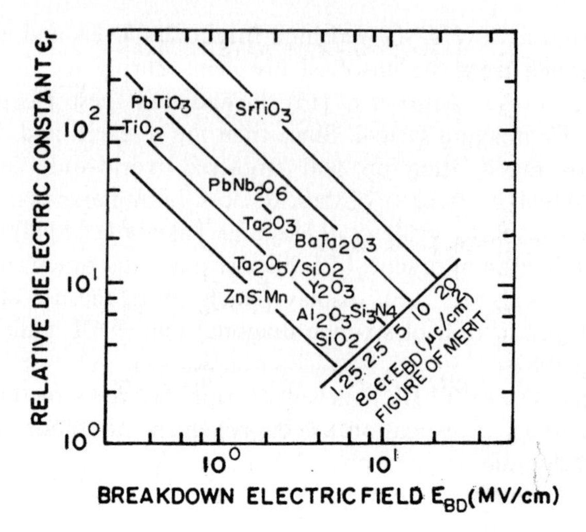

Figure 1.13. Relative dielectric constant versus breakdown electric field (ε_r–E_{BD}) characteristics of typical dielectric materials used for a double-insulating thin-film EL device [48].

1.5.2.3 Methods for phosphor deposition

There are many phosphor deposition techniques investigated to increase the brightness levels of the EL displays: sputtering method (SP) [49], atomic-layer epitaxy (ALE) [50], electron-beam deposition (EBD) [43, 44], multisource deposition (MSD) [51], metal–organic chemical vapour deposition (MOCVD) [52] and low-pressure hydride-transport chemical vapour deposition (HT-CVD) [53]. Of these, the EBD and ALE methods are used to fabricate commercially available TFEL displays. The SP method is extensively studied in making thin-films of ZnS:Tb^{3+},F phosphor.

1.5.2.4 Electrical characteristics

TFEL devices exhibit current density versus voltage (I–V) and phase difference versus voltage (ϕ–V) characteristics with 1 kHz sinusoidal wave drive conditions, as shown in figure 1.14. The I–V profile gives two straight lines linked with a kink at the threshold voltage (V_{th}). The slope of the I–V curve below V_{th} corresponds to the total capacitances of the insulating layer, the phosphor layer, and the second insulating layer in series. On the other hand, the slope above V_{th} corresponds to the total capacitances of only the insulating layers in series [54]. This indicates that the phosphor layer changes its capacitive characteristics into resistive characteristics. This behaviour is further confirmed by the ϕ–V characteristics.

In the a.c.-driven TFEL devices, above threshold voltage, one observes a charge transfer whose value increases with increasing voltage. The increase of charge is linear with voltage. With further increase in V the destructive

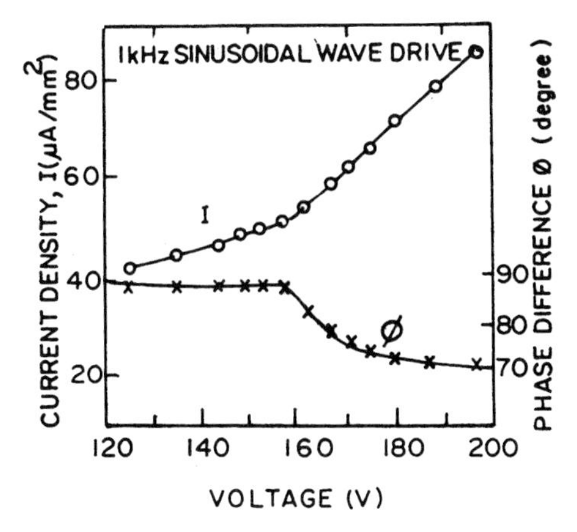

Figure 1.14. Current density versus voltage (I–V) and phase difference versus voltage (ϕ–V) characteristics of a double-insulating thin film EL device [54].

breakdown occurs in the phosphor layer. Below a certain threshold voltage, the device is assumed to be a perfect capacitor. Under the action of the electric field, the charge carriers cross the phosphor layer and are captured either in bulk or at the opposite insulator–semiconductor interface. The accumulation of charge at the opposite phosphor–insulator interface leads also to a polarization or memory effect. Carrier injection into the conduction band of the phosphor layer is generally thought to result from field emission or tunnelling from interface levels. In general, the bulk is not necessarily neutral owing to stored charge, and the electric field is no longer uniform over the thickness of the phosphor film. The electric field profile will depend on the bias polarity. For more details regarding the electrical behaviour of ACTFEL devices, the reader is referred to section 1.3.2.2 and references [23, 55–59], and the references therein. For physics of ACTFEL devices concerning electrical conduction, high-field electron transport, excitation of luminescent centres, concentration effects, optical output, colour, efficiency etc. the reader is referred to the section 2.3.2.2 of Chapter 2.

For example, in ZnS phosphor, with an increase of driving voltage, the average electric field in the phosphor layer at V_{th} reaches a value of 1 to $2 \times 10^6\,\mathrm{V\,cm^{-1}}$ and remains at that value above V_{th}. From the above results, an equivalent circuit model [54] for TFEL devices can be represented as shown in figure 1.15.

1.5.2.5 Electro-optical properties

When the phosphor layer becomes conductive above V_{th}, an in-phase conduction current starts flowing through the phosphor layer, producing

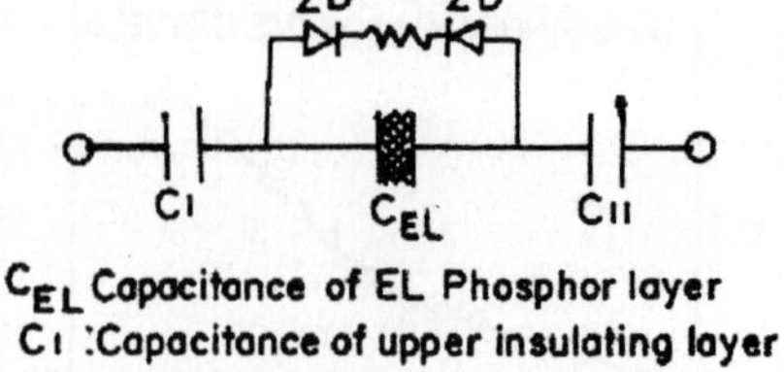

Figure 1.15. Equivalent circuit model of a double-insulating layer a.c. thin-film EL device [54].

EL emission. The brightness is proportional to the in-phase conduction current [54]. Figure 1.16 shows the schematic characteristics of the voltage waveform (V), the current waveform (I) and the EL emission waveform (L) under triangular-wave drive conditions.

The rise and decay times in the case of ZnS:Mn^{2+} electroluminescent devices are of the order of several microseconds and several milliseconds, respectively. The above times indicate that the brightness is directly proportional to the driving frequency, up to frequencies of several kHz. The same proportionality is expected to continue for frequencies up to the order of several hundred kHz. However, consideration must be given to the influence on the time constant by the resistivity of the ITO, the output impedance of the IC driver and the capacitance of the TFEL display panel.

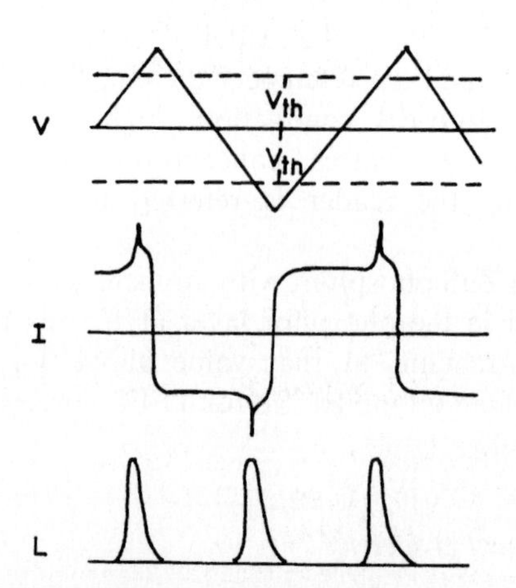

Figure 1.16. Schematic characteristics of voltage waveform V, the current waveform I, and the EL-emission L, under triangular-wave drive conditions [54].

1.5.2.6 Reliability

In the development of TFEL devices, reliability is one of the important factors to be considered. Care is to be taken to have an even surface with no pinholes during the constituent layer formation. An ageing process is needed to stabilize the brightness–voltage characteristics since the *B–V* characteristics of the as-fabricated EL devices show a shift towards higher voltages. Moisture penetrating the phosphor–insulating layer interface through the small dielectric breakdown holes might cause peeling of thin films. Serious attention must be paid to sealing techniques in order to prevent moisture penetration from the atmosphere. In commercial TFEL devices, silicon oil and silica gel powders are placed in the glass-sealed EL device. With this technique, a lifetime of more than 30 000 h has been achieved by various researchers [60, 61].

1.6 Recent developments in TFEL displays

The first commercial monochrome TFEL display was introduced in 1983. Since then significant progress and variety of products appeared in the market; for example, in the case of $ZnS:Mn^{2+}$ monochrome EL panels, high brightness levels of around $200\,cd\,m^{-2}$, high-resolution devices with 1.3 million pixels, low-power consumption devices of $<5\,W$ in a 10.4-inch diagonal EL display and high-contrast devices of $>50:1$ at 500 lux are commercially available [62]. In the case of colour TFEL panels, green/red multicolour TFEL devices are now available [63]. These TFEL displays have found strong acceptance in medical and industrial control applications, where the need for wide view angle, longer life, wide temperature ranges, and fast response time are critical. Other recent developments in TFEL displays include high-contrast monochrome, multicolour, and full-colour displays for futuristic devices.

References

[1] Destriau G and Mattler J 1946 *J. Phys. Rad.* **7** 259
[2] Williams F E and Cusano D A 1956 *J. Phys. Rad.* **17** 742
[3] Hershinger L W, Daniel P J, Schwarz R F and Laseer M E 1958 *Phys. Rev.* **111** 1240
[4] Tanaka I, Izumi Y, Tanaka K, Inoue Y and Okamoto S 2000 *J. Lumin.* **87–89** 1189
[5] Gudden B and Pohl R W 1920 *Z. Phys.* **2** 192
[6] Destriau G 1936 *J. Chim. Phys.* **33** 620
[7] Henish H K 1962 *Electroluminescence* (Oxford: Pergamon Press)
[8] Ivey H F 1966 Electroluminescence and related effects, in *Advances in Electronics and Electron Physics*, Suppl. I (New York: Academic)

[9] Pankove J I 1980 Electroluminescence, in *Topics in Applied Physics*, vol 17 (Berlin: Springer)

[10] Vlasenko N A and Popkov Z 1960 *Opt. Speckrosk.* **8** 39

[11] Vecht A, Werring N J, Ellis R and Smith J F 1973 *Proc. IEEE* **61** 902

[12] Russ M J and Kennedy D I 1967 *Electrochem. Soc.* **114** 1066

[13] Kahng D 1968 *Appl. Phys. Lett.* **13** 210

[14] Igonuchi T, Takeda M, Kakihara Y, Nakata Y and Yoshida M 1974 *SID '74 Digest* 84

[15] Mito S, Suzuki C, Kanatani Y and Ise M 1974 *SID '74 Digest* 86

[16] Barrow W A, Coovert R C, Dickey E, Flegal E, Fullinan M, King C N and Laakso C 1994 *Proc. 14th Inter. Display Res. Conf.*, Monterey, p. 448

[17] Bringuir E 1994 *J. Appl. Phys.* **75**(9) 4291

[18] Chander H, Shanker V, Haranath D, Dudeja S and Sharma P 2003 *Mater. Res. Bull.* **38** 279

[19] Zalm P 1956 *Philips Res. Rep. No. 11* **353** 417

[20] Fischer A G 1962 *J. Electrochem. Soc.* **109** 1043

[21] Fischer A G 1963 *J. Electrochem. Soc.* **110** 733

[22] Fischer A G 1966 EL in II–VI compounds, in *Luminescence of Inorganic Solids* ed. P Goldberg (New York: Academic Press) ch. 10

[23] Ono Y A 1995 *Electroluminescent Displays* (Singapore: World Scientific)

[24] Shionoya S and Yen W M (eds) 1998 *Phosphor Handbook* (Boca Raton, FL: CRC Press) p. 136

[25] Baraff G A 1964 *Phys. Rev.* **133** A26

[26] Van Gool W and Cleiren A P 1960 *Phil. Res. Rep.* **15** 238

[27] Kroger F A and Dikhoff J 1950 *Physica* **16** 297

[28] Curie D 1963 *Luminescence in Crystals* (New York: Wiley) p. 241 (translated G F J Garlick)

[29] Antonov-Romanovsky V V 1959 *Czech. J. Phys.* **9** 146

[30] Desriau G 1955 *Br. J. Appl. Phys.* **4** Suppl. S2.9

[31] Matossi F 1955 *Phys. Rev.* **98** 546

[32] Tanaka S 1988 *J. Lumin.* **40/41** 20

[33] Rennie J and Sweet M A S 1987 *Cryst. Res. Technol.* **2** K119

[34] Wager J F, Hitt J C, Baukol B A, Bender J P and Keszler D A 2002 *J. Lumin.* **97** 68

[35] Kirton J 1981 *Handbook on Semiconductors*, Vol. 4, *Devices Physics* (North-Holland: Amsterdam) ch. 5C

[36] Waite M S and Vecht A 1971 *Appl. Phys. Lett.* **19** 471

[37] Higton M, Vecht A and Mayo J 1978 *Digest 1978 SID Int. Symp. Soc. Inform. Display*, Los Angeles, p. 136

[38] Haynes J R and Briggs H B 1952 *Phys. Rev.* **99** 1892

[39] Lehovec K, Accaardo C A and Jamgochian E 1951 *Phys. Rev.* **83** 603

[40] Chase E W, Hoppelwhite R T, Krupka D C and Kalong D 1969 *J. Appl. Phys.* **40** 2512

[41] Beale M 1993 *Philos. Mag.* **B68** 573

[42] Inoguchi T and Mito S 1977 Electroluminescence, in *Topics in Applied Physics* vol. 17 (Berlin: Springer) p. 196

[43] Uede H, Kanatani Y, Kishishita H, Fujimori A and Okamo K 1981 in *Digest 1981 SID Int. Symp., Society for Information Display*, p. 28

[44] Takeda M, Kanatani Y, Kishishita H, Inoguchi T and Okano K 1981 *Proc. Society Information Displays* **22** 57

[45] Takeda M, Kanatani Y, Kishishita H and Uede H 1980 *Advances in Display Technology III, SPIE Proceedings* **386** 34

[46] Okamoto K, Nasu Y and Hamakawa Y 1980 in *Digest, Biennial Disp. Res. Conf.* p. 143

[47] Marello V and Onton A 1980 *IEEE Trans. Electron. Devices* **ED-27** 1767

[48] Fujita Y, Kuwata J, Nishikawa M, Tohda T, Matuoka T, Abe A and Nitta T 1984 *Proc. Soc. Inf. Display* **25** 177

[49] Ohnishi H, Yamasaki Y and Iwase R 1987 *Proc. Soc. Inf. Display* **28** 345

[50] Suntola T, Anston J, Pakkala A and Linfors S 1980 *Digest 1980 SID Int. Symp. Soc. Inf. Display* p. 108

[51] Nire T, Watanabe T, Tsurumaki N, Miyakoshi A and Tanda S 1989 Electroluminescence in: *Springer Proc. in Phys.* **38** eds H Shionoya and H Kobayashi (Heidelberg: Springer)

[52] Cattel A R, Cockanyne B, Dexter K, Kirton J and Wright P J 1983 *IEEE Trans. Electron Devices* **ED-30** 471

[53] Mikami A, Terada K, Okibayashi K, Tanaka K, Yoshida M and Nakajima S 1991 *J. Cryst. Growth* **110** 381

[54] Inoguchi T and Suzuki C 1974 *Nikkei Electronics* **84** (in Japanese)

[55] Neyts K and Visschere P De 1992 *Solid State Electr.* **35** 933

[56] Bringuier E and Geoffroy A 1992 *Appl. Phys. Lett.* **60** 1256

[57] Allen J W 2000 *J. Lumin.* **87–89** 1189

[58] Krasnov A N and Hofstra P G 2001 *Prog. Cryst. Growth and Charact.* **42** 65

[59] Gumlich H-E, Zeinert A and Mach R 1998 Electroluminescence in: *Luminescence of Solids* ed D R Vij (New York: Plenum Press)

[60] Rack P D, Naman A, Holloway P H, Sun S-S and Tuenge R D 1996 *MRS Bulletin* **21** 49

[61] Okamoto K, Wakitani M, Sato S, Miura S, Andoh S and Umeda S 1983 *Digest 1983 SID Int. Symp. Information Display* p. 16

[62] Mikami A 1994 *Tech. Digest, Int. Symp. Inorganic and Organic Electroluminescence* p. 17

[63] King C N 1994 *Proc. Int. Disp. Res. Conf.* p. 69

PART 1

II–IV GROUP MATERIALS

Chapter 2

Zinc sulphide

Nigel Shepherd and Paul H Holloway
Department of Materials Science and Engineering, University of Florida, Gainesville, USA

2.1 Introduction and historical perspective

Zinc sulphide (ZnS) is *the* classic electroluminescent (EL) phosphor, and is the workhorse of the thin-film electroluminescence (TFEL) sector. It has been used extensively in every phase of EL phosphor research and development. This chapter addresses the fundamental processes that lead to light emission in ZnS phosphors. Likewise, the basic processes of sulphide-based EL devices will be addressed. The discussion will be presented within the frame of solid state and semiconductor physics.

EL phosphors are dilute solid solutions. In this case the ZnS component is the solvent, which in EL nomenclature is known as a host. Luminescent centres (transition metal and rare-earth ions), also known activators, constitute the solute component of ZnS phosphors. The electrical and electro-optical characteristics of phosphor EL devices are determined primarily by the host, whereas colour or emission wavelength is determined by the activator. Emission wavelength is determined by the activator because electroluminescence in ZnS is generated from impact excitation and/or impact ionization of luminescent centres/activators by high-energy electrons. In other words, the electron excitation and radiative relaxation is a transition *localized* at the luminescent centre, i.e. an atomic transition. This light emission mechanism is also commonly referred to as high-field electroluminescence. It is in contrast to the second class of EL devices (light emitting diodes and semiconductor lasers), in which light emission results from recombination of electron–hole pairs.

Zinc sulphide is widely used in all types of high-field electroluminescent (EL) devices. Specifically, it is used in a.c. (alternating current) powder EL devices, d.c. (direct current) powder EL devices, d.c. thin-film EL (TFEL) devices, and a.c. thin-film EL (ACTFEL) devices. The first report of electroluminescence in ZnS is often attributed to Destriau [1]. In 1936, he reported that upon application of a high field to a suspension consisting of ZnS:Cu:oil in a device that resembles a modern powder EL cell, light emission resulted. In the period immediately following Destriau's publication not much is reported on EL in the literature. This was at least in part due to the lack and immaturity of fabrication techniques for transparent electrically conductive films. The development of ZnO in 1950 triggered worldwide EL research and development. At the time these efforts focused on powder phosphors for general lighting applications. Indeed, GTE Sylvania obtained a patent for an a.c. EL powder lamp in 1952 [2]. However, it soon became clear that so-called cold lamps had very limited lifetimes, which strongly depended on the drive level and light output required. Simply put, they were unable to provide adequate sustained brightness over the minimum (500 h) commercially acceptable lifetimes. As a result, interest in the area diminished.

Many early powder phosphor devices were d.c. driven. The associated ionic drift and material diffusion was one of the main reasons for early failures. Another reason for these early failures was the permeation of moisture into the device structure. The absorbed water reacted with the EL phosphor, effectively quenching the luminescence. In the 1980s and 1990s, with a resurgence of interest in powder EL, these two failure modes were addressed by utilizing an a.c. driving voltage and glass encapsulation techniques to minimize moisture-induced degradation of the EL phosphor. Today, a.c. powder EL devices find application mainly in speciality lighting. A few common examples of this application are the backlighting of liquid crystalline displays (LCDs) found in wrist watches and PDAs, night lights for our homes, and emergency lighting in airplanes and cinemas. It should be noted, however, that d.c. powder panels are also available commercially. The peculiarities of powder EL (and TFEL) will be addressed in detail in sections that follow.

While powder EL is used mainly for lighting applications, thin-film EL is a flat panel display (FPD) technology. The term flat panel displays encompasses all direct-view, non-cathode ray tube (CRT) displays. In addition to thin-film electroluminescent (TFEL) displays, the flat panel group includes other technologies such as liquid crystal displays (LCD), plasma displays (PD) and field emission displays (FED). Though liquid crystal displays dominate the FPD sector ($\sim$85%), thin-film EL displays offer several advantages. These include wide viewing angle (>160°), high contrast in high ambient illumination conditions, broad operating-temperature range ($-60\,^{\circ}$C to $+100\,^{\circ}$C), fast response times, and the capability of very high resolution and legibility. Furthermore, their simple all solid-state construction results in inherent ruggedness and makes them insensitive to shock and vibration. TFEL displays (a.c.-driven) have become the most reliable, longest running (>50 000 h) devices in the market [3]. These attributes make TFEL displays ideal for military, avionics, medical and industrial applications.

Electroluminescence in thin-film ZnS:Mn was described as early as 1960 by Vlasenko and Popkov [4]. The reports by these authors and co-workers [4–7] were centred on the physics of the EL mechanism, with little emphasis on prolonging lifetime. The late 1960s saw revived interest in EL research and development and, though Vecht *et al* demonstrated the first d.c. powder EL panel in 1968 [8], the majority of development efforts focused on thin-film EL. The optically active layer in Vecht's device consisted of ZnS powder particles coated with copper ions, which were then embedded into a binder. In 1968, Kahng [9] employed an a.c.-driven device using Lumocen (luminescence from molecular centres), and proposed that impact excitation of luminescent centres by hot electrons was the EL emission mechanism. Here the optically active layer consisted of thin-film ZnS doped with rare-earth fluorides (as molecular centres). In the 1960s, Soxman and Ketchpel [10] also demonstrated TFEL devices with acceptable lifetimes and multiplexing capabilities, though reliability remained unacceptable. These early successes would not have been realized without the developments in thin-film processing, and advances in materials science and physical electronics that took place in the 1960s.

The insulating layer/phosphor/insulating layer sandwich structure first proposed by Russ and Kennedy [11] in 1967 was a major milestone in the development of TFEL display technology, though its benefit was not immediately realized by researchers in the area. Utilizing this double-insulating thin-film structure with a ZnS:Mn phosphor layer and a.c. driving, in 1974 Inoguchi *et al* [12] from Sharp Central Research Labs reported stable, high-luminance EL panels which had excellent lifetimes. Inoguchi's work described the basic phenomena and is often cited as the birth of current TFEL technology. Mito *et al* [13] showed that these EL panels could be used as a TV imaging system at the same conference at which Inoguchi's work was presented. Since then, a.c. thin-film EL devices (ACTFELDs) have been the most intensely investigated of the four types of devices because it has the highest possibility for successful practical application. As a result of these efforts, TFEL is now one of the three main flat panel display technologies. Uede, Takeda [14–16] and their co-workers reported on the production of practical EL ZnS:Mn display units in 1981 and in 1983 Sharp introduced the first commercial ZnS:Mn thin-film display [17]. By the mid 1980s, ZnS:Mn thin-film displays were also introduced by Planar Systems and Finlux (Planar International), and a variety of EL products began appearing on the market. ZnS doped with manganese is yellow emitting. As a result, early thin-film products were monochrome, yellow emitting. At present, both Sharp and Planar utilize colour filtering to obtain red and green from yellow, in order to manufacture multicolour displays based on ZnS:Mn. There is also considerable effort towards developing

ZnS doped with rare earth ions thin-film phosphors that emit the three primary colours, i.e. red, green and blue for full-colour displays. Work in this area was pioneered by Kahng [9] and Chase *et al* [18] in the late 1960s. At present the luminance of blue emitting ZnS phosphors is insufficient for practical application. Commercial production of full-colour thin-film displays began in 1994. These were based on the work by Barrow [19, 20] and his co-workers. They used a stacked structure with the red and green phosphors on one substrate, and blue emitting cerium-doped calcium thiogallate phosphor on a separate substrate. The details regarding the physics and structure of this and other EL devices will be addressed in later sections. Table 2.1 lists the major milestones in the development of ZnS based EL.

Today ZnS and other EL phosphors continue to be extensively researched. Most of this effort is geared towards developing more efficient full-colour (red, green, blue) a.c. thin-film EL displays. Advances in TFEL

Table 2.1. The development high-field electroluminescence in ZnS.

1936	High-field electroluminescence from ZnS discovered by Destriau [1].
1950	Transparent electrically conductive zinc oxide developed.
1950s	Development and basic studies of a.c. powder EL devices. GTE Sylvania obtains a patent for a powder EL lamp. Problems include low brightness, short lifetimes, poor contrast, and high operating voltage among others. Research and commercial interest eventually fades.
1960	Electroluminescence in thin-film ZnS:Mn described by Vlasenko and Popkov [4].
1960s	Thin-film EL research by Soxman and Ketchpel [10]. Reported in 1972.
1967	Breakthrough double-insulating-layer a.c. thin-film device structure proposed by Russ and Kennedy [11].
1968	Pioneering work by Kahng reported high luminance EL from ZnS:TbF_3 thin-film phosphors, and described impact excitation. He called it luminescence from molecular centres (Lumocen) [9].
1968	Vecht demonstrates the first d.c.-driven powder EL panel [8]
1974	Inoguchi and his co-workers reports first ZnS:Mn high-luminance, long-lifetime a.c. thin-film EL panels [12]. Device structure was proposed earlier by Russ and Kennedy [11].
1974	Mito *et al* [13] showed that thin-film EL panels could be used as a TV imaging system.
1981	Uede, Takeda [14–16] and their co-workers reported on the production of practical ZnS:Mn a.c. thin-film EL display units.
1983	Sharp introduces the first commercial ZnS:Mn a.c. thin-film monochrome display [17].
1988	Prototype full-colour thin-film EL display demonstrated by Barrow *et al* [20].
1990s	Alternating-current powder EL devices developed for speciality lighting.
1993	First commercial multicolour (red/green/yellow) ZnS:Mn a.c. thin-film EL display introduced by Cramer *et al* [21].
1994	First commercial full-colour a.c. thin-film EL display panel [19].

technology is occurring rapidly due to an increasing knowledge base coupled with the interplay of developments in materials, deposition techniques and device structures. On the commercial side, ongoing reductions in production costs will eventually make the price of TFEL displays more competitive with the more mature liquid crystal display (LCD) technology, which holds about 85% of the flat-panel market. Considering these factors together with the earlier mentioned advantages of thin-film EL displays over LCDs, the position and growth of electroluminescent displays in the marketplace seems assured.

2.2 Phosphor fundamentals

As mentioned earlier EL phosphors have two constituents: host, and luminescent centre(s). The electrical and electro-optical characteristics of EL phosphor devices are determined primarily by the host, whereas colour or emission wavelength is determined by the luminescent centre. The four basic processes leading to light emission are as follows.

1. Electronic charge carriers are injected into the phosphor layer.
2. Electrons in the phosphor are accelerated to high energies.
3. Energy is transferred from ballistic energy electrons to luminescent centres promoting electrons on the latter into excited states, i.e. impact excitation of the luminescent centre.
4. Radiative relaxation of the luminescent centre resulting in light emission.

In other words, photon emission is due to an *atomic* transition, i.e. an electronic transition localized at the luminescent ion. Before proceeding with a detailed look at the different physical processes and device structures of high-field electroluminescence, a closer look at some of the fundamental concepts applicable to all EL phosphors is warranted.

2.2.1 Fundamentals of atomic transitions

I. Selection rules for optical transitions

The luminescence properties of atoms (ions) are determined by their quantum mechanical properties. In an atom, the energy states of its electrons are characterized by three main quantum numbers. The principal quantum number n essentially characterizes the radius of the electron distribution, and can take the values $n = 1, 2, 3, \ldots$. An electron with $n = 1$ is in the K shell, that with $n = 2$ is in the L shell, that with $n = 3$ the M shell, and so on. The second quantum number, l, is a measure of the angular momentum of the electron, and can assume the values $l = 0, 1, 2, \ldots, n - 1$. Its magnitude is defined as $|\boldsymbol{l}| = [l(l+1)]^{1/2}\hbar$. Since l can range from 0 to $n - 1$, there are n subshells for a shell with principal quantum number n. Subshells with $l = 0$ are referred to as s subshells, those with $l = 1$ as p, those with $l = 2$ as d, and so forth. Thus for example, a 1s state is one with $n = 1$ and $l = 0$, a 2s state is one with $n = 2$ and $l = 0$, a 2p state is one with $n = 2$ and $l = 1$, and a 4d state is one with $n = 4$ and $l = 2$. The magnetic quantum number $m_l = 0, \pm 1, \pm 2, \ldots, \pm l$, characterizes the quantization of the possible orientations (z component) of the angular momentum vector with respect to an external magnetic field applied in the z direction.

Electronic transitions that involve a change of principal quantum n are known as intershell transitions. Those that occur without a change in n are known as intrashell transitions. The radiative relaxation of excited electrons via both intershell and intrashell transitions is exploited in the ZnS type electroluminescence. These transitions are driven by an electric dipole moment. The transition moment integral is defined as

$$M = \int \Psi' \hat{\mu} \Psi \, \mathrm{d}\tau \tag{2.1}$$

where the prime denotes the final (excited) state wavefunction, Ψ is the initial (ground) state wavefunction, $\hat{\mu}$ is the dipole moment operator, and integration is performed over all space. The electric dipole moment (x component) is generally written as

$$\mu = \sum_i e_i \bar{x}_i \tag{2.2}$$

where $\bar{x}_i$ is the electron coordinate vector. Likewise, the total wavefunction Ψ can be written as

$$\Psi = \Psi_e \Psi_s \Psi_\nu \tag{2.3}$$

where Ψ_e the electronic orbital wavefunction, Ψ_s is the electron spin wavefunction and Ψ_ν is the nuclear vibrational wavefunction. It can be shown (see [22], sections 3.6 and 5.4) that the transition moment integral can be written as

$$M = \int \Psi'_e \mu_e \Psi_e \, \mathrm{d}\tau_e \int \Psi'^*_s \Psi_s \, \mathrm{d}\tau_s \int \Psi'^*_\nu \Psi_\nu \, \mathrm{d}\tau_n . \tag{2.4}$$

The intensity of a transition is related to the transition moment integral through the oscillator strength, which is a measure of the dipole moment associated with the shift of charge that occurs when electron redistribution takes place during an electronic transition. The oscillator strength is defined as

$$F = (8\pi^2 m_e \nu / 3he)^* M^2 . \tag{2.5}$$

Clearly, if any of the integrals in equation (2.4) is zero, M and F are zero. Such transitions are formally forbidden. Selection rules are derived from equation (2.4) and are a set of conditions for which $M \neq 0$. The first integral in equation (2.4) is the basis for orbital (parity) selection rules, and the second integral is the basis for spin selection rules. The square of the third integral represents the well known Franck–Condon factor, which in essence is an overlap integral between the initial and final vibrational state wavefunctions. It indicates that the intensity of a particular simultaneous electronic and vibrational transition is directly proportional to the vibrational overlap. The selection rules for optical transitions are as follows.

1. Spin selection rule. Inspection of equation (2.4) shows that the integral $\int \Psi_s'^* \Psi_s \, d\tau_s$ must be nonzero if a transition is to be allowed ($M \neq 0$). Because of the orthogonality of the spin wavefunctions, if $\Psi_s' \neq \Psi_s$, the integral $\int \Psi_s'^* \Psi_s \, d\tau_s$ is zero. Therefore, a transition is spin-allowed only if the multiplicities (see part II of this subsection) of the two states involved are identical. Thus, for example, singlet → singlet and doublet → doublet transitions are allowed, but singlet → triplet and quartet → doublet transitions are forbidden. The spin selection rule can be summarized as
$$\Delta s = 0 \tag{2.6}$$
where s is the electron spin quantum number (i.e. $\pm\frac{1}{2}$).
2. Angular momentum selection rule.
$$\Delta l = \pm 1. \tag{2.7}$$
It reflects the fundamental physical nature of photons which are bosons with unit spin ($s = 1$). Due to their unit spin, photons have unit intrinsic spin angular momentum. This means that if a photon is generated by an electron undergoing a transition, the angular momentum of the electron must change to account for the angular momentum carried away by the photon as its spin. The same argument holds for electronic transitions where photons are absorbed rather than generated.
3. Parity selection rule. This rule states that the only allowed transitions are those accompanied by a change of parity. This condition is derived directly from the first integral of equation (2.4), i.e. $\int \Psi_e' \mu_e \Psi_e \, d\tau_e$. The dipole moment operator μ_e is an odd function (will change sign in the case that the electronic coordinates are replaced by their negatives). Thus, the entire function $\Psi_e' \mu_e \Psi_e \, d\tau_e$ will be odd if the initial and final state wavefunctions are both even or both odd. Integration of an odd function over all space gives zero, because the partial integral from 0 to $+\infty$ exactly cancels the integral from 0 to $-\infty$. When the integral $\int \Psi_e' \mu_e \Psi_e \, d\tau_e$ vanishes, the transition dipole moment and consequently the oscillator strength is zero. Therefore, the function $\Psi_e' \mu_e \Psi_e \, d\tau_e$ must be even as a whole to obtain $M \neq 0$, which leads to the condition that transitions accompanied by the emission or absorption of dipole radiation can only occur between even and odd states (an even state being one represented

by an even wavefunction, and an odd state one represented by an odd wavefunction). The reader is referred to [22, 23] for a deeper understanding of the quantum mechanics of electronic transitions in atoms.

These selection rules are general and apply to all optical transitions. However, selection rules are not absolute and there are several conditions that relax them. In the ZnS phosphor system specifically, parity forbidden *d–d*, and *f–f* transitions of practical importance are observed. Selection rules can be relaxed be placing the luminescent centre in a lattice site without a centre of symmetry (site with uneven crystal field components), spin–orbital coupling, and electron–vibration coupling etc., due to the admixture of suitable wavefunctions into the original, unperturbed wavefunctions. For details see [24–26].

II. Term symbols

Term symbols or spectroscopic notation are often used to represent the energy states of luminescent ions. Understanding this representation is therefore useful. A term symbol has the general form

$$^{2S+1}L_J, \tag{2.8}$$

which conveys three pieces of information succinctly, as follows.

1. The letter L indicates the total orbital angular momentum. It reflects how the individual orbital angular momenta of the several electrons that are present in an unfilled shell add together, or oppose each other. L is obtained by adding the individual orbital angular momenta using the Clebsch–Gordon series:
$$L = l_1 + l_2, l_1 + l_2 - 1, \ldots, |l_1 - l_2|. \tag{2.9}$$
It is quantized and can have $2L+1$ possible orientations. Considering two p electrons (for which $l_1 = l_2 = 1$) as an example, then $L = 2$, 1, 0. The code for converting the value of L into a letter is: $L = 0$(S), 1(P), 2(D), 3(F), 4(G), 5(H), 6(I) etc.
2. The top left superscript $2S+1$ is the multiplicity of the term, where S is the total spin angular momentum. Noting that each electron has $s = \frac{1}{2}$ and using equation (2.9), for two electrons one finds that $S = 1, 0$ (depending on whether their spins are paired or not). Closed shells for example have all their electrons paired, and therefore no net spin. In such a case $S = 0$, and the multiplicity $2S+1 = 1$ is designated a singlet, such as ^{1}S. When there are two unpaired electrons, $S = 1$ and $2S+1 = 3$ give a triplet, such as ^{3}D. For a single electron, $S = s = \frac{1}{2}$ and $2S+1 = 2$, which gives a doublet term. If there are three electrons, the total spin angular momentum is obtained by coupling the spin of the third electron to each of the values of S for the first two, resulting in $S = \frac{3}{2}, \frac{1}{2}$.

3. The right subscript on the term symbol, J, is the total angular momentum quantum number, and is also determined using the Clebsch–Gordon series, i.e. $J = L + S,\ L + S - 1, \ldots, |L - S|$. Thus for example, ^{3}D has $S = 1$ and $L = 2$ which results in $J = 3,\ 2,\ 1$ and so the term has three levels $^3D_3$, $^3D_2$ and $^3D_1$.

The selection rules in term symbol notation are:

$$\Delta S = 0,$$
$$\Delta L = 0, \pm 1 \text{ with } \Delta l = \pm 1,$$
$$\Delta J = 0, \pm 1.$$

The rule about $\Delta L = 0,\ \pm 1$ with $\Delta l = \pm 1$ indicates that the orbital angular momentum of an individual electron participating in an optical transition must change ($\Delta l = \pm 1$), but whether or not this influences the overall total orbital momentum of the system depends on the *jj* coupling between electrons.

2.2.2 Luminescent centres

When considering luminescent centres for electroluminescence applications, there are three properties that are of primary importance: radiative kinetics, the cross-section for impact excitation, and radiative transition energy (colour).

2.2.2.1 Radiative kinetics

When a luminescent centre is made to be in an excited state, it must eventually dissipate the absorbed energy and return to the ground state. The relaxation process can occur via (a) emission of a photons (radiative relaxation), (b) non-radiative relaxation such as phonon emission to the lattice, or (c) energy transfer to another luminescent centre. Absorbed energy which does not participate in the radiative processes is dissipated to lattice via non-radiative processes.

The rate of spontaneous return from the excited state to the ground state for a simple two-level system is ([27], p. 38)

$$\frac{dN_e}{dt} = -N_e P_r \tag{2.10}$$

where N_e is the number of electrons in the excited state, t is time, and P_r is the probability for spontaneous emission with relaxation from the excited to the ground state, and is in essence the radiative relaxation rate. Solving for N_e one obtains

$$N_e = N_e(0)\exp(-P_r t) = N_e(0)\exp(-t/\tau_r) \tag{2.11}$$

where $\tau_r = P_r^{-1}$ is the radiative decay time, and is defined as the time for the excited population to decrease to $1/e$ (or 37%) of its original value. Transitions which are allowed by selection rules are characterized by short decay times in the nanosecond range, whereas the decay time for forbidden transitions are typically in the millisecond range (or longer). Though the decay time is an important characteristic for many applications, systems with long decay times are not necessarily poor EL phosphors.

The internal radiative quantum efficiency or yield is defined as the fraction of radiative relaxation to the total relaxation,

$$\eta_{iq} = \frac{R_r}{R_r + R_{nr}}, \tag{2.12}$$

where R_r is the radiative relaxation rate, and R_{nr} is the non-radiative relaxation rate. R_{nr} is a function of temperature with the following dependence:

$$R_{nr} = R(0) \exp\left(\frac{-E}{K_B T}\right) \tag{2.13}$$

where T is temperature, $R(0)$ is on the order of the lattice vibrational frequency, E is an activation energy, and K_B is Boltzmann's constant. From equations (2.12) and (2.13) it follows that when T increases the non-radiative rate increases rapidly, and efficiency decreases. This phenomenon is referred to as thermal quenching ([27], ch. 4, and [28]).

2.2.2.2 *Cross-section for impact excitation*

As alluded to earlier, impact excitation is believed to be the primary process involved in the ZnS type electroluminescence [28, 29]. The cross-section for excitation is related to the probability that a high energy electron will impact excite a luminescent centre in the EL phosphor material. As a first-order approximation, it is proportional to the geometric cross-section of the ionized luminescent centre when substituted in the host matrix [30]. Table 2.2 gives the ionic radii of several luminescent centres together with that of Zn for comparison. The average impact length is found from

$$l_I = (\sigma N)^{-1} \tag{2.14}$$

where σ is the impact cross-section in cm^2 and N is the concentration of centres in cm^{-3}.

2.2.2.3 *Radiative transition energy*

In material systems where electroluminescence is generated by injected electron–hole recombination processes (see figure 2.1), at high fields (>1 MV/cm), electron and hole generation due to the ionization of dopants is followed by the spatial separation of charge [30, 34]. That is,

Table 2.2. Ionic radii of Zn and the more common luminescent centres [31–33].

Ion	Radius (Å)	Ion	Radius (Å)
Zn^{2+}	0.74	Mn^{2+}	0.80
Ce^{3+}	1.034	Gd^{3+}	0.938
Pr^{3+}	1.013	Tb^{3+}	0.923
Nd^{3+}	0.995	Dy^{3+}	0.908
Sm^{3+}	0.964	Ho^{3+}	0.894
Eu^{3+}	0.950	Er^{3+}	0.881
Er^{2+}	1.09	Tm^{3+}	0.869

due to the aforementioned high fields, electrons move towards the anode and holes move towards the cathode. This spatial separation reduces the probability of electron–hole recombination and is known as field quenching [30]. It renders efficient cathodoluminescent phosphors ineffective as EL phosphors. Thus, high-field electroluminescence makes use of intershell and intrashell transitions of luminescent ions, which are well shielded against external fields. The emitted photon energy (or colour) is a measure of the energy separation between the initial and final states of the radiative recombination event.

In general, two types of luminescent ion are employed in ZnS electroluminescence: transition metal ions such as Mn^{2+}, and rare-earth (lanthanide) ions mainly as RE^{3+}. Intrashell *d–d* transitions are exploited in transition metal ions, whereas both parity allowed 5*d*–4*f* (e.g. Ce^{3+}) and parity forbidden 4*f*–4*f* (e.g. Tb^{3+}) occur in rare-earth ions. Theoretically, parity forbidden transitions have small probabilities and based on equation (2.11)

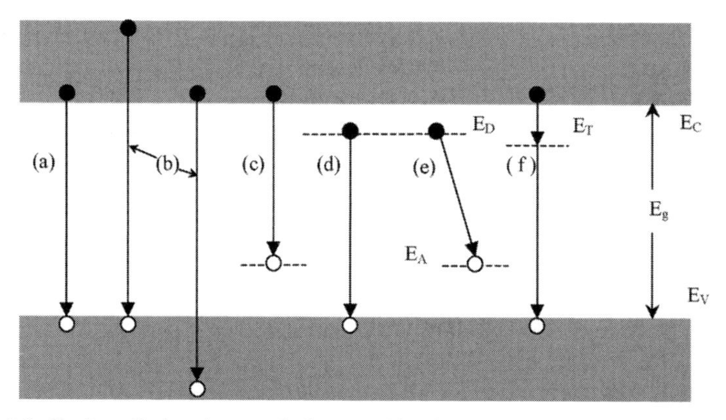

Figure 2.1. Basic radiative electron–hole recombination transitions. ●, electrons; ○, holes. E_C: conduction-band edge; E_V: valence-band edge; E_G: band-gap; E_D: donor level; E_A: acceptor level; E_T: trap level [35].

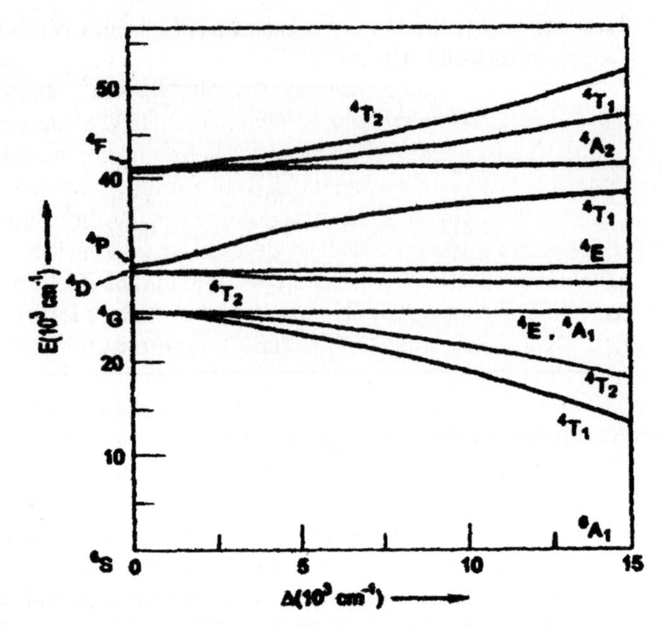

Figure 2.2. Energy level for Mn^{2+} as a function of octahedral crystal field [27].

are usually long-lived. Yet yellow emitting ZnS:Mn is the brightest (>600 candelas/m^2 at 60 Hz, [36]) high-field EL phosphor, and ZnS:Tb with a brightness greater than 120 candelas/m^2 at 60 Hz [37] is the best green emitting phosphor. Thus, as alluded to earlier, selection rules are not absolute and material systems with long decay times are not necessarily poor EL phosphors.

ZnS:Mn is the best understood and most commercialized high-field electroluminescence material system. In general, transition metal ions have an incompletely filled d shell (d^n, $0 < n < 10$). Specifically, manganese as Mn^{2+} has five electrons in its 3d shell. At fields of about 1.5 MV/cm, ballistic conduction electron impact excites one of the five d electrons into an excited state within the same ion. Radiative relaxation to the ground state is responsible for the yellow (585 nm) EL emission of ZnS:Mn. The nature of the excited state of Mn^{2+} has been well characterized by Busse and co-workers [38, 39], and figure 2.2 depicts the Mn^{2+} energy level diagram. The free ion energy levels are on the left of this figure. The notation to the right of figure 2.2 are crystal-field (the electric field at the site of the ion under consideration due to its surroundings) levels marked as $^{2S+1}$X, where X may be A (no degeneracy), E (two-fold degeneracy) or T (three-fold degeneracy), and S is the total spin quantum number. The subscripts indicate certain symmetry properties [27]. The abscissa of figure 2.2, Δ, indicates the value of the crystal field.

The yellow emission of the Mn^{2+} ion is a $^4T_1(3d) \rightarrow {}^6A_1(3d)$ transition. Note that $3d$–$3d$ transitions are parity forbidden because the principal quantum number does not change and, as a result, the Mn^{2+} excited state is long-lived ($\sim$1.5 ms). However, parity selection rules are relaxed if the luminescent centre is placed in a non-centrosymmetric lattice site; that is, a lattice site without a centre of symmetry or inversion [40]. Indeed in ZnS, Mn^{2+} and other luminescent ions substitute directly on to Zn^{2+} lattice sites that exhibit tetrahedral symmetry. Tetrahedral lattice sites lack a centre of symmetry, as can be ascertained from inspection of the ZnS zincblende crystal structure (see section 2.2.3). The parity selection rule is also relaxed by the coupling of electronic transitions to vibrational transitions of suitable symmetry [40].

By considering the electron configuration of Mn^{2+} ($1s^2 2s^2 2p^6 3s^2 3p^6 3d^5 4s^2$) and the Aufbau building-up principle [41, p. 367], it is clear that the $3d$ orbitals are not shielded from the host lattice by any occupied orbitals. Thus, the energy difference between the excited state and ground state is crystal-field dependent (see figure 2.2), and can therefore be modified by the host lattice. The details surrounding luminescent centre–ZnS host interactions are more fully addressed in section 2.2.4. The width of the ZnS:Mn emission band (figure 2.3) is due to electron–lattice coupling and

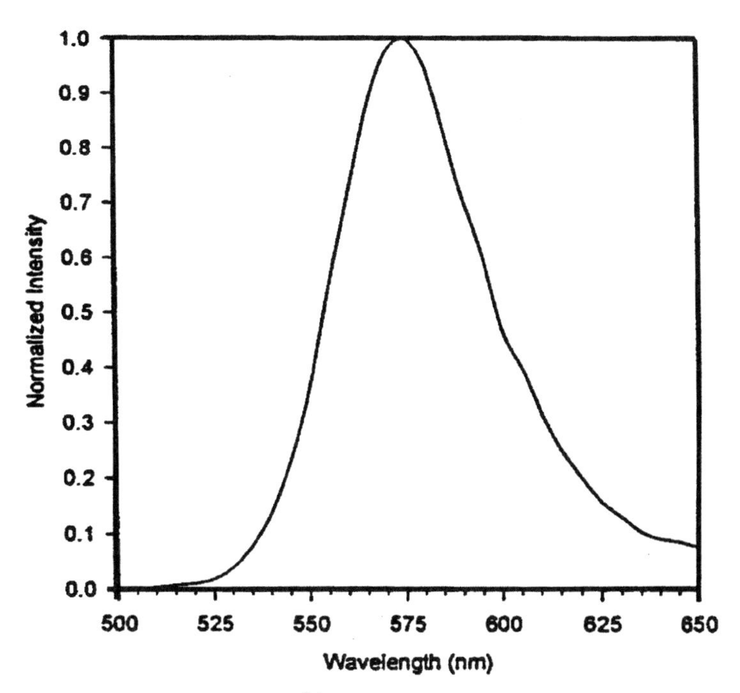

Figure 2.3. EL spectrum for ZnS:Mn^{2+}.

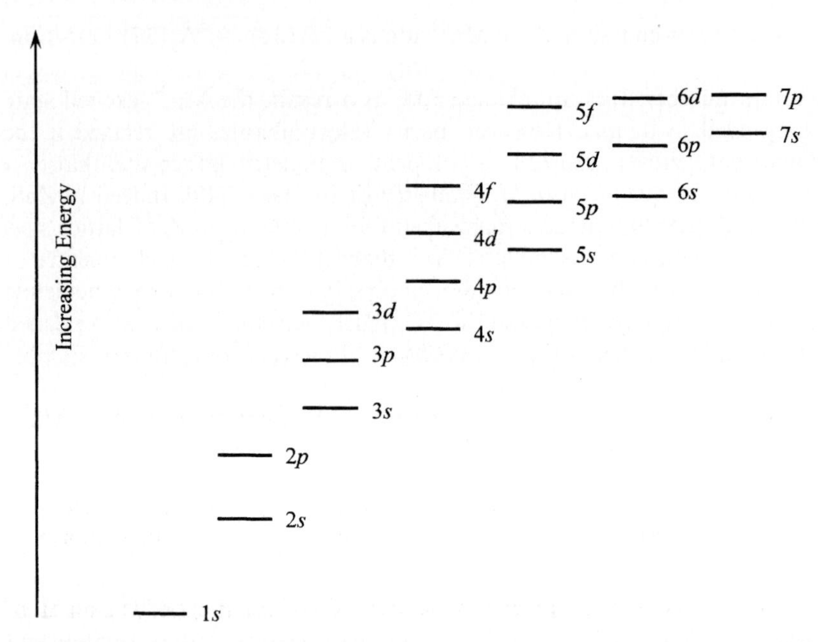

Figure 2.4. The relative energy levels of the atomic orbitals in multi-electron atoms [42]. Orbitals fill up in order of increasing energy as shown.

is described well by Huang–Rhys theory ([27], ch. 3). We shall return to the discussion of the important ZnS:Mn phosphor system in later sections.

Rare-earth (RE) or lanthanide ions are the second type of luminescent centres that are employed in zinc sulphide phosphors. These are typically used as RE^{3+}, which in general, except for Ce^{3+}, exhibit parity forbidden, long-lived $4f$–$4f$ transitions. Rare-earth ions have an electron configuration of $1s^2 2s^2 2p^6 3s^2 3p^6 4s^2 3d^{10} 4p^6 5s^2 4d^{10} 5p^6 6s^2 4f^x$, and the energy levels of the atomic orbitals fill up in order of increasing energy, as is shown in figure 2.4. Thus, the $5s$, $5p$ and $6s$ orbitals fill up before the $4f$ orbital, even though the radial probability distribution of the $4f$ orbital has its maximum closer to the nucleus. As a result, the emission spectra of transitions which arise from the $4f^n$ configuration appear as narrow lines because the $4f$ electrons are well shielded from their environment by the $5s$ and $5p$ electrons. Figure 2.5 is a schematic representation of this shielding effect in RE^{3+} lanthanide ions.

In cases where emission is a result of $5d$–$4f$ transitions (e.g. Ce^{3+} and Eu^{2+}) the emission spectra are broader because the $5d$ electrons are unshielded and are therefore influenced by their surroundings, i.e. the excited state is influenced by the crystal field, the nephelauxetic effect and other luminescent centre–host lattice interactions. Since $5d$–$4f$ transitions are parity allowed, they are faster than the parity forbidden $4f$–$4f$ transitions. The book by

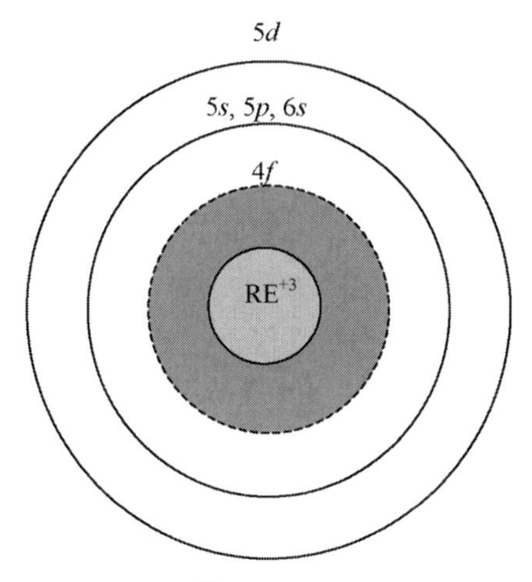

Figure 2.5. Schematic diagram of a RE^{3+} ion showing the radial position of the outer orbitals. Note that 4*f* electrons are shielded, but 5*d* electrons are exposed to the environment.

Blasse and Grabmaier [27] provides excellent details of these effects on luminescent ions.

Figure 2.6 shows the partial energy level diagram for several rare-earth ions. Among these, green emitting Tb^{3+} is the most efficient in ZnS. Red emitting ZnS:Sm,Cl and blue emitting ZnS:Tm,F are currently being developed. We shall elaborate on these and other ZnS phosphors in the device section of this chapter. Many lanthanide ions exhibit transitions in the near, mid and far infrared regions of the electromagnetic spectrum. The infrared properties of rare-earth ions are also being actively researched for several applications including optical telecommunications, opto-electronic integrated circuitry, lasers and sensors, to name a few [43].

2.2.3 Zinc sulphide—the host

As mentioned earlier all phosphors consist of a host material, and a luminescent centre or light-emitting dopant. Before proceeding with a detailed look at the ZnS host system, it is worth reviewing the properties that are desirable of all high-field electroluminescence hosts.

1. The band-gap must be sufficiently large, such that light emitted from the luminescent centre will not be absorbed (the host should be transparent in the range of the electromagnetic spectrum that is of interest). Complete visible transmission requires a band-gap of at least $\sim$3.1 eV (400 nm). The equation for converting between energy, wavelength, and wavenumbers is

$$\lambda = 1240/E, \quad \lambda = 1/\text{Å} \qquad \text{where } \lambda = [\text{nm}], E = [\text{eV}]. \tag{2.15}$$

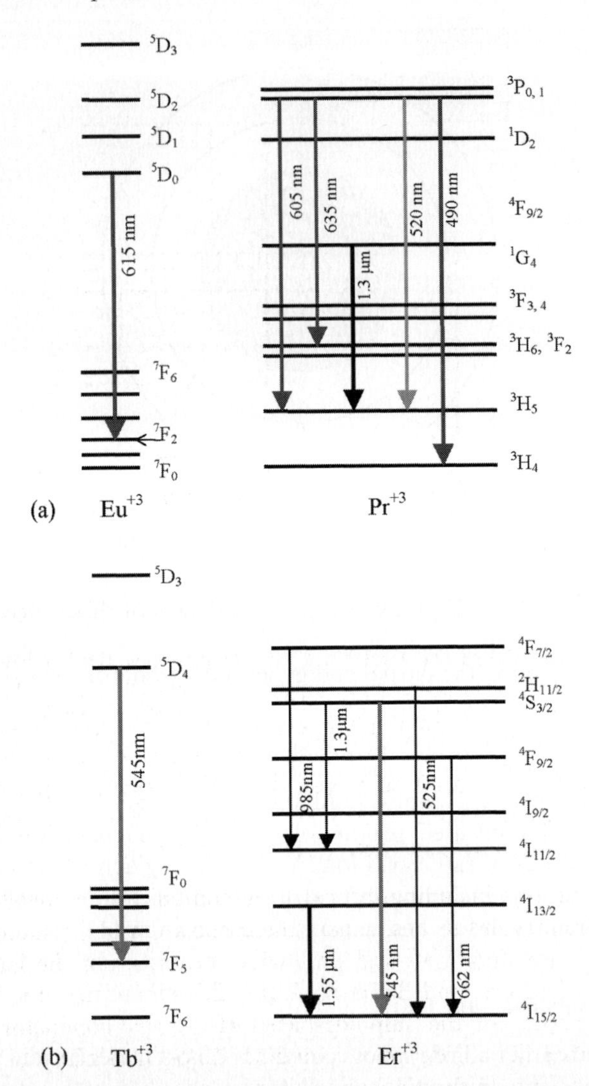

Figure 2.6. Partial schematic energy level diagram for selected rare-earth ions which emit (a) red, (b) green and (c) blue light. Note that several emit in the near infrared as well. Pr^{3+}, Dy^{3+} and Tb^{3+} show emissions in the mid-infrared [44]. Ho^{3+} emits at wavelengths ranging from 550 nm to 3.9 μm.

2. Hosts must be non-conducting below the EL threshold. This is required for a voltage drop and subsequent high electrical field across the phosphor, and leads to capacitive sub-threshold behaviour.
3. The host must have a high breakdown strength to allow for efficient acceleration of electrons. The breakdown field of the host must be at least 1 MV/cm.

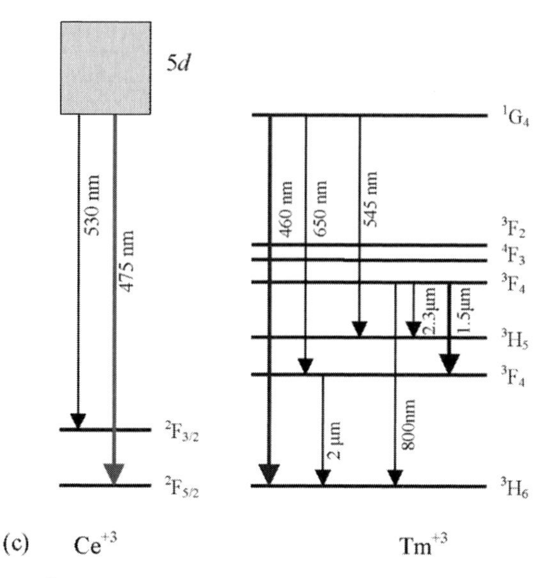

Figure 2.6. *Continued*

4. The host must have the best possible crystallinity and a low phonon-coupling constant in order to minimize electron scattering. To maintain good crystallinity at the dopant levels (~1 at%) typically used in EL phosphors, it is desirable to match the geometric size as well as the valence of the host cation with that of the luminescent centre [45, 46].
5. The host must provide a substitutional lattice site of appropriate symmetry for the luminescent centre (see section 2.2.4).

Sulphide-based II–VI compound semiconductors and thiogallates have empirically been found to best satisfy these requirements. Examples of the former group are ZnS, CaS and SrS, while examples of the latter group are $CaGa_2S_4$, $SrGa_2S_4$ and $BaGa_2S_4$. Table 2.3 summarizes the important physical properties of the sulphide-based II–VI semiconductor hosts. It should be noted that oxide hosts including Zn_2SiO_4, $ZnGa_2O_4$, Zn_2GeO_4 and Ga_2O_3 are also of practical importance.

2.2.3.1 Physical properties of ZnS

ZnS is a IIb–VIb compound semiconductor. Figure 2.7 is a schematic band diagram of this material [47]. As can be observed, the maximum of the valence band lines up in **k** (momentum) space with the minimum of the conduction band at the Γ point. Thus, ZnS is a direct band-gap semiconductor, i.e. momentum is conserved during transitions between the valence and conduction bands, without the participation of phonons. The 3.8 eV band gap is wide enough to be transparent to the whole visible spectrum, yet is not so

Table 2.3. Physical properties of ZnS, CaS and SrS [45, 46, 48, 52–56].

Property	ZnS (IIb–VIb)	CaS (IIa–VIb)	SrS (IIa–VIb)
Melting point (°C)	1800–1990	2400	>2000
Band gap	3.6	4.4	4.3
Transition type	Direct	Indirect	Indirect
Crystal structure	Zincblende/hexagonal	Rock salt	Rock salt
Lattice constant (Å)	5.409	5.697	6.019
Dielectric constant (static)	8.3	9.3	9.4
Dielectric constant (high frequency)	3.6		
Ionicity	0.623	>0.785	>0.785
Impact ionization threshold energy (eV)	3.60*		
Heat of formation (kJ/mol)	477 (Zincblende), −206 (Hexagonal)		
Electron effective mass, m^* for the various valleys (see figure 2.7)	$0.27m$ (Γ) $0.222m$ (L) $1.454m$ (X)		
Electron mobility ($cm^2/V\,s$)	<180		
Hole mobility ($cm^2/V\,s$)	5		
Polar optical-phonon energy (eV)	0.044 (357 cm^{-1})		
Frolich's constant, α, electron-optical phonon	0.6		

* This value was given by [48] and [49]. Other sources [50, 51] give this value as $\frac{3}{2}E_g$, or ~5.4 eV.

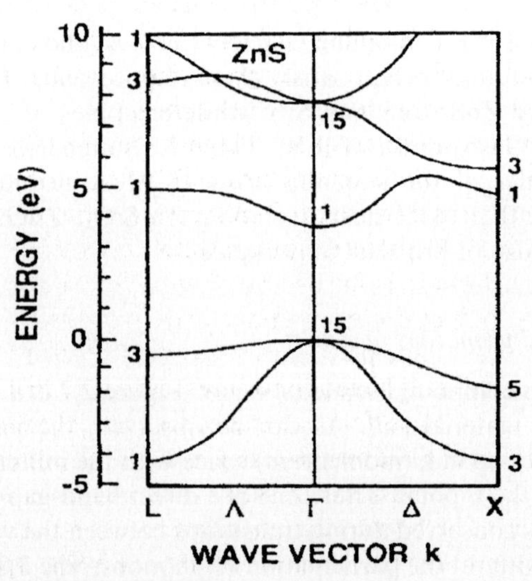

Figure 2.7. Band structure of ZnS [47].

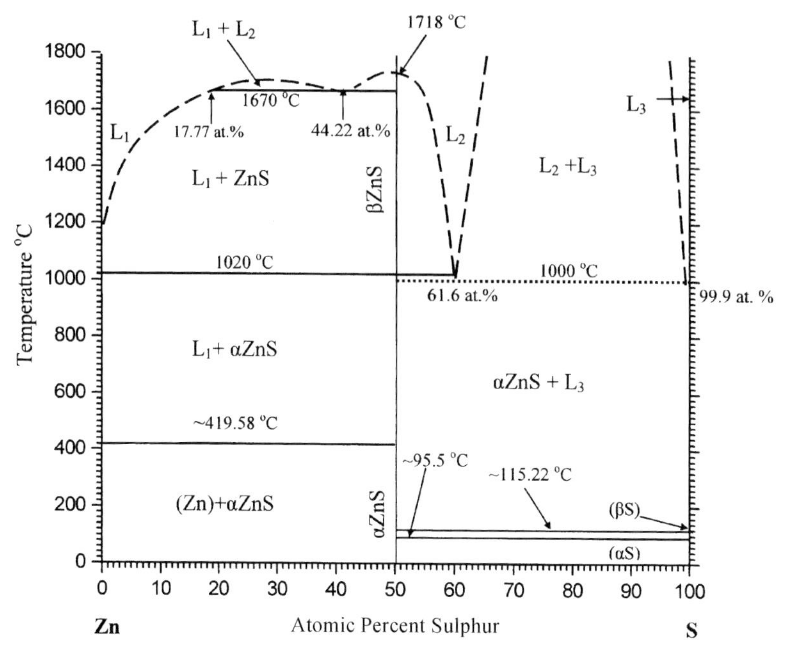

Figure 2.8. Binary phase diagram for the Zn–S system.

large that it limits the electron multiplication process and restricts the number of electrons available for impact excitation of the activator. As the reader may ascertain from table 2.3, zinc sulphide has a small effective electron mass, a suitably large dielectric constant, low phonon energy and a small electron–phonon coupling constant. The combination of the 3.8 eV band gap and the electrical properties given above, makes ZnS an excellent host for high-field electroluminescence phosphors.

Figure 2.8 shows the binary phase diagram of ZnS [57]. It has the cubic zincblende (sphalerite, αZnS) crystal structure when processed at low temperatures or the hexagonal (wurtzite, βZnS) crystal structure when processed at high temperatures. At room temperature the zincblende phase is stable, while the hexagonal phase is metastable. Figure 2.9 shows the two crystal structures of ZnS. A unit cell of the zincblende structure (figure 2.9a) is an FCC lattice with two atoms per lattice site: one at each FCC position and another at $a_0\sqrt{\frac{3}{4}}$ in the $[11\bar{1}]$ direction with respect to the first lattice position. It is an FCC unit cell with half of its tetrahedral sites occupied. Here sulphur is in the face and corner positions, and zinc occupies the interior, tetrahedral positions. An equivalent structure results if the positions of the S and Zn atoms are reversed. The tetrahedral coordination of the Zn atoms is more clearly shown in figure 2.9(b), and is an important point. Specifically, tetrahedral sites lack a centre of symmetry or inversion, and this affords the

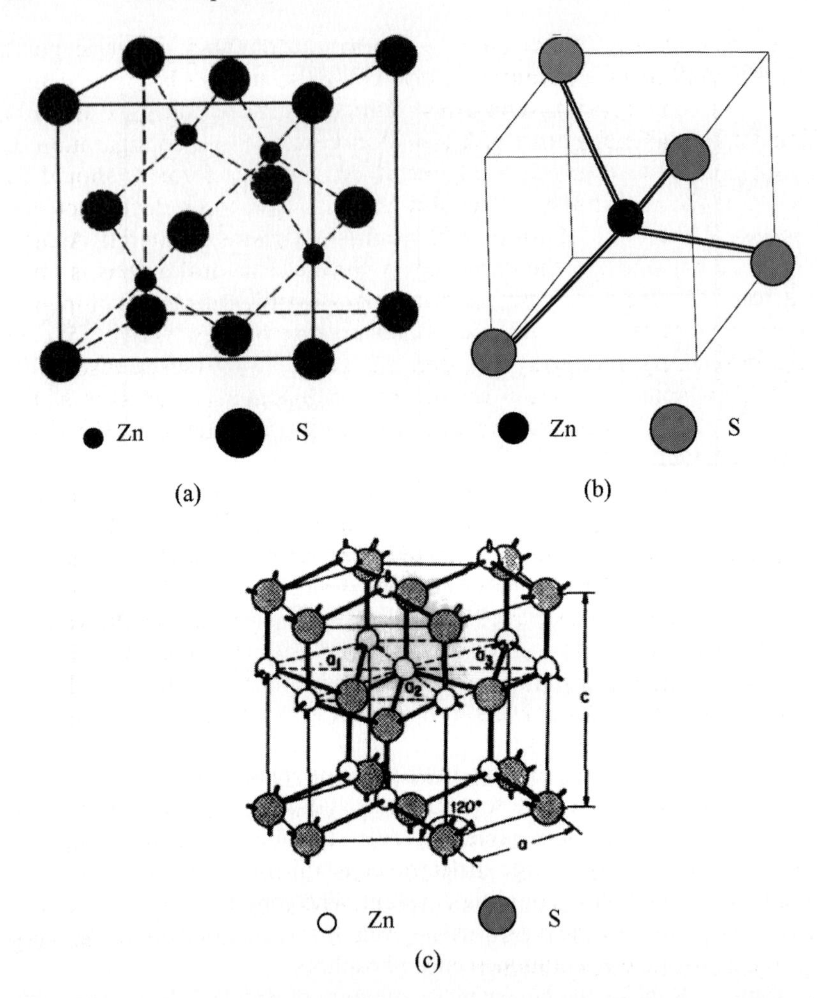

Figure 2.9. (a) Zincblende unit cell for ZnS and (b) tetrahedral coordination of Zn lattice sites. Notice the lack of a centre of symmetry or inversion. (c) Hexagonal unit cell for ZnS.

relaxation of parity selection rules for *d–d* and *f–f* intraband transitions of luminescent ions, when such ions are substituted on to Zn lattice sites (see section 2.2.4).

2.2.3.2 Point defects in ZnS

I. Intrinsic point defects. Point defects in ZnS are either intrinsic, i.e. associated with the Zn or S sublattices, or extrinsic, i.e. associated with an impurity or intentional dopant. Defects are widely believed to influence the optical and electrical properties of semiconductors, and thus their nature and concentration is important. As a binary compound, intrinsic point defects in ZnS include vacancies and interstitials on or in both the cation (Zn^{2+}) and anion (S^{2-}) sublattices. Anti-site defects, such as when a sulphur atom takes the place of a zinc atom resulting in a configuration of sulphur bonded to four other sulphur atoms, can also form. It should be noted that the equilibrium defect concentration and types depend on the deposition technique, deposition conditions and substrate material. Assignment of energy levels in the band gap to specific structural defects is often nontrivial [58], and the electronic properties of point defects in compound semiconductors are not fully understood because of their variety [59]. In addition to being electrically and optically active, point defects also affect the diffusion characteristics of the material. Self-compensation such as the pairing of cation vacancies with donors is common throughout the II–VI compounds [60].

The doubly ionized Zn/cation vacancy (V_{Zn}^{2+}) is the most well characterized point defect in ZnS, and is reported to create a defect level $\sim$1.0–1.1 eV above the ZnS valence band edge [61–63]. This value is close to the hole-trap depth of 1.2 eV reported by Hitt *et al* [50]. A singly ionized V_{Zn}^{+} vacancy is reported to create a defect level in the range 0.2–0.6 eV above the valence band, depending on the author [62, 63]. An estimated concentration of $7 \times 10^{16}\,cm^{-1}$ hole traps tentatively attributed to zinc vacancies and their complexes has been determined by electrical characterization of ZnS:Mn a.c. thin-film EL devices [64].

Less is known about the electrical characteristics of anion vacancies (also known as F-centres [58]) in II–VI compounds [61, 65]. Impurity absorption and photoconductivity measurements conducted by Georgobiani *et al* [62] on chlorine-doped ZnS revealed donor levels at 0.515, 0.615 and 1.08 eV below the conduction-band edge, though these were not assigned to any specific structural defect. Figure 2.10 is a representation of the deep levels reported by Georgobiani.

II. Extrinsic point defects. The second type of point defects in ZnS are extrinsic point defects, or those that result from the presence of impurities or intentional dopants in the lattice. Much of the interest in extrinsic defect levels in ZnS and other II–IV compounds is related to the doping of these materials, *n* and *p* type, in order to exploit electron–hole injection and subsequent radiative recombination (injection type electroluminescence) [60, 66–70]. However, to date *p*-type ZnS has not been demonstrated [69] and, as a result, little has been reported regarding defect levels due to extrinsic acceptors in ZnS.

As discussed earlier, many of the rare-earth ions are employed in their 3+ oxidation state (see subsection 2.2.2.3 and table 2.2), and it is desirable that the luminescent ion substitutes on to tetrahedrally coordinated Zn lattice sites. The valence mismatch between Zn^{2+} and RE^{3+} is typically

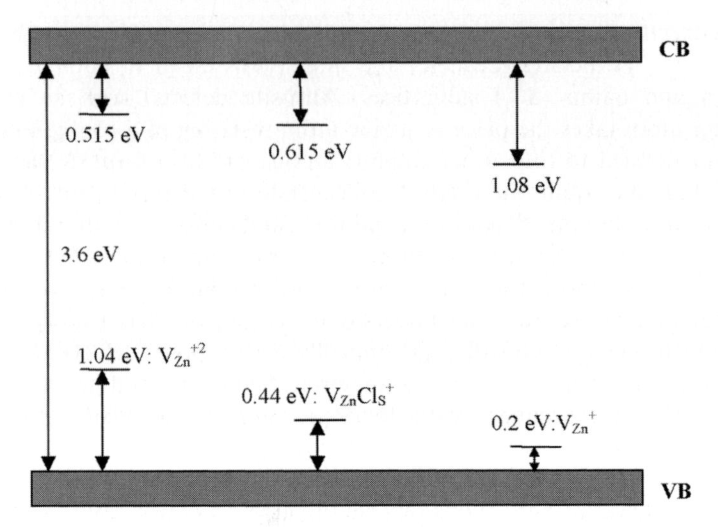

Figure 2.10. Schematic band-gap diagram of ZnS:Cl showing deep defect levels [62].

neutralized by co-doping with charge compensators such as chlorine and fluorine. Chlorine as an anion substitute in ZnS has been shown to create a donor level at 0.25 eV below the conduction band by photoconductivity measurements [71]. Georgobiani *et al* [62] have reported that, in addition to the Cl donor level that results during ZnS:Cl single crystal growth, an ionized acceptor level appears from a self-compensation mechanism. These acceptor levels are thought to be associated with induced, compensating isolated zinc vacancies.

V_{Zn}–Cl_S complexes, also know as A-centres, are also thought to exist [61–63] in ZnS. Vlasenko *et al* [63] and Lewis *et al* [36] assigned a level at 1.8 eV below the conduction band to the V_{Zn}–Cl_S complex, whereas Neumark [58] assigned a level 0.95 eV below the conduction band to the same structural defect. In contrast, Georgobiani *et al* [62] assigned an acceptor level at ~0.44 eV above the valence band to the V_{Zn}–Cl_S complex.

The foregoing shows that the topic of point defects in ZnS (and compound semiconductors in general) is a complex one, and there is ongoing active debate about the assignment of defect levels in the band gap to specific structural defects. The problem is compounded by the variety of defects, and more scientific data on the subject would be welcomed. Two-dimensional defects in ZnS including dislocations and stacking faults have been reviewed by Zhai [72].

2.2.4 Luminescent centre–host lattice interactions

The electrons of isolated (i.e. uninfluenced by any electric or magnetic fields) luminescent ions arrange themselves in a manner such that the total energy of the ion is minimized. However, when an electric or magnetic field is present, the arrangement of the energy levels of the previously isolated ion is perturbed. Core-level electrons are tightly bound to the nucleus, and in addition are shielded by outer shell electrons. These two factors combine to negate the influence of external field on core-level electrons. Outer shell electrons, however, are not as tightly bound to the nucleus and are unshielded. As a result, the energy levels associated with outer shells can be significantly influenced by external fields. This is particularly true for ions which exhibit *d–d* and *d–f* radiative transitions. Thus, while high-field EL exploits radiative transitions localized at the luminescent centre, the position of the final energy levels associated with the light emission from transition metal and some lanthanide activators is influenced by the host.

The outer energy levels of ions, bound together to form a solid, are significantly influenced by their surroundings. In an ideal ionic solid, cations and anions are treated as positive and negative point charges, and the atomic energy levels are modified according to the coulombic forces which exist between oppositely charged ions. The coulombic forces induce a crystal field in the solid that is a function of symmetry (ion arrangement), bond length, and valence or charge of the constituent ions. Real solids, however, deviate from ideal ionic behaviour because nearest-neighbour electron clouds overlap and intermix to form partially covalent bonds. In addition to crystal field effects, the atomic energy level positions are a function of the covalency/ionicity of the luminescent ion–ligand bond. The shift of emission energy of luminescent centres to lower values with increasing covalency of the host (i.e. reduced electronegativity difference between the constituent ions) of the host is known as the nephelauxetic effect ([27], section 2.2). Rack and Holloway ([73], section 7.1) have treated luminescent centre–host lattice interactions in detail, as have Blasse and Grabmaier [27]. In this section, we summarize the main factors which affect the final energy level positions of the luminescent centre, namely, lattice symmetry, ion spacing (bond length) and molecular orbital overlap between cations and anions [74, 75].

2.2.4.1 Symmetry

As discussed earlier, transition metal or rare-earth luminescent ions substitute on cation (Zn^{2+}) lattice sites. The arrangement of nearest-neighbour anions with respect to the cation (luminescent centre) site in ZnS is tetrahedral, though octahedral coordination (FCC lattice with two atoms per lattice point: one ion at each FCC position and another at $a_0/2$ in the [100] direction with respect to the first) is observed in some metal sulphide hosts.

The tetrahedral coordination of luminescent ions which exhibit parity-forbidden intraband (*f–f* or *d–d*) transitions in ZnS is an important point

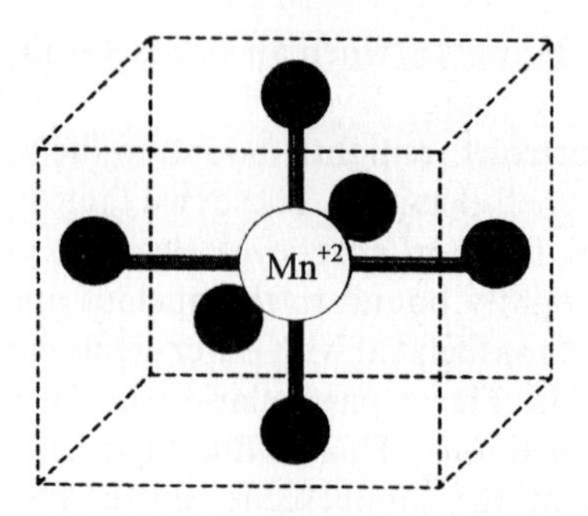

Figure 2.11. Octahedral coordination of a central luminescent ion.

and deserves elaboration. Tetrahedral lattice sites are lacking a centre of symmetry/inversion as can be seen from examination of figure 2.9(b). When the $4f$–$4f$ rare-earth luminescent centre occupies a lattice site without inversion symmetry, it is in a crystallographic position where uneven components of the crystal field are present ([27], section 2.3.3). The uneven components mix small amounts of opposite-parity wavefunctions (e.g. $5d$) into the $4f$ wavefunctions, which results in the transition moment integral and therefore the oscillator strength having some intensity. There would be no colour television (cathodoluminescence) or energy-saving phosphorescent lamps, for example, without uneven crystal fields. The discussion, though seemingly academic, has serious implications for practical applications. For details the reader is referred to [25, 76–78].

Though the uneven crystal field of non-centrosymmetric host sites has the effect of relaxing the parity selection rule for forbidden $4f$–$4f$ transitions, the energy levels associated with $4f$ electrons are well shielded by the $5s$ and $5p$ orbitals and are essentially uninfluenced by the host. In contrast to $4f$ electrons, $3d$ and $5d$ electrons are unshielded, and the associated energy levels are significantly affected by the host and crystal field. Figure 2.2 shows the emission energy of Mn^{2+} ($3d$–$3d$) as a function of octahedral crystal field. In order to understand the effect of the host on $3d$ and $5d$ electrons, the shape and symmetry of the d orbitals must be considered. Take the $3d$ orbital of Mn^{2+} ions in octahedral coordination (figure 2.11) as an example. The principal quantum number is $n = 3$, and l, the angular momentum quantum number, can assume the values $l = 0, 1, 2$. Its magnitude is defined as $|\boldsymbol{l}| = [l(l+1)]^{1/2}\hbar$. Since l can range from 0 to $n - 1$, there are n subshells for a shell with principal quantum number n. In this case the magnetic quantum number m_l can assume the values $0, \pm 1, \pm 2$. This gives one $3s$ orbital, three $3p$ orbitals and five $3d$ orbitals. In the host, the environment of luminescent ion is no longer spherical as is the case of the isolated ion, and the d orbitals are no longer degenerate. The shapes (probability distributions) of the five d orbitals are shown in figure 2.12. Here, one should imagine Mn^{2+} ions at the centre of axes, octahedrally coordinated by six

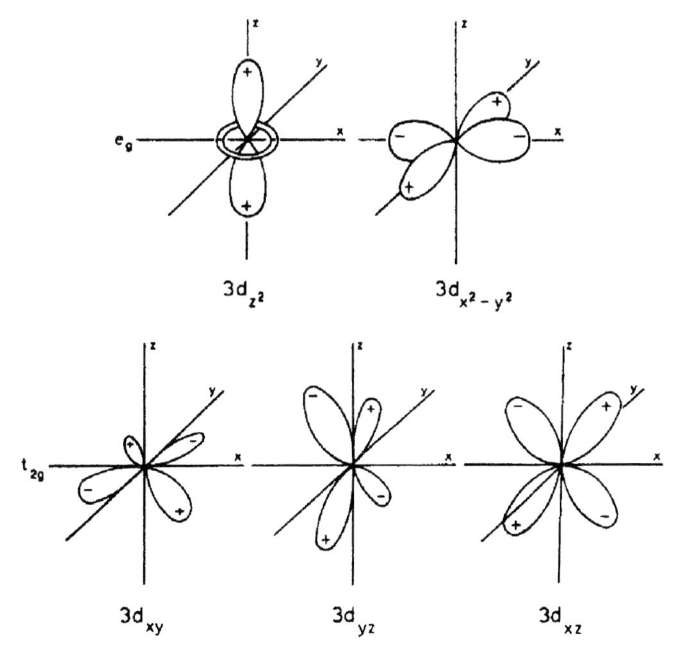

Figure 2.12. The shape and symmetry of the five *d* orbitals of the luminescent centre subject to octahedral coordination in the host ([42], section 6.2).

S^{2-} ligands which lie along the *x*, *y* and *z* axes. The doubly-degenerate *d* orbitals labelled e_g (the $3d_{z^2}$ and $3d_{x^2-y^2}$ components) point at the ligands and an electron in one of these *d* orbitals must come close the ligand's unpaired electrons. The resulting repulsion will raise the energy of the *d* orbital electron. In contrast, the triply-degenerate orbitals labelled t_{2g} point between the ligands, and so the energy of electrons in those orbitals will not be greatly affected by the presence of ligands. The net effect is a lifting of the degeneracy of the 3*d* energy levels, with the e_g levels being higher in energy than the t_{2g} levels, as is shown in figure 2.13. For tetrahedral coordination, the *e* orbitals lie below the t_2 orbitals [41, section 17.3]. The foregoing discussion demonstrates how the host affects the energy levels of unshielded *d* electrons, via crystal field splitting (Δ). In general, the *d* orbital crystal field splitting of tetrahedral symmetry is $\frac{4}{9}$ the magnitude of octahedral symmetry ($\Delta_{tet} = \frac{4}{9}\Delta_{oct}$) [79].

2.2.4.2 *Bond length*

In addition to symmetry of the site occupied by the luminescent ion, the energy level positions of the luminescent centre are also influenced by the bond length or ion spacing of the host material. As discussed earlier, the energy levels associated with 4*f*–4*f* transitions are well shielded by the

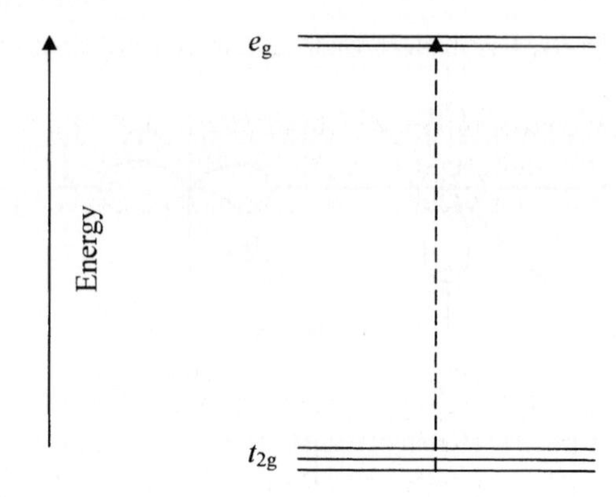

Figure 2.13. Schematic representation of the crystal field splitting (lifting of degeneracy) of octahedrally coordinated, unshielded *d* orbitals.

5*s* and 5*p* orbitals and are essentially uninfluenced by the host. Hence, the discussion here is limited to the influence of bond length on the unshielded *d* orbitals of luminescent centres which exhibit *d*–*d* or *d*–*f* radiative transitions.

The influence of bond length on the *d* orbitals can be easily understood by treating the constituent ions as point charges and considering Coulomb's law, which describes the electrical force that exists between two charged particles. Given two particles with charge q_1 and q_2 respectively, Coulomb's law can be written as

$$F = \frac{q_1 q_2}{4\pi\varepsilon r^2} \tag{2.15}$$

where r is the distance between the point charges, and ε is the dielectric permittivity of the medium between the two charges, i.e. the host. The electrostatic force on a point-charge q, located at a point in space at which the electric field is E, is given by

$$F = qE \tag{2.16}$$

regardless of what arrangement of charges may have established E. From this it follows that the local electric field due to charge q is

$$E_q = \frac{q}{4\pi\varepsilon r^2}. \tag{2.17}$$

Equations (2.15) and (2.17) demonstrate the $1/r^2$ dependence of the electrostatic force and electric field between two charged particles. As the bond length r decreases, the electrostatic force and field increase. The converse is true when r increases. Applying this concept to ions in a solid, the crystal field in the material increases as the bond length or ion spacing decreases, and cations and anions come closer together. Conversely, when r is large, the crystal field decreases and the d orbital energy level states become more like the free ion case.

Using the point charge model which represents ions as point charges and takes into account the radial nature of the d orbitals, Gerloch ([80], pp. 31–36) has shown that the energy level splitting dependence on bond length is ([80], p. 36)

$$D_q = \frac{1}{6} ze^2 \frac{a^4}{r^5} \tag{2.18}$$

where D_q is a measure of the energy level separation (splitting), z is the charge or valence of the anion, e is the electron charge, a is the radius of the d wave-functions, and r is the bond length. Equation (2.18) shows that for transitions which involve outer unshielded d orbitals ($3d$ or $5d$), the final energy level positions and therefore the emission energy can vary depending on the bond length of the host material. The magnitude of the d orbital splitting is $10D_q$ for octahedral symmetry, and $\frac{4}{9} \times 10D_q$ for tetrahedral symmetry.

2.2.4.3 *Molecular orbital (MO) overlap*

The point charge model of ions ignores the true wave nature of its electrons. The quantum mechanical probability of locating an electron in space is given by

$$\int \Psi\Psi^* \, \mathrm{d}\tau = 1 \tag{2.19}$$

where Ψ is the electron wavefunction, Ψ^* is the complex conjugate of the wavefunction, $\mathrm{d}\tau = \mathrm{d}x\,\mathrm{d}y\,\mathrm{d}z$, and integration is performed over all space. The implication of this statement is that when ions combine to form a solid, their molecular orbitals (where the electrons reside) overlap, intermix and hybridize. Thus, in addition to the symmetry and bond length effects of the host, the final energy level positions associated with light emission are influenced by molecular orbital overlap.

Though the quantum mechanics of MO theory is quite involved and beyond the scope of this chapter (the reader interested in molecular orbital theory is referred to [81–83]), we shall proceed with a brief overview of the underlying concept. Simply put, one is required to solve the Schrödinger equation

$$\hat{H}\Psi = E\Psi \tag{2.20}$$

for a multi-particle system. Here $\hat{H}$ is the Hamiltonian operator and E is the total energy of the system. The Hamiltonian reflects all the electron–nuclei (attractive), electron–electron and nucleus–nucleus (repulsive) interactions. It is the sum of kinetic energy operators for the nuclei and the electrons, and the potential energy terms which represent the various coulombic interactions. It must also reflect the influence of external electric or magnetic fields. In short, the Hamiltonian must contain all the interactions which shift, or split the electron energy levels. Therefore, to solve the Schrödinger equation one must (a) set up the Hamiltonian matrix and (b) perform diagonalization of the Hamiltonian matrix to obtain the eigenvalues, or allowed energy level positions.

To demonstrate the complexity of the multi-body problem, consider a closed system with N nuclei and n electrons. One is then required to solve

$$\hat{H}^{\text{total}}\Psi_i = E_i\Psi_i \tag{2.21}$$

where Ψ_i are the eigenfunctions and E_i are the eigenvalues. In this case the many-particle Hamiltonian operator is [81]

$$\begin{aligned} &\hat{H}^{\text{total}}(1,2,3,\ldots,N;1,2,3,\ldots,n) \\ &\quad = -\frac{h^2}{8\pi^2}\sum_{\text{A}}^{N} M_{\text{A}}^{-1}\nabla_{\text{A}}^2 + \sum_{\text{A}<\text{B}} e^2 Z_{\text{A}} Z_{\text{B}} r_{\text{AB}}^{-1} \\ &\qquad - \frac{h^2}{8\pi^2 m}\sum_{p}^{n}\nabla_p^2 - \sum_{\text{A}}\sum_{p} e^2 Z_{\text{A}} r_{\text{A}p}^{-1} + \sum_{p<q} e^2 r_{pq}^{-1} \end{aligned} \tag{2.22}$$

where M_{A} is the mass of nucleus A, m and e are the electronic mass and charge, $Z_{\text{A}}e$ is the charge on nucleus A, r_{ij} is the distance between particles, $\nabla^2 = \partial/\partial x^2 + \partial/\partial y^2 + \partial/\partial z^2$ and summation involving indices A and B are over the atomic nuclei, while those involving p and q are over electrons. As usual the $-(h^2/8\pi^2 m)\nabla^2$ terms refer to kinetic energy and the $e^2 Zr^{-1}$ terms refer to potential energy. The Hamiltonian becomes even more complex if external magnetic or electric fields are present. The Schrödinger equation for the entire system is thus

$$\begin{aligned} &\hat{H}^{\text{total}}(1,2,3,\ldots,N;1,2,3,\ldots,n)\Psi(1,2,3,\ldots,N;1,2,3,\ldots,n) \\ &\quad = E\Psi(1,2,3,\ldots,N;1,2,3,\ldots,n) \end{aligned} \tag{2.23}$$

where Ψ is a complete wavefunction for all of the particles in the system, and E is the total energy. The resulting large number of equations is almost impossible to solve simultaneously and *ab initio* or first principles calculations require enormous computing resources. When one considers that each nucleus or electron is characterized by three cartesian coordinates, the situation becomes even more complex because equation (2.23) becomes a partial differential equation with an additional $3N + 3n$ variables.

More often, semi-empirical techniques are used by theorists involved with electronic structure calculations based on MO theory. This approach makes use of empirical data in conjunction with a less rigorous theoretical treatment. With regards to the latter, a simplified Hamiltonian operator is used. Simplification of the Hamiltonian is facilitated by the Born–Oppenheimer approximation [84] which in essence states that due to the large mass of nuclei compared with that of electrons, nuclei are essentially immobile (too heavy to respond quickly enough) during the electron charge redistribution that characterizes electronic transitions in ions. This means that elements of the Hamiltonian which relate to the momenta of nuclei can be negated and the effective Hamiltonian is essentially the *electronic* Hamiltonian operator:

$$\hat{H}^{\text{total}}(1, 2, 3, \ldots, N; 1, 2, 3, \ldots, n)$$

$$\cong \hat{H}^{\text{electronic}} = -\frac{h^2}{8\pi^2 m}\sum_{p}^{n} \nabla_p^2 - \sum_{\text{A}}\sum_{p} e^2 Z_{\text{A}} r_{\text{A}p}^{-1} + \sum_{p,q} e^2 r_{pq}^{-1}. \quad (2.24)$$

Though the Born–Oppenheimer approximation simplifies the Hamiltonian, analytical solution to the Schrödinger equation even by semi-empirical methods is troublesome because of the many electron–electron interaction terms: $\sum_{p,q} e^2 r_{pq}^{-1}$ where r_{pq} is the separation between electrons p and q, and summation is over all pairs of electrons. This problem is reduced by neglecting some of the overlap integrals characterizing the electron–electron interactions. Depending on the degree to which those overlap integrals are neglected the approach is known as complete neglect of differential overlap (CNDO), or intermediate neglect of differential overlap (INDO).

Currently, the numerical solutions for eigenfunctions and eigenvalues obtained by the semi-empirical computational technique introduced by Hartree [85] and later modified by Vladimir Fock [86] are widely accepted as reliable. The following is a synopsis of how the technique works. Consider a $3d$ electron: the Schrödinger equation with the Hamiltonian given by equation (2.25) can be written as

$$-\frac{h^2}{8\pi^2 m}\nabla^2 \Psi_{3d} - Ze^2 \Psi_{3d} + V_{pq}\Psi_{3d} = E\Psi_{3d} \quad (2.25)$$

where V_{pq} is an average electron–electron repulsion term and is dependent on the wavefunctions of all the other electrons. Equation (2.25) may be solved for Ψ_{3d} by numerical integration, and the obtained solution will be different from that originally guessed. The process is then repeated for another orbital such as a $4s$, using the improved electron–electron repulsion term obtained from the Ψ_{3d} calculation. This procedure is repeated for other orbitals, each time using the improved orbitals found at the earlier stage. The whole procedure is repeated using the improved orbitals, and a second set of improved orbitals is obtained. The recycling is continued until the orbitals and energies obtained are the same as those used at the beginning of the latest cycle. Such solutions are termed *self-consistent*, and are accepted as solutions of the problem. Due to the enormous advances of computer technology, several companies now offer software packages for use with ordinary desktop PCs that calculate the electronic structure of small clusters based on MO theory.

The type of bonding (ionic versus covalent) is a measure of the degree of MO overlap in materials. Increasing covalency means that the electronegativity difference between constituent ions is decreasing. An estimation of the fraction of covalency for a given bond is expressed by ([87], p. 43)

$$f_c = \exp[(-0.25)\Delta EN^2] \tag{2.26}$$

where ΔEN is the electronegativity difference between the anion and cation. The ionicity is

$$f_i = 1 - f_c. \tag{2.27}$$

When bonding is primarily ionic, MO overlap is small. If the bond is primarily covalent, MO overlap is large because the electrons spread out over wider orbitals. This results in a decreased interaction between electrons. Electronic transitions between energy levels strongly influenced by this type of interaction shift to lower energy for increasing covalency of the host. The red shift in emission energy associated with increasing covalency of the host is known as the nephelauxetic effect ([27], section 2.2).

2.3 Zinc sulphide electroluminescent devices

Section 2.2 was a summary of the quantum mechanical concepts applicable to all EL phosphors including ZnS doped with transition metal or rare-earth ions. In this section, we examine the various device structures and fundamental physical mechanisms that lead to electroluminescence in ZnS phosphors. With regards to the basic physical mechanisms that lead to light emission, we are concerned with:

I. the injection of electrical charge into conduction band of the ZnS phosphor layer,
II. high-field transport of electrical charge across the phosphor layer which facilitates acceleration of a fraction of the injected electrons to ballistic energies,
III. impact excitation efficiency, which governs how many luminescent ions have electrons raised from the ground into an excited state via inter- or intraband transitions due to impact excitation, and
IV. radiative decay, resulting in light emission, i.e. electroluminescence.

Radiative decay and the related basic quantum mechanical underpinnings and phenomena were discussed in section 2.2.

Though only a.c. thin-film and powder devices are commercially produced, we shall review all types of ZnS EL devices, namely d.c. thin-film EL (TFEL) devices, a.c. thin-film EL (ACTFEL) devices, a.c. powder EL devices, and d.c. powder EL devices. We begin with d.c. thin-film EL (TFEL) devices because they have the simplest structure, and provide the clearest picture of electrical conduction in ZnS phosphor films. The concepts introduced here are applicable with minor modifications to the more complex, a.c. thin-film structures.

2.3.1 Direct current thin-film devices

Direct current (d.c.) electroluminescence in thin-film ZnS:Mn was described as early as 1960 by Vlasenko and Popkov [4]. The research of these authors [4–7] was centred on exposing the physics of the EL mechanisms, as is most research dealing with d.c. high-field electroluminescence. Though d.c. electroluminescent devices provide useful information about the electrical characteristics of EL phosphors, they tend to fail catastrophically after short operation times.

A cross-sectional view of the d.c.-driven device is given in figure 2.14. The transparent conductor is typically indium/tin/oxide (ITO), which is a degenerately doped n-type semiconductor. Thus, for all practical purposes the d.c. device is a metal/insulator/metal (MIM) structure, with the ZnS phosphor (insulator) sandwiched between two metal layers.

The bulk resistivity (ρ) of ZnS is $\geq 10^{13}\ \Omega$ cm [88], which is not surprising when one remembers that the material is a wide band-gap, undoped semiconductor. Thus, at low applied fields, the d.c.-driven device is a leaky capacitor (i.e. a perfect capacitor connected in parallel with a resistor). In the low field regime electron transport is linear with a drift velocity:

$$v_{\mathrm{d}} = \mu F \tag{2.28}$$

where μ is the electron mobility and F is the applied field. The mobility is primarily limited by scattering with polar optical phonons [55]. Zinc sulphide phosphors abound with defects and impurities, and as a result the mobility is low at $\sim 60\ \mathrm{cm^2/V\,s^{-1}}$. The zero-field free carrier density (n) can be estimated from the bulk conductivity as

$$\sigma = en\mu, \tag{2.29}$$

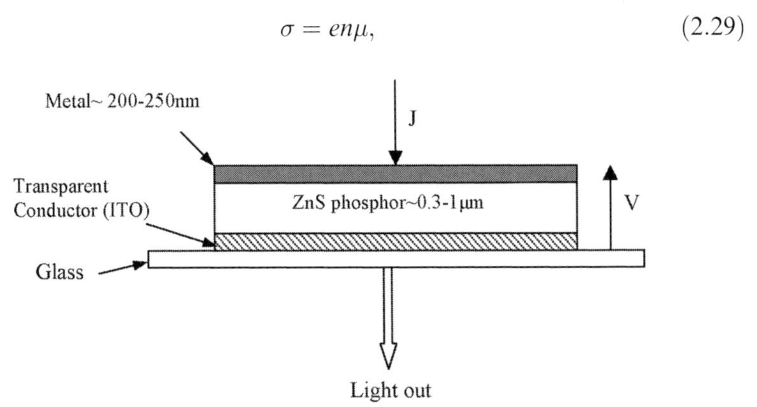

Figure 2.14. Cross-section of the d.c.-driven MIM thin-film electroluminescent device.

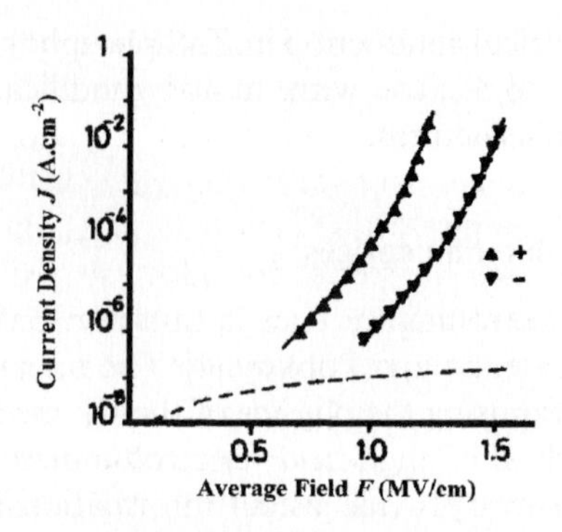

Figure 2.15. Current–voltage characteristic of a d.c.-driven ZnS MIM EL device at 300 K, for both polarities [88]. The dashed line shows Ohm's law $J = (10^{-13}\,\Omega^{-1}\,\text{cm}^{-1})F$.

giving a value of $n \approx 10^4\,\text{cm}^{-3}$, where $e = 1.602 \times 10^{-19}\,\text{C}$ $(\sigma = \rho^{-1} = 10^{-13}\,\Omega^{-1}\,\text{cm}^{-1})$. The free-carrier density is related to the current density via

$$J = env_{\text{d}} \tag{2.30}$$

and a plot of the current density (J) versus applied field should yield a slope of $10^{-13}\,\Omega^{-1}\,\text{cm}^{-2}$ in accordance with Ohm's law:

$$J = \sigma F. \tag{2.31}$$

This low-field, linear transport regime is represented by the region up to ~0.2 MV/cm on the dashed curve of figure 2.15. The sub-linear J–F character of this curve (Ohm's law) at fields greater than 0.2 MV/cm is due to electron velocity saturation ($v_{\text{s}} = 10^7$ cm/s) at high fields. One may estimate the maximum free-carrier concentration at high fields by substituting v_{s} for v_{d} in equation (2.30), and assuming $J = 1\,\text{A}\,\text{cm}^{-2}$. In that case one obtains $n_{\text{max}} = 10^{12}\,\text{cm}^{-3}$.

Inspection of figure 2.15 reveals that the J–F characteristic of d.c. MIM ZnS devices is actually *super-linear* in the high field region. The passage of such large current densities by an 'insulator' is due to a field-dependent carrier density, and quantum mechanical or Fowler–Nordheim [89] tunnelling.

Before proceeding with a discussion of Fowler–Nordheim tunnelling proper, a few more comments on the electrical properties of ZnS are necessary. The zero field carrier density ($n \sim 10^4\,\text{cm}^{-3}$) can be used to determine the position of the Fermi level relative to the conduction band edge, from

$$n = N_{\text{c}} \exp[-(E_{\text{c}} - E_{\text{F}})/K_{\text{B}}T] \tag{2.32}$$

where N_{c} is the effective density of conduction band states, E_{c} is the conduction band edge, E_{F} is the Fermi level, K_{B} is Boltzmann's constant (8.616×10^{-5} eV/K) and T is temperature (300 K in this case). N_{c} is around $10^{19}\,\text{cm}^{-3}$ at 300 K for ZnS [28]. Using the above given values and solving with respect to E_{F} one obtains

$$E_{\text{F}} = E_{\text{c}} - 0.9\,\text{eV}, \tag{2.33}$$

which means that the material is significantly *n*-type. The *n*-type conductivity is believed to result from the large number of unintentional dopants and impurities previously mentioned. Photoconductivity [88] and optical absorption [90] measurements confirm that the ZnS band gap contains a broad distribution of energy states, such that the highest filled levels lay at $\sim E_F = E_c - 0.9$. The total density of forbidden gap states (integrated over energy) is expected to be at least 10^{18} cm^{-3}, as hole creation through acceptor doping at these concentrations has always been compensated by unidentified donor-like 'flaws'. Due to this self-compensation mechanism, ZnS has so far proven to be *p*-undopable. We shall return to the influence of these donor-like 'flaws' on the electrical properties of ZnS thin films shortly.

2.3.1.1 Carrier sourcing: field emission or Fowler–Nordheim tunnelling

Figure 2.16 is a schematic band diagram of a metal–insulator contact. Electrons from the metal face a high energy barrier at the interface, and as a result such contacts are known as 'blocking' or Schottky contacts. This is commonly the case with most metal/wide band-gap semiconductor contacts. The basic concept of Fowler–Nordheim tunnelling [89] is that the wavefunctions of electronic states in metal penetrate into the classically forbidden barrier, where they overlap with empty conduction-band states of the ZnS. The electric potential at the interface brings into coincidence the energy of the ZnS conduction band states with the metal Fermi level, at which point carrier injection from the cathodic metal into the ZnS conduction band can occur by tunnelling. Simply put, the current in d.c.-driven ZnS devices is sustained by the injection of electrons via tunnelling, from the negatively

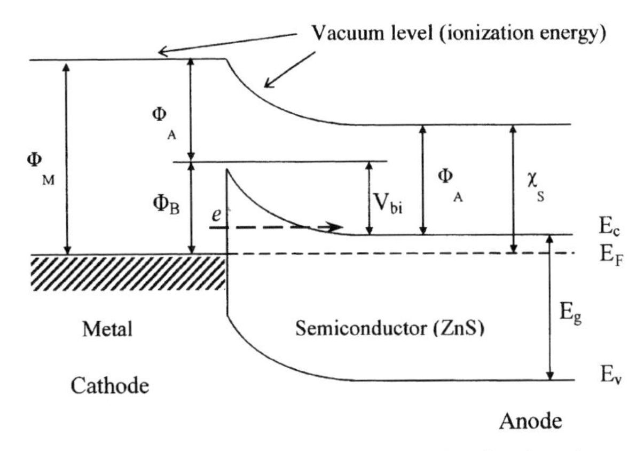

Figure 2.16. Schematic of the metal–ZnS Schottky barrier showing electron injection by Fowler–Nordheim tunnelling from the metal. The labels are as follows: Φ_M = metal work function, Φ_A = electron affinity, χ_S = semiconductor work function, E_F = Fermi level, E_c = conduction band edge, E_v = valence band edge, $E_g = E_c - E_v$ = band-gap energy, $V_{bi} = \Phi_M - \chi_S$ = built-in interface potential, $\Phi_B = \Phi_M - \Phi_A$ = barrier height.

biased metal (this does *not* occur in a.c.-driven devices where the d.c. path is blocked!).

The tunnelling current is obtained from the product of the carrier charge, velocity and density, multiplied with the tunnelling probability:

$$J_{\mathrm{T}} = e v_{\mathrm{R}} n \Theta \tag{2.34}$$

where the velocity is taken as the average velocity with which carriers approach the barrier, i.e. the Richardson velocity ($v_{\mathrm{R}} = [K_{\mathrm{B}} T / 2\pi m^*]^{1/2}$), and the tunnelling probability (Θ) is determined from the time-independent Schrödinger equation,

$$-\frac{\hbar^2}{2m^*}\frac{\mathrm{d}^2\Psi}{\mathrm{d}x^2} + V(x)\Psi = E\Psi, \tag{2.35}$$

which can be rewritten as

$$\frac{\mathrm{d}^2\Psi}{\mathrm{d}x^2} = \frac{2m^*}{\hbar^2}[V(x) - E]\Psi. \tag{2.36}$$

For a triangular potential barrier of the type shown in figure 2.16, $V(x) - E = -e\Phi_{\mathrm{B}}$ where Φ_{B} is the interface barrier height. If it is assumed that $V(x) - E$ is independent of position in a region between x and $x + \mathrm{d}x$, equation (2.36) can be solved, yielding

$$\Psi(x + \mathrm{d}x) = \Psi(0)\exp(-k\,\mathrm{d}x) \quad \text{with } k = \frac{\sqrt{2m^*[V(x) - E]}}{\hbar} \tag{2.37}$$

where the minus sign indicates that the electron (wavefunction) is moving from left to right. For a slowly varying potential, the amplitude of the wavefunction at $x = L$ can be related to the wavefunction at $x = 0$ through the WKB (Wigner, Kramers, Brillouin) approximation. Namely:

$$\Psi(L) = \Psi(0)\exp\left(-\int_0^L \frac{\sqrt{2m^*[V(x) - E]}}{\hbar}\right). \tag{2.38}$$

The tunnelling probability is the ratio of the amplitude of the wavefunction at the right side of the barrier (in the metal) to amplitude on the left side (in the ZnS). That is

$$\Theta = \frac{\Psi(L)\Psi^*(L)}{\Psi(0)\Psi^*(0)} = \exp\left(-\frac{4}{3}\frac{\sqrt{2em^*}}{\hbar}\frac{\Phi_{\mathrm{B}}^{3/2}}{E}\right) \tag{2.39}$$

where the electric field $E = \Phi_{\mathrm{B}}/L$. The field emission or Fowler–Nordheim tunnelling current is then

$$J \approx E^2 \exp\left[-\frac{4\sqrt{2m^*}}{3e\hbar E}(e\Phi_{\mathrm{B}})^{3/2}\right]. \tag{2.40}$$

For an exact derivation of equation (2.40) and more details surrounding field emission, the reader is referred to [89, 91–97].

It should be noted that at high fields (>1 MV/cm) the ZnS conduction band is also populated by field emission from deep levels in the band-gap. At low fields, such levels may populate the conduction band by thermal excitation. These deep levels also act as carrier sinks. The reader may have observed that since the trap or donor-like 'flaws' density is greater than 10^{18} cm^{-3} (i.e. is several orders of magnitude higher than the largest free-carrier density achievable, $\sim 10^{12}$ cm^{-3}; see section 2.3.1), the bulk is an efficient electron source at high fields.

2.3.1.2 Space charge effects in d.c. devices

A one-carrier injection current consisting exclusively of injected electrons is necessarily a space-charge current, since each injected carrier contributes one excess electron to the insulator. Beale and Mackay have shown [98] that when a d.c. device is biased into steady-state conduction at a given voltage (V), a small change in the applied voltage $\mathrm{d}V$ resulted in a corresponding change of the current $\mathrm{d}I = I(V + \mathrm{d}V) - I(V)$, only after some time had elapsed. The delay needed to attain steady-state conduction at $I(V + \mathrm{d}V)$ was inversely proportional to the current I. These findings are indicative that the bulk of the film contains some space charge, which varies with applied voltage at a rate proportional to the passage of charge. For details see [88, 98].

According to Gauss's law, space charge results in a distorted or non-uniform electric field. This distortion is expected to be severe in ZnS films as the number of native point defects and complexes exceed 10^{18} cm^{-3}. Since the distribution of these defects is likely to be non-uniform over the thickness of the ZnS film, the electric field profile will depend on the bias polarity. Thus, the asymmetry of the current–voltage characteristic with regards to polarity depicted in figure 2.15 is not only due to the difference in barrier height at the metal–ZnS and ITO–ZnS interfaces, but is related to the features of the bulk as well. The book by Lampert and Mark [95] is devoted to carrier injection in insulators (wide band-gap semiconductors), and reviews space charge effects in detail. Discussion of other fundamental processes that lead to EL in high-field devices is given below.

2.3.2 Alternating current thin-film devices

Of the various kinds of high-field EL devices, the a.c. thin-film variety is the most widely researched, and due to these efforts is the most successful in the commercial marketplace. The research and commercialization of these devices are primarily focused on their application as a flat-panel display technology, though new and innovative applications are emerging. ACTFEL devices possess several attractive advantages over other FPD technologies. These include wide viewing angle (>160°), high contrast in high ambient illumination conditions, broad operating-temperature range (−60 °C to +100 °C), fast response times, the capability of very high resolution and legibility, excellent reliability (>50 000 h) [3] and in addition, they are insensitive to shock and vibration due to their simple all-solid-state rugged construction. These attributes make TFEL displays ideal for military, avionics, medical and industrial applications, and are undoubtedly responsible for their commercial success. Several comprehensive reviews of ACTFEL devices have been published including the work by Ono [29, 31], Rack and Holloway [73], and Mueller-Mach and Mueller [99].

2.3.2.1 ACTFEL device structure and material requirements

The work of Inoguchi *et al* is often cited as the birth of current ACTFEL technology [12], though Russ and Kennedy [11] must be acknowledged for being the first to propose the double-insulator structure. Figure 2.17(a) is a schematic cross-section of the double insulator a.c. device structure. In

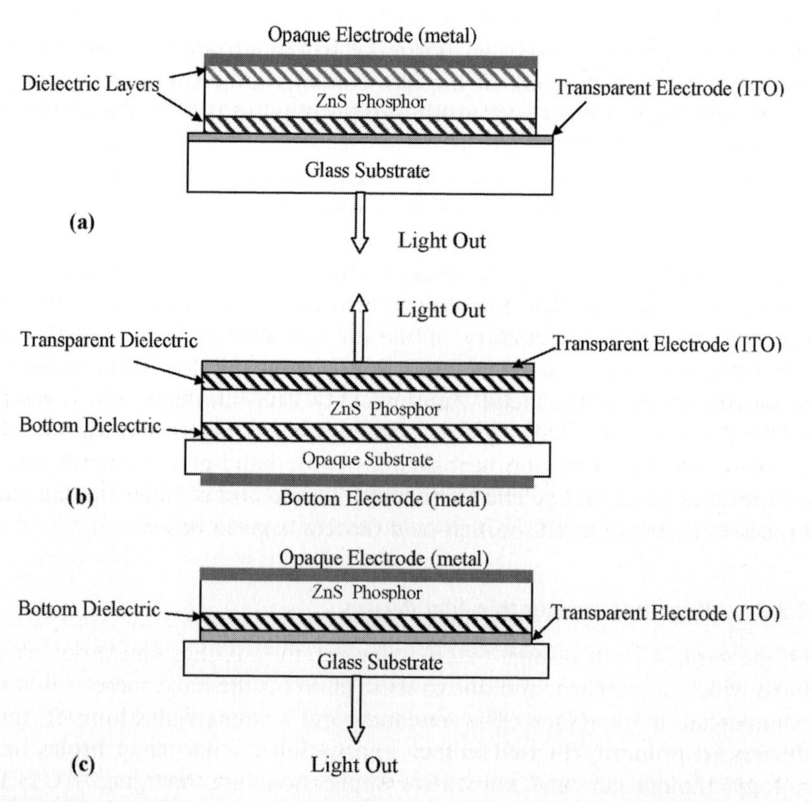

Figure 2.17. Schematic cross-section of (a) the standard double insulator structure, (b) inverted structure and (c) half-stack structure.

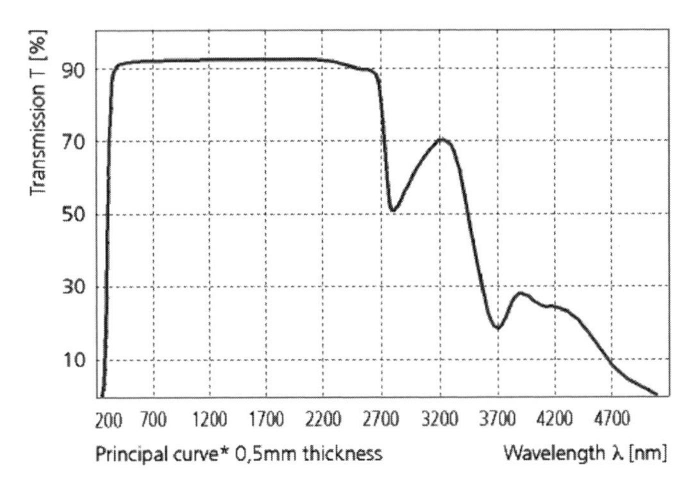

Figure 2.18. Transmission spectrum of 7059 glass (www.pgo-online.com).

physical terms, it is two back-to-back metal/insulator/semiconductor (MIS) junctions. Variations of this structure are the inverted structure and half-cell structure shown in figure 2.17(b) and (c). We have discussed the properties of the ZnS phosphor layer above, with the exception of specific ZnS:activator combinations for particular emission wavelengths. This section is devoted to the properties required of the various other layers in a.c. double insulating device structure for optimum reliable performance.

2.3.2.1.1 Glass substrates. Glass substrates are typically used in ACTFEL devices for flat-panel display applications, and influence their properties and reliability. Glass substrates must have a high transmission coefficient for visible light, a thermal coefficient of expansion that is close to that of the deposited films, a high softening temperature to accommodate process temperatures greater than 650 °C, and low alkali metal content (diffusion of metal ions from the glass into the phosphor layer results in deteriorated performance). Corning 7059 non-alkaline, phosphosilicate glass is relatively inexpensive and is commonly used even though its softening temperature is only 600 °C. Figure 2.18 shows the transmission spectrum of this glass.

2.3.2.1.2 Insulator layers. In the double-insulator device structure, the d.c. path is blocked. The insulator thickness is usually in the 250 nm range, which is too thick for electrons to tunnel through from the metal under cathodic bias. The device is therefore only capacitively coupled to the externally applied field. The main function of the insulator layers is to protect the ZnS phosphor from electrical breakdown (runaway avalanche breakdown eventually leading to destructive breakdown) at high fields (greater than 2×10^6 V/cm). The properties of the insulator layers are therefore critical to device performance. The most important requirements [29, 31–33, 100–103] of these layers are as follows.

1. High dielectric constant.
2. High dielectric breakdown electric field strength.
3. Must provide interface states at the insulator–phosphor interface, from which electrons can tunnel into the phosphor conduction band under the influence of an applied field above a certain threshold value. We shall examine this requirement in detail when we look at the electrical model for ACTFEL devices.
4. A small number of pinholes and defects. Pinholes and defects offer sites for local field enhancement and consequent premature dielectric breakdown.
5. Good mechanical adhesion and stress accommodation over process temperatures which range up to 600 °C (the coefficients of thermal expansion varies widely for the different layers).
6. Must act as a barrier to metal-ion diffusion into the phosphor layer.

As we will discuss in detail shortly, above a certain applied threshold voltage carrier injection into the phosphor layer conduction band occurs due to field emission from interface states. Below the threshold field, the double insulator device structure is a perfect capacitor. Maxwell's equations then apply, and impose the following boundary conditions at the insulator–phosphor interfaces:

$$\varepsilon_{\mathrm{i}}^{\mathrm{t,b}} E_{\mathrm{i}}^{\mathrm{t,b}} = \varepsilon_{\mathrm{p}} E_{\mathrm{p}} \tag{2.41}$$

where ε is the dielectric constant at the frequency of operation, E is the electric field, the subscripts i and p refer to the insulator and phosphor respectively, and the superscripts t and b refer to the top and bottom insulator. The total applied voltage is divided between each layer according to the following:

$$V_{\mathrm{app}}^{\mathrm{tot}} = E_{\mathrm{i}}^{\mathrm{t,b}} d_{\mathrm{i}}^{\mathrm{t,b}} + E_{\mathrm{p}} d_{\mathrm{p}} \tag{2.42}$$

where d_{i} and d_{p} are the thicknesses of the insulator and phosphor layers respectively.

From equations (2.41) and (2.42) the fraction of the total applied voltage that appears across the phosphor layer can be calculated as ([29], p. 63)

$$V_{\mathrm{p}} = E_{\mathrm{p}} d_{\mathrm{p}} = \frac{\varepsilon_{\mathrm{i}} d_{\mathrm{p}}}{\varepsilon_{\mathrm{i}} d_{\mathrm{p}} + \varepsilon_{\mathrm{p}} d_{\mathrm{i}}} \times V_{\mathrm{app}}^{\mathrm{tot}}. \tag{2.43}$$

As equation (2.43) shows, the dielectric constant of the insulating layer (ε_{i}) must be as large as possible and its thickness (d_{i}) as small as device reliability will permit, in order to maximize the voltage drop across the EL phosphor layer. Another advantage of insulators with a high dielectric constant is better device reliability due to lower operating voltages, because a proportionally smaller voltage will appear across the insulator layers.

The threshold field for tunnel emission from interface traps in ZnS is $\sim 1.8 \times 10^6$ V/cm ([29], p. 64). According to equation (2.43), the minimum field that the insulator layer must sustain without breakdown in order to perform its function properly is

$$E_i^{t,b} = \frac{\varepsilon_p E_p}{\varepsilon_i^{t,b}} = \frac{8.3 \times 1.5 \times 10^6 \text{ V/cm}}{\varepsilon_i^{t,b}} \tag{2.44}$$

where ε_p (ZnS) is taken as 8.3. This equation shows that high breakdown field and high dielectric constant are inversely proportional, and this has been verified experimentally for several insulators. As high dielectric constant and high breakdown field are difficult to satisfy simultaneously, a figure of merit introduced by Howard [101] is used to rate insulators for ACTFEL devices. It is simply the product of the dielectric constant and the electrical breakdown field (E_{DB}), and is a measure of the maximum trapped charge ($\mu C/cm^2$) density at the insulator–phosphor interface. Table 2.4 presents a comparison of the important properties of several insulators. Unfortunately the insulators with the highest figures of merit

Table 2.4. Typical insulator materials with relative dielectric constant, dielectric breakdown field, figure of merit (maximum trapped charge density at insulator–phosphor interface), breakdown mode and deposition technique. SHB = self healing breakdown; PB = propagating breakdown; PECVD = plasma enhanced chemical vapour deposition; ALE = atomic layer epitaxy; EBE = electron beam evaporation.

Material	Dielectric constant (ε_i)	Breakdown field E_{DB} (MV/cm)	Figure of merit $\varepsilon_i^* E_{DB}$ ($\mu C/cm^2$)	Breakdown mode	Deposition method
SiO_2	4	6	2	SHB	Sputtering
SiON	6	7	4	SHB	Sputtering/PCVD
Al_2O_3	8	5	3.5	SHB	Sputtering
Al_2O_3	8	8	6	SHB	ALE
Si_3N_4	8	6–8	4–6	SHB	Sputtering
Y_2O_3	12	3–5	3–5	SHB	Sputtering/EBE
Ta_2O_5	23–25	1.5–3	3–7	SHB	Sputtering
$BaTiO_3$	14	3.3	4	SHB	Sputtering
$BaTa_2O_6$	22	3.5	7	SHB	Sputtering
$PbTiO_3$	150	0.5	7	PB	Sputtering
TiO_2	60	0.2	1	PB	ALE
$SrTiO_3$	140	1.5–2	19–25	PB	Sputtering
$Sr(Zr,Ti)O_3$	100	3	26	PB	Sputtering

exhibit propagating breakdown which, when initiated, spreads catastrophically and eventually destroys the device.

2.3.2.1.3 Electrodes. Devices for display applications require at least one optically transparent conducting electrode. Indium/tin/oxide (ITO) films are predominantly used. The sheet resistance of ITO is around $5\,\Omega$/sq and the resistivity of the layer increases with thickness. The resistivity of the ITO layer must be less than $10^{-4}\,\Omega\,\text{cm}$ ([29], p. 63) which limits the thickness to around 200 nm (resistivity = sheet resistance × thickness). Though optically transparent, ITO is a degenerately doped *n*-type semiconductor. The *n*-type conductivity of ITO is due to the thermal ionization of shallow donors which arise from the substitution of Sn^{4+} on to In^{3+} lattice sites, and oxygen vacancies [104]. Zinc oxide (ZnO) and zinc oxide-doped Al, In or Ga thin films are also used as transparent conducting electrodes. In the undoped material *n*-type conductivity is due shallow donor levels due to oxygen vacancies, while in the doped system the 3+ valence state of Al, In or Ga substituting on Zn^{2+} lattice sites creates shallow donor levels [104].

The top metal electrode is typically 300–500 nm thick and is the last layer to be deposited during device fabrication. It must have a low resistivity, be resistant to electromigration (metal-ion migration) at high fields, have good adhesiveness to the insulating layers and must possess some ability to prevent breakdown spread when dielectric breakdown of the insulator (double-insulating structure) or phosphor (half-stack) layers occurs. Aluminium is the metal that best satisfies these requirements. The main drawback of Al is its high reflectivity. Mirror-type reflections occur when external light penetrates the devices and this negatively affects the display contrast. Filters are often used to suppress this back reflection so that only light generated by the device is observed by the viewer.

2.3.2.2 *The physics of a.c. devices*

In this section we are concerned with the basic physical processes and their interrelationships which lead to high-field electroluminescence in ZnS devices, namely:

1. the injection of electrical charge into the conduction band of the ZnS phosphor layer,
2. high-field transport of electrical charge across the phosphor layer which facilitates acceleration of a fraction of the injected electrons to ballistic energies, and
3. impact excitation efficiency, which governs how many luminescent ions have electrons raised from the ground into an excited state via inter- or intraband transitions due to impact excitation (or impact ionization).

Radiative decay, resulting in light emission was addressed in section 2.2.

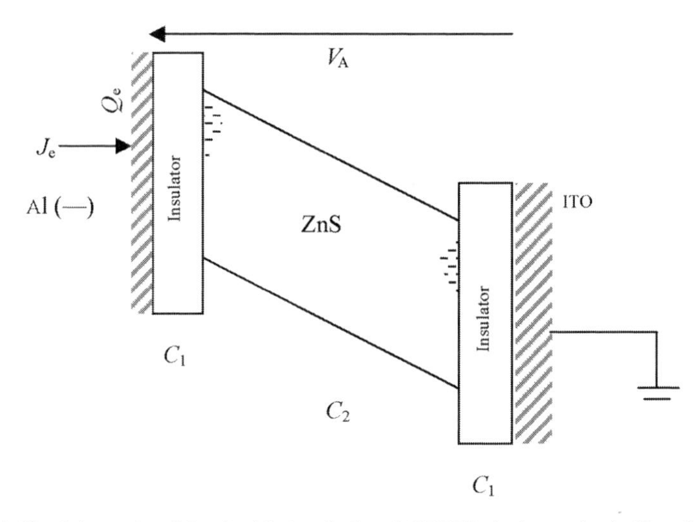

Figure 2.19. Schematic of the double-insulating ACTFEL device under half-cycle bias.

2.3.2.2.1 Electrical conduction in a.c. devices: the relationship between the external voltage and internal charge transfer. Figure 2.19 shows the double-insulating a.c. device structure with one electrode grounded. This is equivalent to half a cycle of the a.c. drive frequency. A single a.c. cycle is completed when the polarity of the applied voltage is reversed. Below a certain threshold voltage, the device is assumed to be a perfect capacitor. If the capacitances per unit area of the top (C_{IT}) and bottom (C_{IB}) insulators are added together in series to give an effective capacitance

$$C_1 = \frac{C_{IT}C_{IB}}{C_{IT} + C_{IB}}, \tag{2.45}$$

then the total series capacitance of the insulator/ZnS/insulator multilayer stack is

$$C = \frac{C_1 C_2}{C_1 + C_2} \tag{2.46}$$

where C_2 is the capacitance of the ZnS phosphor layer. The external current density (J_e) flowing in the external circuit is related to the external charge density (Q_e) on the metal electrode. That is

$$J_e = dQ_e/dt. \tag{2.47}$$

Boundary conditions impose that the charge on the metal electrode (Q_e) be proportional to the oxide field, and in the absence of conduction in the ZnS layer J_e is an ordinary displacement current

$$J_e = C\,dV_A/dt \tag{2.48}$$

where C is given by equation (2.46) and V_A is the applied voltage with respect to ground. The applied voltage is capacitively divided according to

$$V_P = \left[\frac{C_1}{C_1 + C_2}\right] \times V_A \tag{2.49}$$

$$V_I = \left[\frac{C_2}{C_1 + C_2}\right] \times V_A \tag{2.50}$$

where V_P and V_I are the portions of the applied voltage that drop across the phosphor and insulator layers, respectively.

Above the threshold voltage, electrical conduction in the ZnS layer occurs. Carrier injection into the ZnS conduction band is generally thought to result from field-emission or tunnelling from interface levels. The field-emission current is described by equation (2.40) with the barrier height Φ_B replaced by an effective interface trap depth.

When electrical conduction in the ZnS layer occurs, the external, observable current (J_e) increases. This comes about because the voltage drop across the phosphor must be compensated by an increase in the voltage across the insulator layers in order that V_A may remain constant. This causes an increase in the rate of change of Q_e with respect to time, compared with that needed to build the field in the non-conducting insulator/ZnS/insulator stack (boundary conditions impose that Q_e be proportional to the oxide field). Equation (2.48) then becomes

$$J_{e,c} = C\,\mathrm{d}V_A/\mathrm{d}t + J_c \tag{2.51}$$

where J_c is the excess current density due to the increased insulator field. The voltage transferred from the ZnS to the insulator is given by ([29], section 4.2)

$$\Delta V = \left[\frac{C_1}{C_1 + C_2}\right] \times [V_A - V_{th}] \tag{2.52}$$

where V_{th} is the threshold voltage for electrical conduction in the phosphor layer, and highlights the requirement for high dielectric breakdown strength of the insulating layers. The corresponding charge that flows in effecting this voltage transfer is

$$\Delta Q_P = C_1[V_A - V_{th}] = \int_0^\infty J_c\,\mathrm{d}t, \tag{2.53}$$

which relates the internal charge flow (per unit area) in the ZnS layer to the externally applied voltage, and the capacitance per unit area of the insulating layer.

J_c is known as the dissipative current density and characterizes charge transfer in the ZnS layer. Bringuier [28] has shown that the dissipative current (J_c) can be written as

$$J_c = \frac{C_1}{C_1 + C_2} \frac{\mathrm{d}}{\mathrm{d}t}\left(\frac{D}{d}\right) \tag{2.54}$$

where D is the electric dipole moment per unit area of the ZnS film, and d is its thickness. Equation (2.53) shows that the externally observable dissipative current is related to the rate of change of the internal electric dipole moment. This dipole moment arises from the separation of electronic charge from its associated ionized donors at voltages above threshold when electrical conduction in the ZnS layer sets in. Clearly if there is no charge separation, $D = 0$ and therefore the dissipative current is zero. The rate of change of internal electric dipole moment characterizes charge transfer in the ZnS layer. Considering equations (2.53) and (2.54), the charge transferred across the phosphor between the two zeros of the applied voltage is then

$$\Delta Q_\mathrm{P} = -\frac{C_1}{C_1 + C_2}\left(\frac{\Delta D}{d}\right) \tag{2.54a}$$

$$\Delta D = \sum e_\mathrm{i} d_\mathrm{i}. \tag{2.54b}$$

The ideal electrical behaviour of ACTFEL devices is often modelled as a capacitor in parallel with a nonlinear resistor, or two back-to-back Zener diodes. Both equivalent schemes are shown in figure 2.20.

As mentioned above, electrons are sourced by field-emission from interface states near one insulator/sulphide boundary and are sinked by similar states at the opposite sulphide/insulator interface. One imagines the former electrode as a momentary cathode and the latter as a momentary anode. Upon reversal of polarity, their roles are reversed. The transferred charge which accumulates at the opposite interface results in an internal field, which opposes and cancels enough of the externally applied field to reduce the electric field in the ZnS to the threshold field for electrical conduction. This effect is known as *field clamping* and imposes a limit to the field that can be applied to the phosphor. In ZnS phosphors, electron sourcing occurs from a distribution of states and field clamping is observed over a *range* of applied fields. It has been found that the lower the *energy* density of interface states, the larger the range of field to be explored in order to obtain a given transferred charge [28, 105].

The accumulation of charge at the opposite phosphor/insulator interface leads also to a *polarization* or memory effect. This effect is illustrated in figure 2.21 where it is observed that if the polarity of the succeeding pulse is the same, light output is greatly reduced. On the other hand if the polarity of the succeeding pulse is inverted, a large light output is observed. The explanation of this behaviour is related to the long residence time of accumulated electrons at interface states even after the applied electric field is removed, which results in remnant polarization. Consequently, if the

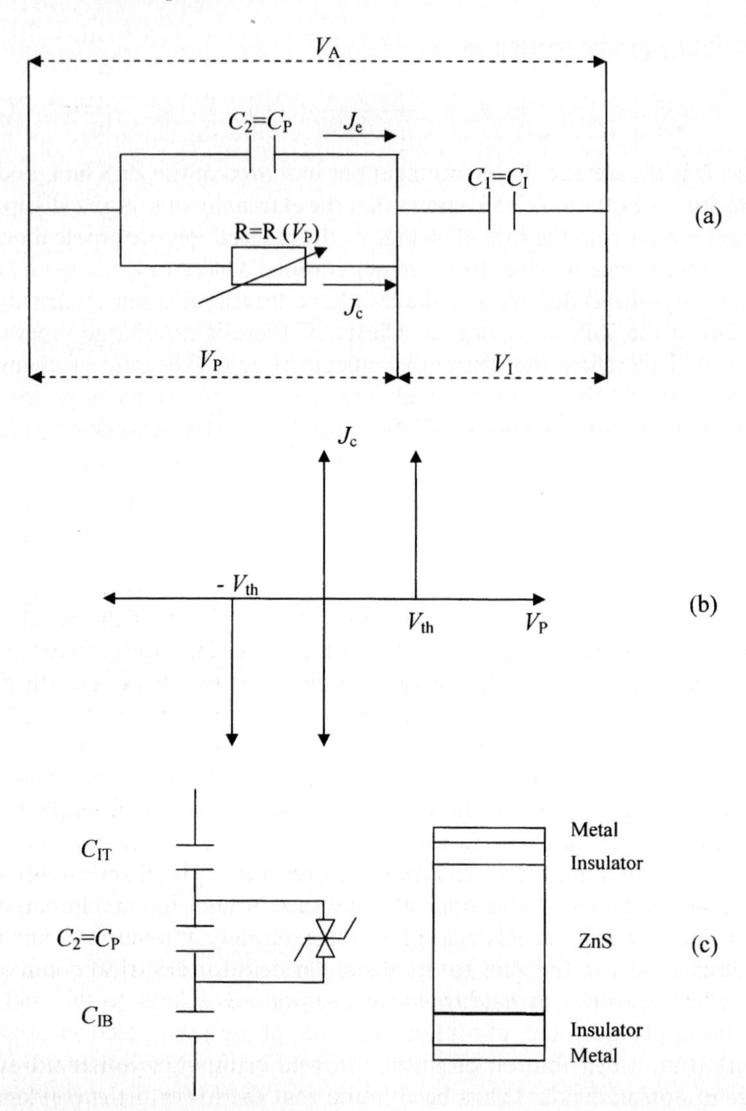

Figure 2.20. (a) The equivalent circuit of an ACTFEL device modelled as a capacitor in parallel with a nonlinear resistor, (b) ideal quasi-static current–voltage characteristic of the nonlinear resistor, and (c) equivalent circuit of an ACTFEL device modelled as two back-to-back Zener diodes.

polarity of the succeeding pulse is the same as that of the first pulse, the effective inner electric field across the ZnS phosphor layer is lowered by the pre-existing counter polarization from the first pulse. The converse is true when sequential pulses have inverted polarity. As discussed in section 2.3.1, ZnS phosphors abound with defects and flaws that act as both

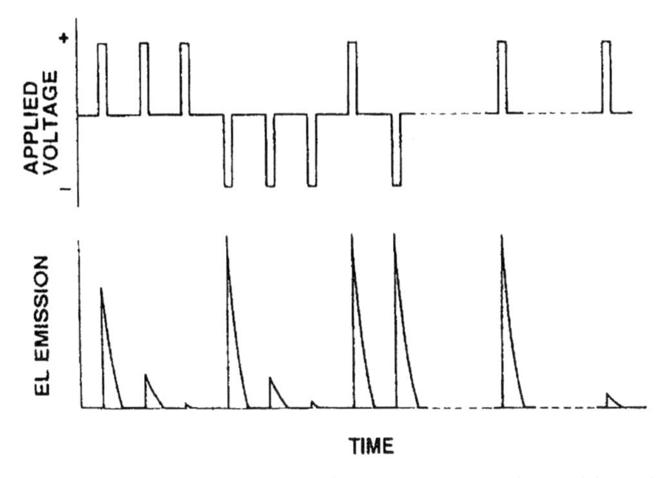

Figure 2.21. Illustration of the dependence of EL emission intensity on drive polarity [107].

carrier sources and sinks. Since the distribution of flaws is unlikely to be uniform over the thickness of the ZnS film, the electric field profile will depend on the bias polarity. In other words, the electrical activity of the bulk ZnS film must also be considered. For more details regarding the electrical behaviour of ACTFEL devices the reader is referred to [22, 29, 106] and references contained therein.

The preceding discussion established the relationship between externally applied fields and internal charge transfer phenomena. A fraction of the electrons that populate the conduction band by field-emission or Fowler–Nordheim tunnelling from interface states (above V_{th}) will be accelerated to energies sufficiently high to effect impact excitation of luminescent centres in the ZnS host. Radiative relaxation of the impact-excited luminescent ions results in electroluminescence. This leads naturally to our next point of discussion, namely, high-field transport.

2.3.2.2.2 High-field electron transport. The objective of high-field transport theories is to predict the fraction of electrons injected into the conduction band, which can attain the specific energy required to impact-excite luminescent centres into an excited state. Subsequent radiative return the ground state results in light emission. The electron energy distribution is a complicated function of high-field scattering mechanisms such as intervalley scattering, polar optical phonon (phonons with energies in the optical range) scattering, acoustic phonon (phonons with energies in the acoustic range) scattering, impact excitation and band-to-band impact ionization. Electron mobility at low at low fields is controlled by ionized impurity scattering [28]. Electrons in the conduction band travel under the influence of a field $F \approx 1$ MV/cm, which in atomic units corresponds to $F \approx 0.01$ V/Å.

This value is much less than atomic fields so the usual band-structure concepts retain their validity [108, 109].

Simulations of high-field electron transport in ACTFEL devices have been investigated by several groups using Monte Carlo techniques. Initially, Brennan [48] used a full-band ZnS structure to calculate the electron energy distribution for electric fields up to 1 MV/cm. For this electric field region, he concluded that very few electrons obtained sufficient energy to excite Mn luminescent centres. Later, a simple parabolic-band model [110] was used to simulate the high field transport properties in ZnS in which the conduction band was described by a single parabolic band. In addition, this parabolic-band model assumed that the electron scattering was dominated by polar optical phonon scattering. These calculations suggested that the electrons in ZnS experienced nearly loss-free transport, which is referred to as electron runaway, and resulted in a very energetic electron energy distribution. Later, a non-parabolic conduction band was implemented by Bhattacharyya *et al* [55], where they included scattering due to polar optical phonons, acoustic phonons, intervalley scattering, and ionized and neutral impurities. In this case, electron runaway was no longer observed and the electron energy distribution became stable. It was determined that non-polar interactions and conduction-band non-parabolicity stabilized the electron distribution by increasing the electron–phonon scattering rates.

Figures 2.22–2.24 summarize the main results from these numerical calculations for ZnS. Figure 2.22 shows scattering rates as a function of energy for the Γ valley at 300 K. These data indicate that for energies less

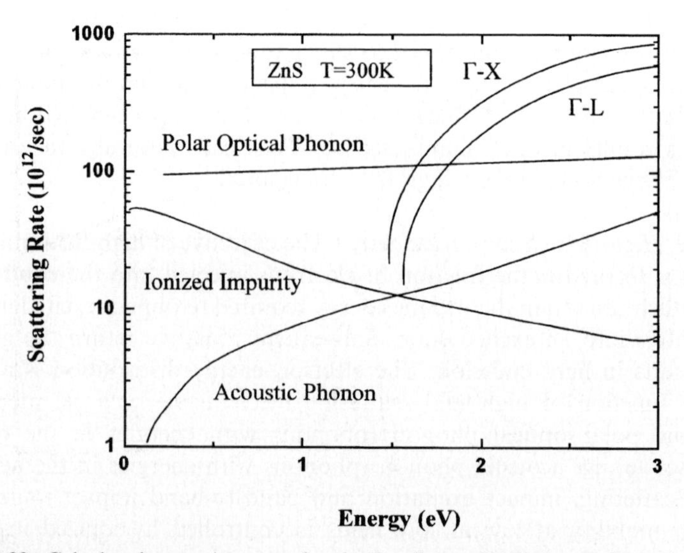

Figure 2.22. Calculated scattering rates for the Γ valley at 300 K due to the various scattering mechanisms as a function of energy for ZnS [55].

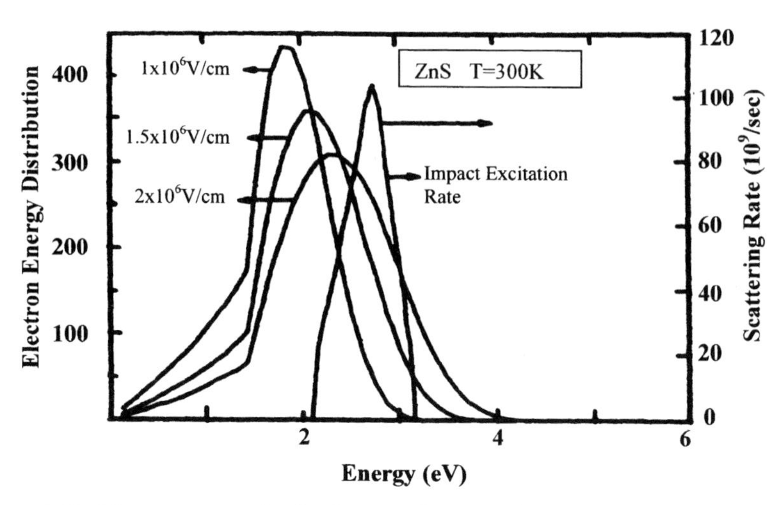

Figure 2.23. Electron energy distribution as a function of energy for different phosphor fields using the non-parabolic band structure model [55]. The impact excitation rate as a function of energy is also included.

than 1.7 eV, electron mobility is determined by polar optical phonon scattering. Above 1.7 eV, mobility is dominated by intervalley scattering. figure 2.23 shows the electron energy distribution ($n(E)$) as a function of electron energy for different phosphor fields, and the impact excitation rate for Mn^{2+} as a function of energy obtained from the nonparabolic band model [55]. The

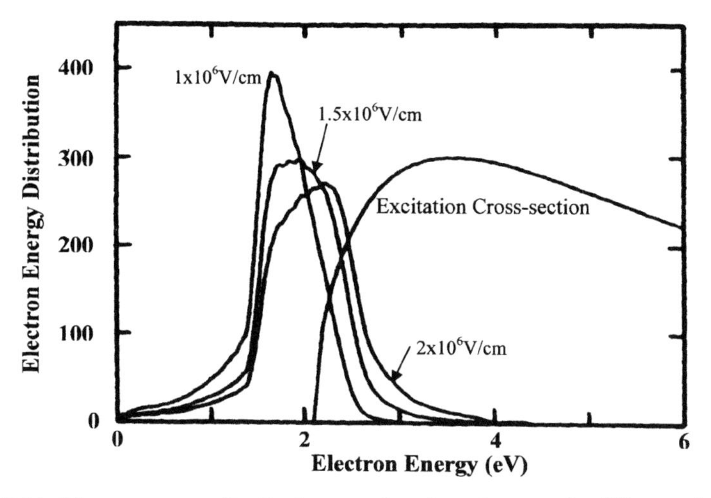

Figure 2.24. Electron energy distribution as a function of energy for different phosphor fields using the full-band structure model [55]. The impact excitation cross-section as a function of energy is also included.

electron energy distribution ($n(E)$) as a function of electron energy using the full-band model is shown in figure 2.24. Also shown here is the Mn^{2+} impact excitation cross-section as a function of electron energy. The data indicate that a majority of electrons are able to attain energies of around 2.12 eV, the minimum energy required to excite Mn^{2+}.

The overlap between the electron energy distribution and the impact excitation rate (or impact excitation cross-section) is a direct measure of the quantum efficiency of the device.

Another model, the so-called 'lucky drift' model, is a simpler analytical model which uses the electron mean free path to describe the high-field transport properties [111]. This approach does not require extensive computing resources as do Monte Carlo simulations. The mean free path is characterized by an electron–phonon interaction. Specifically, the energy exchange between electrons and phonons is described by an electron–phonon interaction Hamiltonian, where the electrons can emit or absorb one phonon at a time. Staying with the quantum mechanical description, phonons are bosons and therefore the phonon occupation number $n(\omega)$ at temperature T is described by Bose–Einstein statistics:

$$n(\omega) = [\exp(\hbar\omega/K_B T)]^{-1}. \tag{2.55}$$

For ZnS at 300 K, $n(\omega) = 0.24$ [28] [$n(\omega) = 0.223$ ([29], p. 60)]. At high fields, the lucky-drift model considers scattering to be dominated by polar optical phonons. The energy, $\hbar\omega$, of polar optical phonons in ZnS typically around 0.044 eV. The probability of the phonon occupation number changing from n to $n+1$ is proportional to $n+1$, and that from n to $n-1$ is proportional to n. The ratio of the phonon emission rate $r_e(n \to n+1)$ to the phonon absorption rate $r_a(n \to n-1)$ is therefore

$$\frac{r_e}{r_a} = \frac{n+1}{n}. \tag{2.56}$$

Because $r_e > r_a$, a net energy transfer from the electron to the lattice results. This electron energy loss stabilizes its drift velocity, and prevents electron runaway. The electron–phonon coupling is characterized by an electron–phonon scattering rate, which is determined using the golden rule

$$\frac{1}{\tau} = \frac{2\pi}{\hbar} |\langle \Psi_f | H_{e\text{-}ph} | \Psi_i \rangle| N(E) \tag{2.57}$$

where τ is the electron mean-free time, Ψ_f and Ψ_i are the final and initial state wavefunctions, $H_{e\text{-}ph}$ is the electron–phonon Hamiltonian operator and $N(E)$ is the density of conduction band states at energy E. The electron–phonon scattering rate $1/\tau$ is related to the phonon emission and absorption rate through

$$1/\tau = r_e + r_a. \tag{2.58}$$

Considering equations (2.56) and (2.58), the average electron energy loss per unit time is then

$$\hbar\omega(r_e - r_a) = \hbar\omega/[2n+1]\tau = eFv_d \approx 10^{13}\ \text{eV/s} \tag{2.59}$$

where $v_d = v_d(E)$ is the average saturation drift velocity over the electron energy distribution, and is determined by [111]

$$v_d = \left[\frac{\hbar\omega}{(2n+1)m^*}\right]^{1/2}. \tag{2.60}$$

It can be seen from equation (2.59) that in the high-field electron velocity saturation regime, F, the applied field strength is a direct measure of the electron–phonon inelastic collision rate. The energy balance condition, that is, the balance between energy gain from the applied field and energy loss to phonons, is determined from

$$\frac{dE}{dt} = eFv_d - \frac{\hbar\omega}{[2n(\omega)+1]\tau(E)} = 0. \tag{2.61}$$

Having discussed the energy loss (stabilization) mechanism, the essence of the 'lucky drift', electron mean-free path model can be summarized as follows [111, 112]: the electron *momentum* relaxes over a mean-free path given by $\lambda_m = v_g\tau_m$, while the electron *energy* relaxes over the energy relaxation length $\lambda_E = v_d\tau_E$. Here $v_g = v_g(E)$ is the electron group velocity and v_d is the field dependent saturation drift velocity. At high fields λ_E is much greater than λ_m and the fraction of electrons which can attain a given energy E is not proportional to $\exp[-E/eF\lambda_m]$ but rather to $\exp[-E/eF\lambda_E]$, which is much larger. Table 2.5 shows the fraction of electrons that attain energies of at least 2.12 eV (minimum energy required for impact excitation of Mn^{2+} ions) according to the different models for

Table 2.5. Fraction of electrons beyond 2.12 eV in ZnS at 300 K for $F = 1$ MV/cm and $F = 2$ MV/cm, according to the different models.

Reference	Brennan 1988 [48]	Bringuier 1991 [111]	Bhattacharyya *et al* 1993 [55]
Calculation type	Numerical/ empirical pseudo-potential band-structure	Analytical/ mean-free path	Numerical/ non-parabolic multivalley band-structure
% of electrons above 2.12 eV at $F = 1$ MV/cm	∼1%	27%	26%
% of electrons above 2.12 eV at $F = 2$ MV/cm	∼50%	72%	65%

ZnS. Good agreement is observed between Bhattacharyya's numerical approach and Bringuier's analytical approach.

2.3.2.2.3 Excitation of luminescent centres: impact excitation and impact ionization. As mentioned earlier, the emission wavelength of ZnS phosphors is determined by the internal energy levels of luminescent centres embedded in the ZnS host. There are two electrical (exciting energy brought by electrical charge) mechanisms whereby luminescent centres are brought to an excited state. The first of these mechanisms is direct impact excitation. In this case inelastic collisions of conduction-band electrons with luminescent ions excite their electrons to higher energy states *localized* at the light emitting centres. That is, the centres are raised to an excited state, *without a change in their charge or oxidation state*. The second mechanism is impact ionization. Here the energy transfer from conduction-band electrons is sufficient to effect ionization of the luminescent centre. That is, an electron formerly bound to the luminescent ion is promoted to the conduction band of the host. The ionized centre may then bind with a free conduction-band electron, and the subsequent relaxation may be radiative or not. The light emission process is two staged: ionization, followed by recombination with a conduction electron. This mechanism is often referred to as 'delayed recombination' [113] because the two stages are separated in time. The occurrence of a particular mechanism depends on the position of the ground or excited-state energy levels of the luminescent ion with respect to the host band-edges. For example, the excited state of Eu^{2+} lies at $E_c - 0.13\,eV$ in CdS [114, 115]. The excited state of Eu^{2+} in CdS is therefore shallow, and can release an electron to the host conduction-band either thermally or by tunnelling at fields well below 1 MV/cm [116, 117]. As a result the electroluminescence in CdS:Eu^{2+} is generated by the impact ionization–delayed recombination mechanism [114–117]. In contrast, the ground-state of Mn^{2+} is believed to lie at $\sim$0.9 eV below the valence-band edge of ZnS and the first excited state at $E_c - 2.5\,eV$. The excited state lies deep in the band-gap, is therefore localized, and the electroluminescence mechanism is direct impact-excitation. It should be noted that the impact ionization–delayed recombination excitation mechanism is generally a low probability process because, once in the conduction-band, free electrons are immediately swept towards the momentary anode. The spatial separation of electrons from ionized luminescent centres reduces the probability of electron recapture by the centre. Therefore, we shall begin the discussion by looking at the impact excitation mechanism, after which we shall examine the difference between impact excitation and impact excitation as they relate to ZnS.

I. Impact excitation rate. Under the influence of the applied electric field, drifting electrons spend part of their time above the impact excitation threshold energy, E_e. In general, the impact excitation threshold is equal to the luminescent transition energy plus the Stokes shift. According to the ergodic principle, this fraction of time is equal to the fraction of carriers $P(E_e)$, which overtake energy E_e. The ergodic principle is a basic assumption of statistical physics which states that the mean over the ensemble is equal to the mean over time; that is

$$\langle x^2 \rangle_e = \langle x^2 \rangle_t. \tag{2.62}$$

For ZnS:Mn, E_e is 2.24 eV while the emission transition energy is 2.12 eV due to the Stokes shift [118]. If N [cm^{-3}] is the concentration of excitable centres and we suppose that the impact excitation cross-section σ_0 [cm^2] is a constant above the threshold field, the impact exciting probability per unit length is

$$\sigma_0 N \quad [\text{cm}^{-1}]. \tag{2.63}$$

Equation (2.63) must be weighted by the fraction of *time* that electrons spend above E_e or equivalently, according to equation (2.62), by the fraction of electrons $P(E_e)$ which have energies greater than or equal to the impact excitation threshold E_e in a statistical ensemble. If we define the impact excitation rate α_e as the number of impacts per unit length *drifted*, we must be aware that the path actually travelled by the electron is v_g/v_d times the drift distance. As before, v_g is the electron group velocity and v_d is the field-dependent saturation drift velocity. Hence, the probability of a drifting electron colliding with a luminescence centre is increased by a factor of $\sim$10 for drifting electrons [111]. For ballistic (collisionless) electrons v_g/v_d is unity since in that case the group and drift velocity are the same. With the above considerations the impact excitation rate can then be written as

$$\alpha_e = \sigma_0 N (v_g/v_d) P(E_e), \tag{2.64}$$

which shows that α_e is dependent on luminescent centre factors (σ_0 and N) as well as high-field transport related factors ((v_g/v_d) and $P(E_e)$). In reality both σ_0 and v_g/v_d are energy dependent, i.e. $\sigma = \sigma(E)$ and $(v_g/v_d) = (v_g/v_d)(E)$. The consideration of a finite σ at E_e is known as the 'hard-threshold' approximation. In actuality, Shen and Xu [119] and Allen [120] have shown that $\sigma(E)$ is expected to increase smoothly from E_e, peak at some value E_m, then vanish. Bringuier [111] has shown that if $\sigma(E)$ exhibits a narrow peak at E_m and $P(E)$ varies little at E_m, then the impact excitation rate is more accurately given by

$$\alpha_e = N\bar{\sigma}\frac{v_g}{v_d} E_m P(E_m) \tag{2.65}$$

where $\bar{\sigma} = (\int_0^{+\infty} \sigma(E)\, \mathrm{d}E)/E_m$ is an energy-averaged effective cross-section for impact excitation. If $\sigma(E)$ is not sharply peaked, the *effective* threshold for impact excitation lies somewhere between E_e and E_m, and the impact rate vanishes if the energy of conduction electrons is beyond the $\sigma(E)$ window. Equation (2.65) shows that the effective energy threshold is E_m

rather than E_e, and E_e is termed a 'soft threshold' in this scenario. According to Bringuier [111] and references contained therein, the impact excitation cross-section for Mn^{2+} should not exceed $\sim 10^{-17}\,cm^2$. The lower excitation efficiency of rare-earth doped ZnS phosphors compared with ZnS:Mn is attributed in part to their smaller excitation cross sections [111, 120, 121].

II. Impact excitation efficiency. The impact excitation efficiency governs the fraction of luminescent ions that have electrons excited from the ground into an excited state via inter- or intraband transitions due to impact excitation. Given the impact excitation rate α_e, the impact excitation efficiency is the number of impacts per unit length or distance (L) travelled by electrons drifting under the influence of the applied field. That is, the impact excitation efficiency is the product

$$\alpha_e L. \tag{2.66}$$

Equivalently, the impact excitation efficiency may be expressed in terms of the fraction of excited luminescent centres, f^{ex}. If the internal transferred charge per electron in the interfacial model is $\Delta Q/e$ [$C\,cm^{-2}/C$], then [28, 111]

$$f^{ex} = (\Delta Q/e)\alpha_e L/(nL) = (\Delta Q/e)\bar{\sigma}(v_g/v_d)E_m P(E_m) \tag{2.67}$$

where n is the number of excitable carriers. In the most favourable case, that is, the case of Mn^{2+} which is the most efficiently excited centre, $\Delta Q/e \approx 5\times 10^{13}\,cm^{-2}$ [111, 122] and f^{ex} is around 1% [111, 122].

It should be noted that impact excitation has a negligible affect on electron transport. A hot electron loses energy by inelastic collisions at a rate of $\sim E_e\alpha_e \approx 2\times 10^4\,eV/cm$ (taking $\bar{\sigma}(v_g/v_d)$ to be $\sim 10^{-16}\,cm^2$ [28, 111], which must be compared with the rate of energy gain from the field which is $eF \approx 2\times 10^6\,eV/cm$. Therefore impact excitation negligibly affects electron transport. As discussed in subsection 2.3.2.2.2, intervalley and polar optical phonon scattering dominates at high field, and scattering due to impurity impact ionization does not affect electron transport. However, impurities and the associated scattering mechanism do modify the phosphor field distribution by introducing space charge.

III. Impact excitation yield. The number of emitted photons is normally different from the number of excited centres (N^{ex}) due to the existence of, and competition with, non-radiative relaxation pathways. We define *yield* as a ratio of the type: response/excitation. The *response* will be the number of excited centres N^{ex}. In a.c.-driven structures the *excitation* is the transferred charge (transient charge) ΔQ_P flowing through the phosphor due to the above-threshold, short, applied voltage pulse. The basic assumption is that excitation is due to the *current* impact exciting or ionizing the luminescent centres. Possible field ionization (Zener breakdown) of the luminescent centres is neglected. Additionally, the relevant transferred charge is taken to be that flowing in one direction, i.e. the integral of the positive pulse of dissipative current (positive means J_c and V of like signs). The definition of the excitation yield is then

$$\frac{N^{ex}}{[\Delta Q_P/e]}. \tag{2.68}$$

That is, the excitation yield is the number of excited centres per transferred charge across the phosphor thickness, and is unitless when all quantities refer to unit surface area.

As we have discussed in subsection 2.3.2.2.1, the internal quantity ΔD associated with ΔQ_P must be weighted by the distance drifted downfield (equations (2.53)–(2.54a)). This turns out to be the best measure of the excitation parameter N^{ex} [28], because the impact excitation (or ionization) efficiency of the ith conduction electron is proportional to the distance d_i, drifted. If α_e [cm^{-1}] denotes the impact rate which we shall assume to be spatially constant, then the number of impacts of electron i is $\alpha_e d_i$, and the number N^{ex} of impacts due to *all* electrons above the threshold energy is

$$N^{ex} = \sum \alpha_e d_i = \alpha_e \sum d_i = \alpha_e \, \Delta D/e. \tag{2.69}$$

Therefore $N^{ex}/\Delta Q_P$ is a direct measure of α_e (see equation (2.54a)). The dimensionless excitation yield is exactly $\alpha_e d$, that is, the number of impacts caused by an electron travelling over the whole phosphor layer thickness if non-uniformities in the field profile are ignored. In a.c. structures where the field varies with time, this has to be time integrated over the dissipative/conduction current duration.

When the field profile is non-uniform, the impact excitation rate α_e is then a function of position x, through the local field $F(x)$. The number of impacts caused by the ith electron is then given by

$$\int_{x_{1i}}^{x_{2i}} \alpha_e(F(x))\,\mathrm{d}x = d_i\bar{\alpha}_e(x_{1i}, x_{2i}) \tag{2.70}$$

where x_{1i} is the point where the charge carrier was sourced, x_{2i} the point where it was sinked, and $\bar{\alpha}_e(x_{1i}, x_{2i})$ is the average value of $\alpha_e(F(x))$ over the interval (x_{1i}, x_{2i}). The excitation yield is then an averaged measure of $\alpha_e(F(x))$ and depends on the field profile $F(x)$. For more details relating to impact excitation yield the reader is referred to [28, 111] and references contained therein.

2.3.2.2.4 Impact excitation versus impact ionization: experimental indicators. In the preceding subsection, we described the electrical difference between the two excitation mechanisms. Direct impact excitation occurs when inelastic collisions of conduction-band electrons with luminescent ions excite their electrons to higher energy states *localized* at the light emitting centre (the centres are raised to an excited state, *without a change in their charge or oxidation state*). Impact ionization occurs when the energy transfer from conduction-band electrons is sufficient to effect ionization of the luminescent centres, and electrons formerly bound to the luminescent ions are promoted to the conduction band of the host. The occurrence of a particular mechanism depends on the position of the ground or excited-state energy levels of the luminescent ion with respect to the ZnS host band-edges. Determination of the luminescent ion levels with respect to the ZnS band edges in principle can be accomplished by XPS. However, the low dopant concentration of luminescent ions required in EL phosphors (and corresponding extremely weak XPS signals) makes this a non-trivial endeavour. We shall elaborate on the specifics of luminescent centre dopant concentrations in the next subsection. Here we will attempt to describe some of the alternative manifestations which may enable one to discriminate between the two mechanisms.

The clearest identification of the excitation mechanism is achieved in d.c.-driven devices. Steinberger *et al* [113] have studied the identification of the actual excitation mechanism in d.c. device structures, using both a constant d.c. bias and pulsed d.c. driving. Following the presentation of Steinberger *et al* [113], in the case of constant bias, the luminescence output EL_{out} (in photons/s) is proportional to the number of excited centres (N^{ex}) while the current I (in electrons/s) is flowing. In this case the yield

$$\frac{EL_{out}}{I} \tag{2.71}$$

is not the *excitation* yield of the carriers (number of centres excited per electron crossing the film) discussed above, because EL_{out} is also sensitive to the *radiative* yield of the centres (see subsection 2.2.2.1). It may, however, in some instances, follow the excitation yield. Then the dependence of the yield defined by equation (2.71) on current I or bias V may provide information on the excitation mechanism. For example, an impact excitation mechanism in a uniform field should give rise to a superlinear $EL_{out}(I)$ relationship. For details of these effects in d.c.-driven devices the reader is referred to [113, 123, 124] and references contained therein.

In the case of pulsed d.c. driving in the form of rectangular voltage pulses, Steinberger *et al* [113] not only observed light during the 'on' period of the voltage pulse, but light was also observed when voltage was switched off. They noted that either excitation mechanism may give rise to a second burst of light. Specifically:

1. If the centres are impact-excited, the second burst should be ascribed to a current flowing through the sulphide film while $V = 0$. This can be checked by measuring I. If $I = 0$, an impact-excitation mechanism is

unlikely, but cannot be ruled out altogether, since the local current may be inhomogeneous throughout the phosphor and average to zero. This requires a non-uniform field, also averaging to zero, but strong enough to lead the electrons to high kinetic energies.
2. If the centres are impact ionized, the second burst can be attributed to delayed recombination of the ionized centres with free electrons. This also requires a current flowing after voltage turn-off, but the local fields due to inhomogeneities need not be high.

The applied waveforms/voltages are typically monitored using an oscilloscope, as is the conduction current behaviour (using sense elements as presented in subsection 2.3.2.3.3 where we discuss input electrical power). The luminescence behaviour and conduction current as a function of applied voltage are monitored on the same oscilloscope. We note that if the luminescence decay curve is fitted to a double exponential decay (two separate decay times) after the voltage pulse has been removed, then one *may* have possible evidence that different (at least two) EL mechanisms are active in the device. Simultaneous inspection of the conduction current is required for proper analysis. In the analysis of luminescence decay behaviour the reader must realize that the slowest component in the measurement setup determines the fastest decay time that can be measured. It is desirable that the slowest component in the measurement setup be much faster than the decay time. For example, if one uses a modern oscilloscope with response times in the GHz range and a photomultiplier tube (PMT) to which a pre-amplifier with a bandwidth of 100 kHz has been attached, the fastest decay time that can be measured will be determined by the 10 μs response of the pre-amplifier, as the response times of PMTs is typically fast and in the nanosecond range. Thus, a measured decay time in the 10 μs range would be unreliable but one in the millisecond range could be reported with confidence.

In a.c. structures the situation is similar [125–128] but is complicated by the capacitive coupling of the device to the drive voltage. In addition, it should be noted that the second light pulse during the voltage 'off' period can be strengthened by space-charge effects. The effective duration of the excitation is determined by the dissipative current, which in some instances exhibits a lasting tail even though the voltage has vanished. For example, Geoffroy and Bringuier [129] observed a reverse current at the end of the 2 μs exciting voltage pulse at high drive levels, thereby bringing new excitations after the voltage was turned off. In other words, even in the absence of a net charge flow, there may be some space-charge rearrangement inside the phosphor causing luminescence centre excitation. Therefore, the observation of luminescence buildup subsequent to excitation does not necessarily invalidate the direct-impact excitation mechanism.

In relation to the impact ionization mechanism, only a few primary carriers (injected or from the bulk) are needed to ionize a fraction of the luminescent centres, which as a consequence liberate secondary electrons, and results in carrier multiplication over the phosphor film thickness. In the presence of ionized impurities the phosphor field will vary significantly over the film thickness. Therefore, the dependence of the threshold voltage on dopant concentration can be used as a test for discriminating between the two mechanisms. Specifically, the threshold voltage is expected to move to lower values as the dopant concentration increases in phosphor systems where the impact ionization mechanism dominates. For example, Ando and Ono [130] have found that Eu-doped and Ce-doped alkaline-earth sulphides exhibit threshold fields of 1.2 and 1.0 MV/cm respectively, which is noticeably smaller than their undoped counterpart which exhibit thresholds of around 1.6 MV/cm.

2.3.2.2.5 Other excitation mechanisms. In addition to the direct electrical excitation mechanisms described above, there may be indirect ones [131–134]. The one most frequently mentioned is energy transfer to nearby luminescent centres from hole–electron pair recombination. This mechanism is potentially efficient if the pairs are numerous enough, and thus depends on the band-to-band impact ionization (electron–hole pair generation) rate. On the other hand, as discussed earlier, spatial separation of charge at high fields reduces the probability of electron–hole recombination processes.

Before ascribing luminescence buildup after the voltage pulse has been removed to an indirect mechanism, great care should be exercised in systematically measuring the dissipative/conduction current in order to make sure that the dissipative current has completely vanished. Observation of the so-called 'intrinsic luminescence' (at 460 nm in ZnS) after the voltage has been removed is a hint to carrier flow lasting after the voltage has been removed [135], since that luminescence usually follows the electric current in time [136, 137].

2.3.2.2.6 Concentration effects. So far we have discussed the optical properties of luminescent ions, electroluminescent phosphors and thin-film EL devices without specific mention of the ideal concentration of luminescent centres necessary for optimal device performance, i.e. maximum optical output. Clearly, this is a critical parameter for luminescent materials. It is desirable to have a maximum number of luminescent centres *within a certain concentration range*, where the host-luminescent ion material system remains a dilute solid solution. Above a critical dopant level, the precipitation of secondary phases and clustering of dopant ions initiates. These microstructural changes degrade crystallinity, which leads to increased electron scattering because the mean free path of electrons decreases. This results in an enormous negative effect on the electrical, optical and electro-optical properties of EL devices, and can lead to a complete quenching of electroluminescence output. In addition to these macroscopic effects, above a critical dopant concentration the electro-optical properties of EL devices are negatively affected on a quantum level, because the wavefunctions of luminescent ions begin to interact when they are in close proximity to each other. We devote the remainder of this subsection to this latter effect that also leads to reduced optical output.

The optimal concentration of luminescent ions in ZnS from empirical observations ranges between 1 and 2 at% depending on the dopant. Above this dopant concentration the spatial separation between ions decreases, and optical energy is transferred between them. The resulting reduction in optical output is then said to be due to concentration quenching. Blasse and Grabmaier ([27], ch 5) have treated concentration quenching in detail for a range of compounds containing rare-earth ions. Here we present an overview of the underlying physical mechanisms. The reader is reminded that a mole of ZnS contains one mole of Zn atoms (6.023×10^{23} atoms) and one mole of S atoms. Because the ZnS host contains n_{Zn} moles of Z and n_{S} moles of S, the mole fraction of the Z_n is $N_{\mathrm{Zn}} = n_{\mathrm{Zn}}/\ (n_{\mathrm{Zn}} + n_{\mathrm{S}})$ and similar expression can be written for the mole fraction of the S. As the components are elemental, the mole fractions and atomic fractions or percentages are equivalent. The system is considered closed, so that the sum of the mole fractions equal unity, i.e. $N_{\mathrm{Zn}} + N_{\mathrm{S}} = 1$.

When luminescent centres are too close to each other, the return to ground state may occur by transfer of the excitation energy from an excited centre A, to another centre B. If the energy transfer is followed by emission from B, A is said to be a sensitizer for B. However, B may also decay non-radiatively and in this case B is said to be a quencher of the A emission. Energy transfer between two centres requires an interaction between them. The energy transfer process is a well understood phenomenon and has been treated in detail by several books [25, 138]. Consider two centres, A and B, separated by a finite distance R in the host lattice. When the distance R is very small, the centres A and B will have a non-vanishing interaction with each other. The interaction may be either an exchange interaction (if we have wavefunction overlap) or an electric or magnetic multipolar interaction. Energy transfer can only occur if the energy differences between the ground and excited states of A and B are equal (resonance condition) and if a suitable interaction between both systems exists. In practice, the resonance condition can be assessed by considering the spectral overlap of the emission and absorption spectra of luminescent ions being discussed. When these conditions are met and if A is in the excited state and B in the ground state, the relaxed excited state of A may transfer its energy to B.

The rate of nonradiative energy transfer processes has been calculated by Forster and Dexter [206, 207] for all interaction types and yields

$$P_{\mathrm{AB}} = \frac{2\pi}{\hbar} |\langle \mathrm{A}, \mathrm{B}^* | H_{\mathrm{AB}} | \mathrm{A}^*, \mathrm{B} \rangle|^2 \int g_{\mathrm{A}}(E) g_{\mathrm{B}}(E)\, \mathrm{d}E \qquad (2.72)$$

where the integral presents the spectral overlap ($g_i(E)$ being the normalized optical line shape function of centre i), the matrix element represents the interaction between the initial state $|A^*, B\rangle$ and the final state $|A, B^*\rangle$, and H_{AB} is the interaction Hamiltonian. The main feature of equation (2.72) is that the transfer rate P_{AB} vanishes for vanishing spectral overlap. The distance dependence of the transfer rate depends on the type of interaction. For electric multipolar interaction the distance dependence is given by R^{-n} ($n = 6$, 8 for electric dipole–electric dipole interactions and electric dipole–electric quadrupole interactions, for example). For exchange interaction the distance dependence is exponential, since exchange interaction requires wavefunction overlap. A high transfer rate, i.e. a high value of P_{AB}, requires a considerable amount of spectral overlap between the A emission band and the B absorption band(s) and interaction, which may be of the multipole–multipole type or of the exchange type. The strength of the electric multipolar interactions is determined by the intensity of the associated optical transitions. High transfer rates involving multipolar interactions can only be expected if the optical transitions involved are allowed electric-dipole transitions. If the absorption strength vanishes, the transfer rate for electric multipolar interaction vanishes too. However, the overall transfer rate does not necessarily vanish, because there may be contributions by exchange interaction. The transfer rate due to exchange interaction depends on wavefunction overlap and therefore spectral overlap, but *not* on the spectral properties of the transitions involved.

The occurrence of nonradiative energy transfer can be detected in several ways. If the excitation spectrum of the B emission is measured, the absorption bands of A will be found as well, since excitation of A yields emission from B via energy transfer. If A is excited selectively, the presence of B emission in the emission spectrum points to $A \rightarrow B$ energy transfer. Finally, the decay time of the A emission should be shortened by the presence of nonradiative energy transfer, since the transfer process shortens the lifetime of the excited state A^*.

The general concept is the same irrespective of whether A and B are the same species or not. It should be noted that there is no reason why the energy transfer should be restricted to a single step, so that the first transfer step may be followed by many others. This process can take the excitation energy far from the site where the absorption took place and is known as energy migration. If the excitation energy reaches a site where it is lost non-radiatively (a killer or quenching site) the luminescence efficiency of the phosphor will be low. This phenomenon is called concentration quenching. This type of quenching will not occur at low concentrations, because then the average distance between the luminescent ions is large, energy migration is hampered, and the killer centres are not reached. Whether or not energy transfer will occur between identical ions depends in the first place on the spectral overlap of their emission and absorption spectra, and the interaction between them. For ions with large Stokes shifts (a large energy separation between the absorption and emission maxima) spectral overlap will be very small or even vanish, so that energy transfer becomes impossible.

The well-shielded character of the $4f$ electrons of rare-earth ions would seemingly suggest weak interaction, and therefore small energy transfer rates between identical rare earth ions. However, although the radiative rates are small, the spectral overlap can be large, which originates from the fact that the absorption and emission lines coincide on a configurational coordinate diagram, i.e. the minimum of the excited state coincides with the minimum of the ground state in vibrational space (see [27], section 2.1). Further, the transfer rate will easily surpass the radiative relaxation rate, since the latter is low.

The critical distance for energy transfer (R_c) is defined as the distance for which the nonradiative energy transfer rate P_{AB} equals the radiative transfer rate, P_S. For $R > R_c$ radiative emission from A prevails and for $R < R_c$ energy transfer from A to B dominates. If the optical transitions of A and B are allowed electric-dipole transitions with a considerable spectral overlap, R_c may be some 30 Å. If these transitions are forbidden as is the case of most rare earth ions, exchange interaction is required for the transfer to occur and this restricts the value of R_c to some 5–10 Å.

Energy migration has been observed in many rare earth compounds, and concentration quenching usually becomes effective for concentrations of a few atomic percent of dopant ions. Energy transfer over distances of up to some 10 Å is possible. As an example, the transfer rate between Eu^{3+} ions may be of the order of $10^7\ s^{-1}$ if the distance between ions is 4 Å or shorter. This compares with a radiative relaxation rate of 10^2–$10^3\ s^{-1}$. Consequently the excitation energy may be transferred more than 10^4 times during the lifetime of the excited state. If the Eu–Eu distance is larger than 5 Å exchange interaction is ineffective, and only multipolar interactions (which will be weak anyway) need be considered. It has been demonstrated that for compounds of Eu^{3+}, Gd^{3+} and Tb^{3+}, concentration quenching of the luminescence is due to energy migration to killer centres ([27], section 5.3.1).

In the case of Sm^{3+}, Pr^{3+} and Dy^{3+}, the quenching of the luminescence occurs in ion pairs and not by energy migration. This mechanism is known as cross-luminescence and occurs when only a part of the excitation energy is transferred from ion to ion. This process is depicted in figure 2.25 for Pr^{3+}. The reader seeking detailed information on concentration quenching is referred to ([27], ch 5, and [138]) and references contained therein.

It is important to realize that multiphonon emission from energy levels of associated ions also leads to suppressed emission. This mechanism is important only if the energy difference between the levels involved is less than about five times the vibrational frequency and is independent of the concentration of luminescent centres. For ZnS this energy difference

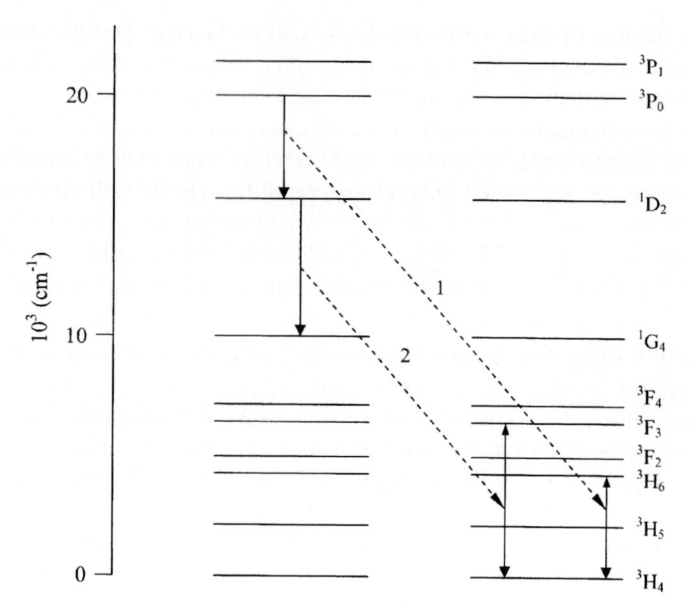

Figure 2.25. Schematic diagram showing how the $^3P_0$ and $^1D_2$ emission can be quenched by cross-relaxation between two Pr^{3+} ($4f^2$) indicated by the arrows labelled 1 and 2.

corresponds to 1785 cm^{-1} or 0.22 eV. Törnqvist [139] has studied the effects of manganese concentration on the electro-optical characteristics of ZnS:Mn ACTFELDs. The devices that were studied consisted of a 0.35 μm thick ZnS:Mn EL phosphor sandwiched between two Al_2O_3 insulator layers. All films were grown by atomic layer epitaxy. The ZnS films had manganese concentrations ranging from 0.8 to 0.15 wt% (Mn/ZnS). Weight percent is converted to atomic percent by:

$$\text{at\% A} = \left| \frac{\dfrac{\text{wt\% A}}{\text{at.wt A}}}{\dfrac{\text{wt\% A}}{\text{at.wt A}} + (\text{wt\% B})/(\text{at.wt B})} \right| \times 100 \qquad (2.73)$$

where A refers to the dopant and B refers to the ZnS host. The ratio of the radiative decay rate to the total decay rate measured at 3500 cd/m^2 was 68% for an Mn concentration of 0.19 wt% or 0.354 at%. The corresponding ratio was only about 50% when the concentration was 0.66 wt%. At 0.66 wt% the luminescence decay rate was faster by a factor of four compared with the rate at 0.19 wt%. Törnqvist proposed that at the higher dopant concentrations, excited Mn^{2+} ions interact by energy transfer to unexcited Mn^{2+}, and that the interaction results in nonradiative decay as is shown in figure 2.26. Sasakura *et al* [140] have studied evaporated ZnS:Mn thin films with Mn concentrations ranging from 0.03 to 10 wt%. They

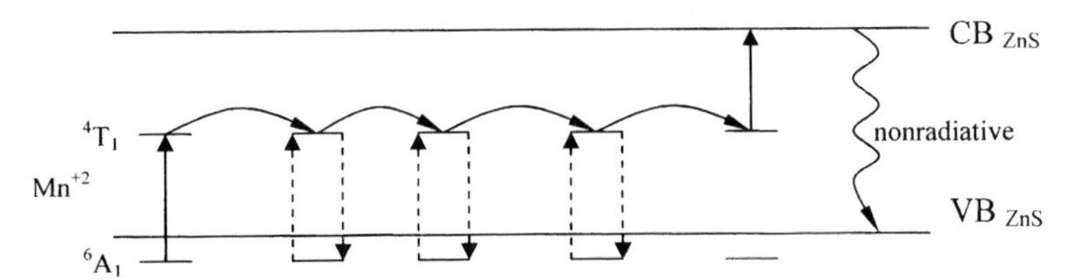

Figure 2.26. Schematic illustration of the energy transfer and luminescence quenching at high Mn^{2+} concentrations in ZnS:Mn.

employed Y_2O_3 as the insulator layers. The highest efficiencies were obtained for phosphor layers with 0.45 wt%. Above this concentration the efficiency and luminescence decay time were observed to decrease rapidly. Ono reports a similar effect at Mn concentrations of 0.5 wt% ([29], p. 23).

It should be expected that the processing of the ZnS will determine the structural properties of the phosphor. Choice of the insulator layers affects the electrical properties of the device via the density and depth of interface states that it provides for field emission. These parametres govern the transport and electrical properties of ACTFELDs and consequently their electro-optical performance. Therefore, depending on deposition technique and choice of insulators, the optimal manganese concentration will vary within the dilute solid-solution limit.

2.3.2.3 Figures of merit for a.c. thin-film EL devices

For practical application, a.c. thin-film EL displays and all other EL devices for that matter must exhibit certain key features. In this section we review these properties and methods for their evaluation.

2.3.2.3.1 Optical output. For practical applications the optical output of EL devices is a critical parameter. Two systems of units for quantifying optical output are used. Radiometric units are optical power units and can be used to quantify all electromagnetic radiation. These units quantify the photon flux density (radiant flux density) at a given wavelength. Optical power or *irradiance* is measured in $W/cm^2 \cdot nm$.

Photometric units can be used to quantify light only in the visible part of the spectrum. The photometric system takes into account the spectral response of the human eye. This response curve is called the luminosity curve for the standard observer (also known as the CIE curve). A brightness scale is the basic measuring element of the standard observer curve. That is, human observers were asked to rank by brightness various wavelengths of light of known flux density. The response curve can be interpreted as follows: a wavelength of 555 nm will appear brighter than any other wavelength of the same radiometric power. A source that can emit the

Table 2.6. Relative luminosity as a function of wavelength: the basis of the CIE or standard observer curve which accounts for the spectral response of the human eye.

Wavelength (nm)	Relative luminosity (η)	Wavelength (nm)	Relative luminosity (η)
410	0.001	560	0.995
420	0.004	570	0.952
430	0.012	580	0.870
440	0.023	590	0.757
450	0.038	600	0.631
460	0.060	610	0.503
470	0.091	620	0.381
480	0.139	630	0.265
490	0.208	640	0.175
500	0.323	650	0.107
510	0.503	660	0.061
520	0.710	670	0.032
530	0.862	680	0.017
540	0.954	690	0.008
550	0.995	700	0.004
555	1	710	0.002
		720	0.001

same radiometric energy at 555 nm and 610 nm will appear only half as bright when operated at 610 nm as it does when operated at 555 nm. This relative brightness is known as the relative luminosity. Brightness is formally referred to as luminance (L = lumens/m^2). Table 2.6 shows the relative luminosity as a function of wavelength. A plot of the relative luminosity versus wavelength yields the luminosity curve for the standard observer or CIE curve. Notice that at 555 nm the relative luminosity is equal to unity, i.e. the human eye is most sensitive to this wavelength. The relationship between luminous flux and radiometric flux is

$$F_v = \phi_r \times 683\,\text{lm/W} \times \eta \tag{2.74}$$

where F_v is the luminous flux (lm), ϕ_r is the radiometric flux (W), 683 lm/W is a physical constant and η is the relative luminosity at the wavelength under consideration. In other words, luminous flux is the optical power (W) weighted against the CIE curve. Intensity in the photometric system is often specified in candelas rather than lumens [1 candela (cd) $\equiv$ 1 lumen/steradian (sr)]. The units of luminance or brightness then becomes cd/m^2 (= lm/m^2 sr).

Irradiance and luminance are measured using a calibrated optical spectrometre. Ideally one uses an integrating sphere for collection of the output photons. It should be noted that the number of photons produced by the EL device is always larger than the number of photons reaching the detector due to imperfect collection optics, and the transmission efficiency of any gratings used. In addition, photon losses occur for emitted light travelling the optical path from the EL device to the light detector. Specifically, the flux density at any radius from a point source is given by the well known simple equation [141]

$$H_{\mathrm{rad}} = \frac{\phi_{\mathrm{r}}}{4\pi R^2} \tag{2.75}$$

where ϕ_{r} is the radiometric flux or power (W), H_{rad} is the radiometric flux density ($\mathrm{W/cm^2}$) and R is the distance from the source. Equation (2.75) indicates that the magnitude of the measured flux density will decrease as the inverse square of the distance from the source, and is known as the *inverse-square law*.

The American National Standards Institute/Human Factors Society (ANSI/HFS) 100-1988 standard is usually used to estimate the required luminance levels needed in subpixels for colour display applications. This standard requires an ambient luminance level of $35\,\mathrm{cd/m^2}$ for white light at 60 Hz. According to the ANSI standard, the ratio for the red:green:blue (R:G:B) components should be 27:65:8. The European Broadcasting Union Standard (EBU) requires a R:G:B ratio of 30:59:11. This results in a required ambient (areal) luminance of 9.3, 23 and $2.7\,\mathrm{cd/m^2}$ for the R:G:B components in the ANSI standard and 10.5, 20.6 and $3.9\,\mathrm{cd/m^2}$ in the EBU systems, respectively. In order to obtain the minimum $35\,\mathrm{cd/m^2}$ of ambient white light, the minimum required colour subpixel luminance at 60 Hz is 42, 104 and $12\,\mathrm{cd/m^2}$ for red, green and blue, respectively, according to the ANSI standard. The corresponding minimum subpixel luminance levels according to the EBU system are 48, 93 and $18\,\mathrm{cd/m^2}$ for red, green and blue. Monochrome, yellow emitting ZnS:Mn easily satisfies these requirements, as does red from filtered yellow ZnS:Mn, and green from ZnS:Tb,F/ZnS:TbOF (see table 2.7). These EL phosphor systems are used in commercial TFEL panels. Though the standard requires an ambient luminance level of $35\,\mathrm{cd/m^2}$ for white light at 60 Hz, in order to remain competitive with other flat panel displays TFEL technologies need to produce luminance levels well above this limit. For example, High-Definition Flat Panel Displays require around $200\,\mathrm{cd/m^2}$ of white ambient luminance. The reader is referred to table 2.7 for a listing of the luminance and colour coordinates of ZnS together with some other phosphors. For comparative assessment, the luminance or brightness value is reported at 40 V above the threshold. That is, the so-called B_{40} value is used. The threshold voltage is defined as the voltage required to produce a luminance output of $1\,\mathrm{cd/m^2}$ ([29], p. 38). In order to properly compare EL luminance data from different devices the drive waveforms (together with pulse width, rise-time and fall-time), drive frequency, phosphor layer thickness and operating temperature must be known.

Table 2.7. Comparison of the visible characteristics of ZnS EL phosphors.

Phosphor material	Emission colour	CIE coordinates	Subpixel luminance at 60 Hz (cd/m^2)	Luminous efficiency (lm/W)
ZnS:Mn	Yellow	$x = 0.5$, $y = 0.5$	600	5
ZnS:Mn/filter	Red	$x = 0.65$, $y = 0.35$	75	0.8
ZnS:Mn/filter	Yellow-green	$x = 0.45$, $y = 0.55$	80	1.3
ZnS:TbOF	Green	$x = 0.30$, $y = 0.60$	125	0.5–1
ZnS:Tb,F	Green	$x = 0.30$, $y = 0.60$	90	0.08
ZnS:Sm,Cl	Red	$x = 0.64$, $y = 0.35$	12	0.05
ZnS:Sm,F	Orange-red	$x = 0.60$, $y = 0.38$	8	<0.01
ZnS:Tm,F	Blue	$x = 0.15$, $y = 0.15$	<1	1.6
ZnS:Mn/SrS:Ce	White	$x = 0.42$, $y = 0.48$	450	0.2
SrS:Ce,Eu	White	$x = 0.41$, $y = 0.39$	36	0.05
CaS:Eu	Red	$x = 0.68$, $y = 0.31$	12	0.1
CaS:Ce	Green	$x = 0.27$, $y = 0.52$	10	0.04
$CaGa_2S_4$:Ce	Blue	$x = 0.15$, $y = 0.19$	13	

2.3.2.3.2 Color. Any colour can be obtained by appropriate combination of the three primary colours i.e., red, green and blue. The standard for colourimetry is the Commission Internationale d'Éclairage (CIE) system which employs a three-dimensional diagram to represent the attributes of colour. The CIE colour coordinates are defined as the ratios

$$x = \frac{X}{X+Y+Z}, \qquad y = \frac{X}{X+Y+Z}, \qquad z = \frac{Z}{X+Y+Z} \tag{2.76}$$

where X, Y and Z are the integrated tri-stimulus values over the entire blue, green and red regions of the electromagnetic spectrum, i.e. over the entire visible spectrum. The green spectrum used in this scheme corresponds to the photopic response of the eye and peaks at 555 nm, as discussed in the preceding subsection. Since the sum of the components x, y and z equals unity, it is sufficient to specify x and y only in order to define a colour. Figure 2.27 is a representation of the two-dimensional CIE colour coordinate system.

I. ZnS:Mn, monochrome yellow and orange yellow. ZnS:Mn is the best understood and most developed TFEL material and technology. ZnS:Mn is used in all EL display types, that is, monochrome, multicolour and full-colour as we shall see shortly. Manganese has both the correct valence and similar ionic radius (only 8% difference) with respect to Zn (see table 2.2). In addition ZnS and MnS have the same crystal structure. This leads to a large solid solubility of Mn in the ZnS host. Manganese can be evenly distributed without the need for charge compensators, and a large impact

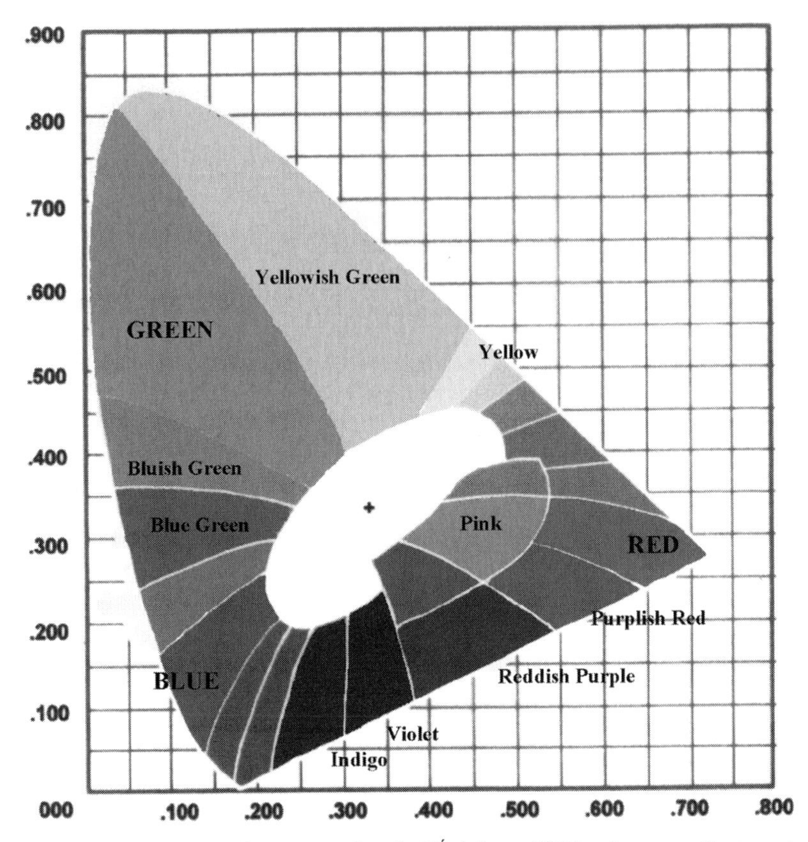

Figure 2.27. The Commission Internationale d'Éclairage (CIE) colour coordinate system.

cross-section of around $2 \times 10^{-16}\,\mathrm{cm}^2$ is realized. As a result, ZnS:Mn ACTFELDs show the highest luminance (optical output) and luminous efficiency compared with other ZnS phosphors. ZnS:Mn exhibits a yellow luminance of $600\,\mathrm{cd/m^2}$ at 60 Hz and $5000\,\mathrm{cd/m^2}$ at 1 kHz. The corresponding luminous efficiency at 1 kHz was 5 lm/W (see table 2.7). Lewis *et al* [36] showed that codoping ZnS:Mn with K and Cl limited the space charge from hole trapping and led to 60% increases in the brightness and efficiency of sputter-deposited EL thin films. As shown in figures 2.2 and 2.3, the yellow (585 nm) emission of the Mn^{2+} ion is a ${}^4T_1(3d) \rightarrow {}^6A_1(3d)$ transition. Though the $3d$–$3d$ transitions are formally parity forbidden because the principal quantum number does not change, uneven crystal field around the Mn^{2+} ions make them partially allowed as was discussed in section 2.2.4.1. The CIE coordinates of this emission are $x = 0.5$ and $y = 0.5$. The emission of ZnS:Mn appears at 585 nm when the material has the zincblende structure and at 580 nm when the crystal structure is hexagonal ([299], p. 85).

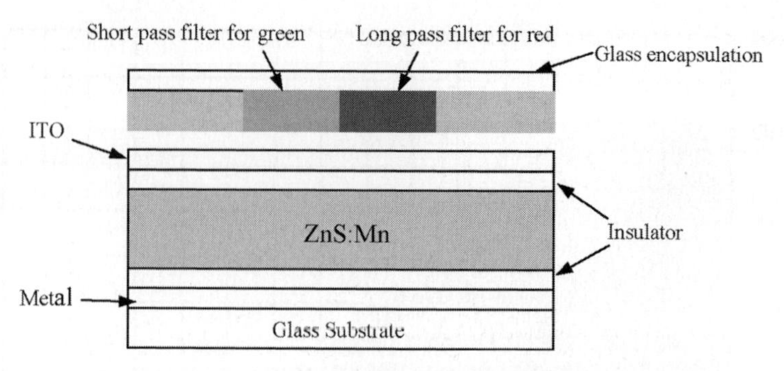

Figure 2.28. Schematic structure of a multicolour ZnS:Mn TFEL display. The red and green segments are filters that allow the passage of only red or green light [143].

II. ZnS:Mn—multicolour, red and green filtered from yellow. Multicolour green/yellow/red emitting devices can also be made from ZnS:Mn. This is afforded by the broad, intense emission of ZnS:Mn which with appropriate filtering provides sufficient photons in the red and green range for practical applications. In 1991 Tuenge and Kane [142] employed a long-wavelength pass filter made of CdSSe (cadmium sulphoselenide) on yellow emitting ZnS and demonstrated red emission with CIE coordinates $x = 0.65$ and $y = 0.35$. The reported brightness (luminance) and luminous efficiency were $75\,\mathrm{cd/m^2}$ and $0.8\,\mathrm{lm/W}$ respectively at 60 Hz. At the same time Okibayashi *et al* [143] obtained yellowish-green emission using a high-pass filter from filtered ZnS:Mn with CIE coordinates of $x = 0.45$ and $y = 0.55$ and an intensity of $80\,\mathrm{cd/m^2}$. In 1991 Sharp and Planar began manufacture of these multicolour EL devices. Figure 2.28 is schematic of a ZnS:Mn multicolour device.

III. ZnS:Tb,F, green. The energy level diagram for the Tb^{3+} ion is given in figure 2.6. The emission spectrum of this ion together with other rare-earth ions which emit in the green region of the visible spectrum is shown in figure 2.29. The optical output of $125\,\mathrm{cd/m^2}$ (at 60 Hz) of this green phosphor with CIE coordinates of $x = 0.32$ and $y = 0.60$ is second only to ZnS:Mn. Fluorine is used in this system as a charge compensator, as is chlorine. table 2.2 indicates that the ionic radius of Tb^{3+} is 25% larger than Zn^{2+}. This size mismatch makes thermal diffusion of the Tb ions to Zn sites in the ZnS lattice difficult. It has been found that sputtering is very effective for the incorporation of Tb^{3+} into ZnS because sputtering imparts kinetic energies of several eV to the nucleating atoms during film growth [144–147]. These researchers proposed that the excess energy enables better substitution of Tb^{3+} on to Zn^{2+} lattice sites, and that F^- ions act as charge compensator for the excess positive charge (of the Tb^{3+}). Fluorine is believed to reside as an interstitial in the lattice.

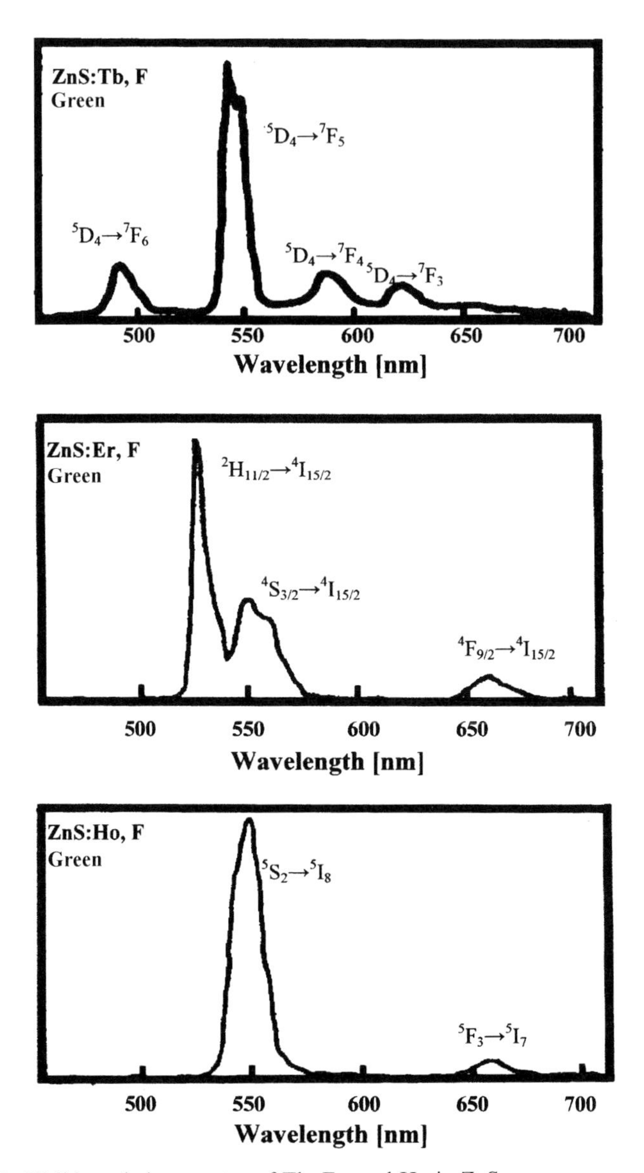

Figure 2.29. Visible emission spectra of Tb, Er, and Ho in ZnS.

In 1971 Krupka and Mahoney showed that the main excitation mechanism in ZnS:Tb thin-film electroluminescence is direct impact excitation of Tb^{3+} ions [131, 148]. Recently, it has been established that there are at least two alternative excitation mechanisms present in devices, namely, resonant energy transfer from electron–hole pairs, and indirect excitation via donor–acceptor pairs or bound excitons which can be generated or trapped around the Tb ion, or Tb related centre [149, 150]. In the latter mechanism, energy from the recombination of the pairs is transferred to the Tb centre from the ZnS host. These alternative mechanism have been shown to exist in ZnS:Mn as well [135].

TbOF complexes have been shown to improve the luminance and luminous efficiency of the green phosphor [151–154]. The pioneering work with molecular complexes was performed by Kahng, Chase and coworkers in the late 1960s [9, 18]. Based on the experimental data obtained, Sohn *et al* [154] proposed that oxygen doping gave rise to an increase luminance due to:

(a) an improvement of crystallinity, film uniformity and grain size,
(b) shallow interface states with low density introduced by oxygen which leads to a decrease in threshold voltage,
(c) oxygen acting as a recombination centre when it substitutes for some of the Zn or S vacancies, and
(d) the suppression of non-radiative energy transfer via grain boundaries and/or vacancies.

Kim *et al* [155] evaluated the effects of O co-doping of ZnS:TbF and showed that at concentrations below the optimum value of $[O]/[Tb] = 1$, the brightness was low due to low transfer charge, presumably due to a low density of interface states. Above the optimum value of O, the brightness again decreased due to lower excitation and radiative recombination efficiencies. Kim *et al* [156] also showed that co-doping ZnS:TbOF with Ce led to energy transfer into the Tb centres and increased Tb EL brightness and efficiency.

IV. ZnS:Sm(F,P,Cl), red. Compared with Zn^{2+}, the ionic radius of Sm^{3+} is 30% larger. In addition, a valence mismatch also exists. These factors negatively influence the luminance output of ZnS:Sm phosphors. Chase *et al* [18] were the first to report reddish emission with colour coordinates $x = 0.6$ and $y = 0.38$ from Sm-doped ZnS, where F was used as the charge compensator. When P [157] and Cl [158, 159] were used as charge compensators, colour coordinates of $x = 0.63$, $y = 0.36$ and $x = 0.64$, $y = 0.35$ were obtained, respectively. These latter two chromaticities are quite close to the standard red used in CRTs. The shifts in emission colour observed with the change in charge compensator indicate that the relative intensities of the transitions shown in figure 2.30 are affected by the local crystal fields. The maximum reported luminance for ZnS:Sm,Cl of $1000\,cd/m^2$ at 5 kHz [158, 159] extrapolates to $12\,cd/m^2$ at 60 Hz. This intensity is several factors lower than that required (as discussed in subsection 2.3.2.3.1) for applications in display technologies.

V. ZnS:Tm,F, blue. ZnS doped with Tm and F as a charge compensator has been studied for a possible blue EL phosphor, albeit with limited

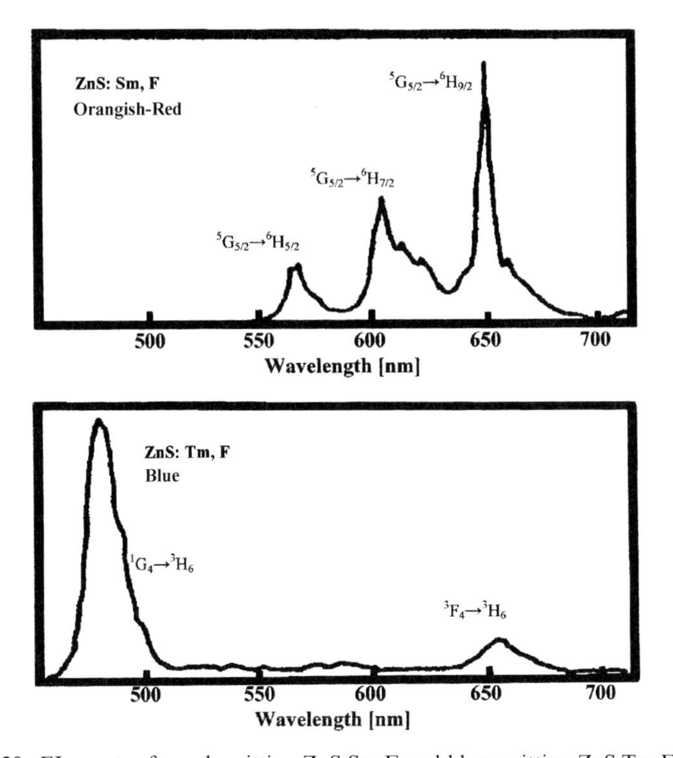

Figure 2.30. EL spectra for red emitting ZnS:Sm,F and blue emitting ZnS:Tm,F.

success. The low luminance value of 0.2 cd/m^2 [73] is due to the ionic-radius and valence mismatch between Zn^{2+} and Tm^{3+}, and makes this system unsuitable for display applications. However, the emission of Tm^{3+} has colour coordinates of $x = 0.15$, $y = 0.15$ and is pure blue.

It has been reported that oxygen co-doping improves the luminance of ZnS:TmF_3 [160]. They proposed that the luminance increase observed with oxygen co-doping was due to improved film crystallinity, and decreased non-radiative energy transfer via sulphur vacancies because oxygen substitutes on to some such sites.

VI. ZnS for full colour (red, green, blue) applications. ZnS:Mn is the phosphor of choice in TFEL display applications where one other colour is needed, as discussed in subsections 2.3.2.3.1 and II above. With very little change to the existing well understood monochrome technology, multicolour red, green, and yellow emitting displays with sufficient luminance could be obtained (see figure 2.28).

Full-colour displays require the three primary colours. As discussed above, red, green and blue emission can be obtained by using the appropriate

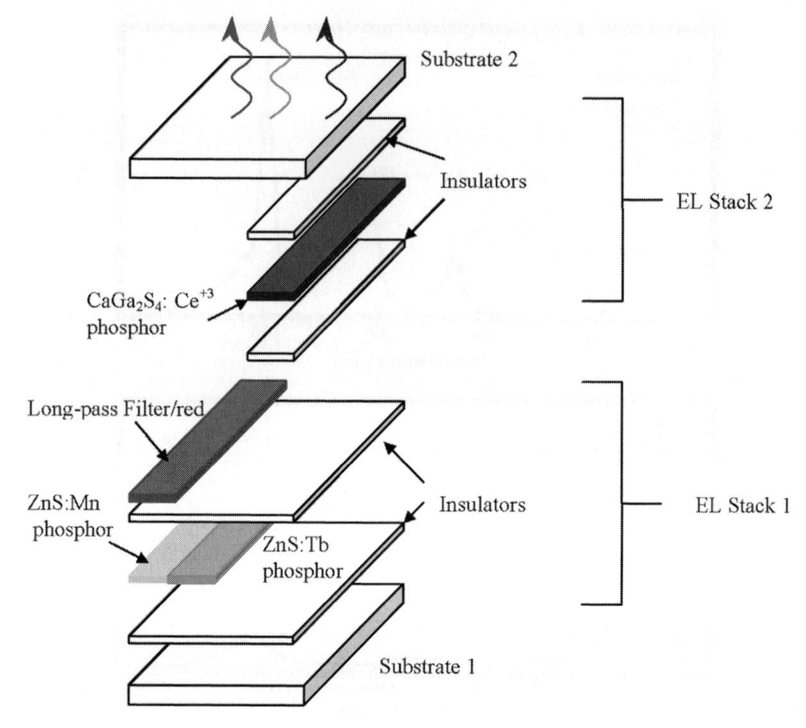

Figure 2.31. Schematic of the dual-substrate display concept. The electrodes have been omitted for clarity. The top electrode of the bottom stack and both electrodes of the top stack must be transparent in the visible part of the spectrum.

rare-earth dopant in ZnS, though the luminance and efficiency of the red and blue emission is poor and insufficient for practical applications. The first approach at full-colour EL displays was the so-called dual-substrate display. It was based on the work of Barrow *et al* [19, 161] and King [162] and was manufactured by Planar [99]. Figure 2.31 shows the principle: green emitting ZnS:Tb and red emitting ZnS:Mn (filtered) are deposited side by side on one substrate. On a second substrate, monochrome blue emitting calcium thiogallate doped with Ce^{3+} ($CaGa_2S_4$:Ce^{3+}) was deposited. Optical contact between the two substrates put face-to-face was established by filling with suitable oil. Of course all of the electrodes of the upper display must be transparent.

In parallel to the above approach, the colour-by-white approach was being developed [163]. The concept (see figure 2.32) is to filter red, green and blue from a white emitting phosphor. The luminance and efficiency of the white phosphor must be good enough to allow for the light absorbing filtering. There is no white emitting phosphor and therefore multilayer stacks of phosphor films are used. Red and green from filtered ZnS:Mn

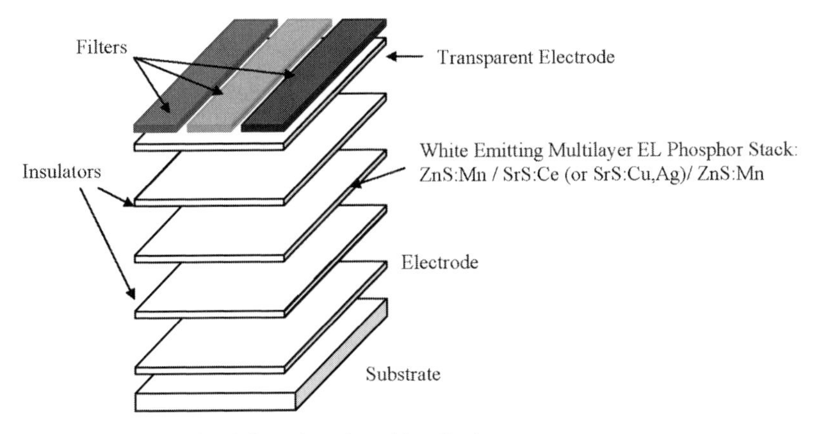

Figure 2.32. Schematic of the colour-by-white display concept.

and blue from strontium sulphide phosphors (SrS:Ce or SrS:Ag,Cu) combine to cover most of the visible spectrum. In 1992 Nire *et al* [164] demonstrated a successful example which employed a composite ZnS:Mn^{2+}/SrS:Ce^{3+}/ZnS:Mn^{2+} multilayer active region. The first colour-by-white full-colour was introduced to the market in 1997 by Planar. The principal device structure and choice of materials was the same as those used by Törnquist [165] and Soininen [166]. After being filtered and recombined, a mixed white with an excellent luminance of 45 cd/m^2 was achieved at 240 Hz drive frequency. A similar product from Westaim exhibits better blue colour coordinates [99]. Note that SrS:Ce^{3+} does not produce a saturated blue (see figure 2.6(c)). SrS:Cu [166] and SrS:Ag,Cu [99, 167] are expected break the 'blue barrier' and produce blue with improved colour coordinates.

2.3.2.3.3 Efficiency. The conversion of electric input into optical output is subject to several electro-optical processes which determine the total optical efficiency of EL devices. The total efficiency of the device can be written as

$$\eta = \eta_{\text{out}}\eta_{\text{rad}}\eta_{\text{hot}}\eta_{\text{exc}} \tag{2.77}$$

where η_{out} is the outcoupling efficiency, η_{rad} is the internal quantum efficiency, η_{hot} is an efficiency related to the fraction of electrons that become sufficiently hot to excite luminescent dopants (see table 2.5), and η_{exc} is the efficiency related to the fraction of hot electrons that *actually* excites dopants.

We begin by looking at the first term in equation (2.75), η_{out}, the outcoupling efficiency. For information display applications photons generated in the phosphor layer must escape in the direction of the viewer. When light travelling in a material of a certain optical density (refractive index) impinges on the boundary with a material of a lower optical density at a certain critical angle, the light will be totally internally reflected. This effect traps the majority of the generated light in the phosphor layer because

$n_{\text{phosphor}} > n_{\text{insulators}} > n_{\text{glass}} > n_{\text{air}}$, where n refers to the refractive index of the respective medium. Furthermore, some light will be trapped in the insulators and glass substrate by the same phenomenon ($n_{\text{insulators}} > n_{\text{glass}} > n_{\text{air}}$), and only a small portion is available to the viewer. The percentage of light that escapes a device can be estimated. Consider Snell's law,

$$n_1 \sin\theta_1 = n_2 \sin\theta_2, \tag{2.78}$$

where n_1 is the refractive index of ZnS (2.4), θ_1 is the angle of incidence, n_2 is the refractive index of air for example ($n_2 = 1$), and θ_2 is the angle of refraction. When the refraction angle is 90° no refraction will occur because the light will be reflected internally. The critical angle for total internal reflection can be obtained from equation (2.78):

$$\theta_1 = \theta_c \arcsin(1/2.4) = 24.6°. \tag{2.79}$$

The optical outcoupling efficiency when all incidence angles smaller than the critical angle are considered is then

$$\eta_{\text{out}} = \int_0^{24.6} \sin\theta \, d\theta \approx 0.1 \tag{2.80}$$

assuming the back (bottom) electrode is completely (100%) reflective. In other words only about 10% of the generated light is seen by the viewer, and 90% travels in the plane of the EL stack. Mach and co-workers have shown [168, 169] that light extraction can be increased to about 40% by using ZnS with roughened surfaces. Rough ceramic reflector substrates have also been shown to significantly improve optical outcoupling [170–172], and is partially responsible for the success of Westaim's full-colour displays. It should be noted, however, that excessive surface roughness reduces contrast because of increased diffuse scattering.

The internal quantum efficiency (η_{rad}) is a measure of the radiative efficiency (see equation (2.12)). It is dependent on the luminescent centre concentration because of concentration quenching, on temperature because the non-radiative recombination rate increases exponentially with temperature (see equation (2.13)), and on processing because the concentration of non-radiative recombination pathways depends on the degree of crystalline perfection. Mueller-Mach and Mueller [99] have reported an internal quantum efficiency of 0.4 for ZnS:Mn.

The efficiency related to the fraction of electrons which become sufficiently energetic (hot) in order to effect excitation of luminescent centres is dependent on the electron energy distribution (see subsection 2.3.2.2.2 and table 2.5).

The fraction of hot electrons that *actually* excites luminescent dopants determines the excitation efficiency, η_{exc}. The impact length for excitation l_{ex} is inversely proportional to the dopant concentration (higher concentrations means less distance between dopants) and excitation cross-section.

That is

$$l_{ex} = [N(\mathrm{cm}^{-3}) \times \sigma(\mathrm{cm}^2)]^{-1}. \tag{2.81}$$

As long as the drift length (l_d) needed to gain sufficient energy for excitation from the field is shorter than the impact length l_{ex}, the excitation efficiency is

$$\eta_{exc} = l_d / l_{ex} = l_d N \sigma. \tag{2.82}$$

The magnitude of σ in ZnS:Mn is 1–4 $\times 10^{-16}$ cm^2 [173, 174] and an optimum dopant concentration of 2×10^{20} cm^{-3} has been reported [99].

A parameter of practical importance is the luminous efficiency. It is simply the ratio of the optical power output (in photometric units = luminance) to the electrical power consumed. Therefore, in order to estimate the luminous efficiency (η_{lum} = lm/W), the input power density (W/cm^2) and luminance (cd/cm^2) must be known. The input power density to an EL panel can be estimated from the area under its charge-voltage (Q-V) plot. That is:

$$P_{in} = \text{drive frequency} \times \text{area under Q-V plot} \tag{2.83}$$

$[\mathrm{Hz} \times \mathrm{C/m^2} \times \mathrm{V} = 1/\mathrm{s} \times \mathrm{A.s/m^2} \times \mathrm{V} = \mathrm{W/m^2}]$.

We have discussed the conduction current and its voltage dependence in detail in subsection 2.3.2.2.1. It was also noted in section 2.3.2.2.4 that careful monitoring of the current during and after the voltage pulse has been applied is required for proper identification of the electroluminescence mechanism. Electrical current is the flow rate of charge, that is, amperes are coulombs per second. Charge is therefore current integrated over time. In this case we refer specifically to the conduction or dissipative current and associated electrical charge. A few words on the measurement of this quantity are therefore in order.

Wager and Keir [175] have developed a technique for the measurement and calculation of conduction current (charge) based on the Sawyer–Tower circuit [176] for arbitrary waveforms. This technique uses the instantaneous values of external voltage and current (or charge), assumes the phosphor and insulators are ideal electrical components, and uses simple circuit equations to calculate the conduction current or charge. Figure 2.33 is a schematic of the scheme.

If the series resistor is used as the sense element, the external charge density is simply

$$Q_{ext}(t) = \int_0^t i(t)\,dt = \int_0^t \frac{V_1(t) - V_2(t)}{R_{series}}\,dt \tag{2.84}$$

where R_{series} is the series resistor. The applied voltage is

$$V_{ap} = V_2(t) - V_3(t) \tag{2.85}$$

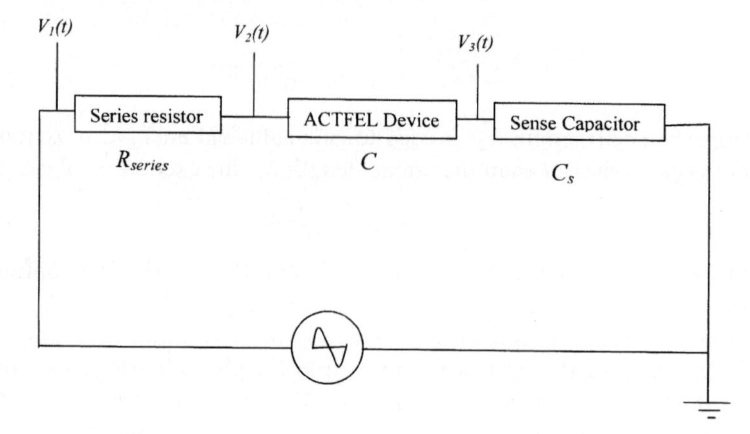

Figure 2.33. Schematic of the Sawyer–Tower circuit for measuring conduction current [176].

and a Q-V plot can be obtained. If the series capacitor is used as the sense element the external charge density is given by

$$Q_{ext} = C_s V_3(t) \tag{2.86}$$

under the condition that the sense capacitance is much larger than the total capacitance of the multilayer stack ($C_s \gg C$, see equation (2.46). Figure 2.34 is the Q-V characteristic corresponding to the ideal model of the double-insulating, a.c. thin-film EL device (back-to-back Zener diodes). As discussed in section 2.3.2.2.1, the EL stack is considered as capacitors connected in series below the threshold voltage. Below the threshold voltage the Q-V plot is a straight line going through the centre of coordinates with slope $C = Q/V$, where C is the total capacitance of the EL stack. This is represented by the dotted arrow in figure 2.34. Above the threshold voltage V_{th}, the Q-V diagram is represented by a parallelogram as shown [177]. The slope of the sides of this parallelogram is the combined capacitance (C_1) of the insulators only (equation (2.45)), as the phosphor experiences 'Zener breakdown' above V_{th}.

In accordance with subsection 2.3.2.2.1 and equation (2.49), $V_{EL,th} = [C_1/C_2 + C_2] \times V_{th}$. Based on equation (2.81) and figure 2.34, the power consumed is

$$P_{in} = 2 \times f \times V_{EL,th} \times \Delta Q. \tag{2.87}$$

Having discussed the measurement of Q-V curves and input electrical power, we can now calculate the luminous efficiency:

$$\eta_{luminous}[\text{lm/W}] = \pi \times (L[\text{lm/m}^2]/P_{in}[\text{W/m}^2]). \tag{2.88}$$

It should be noted that ideal Q-V characteristics are not observed in practice. Figure 2.35 is a more realistic depiction of the Q-V behaviour of an actual

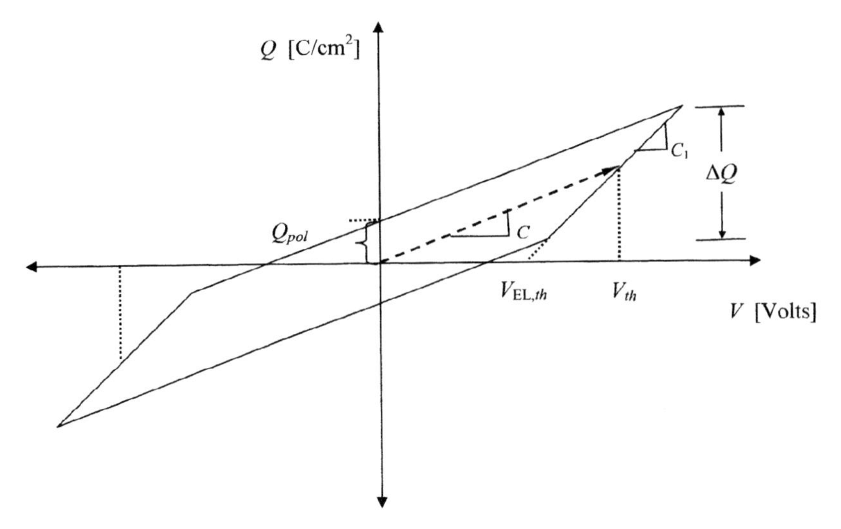

Figure 2.34. Charge-voltage (Q-V) characteristics of an ideal ACTFELD. Here V_{th} is the threshold voltage of the EL device, $V_{\text{EL,th}}$ is the threshold voltage of the EL phosphor layer, ΔQ is the transferred charge, and Q_{pol} is the charge remaining in the EL device after the applied voltage has been reduced to zero (polarization charge).

double-insulating ZnS:Mn EL device under excitation of a trapezoidal waveform (5 μs rise and fall time, and 30 μs dwell time). Note the various charge transfer processes that are shown. Readers seeking more details on a.c. thin-film EL devices are referred to [29, 73, 99].

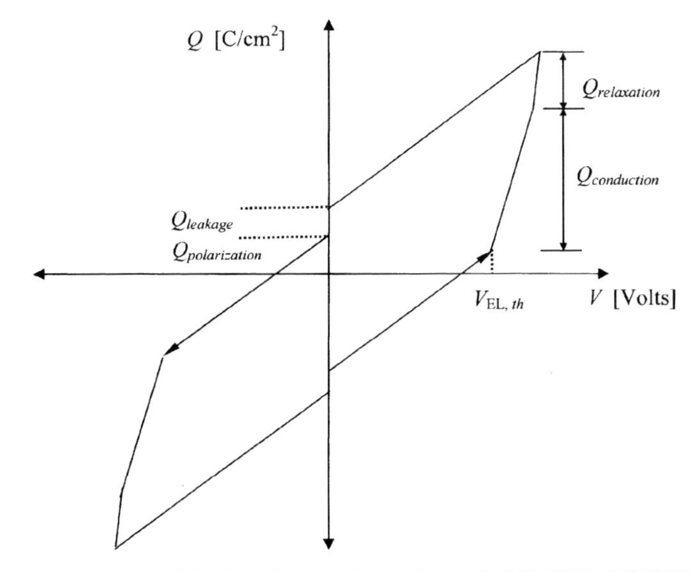

Figure 2.35. Schematic of the Q-V characteristics of a typical ZnS:Mn ACTFEL device.

2.3.2.3.4 Contrast ratio. Electroluminescent displays must exhibit good contrast ratios in order that good colour rendering and sharpness may be achieved. The contrast ratio is essentially the luminance output of the EL device normalized to the ambient illuminance. Illuminance is typically reported in units of lux [lux = lm/m^2]. To measure the contrast ratio, the EL device surface is illuminated by an ambient light source at some incident angle θ from the normal to the EL device surface. In the direction normal to the EL device surface one measures the luminance at the device surface due to illumination, as well as the luminance output due to the EL emitting area. The contrast ratio is defined as the ratio of the luminance of an EL emitting area (pixel) to the luminance of a non-emitting area. From the preceding it is clear that in order to specify a contrast ratio, the incident angle and intensity (in lux = lm/m^2) of the illuminating source must be known, as must the luminance output from the EL emitting area.

A contrast ratio of 20:1 has been reported for commercial multicolour ZnS:Mn devices operating at 60 Hz at 500 lux ambient illumination, and a viewing angle of around 140° [23]. The contrast ratio was 10:1 when the ambient illumination was 1000 lux. For full-colour displays the reported values at 200 lux ambient are 10:1 and 20:1 at 60 and 180 Hz drive frequency respectively [19]. Device engineering is often used to improve the contrast ratio. For example, if black stripes are placed between the colour filters shown in figure 2.32, they will serve to absorb ambient light and as a result the contrast ratio will improve.

2.3.2.4 *Deposition techniques for ACTFEL device materials*

I. Metal electrodes. Metal electrodes are typically deposited by physical vapour deposition, i.e. by thermal evaporation. The vacuum chamber is evacuated to a pressure in the neighbourhood of the vapour pressure of the metal to be deposited (typically Al). A high current is then passed through a high resistivity metallic crucible such as tungsten, on to which the Al is placed. The resulting Joule heating of the crucible and its contents induces the Al to sublimate, and deposit on unmasked areas of the panel being fabricated. Electron beam evaporation (EBE) is also used. This process varies from thermal evaporation in that an electron beam directed at the sample provides the heat required for sublimation.

II. Transparent conducting electrodes (indium/tin/oxide). The transparent conducting electrode normally is ITO. Indium/tin/oxide is a degenerately doped, *n*-type, wide band-gap semiconductor. Magnetron sputtering is now the standard deposition method for high-quality ITO films because of the good characteristics the deposited films and high production throughput [178, 179]. It should be noted that high conductivity in ITO films is due to an oxygen deficiency, and control of the oxygen partial pressure during deposition is important. If the oxygen concentration is too high, the resistivity of the films is higher than the maximum of 10^{-4} Ω cm we discussed in section 2.3.2.1.3. On the other hand, precipitation of metal particles and resultant degradation of transparency occurs when films are oxygen deficient.

Sputtering is ideal for the deposition of films with complex composition. The main features of the sputter deposition can be listed as follows.

1. A voltage is applied across a medium in a 'capacitor', where the target material (ZnS) forms the cathode side of the gap. The medium is typically an inert gas such as argon at a pressure in the 10 millitorr range [1 millibar = 0.75 torr, 1 pascal (Pa) = 7.5 millitorr]. At some voltage level, breakdown of the gas occurs and a glow discharge plasma is initiated.
2. Positive ions from the plasma strike the target (cathode), where energy transfer from incident Ar ions to the target atoms takes place.
3. The transferred energy is sufficient to break bonds and physically sputter target atoms, which subsequently condense on to a substrate.

In contrast to d.c. sputtering, r.f. magnetron sputtering works for insulators because of the r.f. coupling to the externally applied field. The sputtering and deposition rates are also higher for r.f. magnetron sputtering. An intrinsic characteristic of the sputtering process is that when steady-state is established, the composition of the deposited film will be the same as the target material with respect to *concentration*. The crystal structure of the target has no relation to, or impact whatsoever to the crystal structure of the deposited films as the atoms of the deposited films must travel through the plasma, i.e. gas of ions, electrons and molecules of the discharge medium. Sputtered deposited films are typically subjected to a post-deposition anneal in order to improve crystallinity [180]. The reader is referred to the book by edited by Vossen and Kern for explicit details of sputter deposition [181].

III. Insulator layers. Sputtering is also the standard deposition method for oxide and nitride insulating layers as shown in table 2.4. This method is not limited by the melting point of the insulator, and large area deposition with fairly uniform thickness is possible. Sputtering is suitable for industrial in-line processing as the transparent conducting electrode and the insulator layers can be deposited by the same process. As discussed above and below, sputtering is also used for the deposition of ZnS phosphors.

Insulating layers are also deposited by atomic layer epitaxy (ALE). This process produces high quality layers which are free of pinholes and exhibit high breakdown voltages. The obtained layers are typically amorphous as process temperatures and post deposition annealing temperatures are not high enough to induce crystallization. In contrast to thermal evaporation and sputtering, which are source-controlled deposition techniques, atomic layer epitaxy is a surface-controlled growth technique. ALE is used for growth on both amorphous and single-crystalline substrates.

ALE is a novel modification of well-established growth techniques. The process was patented in 1974 [182] by the Finnish Corporation, Nohja. Simply put, atomic layer epitaxy makes use of the difference between chemical absorption (bonding) and physical adsorption. When the first layer of atoms or molecules of a reactive species reaches a solid surface there is usually a strong interaction, i.e. bonding or chemisorption; subsequent layers tend to interact much less strongly (physisorption). If the initial substrate surface is heated sufficiently one can achieve a condition such that only the chemisorbed layer remains attached. Given that, any excess incident molecules or atoms impinging on the film do not stick if the substrate temperature is properly chosen, and one therefore obtains precisely single monolayer coverage in each cycle. ALE is an excellent technique for the preparation of ultra-thin films of precisely controlled thicknesses in the nanometre range. The growth technique is based on chemical reactions at the solid surface of a substrate, to which the reactants are transported alternately as pulses of neutral atoms (either as chopped beams in high vacuum or as switched streams of vapour possibly in an inert carrier gas). The outermost atomic layer of the substrate reacts directly and chemically bonds only with the first monolayer of the incident pulse of atoms. The film therefore grows stepwise: a single monolayer per pulse. That is, at least one complete monolayer of coverage by a constituent element (or of a chemical compound containing it) is formed before the next pulse (of the other constituent element(s)) is allowed to react with the surface. Precaution must be taken to ensure the off-time of a given source beam is sufficiently long (of the order of 1 s) after each monolayer deposition. This complete cycle could be repeated indefinitely, the number of layers grown being determined solely by the number of cycles. An added benefit of this ‘off-time’ is that the surface is allowed to approach thermodynamic equilibrium at the end of each reaction step, a growth condition that is not usually met in other deposition techniques by evaporation. Elemental components of the final compound as well as compounds of the elements can be used as reactants. In the former case ALE is essentially evaporative deposition which relies on heated elemental source materials, and the process is more like molecular-beam epitaxy (MBE). When the reactants are compounds of elements the process is essentially chemical vapour deposition (CVD), relying on sequential surface exchange reactions between compound reactants. Goodman and Messa [183] have reviewed ALE and associated phenomena.

IV. ZnS phosphor layers. Whichever the technique used for the deposition of phosphor layers, the objective is to produce films with excellent stoichiometry and the best possible crystallinity. Crystal imperfections result in enhanced inelastic scattering which limits the maximum acceleration of electrons. Consequently, the number of electrons which gain the minimum energy required for the impact-excitation of luminescent centres is reduced. Therefore, deposition should occur at a temperature high enough to induce large grain growth. This temperature is limited by the dissociation temperature of ZnS. ZnS phosphors are typically subjected to a post-deposition anneal ($\sim$550 °C in vacuum or inert gas) in order to relieve strains in the film, accelerate diffusion of the dopants and further enhance crystallinity.

A number of physical vapour deposition and chemical vapour deposition techniques can be used for the deposition of thin-film ZnS:Mn phosphors. These include electron-beam and thermal evaporation, atomic layer epitaxy, sputtering, metal–organic chemical vapour deposition (MOCVD) and hydride or halogen transport chemical vapour deposition (HT-CVD). Ono has discussed these and other techniques in detail ([29], ch 6).

Of the various deposition methods, ALE and evaporation are used to fabricate commercial ZnS:Mn EL panels ([29], p. 70). The crystallite size of ALE-prepared ZnS:Mn is two to three times larger than that of other deposition methods [184], indicating that thin films prepared by ALE exhibit the best crystallinity. The Zn source is typically $ZnCl_2$, $Zn(CH_3COO)_2$, or $Zn(thd)_2$ where thd is 2,2,6,6,-tetramethyl-3,5-heptanedione [Les92]. The sulphur source is normally H_2S. Manganese compounds such as $MnCl_2$, $Mn(thd)_2$ and $Mn(CO)_5$ have been used as sources for the luminescent centre. There are some advantages in using $ZnCl_2$ and H_2S as source materials. First, the probability of re-evaporation of Zn from the surface is reduced because the surface is covered with a non-metallic layer. Second, molecules containing only one S atom such as H_2S are more reactive than two or poly-atomic forms. The ALE growth process for ZnS is described by the exchange reaction

$$ZnCl_2(ads) + H_2S(g) \rightarrow ZnS(s) + 2HCl(g) \tag{2.89}$$

where ads stands for the adsorbed state, s stands for the solid state, and g stands for gaseous state. $ZnCl_2$ is introduced to the deposition chamber in vapour phase and provided that the substrate temperature is not too high ($\sim$500 °C), a monolayer of $ZnCl_2$ adheres to the surface by chemisorption as discussed in III above. However, the substrate temperature is kept high enough to prevent the formation of zinc-to-zinc bonds on the surface, and therefore only a single layer of $ZnCl_2$ is formed per cycle. Excess $ZnCl_2$ vapour is subsequently flushed away, after which H_2S vapour is introduced to the chamber. Through reaction of H_2S with the adsorbed $ZnCl_2$, a ZnS layer is formed and HCl is liberated. After flushing away excess HCl, the cycle is repeated. Thickness monitoring is automatic as the final thickness of the deposited ZnS film is determined by counting the number of reaction cycles, provided that the dose of reactant in each step is high enough to produce a full monolayer coverage. When elemental sources are used the growth process is the evaporative extreme of ALE as described in III above, and is similar to MBE.

Films deposited by ALE possess high chemical stability as weak chemical bonds are automatically eliminated in each reaction step, which results in highly stable, stoichiometric films. The growth is insensitive to pressure and substrate materials, but the substrate temperature is important, i.e. the temperature must be sufficient for chemisorption but not too high to lead to desorption of the monolayer. As mentioned in III above, ALE produces highly uniform layers even in ultra-thin structures. An additional advantage of the ALE method is the possibility to prepare the whole insulating layer/phosphor layer/insulating layer stack in one continuous process.

Evaporation is also used for the deposition of commercial ZnS:Mn phosphor layers ([29], p. 70). Though the evaporation technique has difficulty producing good films of complex phosphors because of the widely varying vapour pressures and chemical reactivities of the constituent components of these compounds, the classical II–VI compounds form excellent films by evaporation. The source material is placed in a crucible and heated. The strength of the chemical bonding of these II–VI compounds is weak enough that the heat of evaporation is sufficient to largely dissociate the source molecules. The individual atomic species are then transported to the substrate provided that the deposition pressure level is low enough to permit collisionless transport. The atoms then recombine on the substrate to form the original II–VI compound. A feature of this recombination at the substrate is that it can be controlled to produce very stoichiometric films by adjusting the substrate temperature. The mechanism here is that the vapour pressure of the constituent atoms, for example Zn and S, is high enough at substrate temperatures above 200 °C that neither Zn nor S will adhere to other similar atoms. Thus the film growth proceeds by formation of alternate layers of Zn and S atoms and stoichiometry is automatically achieved. Table 2.8 is a comparison of some basic properties of ZnS obtained by various methods.

Sputtering is suitable for in-line processes because metal layers, transparent conducting electrode layers, insulating layers and the phosphor layer can all be grown by the same deposition method. As shown in table 2.8, sputter deposited ZnS:Mn can exhibit an equal luminance output relative to films obtained by other techniques [36]. Without care, the EL brightness and efficiency of sputter-deposited films may be low due to poor crystallinity from bombardment by energetic particles [180]. In addition, sputtered films tend to incorporate some of the discharge gas into the growing film which degrades the crystal structure or crystal perfection. The relatively slow deposition rate allows background gases (such as H_2O) to chemically react with the films during growth, which precludes the possibility of obtaining good films with excellent stoichiometry. Furthermore, II–VI sulphides are difficult to deposit because sulphur is less reactive than

Table 2.8. Basic properties of ZnS as a function of deposition method [36, 187–189].

Deposition technique	Deposition rate (Å/min)	Substrate temperature (°C)	Crystal structure	Emission peak (nm)	Luminance at 1 kHz (cd/m^2)
ALE	10–50	350	Zincblende	585	>3000
		500	Hexagonal	580	>3000
EBE	>1000	200	Zincblende	585	>3000
Sputtering	>100	200	Zincblende	585	>3000 [36]
MOCVD	>100	300–500	Hexagonal	580	>3000
			Zincblende	585	
HT-CVD	>100	500	Hexagonal	580	>3000

oxygen contained in background gases, and therefore it is difficult to avoid forming an oxy-sulphide rather than a pure sulphide. Finally sulphur and rare-earth compounds tend to form negative ions which sputter and damage the depositing film, reducing the deposition rate and affecting EL properties [180, 185]. Oxide formation can be suppressed by sputtering in a partial pressure of H_2S which creates an overpressure of S [186].

In contrast to ZnS:Mn, sputtering is the best technique for the deposition of ZnS doped with rare-earth ions phosphors. The brightest green-emitting EL phosphors, ZnS:Tb,F and ZnS:TbOF, are obtained by sputtering [144, 155, 156, 190]. The 25% ionic radius mismatch between Zn^{2+} and Tb^{3+} makes it very difficult to incorporate Tb ions into the ZnS host lattice without creating lattice defects. This is also true for other rare-earth ions (see table 2.2). The sputtering process is very effective for the preparation of ZnS doped with rare-earth ions because sputtering imparts a kinetic energy of several eV to the nucleating atoms during the film growth. The ion bombardment during sputtering influences both the crystal growth of the ZnS and the Tb dopant incorporation site in a manner such that the Tb dopant is more effectively incorporated into the thin-film phosphor.

2.3.3 Direct current powder electroluminescence from ZnS

Electroluminescence in powdered ZnS was studied and reported long before thin-film processes were developed. As early as 1936 Destriau [1] reported that upon application of a high field to a suspension consisting of ZnS:Cu:oil in a device that resembles a modern powder EL cell, light emission resulted. In the period immediately following Destriau's publication not much is reported on EL in the literature, which was at least in part due to the lack and immaturity of fabrication techniques for transparent electrically conductive films.

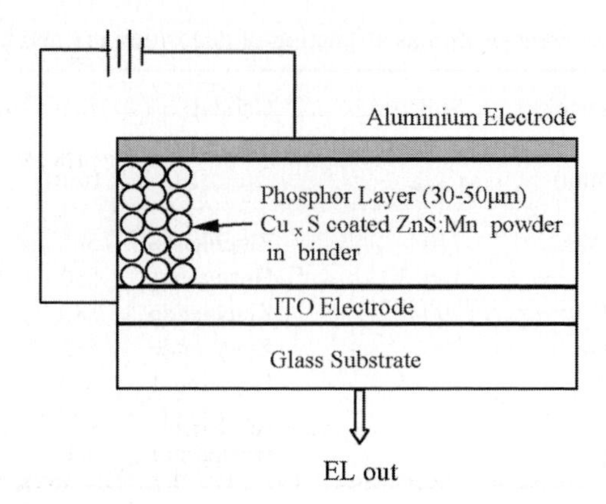

Figure 2.36. Schematic of a d.c. powder EL device.

The pioneering work on 'modern' powder EL panels was performed by Vecht and coworkers who demonstrated the first d.c. powder EL panel in 1968 [8, 191, 192]. Early powder phosphor devices were d.c.-driven. The associated ionic drift and material diffusion resulted in short device lifetimes. Permeation of moisture into the device structure also contributed to premature failure, i.e. absorbed water reacted with the EL phosphor, effectively quenching the luminescence. In the 1980s and 1990s, with a resurgence of interest in powder EL, these two failure modes were addressed by utilizing an a.c. driving voltage, and glass encapsulation techniques to minimize moisture-induced degradation of the EL phosphor. Vecht's publication 'Electroluminescent Displays' [193] reviews both a.c. and d.c. powder EL panels.

Figure 2.36 is a schematic of the d.c. powder EL structure. The phosphor layer is typically 30–50 μm thick, and consists of ZnS-doped Mn powder coated with Cu_2S in a binder. The powder particles must be in the range 0.5–1 μm and only a small amount of binder is necessary. That is, the phosphor-to-binder ratio must be high [193]. The Cu_2S coating is necessary for electron injection into the ZnS particles, and the binder is required to keep the device from becoming too conductive and shorting out.

When a voltage is first applied, a large current flows and no light emission is observed initially. When a critical power of approximately 2 W/cm is attained [193], the layer heats up, and gradually a narrow region ~1 μm thick adjacent to the transparent anode electrode begins to luminesce. This is phenomenon is referred to as the *forming* process. The *formed* region is the region associated with initial light emission and occurs nearest to the positive electrode. The light emission which initially begins adjacent to the transparent anode electrode gradually spreads until the whole thickness of the phosphor becomes light emitting. As the emission of light increases, a

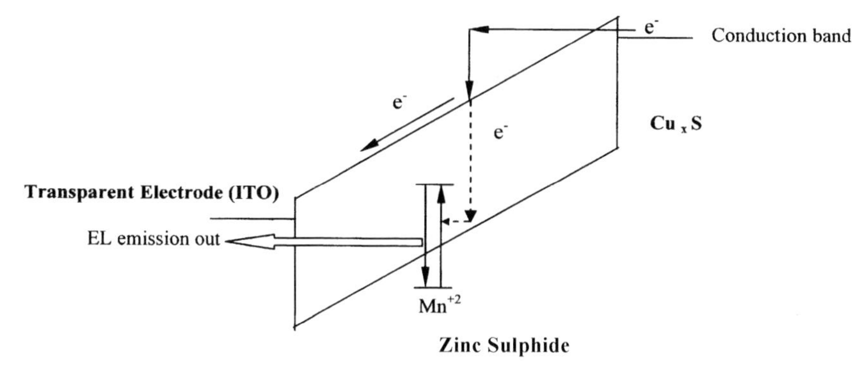

Figure 2.37. Electroluminescence mechanism for d.c. powder EL devices.

gradual drop in device current is observed. At a given voltage, the current will drop by a factor of 5 to 10 [193]. In the forming process electromigration drives all copper out of a region adjacent to the anode, creating a highly resistive copper-free region (*formed region*) near the anode. As a result, a high electric field of around 10^8 V/m is created in this region. Under the influence of this high electric field, electrons tunnel from the Cu_xS at the edge of the formed region, are accelerated, and impact excite luminescent centres. Radiative relaxation of excited electrons to the ground state results in EL emission. Colour tuning is obtained by appropriate choice of the dopant/luminescent centre. The emission of ZnS:Mn,Cu is yellow (~590 nm). When the dopant is Tm^{3+} the emission is blue, whereas Tb^{3+} and Er^{3+} give green emission, and Nd^{3+} and Sm^{3+} give red emission. Figure 2.37 is a schematic of the EL mechanism (for Mn^{2+}). Note the similarity with the EL mechanism of thin film devices.

The luminance output of ZnS:Mn d.c. powder devices is around 500 cd/m^2 at 100 V d.c. Almost the same luminance level is realized as that of the d.c. drive case under pulsed d.c. driving at a frequency of 500 Hz with a 1% duty cycle. The luminous efficiency of such devices is typically in the 0.2–0.3 lm/W range. Encapsulated, moisture-resistant devices exhibit a half-life (time after which the EL output decreases to 50%) of about 1000 h under d.c. drive, and 5000 h under pulsed d.c. driving. D.c. driven powder EL panels are available commercially. Vecht [193] has reported that for optimal device performance the following conditions are required:

(a) uniform particle size (0.5–1 μm),
(b) minimal ionic species which move under high fields,
(c) uniformity with respect to stoichiometry and impurities,
(d) the ability to properly copper coat in order that the desired conductivity could be obtained, and
(e) uniform 'forming' in order to avoid local overheating.

2.3.4 Alternating current powder electroluminescence from ZnS

In contrast to ACTFELDs, which are a display technology, a.c. powder EL devices are a lighting technology. One of the first commercially available EL products was an a.c. EL powder lamp for general lighting applications. GTE Sylvania obtained a patent for an a.c. EL powder lamp as early as 1952 [2]. However, these lamps had very limited lifetimes, which strongly depended on the drive level and light output required. They were unable to provide adequate sustained brightness over the minimum (500 h) commercially acceptable lifetimes. Here too, ionic drift, material diffusion and permeation of moisture into the device structure were the main factors contributing to premature failure. As mentioned above, these failure mechanisms were resolved by utilizing an a.c. driving voltage, and glass encapsulation techniques to minimize moisture-induced degradation of the EL phosphor and as a result, the 1980s and 1990s saw a resurgence of interest in powder EL. Today, a.c. powder EL devices find commercial application mainly in speciality lighting. A few common examples of this application are the backlighting of liquid crystalline displays (LCDs) found in wrist watches and personal digital accessories (PDAs), night lights for our homes, and emergency lighting in airplanes and cinemas.

Figure 2.38 is a schematic of the a.c. powder EL device structure. The phosphor layer is typically 50–100 μm thick and consists of a suitably doped ZnS powder suspended in a dielectric, which acts as a binder as well. The binder is normally an organic material with a large dielectric constant, such as cyanoethylcellulose. The grain size of the powder is typically in the 5–20 μm range. The phosphor layer is sandwiched between two electrodes, one of which must be transparent for optical outcoupling. The electrode/phosphor/electrode stack is supported by a substrate

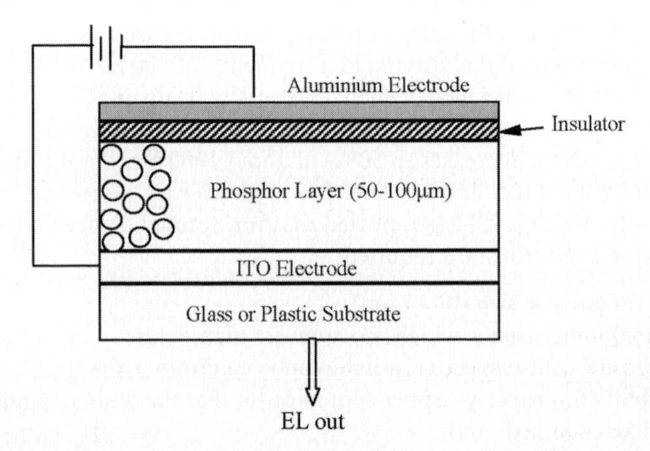

Figure 2.38. Schematic of an a.c. powder EL device.

which may be either glass or flexible plastic. In order to prevent catastrophic dielectric breakdown, an insulating layer may be placed between the phosphor layer and the rear electrode. The critical point is that the ZnS grains should be isolated from each other. If this is achieved, resulting in good isolation, an insulating layer is not necessary.

The EL mechanism in powder EL devices has been studied by Fischer [195–197] and summarized by Ono ([29], ch 3). When the applied electric field exceeds the threshold electric field E_{th}, EL emission begins in the form of pairs of small bright spots, which subsequently elongate to form comet-shaped emissive regions with increased electric fields. As discussed on page 54, ZnS has the cubic zincblende (sphalerite, αZnS) crystal structure when processed at low temperatures or the hexagonal (wurtzite, βZnS) crystal structure when processed at high temperatures. With this in mind, the following explanation for the EL mechanism has been proffered by Fischer [194, 195]: ZnS powders fired at high temperatures have the hexagonal structure but subsequent cooling transforms them into the cubic (zincblende) structure. When this occurs, excess copper (copper exceeding the solubility limit in ZnS) precipitates on defects in the ZnS particles. The result is ZnS particles embedded with Cu_xS conducting needles. That is, ZnS/Cu_xS/ZnS heterojunctions are formed. The Cu_xS conducting needles act as field enhancement tips, and therefore an applied field of 10^6–10^7 V/m can induce a local field of 10^8 V/m or more. This electric field is strong enough to induce tunnelling of holes from one end and electrons from the other end of the Cu needles. The holes are trapped on Cu recombination centres and, upon reversal of polarity, emitted electrons recombine with the trapped holes to produce light. Thus EL emission occurs along the Cu_xS precipitates. Larger particle sizes lead to longer needles and greater field enhancement. The reader should note that this mechanism is similar to the injection type electroluminescence exploited in p–n junction diodes and lasers.

Emission colour is controlled by adding different kinds of luminescent centres. A combination of Cu and Cl (ZnS:Cu,Cl) gives either blue (~460 nm) or green (~510 nm) emission, depending on the relative amount of Cl. This EL emission is caused by the recombination transition of D–A (donor–acceptor) pairs, where Cu constitutes an acceptor and Cl constitutes a donor. The combination of Cu and Al (ZnS:Cu,Al) gives green and the addition of Mn to Cu and Cl (ZnS:Cu,Cl,Mn) gives yelllow (~590 nm). An advantage of this type of EL device is easy control of emission colour by mixing phosphors with different emission colours ([29], ch 3).

Luminance increases proportionally with frequency up to 10 kHz, but lifetime decreases in the same proportion. A luminance level of about 100 cd/m^2 at a voltage of 400 V and frequency of 400 Hz is typical. The luminous efficiency of such devices is about 1 lm/W. The half-life now exceeds 2500 h when driven at a voltage of 200 V and frequency of 400 Hz.

Improvements of phosphor treatment conditions, encapsulation, and drive conditions are expected to yield better lifetimes and luminance output.

2.4 Summary

In the previous sections we reviewed the fundamentals of zinc sulphide phosphors and electroluminescent devices. We saw that the physics involved is complex, and that multiple material requirements and physical phenomena must be simultaneously satisfied in order that practical electroluminescent devices may be realized. The fundamental concepts presented therein are in general applicable to all electroluminescent phosphors. Section 2.2 detailed the quantum mechanics of luminescence from localized quantum states associated with luminescent ions, and the physical properties of the ZnS host. There, it was shown that the emission energy and oscillator strength of luminescent centres can be influenced by their surroundings. Section 2.3 reviewed all types of ZnS electroluminescent devices, the material requirements, and the electrical and electro-optical physical processes that lead to light emission. We noted that the ACTFEL type device is the most developed high-field electroluminescence technology, though powder EL devices for general lighting and backlighting are commercially available. figures of merit for practical applications were discussed and the main limitation of ZnS EL technology was highlighted. With regards to the latter, we specifically refer to the lack of a true red or blue ZnS phosphor which satisfies the minimum required light output levels for practical, full-colour displays. This deficiency is due to the valence and size mismatch of the related luminescent ions (Sm^{3+} and Tm^{3+}) with respect to the zinc substitutional lattice sites, and to the small impact cross-section and oscillator strengths of the luminescent ions being discussed. The emission of ZnS:Tm is further limited by the electron energy distribution as the blue emission of Tm corresponds to 460 nm or 2.7 eV.

Nonetheless, EL output from ZnS phosphors is continually being improved. In this summation we shall review some of the approaches that have resulted in marked improvements in luminance and/or efficiency. We will also look at the position of ZnS displays in a marketplace dominated by liquid crystal displays, and look at possible new applications for ZnS phosphors and devices.

2.4.1 Practical approaches for enhancing luminance in ZnS phosphors

The dual-substrate and colour-by-white approaches have produced acceptable luminance levels and good colour saturation, and are the technologies of choice for commercial full-colour (red-green-blue) electroluminescent displays. However, as mentioned earlier monochrome ZnS:Mn and ZnS:Tb EL phosphors are continually being improved. A combination of advances in material processing, and better understanding of the material and device physics, is at the heart of these improvements. A few examples of research conducted at the University of Florida will be used to illustrate paths for future progress.

I. Codoping of ZnS:Mn and ZnS:TbOF. The effects of K and Cl co-doping on radio-frequency magnetron sputtered ZnS:Mn thin film phosphors have been studied extensively by Lewis *et al* [36], Waldrip and co-workers [180, 197, 198] and Zhai *et al* [199]. K and Cl codoping was achieved by diffusion from an evaporated KCl overlayer during a post-deposition anneal [180, 197]. Lewis *et al* have demonstrated that K and Cl co-doping results in an increase in luminance and device efficiency (at 40 V above the threshold) of 52% and 63% respectively [36] with respect to undoped ZnS:Mn thin films. At 60 Hz driving frequency, the luminance increased from 379 cd/m^2 to 581 cd/m^2, while efficiency increased from 1.18 lm/W to 1.92 lm/W. For comparison with the values presented in table 2.8, we note that 581 cd/m^2 at 60 Hz corresponds to a very impressive 9683 cd/m^2 at 1 kHz (581/60 $\times$ 1000). An earlier publication by Waldrip *et al* [198] had reported improvements of 70% in luminance and 60% in device efficiency from similarly deposited and co-doped films. The results from transient electro-optical measurements [36] showed that the improved EL performance was primarily due to a reduction in static space charge. The reduction of static space charge resulted in a larger average phosphor field, hotter electron energy distribution and increased charge multiplication, all of which contributed to an increase in Mn^{2+} excitation efficiency. The increase in Mn^{2+} excitation efficiency was quantified as 75% [36]. The formation of shallow charge compensating V_{Zn}–Cl_S acceptor complexes and Cl_S donor point defects (see subsection 2.2.3.2) was proposed as the mechanism by which space charge (deep V_{Zn} hole traps) was reduced. Extensive transmission electron microscopy (TEM) and X-ray diffraction (XRD) studies performed by Zhai and co-workers [72, 199] indicate that K and Cl co-doping results in improved crystallinity, with grain sizes growing to larger than 1 μm under unusual circumstances. Zhai's work showed that the reduction in non-radiative recombination due to increased crystallinity was an additional factor, but not the major factor, that contributed to the improved EL performance of K and Cl co-doped ZnS:Mn.

Kim *et al* [156] have demonstrated luminance improvements of 60%, and efficiency improvements of 130% from cerium co-doped ZnS:TbOF thin-film EL phosphors prepared by r.f. magnetron sputtering. Based on the results of luminescence decay measurements, Kim proposed that the improved radiative efficiency was primarily due to energy transfer from Ce^{3+} to Tb^{3+}. Considering that the excited state lifetime of Ce^{3+} is several tens of nanoseconds, then during the 30 μm on-time of the applied voltage pulse, energy transfer could be very efficient assuming that the characteristic time for transfer is smaller than or equal to the excited state lifetime of Ce^{3+}. It was proposed that Ce^{3+} co-doping also leads to a reduction in space charge which, as discussed above, results in improved excitation efficiency. For complete details the reader is referred to [156, 200].

Simultaneous co-doping of ZnS:TbOF with Ag and Cu also resulted in an improvement in luminance [200]. A 22% increase in brightness was reported. Kim [200] attributed the improvement to increased surface roughness (characterized by atomic force microscopy) and consequent improved optical outcoupling, plus improved excitation efficiency due to space charge modification by Ag^{1+} and Cu^{1+}.

II. Interface modification. The objective of interface modification is to enhance carrier injection from interface states into the phosphor layer. This is achieved by the use of new oxides which provide a higher density of interface states, by the use of 'leaky' oxides, and by the use of special electron injector layers. Much of the information regarding the development and use of new and leaky oxides is proprietary. However, Rack and Holloway and their co-workers [73, 201, 202] have demonstrated the effectiveness of electron injector layers for enhancing carrier injection into EL phosphor layers. The specific phosphor studied was blue emitting $Ca_xSr_{1-x}Ga_2S_4$:Ce. Rack *et al* used a 5 nm thick indium electron injector layer, which was deposited on top of the ATO insulator. Unique to the device structure was a 30 nm thick ZnS layer on top of the indium layer, which was added in order to facilitate crystallization of the $Ca_xSr_{1-x}Ga_2S_4$:Ce phosphor. The 'interface' therefore had the following structure: ATO/In/ZnS/$Ca_xSr_{1-x}Ga_2S_4$:Ce. Compared with a similar structure without the In injector layer, a 15 V reduction in threshold voltage and 25% increase in luminance was obtained. The use of electron injector layers in ZnS devices is expected to yield a similar effect and is actively being researched by a number of laboratories.

III. Nanoparticles. The PL and EL properties of Mn doped ZnS nanocrystals prepared by competitive reaction chemistry have been reported by Yang and Holloway [203]. Nanoparticle technology offers numerous opportunities for new device structures, new applications and manipulation of optical properties. It is expected that the large oscillator strengths inherent to quantum confined nanoparticles will lead to enhanced, efficient energy transfer from the ZnS nanoparticle host to embedded luminescent ions. An excellent PL quantum efficiency of ~18 to 25% has already been achieved [204] from ZnS-coated CdS:Mn nanoparticles. Efforts to translate such high PL efficiencies into EL devices are ongoing.

2.4.2 ZnS devices in the FPD marketplace and current trends

The three major companies involved in the production of electroluminescent displays are Planar, Sharp and Westaim (iFire). The corporate research laboratories of Sharp and Planar together with several publicly funded laboratories, consortia and universities (including the University of Florida) have been the major driving force in the development of EL displays, and bringing the technology to the marketplace. iFire, a division of Westaim, is a relative newcomer but has been a significant contributor, particularly in the development of full-colour displays.

The FPDs marketplace is highly competitive and is currently dominated by LCDs and plasma display panels (PDPs). This competition is expected to continue and intensify, which in good for the customer, as the competition will eventually drive prices down and make FPDs accessible to the average household. Industry experts project that the dollar value of FPDs sold in the world will overtake that of CRTs by 2006. The replacement of CRTs alone represents an enormous business opportunity. When other display applications are considered, the business opportunities are even greater. The market for flat panel screens 25 inches and greater in size is expected reach more than US $60 billion in annual sales by 2005. Thus, there is ample room for gaining market share and the growth of full-colour EL panels.

ZnS-based devices is the technology of choice for monochrome and other EL displays where at least one other colour is required. The market for such displays is niche applications where ruggedness, reliability, insensitivity to temperature and readability in high ambient illumination conditions are required. Displays for use in industrial and medical instrumentation are two examples. Monochrome and multicolour devices also find application in military information displays such as in the Abrams tank and avionic displays in jets. The best growth opportunity for these devices appears to be in automotive dashboard displays.

With regards to full-colour ZnS applications, the dual substrate and colour-by-white approaches, though commercialized, have been slow to develop true colour saturation, particularly in the blue. Further, these technologies have been unsuccessful in commercially scaling up to larger screen sizes common to the TV market sector. Thus, full-colour ZnS devices have been excluded from the extremely lucrative 'high definition flat panel TV on the wall' market, and relegated to applications which utilize small display sizes.

An improvement for full-colour EL devices is the recent 'colour-by-blue' technology introduced by iFire. The colour-by-blue technology employs iFire's proprietary thick-film dielectric processing, a new EL phosphor, and a combination of electroluminescence and photoluminescence phenomena to obtain red, green and blue emission. The new EL phosphor is blue emitting $Mg_xBa_{1-x}Al_2S_4$-doped europium. The principle of operation is as follows: blue light from the thin-film EL phosphor interacts with a 'colour correction layer', which tunes the colour to exactly match TV colour requirements and improves contrast. The colour correction layer is a sheet of glass coated with patterned red, green and blue emitting materials. The red and green emitting materials are down-converting *PL* phosphors which absorb blue light from the active $Mg_xBa_{1-x}Al_2S_4$:Eu electroluminescent layer, and re-emits red or green light. iFire reports that replacing the traditional thin film dielectric with its patented thick-film, high-K dielectric processing, combined with the colour-by-blue approach, will require simpler and fewer processing steps for manufacture. Fewer and less expensive manufacturing tools will be required, and these factors should all combine to result in lower production and operation costs, and higher profit margins. The ability to scale to large sizes and achieve high yield and reliability are envisaged, and commercial volume production of 34 inch high-definition television display modules is scheduled to begin in 2005.

2.4.3 New applications for ZnS devices and materials

In the preceding sections we discussed electroluminescence from the localized quantum states of luminescent ions with transition energies in the visible part of the electromagnetic spectrum. However, many lanthanide ions also exhibit transitions in the infrared part of the spectrum. Er^{3+}, Nd^{3+}, Ho^{3+}, Tb^{3+}, Pr^{3+} and Dy^{3+} are examples of lanthanide ions with infrared transitions. Infrared EL phosphors and devices have potential application in consumer electronics, optical communications and opto-electronic integrated circuitry and sensors. Kale *et al* [43] have reported power densities of $\sim 30\,\mu W/cm^2$ for the 1500 nm emission of ZnS:Er and the 910 nm emission of ZnS:Nd. The infrared emission is primarily a result of the properties of the rare-earth activator added to the host. In fact, Kim *et al* [205] have shown both infrared and visible emission from rare earths in sputter-deposited thin films where GaN was the host.

Spintronics is a new area where ZnS:Mn (and alloys) could find potential application. Spintronics research is still in its infancy, but the basic concept is to use electron spin rather than electron charge to achieve functionality. In a spin LED, for example, one seeks to inject electrons with a specific spin alignment. In order that this may be facilitated, the material from which the electrons are injected (the spin injection layer) must be ferromagnetic, i.e. exhibit spin alignment under the influence of an applied magnetic field. Various dilute magnetic semiconductors (DMS) and semi-metals are being researched as spin injection layers. Most of these materials involve the use Mn, which results in ferromagnetic behaviour. The MnS/ZnS:Mn system is a good potential candidate for evaluation.

Finally, the work by Yang and Holloway with respect to nanoparticles involving ZnS was reported above. The emphasis is currently on the basic science of ZnS nanoparticle systems, and it is expected that new functionality will emerge from this deepening understanding. For example, nanoparticles with modified surfaces could be used as biomedical probes to which specific proteins or enzymes would attach, thereby indicating the presence or absence of a specific medical condition. Bio-applications will become increasingly important for the future of this technology area.

Acknowledgment

This work has been supported by a number of agencies including ARO, DARPA and NRL, and most recently by ARO grants DAAD19-00-1-0002 and DAAD19-01-1-0603.

References

[1] Destriau G 1936 *J. Chim. Phys.* **33** 587
[2] *Flat Panel Displays and CRTs* 1985 ed. E Tannas Jr (New York: Van Nostrand Reinhold) p. 620
[3] Rack P D, Naman A, Holloway P H, Sun S-S and Tuenge R T 1996 *MRS Bull.* **21**(3) 49
[4] Vlasenko N A and Popkov Z 1960 *Opt. Speckrosk.* **8** 39
[5] Vlasenko N A 1965 *Opt. Speckrosk.* **18** 260
[6] Vlasenko N A and Yaremko A M 1965 *Opt. Specktrosk.* **18** 265
[7] Vlasenko N A, Zynio S A and Kopytko Yu V 1975 *Phys. Stat. Sol. (a)* **29** 671
[8] Vecht A, Werring N J and Smith P J F 1968 *Brit. J. Appl. Phys.* **1** 134
[9] Kahng D 1968 *Appl. Phys. Lett.* **13** 210
[10] Soxman E J and Ketchpel R D 1972 *Electroluminescence Thin Film Research Reports* JANAIR Report 720903
[11] Russ M J and Kennedy D I 1967 *J. Electrochem. Soc.* **114** 1066
[12] Inoguchi T, Takeda M, Kakihara Y, Nakata Y and Yoshida M 1974 *Digest of 1974 SID International Symposium* p. 84
[13] Mito S, Suzuki C, Kanatani Y and Ise M 1974 *Digest of 1974 SID International Symposium* p. 86
[14] Uede H, Kanatani Y, Kishishita H, Fujimori A and Okano K 1981 *Digest of 1981 SID International Symposium* p. 28
[15] Takeda M, Kanatani Y, Kishishita H, Inoguchi T and Okano K 1980 *Digest of 1980 SID International Symposium* p. 66
[16] Takeda M, Kanatani Y, Kishishita H, Inoguchi T and Okano K 1981 *Proc. SID 22* p. 57
[17] Takeda M, Kanatani Y, Kishishita H and Uede H 1983 *Proc. SPIE 386 Advances in Display Technology III* p. 34
[18] Chase E W, Heppelwhite R T, Krupka D C and Kahng D 1969 *J. Appl. Phys.* **40** 2512

[19] Barrow W, Coovert R E, Dickey E, Flegal T, Fullman M, King C and Laasko C 1994 *Conference Record of the 1994 International Display Research Conference* p. 448

[20] Barrow W, Coovert R E, King C N and Zinchkovski M J 1988 *Digest of 1988 SID International Symposium* p. 284

[21] Cramer D, Haaranen J, Törnqvist R and Pitkanen T 1993 *Application Digest of 1993 SID International Symposium* p. 57

[22] Harris D C and Bertolucci M D 1978 *Symmetry and Spectroscopy: An Introduction to Vibrational and Electronic Spectroscopy* (Oxford: Oxford University Press)

[23] Pauling L and Wilson E B Jr 1935 *Introduction to Quantum Mechanics with Applications to Chemistry* (New York, London: McGraw-Hill)

[24] Sobelman I I 1979 *Atomic Spectra and Radiative Transitions* (Berlin, Heidelberg: Springer) p. 27

[25] Henderson B and Imbusch G F 1989 *Optical Spectroscopy of Inorganic Solids* (Oxford: Clarendon)

[26] DiBartolo B 1968 *Optical Interactions in Solids* (New York: Wiley)

[27] Blasse G and Grabmaier B C 1994 *Luminescent Materials* (Berlin: Springer)

[28] Bringuier E 1994 *J. Appl. Phys.* **75**(9) 4291

[29] Ono Y A 1995 *Electroluminescent Displays* (Singapore: World Scientific)

[30] Mach R and Mueller G O 1991 *Semicond. Sci. Technol.* **6** 305

[31] Ono Y A 1993 in *Encyclopedia of Applied Physics* vol 5 ed. G L Trigg (Weinheim: VCH Publishers; New York: American Institute of Physics) p. 295

[32] Ono Y A 1993 in *SID Seminar Lecture Notes* vol II, F-1, F-1/1

[33] Ono Y A 1995 in *Progress in Information Display Technology* vol 1 ed. H L Ong and S Kobayashi (Singapore: World Scientific) ch 2

[34] Gumlich H E 1981 *J. Lumin.* **135** 795

[35] Ivey H F 1966 *IEEE J. Quantum Electronics* **QE-2**(11) 713

[36] Lewis J S, Davidson M R and Holloway P H 2002 *J. Appl. Phys.* **92** 6646

[37] Kim J P, Davidson M, Moorehead D, Puga-Lambers M, Zhai Q and Holloway P H 2001 *J. Vac. Sci. Technol.* **A19** 2244

[38] Busse W, Gumlich H E, Meissner T and Theiss D 1976 *J. Lumin.* **12/13** 693

[39] Busse W, Gumlich H E, Geoffroy A and Parrot R 1979 *Phys. Stat. Sol.* **93** 591

[40] Cotton F A 1990 *Chemical Applications of Group Theory* 3rd edition (Chichester: Wiley)

[41] Atkins P W 1990 *Physical Chemistry* 4th edition (New York: W H Freeman)

[42] Pickering H S 1978 *The Covalent Bond* (London: Wykeham Publications)

[43] Kale A, Shepherd N, DeVito D, Glass W, Davidson M and Holloway P H 2003 *J. Appl. Phys.* **94**(5) 3147

[44] Shaw L B, Cole B, Thielen P A, Sanghara J S and Aggarwal I D 2001 *IEEE J. Quantum Electronics* **48**(9) 1127

[45] Tanaka S, Shanker V, Shiiki M, Deguchi H and Kobayashi H 1985 *Digest of 1985 SID International Symposium* p. 218

[46] Tanaka S, Shanker V, Shiiki M, Deguchi H and Kobayashi H 1985 *Proc. SID 26* p. 255

[47] Yang Y R and Duke C B 1987 *Phys. Rev. B* **36** 2763

[48] Brennan K 1988 *J. Appl. Phys.* **64**(8) 4024

[49] Dur M, Goodnick S, Pennathur S S, Wager J F, Reigrotzki M and Redmer R 1998 *J. Appl. Phys.* **86**(6) 3176

[50] Hitt J C, Bender J P and Wager J F 2000 *Critical Rev. Solid State Mat. Sci.* **25**(1) 29

[51] Bringuier E 1990 *J. Appl. Phys.* **67**(11) 7040
[52] Tanaka S, Deguchi H, Mikami Y, Shiiki M and Kobayashi H 1986 *Digest of 1986 SID International Symposium* p. 29
[53] Tanaka S, Deguchi H, Mikami Y, Shiiki M and Kobayashi H 1987 *Proc. SID* **28** 21
[54] Thompson T D and Allen J W 1987 *J. Phys. C: Solid State Phys.* **20** L499
[55] Bhattacharyya K, Goodnick S M and Wager J F 1993 *J. Appl. Phys.* **73**(7), 3390
[56] Lide D R and Frederikse H P R eds. 1997 *CRC Handbook of Chemistry and Physics* (Boca Raton: CRC Press)
[57] Sharma R C and Chang Y A 1996 *J. of Phase Equilibria* **17**(3) 261
[58] Neumark G F 1997 *Mater. Sci. Eng.* **R21** 1
[59] Chang L L, Esaki L and Tsu R 1971 *Appl. Phys. Lett.* **19** 143
[60] Watkins G D 1996 *J. Cryst. Growth* **159** 338
[61] Watkins G D 1973 *Solid State Commun.* **12** 589
[62] Georgobiani A N, Maev R G, Ozerbov Y V and Srrumban E E 1976 *Phys. Stat. Sol. (a)* **38** 77
[63] Vlasenko N A, Chumachkova M M, Denisova Z L and Veligura L I 2000 *J. Cryst. Growth* **216** 249
[64] Hitt J C, Keir P D, Wager J F and Sun S S 1998 *J. Appl. Phys.* **82**(2), 1141
[65] Matsuura K, Kishida S and Tsurumi I 1987 *Phy. Stat. Sol. (b)* **140** 347
[66] Yamaga S, Yoshikawa A and Kasai H 1988 *J. Cryst. Growth* **86** 252
[67] Hauksson I S, Simpson J, Wang S Y, Prior K A and Cavenett B C 1992 *Appl. Phys. Lett.* **61**(18), 2208
[68] Hermans J, Woitok J, Geurts J, Sollner J, Heuken M, Stanzl H and Gebhardt W 1996 *J. Cryst. Growth* **159** 363
[69] Faschinger W 1996 *J. Cryst. Growth* **159** 221
[70] Walukiewicz W 1996 *J. Cryst. Growth* **159** 244
[71] Bube R 1978 *Photoconductivity of Solid* (Huntington, New York: Rob Kreiger Publishing)
[72] Qing Zhai 1999 PhD Dissertation, University of Florida
[73] Rack P D and Holloway P H 1998 *Mater. Sci. Eng.* **R21**(4) 171
[74] O'Brien T A, Rack P D, Holloway P H and Zerner M C 1998 *J. Lumin.* **78** 2457
[75] O'Brien T, Zerner M C, Rack P D, and Holloway P H 2000 *J. Electrochem. Soc.* **147** 792
[76] Judd R, 1962 *Phys. Rev.* **127** 750
[77] Ofelt G S, 1962 *J. Chem. Phys.* **37** 511
[78] Carnall W T in *Handbook on the Physics and Chemistry of Rare-Earths,* vol. 3 eds. K A Gschneider Jr. and L Eyring (Amsterdam: North-Holland) ch. 24
[79] Zumdahl S S 1989 *Chemistry* 2nd edition (Lexington: D C Heath) p. 922
[80] Gerloch M 1973 *Ligand-Field Parameters* (London: Cambridge University Press)
[81] Pople J A 1970 *Approximate Molecular Orbital Theory* (New York: McGraw-Hill)
[82] McQuarrie D A 1983 *Quantum Chemistry* (University Science Books)
[83] Hehre W J, Radom L, Schleyer P V and Pople J A 1986 *Ab Initio Molecular Orbital Theory* (Wiley)
[84] Born M and Oppenheimer J R 1927 *Ann. Physik* **84** 457
[85] Hartree D R 1957 *The Calculation of Atomic Structures* (New York: Wiley)
[86] Fischer C F 1977 *The Hartree–Fock Method for Atoms: A Numerical Approach* (New York: Wiley)

[87] Askeland D R 1994 *The Science and Engineering of Materials* (Boston, MA: PWS Publishing)
[88] Beale M 1993 *Philos. Mag. B* **68** 573
[89] Fowler R H and Nordheim L 1928 *Proc. Roy. Soc. London* **A119** 173
[90] Onton A and Marrello V 1982 *Advances in Image Pickup and Display* vol. 5 (New York: Academic) p. 137
[91] Nordheim L W 1928 *Proc. Roy. Soc. London* **A121** 626
[92] Gomer R, 1961 *Field Emission and Field Ionization* (Cambridge, MA: Harvard University Press)
[93] Dyke W P and Dolan W W 1956 in *Advances in Electronics and Electron Physics* vol. 8 ed. L Marton (New York: Academic)
[94] van Oostrom A 1973 *Philips Tech. Rev.* **33** 277
[95] Lampert M A and Mark P 1970 *Current Injection in Solids* (Academic Press)
[96] Stratton R 1955 *Proc. Phys. Soc. London* **B68** 746
[97] Stratton R 1962 *Phys. Rev.* **125** 67
[98] M Beale and Mackay P 1992 *Philos. Mag. B* **65** 47
[99] Mueller-Mach R and Mueller G O 2000 *Semiconductors and Semimetals* vol. 65 (Academic Press) p. 27
[100] Alt P M 1984 *Proc. SID* **25** 123
[101] Howard W E 1977 *IEEE Trans. Electron Devices* **ED-24** 903
[102] Howard W E 1977 *Proc. SID* **18** 119
[103] Tiku S K and Smith G C 1984 *IEEE Trans. Electron Devices* **ED-31** 105
[104] Chopra K L, Major S and Panday D K 1983 *Thin Solid Films* **102** 1
[105] Bringuier E 1989 *J. Appl. Phys.* **66** 1314
[106] Krasnov A N and Hofstra P G 2001 *Progress in Crystal Growth and Charact.* **42** 65
[107] Inoguchi T and Mito S 1977 in *Electroluminescence* ed. J I Pankove *Topics in Applied Physics* vol. 17 ch. 6 (Berlin: Springer) p. 196
[108] Slater J C 1949 *Phys. Rev.* **76** 1592
[109] Bardeen J and Shockley W 1950 *Phys. Rev.* **80** 69
[110] Mach R and Mueller G O 1990 *J. Cryst. Growth* **101** 967
[111] Bringuier E 1991 *J. Appl. Phys.* **70** 4505
[112] Ridley B K 1983 *J. Phys. C* **16** 3373
[113] Steinberger I T, Bar V and Alexander E 1961 *Phys. Rev.* **121** 118
[114] Ando M 1999 in *Proc. 6th International Workshop on Electroluminescence* eds. V P Singh and J C McClure p. 85
[115] Ando M 1992 *Appl. Phys. Lett.* **60** 2189
[116] Yoshiyama H, Sohn S H, Tanaka S and Kobayashi H 1989 in *Proc. 4th International Workshop on Electroluminescence* eds. S Shionoya and H Kobayashi (Heidelberg: Springer) p. 48
[117] Tanaka S, Yoshiyama H, Nakamura K, Wada S, Morita H and Kobayashi H 1991 *Jpn. J. Appl. Phys.* **30** L1021
[118] Bringuier E 1994 *Phys. Rev. B* **49** 7974
[119] Shen M and Xu X 1989 *Solid State Commun.* **72** 803
[120] Allen J W 1986 *J. Phys. C* **19** 6287
[121] Langer J M 1989 in *Proc. 4th International Workshop on Electroluminescence* eds. S Shionoya and H Kobayashi (Heidelberg: Springer) p. 24
[122] Smith D H 1981 *J. Lumin.* **23** 209
[123] Mishima T and Takahashi K 1983 *J. Appl. Phys.* **54** 2153

[124] Mishima T and Takahashi K 1983 *IEEE Trans. Electron Devices* **ED-30** 282
[125] Bringuier E and Geoffroy A 1986 *Appl. Phys. Lett.* **48** 1780
[126] Douglas A A, Wager J F, Morton D C, Koh J B and Hogh C P 1993 *J. Appl. Phys.* **73** 296
[127] Crandall R S 1987 *Appl. Phys. Lett.* **50** 551
[128] Crandall R S and Ling M 1987 *J. Appl. Phys.* **62** 3074
[129] Geoffroy A and Bringuier E 1990 *J. Appl. Phys.* **67** 4276
[130] Ando M and Ono Y A 1991 *J. Appl. Phys.* **69** 7225
[131] Krupka D C and Mahoney D M 1972 *J. Appl. Phys.* **43** 2314
[132] Pankove J I, Lampert M A, Hanak J J and Berkeyheiser J E 1977 *J. Lumin.* **15** 349
[133] Boyn R 1988 *Phys. Stat. Sol. (b)* **148** 11
[134] Panday R and Sivaraman S 1991 *J. Phys. Chem. Solids* **52** 211
[135] Okamoto K and Miura S 1986 *Appl. Phys. Lett.* **49** 1596
[136] Geoffroy A and Bringuier E 1991 *Semicond. Sci. Technol.* **6** A131
[137] Thioulouse P 1985 *J. Cryst. Growth* **72** 545
[138] DiBartolo B (ed) 1984 *Energy Transfer Processes in Condensed Matter* (New York: Plenum)
[139] Törnqvist R 1983 *J. Appl. Phys.* **54** 4110
[140] Sasakura H, Kobayashi H, Tanaka S, Mita J, Tanaka T and Nakayama H 1981 *J. Appl. Phys.* **52** 6901
[141] Mooney W J 1991 *Optoelectronic Devices and Materials* (Englewood Cliffs, NJ: Prentice-Hall)
[142] Tuenge R T and Kane J, 1991 *Digest of 1991 SID International Symposium* 279
[143] Okibayashi K, Ogura T, Terada K, Taniguchi T, Yamashita T, Yoshida M and Nakajima S 1991 *Digest of 1991 SID International Symposium* 275
[144] Ohnishi H, Yamamoto K and Katayama Y 1985 *Conference Record of the 1985 International Display Research Conference* p. 159
[145] Ogura T, Mikami A, Tanaka K, Taniguchi K, Yoshida M and Nakajima S 1986 *Appl. Phys. Lett.* **48** 1570
[146] Miura S, Okamoto K, Sato S, Andoh S, Ohnishi H and Hamakawa Y 1983 *Japan Display '83* 84
[147] Mita J, Koizumi M, Hayashi T, Sekido Y, Kazama M and Nihei K 1986 *Japan Display '86* 250
[148] Krupka D C 1971 *J. Appl. Phys.* **43** 476
[149] Mikami A, Taniguchi K, Yoshida M and Nakajima S 1988 *J. Appl. Phys.* **64** 3650
[150] Swiatek K, Suchocki A and Stapor A 1989 *J. Appl. Phys.* **66** 6048
[151] Okamoto K, Yoshimi T and Miura S 1986 *Appl. Phys. Lett.* **49** 578
[152] Okamoto K, Yoshimi T and Miura S 1989 in *Proc. 4th International Workshop on Electroluminescence* eds. S Shionoya and H Kobayashi (Heidelberg: Springer) p. 139
[153] Yoshino H, Ohura M, Kurokawa S and Ohnishi H 1992 *Japan Display '92* 737
[154] Sohn S H, Hyun D G, Norma M, Hosomi S and Hamakawa Y 1992 *J. Appl. Phys.* **72** 4877
[155] Kim J P, Davidson M R and Holloway P H 2003 *L. Lumin.* Submitted for publication
[156] Kim J P, Davidson M R and Holloway P H 2003 *J. Appl. Phys.* **93** 9597
[157] Tohda T, Fujita Y, Matsuoka T and Abe A 1986 *Appl. Phys. Lett.* **48** 95
[158] Hirabayashi K, Kozawaguchi H and Tsujiyama B 1986 *Japan Display '86* 254
[159] Hirabayashi K, Kozawaguchi H and Tsujiyama B 1987 *Jpn. J. Appl. Phys.* **26** 1472

[160] Sohn S H and Hamakawa Y 1993 *Appl. Phys. Lett.* **62** 2242
[161] Barrow W A, Coovert R E, Dickey E, King C N, Laasko C, Sun S S, Tuenge R T, Wentross R C and Kane J 1993 *Digest of 1993 SID International Symposium* 761
[162] King C N 1994 *1994 SID Seminar Lecture Notes* vol. 1 p. M-9
[163] Tanaka S, Yoshiyama H, Nishiura J, Ohshio S, Kawakami H and Kobayashi H 1988 *Digest of 1988 SID International Symposium*, 293
[164] Nire T, Matsonu A, Wada F, Fuchiwaki K and Miyakoshi A 1992 *Digest of 1992 SID International Symposium* 352
[165] Törnqvist R 1997 *Digest of 1997 SID International Symposium* 855
[166] Soininen E 1998 in *4th International Conference on Science and Technology of Phosphors*
[167] Park W, Jones T C and Summers J C 1999 *Appl. Phys. Lett.* **74** 1785
[168] Mach R, Schrottke L, Mueller G O, Reetz R, Krause E and Hildish L 1988 *J. Lumin.* **40/41** 779
[169] Mach R, Mueller G O, Schrottke L, Benalloul P and Benoit J 1990 in *Proc. 5th International Workshop on Electroluminescence*, Helsinki
[170] Wu X 1996 in *Proc. 8th International Workshop on Electroluminescence* (Berlin: Wissenschaft und Technik)
[171] Doxsee D D and Wu X 1997 in *3rd International Conference on Science and Technology of Phosphors*, Huntington, CA
[172] Lui G, Xiao T, Lobbau K and Wu X 1998 in *Proc. 9th International Workshop on Electroluminescence*, Bend, Oregon
[173] Mach R and Mueller G O 1984 *Phys. Stat. Sol. (a)*, **81** 609
[174] Zeinert A, Benalloul P, Benoit J, Barthou C, Dreyhsig J and Gumlich H E 1992 *J. Appl. Phys.* **71** 2855
[175] Wager J F and Keir P D 1997 *Ann. Rev. Mater. Sci.* **27** 223
[176] Sawyer C B and Tower C H 1930 *Phys. Rev.* **35** 269
[177] Ono Y A, Kawakami H, Fuyama M and Onisawa K 1987 *Jpn. J. Appl. Phys.* **26** 1482
[178] Tueta R and Braguier M 1981 *Thin Solid Films* **80** 143
[179] Shimizu Y and Matsudaira T 1985 in *Conference Record of 1985 International Display Research Conference*
[180] Waldrip K E, Davidson M R, Lee J H, Pathangey B, Puga-Lambers M, Jones K S, Holloway P H, Sun S S and King C N in 1999 *Display and Imaging* eds. H Kobayashi and S Kobayashi **8** 73
[181] Vossen J L and Kern W eds. 1978 *Thin Film Processes* (Academic Press)
[182] Suntola T and Antson J 1974 Finnish Patent 52359; US Patent 4058430 1977; Suntola T, Pakala A and Lindfors S 1983 US Patent 4389973
[183] Goodman C H L and Pessa M V 1986 *J. Appl. Phys.* **60** R65
[184] Theis D, Oppolzer H, Ebbinghaus G and Schild S 1983 *J. Cryst. Growth* **63** 47
[185] Davidson M R, Pathangey B, Holloway P H, Rack P D, Sun S-S and King C N 1997 *J. Electron. Materials* **26** 1355
[186] Abdalla M I, Plumb J L and Hope L L 1984 in *Digest of 1984 SID International Symposium* 245
[187] Ono Y A 1990 in *Proc. 5th International Workshop on Electroluminescence*, Helsinki
[188] Tanninen V P, Oikkonen M and Tuomi T O 1981 *Phys. Stat. Sol. (a)* **67** 573
[189] Tanninen V P, Oikkonen M and Tuomi T O 1983 *Thin Solid Films* **109** 283
[190] Ohnishi H 1994 *in Digest of 1994 SID International Symposium* 129
[191] Vecht A, Werring N J, Ellis R and Smith P J F 1969 *Brit. J. Appl. Phys. Ser.* **22** 953

[192] Vecht A and Werring N J 1970 *J. Phys. D: Appl. Phys.* **3** 105
[193] Vecht A 1973 *J. Vac. Sci. Technol.* **10**(5) 789
[194] Fischer A G 1962 *J. Electrochem. Soc.* **109** 1043
[195] Fischer A G 1963 *J. Electrochem. Soc.* **110** 733
[196] Fischer A G 1971 *J. Electrochem. Soc.* **118** 139C
[197] Waldrip K E, Lewis J S, Zhai Q, Davidson M R, Holloway P H and Sun S S 2000 *Appl. Phys. Lett.* **76** 1276
[198] Waldrip K E, Lewis J S, Zhai Q, Puga-Lambers M, Davidson M R, Holloway P H and Sun S S 2001 *J. Appl. Phys.* **89** 1664
[199] Zhai Q, Lewis J S, Waldrip K A, Davidson M, Evans N, Jones K and Holloway P H 2002 *Thin Solid Films* **44** 105
[200] Kim J P 2001 PhD Dissertation, University of Florida
[201] Rack P D, Holloway P H, Pham L, Wager J F and Sun S S 1995 in *Digest of 1995 SID International Symposium* 480
[202] Rack P D, Holloway P H, Sun S S, Dickey E R, Schaus C F, Tuenge R T and King C N, US Patent 5581150
[203] Yang H and Holloway P H, 2003 *J. Appl. Phys.* **83** 586
[204] Yang H and Holloway P H 2004 Private communication and submitted to *Advanced Materials*
[205] Kim J H, Shepherd N, Davidson M and Holloway P H 2003 *Appl. Phys. Lett.* **83**(4) 641
[206] Foster Th 1948 *Ann. Phys.* **2** 55
[207] Dexter D L 1953 *J. Chem. Phys.* **21** 636

Chapter 3

Zinc selenide and zinc telluride

M Godlewski[1,2]*, E Guziewicz*[1,3] *and V Yu Ivanov*[1]
[1]Institute of Physics PAS, Warsaw, Poland
[2]College of Science, Cardinal S Wyszyński University, Warsaw, Poland
[3]Los Alamos National Laboratory, Los Alamos, New Mexico, USA

3.1 Introduction

Two of the wide band gap II–VI compounds zinc selenide (ZnSe) and zinc telluride (ZnTe) still remain promising materials for opto-electronics applications. ZnSe, in particular, was intensively studied and its several applications in short wavelength devices were demonstrated. These devices are, however, not commercialized, which is due to either too low light emission efficiency or too short device lifetime. In this chapter, we briefly describe various previous efforts to develop light emitting devices, and also analyse some recent ideas, which may result in new applications of ZnSe and ZnTe in light emitting devices in visible and infrared spectral region.

The chapter is organized as follows. First, we briefly describe the history of the development of semiconductor-based light emitting diodes (LED) and recent studies towards a new generation of efficient white light sources. Then we mention the possible role of ZnSe and ZnTe as materials for such devices. However, it seems at the moment that both these II–VI compounds will have limited applications in LED and also laser diode (LD) devices for visible light. Thus, the main part of this chapter is devoted to the discussion of new ideas and a demonstration of new possible applications of ZnSe and ZnTe as electroluminescence (EL) materials in thin-film displays, as nonlinear energy up-converting optical materials, and as tunable mid-infrared lasers.

3.2 Semiconductor-based monochromatic LEDs

3.2.1 Light emitting diodes

Intensive studies of p–n junctions led to the introduction of transistor structures and in the following revolution in electronics. A by-product of these studies is the semiconductor-based light emitting diode. Holonyak and Bevaqua [1] from General Electric were the first to demonstrate light emission from a p–n junction of a semiconductor material. This observation was reported in 1962, i.e. in the early stages of p–n junction research. It took several years to introduce the first practical LED devices. Surprisingly, the first generation of commercialized LEDs was based on p–n junctions of bulk GaP, i.e. material with an indirect forbidden energy band gap. Efficient radiative decay was obtained by using excitonic recombination process of excitons bound at isoelectronic (IBE) centres and at neutral complexes. Strong localization of a primary carrier at an isoelectronic (isovalent) centre relaxes selection rules for momentum conservation and, thus, promotes efficient radiative decay [2].

High efficiency of this recombination channel relates also to the fact that IBE recombination is not affected by competing Auger-type non-radiative recombination transitions, which are destructive in the case of other bound excitonic recombination channels [2]. Consequently, isovalent nitrogen or zinc and oxygen, forming a Zn–O neutral complex in GaP, were used as dopants in the first commercial LEDs. Green colour emission was achieved by nitrogen doping, whereas zinc and oxygen were used as dopants in red colour LEDs. Yellow colour emission from diodes was achieved by introducing nitrogen to ternary GaAsP.

The first commercialized LEDs were used as indicators replacing so-called Nixie tubes. Other applications were unlikely, which was due to a too-low light emission efficiency of 0.001–1 lm/W. Consequently, first-generation LEDs were not suitable for overhead illumination and for monochromatic light sources in traffic signals. Efficiencies higher by at least one order in magnitude were required for these applications.

Fortunately, very rapid progress occurred in the next decades, which resulted in the introduction of efficient generations of LEDs. A breaking point was reached around 1985 when quantum structures, grown by metal organic chemical vapour deposition (MOCVD) or by molecular beam epitaxy (MBE), were mastered. Mostly due to this fact, flux (measured in lumens) doubled every 18–24 months, following the so-called Haitz low [3].

The first LEDs using low dimensional structures were based on GaAs/AlGaAs heterostructures, with GaAs quantum wells (QWs) and AlGaAs barriers. Within the next few years, the efficiency of QW-based LEDs surpassed that of red-filtered incandescent bulbs. Moreover, the cost of diodes, and of the required energy to drive them, was continuously going down, from about US $100 per lumen to about US $0.01 per lumen at present.

The first very bright LEDs were commercialized in mid 1990s. These III–V-based diodes emit in red-orange spectral range. They were introduced to the market by Hewlett-Packard's Optoelectronics Division (now Lumi-Leds Lightning) [4] and by Toshiba [5]. These highest efficiency LEDs are based on MOCVD-grown quaternary aluminium indium gallium phosphide grown on gallium arsenide. Their internal quantum efficiency is very high and currently reaches nearly 100%, i.e. internal quantum efficiency is close to the maximal. Thus, the main efforts are concentrated on improvement of external quantum efficiency, i.e. on the optimization of light emission output from the device. The external quantum efficiency can be considerably improved by applying specially shaped diodes. For example, the use of a truncated inverted pyramid shape minimized internal absorption in the LED.

The next important breakthrough was achieved by etching away the GaAs substrate, which eliminated most of the internal losses. GaAs was replaced with GaP, by the wafer bonding method [3]. GaAs is essential to obtain high structural quality of the LED structure, by growing the active part of the device on a lattice-matched substrate. However, for the device operation GaAs is detrimental and introduces large internal losses. This happens since the band gap of GaAs is smaller than the energy of emitted light by the diode, i.e. photons emitted towards the substrate will be absorbed and thus lost. This is why etching away the GaAs and bonding with GaP is essential. In this way the substrate, which absorbs emitted light, is replaced with the transparent one. The AlInGaP/GaP LED structures thus achieved efficiencies of nearly 100 lm/W, which is better than those of unfiltered incandescent and halogen lamps, and are reaching the efficiencies of fluorescent lamps. Further progress is still possible, and soon expected, which will result in development of a next generation of highly efficient LEDs with fluxes larger than 100 lm.

New applications become possible and economically justified as a result of a rapid progress in LED performance. Applications such as red stoplights in cars, red traffic lights, etc. [6] are now widespread. For example, conventional red traffic lights are at present replaced with a matrix of red LEDs producing 200 lm at a power consumption of 14 W, with lifetimes exceeding several times that of conventional traffic lights (table 3.1). Due to very reduced power consumption costs of replacement are repaid in about 1 year, despite the fact that LEDs are at present more expensive than commercial incandescent light bulbs.

It was clear that for further applications equally efficient green- and blue-colour diodes should be introduced. The first blue LEDs were based on SiC and were fairly inefficient. A breakthrough came with the introduction of GaN-based diodes with ternary InGaN quantum wells [7]. This occurred very recently (in 1996) and resulted in extremely fast progress in

Table 3.1. LEDs versus filtered long-life incandescent light bulbs for traffic light applications.

Color	Filtered 140 W incandescent light bulb Efficiency (lm/W)	LEDs (2000 year state of art) Efficiency (lm/W)
Red	1–6	16
Yellow	4–8	10
Green	3–10	48

the development of short wavelength LED devices. At present red, green and blue LEDs with fluxes above 50 lm in red, 30 lm in green and 10 lm in blue are available, as specified in table 3.1, prepared based on the data given in the Optoelectronic Industry Development Association (OIDA) (http:/www.OIDA.org) technology roadmap update 2002. Further progress is envisaged leading to a similar industrial revolution to the one which occurred after the introduction of transistors in electronics and then of integrated circuits.

3.3 White light sources

3.3.1 Principles of white light sources

The most challenging task remains the commercialization of semiconductor-based white light LEDs, which will replace inefficient incandescent light bulbs for overhead illumination, cold cathode fluorescent lamps as back-lightning for LCD displays, etc. To indicate the scale of this task one should notice that at present nearly 20% of the produced electricity is used for illumination purposes. Such a scale of energy consumption is not economically justified and reduction by nearly a half is presently envisaged. High losses of energy are mostly due to a very low light emission efficiency of massively used incandescent bulb lamps. Their quantum efficiency is about 3–4%, i.e. most of the supplied energy is converted into heat. Also other light sources, which are presently used for overhead illumination, are not very efficient. A large proportion of the produced electrical energy can thus be saved once new and more efficient light sources are introduced.

Historically, all first light sources used a concept of a broad-band black body light emission, with emission extending from a very short to a very long wavelength. Then, in most cases, a significant part of the emission is outside of a visible light spectral region and is lost. Light generated by burning or from incandescent light sources has such a property, i.e. is inherently inefficient. To improve light emission efficiency, light sources should thus generate light with the spectral distribution comparable with the spectral sensitivity of a human eye. This observation led to the formulation of large national research programmes for developing new light sources and to spectacular progress in the field of opto-electronics, which occurred in the past few years.

There are three alternative approaches to achieve white emission from semiconductor-based LED light sources, as recently reviewed in the OIDA technology roadmap update 2002. The first is the wavelength conversion approach, similar to the concept used for example in fluorescent lamps. In this approach narrow band gap emission from an active part of the diode is down-converted by phosphor material deposited on top of an LED.

The second approach uses the concept of colour mixing. White is achieved by mixing of red, green and blue emitted by three independent LED devices [8]. This is most likely the most efficient way to produce white light emission, but may be at the same time the most expensive one. However, the possibility of independent control of each of three diodes may result in a fine-tuning of the colour of an emission and may result in some new attractive applications.

The third approach, the only one which is used at present in practice, is a hybrid approach, in which short wavelength emission from a blue InGaN-based LED is mixed with the yellow emission of YAG:Ce phosphor. By a hybrid approach we mean here that primary blue emission of a diode is mixed with a phosphor emission, excited by a primary light. This concept is by definition less efficient than the second one, since part of a primary light is absorbed by a phosphor and is down-converted into energy.

3.3.2 Basics of colour mixing

All three approaches discussed above use a principle of colour mixing. Before discussing present white light sources, and possible applications of ZnSe-based structures, we will shortly describe fundamentals of colour mixing which were discovered by Sir Isaac Newton. He showed, by passing sunlight through a prism, that white light separates into a spectrum of monochromatic components. Moreover, mixing of a spectrum of monochromatic components with another prism gives back a white light.

Fortunately, the sensation of white can also be achieved by mixing three primary hues. We thus do not need mixing of a full spectrum to achieve white light. For example, mixing of red, green and blue (RGB) gives the impression of white, if intensities of primary colours are chosen in the proper way. Moreover, it has been known for nearly 200 years that we can duplicate any colour by combining in the proper proportions the light from three primary sources. Additive mixing of primary colours

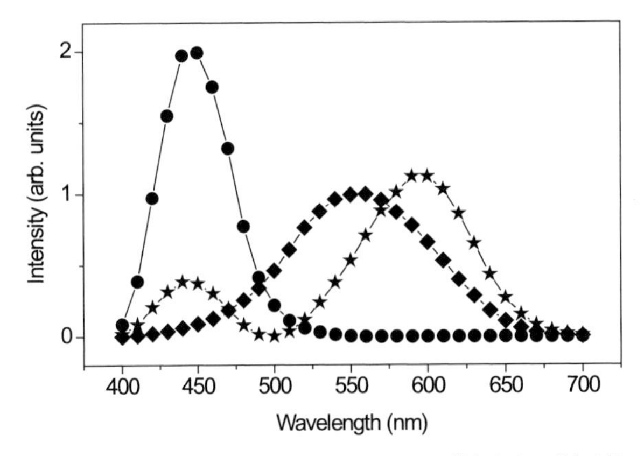

Figure 3.1. CIE 1931 standard colour-matching curves $\bar{x}(\lambda)$ (★), $\bar{y}(\lambda)$ (◆) and $\bar{z}(\lambda)$ (●).

results in huge numbers (about 10 000) of non-monochromatic colours, which can be perceived by the human eye. Sometimes we need only two primary colours; e.g. pink or brown are obtained in this way. Additive mixture of colours and also a subtractive method found several practical applications in light-emitting devices. For example, red, yellow and green filters are used in traffic lights, to select a required colour from a wide band emission of incandescent lamps.

Red, green and blue were selected by the Commission Internationale d'Éclairage (CIE) in 1931 as additive primaries because they provide the greatest colour gamut in mixtures [9–11]. However, other primaries are also possible, as e.g. red-absorbing cyan with green-absorbing magenta and blue-absorbing yellow can be equally good. For example, they are often used in colour printers to obtain full colour printing.

The CIE has adopted a standard of colorimetry, in which a given colour is represented in a three-dimensional diagram. The coordinates of this diagram are derived from three idealized primaries, which are shown in figure 3.1. The primaries, i.e. the standard colour matching curves, $\bar{x}(\lambda)$, $\bar{y}(\lambda)$ and $\bar{z}(\lambda)$, are used to calculate the so-called *tristimulus* X, Y and Z values for a given light source [9–11] which, for a given non-monochromatic light source, are calculated from the following integrals over the entire visible spectrum:

$$X = \int \Phi(\lambda)\bar{x}(\lambda)\,\mathrm{d}\lambda$$

$$Y = \int \Phi(\lambda)\bar{y}(\lambda)\,\mathrm{d}\lambda$$

$$Z = \int \Phi(\lambda)\bar{z}(\lambda)\,\mathrm{d}\lambda.$$

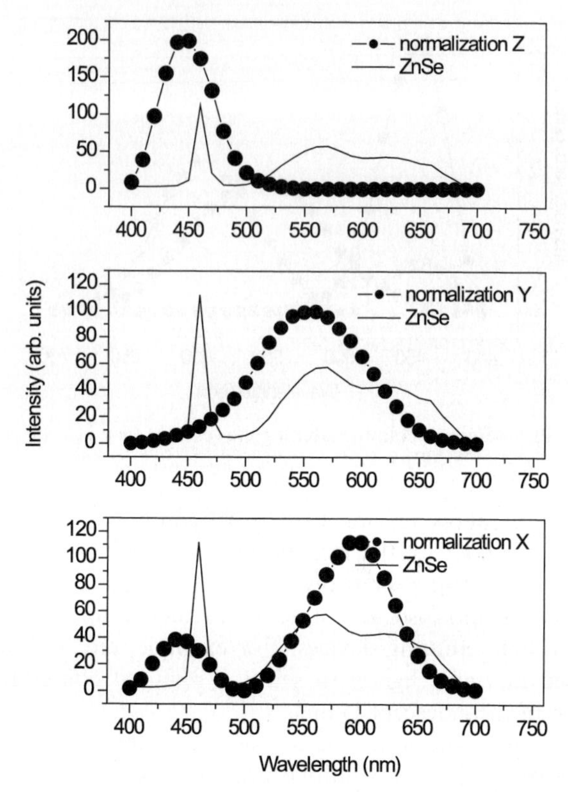

Figure 3.2. Calculation of tristimulus X, Y and Z values performed for a thin film of ZnSe emitting in blue, green and red spectral regions (E Guziewicz *et al*, unpublished results).

The example of such calculations, for a tricolour light emission from thin ZnSe films, is shown in figure 3.2.

The evaluation of the colour (chromacity) is accomplished by defining three new quantities called chromaticity coordinates:

$$x = X/(X + Y + Z)$$
$$y = Y/(X + Y + Z)$$
$$z = Z/(X + Y + Z).$$

Only two of these are independent, because $x + y + z = 1$. Hence, to specify the chromaticity of light it is necessary to give the values of two of the three coordinates; x and y have been selected for this purpose.

The colour map may be expressed as a two-dimensional projection into the xy plane. In that way we obtain the standard chromaticity diagram shown in figure 3.3. The upper part of the diagram is the locus of saturated colours. All the colours resolved by human eye are enclosed between the area

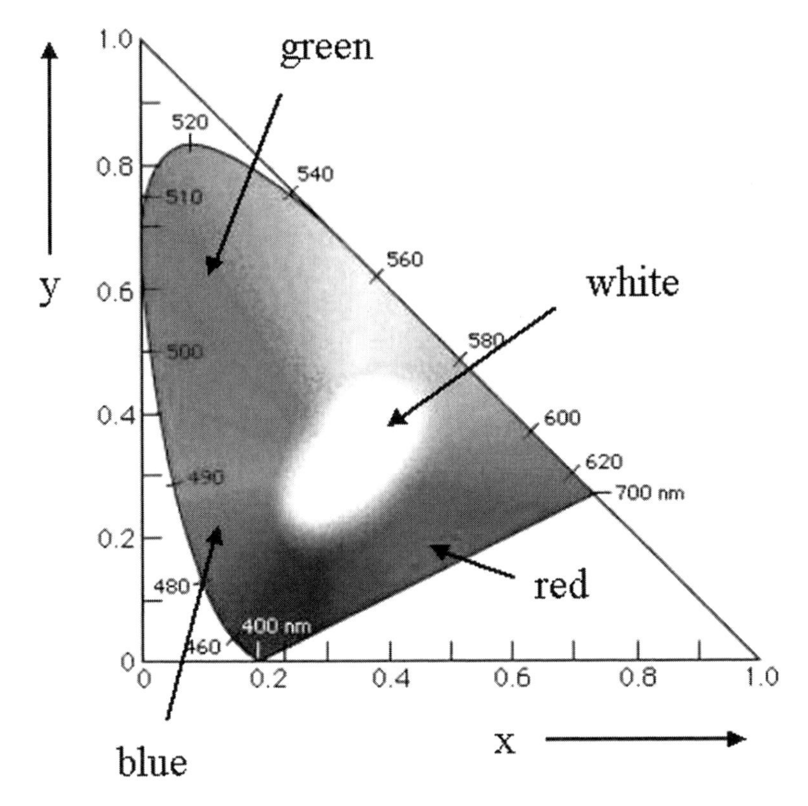

Figure 3.3. The CIE standard chromaticity diagram.

of saturated colours and the straight bottom line called 'magenta'. The white colour appears in the central region of the standard chromaticity diagram. The achromatic point ($x = y = z$), sometimes described as a white point, is located at the centroid of the diagram. The Planckian radiator appears at different temperatures as various shades of whiteness.

It should be noticed that the same chromaticity coordinates can be obtained from different X, Y and Z values. This means that x, y and z describe colour but do not describe the lightness of a given light source. Seven years before the 1931 meeting, the CIE obtained agreement concerning the relative luminances, which are called luminocities. From that time, the relative luminance is indicated directly by the value of Y, on the scale that represents an absolute black by 0 and a perfect white by 100.

Colour temperature is also used to describe the quality of the light source. For example, an incandescent lamp has a temperature of about 2850 K, noon sunlight has a colour temperature of about 4800 K, whereas an average daylight has a colour temperature of about 6500 K.

3.4 Semiconductor-based white light sources

3.4.1 III–V based white LEDs

There have been several attempts to solve the problem of producing semiconductor-based white light sources. At present only one approach has turned out to be good enough for practical applications. A commercial InGaN-based white-light emitting LED [12, 13] uses the concept of a hybrid light mixing source. This white LED consists of an InGaN quantum well emitting a blue/violet colour and YAG:Ce phosphor emitting a yellow colour. Due to the use of a light-converting phosphor and hybrid light-mixing approach this device has at present a rather low luminous efficiency, which is about 10 lm/W. This is more or less equal to that of an incandescent bulb and is well below the efficiency of competing fluorescent lamps. However, this design is already attractive for some specialized applications, such as back lighting in LCD displays [14] or for some specialized lamps.

Another approach includes AlGaInN single QW LED structure with tricolour RGB converters [15]. These white LEDs are less efficient than fluorescent lamps. However, rapid progress in their development ensures that white LEDs can soon reach the light emission efficiencies of the present light sources for overhead illumination.

3.4.2 ZnSe and/or ZnTe?

It is believed at present that III–V-based LED and LD structures are superior as compared with those which are II–VI-based. All commercialized LEDs and LDs are based on heterostructures of III–V compounds. This is why we will describe briefly the earlier research concentrating on monochromatic, visible light LEDs based on ZnSe and ZnTe. The efforts will be directed towards constructing green (ZnSe-based) and red (ZnTe-based) diodes.

3.4.2.1 ZnSe

At present all commercialized light emitting devices are based on III–V semiconductors. Developed in the early 1990s, ZnSe-based LEDs and LDs [16, 17] failed to reach parameters required by industry, mostly because of their too-short lifetimes. This happened despite very active research. Also the following attempts to introduce beryllium chalcogenides [18] into ZnSe or to apply high-quality homo-epitaxial LD structures, i.e. use LD structures grown on bulk ZnSe substrates [19], were not successful. During the past few years, research on ZnSe-based devices has been less concentrated and practically no progress is observed. The results presented a few years ago are still not improved.

The extensive reviews of properties of ZnSe-based monochromatic LEDs and LDs can be found in materials of two Japanese–German workshops on green-blue opto-electronics [20, 21]. Since the recent progress in this field is slow, if any, we will not repeat information given there.

At present only homo-epitaxial LDs are considered as promising. In these diodes one can likely omit strain-related problems and limit the role of misfit dislocations, which are responsible for the formation of so-called dark-line defects in an active part of lasers [22]. Already the first attempts have led to fairly good quality LDs, but their performance was still not satisfactory. Further progress requires availability of ZnSe substrates of a better structural quality. This fact led to very concentrated efforts to produce high-quality ZnSe and also ZnTe (for ZnTe-based diodes) substrates, as recently reviewed by Mycielski *et al* [23, 24]. In consequence of these concentrated efforts, high-quality ZnSe crystals and ZnSe-based solid solutions were obtained using different growth methods from a vapour phase [25].

It is important that the above-mentioned homo-epitaxial ZnSe-based devices show another very attractive property. Homo-epitaxial ZnSe-based LEDs [26, 27] combine blue-green emission from ZnCdSe QWs with a yellow emission from the ZnSe substrate. As a consequence, a bright white emission is observed with good colour parameters. Full-colour light emission could also be achieved by selecting appropriate composition fractions in ternary and quaternary compounds forming quantum well and barrier materials, and by selecting appropriate QW width. As a consequence, white emission was achieved from ZnCdMgSe/ZnCdSe light-emitting diodes [28].

Both these approaches are, however, hampered by the too-short lifetime of ZnSe-based LED devices. Further research is required to solve this deficiency. At present it seems to be very unlikely that these LEDs can compete with those based on GaN for emission in violet and blue. Also, white diodes based on GaN are superior. Some hopes remain for green LD applications. Green LDs are still not produced based on GaN structures.

3.4.2.2 ZnTe

ZnTe with its room temperature band gap of 2.27 eV was considered as a good candidate for green LEDs. As mentioned above, GaN-based (with InGaN QWs) LDs emit in the violet/blue spectral region, but not in the green. This fact stimulated concentrated research in this area. The main obstacle, also for other wide band gap II–VI compounds, was the problem of efficient n- or p-type doping. p-Type doping turned out to be difficult in the case of ZnSe, whereas difficulties in obtaining good quality n-type doped layers hampered any progress in the case of ZnTe. For ZnTe, p-type doping is simple and can be done by introducing either nitrogen or phosphorus. At present, nitrogen is the most efficient p-type dopant, not only of ZnSe, but also of ZnTe.

A breakthrough came with introduction of modern epitaxial techniques. Several groups reported not only growth of good quality ZnTe epilayers by MBE or MOCVD [29–31], but also were successful in n-type doping of these layers. Aluminium was successfully used as an n-type dopant in MBE [31]. Free electron concentration in excess of 10^{18} cm^{-3} was achieved, but at the cost of a rather low electron mobility. Progress in n-type doping of binary ZnTe layers led to the achievement of heavily doped n-type quaternary ZnMgSeTe layers [32] and consequently ZnCdTe/ZnMgSeTe light emitting diodes. These LEDs consist of ZnTe substrate, ZnTe homo-epitaxial buffer layer and p-type ZnMgSeTe cladding layer (30% of Mg). The active part of the diode consists of multiple ZnCdTe quantum wells imbedded between ZnTe barrier layers. The structure was completed with n-type ZnMgSeTe (Al doped) cladding layer, n-type ZnTe and thin n-type ZnSe as contact layers. Bright LED emission at either 2.04 or 2.19 eV, depending on the Cd fraction in QWs, was achieved [33]. It is, however, too early to predict further development of these ZnTe-based devices. At the moment they are less efficient than competing III–V-based LEDs. Also, the lifetime of about 1000 h is too short for device commercialization. Further progress is thus essential.

Sato *et al* [34] demonstrated recently a very original and promising approach. They proposed new method for constructing ZnTe-based diodes. These authors achieved pure green emission from ZnTe plates with an intrinsic p–n junction. Plates were p-type doped with phosphorus using a simple diffusion process. In this way an intrinsic p–n junction can be formed and a LED structure is achieved. Bright LED emission was observed with a lifetime exceeding 1000 h. For this approach it is crucial to use high quality material with a low dislocation density, and with compensating point defects suppressed by low-temperature annealing.

3.5 Niche applications

As already mentioned, it is claimed at present that III–V-based devices are good for opto-electronics applications, whereas II–VI-based devices are only good for some basic research. This statement was justified by the fact that all previous attempts to commercialize ZnSe- or ZnTe-based LEDs or LDs failed. This may change if ZnSe or ZnTe is used for some niche applications. Such new possible applications of ZnSe or ZnTe are described below.

3.5.1 White emission from thin films of ZnSe

Several wide band gap II–VI compounds have been used to produce thin film electroluminescence (TFEL) displays [35–38]. Atomic layer deposition

(ALD), often referred to as atomic layer epitaxy (ALE) [35], was used for manufacturing these devices. Their construction is very simple. A thin film of a transparent and conductive electrode is first deposited on a glass substrate. Then, this electrode is covered with a thin film of a given active material embedded between two layers of dielectric films. The bottom electrode typically uses aluminium [37].

An electric field is applied to the electrodes, which is enhanced by dielectric films to 10^4–10^5 V/cm, and injects high-energy electrons to a semi-conductor film. These electrons, called hot electrons, are accelerated by a strong electric field and reach high energies, in excess of a few eV. Hot carriers are essential, since electroluminescence is excited by the interaction of hot carriers with emission activators. Emission activators are due to transition metal (TM) or rare earth (RE) ions, which are introduced to a thin film of a II–VI semiconductor during a growth. Two mechanisms of TM or RE excitation are active in EL devices. The emission activator is excited either by impact excitation or by impact ionization [38]. In both cases, an intra-shell emission of TM (red/orange/yellow emission of Mn^{2+}) or RE (green emission of Tb^{3+}, red emission of Eu^{3+}, blue-green emission of Ce^{3+} etc.) is excited, which is used as EL from the device.

The first monochromatic TFEL displays were designed using orange-yellow (ZnS:Mn) or green (ZnS:Tb) colour emission by activators. Later, full-colour TFEL displays were introduced [37]. Current technology of full-colour TFEL displays requires the use of two different materials (e.g. ZnS:Mn, SrS:Ce, with an efficiency of 1.3 lm/W – see [36–38] and references therein), which are fabricated separately and then combined into one display unit. The appropriately doped ZnS and SrS provide wide band gap light emissions in two spectral regions of the visible spectrum. Mixing of the two bands gives the impression of white light. Colour filters are then used to allow a selection of colour emission in a given spectral region over the whole visible spectrum. Thus the device uses the concept of colour mixing discussed above in the case of LED structures.

An alternative and simpler approach might be to use two or three emissions coming from one material. This approach has still not been successfully put into practice. Previous attempts using SrS doped with Mn, and Pb- or Pr-doped SrS as active layers of TFEL displays failed due to too-low light emission efficiency [36–38].

Recently, we have observed multicolour emission coming from thin films of ZnSe excited with either a beam of electrons in cathodoluminescence (CL) or by light excitation with above band gap excitation [39]. Below we briefly describe light emission properties from thin films of ZnSe grown by ALD (ALE), and discuss light tuning possibility.

In figure 3.4 we show a cathodoluminescence (CL) spectrum of a 1.1 μm thick layer of ZnSe grown by ALD (ALE) on a GaAs substrate. CL was measured at room temperature using a 20 kV accelerating voltage.

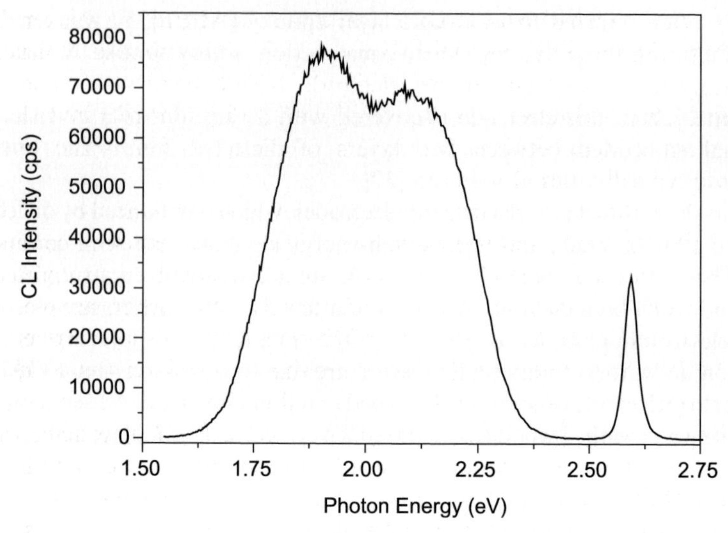

Figure 3.4. CL spectrum of 1.1 μm thick ZnSe film observed at room temperature at 20 kV accelerating voltage (after E Guziewicz *et al* 2003 *Thin Solid Films*).

The observed CL emission bands were identified using high-resolution photoluminescence measured at low temperatures. In figure 3.5 we show low-temperature PL spectra of two ZnSe films of 0.57 and 1.1 μm thickness. Similar spectra were taken for a wide range of samples with thickness varied from 0.04 to 5.5 μm.

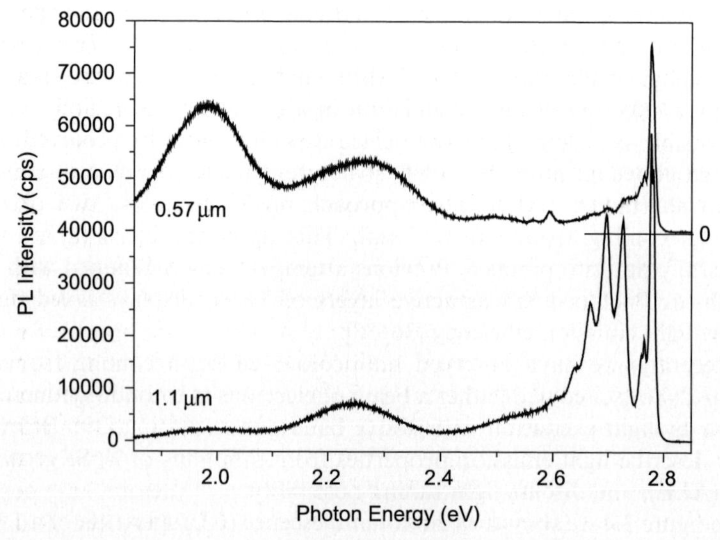

Figure 3.5. Low temperature photoluminescence spectra of two ZnSe films grown by ALD on GaAs substrate (after E Guziewicz *et al* 2003 *Thin Solid Films*).

The most intense 'edge' CL/PL emission is observed for thicker films. For these films 'edge' emission shows a rich structure and consists of several sharp excitonic emissions (see figure 3.5). This emission is due to an overlap of several bound excitonic emissions, and of a structured shallow donor–acceptor pair (DAP) emission at about 2.7 eV. No excitonic emission is observed for the thinnest film.

X-ray diffraction indicates that thicker films are practically strain relaxed. For them the spectral positions of the relevant emission bands are not affected by strain conditions. One can identify their origin by comparing PL spectra from thin films with those known for bulk ZnSe samples. The dominant PL, observed at 2 K at 2.7954 eV, can be related to exciton bound at neutral donor (DBE PL). 2.7958 eV PL was related to neutral donor bound excitons at either Cl, Ga or In [40].

Chlorine is the most likely contaminant of the ALD films since $ZnCl_2$ was often used as a source of zinc in the ALD process. Ga contamination is also possible, since Ga can diffuse into ZnSe from underlying GaAs substrate. It is possible that both Cl and Ga are present in the films, since two DBE PLs were resolved. PL at 2.7903 eV at 2 K is related to radiative recombination of acceptors bound at neutral acceptor centres (ABE PL). The 2.7902 eV position of the ABE PL was observed previously and related to Na contamination [41]. Additional sharp PL at 2.7767 eV (at 2 K) is observed. The origin of this PL remains unknown. One can tentatively relate this band to another ABE PL emission.

In addition to a rich 'edge' emission, red and green PL bands, with maxima at about 2.02 and 2.25 eV, are also observed. We relate these two PL emissions to the donor–acceptor pair (DAP) transitions. The DAP origin of the red and green 'deep' PL emissions of ZnSe was confirmed by the optically detected magnetic resonance (ODMR) study showing magnetic resonance signals of shallow ZnSe donors and deep acceptors [42].

For any practical application the emission should be observed up to room temperature. Thus, the temperature dependence of relevant emission bands was followed. The relative intensity of blue-colour 'edge' PL and red and green PL bands strongly varies with a change of temperature and also film thickness. Deep DAP PLs dominate in thin layers and also become relatively more intensive at higher temperatures. The overall PL intensity drops with increasing temperature by about two orders of magnitude for thin layers and by about one order of magnitude for thicker films, indicating an enhanced rate of non-radiative recombination in structures of a lower structural quality. We estimate that the external quantum efficiency of the PL of thicker films drops to about 1% at room temperature. It has rather low efficiency and should be improved before any possible application. This fortunately can be done by improving the structural quality of the ZnSe films, i.e. by eliminating active channels of non-radiative recombination.

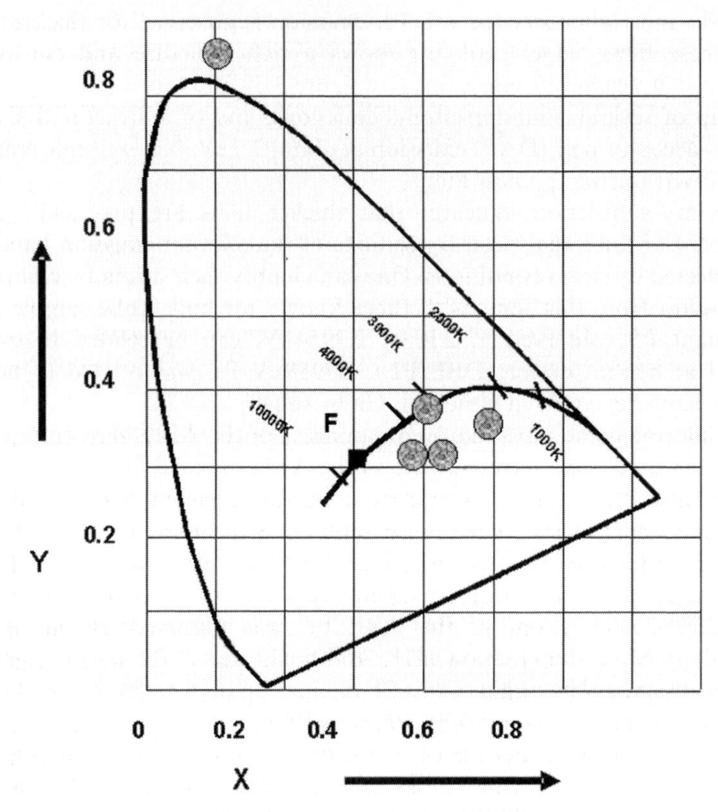

Figure 3.6. CIE (x, y)-chromaticity diagram showing Planckian radiator curve. The point F represents chromaticity coordinates of an equal-energy radiator. The shaded area shows the region of chromaticity coordinates of investigated ZnSe layers (after E Guziewicz *et al* 2003 *Thin Solid Films*).

For possible applications, it is also important that all three emissions are also seen from thin films grown on a glass substrate, i.e. substrates used in the TFEL devices. The blue 'edge' PL emission is accompanied by the green and red PL bands in a full temperature range. Their mixing gives an impression of a pure white. The best impression of a white emission is observed at room temperature, for which 'edge' PLs are partly deactivated (e.g. shallow DAP PL is not observed) whereas deep DAP PL emissions are less affected by an increase in temperature.

As mentioned above, there exists only one 'white point', which is situated at the centroid of the CIE chromaticity diagram at $x = y = z = 1/3$. This point has been marked with a black square in figure 3.6.

In colorimetry, however, light sources whose coordinates are the same as for a Planckian radiator are also regarded as white light sources. Their colorimetric coordinates are situated on the Planckian line, shown as a

Table 3.2. Colorimetric parameters of ZnSe/GaAs layers grown by ALD, calculated using the approach described in section 3.3.2. The sum of x, y and z equals 1 within experimental error.

Colorimetric parameters	Layer thickness (μm)				
	0.04	0.2	0.4	0.57	1.1
x (red)	0.111	0.381	0.432	0.483	0.399
y (green)	0.845	0.316	0.317	0.359	0.373
z (blue)	0.045	0.302	0.250	0.158	0.227
T_c (K)	—	4500	4000	2400	3500

solid line in figure 3.6. Good quality white light is also perceived for light sources with x, y parameters in the white area in figure 3.3.

In table 3.2 we give x, y and z parameters calculated for thin films of ZnSe of different thicknesses. From table 3.2 and figure 3.6 (shaded area) it appears that chromaticity parameters 'oscillate' around 1/3, i.e. the so-called white point. Emission parameters are thus very good. A promising property of the system is also the fact that chromaticity parameters improve with increasing temperature, due to the relative change of intensity of relevant emission bands.

The required colour temperature of light sources depends on their application. Light sources for indoor overhead illumination should have a colour temperature ranging from 3000 to 4000 K ('warm light'), whereas, for example, metal-halide discharge lamps for stadium lighting imitate sunlight with a corresponding T_c between 5500 and 6500 K ('cold light'). For thin films of ZnSe grown by ALD, colour temperatures range from 2400 up to 4500 K, i.e. they fall in a required range of colour temperatures.

PL and CL studies show that the intensity of emission bands varies with the layer thickness. This indicated the possibility of colour adjustment and tuning in devices based on thin films of ZnSe. Depth-profiling CL investigations allowed estimation of in-depth distribution of various emission bands and to explain the observed thickness dependence of emission bands. A complicated thickness distribution of RGB tricolour PL from ZnSe epilayers is shown in figures 3.7 and 3.8.

At 6 keV electron beam energy, which is the optimal condition for a surface-close excitation (see [43] for explanation), blue PL (of excitonic origin) dominates the spectra. At larger beam energies (more than 18 keV) CL emission is excited mostly from the ZnSe/GaAs interface region. Then, the edge PL becomes weaker and the red and green DAP bands dominate. This means that excitonic PL emissions come from the ordered crystalline layer of the film, whereas the green and red DAP emissions mostly come from the disordered part of the film close to the ZnSe/GaAs interface,

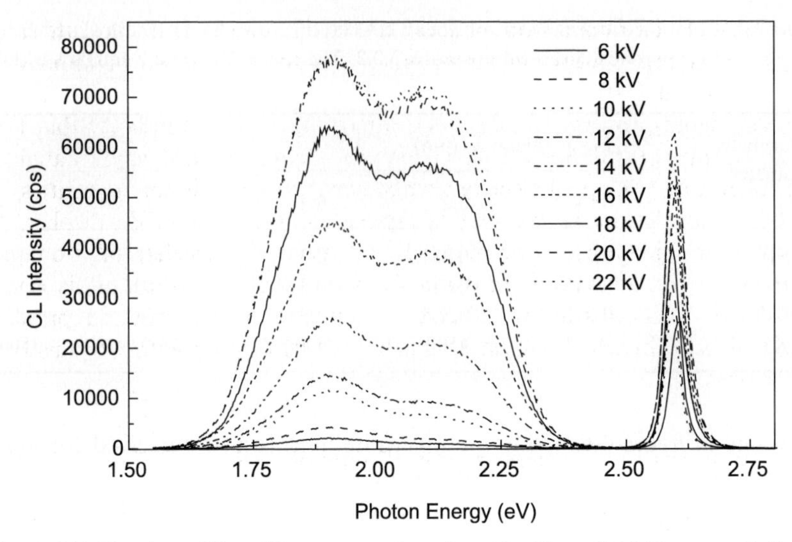

Figure 3.7. Depth-profiling CL spectra taken for thin film of ALD-grown ZnSe at accelerating voltages varied between 6 and 22 kV (after E Guziewicz *et al* 2003 *Thin Solid Films*).

which is due to an increased defect concentration in this region of the samples. In fact, high p-type conductivity of the ZnSe/GaAs interface region was reported in the past [44, 45]. We can thus conclude that in-depth evolution of the PL bands explains their thickness dependence. In

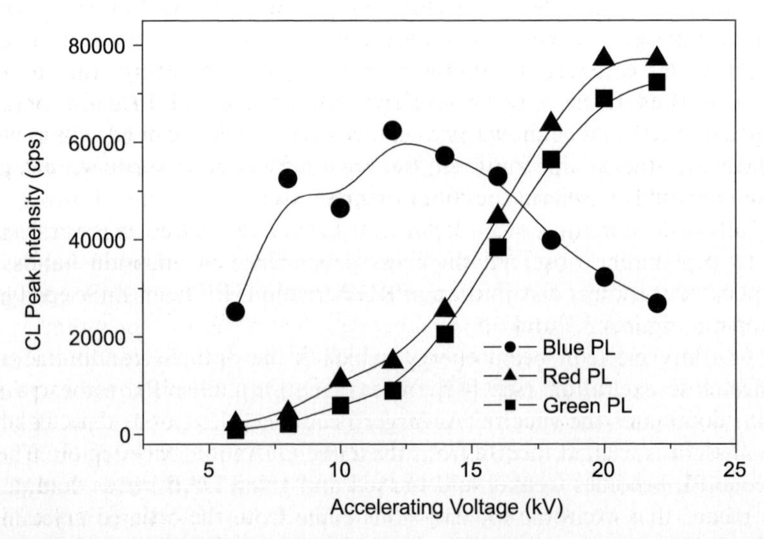

Figure 3.8. Voltage dependence of the CL intensity. The CL data were taken for the ZnSe/GaAs layer with a thickness of 1.1 μm (after E Guziewicz *et al* 2003 *Thin Solid Films*).

thin films the dominant contribution to the PL comes from an interface region. This contribution becomes minor for thick films of an improved structural quality.

The depth-profiling CL data (see figures 3.7 and 3.8) indicate that the colour balance of emission can be varied in a straightforward way. Changing an accelerating voltage of electrons we can change the emission intensities of the CL bands. This can be a very attractive property in a TFEL display. A multi-colour emission can be induced by varying the accelerating voltage, rather than by mixing the emission from different activators, as is done at present. Such possibility of colour tuning does not exist in present TFEL devices, in which emission is activated by doping with various TM or RE ions.

3.5.2 Blue anti-Stokes emission from TM-doped ZnSe

3.5.2.1 ASL in ZnSe

In some cases PL emission is observed for light energies larger than the excitation energy. Such a PL is called an anti-Stokes luminescence (ASL) or energy up-conversion emission. ASL can often exceed the excitation energy by more than the vibrational energy of the lattice and can find several practical applications. For example, investigations of the ASL in dielectric crystals activated with RE ions, carried out in 1960s, led to the introduction of the anti-Stokes luminophors. Energy up-conversion is also widely used in different RE-doped solid laser materials.

Difficulties in achieving efficient short wavelength emission from semiconductor-based LED devices motivated an alternative approach, in which blue or violet emission is obtained under optical pumping, due to the energy up-conversion. Use of efficient red GaAs-based LDs for optical pumping is preferential and ensures that compact devices can be constructed for memory storage applications. For such applications short wavelength emission should be efficient at room temperature.

Regarding energy up-conversion efficiency, light powers of several mW are required in most applications. At present, efficient semiconductor-based red LDs, with about 1 W optical power, are available. This means that light conversion efficiency should be in range of a few times 10^{-3} or more.

Two mechanisms were proposed to explain the ASL in semiconductors. The first of them explains excitation of the ASL by a multi-photon absorption, which may occur with a high density of excitation. The second, expected also for a low and moderate density of excitation, assumes a two-step electronic transition of a carrier from a valence band (VB) to a conduction band (CB) proceeding via some deep impurity (defect) level. In this case, free electrons and holes are created in the CB and VB, respectively, in two complementary photo-ionization transitions (figure 3.9). The ASL can be observed if at least

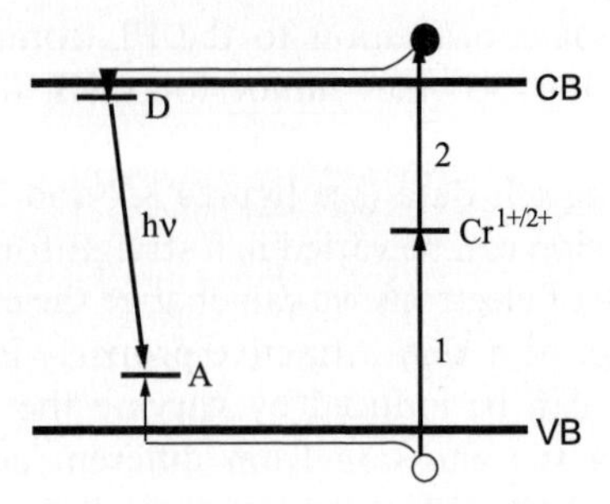

Figure 3.9. Model of energy up-conversion in Cr doped ZnSe (after V Yu Ivanov *et al* 2003 *Acta Physica Polonica A* **103** 695).

part of the photo-generated carriers is not retrapped by an ionized centre. If so, the ASL emission can consist of band-to-band, excitonic and/or donor–acceptor pair (DAP) recombination transitions [46, 47].

The main difference between the two above-mentioned energy up-conversion mechanisms relates to the appearance of impurity (defect)-related band gap level in the second process. If light induced population of this level is large, the following complementary excitation process (moving electron form the level to the conduction band) can be very efficient, as will be demonstrated below.

There are several further requirements for high efficiency of such two step transitions via impurity (defect)-related level. First, the photo-excited, impurity-related level should be close to the mid band gap position. Second, both complementary photo-induced transitions (photo-ionization and photo-neutralization) must be of a similar and large probability, i.e. these transitions should be parity and spin allowed transitions

The above conditions are not sufficient to result in a highly efficient energy up-conversion. The photo-generated free holes (in the valence band) and free electrons (in the conduction band) cannot be efficiently retrapped by the mid band gap impurity (defect) level. They must be trapped by centres active in band-gap-close ('edge') optical recombination transitions, resulting in energy up-conversion. Efficient retrapping of free carriers by photo-ionized (neutralized) centres kills energy up-converted recombination transitions. The remaining condition is that energy transfer processes from centres active in the 'edge' recombination to deep impurity (defect) centres should be inefficient.

ZnSe is an attractive candidate for the energy up-conversion material. ZnSe has an appropriate band gap, with 'edge' PL emission in the blue spectral region. Moreover, this PL emission is observed up to room temperature, which is due to relatively large exciton binding energies and shallow centre (acceptor) ionization energies in ZnSe. Unfortunately, two-photon excitation is inefficient in the case of undoped ZnSe. The quantum efficiency of the process is too low (in the range of 10^{-6}) for any practical application. A

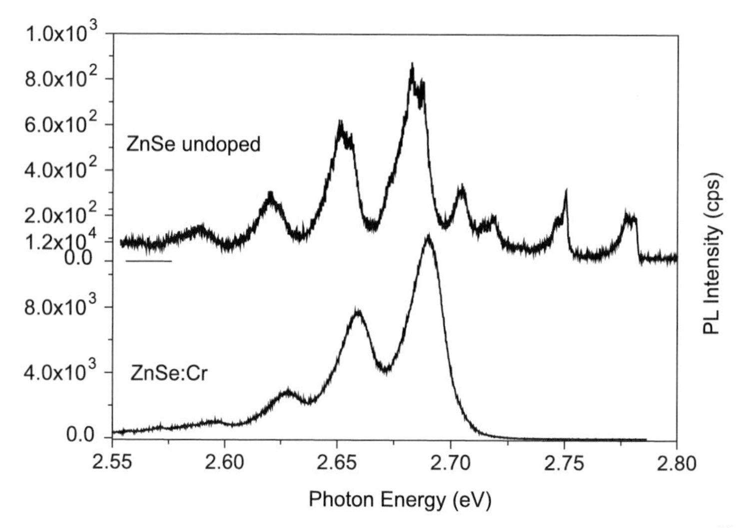

Figure 3.10. Photoluminescence spectra of undoped ZnSe and ZnSe:Cr ($4 \times 10^{18}\,cm^{-3}$) measured at 4.2 K under 2.41 eV photo-excitation [47].

very different situation occurs in the case of Cr-doped samples. The ASL spectra of ZnSe and ZnSe:Cr measured at 4.2 K are shown in figure 3.10.

The PL spectrum under band-to-band excitation ($E_G = 2.82\,eV$) and ASL emission of undoped samples consists of three sharp emission lines at 2.7977, 2.7930 and 2.7829 eV, which are due to a radiative recombination of DBE [48] and two ABE emissions at 2.7930 eV (Na? [48]) and at 2.7829 eV [48]. A structured PL between 2.62 and 2.75 eV is due to the superimposed free (electron)-to-bound (shallow acceptor) (FB) and shallow donor–shallow acceptor pair transitions with zero phonon lines at about 2.720 and 2.692 eV and their LO phonon replica. The 2.692 eV PL can be attributed to the radiative decay of an electron on a shallow donor (Ga or Al) and a hole on a shallow acceptor (Li [49, 50]). This bright blue and structured PL dominates in ASL in Cr-doped ZnSe. The ASL is very bright at helium temperature (see figure 3.11) showing bright blue emission from a cryostat containing a ZnSe:Cr sample observed under green light excitation.

The excitation spectrum of the ASL emission in ZnSe is shown in figure 3.12. The blue ASL is observed for excitation with photon energies larger than 2.0 eV, i.e. smaller energies than the emission energy. Its spectral position and shape of the ASL PL is not dependent on the excitation energy.

Figure 3.12 shows that the excitation spectrum of the ESR signal of the Cr^{1+} exactly correlates with the ASL excitation spectrum. Also, the photoconductivity spectrum of ZnSe:Cr, related to Cr photoionization, correlates with the ASL excitation [47]. Thus, the appearance of the ASL can be explained by efficient photo-generation of free electrons and free holes in two-photon transitions via the deep chromium-related centre.

Figure 3.11. Bright blue anti-Stokes emission of ZnSe:Cr measured at 4.2 K under the 2.41 eV photo-excitation (V Yu Ivanov, unpublished results).

Two complementary Cr ionization transitions, identified by ESR study, are

$$Cr^{2+} + \text{photon} \rightarrow Cr^{1+} + \text{hole in the valence band}$$

$$Cr^{1+} + \text{photon} \rightarrow Cr^{2+} + \text{electron in the conduction band}$$

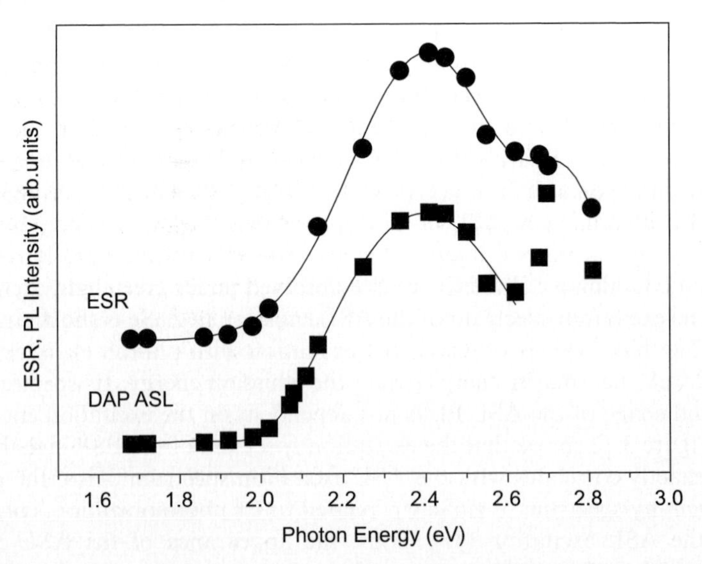

Figure 3.12. Low-temperature excitation spectrum of the ASL (lower curve) and photo-excitation spectrum of the Cr^{1+} ESR signal (upper curve) in ZnSe:Cr sample [47].

The ASL is excited by the 2+ → 1+ photo-ionization transition of chromium, which occurs for the photon energies larger than about 2 eV. Such identity of the ESR excitation band was proved in the previous ESR experiments [51, 52]. The 2+/1+ energy level of Cr lies at about 2 eV above the VB edge of ZnSe [52]. Once the Cr^{1+} state is populated, the complementary 1+ → 2+ transition can take place. ESR studies of Cr-doped ZnSe [52], ZnS [53] and Fe-doped ZnSe [54], and ZnS [55] confirmed relatively high efficiency of such TM-related complementary ionization transitions. Moreover, it was observed that two-step ionization transitions of these two ions result in a population of both shallow donors and shallow and deep acceptors [56]. In the 2+ → 1+ transition free holes (first step) and in the 1+ → 2+ transition free electrons (second step) are created. These free carriers, if not retrapped by chromium, can then participate in the PL recombination transitions resulting in the appearance of the ASL.

For the two-quanta excitation mechanism, via a midgap Cr level, one formally expects a quadratic dependence of the ASL intensity on excitation intensity. In fact, such quadratic dependence was observed, e.g. for undoped ZnSe crystals, but not for ZnSe:Cr (see figure 3.13). For ZnSe:Cr, the efficient ASL process shows, in a large range of excitation densities, a linear dependence of the ASL on excitation intensity. This can be explained by simple kinetics equations, given below, describing the dependence of the

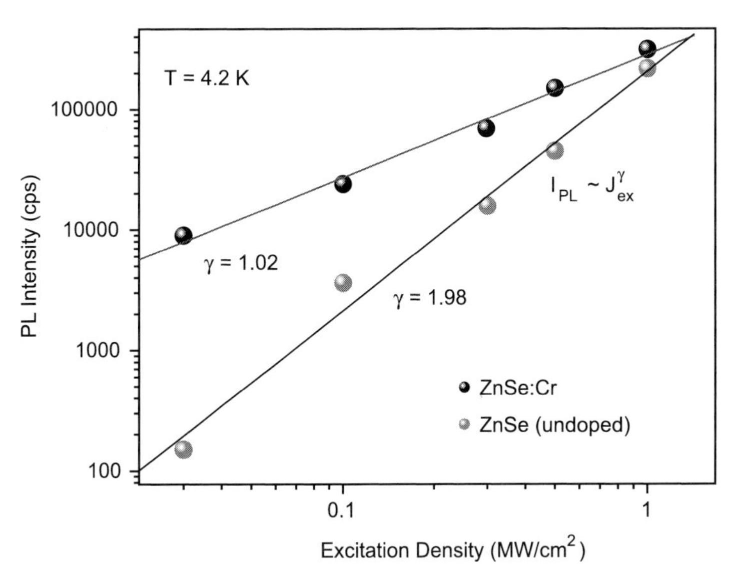

Figure 3.13. Excitation power dependence of the up-converted energy for undoped and chromium doped ZnSe (after V Yu Ivanov *et al* 2003 *Acta Physica Polonica A* **103** 695).

ASL intensity (at low temperature) on excitation density [47]:

$$\frac{dn_{Cr}}{dt} = I\sigma_{OV}(N_{Cr} - n_{Cr}) - I\sigma_{OC}n_{Cr} + n(N_{Cr} - n_{Cr})c^{e}_{Cr} - pn_{Cr}c^{h}_{Cr}$$

$$\frac{dn}{dt} = I\sigma_{OC}n_{Cr} - n(N_D - n_D)c^{e}_{D} - n(N_{Cr} - n_{Cr})c^{e}_{Cr}$$

$$\frac{dp}{dt} = I\sigma_{OV}(N_{Cr} - n_{Cr}) - p(N_A - n_A)c^{h}_{A} - pn_{Cr}c^{h}_{Cr}$$

$$\frac{dn_A}{dt} = p(N_A - n_A)c^{e}_{D} - \beta_{DAP}n_A n_D$$

$$\frac{dn_D}{dt} = n(N_D - n_D)c^{e}_{D} - \beta_{DAP}n_A n_D$$

where n_{Cr} and N_{Cr} are the concentration of the Cr in the 1+ charge state and the total concentration of Cr in the sample, respectively. N_A and N_D are the total concentrations of acceptors (A) and donors (D) in the sample, and n_A and n_D are the concentrations of populated (neutral) acceptors and donors. n and p are the concentrations of free electrons in the CB and free holes in the VB. σ_{OC} and σ_{OV} are the optical ionization rates for the two complementary ionization transitions of Cr. By c^{e}_{D}, c^{h}_{A}, c^{e}_{Cr} and c^{h}_{Cr} we denote capture rates of electrons (e) by ionized donors, holes (h) by ionized acceptors and electrons by Cr^{2+} and holes by Cr^{1+}. By β_{DAP} we describe an average rate of the DAP recombination, β_{ACr} denotes Cr^{1+}-acceptor tunnelling and I stands for the light intensity of the photo-excitation.

Introducing the neutrality condition, finding solution at equilibrium and assuming that concentration of free carriers is low at low temperatures we can obtain the following formula describing intensity of the DAP ASL:

$$I_{PL} \approx \text{const} \cdot I \cdot N_{Cr}$$

if shallow acceptors dominate over Cr^{1+} in hole capture processes. The latter means relatively long lifetime of midgap state, i.e. that the photo-excited 1+ charge state of chromium is metastable. Photo-ESR experiments confirm a metastable population of photo-excited Cr^{1+} centres, with a decay time often of 1000 h at low temperatures.

This observation means that free holes, created in the valence band under 2+ to 1+ chromium ionization, can be efficiently trapped by shallow and deep acceptor centres, active in the ASL and deep DAP PL processes of ZnSe. Thus, holes are not efficiently retrapped by Cr^{1+} centres. Then, the second photon, which is of the same energy as the first one, can be absorbed by the photo-generated Cr^{1+} state. This induces a complementary ionization transition to the conduction band and creates free electrons in the conduction band.

The photo-ESR investigations indicate that a relatively high quantum efficiency of the ASL in ZnSe:Cr (few times 10^{-3}) and a linear dependence

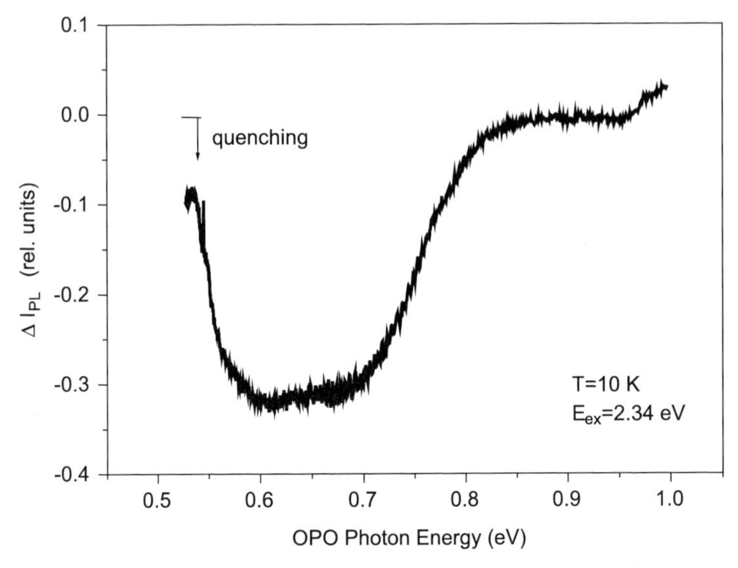

Figure 3.14. Photo-quenching of the ASL emission in ZnSe:Cr, observed in the two-colour experiment with the OPO system and under the above band gap excitation (after V Yu Ivanov *et al* 2002 *Proceedings ICPS'02*, Edinburgh).

of the ASL intensity on excitation density are related to a metastable population of the Cr^{1+} state. Two-colour experiments shown in figure 3.14 confirm metastability of the photo-excited Cr^{1+} state and its influence on the ASL efficiency. In the two-colour experiments two light sources were simultaneously applied to excite and quench the ASL from the sample. First the photon, with energy larger than the ZnSe band gap energy or from the energy range of the Cr 2+ to 1+ ionization transition, excites the blue ASL emission. The second photon, selected from the range of ionization transitions of ZnSe acceptors, was obtained from either the OPO system or the free electron laser (FEL) mid-infrared system. We studied the influence of this second illumination on the intensity of the ASL emission.

For the above-band gap excitation and for infrared illumination with the OPO system, ionization of deep ZnSe acceptors enhances the ASL emission, i.e. ionization of deep acceptors enhances the recombination rate of free holes via the shallow acceptors, which are active in the ASL emission. A different response (see figure 3.14) is observed in the experiment with the ASL excitation (2.34 eV) and with second photon illumination from OPO or FEL, within the range of ionization transitions of ZnSe acceptors. Such two-photon illumination quenches the ASL intensity (figure 3.14). Photo-ionization of shallow acceptors reduces also the population of the Cr^{1+} charge state. This is observed as a rapid photo-induced quenching of the Cr^{1+} signal in the photo-ESR study.

The above observation confirms a crucial role of a metastable occupation of the midgap Cr^{1+} state for obtaining the efficient ASL emission. The photo-excited Cr^{1+} state must be occupied long enough to allow for an efficient second (complementary) 1+ to 2+ Cr ionization transition. Only then are free carriers efficiently photo-excited in both the valence and the conduction band and, in consequence, the efficient ASL emission is observed.

It may be pointed out that metastable occupation of the Cr^{1+} charge state of chromium ion makes ZnSe:Cr also suitable for optical memory applications, as was demonstrated in the case of metastable EL2 defect in GaAs [57, 58].

3.5.2.2 ASL in ZnSe:Fe and in solid solutions of ZnCdS and ZnSSe

As mentioned above, the ASL should be efficient at room temperature and preferably should be pumped with a semiconductor-based red LD. Both these requirements are not fully realized for ZnSe:Cr. Only at low temperatures is the quantum efficiency of the ASL in ZnSe:Cr is high enough for practical applications. Moreover, the most efficient optical pumping occurs with the green colour line of an argon laser and excitation efficiency decreases fast at longer wavelengths.

Two approaches were tried to increase the ASL efficiency and to shift pumping energy into a required spectral range. Chromium was first replaced with iron. The 2+ charge state of iron introduces a mid gap level in ZnSe, i.e. the energy position of this level is optimized for optical pumping with a red LD emission.

In figure 3.15 we show the ASL observed in Fe-doped ZnSe. Weak and broad emission has been observed. Its efficiency is far too low to be of any practical interest. The quantum efficiency of energy up-conversion in ZnSe:Fe has been estimated to be two orders of magnitude smaller than that for chromium doped ZnSe.

Two mechanisms were found to account for the observed low efficiency of the ASL in ZnSe:Fe. Photo-ESR investigations indicate that the photo-excited Fe^{3+} charge state of iron decays quickly even at liquid helium temperature. The midgap level does not appear to be metastably occupied, as in the case of chromium-doped ZnSe. This relates to a very high efficiency of the bypassing process in Fe-doped ZnS and ZnSe [59–61]. The photo-generated Fe^{3+} state has too large a cross section for retrapping of free electrons from the conduction band to be metastably occupied after the photo-generation.

The second mechanism turned out to relate to very efficient Auger-type non-radiative recombination processes observed in Fe-doped wide band gap II–VI compounds (see [62] and references therein). In the Auger process the excited donor–acceptor pair decays non-radiatively by energy transfer to a nearby iron centre, which is ionized.

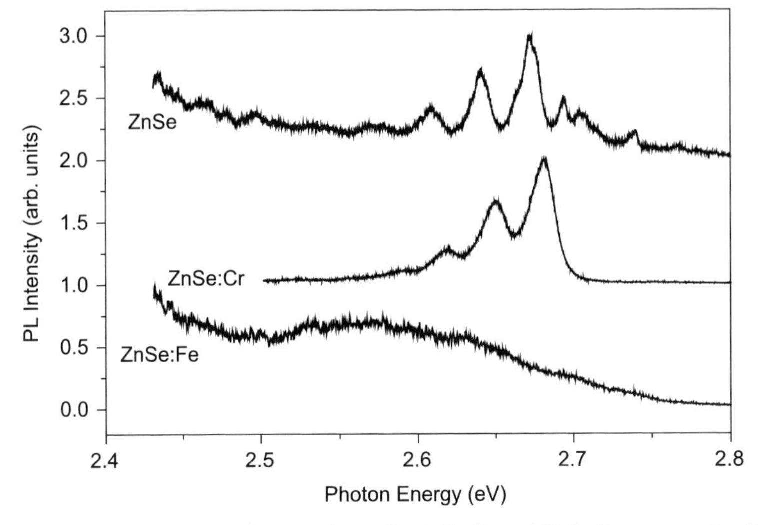

Figure 3.15. Band edge part of the PL in ZnSe, ZnSe:Cr and ZnSe:Fe measured at liquid helium temperature under the anti-Stokes excitation (after V Yu Ivanov *et al* 2003 *Acta Physica Polonica A* **103** 695).

In the case of Fe-doped ZnS and ZnSe this process is efficient, which results in efficient quenching of DAP emissions and in the shortening of DAP decay time, as shown in figure 3.16 for band edge DAP emissions. 'Edge' DAP emission in ZnSe decays much faster than the relevant DAP emissions in undoped ZnSe and in ZnSe:Cr. This is evidence of a very efficient and competing channel of non-radiative recombination in ZnSe:Fe.

Auger-type energy transfer could also be efficient in Cr-doped ZnSe. If so, the quantum efficiency of the ASL is limited by this process and cannot be further improved. Then, the system will never be suitable for efficient light up-conversion material. ODMR investigations were performed to determine the Auger recombination rate in ZnSe:Cr. In the experiment, detection was set at one of the Cr-related intra-shell infrared emissions. The fact was used that Cr ionization (expected for the Auger mechanism), when followed by retrapping of a hole, induces intra-shell transitions of chromium 2+.

Three magnetic resonance signals were detected via an increase in the intensity of Cr^{2+} infrared emissions (figure 3.17). The sharpest one, with a g factor of about 1.1, is due to magnetic resonance at shallow donors in ZnSe. This signal is accompanied by two acceptor-related magnetic resonances, observed at lower magnetic fields, which are due to shallow acceptors. By flipping the spin of either donor or acceptor we enhance DAP transition, which in turn enhances chromium transitions, since relevant resonance signals are detected as an increase in intensity of the Cr^{2+} intra-shell emission. The mechanism of the PL increase must be indirect, i.e. such data indicate that there is an energy transfer from ZnSe DAP PL emissions to chromium ions.

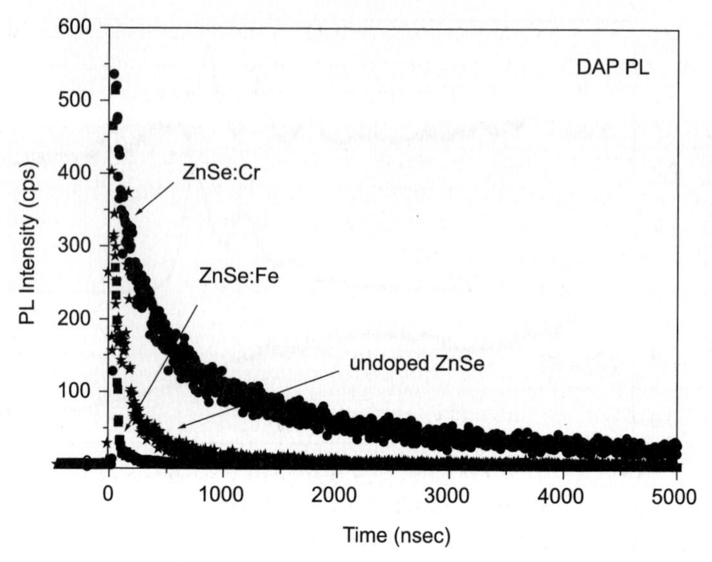

Figure 3.16. PL kinetics of the ASL emission in ZnSe, ZnSe:Cr and ZnSe:Fe (after V Yu Ivanov *et al* 2003 *Acta Physica Polonica A* **103** 695).

The Auger mechanism of this transfer is the most likely explanation of the ODMR data, considering energy overlap between DAP emissions and the Cr ionization transition. Such an overlap is required for the energy transfer to take place.

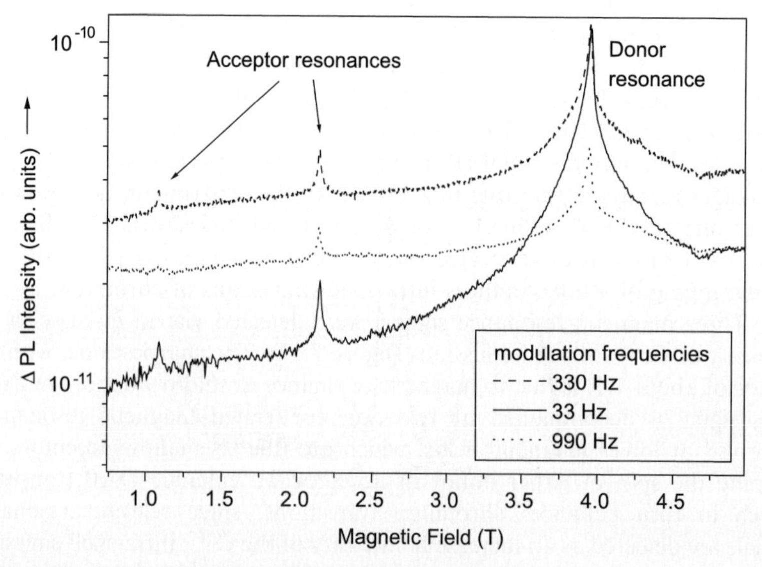

Figure 3.17. 60 GHz ODMR signal detected via increase in the intensity of the Cr-related near-infrared PL emissions (after V Yu Ivanov *et al* 2003 *Acta Physica Polonica A* **103** 695).

The efficiency of this energy transfer was evaluated by performing ODMR experiments at different on–off modulation frequencies of microwaves (see figure 3.17). The modulation frequency of microwaves, when correlated with rate of a given recombination/transfer process, gives the largest magnitude of the relevant ODMR signal. For very efficient transfer/excitation processes the ODMR signal is optimized at a large modulation frequency. In turn, at low efficiency of the transfer/excitation processes, a low modulation frequency of microwaves is needed.

For ZnSe:Cr the ODMR signals are optimized at fairly low (below 100 Hz) modulation frequency, which indicates that the DAP-to-Cr energy transfer process is rather inefficient in quenching DAP transitions of ZnSe. The most likely explanation of this fact is that hole retrapping by Cr^{1+} is needed to excite Cr^{2+} intra-shell emission. If holes are retrapped by acceptors, DAP emissions will not be quenched. Low-temperature metastability of Cr^{1+} is thus also crucial to minimize competing Auger-type non-radiative recombination of DAP transitions.

Since the approach with ZnSe:Fe failed, we tried tuning of the excitation energy of the ASL by introducing chromium into solid alloys of ZnCdS and ZnSSe. Also in these alloys, chromium introduces an excited charge state within the band gap, so alloys were suitable as ASL materials. The ASL emission in two of the II–VI solid alloys is shown in figure 3.18. ASL in Cr-doped ZnCdS and ZnSSe is equally efficient, as in the case of ZnSe:Cr.

There are two further important consequences of using alloys. Both excitation and emission energy (see figure 3.18) could be easily tuned.

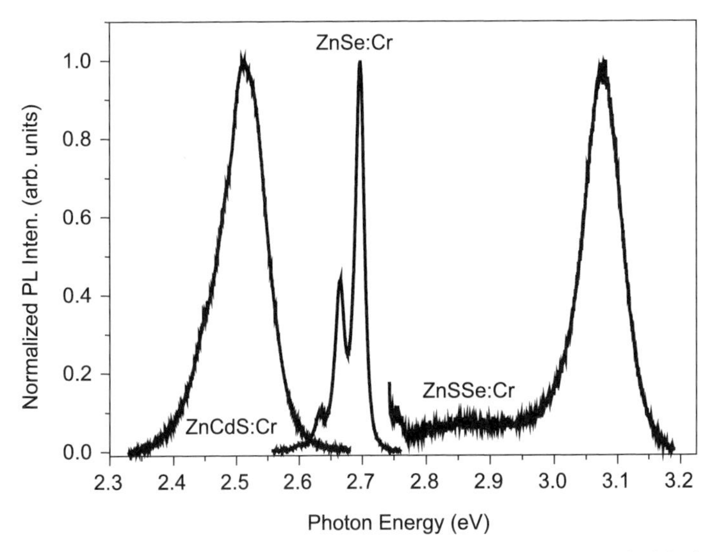

Figure 3.18. ASL emission in solid solutions of ZnCdS and ZnSSe doped with chromium (after V Yu Ivanov *et al* 2002 *Physica B* **308/310** 962).

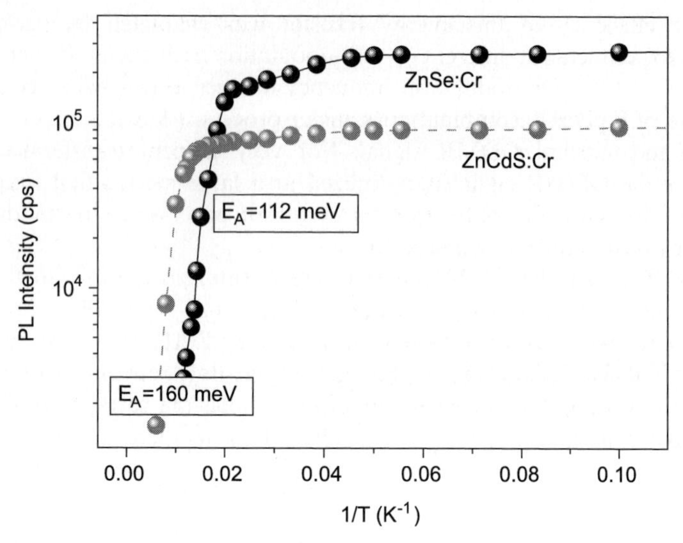

Figure 3.19. ASL intensity versus reciprocal temperature in ZnSe:Cr and ZnCdS:Cr (after V Yu Ivanov *et al* 2002 *Physica B* **308/310** 962).

Moreover, a weaker temperature dependence of the ASL intensity is observed for chromium in the alloys studied (figure 3.19). This is very advantageous, since in the case of ZnSe, the quantum efficiency of the ASL was reduced to about 1% at room temperature, as a result of thermal ionization of shallow acceptors.

In figure 3.19, we demonstrate that the ASL is in fact less affected by an increase in temperature in ZnCdS:Cr than in ZnSe:Cr. This is due to the fact that acceptors in ZnCdS are deeper than those in the ZnSe lattice. Respective acceptor ionization energies are about 110 meV in ZnSe and about 160 meV in ZnCdS. The most promising in this sense is using solid alloys of ZnSSe, as is indicated by our first experimental results. For these alloys acceptors active in the ASL processes are deeper.

3.5.3 Infrared intra-shell emission in ZnSe:Cr

The efficiency of the ASL emission decreases with increasing temperature (figure 3.19). Photo-generated free carriers are no longer efficiently trapped by centres active in the ASL. Shallow donor and acceptor are thermally ionized, thus the ASL is temperature deactivated. For increased temperature DAP PL is first replaced by the FB PL, due to a radiative recombination of free electrons with holes bound at shallow acceptors. Then, once shallow acceptors are thermally ionized at temperatures above 50 K, the DAP PL is replaced by a featureless and weak PL, which we relate to a radiative recombination of free carriers. The deactivation of the ASL is thus characterized by the deactivation energy related to a thermal ionization of shallow acceptors in ZnSe.

Importantly, decreasing efficiency of the ASL process correlates with shortening of lifetime of the photo-excited Cr^{1+}. Our PL experiments indicate that hole retrapping by the 1+ charge state of chromium proceeds via excited intra-shell states of chromium Cr^{2+}. Thus, the excitation of intra-shell emissions of Cr^{2+} becomes more efficient at higher temperatures when photo-generated holes are very efficiently retrapped by ionized chromium centres [63]. Consequently, the concentration of the Cr^{1+} state, an intermediate state in the two-quanta excitation transition, decreases. Population of this state is now no longer metastable and a quadratic dependence on light density is observed for the ASL.

PL experiments indicate that, at increased temperatures, the first step in the ASL excitation is

$$Cr^{2+} + \text{photon} \longrightarrow Cr^{1+} + \text{hole in the valence band},$$

i.e. the Cr 2+ to 1+ ionization transition is followed by retrapping of free holes via intra-shell excited levels of Cr^{2+} ions, which results in two infrared emissions of Cr^{2+} ions:

$$\begin{aligned} Cr^{2+} + \text{hole in the valence band} &\longrightarrow Cr^{2+}(\text{excited state}) \\ &\longrightarrow Cr^{2+}(\text{ground state}) + \text{infrared PL} \end{aligned}$$

Thus, the ASL and near-infrared PLs have the same excitation bands and the two channels of recombination compete. At low temperature the ASL emission dominates; at increased temperatures the infrared emission becomes more pronounced and dominant at room temperature. The competition of two recombination channels is confirmed by the data shown in figure 3.20. Thermally induced decrease of the ASL is accompanied by an increase of the infrared emission. Similar anti-correlation is found in time-resolved experiments. When the decay time of ASL is shortened, it increases for the infrared emission (see figure 3.20).

Typically, photo-ionization bands of transition metal ions are characterized by large oscillator strengths, i.e. the resulting photo-excitation is far more efficient than the intra-shell excitation and, thus, can result in a very efficient optical pumping of laser emission. This fact is important when considering intensive studies on laser action on Cr 2+ intra-shell transitions. Cr-related near-infrared PL bands observed under photo-ionization excitation of chromium ions are shown in figure 3.21. Two infrared emissions, with maxima at about 0.95 and 2.4 μm, are observed. Both these infrared bands are relatively broad, which can result in optically pumped and tunable laser emission with wavelength of emission being suitable for some opto-electronic applications, such as surgery lasers, for remote sensing, gas sensing etc.

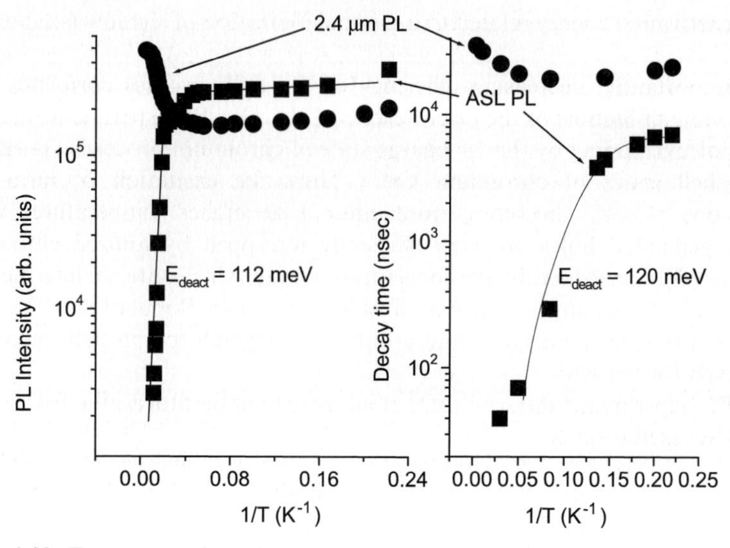

Figure 3.20. Temperature dependences of the PL intensity (left curve) and of the PL decay kinetics (right curve) of the ASL (■) and Cr-related $^5E \longrightarrow {}^5T_2$ intra-shell emission (●) in ZnSe (after V Yu Ivanov *et al* 2002 *Proceedings ICPS'02*, Edinburgh).

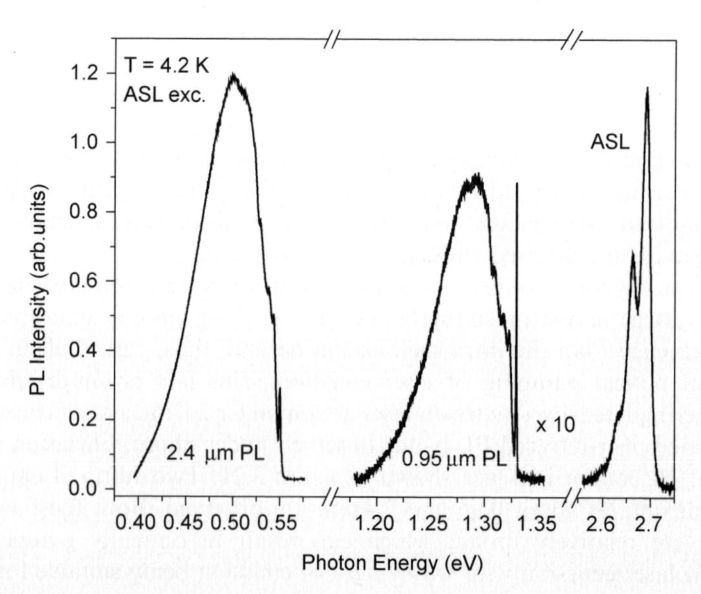

Figure 3.21. Two infrared intra-shell emissions of chromium 2+ in ZnSe shown together with the ASL emission (after V Yu Ivanov *et al* 2002 *Proceedings ICPS'02*, Edinburgh).

Following this idea, mid-infrared lasing action at about 2.5 μm was achieved for several wide band gap II–VI compounds doped with chromium [64–68]. Several wide band gap II–VI compounds were used as host materials. At present, ZnSe is the most promising host material for a mid-infrared laser. For ZnSe:Cr, pulsed and also continuous-wave laser emissions under optical pumping were achieved with power of emitted light above 1 W. This opens a wide range of possible applications for this new mid-infrared laser system.

3.6 Summary

During the last decade of the 20th century and the first years of the 21st century, a revolutionary development in semiconductor-based light emitting devices has been observed, and new generations of highly efficient LED and LD devices were introduced. Soon, semiconductor-based LED devices will be good enough to replace incandescent bulb lamps and fluorescent lamps. All commercialized LEDs and LDs are III–V-based. ZnSe, ZnTe and other wide band gap II–VI compounds are still waiting for technology break-through to start competition with dominant III–V-based LEDs and LDs. It is very unlikely that they will replace III–V compounds in light emitting diodes. Fortunately, some hopes still remain for LD applications. Moreover, there are several possible niche applications for which these two II–VI compounds can be suitable. The most promising ones are described in this chapter.

References

[1] Holonyak N Jr and Bevaqua S F 1962 *Appl. Phys. Lett.* **1** 82
[2] Dean P J and Herbert D C in *Excitons, Topics in Current Physics*, vol. 14, ed. K Cho (Berlin: Springer) p. 55
[3] Martin P 2001 *Laser Focus World*
[4] Kuo C P, Fletcher R M, Osentowski T D, Lardizabal M C, Craford M G and Robbins M 1990 *Appl. Phys. Lett.* **57** 2937
[5] Sugawara H, Ishikawa M and Hatakoshi G 1991 *Appl. Phys. Lett.* **58** 1010
[6] Craford M G 1997 in *High Brightness Light Emitting Diodes* eds. M G Craford and G B Stringfellow (San Diego: Academic Press)
[7] Nakamura S, Senoh M, Iwasa N, Nagahama S, Yamaka T and Mukai T 1995 *Jpn. J. Appl. Phys.* **34** L1332
[8] Koike M, Shibata N, Kato H and Takahashi Y 2002 *IEEE J. Selected Topics in Quantum Electronics* **8** 271
[9] MacAdam D L 1985 in *Color Measurements* 2nd edition (Berlin, Heidelberg: Springer) p. 4
[10] Wyszecki G and Stiles W S 1982 in *Color Science, Concepts and Methods* 2nd edition (New York: Wiley) p. 117

[11] Arens H in 1967 *Color Measurement* (Leipzig: Leipziger Druckhaus) p. 29
[12] Nakamura S 1997 *Proceedings of SPIE* Vol. 3002 (Bellingham: SPIE) p. 26
[13] Schlotter P, Schmidt R and Schneider J 1997 *Appl. Phys. A* **64** 417
[14] Bernard E 1996 *Information Display* **12** 16
[15] Kaufmann U, Kunzer M, Kohler K, Obloh H, Pletschen W, Schlotter P, Schmidt R, Wagner J, Ellens A, Rossner W and Kobusch M 2001 *Phys. Status Solidi* (*a*) **188** 143
[16] Walker C T, DePuydt J M, Haase M A, Qiu J and Cheng H 1993 *Physica B* **185** 27
[17] Nurmikko A V, Gunshor R L, Otsuka N and Kobayashi M 1992 *Int. Conf. on Solid State Devices and Materials*, Tsukuba 1992, extended abstracts, p. 342
[18] Landwehr G, Fischer F, Baron T, Litz T, Waag A, Schull K, Lugauer H, Gerhard T, Keim M and Lunz U 1997 *Phys. Status Solidi* (*b*) **202** 645
[19] Ohkawa H, Behringer M, Wenisch H, Fehrer M, Jobst B, Hommel D, Kuttler M, Strassburg M, Bimberg D, Bacher G, Tonnies D and Forchel A 1997 *Phys. Status Solidi* (*b*) **202**, 683
[20] Special issue of 1994 *Phys. Status Solidi* (*b*) **187**(2)
[21] Special issue of 1997 *Phys. Status Solidi* (*b*) **202**(2)
[22] Yu Z, Eason D B, Boney C, Ren J, Hughes W C, Rowland W H, Cook J W, Schetzina J F, Cantwell G and Harsch W C 1995 *J. Vac. Sci. Technol.* **13** 1694
[23] Mycielski A, Szadkowski A, Kaliszek W and Witkowska B 2001 *Proc. SPIE* **4412** 38
[24] Mycielski A, Szadkowski A J, Kowalczyk L, Zielinski M, Łusakowska E, Witkowska B, Kaliszek W, Jędrzejczak A, Adamczewska J, Kaczor P and Chernyshova M 2002 *Phys. Status Solidi* **229** 189
[25] Yakimovich V N, Levchenko V I, Yablonski G P, Konstantinov V I, Postnova L I and Kutas A A 1999 *J. Cryst. Growth* **198/199** 975
[26] Katayama K, Matsubara H, Nakanishi F, Nakamura T, Doi H, Saegusa A, Mitsui T, Matsuoka T, Irikura M, Takebe T, Nishine S and Shirikawa T 2000 *J. Cryst. Growth* **214/215** 1064
[27] Wenisch H, Fehrer M, Klude M, Ohkawa K and Hommel D 2000 *J. Cryst. Growth* **214/215** 1075
[28] Tamargo M C, Lin W, Guo S P, Guo Y, Luo Y and Chen Y C 2000 *J. Cryst. Growth* **214/215** 1058
[29] Tao I W, Jurkovic M and Wang W I 1994 *Appl. Phys. Lett.* **64** 1848
[30] Ogawa H, Irfan G S, Nakayama H, Nishio M and Yoshida A 1994 *Jpn. J. Appl. Phys.* **33** L980
[31] Chang J H, Cho M W, Wang H M, Wenisch H, Hanada T, Sato K, Oda O and Yao T 2000 *Appl. Phys. Lett.* **77** 1256
[32] Chang J H, Wang H M, Cho M W, Makino H and Yao T 2000 *J. Vac. Sci. Technol. B* **13** 1530
[33] Chang J H, Takai T, Koo B H, Song J S and Yao T 2000 *IEEE Proceedings*, p. 61
[34] Sato K, Hanafusa M, Noda A, Arakawa A, Asahi T, Uchida M and Oda O 2000 *IEICE Trans. Electronics* **R83-C** 579
[35] Suntola T 1994 in *Handbook of Crystal Growth*, Vol. 3B *Thin Films and Epitaxy* ed. D T H J Hurle (Amsterdam: North-Holland) ch. 14, p. 601
[36] Ritala M and Leskelä M 2002 *Handbook of Thin Film Materials* ed. H Nalwa, vol. 1 *Deposition and Processing of Thin Films*, Chapter 2 Atomic Layer Deposition (Academic Press) p. 103
[37] Ono Y A 1995 in *Electroluminescence Displays* (Singapore: World Scientific)

[38] Godlewski M and Leskelä M 1994 *CRC Critical Reviews in Solid State and Materials Sciences* **19** 199
[39] Godlewski M, Guziewicz E, Kopalko K, Łusakowska E, Dynowska E, Godlewski M M, Goldys E M, Phillips M 2002 *Proc. ICL-02*, Budapest, 2003, *J. Luminescence* **102/103** 455
[40] Hitier G 1980 *J. Phys.* (*Paris*) **41** 443
[41] Swaminathan V and Greene L C 1976 *Phys. Rev.* **B14** 5351
[42] Godlewski M, Lamb W E and Cavenett B C 1981 *J. Luminescence* **24/25** 173
[43] Godlewski M, Goldys E M, Phillips M R, Pakula K and Baranowski J M 2002 *Applied Surface Science* **177** 22
[44] Stuecheli N and Bucher E 1989 *J. Electron. Mater.* **18** 105
[45] Pages O, Renucci M A, Briot O and Aulombard R L 1995 *J. Appl. Phys.* **77** 1241
[46] Carlone C, Beliveau A and Rowel N L 1991 *J. Luminescence* **47** 309
[47] Ivanov V Yu., Semenov Yu. G, Surma M and Godlewski M 1996 *Phys. Rev.* **B54** 4696
[48] Dean P J, Herbert D C, Werkhoven C J, Fitzpatrick B J and Bhargava R N 1981 *Phys. Rev.* **B23** 4888
[49] Swaminathan V and Green L C 1976 *Phys. Rev.* **B14** 5351
[50] Merz J L, Nassau K and Shiever J W 1973 *Phys. Rev.* **B8** 1444
[51] Godlewski M and Zakrzewski A 1993 in *II–VI Semiconductors* ed. M Jain (Singapore: World Scientific) p. 205
[52] Godlewski M and Kaminska M 1980 *J. Phys. C* **13** 6537
[53] Godlewski M, Wilamowski Z, Kaminska M, Lamb W E and Cavenett B C 1981 *J. Phys. C* **14** 2835
[54] Surma M, Godlewski M and Surkova T P 1994 *Phys. Rev. B* **50** 8319
[55] Godlewski M and Zakrzewski A 1985 *J. Phys. C* **18** 6615
[56] Godlewski M 1985 *Phys. Rev. B* **32** 8162
[57] Alex V and Weber J 1997 *Mat. Sci. Forum* **258/263** 1009
[58] Alex V and Weber J 1998 *Appl. Phys. Lett.* **75** 1820
[59] Zakrzewski A and Godlewski M 1991 *Appl. Surf. Science* **50** 257
[60] Zakrzewski A and Godlewski M 1990 *J. Appl. Phys.* **67** 2457
[61] Surma M and Godlewski M 1997 *Acta Physica Polonica A* **92** 1017
[62] Godlewski M, Zakrzewski A J and Ivanow V Y 2000 *J. Alloys Compounds* **300/301** 23
[63] Godlewski M 1985 *Phys. Status Solidi* (*a*) **90** 11
[64] Page R H, Schaffers K I, DeLoach L D, Wilke G D, Patel F D, Tassano J B Jr, Payne S A, Krupke W F, Kuo-Tong Chen and Burger A 1997 *IEEE J. Quantum Electronics* **33** 609
[65] Bhaskar S, Dobal P S, Rai B K, Katiyar R S, Bist H D, Ndap J-O and Burger A 1999 *J. Appl. Phys.* **85** 439
[66] Burger A, Chattopadhyay K, Ndap J-O, Ma X, Morgan S H, Rablau C I, Su C-H, Feth S, Page R H, Schaffers K I and Payne A 2001 *J. Cryst. Growth* **225** 249
[67] Podlipensky A V, Shcherbitsky V G, Kuleshov N V, Levchenko V I, Yakimovich V N, Mond M, Heumann E, Huber G, Kretschmann H and Kuck S 2001 *Appl. Phys. B* **72** 253
[68] Sorokin E and Sorokina I T 2002 *Appl. Phys. Lett.* **80** 3289

Chapter 4

Cadmium chalcogenide nanocrystals

***Stephen M Kelly*[1], *Mary O'Neill*[2] *and Tom Stirner*[1,2]**
[1]Department of Chemistry, University of Hull, Hull, UK
[2]Department of Physics, University of Hull, Hull, UK

4.1 Introduction

Electroluminescent nanocrystals exhibit a unique combination of physical properties for applications in electro-optic devices [1–16]. Quantum effects enable the colour of emission to be tuned across the visible spectrum and into the near infrared, e.g. the band gap in CdS is tunable from 4.5 to 2.5 eV as the size is changed from the molecular scale to the bulk semiconductor crystal. The synthesis and evaluation of the physical properties of semiconductor nanocrystals have evolved into a new area of scientific research and development over the past two decades. The concept of quantum confinement in semiconductor nanocrystals and their electroluminescence (EL) and photoluminescence (PL) are covered in other chapters of this handbook and described in a number of excellent and comprehensive review articles [1–16]. However, the area of nanotechnology concerned with the synthesis, processing and consequent incorporation of electroluminescent semiconductor nanocrystals in practical EL devices is not very advanced. Therefore, this chapter is focused on some of the most important issues of relevance to hybrid inorganic/organic devices making use of the quantum-confined EL and charge-transport properties of nanocrystals in general and cadmium chalcogenides in particular.

An optimization of the synthesis, processing and assembly of inorganic semiconductor nanocrystals is required in order to develop a whole new area of electro-optical applications, such as hybrid inorganic/organic light-emitting diodes (HIOLEDs) and photovoltaic devices. Solubility in organic solvents allows semiconductor nanocrystals to be processed by standard procedures for semiconductor and LCD technologies such as spin coating from organic solvents to form uniform thin films. Water solubility allows thin films of semiconductor nanocrystals to be fabricated by electrolytic step-by-step self-assembly processes from aqueous solution. Inorganic II–VI semiconductors such as the cadmium chalcogenides, CdS, CdSe and CdTe nanocrystals are especially suited to electroluminescent electro-optic applications in the visible and near infrared regions of the electromagnetic spectrum. This is due to an advantageous combination of appropriate valence and conduction energy levels and the resultant bandgap, as well as the ease of synthesis and processing from solution to form electroluminescent films. Therefore, the remainder of this chapter will focus on the theory of quantum confinement in cadmium telluride nanocrystals to illustrate general concepts by one particular example, then the synthesis of processable cadmium chalcogenide nanocrystals and finally their assembly and use as the emissive material in flat panel electroluminescent devices.

4.2 Theory

In this section we focus on a theoretical description of the electronic properties of CdTe nanocrystals as typical of the behaviour of cadmium chalcogenides in general. However, since we are mainly interested in the effects of quantum confinement, we ignore here the influence of defect or trap states. Energy levels and exciton energies in quantum dots and nanocrystals have been investigated utilizing the effective-mass theory [17, 18], multi-band $\mathbf{k}\cdot\mathbf{p}$ methods [19], empirical pseudopotential calculations [20, 21] and tight-binding calculations [22]. The influence of many-body effects, the Coulomb interaction and strain have all been taken into account (see e.g. [23, 24]). The effective-mass approximation has been applied to semiconductor nanocrystals and has been shown to seriously overestimate the excitonic transition energies for small nanospheres [20, 25]. Calculations based on the empirical pseudopotential theory [20, 21, 24] have been shown to give much better agreement with experiment [20]. We will therefore employ the pseudopotential method in section 4.3 to investigate CdTe nanospheres and nanorods.

The importance of diffusion as a tool for post-growth tuning of energy levels has been realized for a long time [26, 27] and, using rapid thermal annealing techniques, has been successfully demonstrated in semiconductor quantum-well structures [28]. More recently, such experimental diffusion studies have also been extended to quantum dots [29]. Similarly, the influence of interdiffusion on electron states in InGaAs/GaAs quantum dots has been investigated theoretically by Barker and O'Reilly [30]. One of the advantages of the effective-mass theory is its applicability to semiconductor structures with a position-dependent band-edge profile (e.g. as induced by diffusion). We will therefore employ the effective-mass theory in section 4.4 to investigate the post-growth tuning of energy levels in a relatively large CdTe nanosphere, and we will also use the effective-mass and envelope-function approximations to investigate the possibility of obtaining polarized EL from CdTe nanostructures.

4.3 Empirical pseudopotential method

4.3.1 Theory

The pseudopotential technique is well documented in the literature [21, 31–34]. In short, the bulk eigenfunctions $\phi_{n,\mathbf{k}}$ appearing in the Schrödinger equation

$$H\phi_{n,\mathbf{k}} = E_{n,\mathbf{k}}\phi_{n,\mathbf{k}} \tag{4.1}$$

are expanded in terms of a linear combination of plane waves

$$\phi_{n,\mathbf{k}} = \frac{1}{\sqrt{\Omega}}\sum_{\mathbf{G}} a_{n,\mathbf{k}}(\mathbf{G})\exp[\mathrm{i}(\mathbf{G}+\mathbf{k})\cdot\mathbf{r}], \tag{4.2}$$

where the Hamiltonian H is given by

$$H = -\frac{\hbar^2}{2m_0}\nabla^2 + \sum_{\mathbf{r}_\mathrm{a}} V_\mathrm{a}(\mathbf{r}-\mathbf{r}_\mathrm{a}). \tag{4.3}$$

Here Ω denotes the unit cell volume, $\mathbf{G}$ the reciprocal lattice vectors, and the crystal potential is approximated by a spherically symmetric atomic potential V_a situated at every lattice site $\mathbf{r}_\mathrm{a}$. The local part of the pseudopotential can be written as

$$V_p(\mathbf{r}) = \sum_{\mathbf{G}}[V_S(\mathbf{G})\cos(\mathbf{G}\cdot\mathbf{T}) + \mathrm{i}V_A(\mathbf{G})\sin(\mathbf{G}\cdot\mathbf{T})]\exp(\mathrm{i}\mathbf{G}\cdot\mathbf{r}), \tag{4.4}$$

where the vector $\mathbf{T} = (1,1,1)a_0/8$ denotes the displacements of the anion and cation by $\pm\mathbf{T}$ at the basis of each lattice point with a bulk CdTe lattice constant of $a_0 = 6.481$ Å. The zincblende CdTe pseudopotential form factors, V_S and V_A [35, 36], employed in the present calculations are summarized in table 4.1. The spin–orbit interaction is included using a model described in [37], where the spin–orbit interaction parameters, λ_S and λ_A [37, 38], were determined by fitting the spin–orbit splitting energy $\Delta = 0.927$ eV and found to be 0.025 and 0.009, respectively [35].

The eigenvalues $E_{n,\mathbf{k}}$ can be obtained as a function of the electron wave vector $\mathbf{k}$ by repeated diagonalization of $H(\mathbf{k})$. The resulting bulk CdTe band structure is shown in figure 4.1 where we have used the usual group theory notation for the conduction band (Γ_6), the light- and heavy-hole valence bands (Γ_8) and the split-off band (Γ_7). As can be seen from this figure, for bulk material the eigenvalues $E_{n,\mathbf{k}}$ form continuous bands, the valence band is four-fold degenerate at the Γ point (spin-degeneracy) and the spin–orbit coupling lifts the otherwise six-fold degeneracy of the valence band.

Table 4.1. CdTe pseudopotential form factors in Rydberg [35].

G^2	V_S	V_A
0	0.0	0.0
3	−0.230	0.120
4	0.0	0.0715
8	−0.005	0.0
11	0.081	0.0385

4.3.2 CdTe nanospheres

In stark contrast to the situation for bulk material, in semiconductor nanocrystals the eigenenergies are quantized and hence do not form continuous bands. However, we can define the 'band structure' of nanocrystals as the collection of all discrete eigenenergies in the first Brillouin zone. The corresponding **k** vectors can be obtained by imposing the appropriate boundary conditions on the wave functions. For example, for spherical nanocrystals of radius R with infinite potentials at the boundaries the wave vector, projected on to each Cartesian axis with equal magnitude, is given by [20, 39]

$$\mathbf{k} = \frac{\pi}{\sqrt{3}R}(n_x, n_y, n_z), \tag{4.5}$$

where n are the quantum numbers of the particle. This technique, also referred to as the 'single-band truncated-crystal' approximation, was described in detail by Franceschetti and Zunger [40] for the case of GaAs quantum structures.

Figure 4.2 shows the discrete energy levels calculated for a CdTe nanosphere with radius $R = 1.5$ nm (the lines connecting the eigenenergies are a guide to the eye only). The energy levels were evaluated in the first Brillouin

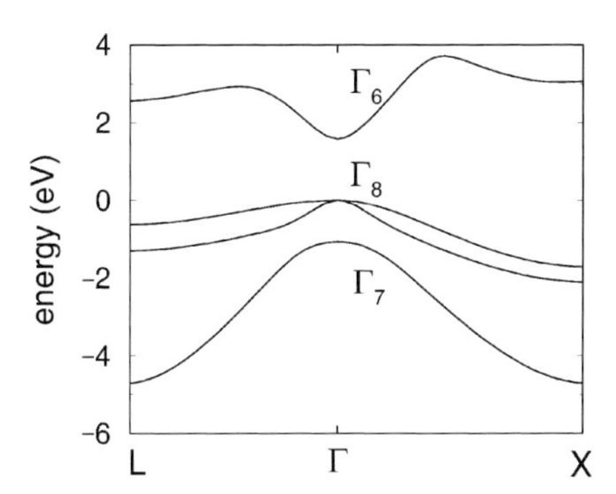

Figure 4.1. Band structure of bulk CdTe.

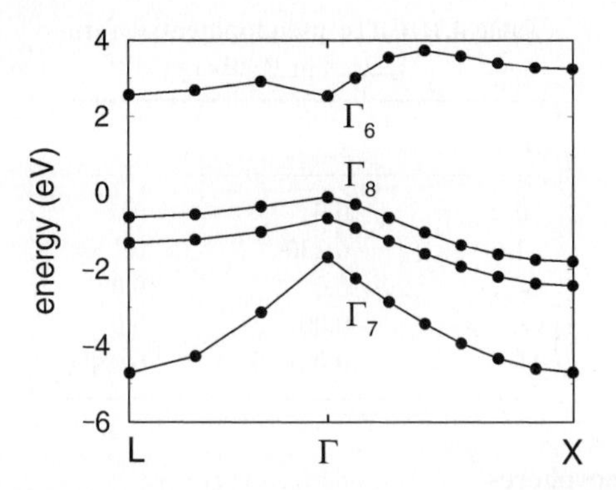

Figure 4.2. Calculated energy level structure of a $R = 1.5$ nm CdTe nanosphere in the first Brillouin zone (the lines are a guide to the eye only) [41].

zone between the high symmetry points along the Γ–X and Γ–L directions. 169 plane waves of different **G** were employed in the calculations in order to converge the eigenenergies to ~1 meV. The eigenenergy structure shown in figure 4.2 is very similar to the band structure of bulk CdTe (see figure 4.1). Consequently, we have labelled the resulting 'bands' Γ_6 (conduction band), Γ_8 (heavy-hole/light-hole valence band) and Γ_7 (split-off band). The two main features that distinguish the nanocrystal 'band' structure from the bulk (in addition to the discretization of the energy levels) are the increase in the fundamental band gap, E_g (i.e. the Γ_6–Γ_8 energy separation at Γ), due to the quantum confinement effect, and the removal of the light-hole/heavy-hole Γ_8 valence band degeneracy at the Γ point.

For spherical CdTe nanocrystals with $R \gtrsim 1$ nm (figure 4.2), the top of the valence band (both for the heavy- and light-hole) and the bottom of the conduction band lie at the Γ point given by the **k** vector at $n_x = n_y = n_z = 1$. For $R \lesssim 1$ nm (not shown), the minimum energy of the conduction band moves from Γ to L [41]. However, as far as EL is concerned, only the transitions at the Γ point (i.e. Franck–Condon excitons) are of interest. The band gap E_g was evaluated at this symmetry point as described above as a function of nanosphere radius R. The Coulomb and correlation energies were calculated as described in [20] for a CdTe relative permittivity value of $\varepsilon_r = 10.6$ [42]. The total exciton transition energy (in SI units) can then be expressed as

$$E = E_g - \frac{1.786\, e^2}{4\pi\varepsilon_r\varepsilon_0 R} - 0.248 E_R \tag{4.6}$$

with

$$E_R = \frac{\mu e^4}{2(4\pi\varepsilon_r\varepsilon_0)^2\hbar^2},$$

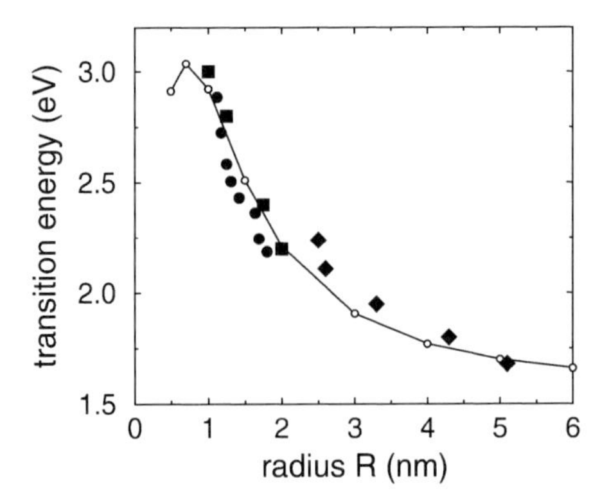

Figure 4.3. Experimental and calculated heavy-hole exciton transition energies of zinc-blende CdTe nanospheres. The empty circles (○) represent the energies calculated by the empirical pseudopotential method (the full line is a guide to the eye only). The symbols represent the experimental data points taken from [43] (◆), [25] (■) and [44] (●).

where e is the electron charge, $\hbar$ is the Planck constant divided by 2π and μ is the exciton reduced mass. Franceschetti and Zunger [24] have demonstrated that, in particular for small nanocrystals (i.e. $R \lesssim 2\,\text{nm}$), the effective-mass approximation (embodied in equation (4.6)) significantly overestimates the Coulomb energy. We have therefore adjusted the Coulomb energy term in equation (4.6) by the factors given in table II of [24]. This correction is a small effect and only influences the calculated transition energies for nano-crystal radii below $\sim$1 nm; however it improves the agreement between theory and experiment for these small nanospheres [41].

The resulting transition energy of the heavy-hole exciton as a function of nanosphere radius R is shown by the full line in figure 4.3 (the empty circles represent the calculated data points). The theoretical curve exhibits a maximum transition energy at $R \approx 0.7\,\text{nm}$ and decreases for very small nano-spheres. This was also observed in similar calculations for the GaP and GaAs material systems [20]. For large nanospheres the transition energy approaches the value of the bulk band gap of 1.56 eV [25]. Also shown in figure 4.3 are the experimental results obtained by Masumoto and Sonobe [43] (◆), Rajh *et al* [25] (■) and Zhang and Yang [44] (●). As can be seen, within the uncertainties of the theory and the experimental error, there is reasonably good agreement between our calculations and the three independent sets of absorption measurements.

Figure 4.3 demonstrates that, by varying the nanosphere radius, the transition energy can be tuned over the whole visible spectral range, i.e. from the red ($\sim$1.8 eV) to the blue ($\sim$2.6 eV) end of the spectrum. In addition, however, a knowledge of the position of the electron energy level in the CdTe nanospheres with respect to the vacuum level is also of importance to

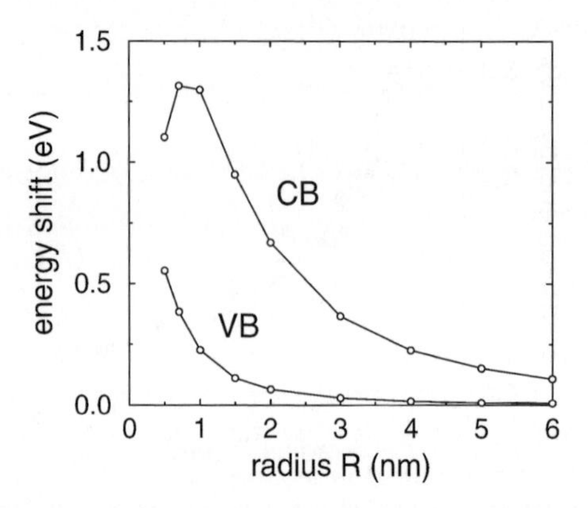

Figure 4.4. Magnitude of the calculated energy-level shifts for both the electron in the conduction band (CB) and the heavy-hole in the valence band (VB) with respect to the bulk CdTe band edge as a function of nanosphere radius (the lines are a guide to the eye only).

electroluminescent devices, since the CdTe nanoparticles act as electron transporting material, and the energy difference between the electron energy level and the work function of the electrode determines the electron injection efficiency [45]. Similarly, for CdSe-PPV composites it was found [46] that CdSe nanocrystals with transition energies larger than the band gap of PPV produced no EL.

Figure 4.4 shows the magnitude of the calculated energy-level shifts for both the electron in the conduction band (CB) and the heavy-hole in the valence band (VB) with respect to the bulk CdTe band edge as a function of nanosphere radius, i.e. the CB level is measured towards and the VB level away from the vacuum level. As can be seen, the CB level shifts to higher energies for decreasing nanosphere radii, while at the same time the VB level moves to lower energies (although the latter effect is only of importance for very small nanosphere sizes). The pseudopotential calculations also indicate that the shift of the CB level is a non-monotonic function of the nanosphere radius. However, as was pointed out by Franceschetti and Zunger [24], such a 'quantum deconfinement' effect may be an artefact of the single-band truncated-crystal approximation. Nevertheless, it is clear that the quantum confinement effect gives rise to a significant shift of the CB level, which can seriously impede the electron injection efficiency for certain electrode metals. For example, if we take the work function of a magnesium electrode as 3.7 eV [47] and the electron affinity of CdTe to be 4.2 eV [48], then figure 4.4 indicates that the lowest CB level of the CdTe nanospheres lies higher than the Mg work function for $R \lesssim 2.5\,\text{nm}$. For aluminium electrodes with a work function of $\sim$4.3 eV [47] the situation would be even worse.

4.3.3 CdTe nanorods

We will now consider cylindrical CdTe nanorods of radius R. For nanorods with infinite potentials at the boundaries the wave vector, projected with equal magnitude on to each of the two Cartesian axes perpendicular to the nanorod axis z, is given by

$$\mathbf{k} = \frac{\alpha_0}{\sqrt{2}R}(n_x, n_y), \tag{4.7}$$

where $\alpha_0 = 2.4048$ is the first zero of the Bessel function.

Figure 4.5 shows the energy level structure calculated for CdTe nanorods with (a) $R = 4$ nm and (b) $R = 1.5$ nm, respectively, in the first Brillouin zone, where we have selected the nanorod (z) axis to lie along the [100] direction. (The dashed lines connecting the eigenenergies in figure 4.5(b) are a guide to the eye only.) As can be seen, for nanorods the energy levels are quantized only along the Γ–L direction (corresponding to [111]), whereas the Γ–X direction coincides with [100] (i.e. is perpendicular to the quantization axes)

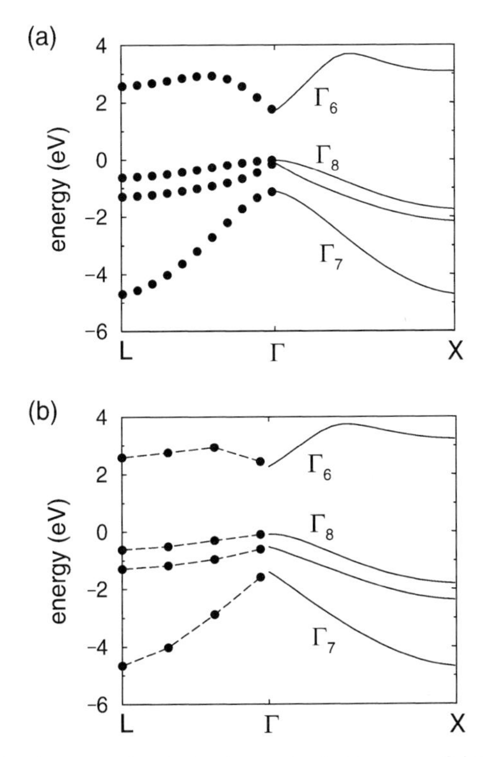

Figure 4.5. Calculated energy level structure of (a) $R = 4$ nm CdTe nanorod and (b) $R = 1.5$ nm CdTe nanorod in the first Brillouin zone (the dashed lines are a guide to the eye only; $z \parallel [100]$).

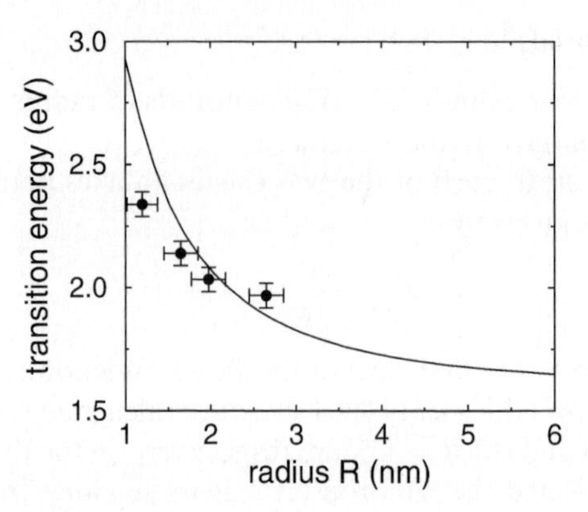

Figure 4.6. Calculated electron–heavy-hole transition energies of zincblende CdTe nanorods. The full line represents the empirical pseudopotential calculations. The full circles with the error bars are experimental data points taken from [54].

and the energy bands remain therefore continuous. Consequently, the features characterizing the nanorod band structure are a hybrid between bulk material and nanospheres, i.e. the fundamental band gap still increases with decreasing nanorod radius and also the Γ_8 valence band degeneracy at the Γ point is removed; however, these effects are slightly less pronounced compared with the nanosphere case.

Figure 4.6 shows the electron–heavy-hole transition energy for a CdTe nanorod as a function of nanorod radius R. In figure 4.6 we have ignored the Coulomb energy between the electron and the hole, which leads to an overestimation of the transition energy in particular for small radii (binding energies of up to 400 meV were calculated for a CdS nanorod of $R \approx 1$ nm and were found to decrease rapidly with increasing radius [49] using the method described in [50–52], we estimate the binding energy for a $R = 1$ nm CdTe nanorod to be $\sim$100 meV [53]). Also shown in figure 4.6 are the experimental PL results obtained by Tang *et al* [54] (the full circles with the error bars). As can be seen, the agreement between our calculations and the experimental data of Tang *et al* [54] is rather qualitative. The discrepancies could, in part, be due to the above-mentioned neglect of the Coulomb energy, the Stokes shift and the occurrence of the wurtzite structure in the CdTe nanorods [54].

So far we have taken no cognizance of the fact that the actual shape of the nanocrystals could be different from a perfect sphere or cylinder. In fact, quantum dots produced in the Stranski–Krastanow epitaxial growth mode are often lens-, pyramid- or truncated-pyramid-shaped [29, 55–58]. Similarly, nanospheres and nanorods could have mesoscopic geometries which reflect the underlying tetrahedral T_d crystal point-group symmetry. For the purpose of illustration we will investigate, in what follows, a nanobox of square cross section with side length L and infinite potential barriers. The

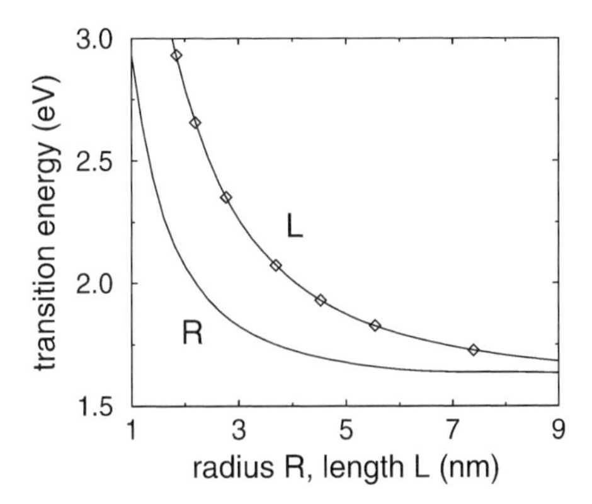

Figure 4.7. Calculated electron–heavy-hole transition energy for the CdTe nanobox (L) as a function of nanobox side length L together with the corresponding curve for nanorods (R) of radius R (see also figure 4.6). The $\diamond$ correspond to the transition-energy curve for the nanorods where the nanorod radius has been scaled by a factor of $\sqrt{2}\pi/\alpha_0$.

wave vector for this case is given by

$$\mathbf{k} = \frac{\pi}{L}(n_x, n_y), \tag{4.8}$$

where again the two quantization axes, x and y, are perpendicular to the nanobox axis z.

Figure 4.7 shows the calculated electron–heavy-hole transition energy for the CdTe nanobox (L) as a function of nanobox side length L together with the corresponding curve for nanorods (R) of radius R (see also figure 4.6). As can be seen, on this scale the transition energy for a nanobox is larger than that for the corresponding nanorod. This is mainly due to the cross-sectional area for the nanorod of radius $R = L$ being larger (by a factor of π) than that of the nanobox. However, if we scale the nanrod radius by a factor of $\sqrt{2}\pi/\alpha_0$ we arrive at the curve depicted by $\diamond$ in figure 4.7. This curve coincides perfectly with the transition-energy curve for the nanobox. Returning now to our initial discussion about the unknown shape of the nanocrystals, we can see from figure 4.7 that for the present theoretical model the transition energy is a universal curve, and can therefore be scaled (according to the ratio of the wave vectors) irrespective of the exact shape of the nanocrystals.

4.4 Effective-mass approximation

4.4.1 Theory

As was demonstrated in the previous section, the empirical pseudopotential method allows a reasonably accurate evaluation of the transition energy of semiconductor nanocrystals. Within the single-band truncated-crystal approximation, its advantage is the computationally efficient calculation of energy levels taking full account of the non-parabolic energy dispersion of the band-edge states. However, in addition to the neglect of interband coupling, the single-band truncated-crystal approximation does not allow the calculation of energy levels in nanocrystals with either finite potential barriers or a position-dependent confinement potential (although the extended and finite-well single-band truncated-crystal approximations [40] partially remedy the first two drawbacks). The effective-mass approximation described in the present section can deal with both of these scenarios and is ideal for investigating potential profiles of arbitrary shape.

The effective-mass Hamiltonian H can be written as

$$H = -\frac{\hbar^2}{2m^*}\nabla^2 + V(r), \tag{4.9}$$

where m^* is the effective mass of the particle and $V(r)$ is the confining potential. Using this Hamiltonian in the framework of the envelope-function approximation [59] has the advantage that the time-independent Schrödinger equation

$$H\psi(r) = E\psi(r) \tag{4.10}$$

can be solved numerically (in the present case for angular momentum and magnetic quantum numbers $l = m = 0$) using the finite differences method. Both the single-particle wave function ψ and energy E can thus be obtained from which, in turn, the Coulomb interaction energy of the exciton can be evaluated.

To first approximation, the Coulomb interaction energy of a 1s-exciton can be written as [60]

$$E_{\mathrm{C}} = -\frac{e^2}{4\pi\varepsilon_r\varepsilon_0}\iint \mathrm{d}r_{\mathrm{e}}\,\mathrm{d}r_{\mathrm{h}}\, r_{\mathrm{e}}^2 r_{\mathrm{h}}^2 \frac{|\psi_{\mathrm{e}}(r_{\mathrm{e}})|^2|\psi_{\mathrm{h}}(r_{\mathrm{h}})|^2}{\max(r_{\mathrm{e}}, r_{\mathrm{h}})}, \tag{4.11}$$

where ε_r is the dielectric constant of the material. The exciton transition energy can now be calculated as

$$E_{\mathrm{t}} = E_{\mathrm{g}} + E_{\mathrm{e}} + E_{\mathrm{h}} + E_{\mathrm{C}}, \tag{4.12}$$

where E_{g} is the bulk band-gap energy of CdTe (1.56 eV [61]), which has to be added since the single-particle electron and hole energies (E_{e} and E_{h}) are calculated relative to E_{g}. For the calculations presented below, the CdTe effective mass of the electron (heavy hole) was taken as 0.096 (0.6) [62] (in units of the free electron mass), the valence band offset was taken as 40% of the total band offset [63] and the value of the dielectric constant was taken as 10.6 [42].

Before we present the results of the effective-mass calculations in the following section, it is worth considering the applicability and validity of

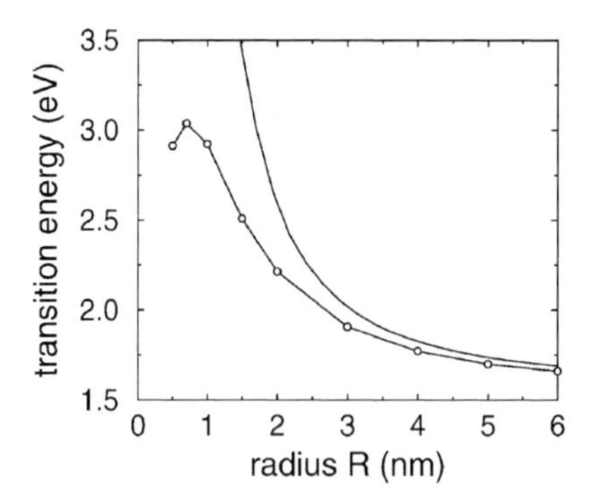

Figure 4.8. Ground-state exciton transition energy for CdTe nanospheres as a function of sphere radius. The full line represents the effective-mass-type calculations, and the line with the empty circles represents the empirical pseudopotential calculations (see figure 4.3). Both sets of calculations assume infinite potential barriers and take the Coulomb energy into account [64].

the present approach, since it is well known that effective-mass-type calculations break down for very small semiconductor structures [20]. In figure 4.8 we compare the empirical pseudopotential calculations for CdTe nanospheres described in the previous section (see figure 4.3) with the results of the present effective-mass-type calculations. Both sets of data assume infinite potential barriers at the nanosphere surface and include the Coulomb energy. As can be seen, the discrepancy in the calculated transition energies between the effective-mass and empirical pseudopotential theory is considerable, in particular for small nanospheres. However, the discrepancy between the two theories decreases with increasing nanosphere radius (e.g. for R larger than $\sim$3 nm the discrepancy is less than 10%). As has been pointed out by Schooss *et al* [60], this discrepancy decreases if finite barriers are assumed in the effective-mass-type calculations. Nevertheless, in what follows, we will only consider a nanosphere of radius 8 nm. It has also been shown [24] that the exchange energy can be neglected for nanospheres of this size.

4.4.2 Post-growth tuning of energy levels

We now consider a nanosphere with a CdTe core of radius $R = 8$ nm in more detail. Since we are ultimately interested in increasing the transition energy by diffusion (e.g. via the rapid thermal annealing technique) we need to embed the CdTe core in a material of larger band gap with a sizable interdiffusion coefficient [65, 66]. Possible 'shell' materials could be ternary compounds, such as $Cd_{1-x}Mn_xTe$, or other binary semiconductors. For the purpose of illustration only, we have chosen the large band-gap semiconductor ZnTe ($E_g = 2.3$ eV) as shell material, where we have assumed that the diffusion takes place on the cation sublattice and the band gap

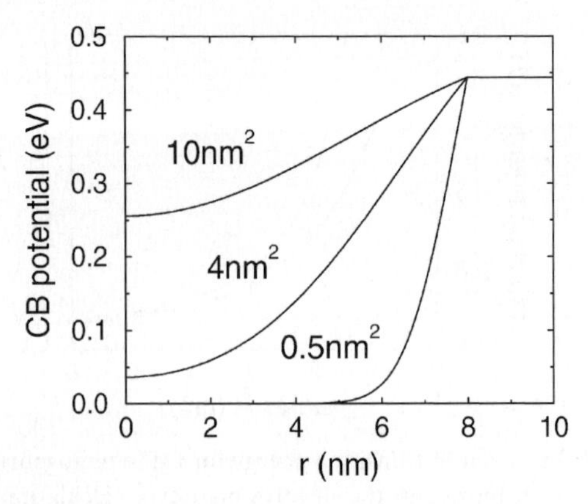

Figure 4.9. Conduction band (CB) potential profile of a nanosphere with $R = 8$ nm CdTe core material surrounded by ZnTe for varying amounts of diffusion (the curves are labelled with the chosen values of Dt).

depends linearly on the Zn concentration. We will also assume that the Zn composition in the shell is constant. This will overestimate the effects of diffusion slightly, but it will not change the essential physics. The concentration profile $x(r)$ due to diffusion of Zn from the shell into the core can be expressed as the series expansion [67]

$$x(r) = 1 + \frac{2R}{\pi r}\sum_{n=1}^{\infty}\frac{(-1)^n}{n}\sin\left(\frac{n\pi r}{R}\right)\exp\left(\frac{-Dn^2\pi^2 t}{R^2}\right), \tag{4.13}$$

where D is the diffusion constant and t the diffusion time. The diffusion profiles are universal in the sense that they depend only on the product Dt, i.e. the values of diffusion constant and time can be interchanged. Figure 4.9 shows the resulting conduction band (CB) profiles of the CdTe nanosphere for varying amounts of diffusion. The curves in this figure are labelled with the corresponding Dt values.

The calculated energy of the e1–hh1 exciton transition in the CdTe–ZnTe core-shell nanosphere corresponding to the system of figure 4.9 is shown in figure 4.10 as a function of the diffusion time. The value of the diffusion constant was chosen as $D = 0.01\ \mathrm{nm^2\,s^{-1}}$. As can be seen from this figure, an increasing amount of diffusion results in a blue shift of the transition energy which saturates for large diffusion times (at which the transition energy approaches the band-gap energy of ZnTe). For the present core–shell system, this diffusion-induced blue shift allows a tuning of the transition energy from the infrared almost to the green end of the spectrum. This effect, which is well known in quantum wells [27], was recently also observed in InGaAs/GaAs [29] and InAs/GaAs quantum dots [57]. For larger nanospheres it was found that the rate of increase in the transition

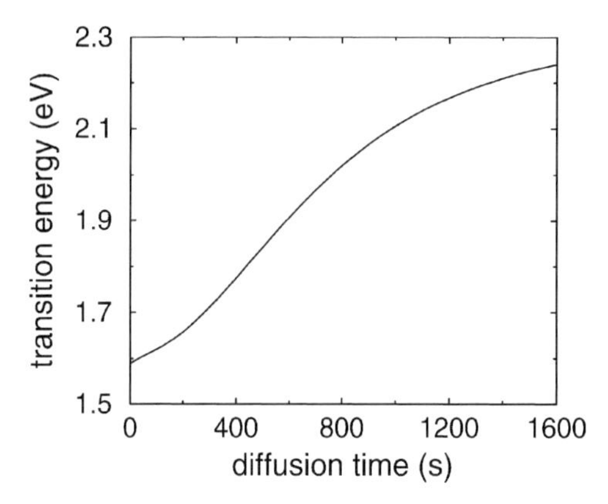

Figure 4.10. Ground-state exciton transition energy for a CdTe–ZnTe core–shell nanosphere as a function of diffusion time. The CdTe core radius is 8 nm ($D = 0.01\,\text{nm}^2\,\text{s}^{-1}$).

energy with increasing diffusion time is smaller [64]. This rate is additionally reduced by the increase in the Coulomb energy E_C for intermediate diffusion times.

The latter (i.e. E_C) is shown in figure 4.11, corresponding to the CdTe–ZnTe core–shell nanosphere of figure 4.10, as a function of the diffusion time. As can be seen from this figure, the Coulomb energy for this core–shell nanosphere is larger than the value of kT at room temperature ($\sim$25 meV) even for large amounts of diffusion. Small amounts of diffusion can be utilized to enhance the Coulomb energy of the 1s-exciton. Eventually, however, large amounts of diffusion will lead to a reduction in the Coulomb energy (towards the value of the free exciton in bulk material).

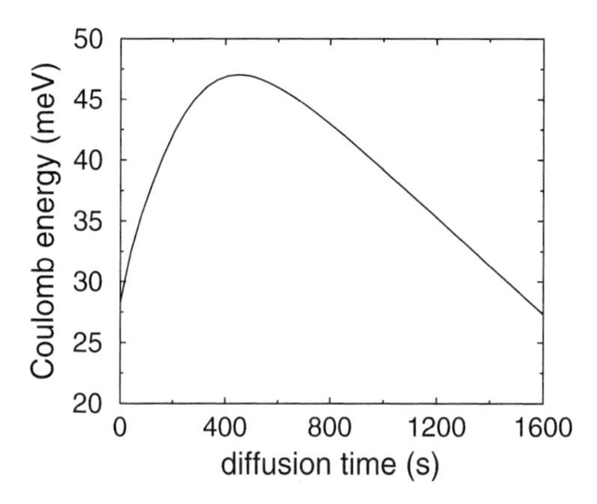

Figure 4.11. Coulomb interaction energy corresponding to the diffused core–shell nanosphere of figure 4.10.

4.4.3 Polarized emission

We will now utilize the envelope-function approximation to investigate the possibility of obtaining polarized emission from CdTe nanocrystals. From a device point-of-view, both polarized and unpolarized emission/absorption are of interest, e.g. in edge-emitting quantum-well lasers [68], polarization-insensitive optical amplifiers [69] or as backlighting for liquid crystal displays. In what follows we will outline theoretically how CdTe nanorods can be utilized to obtain both polarized as well as unpolarized electroluminescence.

The emission of light from CdTe nano*spheres* embedded in a polymer matrix is necessarily unpolarized. In CdTe quantum well structures, however, the absorption of light is polarization dependent. This can easily be demonstrated by an evaluation of the transition dipole matrix element $\langle c|\mathbf{e}\cdot\mathbf{p}|v\rangle$ between the valence and conduction band states $|v\rangle$ and $|c\rangle$, where $\mathbf{e}$ is a unit vector in the direction of the optical electric vector and $\mathbf{p}$ is the momentum operator. In the Luttinger–Kohn model [70] the band-edge Bloch functions can be written as summarized in table 4.2, where the symbol $\uparrow$ ($\downarrow$) means spin up (spin down) and S (X,Y,Z) is a conduction (valence) band function which transforms as an atomic s (p) function under the operations of the tetrahedral group at the Γ point. Consequently, for polarization in the plane of the quantum well (i.e. e_x or $e_y \perp z$) we obtain

$$|\langle\tfrac{1}{2},\tfrac{1}{2}|p_\perp|\tfrac{3}{2},\tfrac{3}{2}\rangle|^2 = |\langle\tfrac{1}{2},-\tfrac{1}{2}|p_\perp|\tfrac{3}{2},-\tfrac{3}{2}\rangle|^2 = \tfrac{1}{2}|p_{cv}|^2$$

$$|\langle\tfrac{1}{2},\tfrac{1}{2}|p_\perp|\tfrac{3}{2},-\tfrac{1}{2}\rangle|^2 = |\langle\tfrac{1}{2},-\tfrac{1}{2}|p_\perp|\tfrac{3}{2},\tfrac{1}{2}\rangle|^2 = \tfrac{1}{6}|p_{cv}|^2$$

and for polarization along the growth axis (i.e. $e_z \parallel z$) we obtain

$$|\langle\tfrac{1}{2},\tfrac{1}{2}|p_\parallel|\tfrac{3}{2},\tfrac{3}{2}\rangle|^2 = |\langle\tfrac{1}{2},-\tfrac{1}{2}|p_\parallel|\tfrac{3}{2},-\tfrac{3}{2}\rangle|^2 = 0$$

$$|\langle\tfrac{1}{2},\tfrac{1}{2}|p_\parallel|\tfrac{3}{2},-\tfrac{1}{2}\rangle|^2 = |\langle\tfrac{1}{2},-\tfrac{1}{2}|p_\parallel|\tfrac{3}{2},\tfrac{1}{2}\rangle|^2 = \tfrac{2}{3}|p_{cv}|^2$$

where the quantization axis lies parallel to the growth (z) axis and $p_{cv} = \langle S|p_N|N\rangle$ with $N = X$, Y or Z (all other matrix elements are zero; we note that the spin and envelope functions give rise to the usual selection rules). We can therefore see that for polarization in the plane of the quantum

Table 4.2. Zincblende basis vectors for the Γ_6 conduction band, and the Γ_8 and Γ_7 valence bands (using the notation of Pidgeon and Brown [71], where $|J,m_J\rangle$ are the band-edge states with total angular momentum J and m_J is the projection of J along z).

Γ_6	$\vert\tfrac{1}{2},\pm\tfrac{1}{2}\rangle$	$\vert S\uparrow\rangle$	$\vert \mathrm{i}S\downarrow\rangle$
Γ_8	$\vert\tfrac{3}{2},\pm\tfrac{3}{2}\rangle$	$\vert\tfrac{1}{\sqrt{2}}(X+\mathrm{i}Y)\uparrow\rangle$	$\vert\tfrac{1}{\sqrt{2}}(X-\mathrm{i}Y)\downarrow\rangle$
Γ_8	$\vert\tfrac{3}{2},\pm\tfrac{1}{2}\rangle$	$\vert\tfrac{1}{\sqrt{6}}[(X-\mathrm{i}Y)\uparrow+2Z\downarrow]\rangle$	$\vert\tfrac{\mathrm{i}}{\sqrt{6}}[(X+\mathrm{i}Y)\downarrow-2Z\uparrow]\rangle$
Γ_7	$\vert\tfrac{1}{2},\pm\tfrac{1}{2}\rangle$	$\vert\tfrac{\mathrm{i}}{\sqrt{3}}[-(X-\mathrm{i}Y)\uparrow+Z\downarrow]\rangle$	$\vert\tfrac{1}{\sqrt{3}}[(X+\mathrm{i}Y)\downarrow+Z\uparrow]\rangle$

well (e.g. normal incident radiation) the optical absorption associated with the heavy-hole–electron transition is three times stronger than that associated with the light-hole–electron transition, whereas for polarization parallel to the growth direction (i.e. TM mode) the heavy-hole–electron transition is absent and only the light-hole–electron transition is present.

In nano*rods* the light- and heavy-hole states are mixed [72] thus giving rise to a decrease in the polarization anisotropy with increasing band mixing. Such a decrease in the polarization anisotropy was recently observed by Notomi *et al* [73] and Lehr *et al* [74] in InGaAs/InP and InGaAs/In(GaAs)P quantum wires of decreasing wire width. However, for nanorods embedded in a matrix with a large dielectric contrast, the electric-field modulation can give rise to a luminescence anisotropy [75, 76]. The electric-field component inside the nanorod for **e** perpendicular to the wire axis z is reduced by a depolarization factor δ, where

$$\delta = \frac{2\varepsilon_I}{\varepsilon_I + \varepsilon_S} \tag{4.14}$$

and ε_S (ε_I) is the relative permittivity of the semiconductor (matrix), whereas the electric-field component along the z-axis is unchanged. Since, in the electric-dipole approximation, the absorption/emission is proportional to the scalar product of the local electric field $\mathbf{E}$ and the interband dipole moment $\mathbf{d}$, the polarization anisotropy σ can be expressed as

$$\sigma = \frac{\alpha_{\parallel} - \alpha_{\perp}}{\alpha_{\parallel} + \alpha_{\perp}} = \frac{\langle c|\mathbf{E}_{\parallel} \cdot \mathbf{d}|v\rangle^2 - \langle c|\mathbf{E}_{\perp} \cdot \mathbf{d}|v\rangle^2}{\langle c|\mathbf{E}_{\parallel} \cdot \mathbf{d}|v\rangle^2 + \langle c|\mathbf{E}_{\perp} \cdot \mathbf{d}|v\rangle^2} \approx \frac{1-\delta^2}{1+\delta^2} \tag{4.15}$$

where $\alpha(\omega)$ is the absorption coefficient and where the last expression assumes that the dipole matrix element between the conduction and valence band is isotropic (which has been shown to overestimate the polarization anisotropy [76]). Such a polarization anisotropy, arising from a large dielectric contrast, was recently observed by Chernoutsan *et al* [49] for InP nanorods in chrysotile asbestos and for CdS nanorods in an Al_2O_3 matrix. In addition, it was argued that the Coulomb interaction energy can be enhanced by such dielectric confinement [49].

4.5 Synthesis

Electroluminescence from inorganic semiconductor compounds, in particular ZnS, has been investigated since the 1960s [77]. However, there are several aspects that render hybrid inorganic/organic LEDs using semiconductor nanocrystals, rather than bulk semiconductor materials, of advantage for flat panel displays based on EL devices [78]. Unsurprisingly, there are many practical issues to be resolved before such devices, as described in the following section on HIOLEDs, appear on the flat-panel market place. The synthesis of stable and soluble cadmium chalcogenide nanocrystals, i.e. CdS, CdSe and CdTe nanocrystals, emitting in the visible region with a high quantum yield, is essential for the realization of such EL displays [79]. Cadmium chalcogenide nanocrystals have been used in most prototype HIOLEDs so far (see the section on electroluminescent devices), although nanorods may be of long-term advantage for some HIOLEDs [80]. Cadmium chalcogenide nanocrystals, just like other semiconductor nanocrystals, can be regarded as a continuous transition state between molecules and bulk solids as described above [81]. Very small cadmium chalcogenide nanocrystals (~2–5 nm) emit across the visible spectrum as a consequence of quantum confinement. Therefore, chemical synthesis allows the EL spectrum of cadmium chalcogenide nanocrystals to be tuned in the visible region for use in hybrid inorganic/organic LEDs. It is usually important for EL applications that the nanocrystals are either soluble in the polar organic solvents used to deposit dissolved materials as a uniform thin layer by spin coating, or soluble in water so that they can be deposited as monolayers by electrolytic techniques, so that electroluminescent layers can be built up on a device substrate. There are many general methods for the synthesis of many different kinds of nanocrystals [10, 12, 82]. The binary II–VI semiconductors, such as metal chalcogenides and especially cadmium chalcogenides, are particularly suited for nanocrystal synthesis due to their high tendency to form regular crystal structures even with a small number of atoms. There is a small but significant number of reports of the synthesis and processing of organically soluble cadmium chalcogenide nanocrystals specifically for use in LEDs. The most common cadmium chalcogenide for this application is CdSe since its emission spectrum can be varied across a large section of the visible spectrum by chemically manipulating the size and size dispersion of the CdSe nanocrystals. Until recently there were few reports of the synthesis of similar CdTe nanocrystals for LEDs using wet chemistry. However, most cadmium chalcogenide nanocrystals have been prepared for other applications.

Colloidal dispersions or suspensions of semiconductor nanocrystals such as CdS were first synthesized in the early 1980s as potential catalysts in photochemical reactions, especially the photocatalytic splitting of water into hydrogen and oxygen [83–88]. This was in order to make use of the large active surface area compared with the bulk material of a nanocrystalline catalyst. Such nanocrystals absorb in the near ultraviolet, thereby generating an exciton on the surface of the nanocrystal, which is then available to take part in a photochemical reaction. Reverse micelles have also been used to control the size of platinized CdS nanocrystals for similar catalytic applications [89]. ZnS nanocrystals have also been used as catalysts in the reduction of ketones to alcohols in a similar manner [90]. It was then realized that such small semiconductor nanocrystals, where the diameter is smaller than the Bohr radius of the exciton, should exhibit quantum effects in PL and El as well as large nonlinear optical effects [91–99]. The discovery of quantum confinement effects in small nanocrystals (~2–5 nm) stimulated the development of a range of new synthetic methods specifically designed for the chemical preparation, purification, characterization and processing of semiconductor nanocrystals of a controlled size, shape and constitution. The first dispersions of very small semiconductor nanocrystals prepared by these new methods were used to demonstrate for the first time the dependence of the semiconductor bandgap on particle size. Unfortunately these colloidal dispersions were often found to be unstable and often the nanocrystals could not be characterized adequately.

Nanocrystals of good optical quality were then synthesized within solid inorganic porous media, such as zeolites [100–103], clays [104] and porous glasses [43, 105–111], which served to control the size of the nanocrystals within the nanometre-sized pores. The most common method of this type is the diffusion-controlled growth in porous silicate and borosilicate glasses. The development of this synthetic approach was partly attributable to the discovery that the blue-shift in the colour of some soda-lime glasses investigated in the early and mid twentieth century, and attributed at that time to very small polarized CdS particles [48, 106], was actually due to the quantum confinement of the CdS nanocrystals within the glass [107]. The physical properties of nanoparticles in glasses as a solid matrix were then investigated in great detail over the following years [108–111]. Nanocrystals have also been prepared in porous titanium dioxide layers in a similar fashion [112, 113]. However, a major problem with this general synthetic approach to semiconductor nanocrystals is the fact that the nanocrystals cannot be isolated from these solid inorganic composites. Furthermore, the inorganic composites themselves are not suited for electro-optic devices based on electroluminescence even when they incorporate II–VI semiconductor nanocrystals.

One of the first methods to synthesize and then isolate II–VI semiconductor nanocrystals of a controlled size and size-distribution involved covering them in an ionic polymer coating by preparing them in an aqueous solution containing a sodium polyphosphate derivative, e.g. sodium hexaphosphate $[NaPO_3]_6$ [114, 115]. The cadmium ions form a complex with the polyphosphonate (PP) chains in aqueous solution:

$$n\mathrm{Cd}^{2+} + \mathrm{PP}^{2n-} \longrightarrow \mathrm{Cd}_n\mathrm{PP}.$$

This charged polymer coating prevents agglomeration of the nanocrystals due to electrostatic repulsion and steric effects attributable to the polymer chains. The exact nature of the interaction between the polymer and the nanocrystal surface is not completely understood, although it is clear that the sodium ions serve to bind the polymer to the nanocrystal surface. However, the presence of the polymer can be used to control the size and shape of the nanocrystals formed by then adding a chalcogenide gas to the reaction solution, e.g. by bubbling hydrogen suphide gas through it, which provides the required sulphur S^{2-} anions in this case:

$$Cd^{2+} + S^{2-} \rightarrow CdS.$$

The formation of the nanocrystals and their subsequent growth using this method can be controlled to a large extent by modifying the pH of the reaction solution. Other ionic polymers used in the synthesis of quantum-confined nanocrystals such as CdS include Nafion and Surlyn [116, 117]. The addition of trialkylamines in low concentrations was found to increase the quantum yield of exciton emission of CdS nanocrystals prepared using hexametaphosphates (HMP) as stabilizer [118]. CdS, CdSe and CdTe polymer complexes have also been synthesized with a bidentate phosphine ligand as stabilizer [119]. Decomposition of the complexes in solution, e.g. in pyridine, results in the formation of the corresponding soluble nanocrystals.

Cadmium chalcogenides have also been prepared in the aqueous droplets in reverse micelles in order to create nanocrystals of a defined size and spherical shape. A monolayer of organic solvent inhibits flocculation and aggregation of the nanocrystals [120]. It was found that the addition of more Cd^{2+} and S^{2-} reagents leads to further growth of the nanoparticles in a kind of inorganic living polymerization reaction [120]. Addition of phenyl selenium anion as phenyl(trimethylsilyl)selenium to a CdSe nanocrystal microemulsion led to the coating of the CdSe nanocrystals with a thin layer of Se–Ph (see figure 4.12).

The presence of this monolayer of organic material around these phenyl-capped nanocrystals meant that they could be isolated from solution and

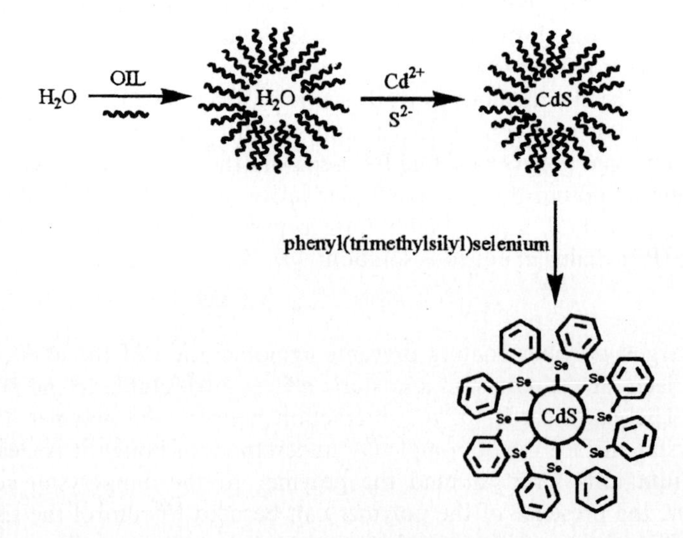

Figure 4.12. Schematic representation of the preparation of inorganic nanocrystals with an organic monolayer coating in reverse micelles using a surfactant (oil) and water.

then re-dissolved in organic solvents, such as pyridine [121]. The surface monolayer did not appear to change the HOMO and LUMO energy levels. A similar phenyl-capping of ZnS(CdSe) core/shell nanocrystals to produce isolatable and re-dispersible nanocrystals with a high quantum yield was also carried out in a reverse micelle medium [122]. Other capped CdS nanocrystals have also been prepared using reverse micelles using a variety of stabilizers such as AOT (sodium dialkylsulphosuccinate) and block copolymers [123–126]. AOT is especially suitable for this application as it can solubilize up to 50 moles of water per mole of surfactant. Reverse micelles of AOT in water have also been used to template the synthesis of CdTe nanocrystals [127]. CdS nanocrystals have also been synthesized in vesicles using dihexadecyl phosphate as the stabilizer [128]. CdS nanocrystals have also been synthesized using water-in-oil microemulsions as a template [129].

The most successful method for the synthesis of nanocrystals is based on the controlled precipitation of nanocrystals in colloidal solutions containing stabilizers [1–16]. These methods are capable of providing nanocrystals as particles of a controlled size, size-distribution and shape and covered by a monolayer of surface stabilizer. The presence of the stabilizers allows the nanocrystals to remain in solution and grow in a controlled fashion by suppressing aggregation of individual nanocrystals to form an inhomogeneous mixture of nanocrystals of different sizes (see figure 4.13). These nanocrystals can then be precipitated out of solution and then isolated as powders. Post-synthesis processing can also provide nanocrystals with an even narrower size distribution and increased quantum yield.

One of the two main methods for the synthesis of semiconductor nanocrystals, such as the cadmium chalcogenides CdS, CdSe and CdTe

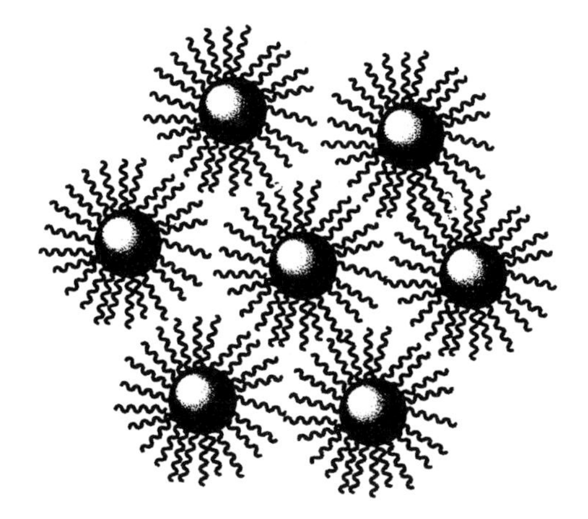

Figure 4.13. Schematic representation of inorganic semiconductor nanocrystals with an organic monolayer coating.

nanocrystals, is the organometallic approach involving the size selective precipitation of nanocrystals in trioctyl phosphine (TOP) and trioctyl phosphine oxide (TOPO) as stabilizers at high temperatures [130]. The phosphine ligands passivate the nanocrystal surface and reduce the number of defects on the surface. The long alkyl chains attached to the phosphorous atoms at the nanocrystal surface convert an inorganic metallic surface into an apolar organic surface and thereby induce solubility of these surface passivated nanocrystals in organic solvents [130–133]. The shape of the nanoparticles can also be controlled using this method, such as CdSe nanodots and nanowires [133]. Attempts have been made to replace some of the reagents, especially the pyrophoric dimethyl cadmium, and modify the reaction conditions of this method in order to make them more environmentally friendly [134–137].

The main alternative method for the synthesis of cadmium chalcogenides involves using aliphatic thiol and aromatic thiophenol stabilizers in colloidal, often aqueous, solutions. It was first noticed that CH_3S groups attached themselves to the surface of CdS nanocrystals, prepared in the presence of other stabilizers, when they were precipitated out of solution in the presence of sodium methylthiolate [138]. It was then found that a variety of other aliphatic thiols performed a similar function when present as additives in the reaction solution itself [138]. A more general method involved preparing the nanocrystals in the presence of long-chain alkylthiols [139]. Aliphatic stabilizers include thioethanol, thioglycerine and thioglycolic acid for the preparation of organically soluble CdS, CdSe and CdTe nanocrystals [140–149]. Aromatic thiols can also be used for this purpose and a whole range of substituted thiophenols have been attached to the surface of organically soluble cadmium chalcogenide nanocrystals [150–159]. The presence of the organic monolayer attached to the nanocrystal surface by the sulphur linkage renders the nanocrystals soluble in organic solvents, stabilizes the surface towards oxidation and lowers the number of traps. Cadmium perchlorate is often used as the source of cadmium cations and hydrogen sulphide gas is a convenient source of sulphide anions as it is commercially available in cylinders.

$$Cd(ClO_4)_2 + H_2S \rightarrow CdS + 2HCl + 4O_2.$$

Other sources of cadmium include cadmium chloride and cadmium nitrate, which also influence the size of the nanocrystals produced. Alternative convenient sources of sulphur are sodium sulphide (Na_2S) and sodium hydrogen sulphide (NaHS). Hydrogen selenide (H_2Se) and hydrogen telluride (H_2Te) are usually generated by addition of acid to aluminium selenide and aluminium telluride.

$$Al_2Te_3 + 3H_2SO_4 \rightarrow 3H_2Te + Al_2(SO_4)_3.$$

The highly reactive surface of nanocrystals with atoms that are not completely coordinated allows a second layer of a different nanocrystalline material to be deposited by epitaxial growth. For example, pre-formed CdSe nanocrystals in solution can be coated with CdS *in situ* to form CdSe(CdS) core/shell nanocrystals [160]. This allows the inner nanocrystal core of these composite core/shell nanocrystals to be more completely and effectively passivated. This results in a higher quantum efficiency and greater resistance towards oxidation than those of normal mono-material nanocrystals due to a lower concentration of surface defects and dangling bonds still present in nanocrystals with an organic passivation layer [160]. The second nanocrystal layer can then be further passivated with an organic monolayer to render the core/shell nanocrystals soluble in aqueous or organic media. Such core/shell CdS nanocrystals have been prepared using most of the known methods so far, e.g. core/shell CdS(ZnS) with an organic thiophenol capping layer have been synthesized in reverse micelles [160]. CdSe(ZnS) core/shell nanocrystals have also been prepared with a phenyl capping layer in reverse micelles. CdTe(HgTe) core/shell nanocrystals have also been prepared with a thiol capping layer [161]. A series of core/shell cadmium chalcogenide nanocrystals, especially with a stable ZnS shell, have been used in composite organic/inorganic LEDs and these are described in more detail in the following section.

The synthesis of cadmium chalcogenide nanocrystals in organic media yields inorganic/organic composites directly. These composite materials, often formed *in situ* as uniform thin plastic films containing appropriate cadmium chalcogenide nanocrystals, are especially suited for electro-optic devices such as LEDs. CdS nanocrystals have been synthesized at the interface between aqueous media and a Langmuir–Blodgett monolayer of organic material [162, 163]. Polymer composites can be formed by polymerizing such organic/inorganic nanocrystal composites where the Langmuir–Blodgett layers contain polymerizable groups such as carbon–carbon triple bonds [162, 163]. A whole range of related but different approaches to polymer networks have been reported [166–169]. Those containing cadmium chalcogenides of particular relevance to LEDs are dealt with in more detail in the section below on devices. Block copolymer micelles have also been used to prepare organic/inorganic polymer composites [170]. Dendrimers can also be used as the organic matrix to synthesize CdS composite materials as a functional organic host intermediate between those represented by organic polymers and small organic molecules [171–173].

4.5.1 Hybrid inorganic/organic LEDs

The market for flat panel displays is very fragmented with a range of different technologies available [174–176]. Liquid crystal displays (LCDs) represent the dominant flat panel display technology at the moment. However, flat panel displays based on electroluminescence have several clear advantages over back-lit LCDs for a range of applications such as mobile telephones and digital cameras where brightness, wide viewing angle, low power consumption and the potential for video-rate addressing are essential [177, 178]. Indeed commercially available bulk semiconductor light emitting diodes (LEDs) already possess low operating voltages, high efficiency, good thermal, mechanical and chemical stability and wide viewing angles due to the Lambertian emission. The colour of emission can be easily tuned by material choice. Furthermore, the emission has a narrow bandwidth, which in theory would facilitate colour mixing for full colour display applications. Unfortunately, it is difficult to fabricate large-area LEDs. Organic light-emitting diodes (OLEDs) represent an emissive and potentially more efficient display technology than either LEDs or LCDs for such applications [177, 178]. OLEDs require materials with appropriate molecular energies for electronic injection, high charge mobility and light emission. Light-emitting polymer devices (PLEDs) combine ease of processing, low voltage operation, fast switching capability and the potential of using flexible substrates. However, the electroluminescence efficiency is still relatively poor due to non-optimized charge injection and transport. Many good hole transporting polymers are known but corresponding electron-transport layers are much rarer. Furthermore, the excited states of polymers absorb some of the emitted light. Therefore, hybrid inorganic/organic light emitting diodes (HIOLEDs) have been investigated in the hope of combining the good optical and electrical properties of semiconductor nanocrystals with the processability of organic polymers and their ability to form thin uniform films on device substrates by standard processing techniques.

Inorganic II–VI semiconductor nanocrystals, and especially cadmium chalcogenide nanocrystals, are particularly suited as the emission layer in such HIOLEDs. The optical properties, including the band-edge of the cadmium chalcogenides, can be tuned simply by changing their size through quantum confinement. Thus they emit across the visible spectrum. This is especially the case for red emission, where few organic emitters are known. The small bandgap required for red emission requires a large number of conjugated carbon–carbon double bonds in organic materials. This renders them susceptible to oxidation and photochemical degradation. The high refractive index of organic semiconductors may contribute to a low external efficiency due to a substantial degree of internal reflection and absorption. The unique optical properties of cadmium chalcogenide nanocrystals can only be made use of if appropriate electrical contact can be established. This can be achieved by incorporating them in hole-transporting host materials, since such nanocrystals exhibit high electron transport. The combination of cadmium chalcogenides as the electron-transporting and light-emitting component of hybrid organic/inorganic OLEDS and processable hole-transporting mainchain polymers appears a particularly attractive approach.

The first HIOLED reported in 1994 used a combination of a layer of PPV as the hole-transport layer and a film of CdSe nanocrystals as the emission and

Table 4.3. Chemical structure of mainchain polymers with hole-transport properties and electron-transport dopants used in hybrid inorganic/organic LEDs.

	Polymer repeat unit/structure	Abbreviation
1	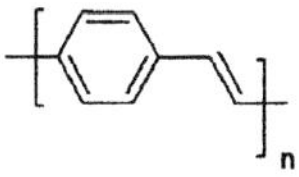	PPV
2	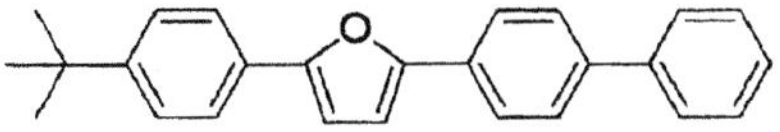	*t*-Bu-PDB
3	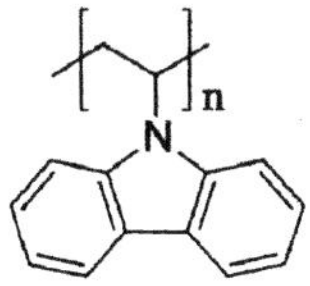	PVK
4	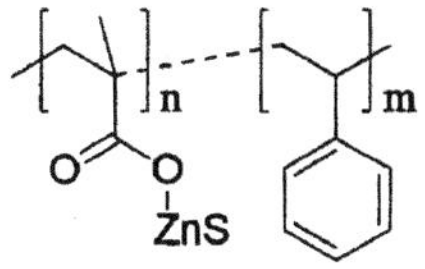	PMMA/PS
5	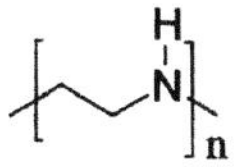	PEI
6	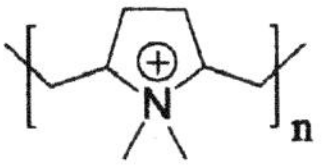	PDDA
7	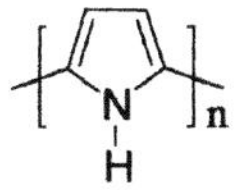	PPy
8	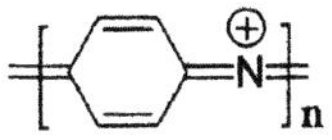	PAni
9	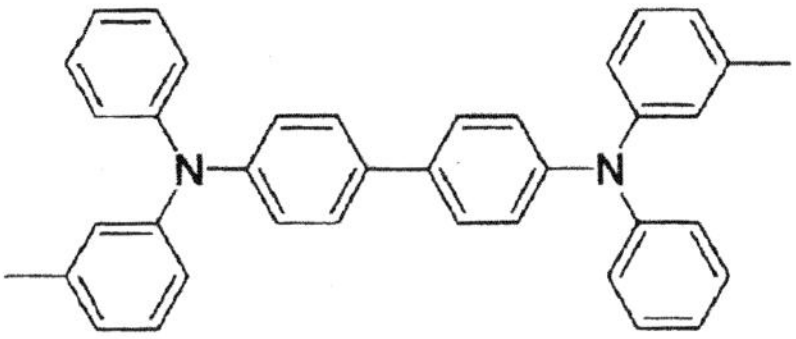	TPD

electron-transport layer [179]. The CdSe nanocrystals, prepared by the TOPO method (see the synthesis section above), were deposited stepwise on to a thin layer (~100 nm) of PPV (**1**) (see table 4.3; bold figures refer to scheme numbers) on an indium tin oxide (ITO) coated glass substrate using a dithiol between the nanocrystal layers to bind them to the surface of the PPV. Alternating layers of nanocrystals and dithiol were built up by this

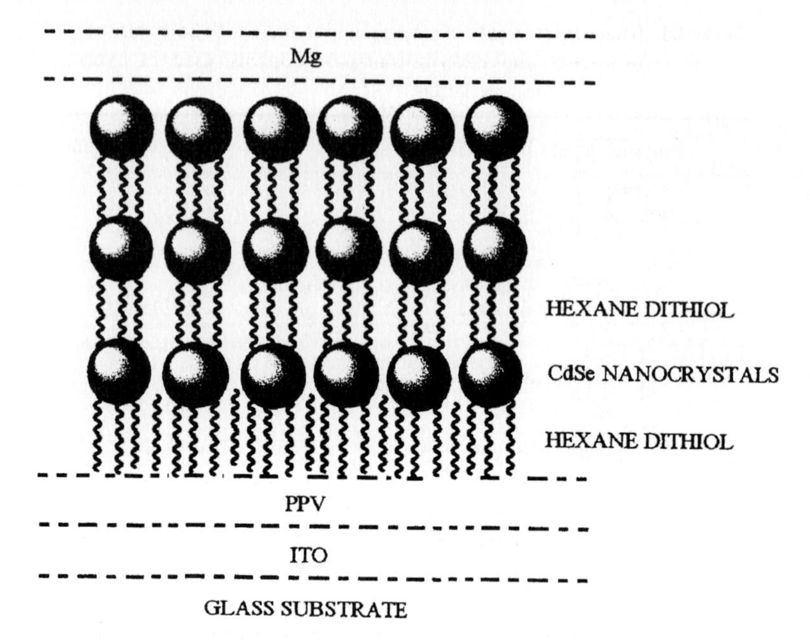

Figure 4.14. Schematic representation of a hybrid inorganic/organic LED. The composite nanocrystal/dithiol material is prepared layer by layer [179].

stepwise process to form a thin nanocrystal layer ($\sim$20–30 nm) and then a magnesium cathode deposited on top, which was then protected in turn with a silver coating (see figure 4.14).

Application of a forward bias between the electrodes with the ITO positively charged led to light emission (100 cd m^{-2}) from the nanocrystals at low turn-on voltage (4–5 V). The colour of emission depends on the size of nanocrystals used (3–5 nm) and was varied from yellow to red. The rectification ratios were low, but electroluminescence was not observed under reverse bias. The energy levels of this hybrid inorganic/organic LED are shown in figure 4.15. The bilayer nature of this device configuration locates the recombination and emission zones at the interface between the hole-transporting PPV (**1**) and the electron-transporting CdSe nanocrystal composite layer. CdSe has a high electron affinity of about 4.6 eV so that electron injection from the magnesium cathode, whose work function is 3.6 eV, is energetically favourable. The high work function of ITO promotes the injection of holes in the layer of PPV (**1**). As figure 4.15 shows, the PPV/CdSe interface shows type-two band alignment with the hole located in the PPV and the electron located in the CdSe. Despite this, intrinsic exciton recombination in both the CdSe nanocrystals and PPV is observed. This suggests that the band alignment is affected by strong electric fields at the interface to enable charge transfer between the two species.

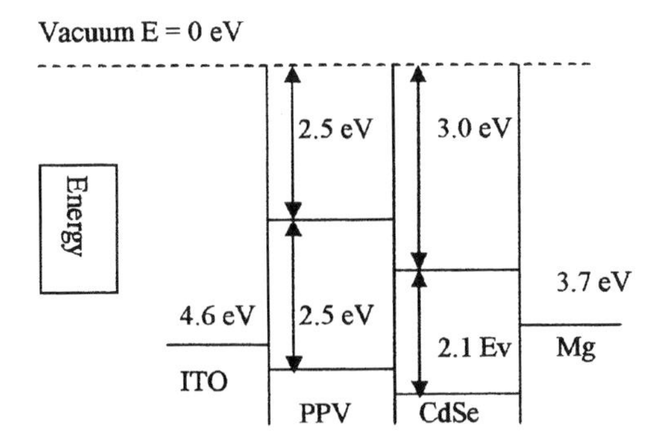

Figure 4.15. Energy level diagram of a hybrid inorganic/organic LED [179, 180].

The quantum efficiency of the hybrid organic/inorganic LED was improved by using core/shell CdSe(CdS) nanocrystals [180]. CdS has a wider bandgap than CdSe and the epitaxial layer of CdS improves the quantum yield and photo-oxidative stability compared with CdSe nanocrystals. The improvement in quantum yield was attributed to carrier confinement with the hole confined to the nanocrystal core and the electron delocalized throughout the structure. The greater stability to oxidation was explained by the low hole-concentration at the surface. The combination of a hole and molecules of oxygen at the surface of the nanocrystal are required for oxidation to take place. The presence of the CdS shell lowers the probability of a hole being present at the nanocrystal surface and consequently reduces the tendency for oxidation. Encapsulation techniques can also reduce the concentration of oxygen and oxygen-containing moisture. The core/shell configuration also reduces non-radiative recombination and results in good electronic access. In this device a thin (25 nm) layer of core/shell CdSe(CdS) nanocrystals was deposited by spin-casting on top of a thin (40 nm) layer of PPV. An ITO anode and a magnesium cathode stabilized by a silver coating were used [180]. The operating voltage is low ($\sim$5 V) at a current density of 10 mA cm^{-2} due to facile charge injection because of the low energy barriers between the cathode and the nanocrystals and the ITO and PPV (**1**). The CdSe nanocrystals have a high electron affinity and are expected to transport electrons well although the thiol stabilizer provides a barrier between individual nanocrystals. PPV has a low ionization potential and transports holes well.

Types of hybrid LEDs using CdSe(CdS) core/shell nanocrystals exhibit a 20 times increase in quantum efficiency ($\sim$0.1–0.2%) and a hundred times increase in lifetime compared to the corresponding LEDs using non-passivated CdSe nanocrystals. For example, the emission of yellow light from one of these bilayer LEDs using CdSe(CdS) core/shell nanocrystals is stable at high brightness (150 cd m^{-2}) over a number of hours (10 h). Other configurations exhibit stable emission, after an initial drop, over considerably longer times (200 h), although the brightness (10 cd m^{-2}) and external efficiency (0.03%) are low.

Similar results were obtained using CdSe nanocrystals with a ZnS coating in a bilayer LED structure with ITO and aluminium electrodes [46]. The hole transport layer (25 nm) is composed of a complicated PPV/sulphonated polystyrene (SPS) PPV/polymethacrylate (PMA) alternating structure deposited by a sequential adsorption technique. The electron-transport and emission layer (50 nm) is formed of CdSe(ZnS) core/shell nanocrystals deposited by spin coating. The photoluminescent yield of the CdSe nanocrystals (2.35 nm) was substantially increased by the core/shell structure. Driving voltages are low (~3.0–3.5 V) and the brightness is also low (120 cd m^{-2}). Surprisingly, the external EL quantum efficiency (~0.1%) is not improved by the addition of the shell structure. Lifetimes of 50–100 h of continuous operation for these prototype displays were reported. Electroluminescence has also been achieved by simply sandwiching a thin layer of nanocrystals between two electrodes [181]. However, the lifetimes of these displays were found to be short.

An early attempt at a multicomponent polymer composite for hybrid organic/inorganic LEDs involved a guest mixture of CdSe nanocrystals (5–10%) and an electron-transporting oxadiazole derivative (*t*-Bu-PDB) (**2**) in a hole-transporting poly(vinylcarbazole) (PVK (**3**)) polymer film as a host matrix [182]. Thin layers (70–120 nm) of this polymer composite were deposited from solution by spin-coating on to a glass substrate. ITO served as the anode and aluminium as the cathode. Emission from the nanocrystals was observed at low voltages and combined emission from the nanocrystals and the host PVK (**3**) at high voltages.

Another approach to creating electroluminescent polymer composites involved growing the nanocrystals in a suitable polymer host. So far this approach has only been used for ZnS nanocrystals [183, 184]. The ZnS nanocrystals were grown in an insulating copolymer matrix (**4**) of polymethacrylate and polystyrene by treating the zinc salt of the polymer with hydrogen sulphide (H_2S) gas. A thin composite film of the ZnS nanocrystal polymer and the hole-transporting component tetraphenylbenzidine (TPB) was deposited from solution by spin coating on to a glass substrate. The single layer device with an ITO anode and an aluminium cathode emits green light (λ_{max} = 520 nm) with a narrow bandwidth (half-width = 120 nm) at a low voltage (2.5 V) [183]. The origin of the emission is not absolutely clear. A similar display without the hole-transport component exhibits blue light (λ_{max} = 435 nm) from the ZnS nanocrystals themselves at a higher turn-on voltage (4.0 V) [184]. Similar behaviour was observed for hybrid inorganic/organic LEDs with Cu-doped ZnS nanocrystals in a comparable polymer matrix [185, 186]. The presence of the copper atoms leads to a significant red shift (25 nm) in the PL emission (λ_{max} = 415 nm) compared with that (λ_{max} = 390 nm) of ZnS nanocrystals of similar size (3.0 nm). The brightness of these displays is low, e.g. 15 cd m^{-2} at a current density of 100 mA cm^{-2} at a driving voltage of 8 V [186]. Impurities lead to a further red shift in EL emission ($\lambda_{max} = 430$ nm).

The use of nanocrystals with a relatively wide particle-size distribution allows the generation of almost white light. Alternating inorganic/organic layers of thiolactic acid stabilized CdSe(CdS) core/shell nanocrystals (core diameter 3.7 nm and shell thickness 0.6 nm) and PPV were assembled from aqueous solution by an electric-field directed polyelectrolytic technique on a glass substrate. A broad emission spectrum (500–800 nm) is observed for turn-on voltages between 3.5 and 5 V for a hybrid organic/inorganic LED with an ITO anode and an aluminium cathode. A poly(ethyleneimine) (PEI (**5**)) monolayer is also present on top of the ITO electrode for synthetic purposes to facilitate the deposition of the layers [187]. The step-by-step fabrication by self-assembly of alternating layers of polymer and nanocrystals is an attractive approach [188, 189] based on a standard procedure [189, 190] to create uniform thin organic/inorganic films in OLEDs. This procedure has been used to fabricate LEDs with CdTe nanocrystals and either poly(diallyldimethylammonium chloride (PDDA (**6**)) [44, 45, 191, 192] or PPV (**1**) [193]. Similarly polypyrrole (PPy (**7**)) [194] and polyaniline (PAni (**8**)) [195] were used to fabricate hybrid organic/inorganic LEDS *via* an electropolymerization technique. A Stokes shift is observed between the ultraviolet–visible absorption edge and the PL emission peak of the CdTe nanocrystals in these devices [196] whereas the EL spectrum of the PDDA composites and the PPV composites virtually coincides with the PL spectrum of the CDTe nanocrystals [191, 193]. The small red-shift of the EL spectrum with respect to the PL spectrum was attributed to the oxidation of CdTe when the nanocrystals are electrically excited in the presence of oxygen [191] and to the involvement of surface trap states [193]. An additional red-shift of the EL peak with increasing applied voltage was observed. This can be partially attributed to the quantum-confined Stark effect at low fields [197], Auger quenching and charge escape from the nanocrystals at high fields [197] as well as a contribution due to oxidation and excitation of trap states [198].

EL has been obtained from LEDs using an unusual composite organic/inorganic hole-transport layer (HTL) as a thin film (20 nm) of polypyrrole (PPy (**6**)) and silicon oxide (SiO_2). A composite PPy (**7**) and silanol film was first produced by an electrochemical process. The silanol was then converted in a second step by an *in situ* thermal process to yield SiO_2 particles in a PPy (**7**) film [199]. The presence of the SiO_2 in the PPy (**6**) film led to an increase in the stability of the HTL.

The brightest HIOLED having an efficiency of 1.6 cd A^{-1} at 2000 cd m^{-2} was recently obtained using a single monolayer of nanocrystals sandwiched between hole and electron transporting organic layers [200]. The core/shell CdSe/ZnS nanocrystals are mixed with the hole-transporting material N,N'-diphenyl-N,N'-bis(3-methylphenyl)-(1,1′-biphenyl)-4,4′-diamine (TPD (**9**)) in solution and deposited by spin-casting on to the substrate. Phase separation of the aliphatic TOPO capped nanocrystals and the TPD results in a nanocrystal monolayer on top of the TPD when an optimized nanocrystal concentration is used.

4.6 Conclusions

Hybrid inorganic/organic light-emitting diodes (HIOLEDs) represent an emissive and potentially very efficient flat-panel display technology for certain applications, such as portable telephones and digital cameras, where brightness, wide viewing angle, low power consumption and the potential for video-rate addressing are essential. This kind of LED requires materials with appropriate energy levels for electronic injection, high charge mobility and light emission. This is especially the case for red emission, where few stable organic emitters are known. Inorganic II–VI semiconductor nanocrystals in general, and cadmium chalcogenides in particular, are especially suited as the emission layer in such hybrid inorganic/organic LEDs.

References

[1] Henglein A 1982 *J. Phys. Chem.* **86** 229
[2] Brus L E 1986 *J. Phys. Chem.* **90** 2555
[3] Fendler J H 1987 *Chem. Rev.* **87** 877
[4] Henglein A 1989 *Chem. Rev.* **89** 1861
[5] Herron N 1991 *ACS Symp. Ser.* **455** 582
[6] Wang Y and Herron N 1991 *J. Phys. Chem.* **95** 525
[7] Brus L E 1991 *Appl. Phys.* **53** 465
[8] Weller H 1993 *Angew. Chem. Int. Ed. Eng.* **32** 41
[9] Alivisatos P 1996 *Science* **271** 933
[10] Alivisatos P 1996 *J. Phys. Chem.* **100** 13226
[11] Zhang J Z 1997 *Acc. Chem. Res.* **30** 423
[12] Fendler J H ed 1998 *Nanoparticles and Nanostructured Films* (Weinheim: Wiley-VCH)
[13] Nirmal M and Brus L 1999 *Acc. Chem. Res.* **32** 407
[14] Xia Y, Gates B, Yin Y and Lu Y 2000 *Adv. Mater.* **12** 693
[15] Alivisatos P 2000 *Pure Appl. Chem.* **72** 3
[16] Eychmüller A 2000 *J. Phys. Chem. B* **104** 6514
[17] Marzin J Y and Bastard G 1994 *Solid State Commun.* **92** 437
[18] Wojs A, Hawrylak A, Fafard S and Jacak 1996 *Phys. Rev. B* **54** 5604
[19] Fu H, Wang L-W and Zunger A 1998 *Phys. Rev. B* **57** 9971
[20] Rama Krishna M V and Friesner R A 1991 *J. Chem. Phys.* **95** 8309; 1991 *Phys. Rev. Lett.* **67** 629
[21] Wang L-W and Zunger A 1994 *J. Phys. Chem.* **98** 2158
[22] Nair S V, Ramaniah L M and Rustagi K C 1992 *Phys. Rev. B* **45** 5969

[23] Lo C F and Sollie R 1991 *Solid State Commun.* **79** 775
[24] Franceschetti A and Zunger A 1997 *Phys. Rev. Lett.* **78** 915
[25] Rajh T, Micic O I and Nozik A 1993 *J. Phys. Chem.* **97** 11999
[26] Homewood K P, Weiss B L and Wismayer A C 1989 *Semicond. Sci. Technol.* **4** 472
[27] Harrison P, Hagston W E and Stirner T 1993 *Phys. Rev. B* **47** 16404
[28] Elman B, Koteles E S, Melman P and Armiento C A 1989 *J. Appl. Phys.* **66** 2104
[29] Leon R, Kim Y, Jagadish C, Gal M, Zou J and Cockayne J H 1996 *Appl. Phys. Lett.* **69** 1888
[30] Barker J A and O'Reilly E P 1999 *Physica E* **4** 231
[31] Cohen M L and Bergstresser T K 1966 *Phys. Rev.* **141** 789
[32] Wang L-W and Zunger A 1994 *J. Phys. Chem.* **98** 2158
[33] Wang L-W and Zunger A 1996 *Phys. Rev. B* **53** 9579
[34] Shaw M J, Briddon P R and Jaros M 1996 *Phys. Rev. B* **54** 16781
[35] Long F, Harrison P and Hagston W E 1996 *J. Appl. Phys.* **79** 6939
[36] Belgoumene B, Kouidri S and Driss Khodja M 1999 *Phys. Lett. A* **261** 191
[37] Bloom S and Bergstresser T K 1968 *Solid State Commun.* **6** 465
[38] Walter J P, Cohen M L, Petroff Y and Balkanski M 1970 *Phys. Rev. B* **1** 2661
[39] Tomasulo A and Ramakrishna M V 1996 *J. Chem. Phys.* **105** 3612
[40] Franceschetti A and Zunger A 1996 *J. Chem. Phys.* **104** 5572
[41] Stirner T, Kirkman N T, May L, Ellis C, Kelly S M, Nicholls J E, O'Neill M and Hogg J C 2001 *J. Nanosci. Nanotech.* **1** 451
[42] Landolt-Börnstein III 17b 1982 *Semiconductors: Physics of II–VI and I–VII Compounds, Semimagnetic Semiconductors* ed. O Madelung (Berlin: Springer)
[43] Masumoto Y and Sonobe K 1997 *Phys. Rev. B* **56** 9734
[44] Zhang H and Yang B 2002 *Thin Solid Films* **418** 169
[45] Gao M Y, Lesser C, Kirstein S, Möhwald H, Rogach A L and Weller H 2000 *J. Appl. Phys.* **87** 2297
[46] Mattoussi H, Radzilowski L H, Dabbousi B O, Thomas E L, Bawendi M G and Rubner M F 1998 *J. Appl. Phys.* **83** 7965
[47] Miyata S and Nalwa H S (eds) 1997 *Organic Electroluminescent Materials and Devices* (Amsterdam: Gordon and Breach)
[48] Walukiewicz 1996 *J. Cryst. Growth* **159** 244
[49] Chernoutsan K, Dneprovskii V, Gavrilov S, Gusev V, Mulijarov E, Romanov S, Syrnicov A, Shalignina O and Zhukov E 2002 *Physica E* **15** 111
[50] Bányai L, Galbraith I, Ell C and Haug H 1987 *Phys. Rev. B* **36** 6099
[51] Ando H, Oohashi H and Kanbe H 1991 *J. Appl. Phys.* **70** 7024
[52] Sugakov V I and Vertsimakha A V 2000 *Phys. Stat. Sol. B* **217** 841
[53] Channon K and Stirner T unpublished results
[54] Tang Z, Kotov N A and Giersig M 2002 *Science* **297** 237
[55] Grundmann M, Stier O and Bimberg D 1995 *Phys. Rev. B* **52** 11969
[56] Cusack M A, Briddon P R and Jaros M 1996 *Phys. Rev. B* **54** R2300
[57] Hsu T M, Lan Y S, Chang W-H, Yeh N T and Chyi J-I 2000 *Appl. Phys. Lett.* **76** 691
[58] Marsal L, Besombes L, Tinjod F, Kheng K, Wasiela A, Gilles B, Rouvière J-L and Mariette H 2002 *J. Appl. Phys.* **91** 4936
[59] Bastard G 1988 *Wave Mechanics Applied to Semiconductor Heterostructures* (Paris: Editions de Physique)
[60] Schoos D, Mews A, Eychmüller A and Weller H 1994 *Phys. Rev. B* **49** 17072

[61] Becker W M 1988 in *Semiconductors and Semimetals* vol 25 (London: Academic Press) ch 2
[62] Dang L S, Neu G and Romestain R 1982 *Solid State Commun.* **44** 1187
[63] Chen P, Nicholls J E, Hogg J H C, Stirner T, Hagston W E, Lunn B and Ashenford D E 1995 *Phys. Rev. B* **52** 4732
[64] Stirner T 2002 *J. Chem. Phys.* **117** 6715
[65] Jamil N Y and Shaw D 1995 *Semicond. Sci. Technol.* **10** 952
[66] Mackowski S, Smith L M, Jackson H E, Heiss W, Kossut J and Karczewski G 2003 *Appl. Phys. Lett.* **83** 254
[67] Crank J 1975 *The Mathematics of Diffusion* (Oxford: Clarendon Press)
[68] Iwamura H, Saku T, Kobayashi H and Horikoshi Y 1983 *J. Appl. Phys.* **54** 2692
[69] Magari K, Okamoto M, Yasaka H, Sato K, Noguchi Y and Mikami O 1990 *IEEE Photon. Technol. Lett.* **2** 556
[70] Luttinger J M and Kohn W 1955 *Phys. Rev.* **97** 869
[71] Pidgeon C R and Brown R N 1966 *Phys. Rev.* **146** 575
[72] Bockelmann U and Bastard G 1992 *Phys. Rev. B* **45** 1688
[73] Notomi M, Okamoto M, Iwamura H and Tamamura T 1993 *Appl. Phys. Lett.* **62** 1094
[74] Lehr G, Harle V, Scholz F and Schweizer H 1996 *Appl. Phys. Lett.* **68** 2326
[75] Ils P, Gréus Ch, Forchel A, Kulakovskii V D, Gippius N A and Tikhodeev S G 1995 *Phys. Rev. B* **51** 4272
[76] Muljarov E A, Zhukov E A, Dneprovskii V S and Masumoto Y 2000 *Phys. Rev. B* **62** 7420
[77] Thornton P R 1967 *The Physics of Electroluminescent Devices* (London: Spon)
[78] Gaponik N P, Talapin D V, Rogach A L, Eychmüller E and Weller H 2002 *Nano. Lett.* **2** 803
[79] Chen L, Klar P J, Heimbrodt W, Brieler F, Froba M, Krug von Nidda H-A and Loidl A 2001 *Physica E* **10** 368
[80] Xiong Y, Xie Y, Yang J, Zhang R, Wu C and Du G 2002 *J. Mater. Chem.* **12** 3712
[81] Ostwald W 1915 in *Die Welt der Vernachlässigten Dimensionen* (Dresden: Steinkopf)
[82] Hu J, Odom T W and Lieber C M 1999 *Acc. Chem. Res.* **32** 435
[83] Darwent J R and Porter G 1981 *J. Chem. Soc. Chem. Commun.* 145
[84] Darwent J R 1981 *J. Chem. Soc. Faraday Trans.* **77** 1703
[85] Meyer M, Wallberg C, Kurihara K and Fendler J H 1984 *J. Chem. Soc. Chem. Commun.* 90
[86] Henglein A 1984 *Pure Appl. Chem.* **56** 1215
[87] Mau A, Huang C-B, Kakuta N, Bard A J, Campion A, Fox M A, White J M and Webber S E 1984 *J. Amer. Chem. Soc.* **106** 6537
[88] Ueno A, Kakuta N, Park K H, Finlayson M F, Bard A J, Campion A, Fox M A, Webber S E and White J M 1985 *J. Phys. Chem.* **89** 3828
[89] Grätzel M 1983 in *Energy Resources Through Photochemistry and Catalysis* (New York: Academic Press)
[90] Yanagida S, Yoshiya M, Shiragami T, Pac C, Mori H and Fujita H 1990 *J. Phys. Chem.* **94** 3104
[91] Rossetti R, Nakahara S and Brus L E 1983 *J. Chem. Phys.* **79** 1086
[92] Brus L E 1983 *J. Chem. Phys.* **79** 5566
[93] Brus L E 1984 *J. Chem. Phys.* **80** 4403
[94] Rossetti R, Ellison J L, Gibson J M and Brus L E 1984 *J. Chem. Phys.* **80** 4464

[95] Steigerwald M L and Brus L E 1990 *Acc. Chem. Res.* **23** 183
[96] Fojitk A, Weller H, Koch H and Henglein A 1984 *Ber. Bunsenges. Phys. Chem.* **88** 969
[97] Weller H, Fojitk A and Henglein A 1984 *Chem. Phys. Lett.* **117** 485
[98] Baral S, Fojitk A, Weller H and Henglein A 1986 *J. Amer. Chem. Soc.* **108** 375
[99] Wang Y 1991 *Acc. Chem. Res.* **24** 133
[100] Wang Y and Herron N 1987 *J. Phys. Chem.* **91** 257
[101] Wang Y and Herron N 1988 *J. Phys. Chem.* **92** 4988
[102] Herron N, Wang Y, Eddy M M, Stucky G D, Cox D E, Moller K and Bein T 1989 *J. Amer. Chem. Soc.* **111** 530
[103] Stucky G D and McDougal J E 1990 *Science* **247** 669
[104] Yoneyama H, Haga S and Yamanaka S 1989 *J. Phys. Chem.* **9** 4833
[105] Kuczynski J, Milosavljevic B H and Thomas J K 1985 *J. Phys. Chem.* **89** 2720
[106] Inman J K, Mraz A M and Weyl W A 1948 in *Solid Luminescent Materials* (New York: Wiley) p. 182
[107] Efros A L and Efros A L 1982 *Sov. Phys. Semicond.* **16** 772
[108] Ekimov A I and Onushchenko A A 1982 *Sov. Phys. Semicond.* **16** 775
[109] Henneberger F, Schmitt-Ring S and Gobel E O (eds) 1993 *Properties of Semiconductor Quantum Dots* (Berlin: Akademie Verlag)
[110] Wogon U 1997 in *Optical Properties of Semiconductor Quantum Dots* (Berlin: Springer)
[111] Gaponenko S V 1998 in *Optical Properties of Semiconductor Nanocrystals* (Cambridge: Cambridge University Press)
[112] Vogel R, Pohl K and Weller H 1991 *Chem. Phys. Lett.* **174** 241
[113] Weller H 1991 *Ber. Bunsenges. Phys. Chem.* **95** 1361
[114] Fojitk A, Weller H, Koch U and Henglein A 1984 *Ber. Bunsenges. Phys. Chem.* **88** 969
[115] Fojitk A, Weller H and Henglein A 1985 *Chem. Phys. Lett.* **117** 45
[116] Wang Y, Suna A, Mahler W and Kasowski R 1987 *J. Chem. Phys.* **87** 7315
[117] Mahler W 1988 *Inorg. Chem.* **27** 435
[118] Dannhauser T, O'Neil M, Johansson K, Whitten D and McLendon G 1986 *J. Phys. Chem.* **90** 6074
[119] Brennan J G, Siegrist T, Carroll P J, Stuczynski S M, Reynders P, Brus L E and Steigerwald M L 1990 *Chem. Mater.* **2** 403
[120] Meyer M, Wallberg C, Kurihara and Fendler J 1984 *J. Chem. Soc. Chem. Commun.* 90
[121] Steigerwald M L, Alivisatos A P, Gibson J M, Harris T D, Kortan R, Muller A J, Thayer A M, Duncan T M, Douglass D C and Brus L E 1988 *J. Amer. Chem. Soc.* **110** 3046
[122] Korton A, Hull R, Opila R, Bawendi M, Steigerwald M, Carroll P and Brus L E 1990 *J. Amer. Chem. Soc* **111** 4141
[123] Lianos P and Thomas J K 1986 *Chem. Phys. Lett.* **125** 299
[124] Lianos P and Thomas J K 1987 *J. Colloid Interface Sci.* **117** 505
[125] Robinson B H, Khan-Lodhi A N and Towey T F 1990 *J. Chem. Soc. Faraday Trans.* **86** 3757
[126] Moffit M, Vali H and Eisenberg A 1998 *Chem. Mater.* **10** 1021
[127] Ingert D, Feltin N, Levy L, Gouzerth P and Pileni M-P 1999 *Adv. Mater.* **11** 220
[128] Youn H-C, Baral S and Fendler J H 1988 *J. Phys. Chem.* **92** 6320

[129] Towey T F, Khan-Lodhi A and Robinson B H 1990 *J. Chem. Soc. Faraday Trans.* **86** 3757
[130] Murray C B, Norris D J and Bawendi, M G 1993 *J. Am. Chem. Soc.* **115** 8706
[131] Dabbousi B O, Bawendi M G, Onitsuka O and Rubner M F 1995 *Appl. Phys. Lett.* **66** 1317
[132] Murray C B, Kagan C R and Bawendi M G 1995 *Science* **270** 1335
[133] Peng X, Manna L, Yang W, Wickham J, Scher E, Kadavanich A and Alivisatos A P 2000 *Nature* **404** 59
[134] Trindale T, O'Brian P and Zhang X-M 1997 *Chem. Mater.* **9** 523
[135] Hambrock J, Birkner A and Fischer R 2001 *J. Mater. Chem.* **11** 3197
[136] Peng Z A and Peng X 2001 *J. Amer. Chem. Soc.* **123** 183
[137] Qu L and Peng X 2002 *J. Amer. Chem. Soc.* **124** 2049
[138] Nosaka Y, Yamaguchi K, Miyama H and Hayashi H 1988 *Chem. Lett.* 605
[139] Fischer Ch-H and Henglein A 1989 *J. Phys. Chem.* **93** 5578
[140] Swayambunathan V, Hayes D, Schmidt K H, Liao Y X and Meisel D 1990 *J. Amer. Chem. Soc.* **112** 3831
[141] Chemseddine A and Weller H 1993 *Ber. Bunsenges. Phys. Chem.* **97** 636
[142] Vossmeyer T, Katsikas L, Giersig M, Popovic I G, Diessner K, Chemseddine A, Eychmüller A and Weller H 1994 *J. Phys. Chem.* **98** 7665
[143] Katari B, Colvin J E and Alivisatos A P 1994 *J. Phys. Chem.* **98** 4109
[144] Rogach A L, Eychmüller A and Weller H 1998 *Macromol. Symp.* **136** 87
[145] Rogach A L, Katsikas L, Korowski A, Dandsheng Su, Eychmüller A and Weller H 1996 *Ber. Bunsenges. Phys. Chem.* **100** 1772
[146] Rogach A L, Koktysh D S, Harrison M T and M T Kotov N A 2000 *Chem. Mater.* **12** 1526
[147] Harrison M T, Kershaw S V, Burt M G, Rogach A L, Kornovski A, Eychmüller A and Weller H 2000 *Pure Appl. Chem.* **72** 295
[148] Eastoe J and Warne B 1996 *Curr. Opin. Coll. Int. Sci.* **1** 800
[149] Sagar W F C 1998 *Curr. Opin. Coll. Int. Sci.* **3** 276
[150] Hayes D, Micic O I, Nenadovic M T, Swayambunathan V and Meisel D 1989 *J. Phys. Chem.* **93** 4603
[151] Dance I G, Choy A and Scudder M L 1984 *J. Amer. Chem. Soc.* **106** 6295
[152] Herron N, Wang Y and Eckert H 1990 *J. Amer. Chem. Soc.* **111** 1322
[153] Ogata T, Hosokawa H, Oshiro T, Wada Y, Sakata T, Mori H and Yanagida S 1992 *Chem. Lett.* 1665
[154] Herron N, Suna A and Wang Y 1992 *J. Chem. Soc. Faraday Trans.* 2329
[155] Veinot J G C, Farah A A, Galloro J, Zobi F, Bell V and Pietro W J 2000 *Polyhedron* **19** 331
[156] Veinot J G C, Ginzburg M and Pietro W J 1997 *Chem. Mater.* **9** 2117
[157] Rogach A L, Kornovski A, Goa M, Eychmüller A and Weller H 1999 *J. Phys. Chem.* **103** 3065
[158] Veinot J G C, Galloro J, Pugliese L, Pestrin R and Pietro W J 1999 *Chem. Mater.* **11** 642
[159] Gaponik N, Talapin D V, Rogach A L, Hoppe K, Shevchenko E V, Kornovski A, Eychmüller A and Weller H 2002 *J. Phys. Chem.* **106** 7177
[160] Peng X, Schlamp M C, Kadavanich A V and Alivisatos A P 1997 *J. Amer. Chem. Soc.* **119** 7019
[161] Qi L, Ma J, Cheng H and Zhoa Z 1996 *Coll. Surf. Physiochem. Eng. Aspects* **111** 195

[162] Yang J, Meldrum F C and Fendler J H 1995 *J. Phys. Chem.* **99** 5500
[163] Roy U N, Mallik K and Kukreja L M 1998 *Appl. Phys. A* **67** 259
[164] Grieser F, Furlong D N, Scoberg D, Ichinose I, Kimizuka N and Kunitake T 1992 *J. Chem. Soc. Faraday Trans.* **88** 2207
[165] Schwerzel R E, Spahr K B, Kurmer J P, Wood V E and Jenkins J A 1998 *J. Phys. Chem. A* **102** 5622
[166] Hilinski E F, Lucas P A and Wang Y 1988 *J. Chem. Phys.* **89** 3435
[167] Gao M, Zhang X, Yang B, Li F and Shen J 1996 *Thin Solid Films* **284/285** 242
[168] Yao H, Takada Y and Kitamura N 1998 *Langmuir* **14** 595
[169] Guo H-Q, Liu S-M, Zhang Z-H, Chen W and Wang Z-G 1999 *Mol. Cryst. Liq. Cryst.* **337** 197
[170] Sookal K, Hanus L H, Ploen H J and Murphy C J 1998 *Adv. Mater.* **10** 1083
[171] Balogh T and Tomalia D A 1998 *J. Amer. Chem. Soc.* **120** 7355
[172] Zhao M, Sun L and Crooks R M 1998 *J. Amer. Chem. Soc.* **120** 4877
[173] Huang J, Sookal K and Murphy C J 1999 *Chem. Mater.* **11** 3595
[174] Kelly S M 2000 *Flat Panel Displays: Advanced Organic Materials RSC Materials Monograph Series* ed. J A Conner (Cambridge: Royal Society of Chemistry)
[175] O'Neill M and Kelly S M 2003 *Adv. Mater.* **15** 1135
[176] Kelly S M and O'Neill M 2004 in *Handbook of Electroluminescent Materials* ed. D R Vij (London: Institute of Physics)
[177] Tang C W and Van Slyke S A 1987 *Appl. Phys. Lett.* **51** 913
[178] Burroughs J H, Bradley D D C, Brown A R, Marks N, Mackay K, Friend R H, Burn P L and Holmes A B 1990 *Nature* **347** 539
[179] Colvin V L, Schlamp M C and Alivisatos A P 1994 *Nature* **370** 354
[180] Schlamp M C, Peng X and Alivisatos A P 1997 *J. Appl. Phys.* **82** 5837
[181] Artemyev M V *et al* 1997 *J. Appl. Phys.* **81** 6975
[182] Dabbousi B O, Bawendi M G, Onitsuka O and Rubner M F 1995 *Appl. Phys. Lett.* **66** 1316
[183] Yang Y, Xue S, Liu S, Huang J and Shen J 1996 *Appl. Phys. Lett.* **69** 377
[184] Yang Y, Xue S, Liu S, Huang J and Shen J 1997 *J. Mater. Chem.* **7** 131
[185] Huang J, Yang Y, Xue S, Yang B, Liu S and Shen J 1997 *Appl. Phys. Lett.* **70** 2335
[186] Que W, Zhou Y, Lam Y L, Chan Y C, Kam C H, Liu B, Gan L M, Chew C H, Xu C Q, Chua S J, Xu S J and Mendis F V C 1998 *Appl. Phys. Lett.* **73** 2727
[187] Gao M, Richter B and Kirstein S 1997 *Adv. Mater.* **9** 802
[188] Cassagneau T P, Sweryda-Krawiec B and Fendler J H 2000 *MRS Bull.* **25** 40
[189] Decher G 1997 *Science* **277** 1232
[190] Iler R K 1996 *J. Colloid Interface Sci.* **21** 569
[191] Gao M Y, Sun J Q, Dulkeith E, Gaponik N P, Lemmer U and Feldmann J 2002 *Langmuir* **18** 4098
[192] Lesser C, Gao M Y and Kirstein S 1999 *Mat. Sci. Eng. C* **8** 159
[193] Chen W, Grouquist D and Roark J 2002 *J. Nanosci. Nanotech.* **2** 47
[194] Gaponik N P, Talapin D V, Rogach A L and Eychmüller E 2000 *J. Mater. Chem.* **10** 2163
[195] Gaponik N P, Talapin D V and Rogach A L 1997 *Phys. Chem. Chem. Phys.* **1** 1787
[196] Harrison M, Kershaw S V, Burt M G, Eychmüller E, Rogach A L and Weller H 2000 *Mat. Sci. Eng. B* **69** 355

[197] Seufert J, Obert M, Scheibner M, Gippius N A, Bacher G, Forchel A, Passow T, Leonardi K and Hommel D 1990 *Appl. Phys. Lett.* **79** 1033

[198] Ioannou-Sougleridis V, Kamenev B, Kouvatsos D N and Nassiopoulou A G 2003 *Mat. Sci. Eng. B* **101** 324

[199] Komaba S, Fujihana K, Osaka T, Aiki S and Nakamura S 1998 *J. Electrochem. Soc.* **145** 1126

[200] Coe S, Woo W-K, Bawendi M and Bulovic V 2002 *Nature* **420** 800

Chapter 5

Alkaline earth sulphides

Virendra Shanker and Harish Chander
National Physical Laboratory, New Delhi, India

5.1 Introduction

Alkaline earth sulphides (AESs) belong to a very old group of phosphors called 'Lenard phosphors' [1]. Luminescence and related properties of AES phosphors have been reviewed from time to time [2–8]. Physical properties of these phosphors, which make them useful for device applications, are determined to a large extent by the interaction of the host lattice and free charge carriers with point defects due to impurity ions (rare earths) or lattice vacancies [9].

Fundamental properties of AES phosphors still remain relatively unknown. Almost all the work to date has been done on thin films and powders, and not on crystals, the reason being the difficulties involved in growing single crystals of these phosphors which would, otherwise, have given clear and fundamental picture about their various defect structures and optical and related properties [10]. Luminescence behaviour of AES had mainly been done on polycrystalline powder materials. Even electroluminescence (EL) studies of AES were carried out on powders [11] under a.c. excitation, probably due to the reason that preparation of thin films of these materials is difficult because of their high melting point and instability in the atmosphere. It was only recently that, due to refinement in the single crystal growth and thin film techniques, the basic and applied aspects of luminescence of these phosphors have been studied.

Thin-film devices using ZnS:Mn phosphor have been already established as flat panel displays [12, 13]. Inoguchi *et al* [14] reported a yellow emitting ZnS:Mn thin-film EL (TFEL) device with double insulating layer structure that had commercially acceptable brightness of more than 3000 cd/m^2 with a lifetime of about 10 000 h and excellent stability. However, the limitation of these ZnS:Mn-based panels is that they only emit yellow-orange light.

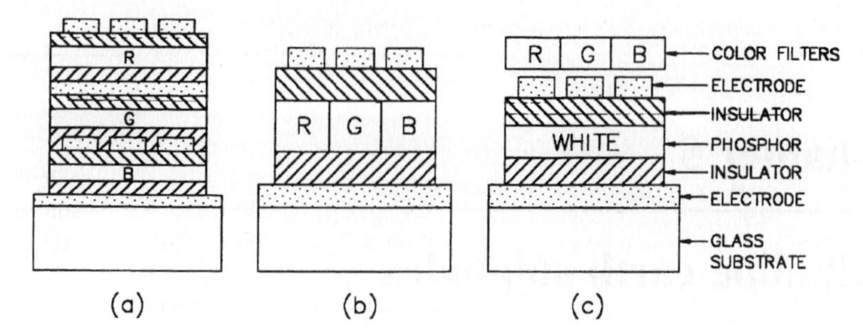

Figure 5.1. Schematic structures of multicolour TFEL panels: (a) stacked-phosphor structure; (b) patterned-phosphor structure; (c) broadband spectrum of white phosphor with patterned colour filters [12].

During recent years, phosphor research has concentrated on the development of the materials for colour TFEL displays. Colour phosphor development has been divided into two different material approaches based on two colour TFEL structures: (i) the patterned phosphor structure, and (ii) the filtered white or colour-by-white structure. In the patterned phosphor structure, three primary colour-emitting phosphors are employed in a pattern or stack form to achieve the true full colour displays. In colour-by-white structure, the phosphor component is either white-emitting or a broad band emitting one and the colours are produced by the use of filters. Broad structure of such device is shown in figure 5.1. The development of new materials based on alkaline earth sulphide thin films at this stage is note-worthy because of their emitting full colour EL.

CaS:Eu, SrS:Ce, SrS:Ce/SrS:Eu, SrS:Ce,Eu, SrS:Ce/CaS:Eu, $SrGa_2S_4$:Ce and $CaGa_2S_4$:Ce are some of the phosphors used for producing full colour emitting EL devices. Today many manufacturers are active in the field of TFEL display devices and are in commercial production of the panels for different applications. The following sections deal with various aspects of electroluminescence in alkaline earth sulphides.

5.2 Physical properties of EL emitting AES compounds

The very first requirement of an electroluminescent material is a large enough band gap so that it can emit visible light from incorporated luminescent centres with least absorption. Further, these materials should be capable of transporting high-energy electrons and must be able to stand a high electric field of the order of 10^6–10^7 V/cm. In addition, the host materials should have a suitable lattice to accommodate the activators and co-activators to form the luminescent centres that emit in the visible. For EL process to be

Table 5.1. Properties of ZnS and alkaline earth sulphides [19].

Material	ZnS	CaS	SrS	BaS
Melting point (°C)	1800–1900	2400	>2000	
Band gap (eV)	3.7	4.4	4.3	3.78
Transition type	Direct	Indirect	Indirect	Indirect
Crystal structure	Cubic zincblende or hexagonal wurtzite	Rock salt (NaCl type)	Rock salt (NaCl type)	Rock salt
Dielectric constant	8.3	9.3	9.4	11.3
Lattice constant (Å)	5.409	5.697	6.019	6.384
Ionic radius (Å)	0.74	0.99	1.13	1.35
Ionicity	0.623	>0.785	>0.785	>0.785

efficient, luminescent centre must have large cross-section for impact excitation and appropriate ionic radius and valency. These centres should be evenly distributed and stable in high field. These requirements put restrictions on the use of materials having band gap less than 2.5 eV. The materials have to be insulators with high dielectric constant. Most alkaline earth sulphides and zinc sulphide meet the above mentioned stringent criteria. The detailed properties of some these are given in table 5.1.

The crystal structure of alkaline earth sulphides is of rock-salt type whereas ZnS has a cubic structure. ZnS has a direct band gap of 3.7 eV. Sulfides of alkaline earth elements have indirect-type band structure with band gap energies in the range of 3.8–4.4 eV, sufficient for all visible radiation to pass through.

Thiogallates of alkaline earths have band gap energy in the range 4.0–4.4 eV, and were known to be efficient as cathodoluminescent materials. These have general formula (alkaline earth)Ga_2S_4 and have been found to give much needed bright blue electroluminescence. Thiogallates have the additional advantage of better chemical inertness. As compared with alkaline earth sulphides, these have relatively low melting point but higher dielectric constant ($\sim$15). Properties of important compounds of this class are given in table 5.2.

Alkaline earth sulphides have ionicity greater than 0.785 with mixed ionic and covalent bonding and are better ionic than ZnS. The ionic radii of rare earths, 1.11 Å for Ce^{3+} and 1.12 Å for Eu^{2+}, are quite close to cations of alkaline earths. In addition, the chemical nature of alkaline earths and that of rare earths is similar. Therefore, the incorporation of rare earths as luminescent centre into the lattice is comparatively easy and effective. As rare-earth elements have larger cross-section for impact

Table 5.2. Important properties of thiogallates of alkaline earths [12].

Host	Crystal structure	Lattice structure			Band gap (eV)	Melting point (°C)	Dielectric constant
		a	b	c			
$Ca–Ga_2S_4$	Orthorhombic	20.09	20.09	12.1	4.2	1150	15
$SrGa_2S_4$	Orthorhombic	20.84	20.49	12.11	4.4	1200	14
$BaGa_2S_4$	Cubic			12.66	4.1	1200	15

excitation, and alkaline earths have a good insulating nature, these make an ideal combination to devise efficient TFEL displays. Table 5.3 gives comparative ionic radii of various luminescent centres, and of important ions belonging to the host materials.

As mentioned earlier, although photoluminescence (PL) of alkaline earth sulphides has been well studied since the beginning of 20th century [1], the pace of development of these phosphors has been slow as these materials absorb moisture and emit foul smelling toxic hydrogen sulphide gas. In turn, luminescence properties are badly affected due to non-stoichiometry and impurities. These materials were explored subsequently for infrared light detectors by photostimulation to meet the requirements of defence in the face of world war threats. The difficulty of reproducing these materials slackened developments for over 30 years. Lehmann [3] gave a synthesis process by which these sulphides were not only prepared

Table 5.3. Comparison of ionic radii of luminescent centres and important host ions.

Ion	Radius Å	Ion	Radius Å
Tm^{3+}	0.869	Ga^{3+}	0.62
Er^{3+}	0.881	Zn^{2+}	0.74
Ho^{3+}	0.894	Mn^{2+}	0.80
Dy^{3+}	0.908	Ca^{2+}	0.99
Tb^{3+}	0.923	Sr^{2+}	1.13
Gd^{3+}	0.938	Ba^{2+}	1.35
Eu^{3+}	0.950	O^{2-}	1.40
Sm^{3+}	0.964	S^{2-}	1.84
Nd^{3+}	0.995	Se^{2-}	1.98
Eu^{2+}	1.090	Ag^{+}	1.15
Pr^{3+}	1.013	Cu^{+}	0.96
Ce^{3+}	1.034	Cu^{2+}	0.72

Table 5.4. Activators and co-activators in CaS and luminescence properties [3].

Activators	Coactivators[a]	Luminescence colour	Luminescence spectrum[b]	Peak (eV)	Type of decay	Decay time constant[c]
O	Nothing	Bluish-green	Band	2.53	Exponential	6.5 μs
P	**Cl**, Br	Yellow	Band	2.13	Hyperbolic	~500 μs
Sc	Cl, Br, Li	Yellowish-green	Band	2.18	—	—
Mn	Nothing	Yellow	Narrow band	2.10	Exponential	4 ms
Ni	Cu, Ag	Red to infrared	Broad band —	—	—	
Cu	**F**, **Li**, **Na**, Rb, P, Y, As	Violet to blue	Two bands	2.10	Hyperbolic	50 μs
Ga	Nothing, or Cu, Ag	Orange, red, and yellow	Broad band	—	—	—
As	F, Cl, Br	Yellowish-orange	Band	2.00	—	—
Y	F, **Cl**, Br	Bluish-white	Broad band	2.8	Hyperbolic	~200 μs
Ag	Cl, Br, Li, Na	Violet	Band	—	Hyperbolic	~1 ms
Cd	Nothing	Ultraviolet to infrared	Very broad band	—	—	—
In	Na, K	Orange	Broad band	—	—	—
Sn	**F**, Cl, Br	Green	Band	2.3	Hyperbolic	~500 μs
Sb	Nothing, or **Li**, **Na**, K	Yellowish-green	Band	2.27	Exponential	0.8 μs
La	Cl, Br, I	Bluish-white	Broad band	—	—	2.55
Au	Li, K, Cl, I	Blue to bluish-green	Two bands	—	Hyperbolic	~10 μs
Pb	**F**, **Cl**, **Br**, I, P, As, Li	Ultraviolet	Narrow band	3.40	Hyperbolic	~1 μs
Bi	**Li**, **Na**, **K**, Rb	Blue	Narrow band	2.77	Hyperbolic	~1 μs

Note: The rare earth elements are listed in table 5.5.
[a] Efficient co-activators are shown in bold letters.
[b] A spectrum changes depending on a co-activator.
[c] The period when the luminescence intensity falls to $1/e$ times the initial value.

reproducibly but were also more inert chemically. After he made many studies on these materials with large numbers of activators, researchers world over started taking interest in the activity. Phosphors of alkaline earth sulphides for cathode ray tube (CRT) applications were developed but these could never be commercialized because of established processing conditions of CRTs. The basic optical properties and band structure of these were studied after growing single crystals of these materials. Calcium sulphide has been the most explored host among alkaline earth chalcogenides. An exhaustive account of activators and co-activators incorporated and their luminescent properties is given in tables 5.4 and 5.5 [3].

Table 5.5. Rare earth activators in CaS and luminescence properties [3].

Ion	Luminescence colour	Luminescence spectrum	Type of decay curve	Decay time constants[b]
Ce^{3+}	Green	Two bands peaked at 2.10, 2.37 eV	Hyperbolic	~1 μs
Pr^{3+}	Pink to green	Lines, green, red, and infrared	Green: exponential	260 μs
Nd^{3+} [a]				
Sm^{3+}	Yellow	Lines, yellow, red, and infrared	Yellow: exponential	5 μs
Sm^{2+}	Deep red (low temperature)	Lines, green, red, and infrared	—	—
Eu^{2+}	Red	Narrow d, peaked at 1.90 eV	Hyperbolic	~1 μs
Gd^{3+}	—	Lines, ultraviolet	Exponential	1.5 ms
Tb^{3+}	Green	Lines, ultraviolet to red	Green: exponential	1.8 ms
Dy^{3+}	Yellow and bluish-green	Lines, yellow, bluish-green, and infrared	Yellow: $(1 + t/\tau)^{-1}$	150 μs
Ho^{3+}	Greenish-white	Lines, blue to infrared	Green: $(1 + t/\tau)^{-1}$	150 μs
Er^{3+}	Green	Lines, ultraviolet, green, and infrared	Green: $(1 + t/\tau)^{-1}$	370 μs
Tm^{3+}	Blue with some red	Lines, blue and red	Blue: exponential	1.05 ms
Yb^{3+}	—	Lines, infrared	—	—
Yb^{2+}	Deep red	Band, peaked at 1.66 eV	Hyperbolic	~10 μs

[a] Luminescence of Nd^{3+} is not identified.

[b] The period when the luminescence intensity falls to 1/e times the initial value. For the type expressed by $(1 + t/, \tau)^{-1}$, the time constant means τ.

5.3 Thin film electroluminescence (TFEL) of alkaline earth sulphides

Alkaline earth sulphide-based phosphors were employed for thin-film EL as ZnS:Mn could only provide orange colour, and even with the use of filters it was not possible to get true multicolours in display devices. In the 1980s, the technological environment was quite suitable for studying the feasibility of CaS/SrS-based phosphors for the EL application as rapid advances in thin-film preparation techniques had been achieved and availability of multicolour photoluminescent phosphors established. Barrow *et al* [15] were the first to report thin-film EL in alkaline earth sulphides with SrS:Ce. It showed improved blue luminance by two orders of magnitude

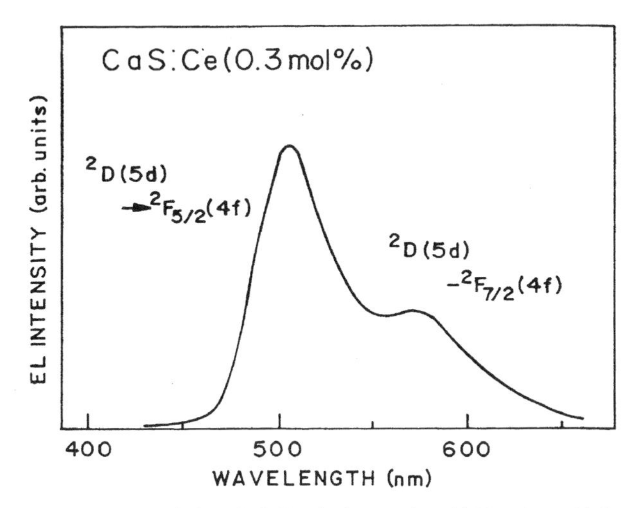

Figure 5.2. EL spectrum of the CaS:Ce device under 1 kHz sinusoidal excitation at 200 V [16].

over the existing ZnS:Tm,F device. This led to fresh interest in research work on alkaline earth sulphides as host for thin-film EL.

Almost simultaneously with the work of Barrow *et al* [15], Shanker *et al* [16] reported a double-insulated CaS:Ce thin-film EL device which was green emitting due to Ce^{3+} luminescent centres characterized by allowed parity of $5d$–$5f$ transitions. The EL spectrum of the CaS:Ce device under 1 kHz sinusoidal excitation at 200 V is shown in figure 5.2. A brightness of 500 cd/m^2 and emission efficiency of 0.11 lm/W was obtained with 5 kHz sinusoidal voltage excitation. In figure 5.3, voltage dependence of brightness (B) and EL efficiency (η) have been shown. CaS:Ce thin film was grown by co-evaporation of CaS and sulphur. The structure of the device fabricated has been shown in figure 5.4.

Kobayashi *et al* [17] extended the work on CaS:Ce and elaborated the role of sulphur co-evaporation and substrate temperature. The brightness was improved to the level of 650 cd/m^2. Substrate temperature was varied up to 400 °C. Crystal morphology of the CaS films so grown was studied by variation in x-ray diffraction patterns (figure 5.5) at different substrate temperatures and with and without sulphur co-evaporation. The changes that take place in crystal morphology as efficiency improves with different process conditions have been shown.

In 1985, researchers started visualizing full thin-film EL display devices. Tanaka *et al* [18] reported multicolour EL in alkaline earth sulphide thin film devices. They proposed and made a CaS:Eu,Cl (0.06 mol%) device for red EL. The EL spectrum consisted of a broad emission band with a peak around 650 nm due to parity allowed transitions of $4f^6(^7F)\ 5d$–$4f^7(^8S_{7/2})$

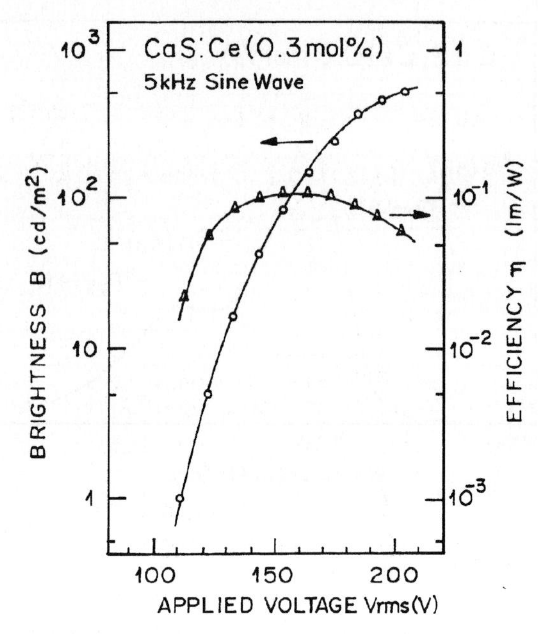

Figure 5.3. Voltage dependence of brightness (B) and EL emission efficiency (η) of a CaS:0.3%Ce device [16].

for Eu^{2+} characteristic emission. The CaS:Ce,Cl (0.1 mol%) device showed an EL spectrum with bright green EL having a strong emission band with a peak of 505 nm and a broad band with a peak around 570 nm due to $5d(^2D)$–$4f(^2F_J)$ $(J = 5/2, 7/2)$ transition for Ce^{3+}. When the lattice is

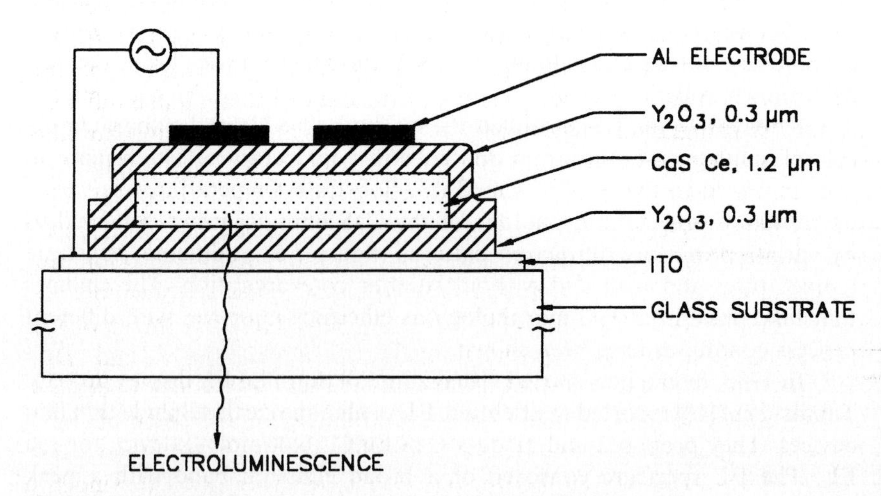

Figure 5.4. Schematic structure of the thin-film CaS:Ce EL device [16].

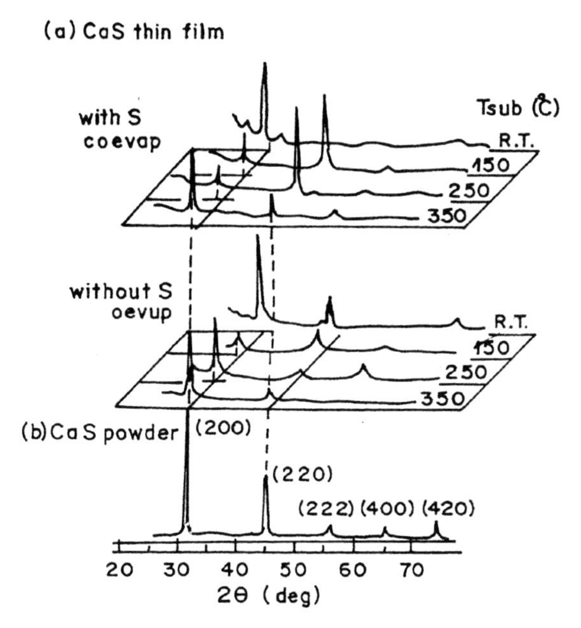

Figure 5.5. X-ray diffraction patterns of (a) CaS thin films grown at different substrate temperatures and of (b) CaS powder [17].

changed from CaS to SrS, the Ce^{3+} ion emission shifts to a shorter wavelength region by about 30 nm compared with that of the CaS:Ce,Cl devices. Due to the fact that the $5d$ state of Ce^{3+} ion is sensitive to crystal field, and its emission strongly depends on host material, devices with SrS:Ce,Cl (0.1 mol%) have a greenish blue emission. The EL spectrum shows a strong emission band with a peak at 475 nm due to the $5d(^2D)$–$4f(^2F_{5/2})$ transition overlapping the $5d(^2D)$–$4f(^2F_{7/2})$ transition of the Ce^{3+} ion. However, the device structure in the case of SrS and BaS is different from CaS. The insulating layer of Y_2O_3 used for sandwiching phosphor film has tendency to react with SrS/BaS. To prevent this, two protective layers of ZnS were employed on either side of the phosphor thin film [18]. The structure of the SrS device is shown in figure 5.6. A CaS:Tb,Cl (0.55 mol%) thin-film EL device fabricated by the group had emission due to both $^5D_3$–^7F_J $(J = 4, 5, 6)$ and $^5D_4$–^7F_J $(J = 3, 4, 5, 6)$ transitions of $(4f)^8$ electron configuration of Tb^{3+} ion. The emission from this device is bluish green due to two prominent transitions, namely $^5D_3$–$^7F_6$ and $^5D_4$–$^7F_5$ occurring at 400 and 550 nm respectively. CaS:Mn (0.52 mol%) devices showed yellowish orange emission resulting from ^4G–^6S transition of $(3d)^5$ electron configuration of Mn^{2+} ions. The EL spectrum has a broad band with a single peak at 580 nm. In the case of the BaS:Ce,Cl (0.3 mol%) device prepared by Tanaka *et al* [18], the EL spectrum shows a very broad

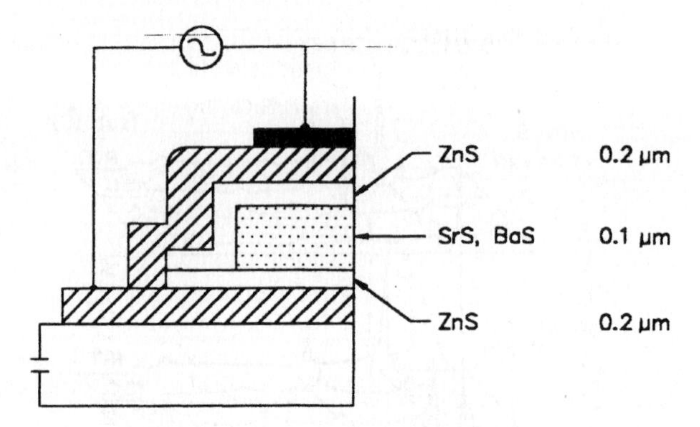

Figure 5.6. Schematic of the thin-film EL device with two ZnS buffer layers [18].

emission band at 550 nm. It was observed that the EL spectra depended on viewing angle due to the higher refractive index of BaS leading to interference. The spectra of all these devices are shown in figure 5.7. The reported efficiency and variation of brightness with applied voltage of all the above devices is given in figure 5.8.

Tanaka [19] extended the work on alkaline earth sulphides with a change in co-activator from chlorine to potassium. He reported nearly the same

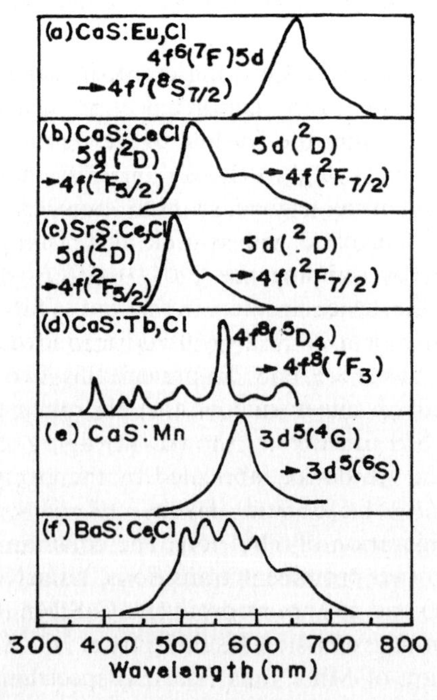

Figure 5.7. The EL spectra of the AES thin film devices. (a) CaS:Eu,Cl, (b) CaS:Ce,Cl, (c) SrS:Ce,Cl, (d) CaS:Tb,Cl, (e) CaS:Mn and (f) BaS:Ce,Cl [18].

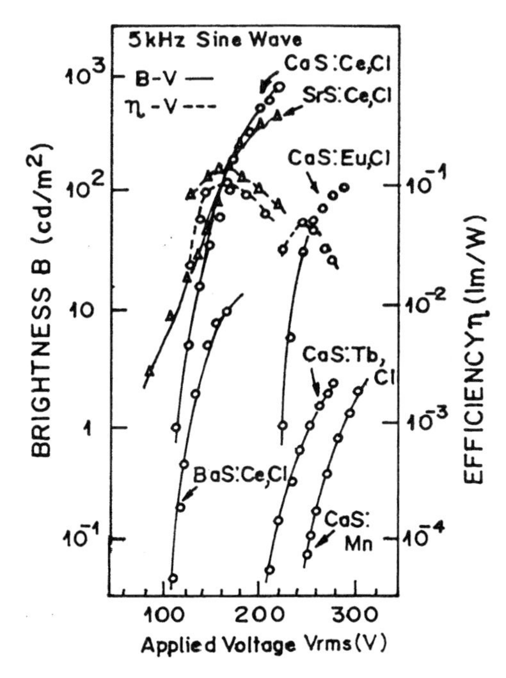

Figure 5.8. Brightness–voltage and efficiency–voltage characteristics of TFEL devices [18].

electroluminescence spectra for Eu and Ce but with periodic oscillations. He explained that these were due to interference effects due to multiple internal reflections. Figure 5.9 show the spectra obtained by him. Studies of the mechanism involved revealed that the Ce^{3+} and Eu^{2+} luminescent centres are first ionized due to impact ionization and/or hole trapping changing into Ce^{4+} and Eu^{3+} ions. Subsequently the centres capture electrons and are converted into the Ce^{3+} or the Eu^{2+} excited state, and finally the centres give rise to luminescence.

At about the same time, Gonzalez [20] reported on the mechanism of a.c. EL in SrS:Ce thin films grown by r.f. sputtering and E-beam evaporation. Studies on time-resolved behaviour of EL intensity and conduction current on the films are given in figure 5.10. Introduction of a ZnS thin film on either side of SrS:Ce film produces higher efficiency electron injection, and brightness is increased by a factor of two over the devices made without ZnS. It is claimed that SrS:Ce thin films grown by the r.f. sputtering technique at a substrate temperature of 350 °C exhibit a brightness of more than 200 cd/m^2 at 1 kHz.

Tanaka [21] reviewed the advances made in the field of alkaline earth sulphides doped with rare earths for TFEL, and the importance of preparation conditions on the characteristics and performance of the thin films produced has been emphasized. Crystallinity of the evaporated thin films has been observed to be strongly dependent on substrate temperature and, with

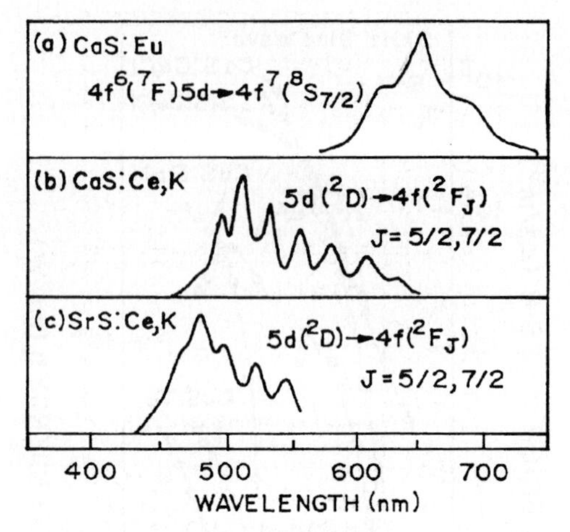

Figure 5.9. EL spectra of (a) CaS:Eu, (b) CaS:Ce,K and (c) SrS:Ce,K TFEL devices [19].

sulphur co-evaporation, the risk of oxygen impurity going into the lattice can be avoided to a large extent. This argument has been supported by Auger electron spectroscopy (AES) measurements. Depth profiles of Sr, S and O atoms in SrS thin films grown with and without sulphur co-evaporation

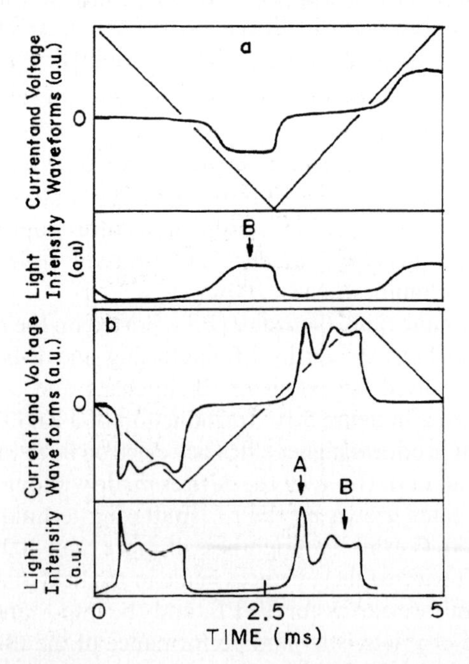

Figure 5.10. (a) Peak B in SrS:Ce device without ZnS. (b) Peak A and Peak B in SrS:Ce device with ZnS layers [20].

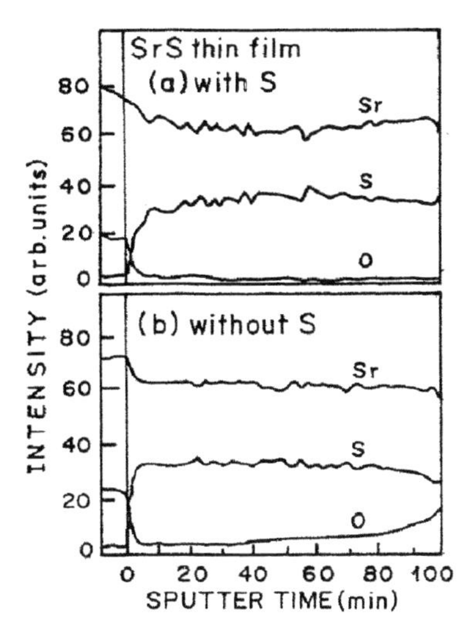

Figure 5.11. Depth profile of Sr, S and O atoms from AES measurements for SrS thin films grown, (a) with and (b) without sulphur co-evaporation [21].

are shown in figure 5.11, which depicts reduction of oxygen concentration with sulphur co-evaporation.

The details of the work on two white emitting TFEL devices based on SrS have also been reported by Tanaka [21]: (1) The SrS:Ce,K,Eu thin film device has EL spectra with emission bands originating from the Ce^{3+} and the Eu^{2+} centres peaking at 480 and 610 nm respectively. Figure 5.12

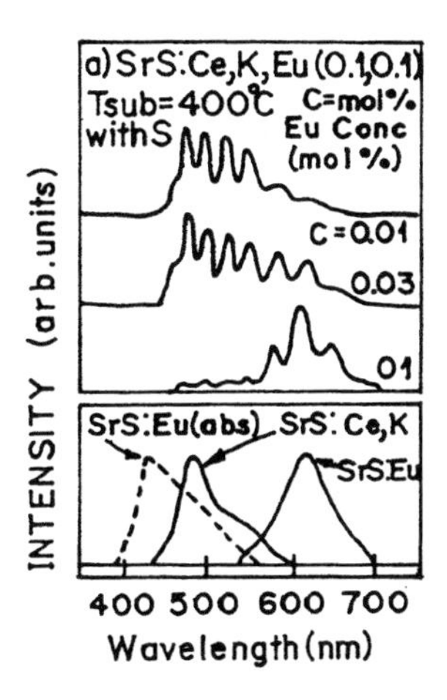

Figure 5.12. (a) EL spectra of SrS:Ce,K,Eu TFEL devices with different Eu concentrations, while Ce concentration was kept constant at 0.1 mol%. (b) PL spectra of SrS:Ce,K and SrS:Eu phosphors (solid lines) and absorption spectrum of SrS:Eu phosphors (dashed lines) [21].

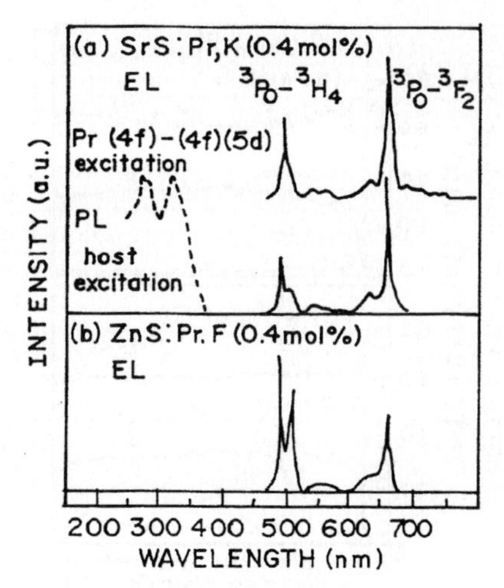

Figure 5.13. (a) PL spectrum of a SrS:Pr,K TFEL device. PL and PL-excitation spectra (dashed line) of SrS:Pr,K thin film are shown. (b) EL spectrum of a ZnS:Pr,K TFEL device [21].

shows EL spectra of the device with varying concentrations of Eu, keeping Ce concentration constant at 0.1%, and PL spectra of SrS:Ce,K and SrS:Eu, and absorption spectrum of SrS:Eu phosphors. It shows how much change in emission colour can be obtained by changing concentration of Eu. (2) In the case of a second device based on SrS:Pr,K, the EL spectra consist of two groups of emission lines. The band with a peak around 490 nm is due to ${}^3P_0$–${}^3H_4$ transition and the band of 660 nm is due to ${}^3P_0$–${}^3F_2$ transition of $(4f)^2$ electron configuration in Pr^{3+} ions. Figure 5.13 shows EL and PL spectra of a SrS:Pr,K device confirming the reported transitions for PL and EL, and the spectrum of ZnS:Pr,F is given for comparison.

The EL excitation mechanism of these devices has been discussed [21] on the basis of the band structure of alkaline earth sulphides and the energy levels of the luminescence centres. Excitation spectra of photo-induced transferred charge and excitation spectra of PL, along with the PL spectra of different TFEL devices, are presented in figure 5.14. The excitation spectra of photo-induced transferred charge of thin-film devices with different rare earth activators have been compared with the PL and PL-excitation spectra [21]. It has been concluded that the luminescent centres are ionized when the centres are excited to the $(4f)^{n-1}(5d)$ excited state.

Like Gonzalez [20], Müller *et al* [22] have also stressed the importance of growing alkaline earth sulphide films by the sputtering technique, and reported efficient ZnS-like EL efficiency. The structure of the devices fabricated has been glass, ITO conducting transparent film, (Al+Ta) oxide, silicon oxide, CaS:Eu,F,Cu,Br, silicon oxide and (Al+Ta) oxide. Efficiencies

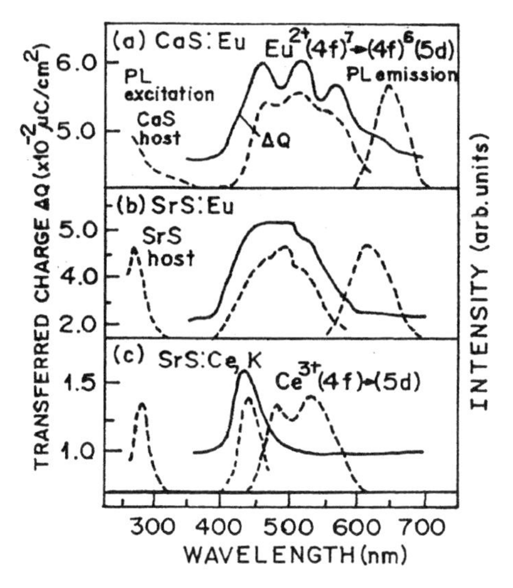

Figure 5.14. Excitation spectra of photo-induced transferred charge of (a) CaS:Eu, (b) SrS:Eu, and (c) SrS:Ce,K TFEL devices. PL and PL-excitation spectra also shown [21].

in the range 0.3–0.6 lm/W have been reported. These have been due to loss-free acceleration of electrons as in the case of ZnS:Mn. The most interesting observation has been the time-resolved spectroscopy of emission spectra of a CaS:Eu device in the time intervals around the pulse edges (figure 5.15). Leading and trailing edge emissions differed significantly, and the effect of polarity change was also quite evident. The explanation is that different centres are involved during the various stages of a.c. field pulse.

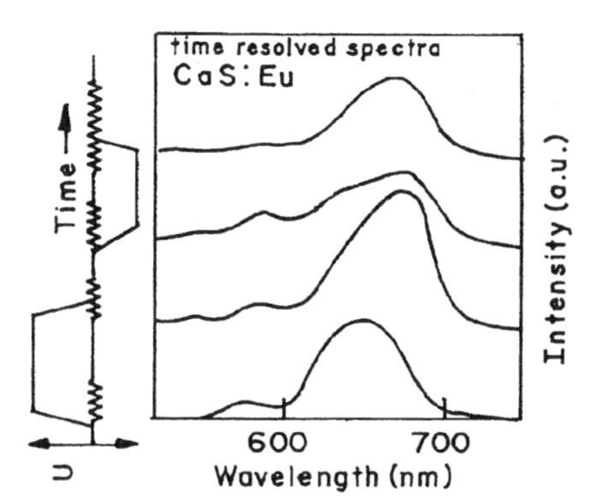

Figure 5.15. Excitation spectra of a CaS:Eu ELD sample, measured during the cross-hatched time intervals of the applied voltage, schematically given at the left side. The zeros of the spectra have been shifted to the parts of the waveform where they were taken [22].

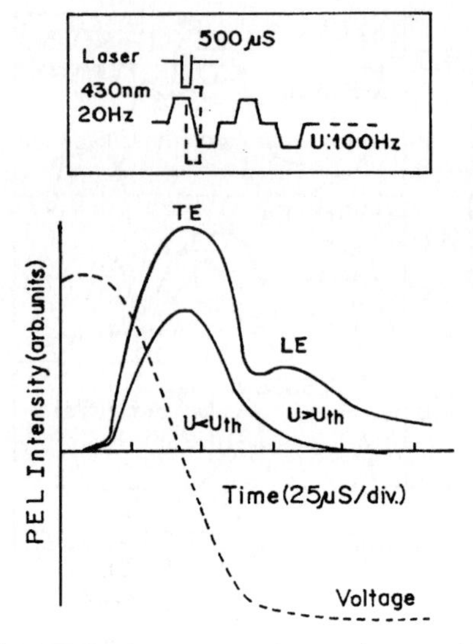

Figure 5.16. Time-resolved PEL response at voltages above and below the EL threshold (TE: trailing edge; LE: leading edge emission) [23].

The observation is important in understanding the phenomena of thin-film electroluminescence.

Towards the understanding of mechanism of TFEL, Troppenz *et al* [23] have reported on the trailing edge phenomena in SrS:$CeCl_3$ TFEL devices. The devices in conventional and multilayer structures were investigated employing time and voltage dependent PL measurements with laser excitation. Interestingly, it was found that photo-induced transferred charge, and also the PEL (photo-induced electroluminescence) is observed quite significantly at much lower voltages than EL threshold voltage at the falling region of voltage pulse, i.e. the trailing-edge (figure 5.16). It has been concluded that the ionization of Ce^{3+} luminescent centres was prerequisite. Further, the PL intensity is seen to decrease with the increased applied voltage. So much so that the PL intensity drops to 65–70% of the value as compared with the value at EL threshold voltage. The PEL was interpreted as a partial regain of the quenched PL. It was proposed that the quenching is caused due to weakly accelerated electrons.

Heikkinen *et al* [24] reported an extensive study on the x-ray photoelectron spectroscopy (XPS) measurements of SrS:Ce TFEL devices. Cerium-doped SrS thin films were analysed using XPS with the aim of obtaining the most efficient doping parameters. Atomic layer epitaxy (ALE) technique was used to deposit the films and β-diketonate complexes of the metals and H_2S were employed as precursors with soda lime glass as substrate. The study was supported with x-ray diffraction (XRD) and

X-ray fluorescence spectroscopy measurements. It was observed that cerium had gone in 3+ form in the lattice. Strontium sulphide was found to be the major phase in the films though traces of strontium as sulphite, carbonate and hydroxide were also present. XRD measurements confirmed that the films were well crystallized cubic SrS.

5.4 Full colour EL phophors

The development of a high efficiency EL phosphor is a combination of many variables at the optimized level. Proper interaction among the host lattice and the rare earth, or transition metal ion dopant, is of utmost importance. Luminance of a device and the emitted colour is the result of this very interaction. To take care of non-radiative losses, and to minimize the concentration of traps, co-activators have to be selected judiciously with an eye on charge compensation. The choice of top and bottom insulating film also helps to increase the performance of the devices produced. Again, luminance output is critically dependent on crystallinity of the thin film. The best films for the purpose are those with columnar grains of about 2 μm, but the films normally fabricated have thicknesses in the range 500–1000 nm. Columnar growth of films facilitates excitation of hot electrons and a low-loss passage to emitted light. The cost and energy economy of processing dictates a low-temperature operation all along the fabrication steps, as costlier substrates are required for higher temperatures. In this direction, some new phosphors based on SrS have been tried.

Summers *et al* [25] have reviewed the recent progress in the development of full colour electroluminescent (FCEL) phosphors with SrS as host. First in the series was SrS:Cu. The work on phosphor had been in progress since 1985. A real break-through was reported by Tong *et al* [26] where an *in situ* vacuum annealing under sulphur flux reduced the native defect density of the device. It greatly improved the morphology and EL intensity. The samples were made at 600 °C and annealed for 30 min at 650 °C under a 2.5 sccm (N_2 equivalent) *t*-butyl mercaptan (sulphur vapour) flow. Blue-emitting devices yielded 40 cd/m^2 ($V_{th} = 40$ and 1 kHz) and colour coordinates of $x = 0.20$ and $y = 0.32$. Figure 5.17 shows SEM micrographs of SrS:Cu thin films depicting the effect of annealing. The report regarding SrS:Mn,Cu and SrS:Cu,Ag also showed much better efficiency. Copper is seen to be playing an important role of flux and sensitizing centre. A greatly improved SrS:Mn,Cu device with an EL performance of 70 cd/m^2 ($V_{th} = 40$, 60 Hz) and an efficiency of 0.43 lm/W, was obtained for a 400 °C deposition, and a post-deposition *in situ* annealing in vacuum. The SrS:Cu,Ag device gave a luminance of 27 cd/m^2 and an efficiency in the range 0.25–0.30 lm/W with purer blue emission of coordinates $x = 0.17$ and $y = 0.15$.

Figure 5.17 SEM micrographs of SrS:Cu thin films: (a) as-grown, (b) vacuum annealed at 650 °C under a *t*-BuSH flow of 2.5 sccm, and (c) a cross-sectional view of annealed SrS:Cu [26].

The effect of external light irradiation on EL properties of alkaline earth sulphides doped with rare earths have been the subject matter of studies to understand the mechanism of generation of dynamic space charge. The report of Tanaka *et al* [27] describes SrS thin films activated with Ce, Pr, Nd, Sm, Gd, Tb, Dy, Ho, Er and Tm. Electroluminescence has been measured for all these films with 340 nm light irradiation having 0.2 mW power, and without any irradiation. EL spectra of SrS activated with different rare earths are shown in figure 5.18. The ratio of change of dynamic space charge due to the light irradiation to the space charge without irradiation for the activators is given in figure 5.19. It has been found that the photo-induced space charge and dynamic space charge is observed with all activators and it seems that it is due to ionization of deep traps in the interfaces between the

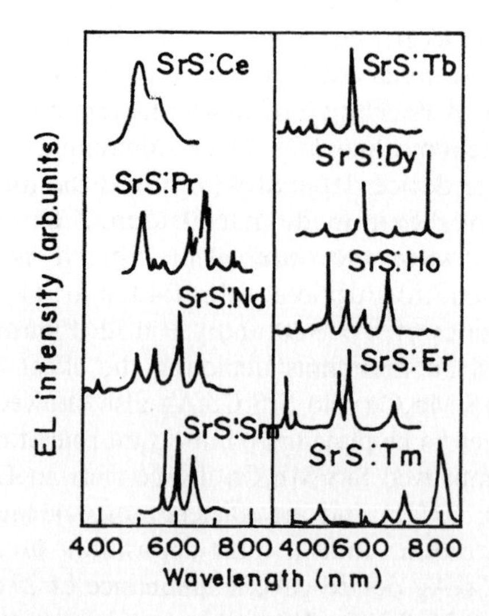

Figure 5.18. EL spectra of SrS activated with Ce, Pr, Nd, Sm, Tb, Dy, Ho, Er, and Tm at room temperature [27].

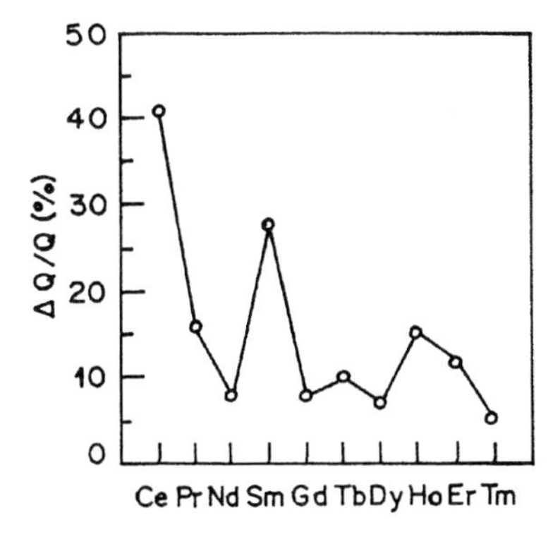

Figure 5.19. Change ratio of the dynamic space charge caused by the external light irradiation in the TFEL devices of SrS activated with Ce, Pr, Nd, Sm, Gd, Tb, Dy, Ho, Er, and Tm [27].

ZnS buffer layer and phosphor layer, or the electron generation in the buffer layers, or both.

The work on TFEL display devices was slowed down because of unavailability of an efficient blue-emitting phosphor material. Cerium-doped strontium gallium sulphide ($SrGa_2S_4$:Ce) and cerium-doped calcium gallium sulphide ($CaGa_2S_4$:Ce) have shown promise. Figure 5.20 shows the spectra of $SrGa_2S_4$:Ce, $CaGa_2S_4$:Ce and $BaGa_2S_4$:Ce [28].

Braunger *et al* [29] have deposited $SrGa_2S_4$:$CeCl_3$ phosphor on a substrate heated at 600 °C by thermal co-evaporation of the three elements of the host material with $CeCl_3$ as the dopant source and LiF as the growth enhancing material. The technique has been known as the reactive multi-source PVD technique. It has been reported that the devices made had highest luminous efficiency. LiF helps in the growth of stoichiometric polycrystalline films, even though the vapours are not in proper ratio, with very large grain diameters compared with film thickness. Thus, LiF has a sort of self-stabilizing effect on the composition of the films. The amount of strontium condensation is related to the film thickness. The resulting devices were analysed by emission spectrum and by studying external charge generated by applied voltage. Transient measurement, including the bridge method, have also been reported.

Another detailed report on $SrGa_2S_4$:Ce has been from Tanaka *et al* [30]. The report describes Ce-doping effects on blue emission for $SrGa_2S_4$:Ce electroluminescent thin films grown by molecular beam epitaxy. The measurements on EL spectra, EL luminance, PL spectra, x-ray diffraction (XRD) intensity and XRD rocking curve were made with varying Ce concentration from 0.87 to 8.37 mol%. For EL brightness, 2.4% Ce is the optimum. The Ce spectrum is only slightly affected by change in concentration of Ce. This insensitivity to Ce-doping may be due to the ionic nature of the Sr site and

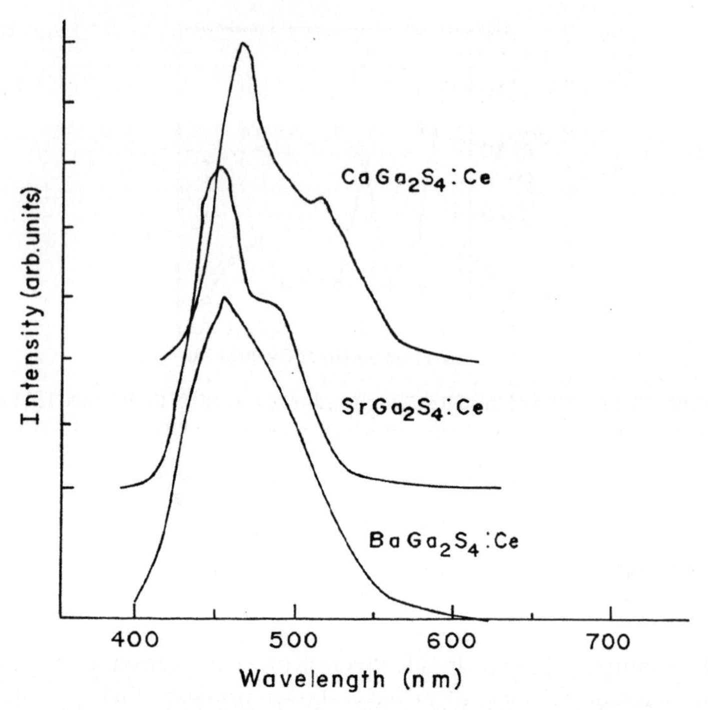

Figure 5.20. EL spectral distributions from cerium-doped MGa_2S_4 where M is Ca, Sr or Ba. The emission near 450 nm results in a blue colour [28].

the weak Ce–Ce interaction originating from $SrGa_2S_4$ crystal structure. EL and PL spectra of $SrGa_2S_4$:Ce films with Ce varying concentration are shown in figure 5.21. EL luminance (figure 5.22(a)), XRD intensity and FWHM of rocking curve of (422) peak versus Ce (figure 5.22(b)) doping concentration are shown in figure 5.22.

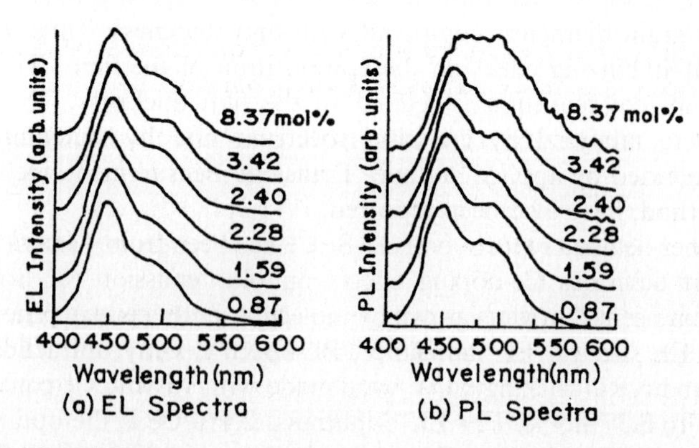

Figure 5.21. EL and PL spectra of $SrGa_2S_4$:Ce thin films grown by MBE with Ce-doping concentrations from 0.87 to 8.37 mol% [30].

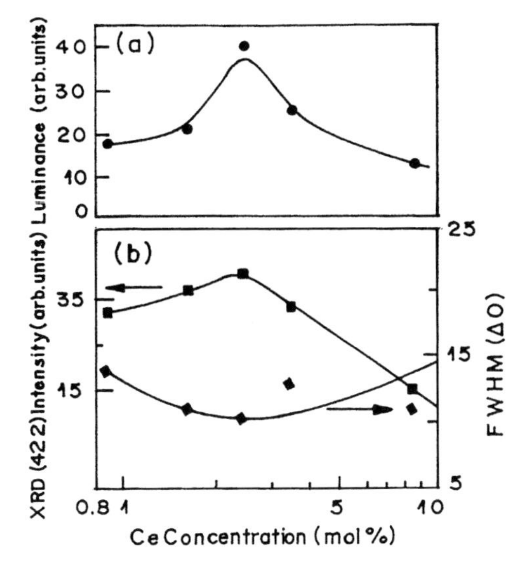

Figure 5.22. (a) EL luminance, (b) XRD intensity and FWHM of rocking curve of (422) peak versus Ce-doping concentration [30].

The reports presented above are by no means exhaustive or can cover the work of all the researchers in the field. It is only a small attempt to say how the field of alkaline earth sulphides has contributed to the science and technology of thin-film electroluminescent display devices. This work will remain incomplete without a summary of the phosphors that have made it to the top and are being used in commercial thin-film EL devices.

5.5 Phosphors of commercial TFEL display devices

Recently, there have been some good reports on the technological status of thin film electroluminescent display devices, namely

1. 'The structure, device physics and material properties of thin film electroluminescent displays' by Philip D Rack and Paul H Holloway published in *Materials Science and Engineering* 1998 **R21** 171–219.
2. The book *Electroluminescent Displays* by Y A Ono published by World Scientific, Singapore, 1995.
3. 'The present and future prospects of electroluminescent phosphors' by Hiroshi Kobayashi and Shosaku Tanaka published in *FIC*, DP 95.
4. 'Electroluminescent Displays' by Christopher N King published as white paper by Planner Systems Inc, USA, in August 2001.
5. *The Quest for the Ideal Inorganic EL Phosphors* by Xingwei Wu published by iFire Technology Inc, Canada, in August 2001.

5.6 Phosphors for monochrome displays

Monochrome thin-film EL displays are available basically in orange-yellow emission colour. This is primarily so because ZnS:Mn is the most efficient luminescent material for TFEL and its peak emission is orange-yellow. Luminescence efficiency is greater than 5 lm/W as compared with the second highest for ZnS:Mn/SrS:Ce multilayer composite, ~1.5 lm/W. The phosphor is the ideal one as the electronic configuration of Zn has a fully occupied $(3d)^{10}$ shell and the Mn atom has a half occupied $(3d)^5$ shell. Hence, Zn^{2+} and Mn^{2+} ions have similar chemical natures, so solid solutions of ZnS and MnS are easily and properly made. The Mn $(3d)^5$ level, which is the ground state of *d–d* transitions of Mn luminescent centres, lies nearly 3 eV below the valence band maximum in the ZnS lattice, so the configuration of Mn^{2+} ions is quite localized and isolated from band electrons. This explains the efficient excitation of Mn^{2+} centres in ZnS via direct impact. The optimum concentration of Mn in ZnS is ~1 mol%.

5.7 Phosphors for colour displays

There are two basic approaches to fabricating full colour thin-film EL, as discussed in section 5.1. Three primary colours (blue, green and red) or white emitting phosphors are required. These will be discussed in the following.

5.7.1 Blue-light emitting phosphors

Early candidates for the efficient blue emitting phosphor, with appropriate colour coordinates of $x = 0.15$ and $y = 0.07$ which has been most difficult to achieve are SrS:Ce, SrS:Ce,Ag, $SrGa_2S_4$:Ce, $CaGa_2S_4$:Ce, $Sr_{0.5}Ca_{0.5}Ga_2S_4$:Ce and ZnS:Tm,F. All these were good candidates and paved the way for better and still better ones. Leading the list at present are SrS:Cu with Ag or Mn as sensitizers and BaAlS:Eu patented by iFire Technology Inc, Canada. It is claimed that SrS:Cu has an efficiency of 0.2 lm/W with coordinates of $x = 0.17$ and $y = 0.16$. For $BaAl_2S_4$:Eu, the coordinates reported are $x = 0.14$, $y = 0.06$ which happen to be the best so far.

5.7.2 Green-light emitting phosphors

Terbium-doped ZnS is the most efficient thin-film EL phosphor and has a luminous efficiency of better than 1 lm/W and a luminance of 100 nits has been reported at 60 Hz drive frequency for sputtered films of ZnS:TbOF. It has colour coordinates of $x = 0.31$ and $y = 0.60$ and emission with a sharp peak at 545 nm. SrS:Ce with a green filter can do the job with coordinates of $x = 0.28$, $y = 0.53$ and luminance of 110 nits. There is a claim for a better green emitting (Zn/Mg)S:Mn phosphor, again by iFire Technology Inc, Canada. It is reported that the phosphor provides an overall gain in brightness and efficiency of a full-colour display, as well as chromaticity.

5.7.3 Red-light emitting phosphors

The best phosphor available for red emission in thin-film EL is ZnS:Mn with a filter. The filter used is a thin film of CdS_xSe_{1-x}. It has a high luminance of 70 cd/m^2 at 40 V above threshold at 60 Hz drive frequency and a luminescence efficiency of 0.8 lm/W. The other phosphors which have been tried are CaS:Eu, CaS:Eu,F, CaS_xSe_{1-x}:Eu and ZnS:Sm,F. Some of these have been used directly, without filter, but efficiency has been lower, by as much as a factor of four.

5.7.4 White-emitting phosphors

White-emitting phosphors are required to create a full colour display by the technique known as 'colour-by-white'. In this only one set/one white emitting phosphor thin film is used and different coloured filters are used for a gamut of colours. Due to this reason, it is a preferred technique. Early developments in white phosphor based on alkaline earth sulphide doped with rare earths have been SrS:Pr,K, SrS:Ce,Eu, SrS:Ce/SrS:Eu and SrS:Ce/CaS:Eu. These could not find commercial acceptance as the luminance level was far too low. The most efficient combination for the purpose has been a set of alternate layer of SrS:Ce and ZnS:Mn. Various groups have fabricated this multilayer stack of the white phosphor. Planar International has reported a best output of 470 nits at 60 Hz drive frequency using the atomic layer epitaxy process to deposit the layers of SrS:Ce/ZnS:Mn phosphors. A SrS:Cu and ZnS:Mn combination has also been used with luminance of 240 nits.

A system of active-matrix EL (AMEL) developed by Planner Systems Inc in August 2001 deserves a mention here as it might change the outlook of display systems in the future. This is a miniature high-resolution display for head-mounted and personal viewer applications. The display diagonal for these applications ranges from 7 to 25 mm. To project the image of such displays, the use of compact optics is required. These devices have high luminance, low weight, compact size and a high resolution of ~2000 lines per inch with high contrast. The devices are fabricated on a silicon wafer substrate using the inverted EL structure with a transparent ITO as the top electrode. The lower electrode is the top of metallization layer of the silicon IC to be used for addressing electronics. The film deposition technique employed for such devices is atomic layer epitaxy (ALE), only because it suits the other operations of the process.

References

[1] Lenard P, Lenard P, Schmidt F and Tomascheck R 1928 *Handbuch der Experimental Physics* vol. 23 (Leipzig: Akademisch Verlags)
[2] Pringsheim P 1949 *Fluorescence and Phosphorescence* (New York: Interscience) p. 595
[3] Lehmann W 1972 *J. Lumin.* **5** 87
[4] Rao R P J 1986 *Mater. Sci. Lett.* **2** 3357
[5] Pandey R and Sivaraman S 1991 *J. Phys. Chem. Solids* **52** 211
[6] Ghosh P K and Ray B 1992 *Prog. Cryst. Growth Charact.* **25** 1
[7] Singh N and Vij D R 1994 *J. Mater. Sci.* **29** 4941
[8] Singh N and Vij D R 1998 *Luminescence and Related Properties of II–VI Semiconductors* ed. D R Vij and N Singh (Nova Science Publishers) ch. 7
[9] Pandey R, Kunz A B and Vail J M 1988 *J. Mater. Res.* **3** 1362
[10] Rennie J, Nakazawa E, and Koda Y 1990 *Jpn. J. Appl. Phys.* **29** 509
[11] Vecht A, Waite M, Higton M and Ellis R 1981 *J. Lumin.* **24/24** 917
[12] Ono Y A 1995 *Electroluminescent Displays* (Singapore: World Scientific)
[13] Gumlich H E, Zeinert A and Mauch R 1998 *Luminescence of Solids* ed. D R Vij (New York: Plenum Press) ch. 6
[14] Inoguchi Y, Takeda M, Kakihara Y, Nakata Y and Yoshida M 1974 *Digest of 1974 SID International Symp.* 84
[15] Barrow W A, Coovert R E and King C N 1984 *Digest of 1984 SID International Symp.* 249
[16] Shanker V, Tanaka S, Shiiki M, Deguchi H, Kobayashi H and Sasakura H 1984 *Appl. Phys. Lett.* **45** 960
[17] Kobayashi H, Tanaka S, Shanker V, Shiiki M and Deguchi H 1985 *J. Crystal Growth* **72** 559
[18] Tanaka S, Shanker V, Shiiki M, Deguchi H and Kobayashi H Proc. 1985 *SID* **26** 255; 1985 *Digest of 1985 SID International Symposium* 218
[19] Tanaka S, 1988 *J. Lumin.* **40-41** 20
[20] Gonzalez C, 1988 *J. Lumin.* **40-41** 771
[21] Tanaka S, 1990 *J. Crystal Growth* **101** 958
[22] Müller G O, Mach R and Selle B 1990 *J. Crystal Growth* **101** 999
[23] Troppenz U, Hûttl B, Velthaus K O and Mauch R H 1994 *J. Crystal Growth* **138** 1017
[24] Heikkinen H, Johansson L-S, Nykänen E and Niinistö L 1998 *Appl. Surf. Sci.* **133** 205
[25] Summers C J, Wagner B K, Tong W, Park W, Chaichimansour M and Xin Y B 2000 *J. Crystal Growth* **214–215** 918
[26] Tong W, Xin Y B, Park W and Summers C J 1999 *Appl. Phys. Lett.* **74** 1379
[27] Tanaka I, Izumi Y, Tanaka K, Inoue Y and Okamoto S 2000 *J. Luminescence* **87–89** 1189
[28] Sun S S, Tuenge R Y, Kane J and Ling M 1994 *J. Electrochem. Soc.* **141** 2877
[29] Braunger D, Oberacker Y A and Schock H-W 1996 *Inorganic and Organic Electroluminescence/EL* **96** Berlin, eds R H Mauch and H E Gumlich (Berlin: Wissenschaft und Technik Verlag) p. 291
[30] Tanaka K, Inoue Y, Okammoto S, Kobayashi K, Tsuchiya Y and Takizawa K 1996 *Inorganic and Organic Electroluminescence/EL* **96** Berlin, eds R H Mauch and H E Gumlich (Berlin: Wissenschaft und Technik Verlag) p. 325

Chapter 6

Zinc oxide

Shashi Bhushan
School of Studies in Physics, Pt. Ravishankar Shukla University, Raipur, India

6.1 Introduction

Zinc oxide (ZnO) is a wide band gap material and presents an interesting system because of its multidirectional applications, e.g. opto-electronic displays [1–3] electrophotography [1], photovoltaic conversion [4] and electro-acoustic transducers [5] etc. Its band gap is 3.2 eV and gives multi-emissions ranging from ultraviolet to infrared regions. It exists in wurtzite structure in which each ion is connected with four counter ions in the tetrahedral bond of sp^3 hybridization [6]. Although tremendous efforts have been made in investigating the origin of different emission peaks/bands emitted by this system, controversies still prevail. This material has been successfully produced in powder, crystal, thin-film and nanocrystalline forms. The present chapter is mainly concentrated on various investigations made on photoluminescence (PL) and electroluminescence (EL) properties. Some important results of some other luminescence phenomena like cathodoluminescence (CL) and thermoluminescence (TL) are also discussed. Preparational methods of materials in different forms along with applications and future scope are also covered.

6.2 Preparation methods

6.2.1 Powders

Lehmann [7] prepared luminescent ZnO phosphors under varied preparative conditions, which gave different emissions, e.g. under reducing condition green emission was found which was enhanced in the presence of sulphur and yellow-orange and red-infrared emissions appeared in the presence of Se and ammonia respectively. Joshi and Kumar [8] prepared ZnO by firing it at 1000 °C for 1 h in air and reducing atmosphere. Lauer [9] heated the high purity ZnO at temperatures between 900 and 1000 °C in an O_2 ambient. When second heat treatment at 1200 °C for 2 h under reducing or inert gas atmosphere was given to ZnO added to various chemicals like Ga_2O_3 and $MnSO_4$, blended and heated at 1400 °C for 3 h, materials were produced suitable for low voltage cathode ray excitation [10]. Jechev *et al* [11] prepared ZnO powder samples by thermal decomposition of ZnC_2O_4 in the presence of oxygen. Bhushan and co-workers [12, 13] prepared ZnO phosphors either by firing ZnO alone or when mixed with impurities at 1050 °C under different atmospheres (O_2, N_2 or vacuum) for 1 h or by burning highly pure zinc metal (99.999%) at 1150 °C for 1 h under the continuous flow of pure oxygen. Wada *et al* [14] obtained ZnO by firing at ~900 °C the finely powdered material under different degrees of reduction. Quang *et al* [15] prepared the Eu-doped ZnO samples in the presence of group V ions (e.g. N) and found sharp transitions in such systems. Singh and Mohan [16] reported the preparation of (ZnO,ZnSe):Cu,Cl phosphors at different concentrations of ZnO and ZnSe by firing at 1000 °C for 30 min in N_2 atmosphere. Bhushan and Diwan [17] prepared (ZnS,ZnO):Cu,Cl electroluminors by firing a mixture of ZnS, ZnO, $CuSO_4$, NH_4Cl and sulphur at 1000 °C for 2.5 h, which showed d.c. EL.

6.2.2 Pellets

Miralles *et al* [18] prepared ZnO varistors by mixing oxides of Zn, Bi, Mn and Co and pressing them into discs followed by sintering at 1250 °C for 1 h. Later, these workers [19] treated the mixture of ZnO with Bi, Mn and Co with acetone and after drying and calcining, pressed them into discs, and then sintered in air up to 1200 °C for 2 h. The presence of Bi acted as a seed and favoured the formation of grains with larger size. Bi- and Mn-doped ZnO ceramics were reported by Garcia *et al* [20] by pressing pure ZnO powder/ZnO powder containing oxides of Bi and Mn and sintering in air for 1 h at 1200 °C. Fernandez *et al* [21] prepared ZnO:Mn ceramics by first mixing high purity ZnO powder with CO_3Mn or Mn_3O_4 in a mortar for 1 h, pressing into discs at 9×10^3 kg cm^{-2} and finally sintering in an oxygen atmosphere for 2 h at 1225 °C. Kossanyi *et al* [22] and Kouyate *et al* [23] obtained pellets by mixing a small amount of ethanol, pressing at 14 tons/cm^2 and then firing at 1100 °C for 5 h at atmospheric pressure. Hsieh and Su [24] prepared ZrO_2 doped with ZnO pellets by firing at 1100 °C.

6.2.3 Thin films

Takata *et al* [25] and Nanto *et al* [26] used an r.f. sputtering technique with a ZnO target (99.999%) in an 80% Ar/20% O_2 gas mixture. Substrates used were Al, quartz, glass, p^+-Si or n^+-Si or sapphire. Nanto *et al* [27] prepared anodized films on high purity zinc plates or on zinc films deposited on conducting solids by using vacuum evaporation and plating techniques. KOH was used as electrolyte and platinum as cathode in this process. Oritz *et al* [28] and Falcony *et al* [29] used the spray pyrolysis technique to prepare Tb-doped ZnO films on Pyrex glass slides by using zinc acetate in a mixture of 3 parts isopropyl alcohol and 1 part de-ionized water and air as a carrier gas. Oritz *et al* [30] also prepared Li-doped ZnO films by the spray pyrolysis technique. The deposited films showed hexagonal polycrystalline structure and the optical transmission depended on substrate temperature which showed an absorption edge shifting to longer wavelengths with higher substrate temperature. Ristov *et al* [31] used the chemical deposition technique to prepare ZnO films through the decomposition of complexes $(NH_4)_2ZnO_2$ and Na_2ZnO_2 etc. in hot water at a temperature between 95 and 100 °C. Glass, quartz and mica were used as substrates. Bala and Firszt [32] prepared ZnSe–ZnO diodes by anodic oxidation of ZnSe crystals. They prepared undoped and doped with Ag, In and Ga films by this method. Sands *et al* [33] prepared $ZnS_{1-x}O_x$:Mn thin-film materials by r.f. sputtering using a solid target of ZnS in Ar/O_2 mixtures. Wenas *et al* [34] prepared ZnO films by first containing diethyl zinc and H_2O in bubblers and then keeping in temperature controlled baths. Glass substrates were used for deposition of films by varying temperature from 100 to 300 °C and keeping the total pressure of 6 torr. Protective ZnO coating was made on chemically deposited CdS/CdSe photoelectrodes by the sol-gel technique by Rincon *et al* [35]. For this, a 14% w/w solution of zinc acetate was prepared in pure methanol. The deposition rate could be varied depending upon the withdrawal velocity of the substrate. After deposition the substrate was heated at ~400 °C to desorb solvents from the sample and to transform the adsorbed zinc complex into ZnO. The thickness of ZnO coating varied from ~350 to 2250 Å for 1 to 10 immersions. Katayama and Izaki [36] prepared ZnO films 1 μm thick by cathodic deposition from 0.1 M zinc nitrate aqueous solution at 335 K. The solution was prepared with distilled water and reagent grade chemicals. A soda-lime glass coated with conductive tin oxide was used as cathode. Prior to electrodeposition, the glass cathodes were rinsed in acetone, anodically polarized in 1 M NaOH aqueous solutions and then rinsed with distilled water. Zinc sheet (99.999%) was used as the active anode and an Ag/AgCl electrode as reference. Song *et al* [37] prepared ZnO:Al films by r.f. magnetron sputtering which gave an interesting transparent conductive oxide quite suitable for electronic devices.

6.2.4 Crystals

Hirose [38] prepared ZnO crystals by vapour phase reaction with ZnI_2 as source by maintaining the growth region between 1150 and 1200 °C. Crystals grew by oxidation. They were mainly needles of 15 mm length or plates of 8 mm^2 area [39]. A number of transport agents like HCl, NH_3, NH_4Cl, H_2, Cl_2 and $HgCl_2$ were used by Shiloh and Gutman [40], of which NH_3, NH_4Cl and $HgCl_2$ were found to be the most efficient. Extremely pure single crystals of 2 mm diameter and 8 mm length were obtained by them. Halogens or halides were used as transport agents by Piekarczyk *et al* [41] for growing such crystals. ZnO crystals were obtained from the vapour phase by Shalimova *et al* [42]. Takata *et al* [43] prepared ZnO single crystals by hydrothermal synthesis. They used these crystals in making an M-S diode by evaporating on Au (or In) dot and In (or Al) on (0001) ZnO surface with Al giving an ohmic contact and In and Au giving rectifying contacts. Matsumoto *et al* [44] prepared ZnO crystals by vapour transport in a closed tube using Zn or $ZnCl_2$ as a transport agent. Crystals used by Bhushan and Chukichev [45] for CL studies were also prepared by the hydrothermal method. Butkhuji *et al* [46] developed a method for the growth of a single crystal layer of ZnO by heat treatment of crystals in an activated oxygen atmosphere. Bala *et al* [47] grew ZnO structures on (111) ZnSe crystals by plasma oxidation.

6.2.5 Nanocrystallites

Suspensions of ZnO nanocrystalline particles were prepared by Smith and Vense [48] in a non-aqueous solvent such as 2-propanol. For this preparation, 25 ml of a 0.02 M NaOH solution was slowly added while stirring to 225 ml of a 0.001 M $Zn(CH_3COO)_2{\cdot}2H_2O$ solution. After both solutions were first cooled to 0 °C, a rapid formation of extremely small ZnO particles was followed by a relatively slow growth of the particles. TEM studies showed a crystalline structure and a mean radius of about 30 Å [49]. Preparation in ethanol was made by Spanhel and Anderson [50] and Meulenkamp [51], by adding 50 ml of 0.14 M $LiOH{\cdot}H_2O$ solution to 50 ml of 0.1 M $Zn(CH_3COO)_2{\cdot}2H_2O$ solution. Again both solutions were first cooled to 0 °C before the hydroxide solution was added slowly to the zinc solution while stirring. This method yields a suspension with a particle concentration about 2 orders of magnitude higher than the 2-propanol preparation. In both cases a rapid formation of extremely small ZnO particles was followed by a relatively slow growth of particles [51, 52]. Nanocrystalline ZnO with different sizes were obtained by first storing the resulting reaction mixture at room temperature and then sampling at regular intervals. Chaparro *et al* [53] prepared ZnO layers with 0.5 μm thickness by reactive magnetron sputtering using Zn–Al alloy as target. Isolating glass was used as the substrate at 300 and 400 °C, resulting in layers with 2×10^3 and 1.5×10^3 Ω cm resistivity. Layers prepared with other targets or lower substrate temperature showed higher resistivity. XRD analysis gave grain sizes of about 20 nm for both ZnO films. Choi *et al* [54] investigated the effects of surfactant types and concentrations on the microstructure and nanostructure of the deposits by depositing ZnO films from a 0.02 M aqueous solution of zinc nitrate mixed with various weight percents of cationic surfactant, cetyl trimethyl ammonium bromide or anionic surfactant sodium dodecyl sulphate.

6.3 Luminescence properties

6.3.1 General

Lehmann [7, 55] found the following four bands in PL emission spectra (figure 6.1) of ZnO under 365 nm excitation: (i) An ultraviolet near edge emission with a peak at 3.18–3.20 eV which appeared at room temperature in useful intensity when materials were n-type doped and prepared under reducing conditions, (ii) a green band which required a slight reduction of firing conditions [56] and was found to be enhanced in the presence of traces of sulphur [57, 58], (iii) an orange band due to Se with peak at 1.95–2.0 eV and (iv) a red-near infrared band with a peak at 1.55 eV observed when ZnO was fired in NH_3 at 1000–1100 °C. Firing in either H_2 or N_2 was ineffective. S and Se are isoelectronic to O in ZnO, may split states out of the valence band, and emission may appear due to transition between such states and the band edges. Red emission similarly was proposed to be due to N–H pairs with N replacing O in ZnO thus forming an acceptor and interstitial H forming a donor. Dingle [59] on the basis of Zeeman and ESR data proposed that a substitutional Cu impurity at a regular zinc site

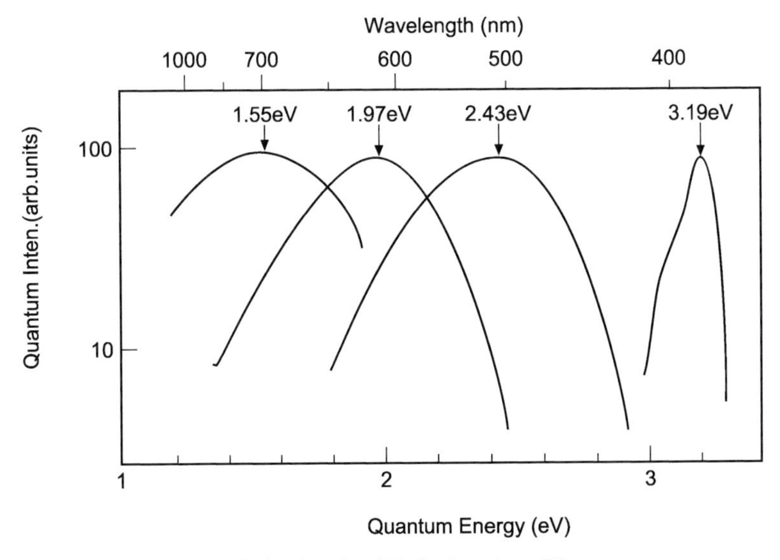

Figure 6.1. Luminescent emission bands of ZnO phosphors [7].

could be a possibility for the green band. Lauer [9] reported a band at 1.7 eV and from the measurements of emission and excitation spectra, intensity and temperature dependence of luminescence, found that the results of this band were similar to those of the yellow (2.02 eV) emission band in ZnO. It was proposed that the luminescence transition between the edge of the conduction band to a hole trapped in the bulk at 1.60 eV above the edge of the valence band was responsible for this emission.

Joshi and Kumar [8] studied the PL spectra of pure and self-activated ZnO phosphors at 77 and 293 K and found three bands in green, yellow and orange regions with peaks at 5100, 5650 and 5850 Å respectively. They associated these emissions with excess of oxygen and some other kinds of structural defects. Kumar [60] found that the intensity of the yellow emission band was reduced in the presence of Au, Tl, Mn and Cu metal ions, while the orange emission band of cation vacancies in the self-activated ZnO phosphor improved in the presence of chlorine ions. Agrawal and Kumar [61] studied the temperature dependence of band width and band shift in self-activated ZnO phosphor, analysed the results by using a configuration coordinate model, and proposed that emitting centres in self-activated ZnO were due to localized levels in energy gap and the emission followed the Prener and Williams model. Jechev *et al* [11] made calculations by using a configuration coordinate model for a centre consisting of Zn ion surrounded by three oxygen ions and an oxygen vacancy. They calculated vibrational frequency, the edge of fundamental absorption and the activation energy for temperature quenching for green emission of ZnO. Cox *et al* [62] found that the yellow PL of Li-doped ZnO gave ESR spectra for shallow donor–Li acceptor pairs, showing that at least a fraction of the yellow emission was a donor–acceptor luminescence. The distribution of separations between donors and acceptors broadened the ESR lines, particularly at high mw power. Bhushan *et al* [12] observed PL and EL spectra of ZnO prepared at a pressure of 1 torr (figure 6.2), in atmosphere of N_2 and open air. They found three peaks in (i) the blue-green region (peak at ~4640 Å), (ii) the green region (peak at 5080 Å) and (iii) the green-yellow region (5490 Å). While the nature of the green-yellow band remained almost the same for the samples prepared in N_2, for that prepared in air its intensity increased and thus this band was associated with centres formed by cation vacancies in the presence of an excess of oxygen. Since the green band was present with materials prepared under reduced pressure, the action of sulphur was considered secondary as it caused reduction of oxygen, forming sulphur dioxide.

Thus, the green band was attributed to stoichiometric excess of zinc, which was created due to the absence of oxygen under reduced pressure. The excess of zinc could be incorporated either in the interstitial positions or at the normal lattice sites with the formation of an equivalent amount of anion vacancies. The blue band was associated with edge emission which shifted to longer wavelength in the presence of some impurities of

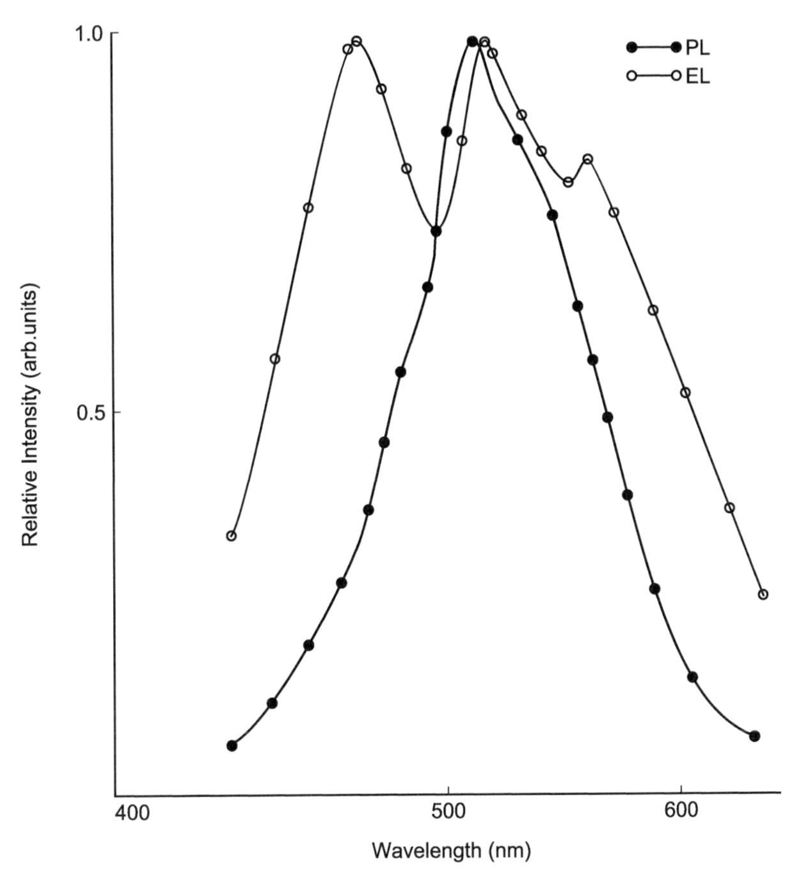

Figure 6.2. PL and EL spectra of undoped ZnO phosphors prepared at a pressure of 1 torr [12].

the order of ppm present in the starting material. Takata *et al* [43] rejected the Dingle model and favoured the model of Bhushan *et al* [12]. Matsumoto *et al* [44] observed green luminescence under ultraviolet excitation of ZnO crystals prepared by the vapour transport method and considered it to appear because of deviation from stoichiometry. In their later work, Bhushan and Asare [13] reported PL and EL of ZnO phosphors prepared by burning 99.999% pure metal zinc at a temperature of 1150 °C for 1 h under the continuous flow of 99.3% pure oxygen. XRD studies showed that the resulting material was ZnO. The particle size of such materials was found to be about 69 μm. The different bands which were found in ZnO were in the green, yellow-orange and orange regions. The peak positions were obtained by resolving the observed curves in terms of Gaussian distribution curves. It was found that in EL the ratios of the intensities of green to yellow-orange band decreased while that of yellow-orange to orange bands increased with grinding time. It was assumed that the centres responsible for the green band, which contributed decisively to the integral intensity of radiation were mainly located near the boundaries of some coherent zones and less inside of them. The centres responsible for the yellow-orange band were assumed to be mainly inside the coherent zones and those for orange emission were situated on boundaries as well as inside the coherent domains. The decrease in green emission supported the stoichiometric zinc model.

Bhushan *et al* [12] also prepared rare earth (RE) doped ZnO and found shifts in peaks of emission bands, on longer as well as shorter wavelength sides, as compared with those observed in undoped ZnO. Based on intensity shift a donor–acceptor model was proposed for such materials in which the RE formed the donor levels while the different centres responsible for the emission of undoped system formed the acceptor levels. Bhushan and co-workers [63, 64] reported brightness waves (BW) of EL in ZnO:La and ZnO:Nd under sinusoidal excitations from liquid nitrogen temperature to above room temperature. The BW at different frequencies (fixed voltage and temperature) and different voltages (fixed frequency and temperature) for ZnO:La are shown in figures 6.3(a) and (b), respectively. It is seen that the action of increasing the frequency is similar to that of lowering the temperature. At very high frequencies (1500 Hz) the secondary waves disappear. With increasing voltage the difference of height between two primary waves decreases and further the secondary waves disappear at very high voltages. A part of the electrons, which are freed from certain centres in the presence of an electric field, become trapped preferentially at the anodic side of a crystal or in lower field regions, and when the external field approaches zero these electrons have a chance to recombine with empty luminescent centres. Thus, there is always a superposition of the external field and a polarization field such that when the external field drops to zero there is still the polarization field, and the number of such recombinations is maximum near the time when the external field passes through zero, and a secondary wave appears clearly from the primary waves. At a very low temperature the release of electrons from traps is small and the secondary wave can be expected near the second primary wave. It should move to the first primary peak upon increasing the temperature since then the polarization field decays more quickly with increasing number of recombination processes. The secondary wave disappears when the temperature is such that traps can hold electrons only to a minor extent during one half cycle. According to Haake [65] the time t_0 at which the maximum light intensity is observed is a function of γ such that γ increases with decreasing time t_0, γ is given by

$$\gamma = p/w = (s/2\pi f)\exp(-E/kT) \tag{6.1}$$

where p is thermal release probability of electrons from traps, s is the frequency of escape, k is the Boltzmann constant and T is the absolute temperature. At constant T, $\gamma \approx 1/f$, i.e. increasing frequency f decreases

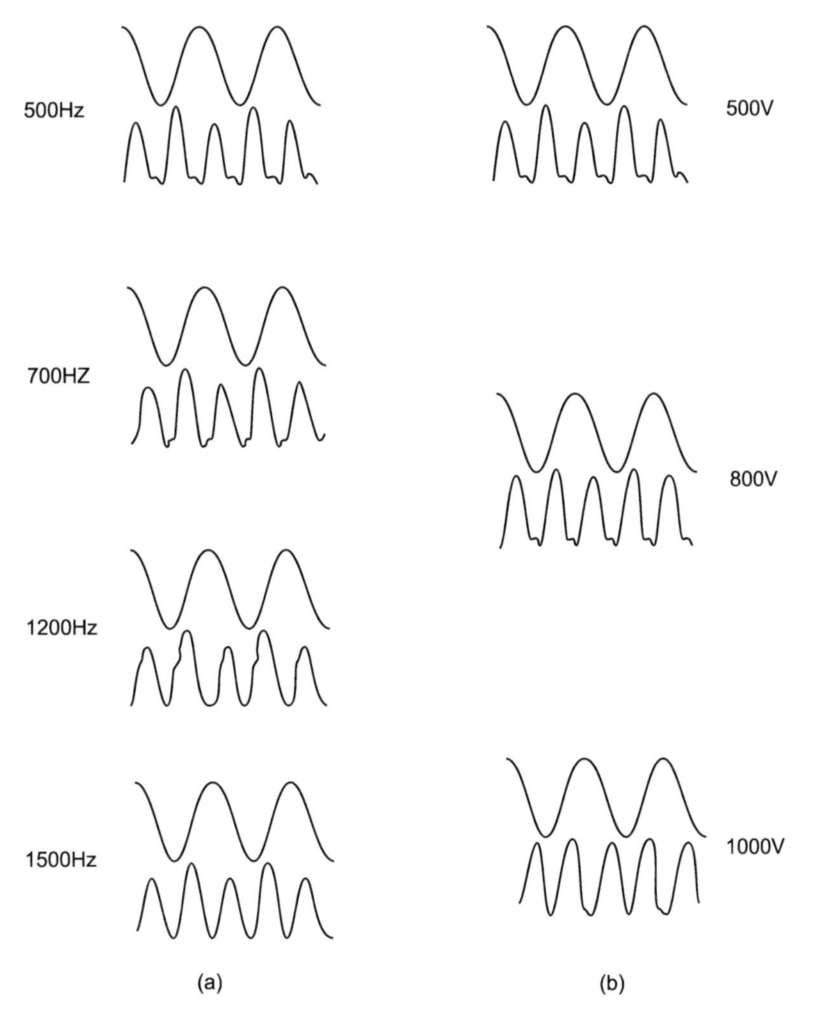

Figure 6.3. Brightness waves of ZnO:La phosphors at (a) different frequencies ($u = 500$ V and $T = 30\,^{\circ}$C) and (b) at different voltages ($f = 500$ Hz and $T = 30\,^{\circ}$C) [63].

γ and hence increases t_0 and the secondary wave creeps forward along the time scale with increasing frequency. At very high frequency, the time becomes so small that the traps cannot hold electrons and therefore secondary waves disappear. According to Patek [66] the light sum emitted in the primary wave satisfies the relation

$$S_p = A\exp(-b/U^{1/2}) \tag{6.2}$$

where A and b are constants. Thus, at higher voltages due to exponential variation the difference between two peaks is eliminated. At very high voltages

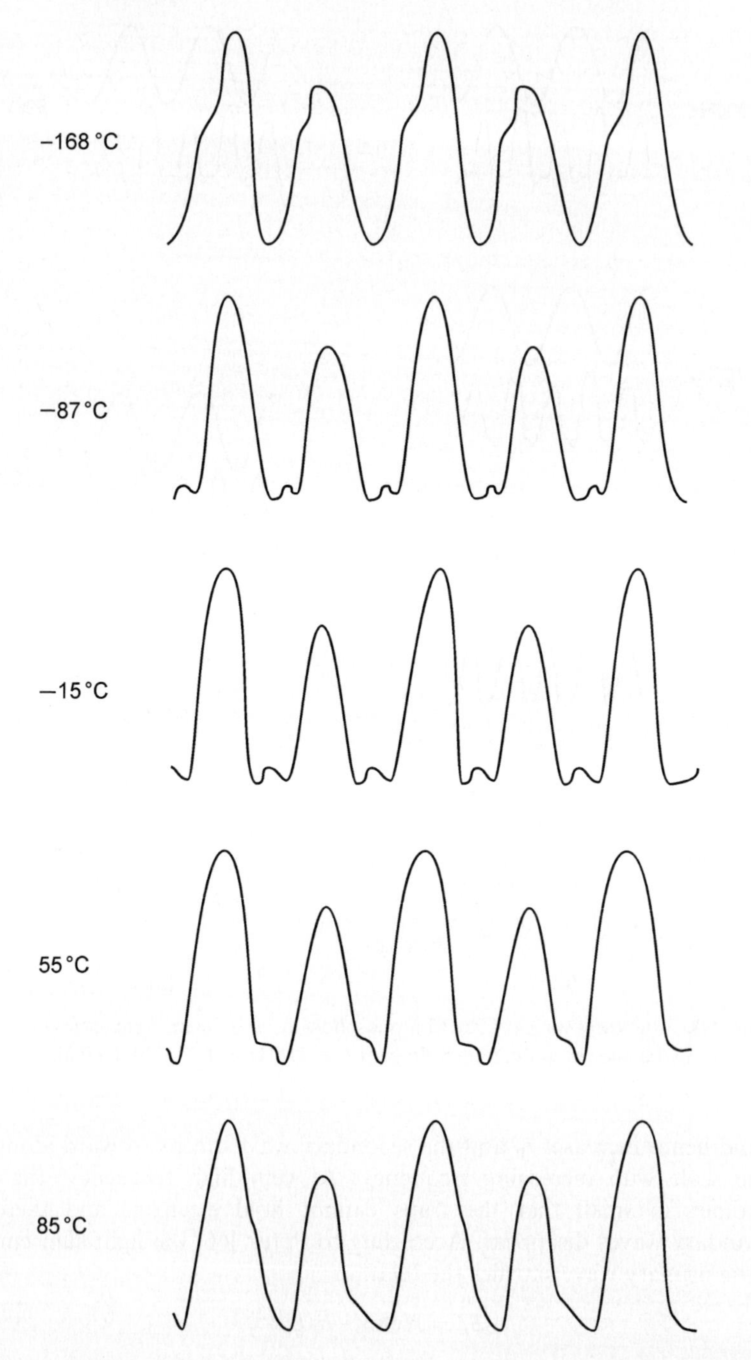

Figure 6.4. Brightness waves of ZnO:La phosphors at different temperatures ($u = 400$ V and $f = 500$ Hz) [63].

(1000 V) the electrons responsible for secondary waves find sufficient energy for acceleration and partake in primary emission and thus secondary waves disappear. Thus, at very high voltages the polarization field is cancelled. The behaviour of BW at different temperatures is shown in figure 6.4.

The secondary peak lies between two primary peaks at 15 °C and is most pronounced at this temperature. The temperature dependence of EL brightness studied between −155 and +20 °C is shown in figure 6.5. It consists of three peaks with highest intensity at −19 °C, a temperature quite close to −15 °C, at which the secondary wave is most pronounced. Thus, the existence of secondary waves is due to traps. In ZnO:Nd the secondary waves grow and disappear twice in the temperature range 93–363 K [64].

Quang *et al* [15] reported sharp emissions at 580, 590, 620 and 650 nm in ZnO:Eu,N (figure 6.6) corresponding to transitions $^5D_0 \rightarrow {}^7F_0, {}^7F_1, {}^7F_2$ and $^7F_3$ respectively along with overlap of broad band background peaking at 2.3 eV probably due to radiative transitions within an RE^{3+} donor–V_{Zn} acceptor pairs. Oritz *et al* [28] and Falcony *et al* [29] reported PL in undoped and $TbCl_3$-doped ZnO films prepared by spray pyrolysis and found that the

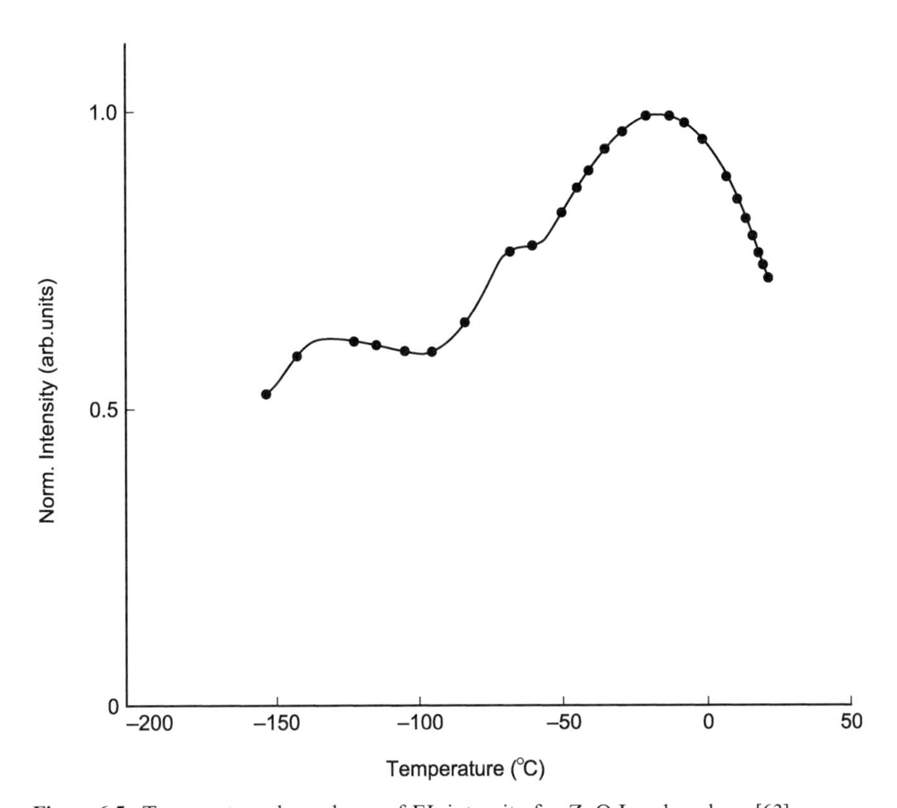

Figure 6.5. Temperature dependence of EL intensity for ZnO:La phosphors [63].

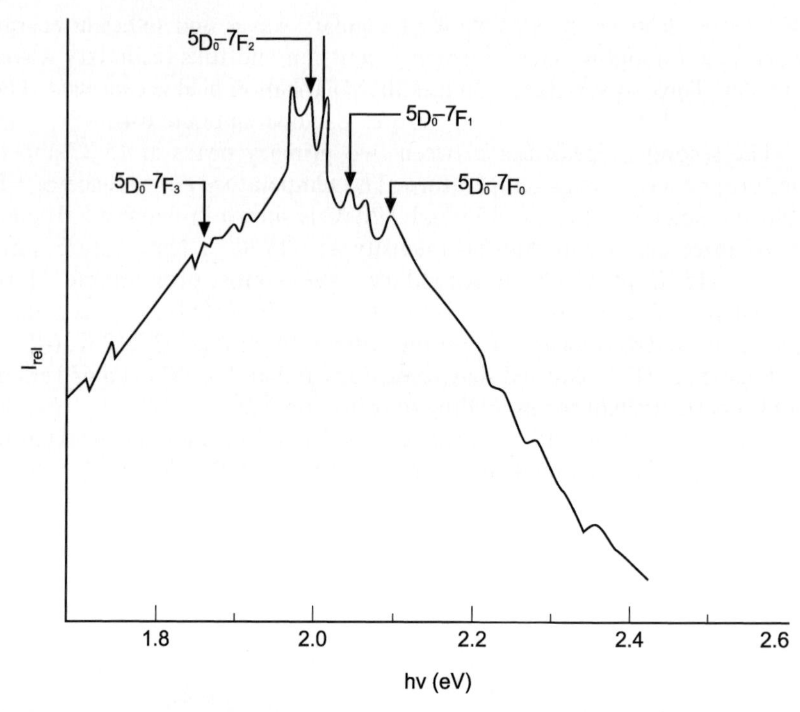

Figure 6.6. PL spectrum of ZnO:Eu,N at 300 K (excitation wavelength 365 nm) [15].

intrinsic films showed emission with a peak at 510 nm, and for doped films emissions with a peak at 550 nm. The blue-green emission was related to a transition within a self-activated centre formed by a doubly ionized zinc vacancy and the ionized interstitial Zn^+ at one and/or two nearest neighbour interstitial sites. The emission peak of doped film was related to the transition $^5D_4 \rightarrow {}^7F_5$ of Tb^{3+} ions, which corresponds to 5425 Å. Kossanyi *et al* [22] and Kouyate *et al* [23] studied the PL of RE-doped ZnO under excitation conditions in which light was primarily absorbed by ZnO. No emission from RE could be observed and only the luminescence of semiconducting substrate, in which RE put the fingerprint of its absorption, was characterized.

Fernandez *et al* [21] found that the emission band of Mn-doped ZnO centred at 528 nm had excitation peaks at 209 and 248 nm, which differed from that of the well known green band of undoped ZnO which was excited by 350 nm radiation. In the two bands observed by Miralles *et al* [19] in the EL spectra of ZnO doped with Co, with peaks at 700 and 900 nm, the former was, however, associated with transitions from conduction band to Co level whereas the latter was associated with transitions from this level to the valence band. The intense white PL observed by

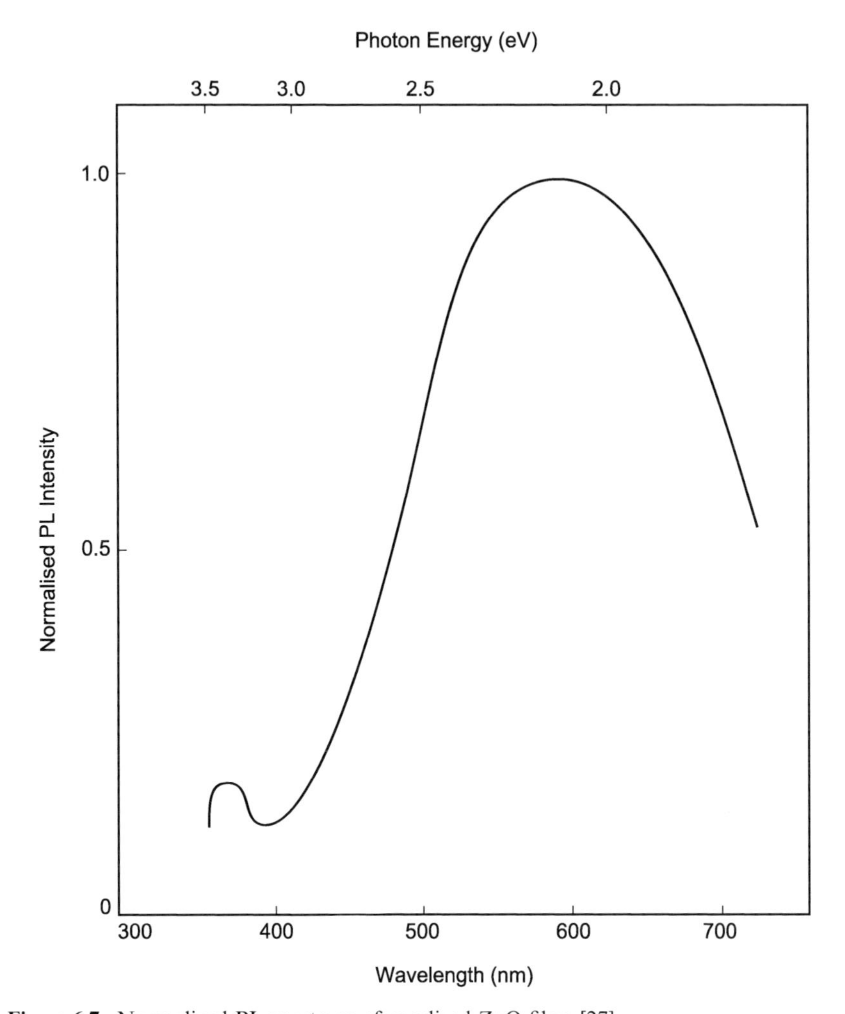

Figure 6.7. Normalized PL spectrum of anodized ZnO films [27].

Nanto *et al* [27] in ZnO films (figure 6.7) formed by anodization was attributed to radiative recombination between localized electron and hole in gap states which were formed by native defects in the films. PL spectra consisted of a sharp emission band at ~380 nm and a very broad visible emission with a peak at ~580 nm. It was guessed that the native defects in ZnO might form energy levels in the band gap and they might act as electron and hole trap centres in the radiative recombination process. Intense white EL and CL were also found in such films.

In the PL spectra of Li-doped ZnO films prepared by spray pyrolysis at low substrate temperature emissions at 420 and 500 nm and those prepared at high substrate temperature, a light emission with a peak at 555 nm was reported by Oritz *et al* [30]. These peaks were associated with blue emission of Pyrex glass substrate, the characteristric blue-green emission of intrinsic ZnO and due to incorporation of Li respectively.

PL, CL and EL and optical absorption spectra were reported by Vavilov *et al* [67]. They proposed a possible mechanism of excitation energy transfer by the electron subsystem to intra-centre states of $3d$ transition element ions. The impact of Pb doping on optical and electrical properties of ZnO powders was investigated by Vanheusden *et al* [68]. They observed a decrease in the 2.26 eV peak and a concomitant smearing of the band edges, narrowing the effective gap of the grains $\sim$2 eV with increasing Pb content. The possibility of a PbO-like phase residing at the grain boundaries was given. The free carrier concentration in grains decreased due to increasing Pb content, which was due to electron transfer from O vacancy donors to substitutional Pb centres, acting as an electron trap. Later Vanheusden *et al* [69] found that green emission intensity was influenced by free carrier depletion at the particle surface and they suggested that the singly ionized vacancy was responsible for the green emission in ZnO, which resulted from recombination of a photo-generated hole with a singly ionized charge state of this defect. Hayashi *et al* [70] reported PL in ZnO:Eu which showed red emission. It was found that the intensity and fine structure of Eu^{3+} emission and their temperature dependence were strongly influenced by doping conditions of Eu. The observed red band was associated with excess oxygen.

The luminescence and non-radiative characterization of excited states involving oxygen defect centres in polycrystalline ZnO were reported by Egelhaaf and Oelkrug [71]. The green luminescence of polycrystalline ZnO was investigated by these workers by studying diffuse reflection, steady state and time resolved PL and was associated to donor–acceptor transition. Infrared emission in ZnO:Cu^{2+} (d^9) was reported by Kimpel and Schulz [72]. $^3E(D) \longrightarrow {}^2T_2(D)$ emission of substitutional Cu^{2+} in ZnO was observed in the 3700–5800 cm^{-1} spectral range under 50 keV e-beam excitation at cryogenic temperature. Yu *et al* [73] made their first report on strong luminescence and stimulated emission from high quality ZnO thin films. Reynolds *et al* [74] found that the green band in ZnO was analogous to the yellow band in GaN. It was likely that respective transitions (donor–acceptor transitions) in two materials were defect related and share common mechanism.

It is worth mentioning some of the more important observations made on CL and TL. Hiraki *et al* [75] found that ZnO phosphors could be used under low voltage CR excitation. Kramer [76] reported blue-green luminescence ($\lambda_{max} \approx 505$ nm) in ZnO during the interaction of low-kinetic energy (20–40 eV) gas-phase+ions. Luminescence was attributed to hole injection followed by radiative recombination. Phosphor-grade ZnO was produced under reducing conditions, containing excess Zn, that served as an activator. The excitation band of the Zn activator overlapped the ultraviolet emission band and an internal conversion of ultraviolet to blue-green occurred. The blue-green luminescence observed during ion excitation agreed with spectral distribution obtained by electron excitation of the ZnO sample. Electron emission of the ZnO sample only produced the ultraviolet band at high current densities and the ultraviolet band was not observed during ion excitation. Wada *et al* [14] reported CL and ESR in ZnO. The g-values of the ESR spectrum measured from the powders at 77 K were $g_{\parallel} = 1.956 \pm 0.01$ and $g_{\perp} = 1.955 \pm 0.001$. The ESR signal of donors arose from the oxygen vacancy. Bhushan and Asare [77] reported results of CL, EL and PL in undoped ZnO, ZnO:Nd and ZnO:Ce phosphors. By resolving the observed spectral curves into Gaussian distribution curves three peaks were found in undoped ZnO and ZnO:Nd whereas in ZnO:Ce four peaks appeared. It was also found that the efficiency of these phosphors respectively were in the ratio 1:0.7787:0.5879. Bhushan and Chukichev [45] found that the temperature dependence of CL spectra studied from liquid He temperature to room temperature showed that the half band width varied according to the relation

$$W = A[\coth(h\nu/2kT)]^{1/2}. \tag{6.3}$$

This showed that the configuration coordinate model could be applied to the green band of ZnO and thus as expressed earlier the green band of ZnO might be considered due to stoichiometric zinc, the levels of which were localized levels located in the energy gap. This result was similar to that obtained by Agrawal and Kumar [61] from PL spectra. The value of vibrational frequency (ν) was found to be $4.579 \times 10^{12}\,\mathrm{s}^{-1}$. Applying a configuration coordinate model to the green centre from PL spectra, Jechev *et al* [11] reported the value of ν to be $1.06 \times 10^{13}\,\mathrm{s}^{-1}$.

De Mauer and Maenhout-vander Vorst [78] resolved the thermal glow curve of heat-treated ZnO observed under the excitation of Wood's light into five singular glow peaks at -161, -144, -129, -117 and $-101\,^{\circ}\mathrm{C}$ showing a first-order process. The peak at $-101\,^{\circ}\mathrm{C}$ emitted 520 nm and the spectrum of the other four peaks showed a maximum at 570 nm. The values of trap depths varied between 0.12 and 0.299 eV and the capture cross-section from $\sim 10^{-19}$ to $10^{-22}\,\mathrm{cm}^2$. TL in single crystals of ZnO under the excitation of 3650 Å was reported by Shalimova *et al* [42] with peaks at 110 and 180 K due to green emission of the crystal and the maximum at 150 K due to yellow emission. Trap depths of the peaks at 110 and 180 K were 0.13 ± 0.03 and 0.35 ± 0.04 eV respectively. The peak at 110 K was not stable and varied from 97 to 120 K and was related to electron traps in the form of oxygen vacancies. The peak at 150 K was due to hole traps, in particular zinc vacancies. TL in RE-doped ZnO under the excitation of ultraviolet, β- and γ-ray irradiation were reported by Bhushan and co-workers [79, 80]. A number of methods, namely isothermal decay, heating rate, initial rise and peak shape methods, were used to determine TL parameters.

Under β-ray excitation undoped ZnO did not show any TL; however, ZnO:Gd showed a peak at $\sim$342 K. ZnO:Gd showed a number of peaks under γ-ray excitation with a higher temperature peak at 620.25 K. Undoped ZnO showed two peaks at 346 and 511.25 K under ultraviolet excitation. Similarly three peaks at 308, 385 and 569.75 K along with a shoulder at 348 K were observed in ZnO:Pr. Higher doses of γ-rays produced a peak at $\sim$608.6 K, along with some other peaks in ZnO:Pr. TL of ZnO:Cu,La consisted of a pronounced peak at 600 K along with some sharp transitions under ultraviolet and γ-ray irradiation. Trap depths were found to be more than 1 eV and the order of kinetics was found to be 2. The existence of higher temperature peaks (above 500 K) indicated that ZnO phosphors could be explored for γ-ray dosimetry.

6.3.2 Electrodes

EL at *n*-ZnO electrodes in electrolytes with emission under cathodic and anodic pulsed polarization was observed by Fichou and Kossamyi [81]. Under cathodic polarization the narrow ultraviolet band-to-band emission with a structure at 384 and 394 nm showed that injection occurred via the valence band; a broad EL band (400–700 nm) was also observed, which presented a 80 nm blue shift, and was considered in terms of a donor–acceptor mechanism. Under anodic polarization the excitation mode, either forward or reverse, was the determinant for EL characteristics of the ZnO/electrolyte junction. Kouyate *et al* [82] found PL and EL in Nd-, Er-, Eu- and Dy-doped electrodes. PL showed a typical emission of ZnO, and no emission of RE could be observed when excited with 385 nm. EL was found to be different depending upon mode of excitation. Under forward polarization, when the electrolyte contained a redox couple capable of injecting holes in the valence band, luminescence of ZnO was observed with only a slight absorption of RE. Under reverse bias, EL resulted from the impact excitation process of RE ion with radiative transitions between shielded 4*f* levels. Direct impact excitation of Ho^{3+}- and Sm^{3+}-doped ZnO were also reported under anodic polarization [83]. Kouyate *et al* [84] observed EL in RE^{3+} (Sm^{3+} and Eu^{3+}) doped ZnO electrodes in contact with aqueous electrolytes under cathodic polarization and the sole luminescence found was due to ZnO. With anodic polarization, characteristic emission of Sm^{3+} and Eu^{3+} was observed. Energy transfer between Sm^{3+} (donor) and Eu^{3+} (acceptor) was also found. Direct impact excitation of thulium(III) luminescence due to hot electrons in polycrystalline ZnO:Tm^{3+} electrodes in contact with an aqueous electrolyte was also reported by Bachir *et al* [85]. EL studied under anodic polarization showed a characteristic emission spectrum of Tm^{3+}. Two main emission peaks of Tm^{3+} found at 480 and 805 nm originated from transitions ${}^1G_4 \rightarrow {}^3H_6$ and ${}^3H_4 \rightarrow {}^3H_6$.

6.3.3 Varistors

The effect of Co on the EL spectra of ZnO varistors was reported by Miralles *et al* [86]. They proposed a model to explain the different intensity ratios found between the bands centred at 700 and 950 nm. Bachir *et al* [87] observed characteristic EL of Ho^{3+} in a ZnO varistor. The presence of hot electrons was found to be responsible for excitation of a trivalent RE ion. Results were found to be consistent with the model of electrical breakdown at the grain boundaries.

6.3.4 Ceramics

PL from electron-irradiated ZnO ceramics was reported by Garcia and Remon [88]. It was found that irradiation caused quenching of the PL green band and a part of this quenching was attributed to damage in the oxygen sub-lattice. Garcia *et al* [20] further studied the effect of addition of Bi and Mn on PL from ZnO ceramics and the effect of the presence of impurities on the green emission band was compared with the effect of oxidizing treatments. A narrow green band was observed in the Mn-doped samples. Radiative recombination in ZnO ceramics was reported by Proskura [89]. Achour *et al* [90] on the basis of CL studies on ZnO ceramics found that impurities like Bi, Mn and Co influenced the green emission of ZnO. It was found that Co induced a low energetic characteristic sharp emission whereas Mn led to a relatively broad one. Bachir *et al* [91] prepared Eu^{3+}-, Tm^{3+}- and Er^{3+}-doped ZnO ceramics and found emission due to RE under the application of electric field. The mechanism of excitation was based on hot electrons impact excitation of trivalent RE ions inserted into the semiconductor lattice.

6.3.5 Hydrogenation

The effect of hydrogenation on the emission of ZnO was reported by Sekiguchi *et al* [92]. It was found that hydrogen plasma treatment strongly passivated the green emission and enhanced the band-edge emission. Band-edge emission showed strong temperature dependence and the intensity increased significantly at lower temperatures. From PL studies Liu *et al* [93] found that water vapour enhanced the green emission of ZnO in the surface and caused a two-band (green and yellow) emission in the bulk. Mn was found to behave as a quencher for both green and yellow emission of ZnO. From positron lifetime measurements, the existence of two distinct vacancy-type defects V_{Zn} and V_O were found. Thus emission was due to interstitial Zn and O.

6.3.6 Excitonic

Iwai and Namba [94] found laser emission from ZnO due to annihilation of a bound exciton at very low temperature and a free exciton with the simultaneous emission of one longitudinal optical phonon at 77 K. Klingshirn [95] found that emission of ZnO observed under high one- and two-quantum excitation could be attributed to interaction of free exciton–phonons, free excitons, and free electrons, and of bound excitons with phonons and free and bound electrons. Miyamoto and Shionoya [96] studied emission spectra of ZnO under N_2 laser excitation at 1.5 K which consisted of two emission lines separated by 3 meV in the low voltage side of the H exciton line. Segawa and Namba [97] reported stress-induced splitting of emission lines from excitonic molecules in ZnO, and exciton–phonon emission was reported by Nikilenko *et al* [98]. These workers found that characteristics of exciton–phonon emission observed at higher temperature was useful to evaluate the degree of perfection of crystal structure and mechanism of radiation of simple crystalline and polycrystalline samples made from ZnO. Savikhin *et al* [99] showed for the first time that a substantial fraction of the exciton emission of ZnO crystals was due to excitons trapped by surface centres. It was suggested that anomalously short lifetimes ($\sim$60 ps) observed for surface excitons were consequences of the Auger process. In their later work [100] on excitons they showed that the major part of the edge emission of ZnO was due to the excitons bound by surface centres. A comparison of temperature of different quasiparticles in ZnO crystals studied under laser excitation of high intensities demonstrated that in the pumped region of the crystal the energy spectrum of the excitons and electrons reflected the temperature of non-equilibrium phonons in specified regions of *K*-space [101]. Butkhuji *et al* [46] studied single crystal layer of ZnO in which, in addition to A-excitons, emission bands corresponding to B- and C-excitons were registered. A violet band at 400 nm associated with a doubly-charged oxygen vacancy and a singly-charged Zn vacancy was dominant in PL spectra of ZnO layers with p-type conductivity. In ZnO with n-type conductivity, green and yellow-orange bands were generated due to oxygen vacancies and alkali metals respectively. Bala *et al* [47] studied the strained ZnO structures grown on (111) ZnSe crystals through plasma oxidation by EL, and PL methods showed lines of heavy and light hole excitons. When single crystals of ZnO were excited by a powerful laser, the mechanism of *e–h* plasma to explain the formation of the long-wave tail in the PL band was discussed by Litovchenko *et al* [102]. A conclusion was made on the predominant contribution of plasma oscillations in a hole subsystem to the losses of energy in the process of *e–h* recombination and long wave tail formation in the PL spectrum. A nonlinear Zeeman effect was reported in the Ni^{3+}-doped ZnO crystal by exciting with photon energies above the excitonic band gap [103]. Controversy about the infrared emission in ZnO was solved by Heitz *et al* [104] by studying the Zeeman spectroscopy of V^{3+} in ZnO. This emission was associated with the ${}^3T_2(F) \rightarrow {}^3A_2(F)$ transition of substitutional V^{3+} on the cation place.

6.3.7 Mixed bases and other forms

Lehmann [105] found ZnO to be a suitable material for contact EL because of its high conductivity. Wachtel [106] observed that addition of ZnO up to about 1 mol% improved the EL of ZnS:Cu,Cl. Bhushan [107] and Bhushan and Das [108] reported binderless EL in (ZnS,ZnO):Cu,Cl after treating with salt solutions of Na_2SiO_3 or $CuSO_4$ and a transport process of Schottky emission type was found to be mainly responsible in such EL. Wave shape and emission spectra of EL from (ZnS,ZnO):Cu,Sn phosphor were reported by Maheshwari and Khan [109]. Light output varied according to the Alfrey–Taylor relation and current varied as $I \approx \exp(aV^{1/2})$ at low voltages and $I \approx \exp(aV)$ at higher voltages. The excitation mechanism of infrared radiation of the ZnO–Cu_2O heterojunction was discussed by Drapak [110]. EL in the ZnO–CdO alloy system was reported by Singh and Mohan [16]. They found that beyond 15% of CdO no EL was found. The peak observed at 5200 Å shifted to the higher wavelength side due to the addition of CdO. Bhushan and Diwan [17] reported d.c. EL in (ZnS,ZnO):Cu,Cl with a burst of emission (figure 6.8) appearing at both switching on and switching off the field and this emission decayed very slowly to a saturation value.

The emission colours of burst and saturated light were different. The intensities of burst depended upon the temperature and the choice of polarity of the electrodes. Bala and Firszt [32] investigated influence of In, Ga and Ag dopants on the EL spectra of ZnSe–ZnO diodes. Minami *et al* [111] observed two broad bands in d.c.-excited Au–ZnO–In M-S diodes located at about 5000 and 9300 Å respectively with total colour to be white. At low current densities, the intensity of the second band was higher; under high current densities that of the first band was higher. The second band was considered to be due to defect and impurity centres, and due to transitions of hot electrons between states in the conduction band. Red emission at about 6500 Å observed in sputtered films of ZnO under N_2 laser excitation by Nanto *et al* [26] was associated with the native defects which were induced in the sputtering process. $ZnS_{1-x}O_x$:Mn films prepared by r.f. sputtering were found to be more useful for d.c. EL by Sands *et al* [33]. Muraleedharan *et al* [112] reported N_2 laser-induced PL in (ZnS,ZnO):Ce phosphor. The emission peak in PL spectra was found to shift towards the longer wavelength side due to an increase in concentration of ZnO.

From the measurement of optical absorption and emission of Pr^{3+} and Nd^{3+} ions in 1ZnO, $2TeO_2$ glasses, measured in visible and infrared regions, oscillator strength, transition probabilities and branching ratios were obtained by Kanoun *et al* [113]. PL in ZnO–CdO–SiO_2 glasses was found by Paje *et al* [114]. A broad band centred between 540 and 560 nm for excitation at 250 nm was observed. Luminescence was found to be dominated by short ordering nuclei and/or crystalline centres and decay curves consisted of three exponentials with lifetimes of 0.025, 0.1 and 0.52 ms. On the basis

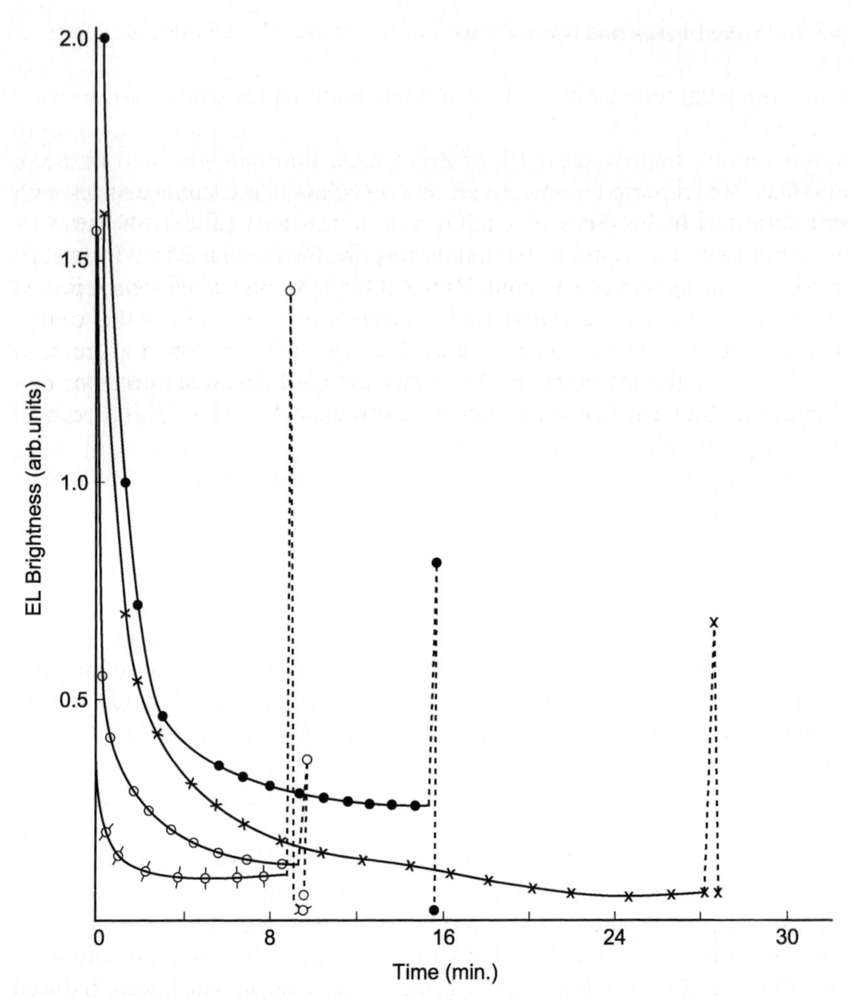

Figure 6.8. Dependence of d.c. EL brightness on time at different temperatures for (ZnS, ZnO): Cu, Cl phosphor —— switching on. ----- switching off. × 27 °C (room temperature), ● 35 °C, ○ 41 °C (conducting glass positive, metal negative); ϕ27 °C (opposite polarity) [17].

of time-resolved spectra, blue emission observed in ZnS:O single crystals under high levels of optical and extreme excitations was associated with recombination of Zn_i donors and O_5–O_5 acceptors [115]. Hsieh and Su [24] used ZnO in ZrO_2 pellets as impurity and observed TL induced by ultraviolet radiation. It was found that a TL peak at 90 °C of ZrO_2 pellets sintered at 1100 °C shifted to 85 °C with intensity increased three times for the addition of 1% ZnO. From the study of emission spectra of TL, the effect of doping of ZnO was found to increase the number of luminescence centres. The *e–h* recombination of Zn–O defects produced intense PL at liquid He temperature in Si [116].

6.3.8 Nanocrystallites

Formation and spectroscopic properties of rock salt ZnO nanoparticles, were reported by Bing-Suo *et al* [117]. The optical and structural properties were found to be similar to exciton-confined Cu_2O nanoparticles. Ruter and Bauhofer [118] found a strongly enhanced luminescence efficiency in zinc borate glasses and glass fibres doped with Eu^{3+} and Tb^{3+}. The enhancement was attributed to a high content of ZnO nanocrystallites which operated as sensitizer for the RE ions. Figure 6.9 shows the emission spectrum of a suspension of ZnO particles upon photo-excitation [3]. This spectrum

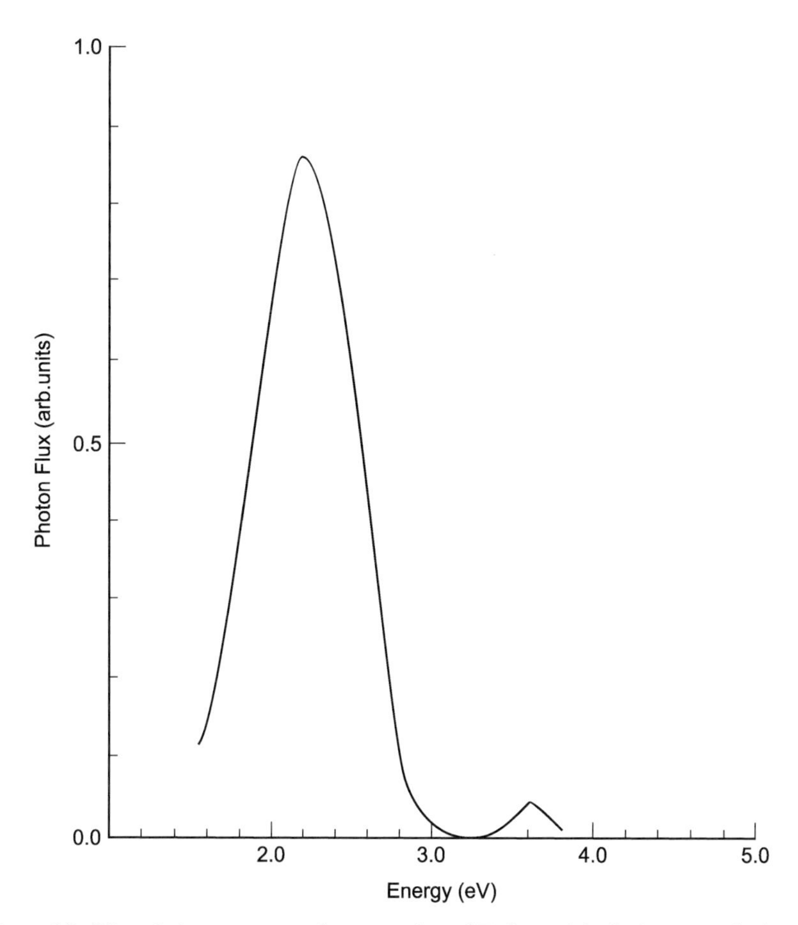

Figure 6.9. PL emission spectrum of a suspension of ZnO particles in 2-propanol taken at room temperature upon excitation with light of 4.4 eV [3].

contains two bands: (i) a broad trap emission band centred at 2.2 eV and (ii) exciton emission band centred at 3.6 eV. The position of an exciton emission band corresponds to the value reported in the literature for the band gap of macrocrystalline ZnO [119, 120]. From the absorption studies an onset 3.38 eV corresponded to a particle radius of 35 Å. The emission spectrum at different times during the growth of ZnO particles in 2-propanol at room temperature consisted of a weak solvent Raman peak at 4 eV, shifted with respect to the excitation energy by about 0.4 eV ($3500\,cm^{-1}$) corresponding to an OH vibration. Previously for ZnO particles with a radius smaller than about 25 Å only a visible emission was observed [48]. On increasing the particle size the ultraviolet emission increases and that of visible emission decreases. The ultraviolet band was associated with annihilation of excitons and the intense visible emission band with a transition of a photo-generated electron from a shallow level close to the conduction band to deeply trapped hole [3].

6.4 Applications and future scope

ZnO is a versatile material because of its usefulness in various devices such as opto-electronic, solar cells, electro-photography and surface acoustic wave devices. It is a promising material for EL devices. The intense emission observed from its nanocrystallites shows good promise for its future. Zn-doped ZnO phosphors have been of worldwide interest for use in low-voltage fluorescent applications [121]. The power efficiencies of the red and blue-green emissions in d.c. EL of annealed thin films of sputtered ZnO were of the order of 10^{-7} and 10^{-6} respectively [25]. Thus it was emphasized that sputtered ZnO films in MIS diode Au/SiO/ZnO/p^+-Si form with highly efficient red and blue-green EL suitable for application as display devices could be obtained by suitable annealing. Nanto *et al* [27] suggested that anodized ZnO films giving white emission might be useful in opto-electronic devices, since the low-temperature anodization process was inexpensive and fast. The brightness in this case was approximately comparable with that of ZnO:Zn phosphors under N_2 laser excitation. ZnO films formed by sputtering giving efficient blue-green and red EL and PL required deposition on high-temperature substrates or post-deposition preparative treatment to achieve efficient emission. Saparin *et al* [122], on the other hand, used dependence of local CL intensity for some materials, one of which was ZnO and a CL contrast of direct writing pattern by means of the scanning electron microscope.

In this chapter various efforts made by different workers have been presented. Attempts should be made to stabilize the conditions under which the different emissions are found with prominence. Devices based on EL emission from ZnO electrodes in contact with electrolytes should be developed for technological applications. Development of LEDs using ZnO also forms an important future problem in this line.

ZnO EL phosphors were also found to be useful in the storage of images. Kazan [123] suggested a new method of image storage using the principle of field effect conductivity control. Such a device was prepared using an EL powder for the generation of the output image and a ZnO powder layer for control and storage purposes. In operation, the ZnO surface was first corona charged to a negative potential to produce its conductivity and erase old information. After this the panel was exposed to an optical image, which discharged local areas producing a stored charged pattern on the ZnO surface. A conductivity pattern in accordance with this charge pattern was created in the ZnO layer, which in turn controlled the luminescence output of the corresponding areas of the adjacent phosphor layer. Stored images having a brightness of 200 foot-lamberts with a maximum contrast ratio of 100:1 were obtained. Kazan demonstrated panels of size $12'' \times 12''$ having a limited resolution between 400 and 800 TV lines. However, this work needs further improvement. Further, the high conductivity of ZnO thin films with a high transparency in the visible spectrum can be used as a transparent electrode in opto-electronic displays [2]. Yu *et al* [73] have already given probably the first report on strong RT emission and stimulated emission from high-quality ZnO thin films. It presents a direction for future investigations.

The nanocrystallites of ZnO show two emission bands, one being an ultraviolet emission band (exciton emission) and the other a visible emission band (trap emission). The visible emission is well known and it is the main reason for the use of ZnO as a phosphor in various applications. The quantum efficiency has been found to lie in the range between about 20% for a particle radius of 7 Å to 12% for a particle radius of 10 Å [3]. Attempts should be made to prepare nanocrystallites of varying band gaps and thus producing desired emissions in the visible region. Along with this, changes due to incorporation of impurities may also be taken as a matter of investigation. In fact, there lies a tremendous future for such materials.

References

[1] Chopra K L, Major S and Pandya D K 1983 *Thin Solid Films* **102** 1
[2] Nanto H, Minami T, Shooji S and Tanaka S 1984 *J. Appl. Phys.* **55** 1029
[3] Van Dijken A 1999 PhD thesis '*Optical properties and quantum confinement of nanocrystalline II–VI semiconductor particles*' Utrecht University p. 33
[4] Tomar M S and Garcia F G 1984 *Sol. Wind. Technol.* **1** 71
[5] Lehmann H W and Widmer R 1973 *J. Appl. Phys.* **44** 3868
[6] Tsukada M, Adachi H and Satoko C 1983 *Prog. in Surf. Sci.* **14** 113
[7] Lehmann W 1968 *J. Electrochem. Soc.* **115** 538
[8] Joshi J C and Kumar R 1973 *Ind. J. Pure Appl. Phys.* **11** 422

[9] Lauer R B 1973 *J. Phys. Chem. Solids* **34** 249

[10] Hiraki M, Kagami A, Hase T, Narita K and Mimura Y 1974 *J. Lumin.* **12/13** 941

[11] Jechev N I, Gochev D K and Kynev K D 1975 *Bulg. J. Phys.* **11** 247

[12] Bhushan S, Pandey A N and Kaza B R 1979 *J. Lumin.* **20** 29

[13] Bhushan S and Asare R P 1981 *Geeh. J. Phys. B* **31** 331

[14] Wada T, Kikula S, Kiba M, Kiyozumi K, Shimojo T and Kakehi M 1982 *J. Crystal Growth* **59** 363

[15] Quang V X, Liem N Q, Thanh N C, Chuong T V and Thanh T L 1983 *Phys. Stat. Sol. (a)* **78** K161

[16] Singh V B and Mohan H 1968 *Z. Physik* **208** 441

[17] Bhushan S and Diwan D 1986 *Cryst. Res. Technol.* **21** 161

[18] Miralles A, Cornet A and Morante J R 1986 *Semiconduct. Sci. Technol.* **1** 230

[19] Miralles A, Cornet A and Morante J R 1989 *Mat. Sci. Forum* **38–41** 567

[20] Garcia J A, Remon A and Piquears J 1987 *J. Appl. Phys.* **62** 3058

[21] Fernandez P, Remon A, Garcia J A, Llopis J and Piqueras J 1988 *Appl. Phys.* **A46** 1

[22] Kossanyi J, Kouyate D, Pouliquen J, Ronfard-Haret J C, Valat P, Oelkrug D, Mammel U, Kelly G P and Wilkinson F 1990 *J. Lumin.* **46** 17

[23] Kouyate D, Ronfard-Haret J C, Valat R, Kossanyi J, Mammel U and Oelkrug D 1990 *J. Lumin.* **46** 329

[24] Hseih W C and Su C S 1994 *Appl. Phys. A: Solids Surf.* **A58** 459

[25] Takata S, Minami T and Nanto H 1981 *Jap. J. Appl. Phys.* **20** 1759

[26] Nanto H, Minami T and Takata S 1981 *Phys. Stat Sol. (a)* **65** K131

[27] Nanto H, Minami T and Takata S 1983 *J. Mat. Sci.* **18** 2721

[28] Oritz A, Faleony C, Garcia M and Sanchez A 1987 *J. Phys. D: Appl. Phys.* **20** 670

[29] Falcony C, Oritz A and Garcia M 1988 *J. Appl. Phys.* **63** 2378

[30] Oritz A, Garcia M and Faleony C 1990 *Mater. Chem. Phys.* **24** 383

[31] Ristov M, Sinadinovski G J, Grozdanov I and Mitreski M 1987 *Thin Solid Films* **L49** 65

[32] Bala W and Firszt F 1987 *Acta Phys. Pol.* **A71** 469

[33] Sands D, Brunson K M, Cheung C C and Thomas C B 1988 *Semicond. Sci. Technol.* **13** 816

[34] Wenas W W, Yamada A and Jakahashi K 1991 *J. Appl. Phys.* **70** 7119

[35] Rincon M E, Sanchez M, Ruiz-Garcia J 1998 *J. Electrochem. Soc.* **145** 3535

[36] Katayama J and Izaki M 2000 *J. Appl. Electrochem.* **30** 855

[37] Song D, Widenborg P, Chin W and Aberle Armin G 2002 *Solar Energy Materials and Solar Cells* **73** 1

[38] Hirose M 1971 *Jpn. J. Appl. Phys.* **10** 401

[39] Hirose M and Furuya Y 1972 *Jpn. J. Appl. Phys.* **11** 423

[40] Shiloh M and Gutman J 1971 *J. Cryst. Growth* **11** 105

[41] Piekarczyk W, Gazda S and Niemysky T 1972 *J. Cryst. Growth* **12** 272

[42] Shalimova K V, Nikitenko V A and Pasko P G 1975 *Opt. and Spect.* **39** 332

[43] Takata S, Minami T, Nanto H and Kawamura T 1981 *Phys. Stat. Sol. (a)* **65** K83

[44] Matsumoto K, Konemura K and Shimova G 1985 *J. Crystal Growth* **71** 99

[45] Bhushan S and Chukichev M V 1988 *J. Mat. Sci. Lett.* **7** 319

[46] Butkhuzi T V, Bureyev A V, Georgobiani A N, Kekelidze N P and Khulordaka T G 1992 *J. Cryst. Growth* **117** 366

[47] Bala W, Firszt F, Legowski S and Mecezynska H 1992 *Acta Phys. Pol. A* **82** 896

[48] Smith J M and Vense W E 1970 *Phys. Rev. A* **31** 147

[49] Van Dijken A, Jansen A H, Smitsmans M H P, Vanmackelbergh D and Meijerink A 1998 *Chem. Mater.* **10** 3513
[50] Spanhel L and Anderson M A 1991 *J. Am. Chem. Soc.* **113** 2826
[51] Meulenkamp E A 1998 *J. Phys. Chem.* **B102** 5566
[52] Wong E M, Bonevich J E and Searson P C 1998 *J. Phys. Chem.* **B102** 7770
[53] Chaparro A M, Ellmer K and Tributsch H 1999 *Electrochimica Acta* **44** 1655
[54] Choi K S, Liehtenegger H C and Stucky G D 2002 *J. Am. Chem. Soc.* **124** 12402
[55] Lehmann W 1966 *Solid State Electronics* **9** 1107
[56] Leverenz H W 1950 *An Introduction to Luminescence of Solids* (New York: Wiley) p. 67
[57] Smith L 1952 *J. Electrochem. Soc.* **99** 155
[58] Thomson S M 1950 *J. Chem. Phys.* **18** 770
[59] Dingle R 1969 *Phys. Rev. Lett.* **23** 579
[60] Kumar R 1977 *Ind. J. Pure Appl. Phys.* **15** 55
[61] Agrawal R K and Kumar R 1976 *Ind. J. Pure Appl. Phys.* **14** 931
[62] Cox R T, Block D, Herve A, Picard R and Santier C 1978 *Solid State Commun.* **25** 77
[63] Bhushan S, Pandey A N and Kaza B R 1978 *Phys. Stat. Sol. (a)* **46** K123
[64] Bhushan S, Kaza B R and Pandey A N 1979 *Pramana* **12** 159
[65] Haake C H 1957 *J. Appl. Phys.* **28** 117
[66] Patek K 1960 *Czech. J. Phys.* **B10** 452
[67] Vavilov V S, Chukichev M V, Rezvanov P P, Ben P V, Sokolov V I, Surkova T P and Naumov A Yu 1992 *Bull. Russ. Acad. Sci.* **56** 266
[68] Vanheusden K, Warren W L, Voigt J A and Seager C H 1995 *Appl. Phys. Lett.* **67** 1280
[69] Vanheusden K, Seager C H, Warren W L, Tallant D R and Voigt J A 1996 *Appl. Phys. Lett.* **68** 403
[70] Hayashi Y, Narahara H, Uchida J, Noguchi T and Ibuki S 1995 *Jpn. J. Appl. Phys. 1, Regul. Pap. Short notes* **34** 1878
[71] Egelhaaf H J and Oelkrug D 1996 *J. Cryst. Growth* **161** 190
[72] Kimpel B M and Schulz H J 1991 *Phys. Rev. B Condens. Matter* **43** 9938
[73] Yu P, Tang Z K, Wong G K L, Segawa Y and Kawasaki M 1996 *QUELS'96 (Quantum Electronics Technical Digest Series), Conf. Edition (IEEE 96 CH 35902)* **10** 102
[74] Reynolds D C, Look D C, Jogai B and Morkos, H 1997 *Solid State Commun.* **101** 693
[75] Hiraki H, Kagami A, Hase T, Narita K and Mimuri Y 1976 *J. Lumin.* **12/13** 941
[76] Kramer J 1976 *J. Appl. Phys.* **47** 1719
[77] Bhushan S and Asare R P 1987 *Cryst. Res. Technol.* **22** K23
[78] De Mauer D and Maenhout-vander Vorst W 1968 *Physica* **39** 123
[79] Bhushan S, Diwan D and Kathuria S P 1984 *Phys. Stat. Sol. (a)* **83** 605
[80] Diwan D, Bhushan S and Kathuria S P 1984 *Cryst. Res. Technol* **19** 1265
[81] Fichou D and Kossanyi J 1986 *J. Electrochem. Soc.* **133** 1607
[82] Kouyate D, Ronfard-Haret J C and Kossanyi J 1991 *J. Lumin.* **50** 205
[83] Ronfard-Haret J C, Kouyate D and Kossamyi J 1991 *Solid State Commun.* **79** 85
[84] Kouyate D, Ronfard-Haret J C and Kossanyi J 1993 *J. Lumin.* **55** 209
[85] Bachir S, Ronfard-Haret J C, Azuma K, Kouyate D and Kossanyi J 1993 *Chem. Phys. Lett.* **213** 54
[86] Miralles A, Cornet A and Morante J R 1989 *Mater. Sci. Forum* **38–41** 567

[87] Bachir S, Kossanyi J and Ronfard-Haret J C 1994 *Solid State Commun.* **89** 859
[88] Garcia J A and Remon A 1987 *Appl. Phys. A* **42** 297
[89] Proskura Al 1990 *J. Appl. Spectrosc.* **52** 368
[90] Achour S, Benlahrache M T and Harrabi A 1994 *Electroceramics IV, Aachen, Germany* **1** 631
[91] Bachir S, Sandouly C and Kossanyi J 1996 *J. Phys. Chem. Solids* **57** 1869
[92] Sekiguchi T, Ohashi M and Terada Y 1997 *Jpn. J. Appl. Phys. Lett.* **36** L289
[93] Liu M, Kitai A H and Mascher P 1992 *J. Lumin.* **54** 35
[94] Iwai S and Namba S 1971 *Sci. Papers IPCR* **65** 1
[95] Klingshirn C 1975 *Phys. Stat. Sol.* **B71** 547
[96] Miyamoto S and Shionoya S 1976 *J. Lumin.* **12/13** 557
[97] Segawa Y and Namba S 1976 *J. Lumin.* **12/13** 569
[98] Nikilenko V A, Tereschenko Al, Kucheruk V P, Naumo N P and Paslko P G 1989 *J. Appl. Spectrosc.* **50** 475
[99] Savikhin S F, Freiberg A M and Tranikov V V 1989 *JETP Lett.* **50** 122
[100] Travnikov V V, Freiberg A and Savikhin S F 1990 *J. Lumin.* **47** 107
[101] Zukauskas A, Kriukova I and Latinis V 1990 *Lith Phys. J.* **30** 93
[102] Litovchenko V G, Korbutyak D V and Kryuchenko Yu V 1995 *Ukr. Fiz. Zh.* **40** 1087
[103] Thurian P, Heitz R, Hoffmann A and Broser I 1992 *J. Cryst. Growth* **117** 727
[104] Heitz R, Hoffmann A, Hausmann B and Broser I 1991 *J. Lumin.* **48/49** 689
[105] Lehmann W 1957 *J. Electrochem. Soc.* **104** 45
[106] Wachtel A 1960 *J. Electrochem. Soc.* **107** 602
[107] Bhushan S 1971 *Ind. J. Pure Appl. Phys.* **9** 1065
[108] Bhushan S and Das R 1993 *J. R. S. Univ.* **6B** 67
[109] Maheshwari R C and Khan M H 1975 *J. Lumin.* **11** 107
[110] Drapak I T 1976 *I Z V. Vuz. Fiz.* **6** 45
[111] Minami T, Takata S, Yamanishi M and Kanamura T 1979 *Jpn. J. Appl. Phys.* **18** 1617
[112] Muraleedharam R, Khokhar M S K, Namboodirni V P and Girigavallabham C P 1996 *Mod. Phys. Lett. B* **10** 883
[113] Kanoun A, Alaya S and Marref H 1990 *Phys. Stat. Solidi B* **162** 523
[114] Paje S E, Llopis J, Zayas M E, Rivera E, Elark A and Rincon J Ma 1992 *Appl. Phys. A: Solids Surf.* **A54** 239
[115] Gurskii A L, Lutsenko E V, Morozova N K and Yablonskii G P 1992 *Sov. Phys. Solid State* **34** 1890
[116] Henry M O, Campion J D, McGuigan K G, Lightowlers E C, Carmo Mc dO and Nazare M H 1994 *Semicond. Sci. Technol.* **9** 1375
[117] Bing-Suo Z, Bin W, Guo-Qing T, Gui-Lan Z and Wenju C 1994 *Chinese Sci. Bull* **39** 1171
[118] Ruter D and Bauhofer W 1996 *Appl. Phys. Lett.* **69** 892
[119] Henglein A 1988 *Top. Curr. Chem.* **143** 113
[120] Bahnemann D W, Kormann C and Hoffmann M R 1987 *J. Phys. Chem.* **91** 3789
[121] Bylander E G 1978 *J. Appl. Phys.* **49** 1188
[122] Saparin G V, Obyden S K, Chukichev M V and Popov S I 1984 *J. Lumin.* **31/32** 684
[123] Kazan B 1960 *Proc. IEEE* **56** 285

PART 2

III–V GROUP MATERIALS

Chapter 7

Gallium arsenide and its ternary alloys (self-assembled quantum dots)

D Wasserman and S A Lyon
Princeton University, USA

7.1 Introduction

Tremendous progress has been made over the past two decades in the field of semiconductor growth. This progress has enabled significant strides in the development of semiconductor device technology, and at the same time has also given researchers the means to study fundamental physics at a level which could only have been dreamed of in the past. Much of the growth in these fields was made possible by the development of technology which allows scientists to fabricate defect-free interfaces between different semiconductors. Such interfaces are known as heterojunctions, and their development has led to an explosion in semiconductor research. Most heterojunctions are grown using either molecular beam epitaxy (MBE) or metal-organic chemical vapour deposition (MOCVD) machines, which allow growth with atomic monolayer resolution.

One of the many areas of research which has sprung from the development of heterojunctions is the field of quantum dots (QDs). A quantum dot, most simply defined, is a structure which can confine a charge carrier in all three dimensions. More practically, this confinement is of little interest unless the carrier is confined to a volume which is small enough to allow the carrier to exhibit quantization effects. Although there is no universally accepted definition of a 'quantum dot', a device which confines carriers to the point of quantization is, in general, a good rule of thumb. Because the confining potential seen by an electron in a quantum dot is similar to the potential acting upon an electron in an atom, quantum dots are often referred to as 'artificial atoms'.

Dots are fabricated in numerous ways. One method involves the growth of a quantum well (QW), which confines carriers in one dimension, and then the use of lithographical techniques to control the confinement within the plane of the quantum well. These lithographically-defined quantum dots allow researchers a fair amount of control over location and size of the dots, though making large arrays of such dots is difficult and their optical properties are less than ideal, due to fabrication-related damage. Quantum dots exhibiting excellent optical properties have also been made in solution. However, it is extremely difficult to make electroluminescent devices out of such dots, as making electrical contact to the dots themselves presents something of a challenge. The most common method of making quantum dots is by self-assembled epitaxial growth, and for the remainder of this chapter, it will be these dots we are referring to when we speak of 'quantum dots'.

Three-dimensional confinement of electrons quantizes the allowed energies of the electrons (the electron density of states) into discrete levels. Figure 7.1 illustrates the electron density of states as confinement is increased. The discrete density of states electrons are believed to exhibit in quantum dots has led to numerous proposed (and realized) applications for dots, as follows.

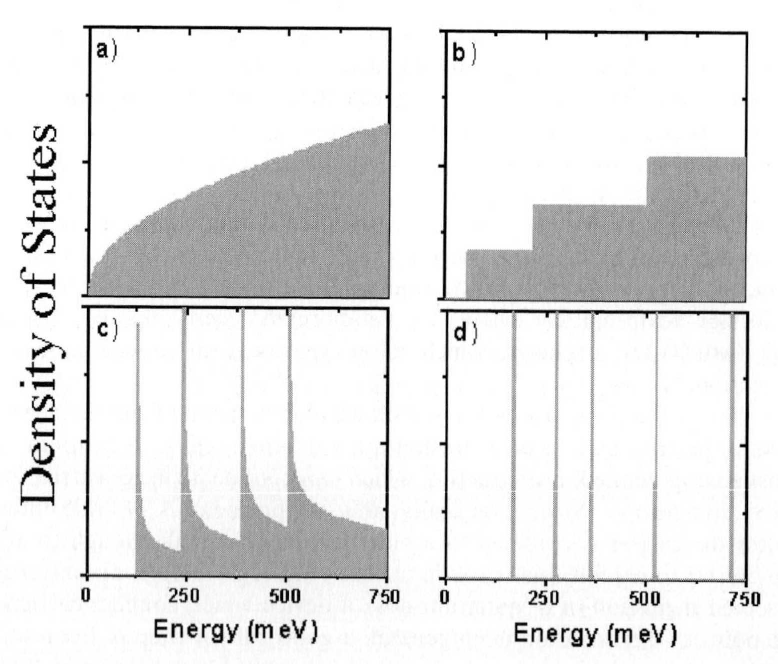

Figure 7.1. Normalized electron density of states for (a) a bulk semiconductor, (b) a 10 nm infinite quantum well, (c) an 8 nm × 12 nm infinite quantum wire and (d) a quantum box or quantum dot.

Quantum computing: Because the separation between energy levels in QDs can be significantly larger that the thermal energy of carriers in the dot (kT), it has been suggested that scattering between energy levels is suppressed. A quantum computing device based on electron spin requires long spin coherence times, which could be achieved if, in fact, thermal transitions between dot states are negligible, allowing the electron to remain in a given state for an extended period of time.

Single photon generation: This would require the ability to isolate, both electrically and optically, a single quantum dot and be able to control carrier injection to this dot. This would allow for the possibility of generating single photons, which is a prerequisite for many quantum communication proposals.

Light detection: Quantum dots are expected to have increased oscillator strength for optical transitions within the dots. Not only would this make them efficient emitters, but also efficient detectors. An incident photon on a quantum dot detector, depending on its energy, can excite an electron hole pair or a carrier from a state in the dot to an excited state. Both transitions would be measured by a corresponding current associated with the absorption of a photon.

Optical emitters: From an opto-electronic standpoint, the discrete density of states in QDs is ideal for devices such as lasers. A typical semiconductor laser uses a quantum well as its active region. Electrons in quantum wells can exist in a continuum of energies, and thus recombine from a wide range of states, which leads to a wide gain spectrum for a quantum well laser. If the quantum well active region is replaced by an array of identical QDs, then electrons can only exist at, and thus recombine from, certain discrete energy levels, which would lead to significantly lower threshold currents for semiconductor laser devices. In addition, because carriers in QDs are subject to confinement potentials larger than those in other semiconductor light-emitting devices, carriers in dots are less likely to escape the active region (quantum dot) than they would be if less tightly confined. A related advantage is a result of the energy spacing between states in these dots being significantly greater than the thermal energy kT of carriers in device structures. Because of this, QDs are expected not to suffer the energy broadening associated with this thermal energy which is present in all other semiconductor devices. Thus, dot-based laser devices, when first proposed, were expected to have improved temperature characteristics in addition to their increased optical efficiency [1].

7.2 Quantum dot growth

Quantum dots grown epitaxially on GaAs or one of its ternary compounds are most often formed by a self-assembly process known as Stranski–Krastanow (SK) growth [2]. In Stranski–Krastanow growth, QDs are

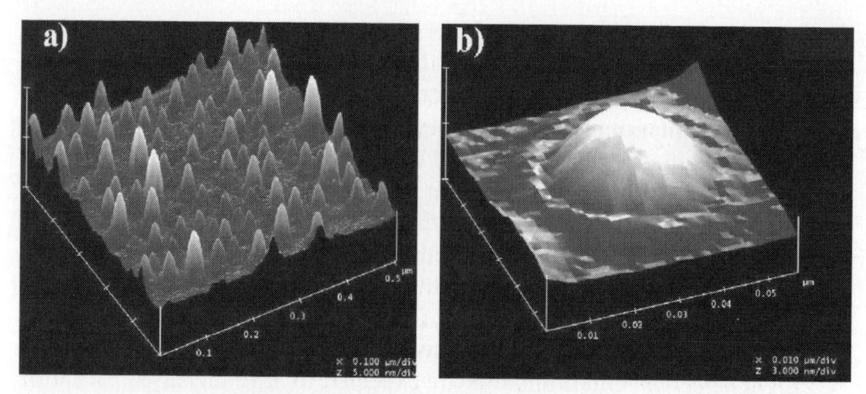

Figure 7.2. AFM image of (a) self-assembled InAs quantum dots on GaAs (image scale 500 nm × 500 nm), (b) close-up of individual quantum dot.

formed due to the lattice mismatch between adjacent epitaxial layers. For the example of InAs QDs on GaAs, InAs (which has a bulk crystal lattice constant of 6.04 Å), is grown upon GaAs (which has a bulk crystal lattice constant of 5.65 Å). The 7% lattice mismatch between the two layers causes the InAs to initially grow with compressive strain. After deposition of ~1.7 monolayers (MLs) of InAs, the compressive strain in the InAs layer is accommodated by the formation of nano-scale InAs islands. The final form of the QD layer includes the nano-scale islands, as seen in figure 7.2, and usually a thin layer, approximately 1 ML thick, of the QD material referred to as the wetting layer (WL).

Quantum dot growth is not limited to InAs on [100] GaAs substrates, or even to the III–V material system. Early examples of epitaxially grown QDs include Ge dots in a Si matrix [3]. In addition, QDs can be grown using $In_xGa_{1-x}As$, InAlGaAs, InAlAs, InSb, GaSb, AlSb and InP as the dot material [4–8], while in addition to [100] GaAs, the surrounding matrix can be $In_xGa_{1-x}As$, $Al_xGa_{1-x}As$, InP or [311] GaAs to mention only a few [9–12].

As mentioned earlier, three-dimensional confinement of electrons in semiconductors can be achieved by other means, though these devices can suffer from problems such as etch damage, low density of dots, fabrication complexity, and difficulty in accessing the dots electrically. Stranski–Krastanow dots, on the other hand, grow dislocation-free at densities of approximately 10^{10} to 10^{11} dots cm^{-2}. In addition, self-assembled QDs tend to grow with fairly good size uniformity, with typical size variations reported to be approximately ±10% [13–15].

In most cases, the QD material has a lower bandgap than the surrounding matrix. An electron electrically injected or photo-excited into the conduction band of a material containing quantum dots will 'fall' into the lower energy wetting layer, and from there into an even lower energy state in the QD. Holes behave similarly in the valence band. If a QD contains both a hole and an electron, the two will often recombine and emit a photon of energy equal to the energy spacing between the QD energy levels of the hole and electron. For this reason, photoluminescence (PL) and electroluminescence (EL) experiments are valuable tools for exploring the electronic structure of QDs.

The peak of a QD luminescence spectrum depends on a number of factors, many of them controllable during the growth of the dots. The most important parameters determining this recombination energy are the size and composition of the dot, and the makeup of the surrounding matrix. For instance, carriers in InAs QDs of equal size and shape will recombine at different energies depending on whether the dots are embedded in an $Al_xGa_{1-x}As$ matrix or a GaAs matrix. Growing the dots in $Al_xGa_{1-x}As$ will increase the size of the potential barriers seen by the electron and hole in the dot, and, due to alloying between the dot layer and the surrounding matrix, actually change the composition of the quantum dots. Both of these effects, by increasing the quantum confinement and alloying Al into the dot layer, increase the energy level separation between electrons and holes in the dot [16, 17]. Similarly, a change in the material composition of the deposited quantum dot layer, from, for example, InAs to $In_xGa_{1-x}N_yAs_{1-y}$, InN_xAs_{1-x}, or $In_xAl_{1-x}As$, can shift the emission spectra to lower or higher energies, depending on the material grown [6, 18]. In addition, substrate temperature during dot growth can affect the size, shape and density of dots grown [19] (figure 7.4(b)). Finally, varying the amount of the dot material deposited can also significantly alter the dot energy level spacing and dot density, as can be seen in figure 7.3(a) and figure 7.4(a), respectively [20, 21].

In addition to the *in situ* methods of controlling the electronic properties of dots, researchers have been able to affect electronic properties of the dots *ex situ*, most notably by performing high-temperature anneals which, by alloying the dots with the surrounding matrix, raises (lowers) the QD electron (hole) ground state in energy and decreases the energy spacing between successive electron or hole states [22–24].

The above methods of controlling the electronic structure of QDs allow growers to adjust the energy spacing between electron and hole states and exert some control over dot size and density. However, there remain significant obstacles to the full realization of quantum dots' technological potential due to the inherent difficulty in controlling self-assembled dot growth. As can be seen in figures 7.2 and 7.4, QDs do not form in any sort of ordered array and, in addition, can vary in size and even shape in a single layer. In order to realize many of the potential applications possible with QDs, one must be able to grow ordered dot arrays, with dots formed in predetermined locations so that they may be accessed electrically. Other applications, such as QD lasers, do not require ordered dot growth, but would benefit greatly from the ability to grow dots of uniform size and shape.

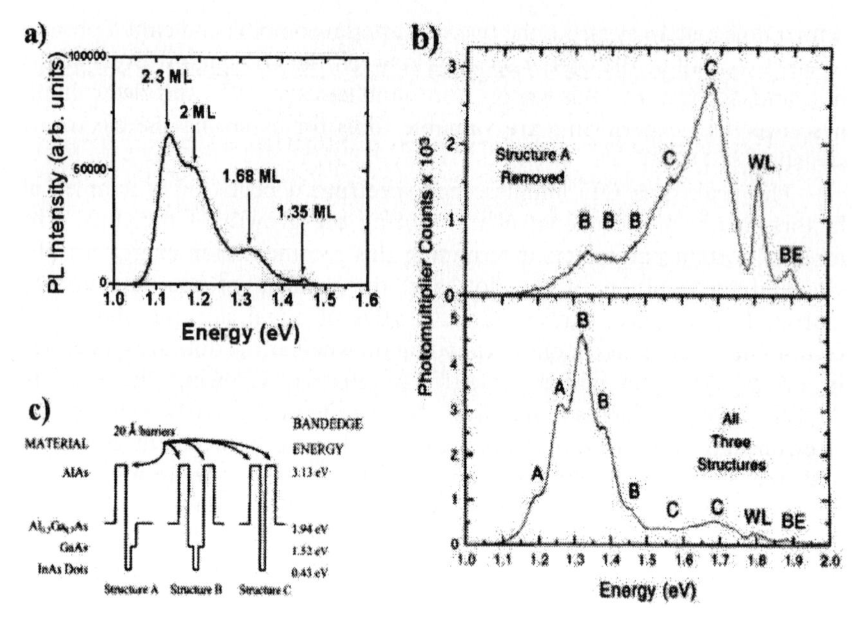

Figure 7.3. (a) PL spectra from InAs quantum dot layers grown with varying InAs deposition times [20]. (b) PL spectra from InAs quantum dots in three different cladding structures, (c) schematics of three structures used in (b) [17]. Peaks in all three spectra were determined by etching samples layer by layer and noting changes in PL spectra.

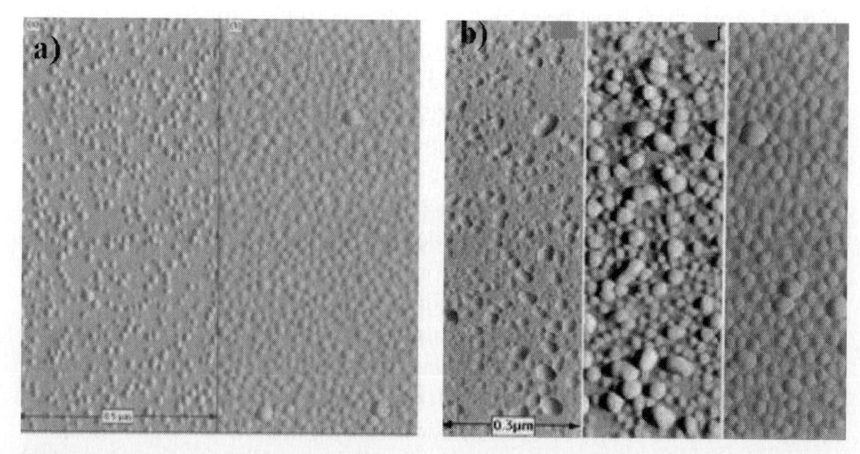

Figure 7.4. Examples of the effects of growth parameters on self-assembled quantum dot formation (a) from left to right, coverage of 2 and 3 ML [21], (b) from left to right, substrate growth temperature of 425 °C, 500 °C and 540 °C [19].

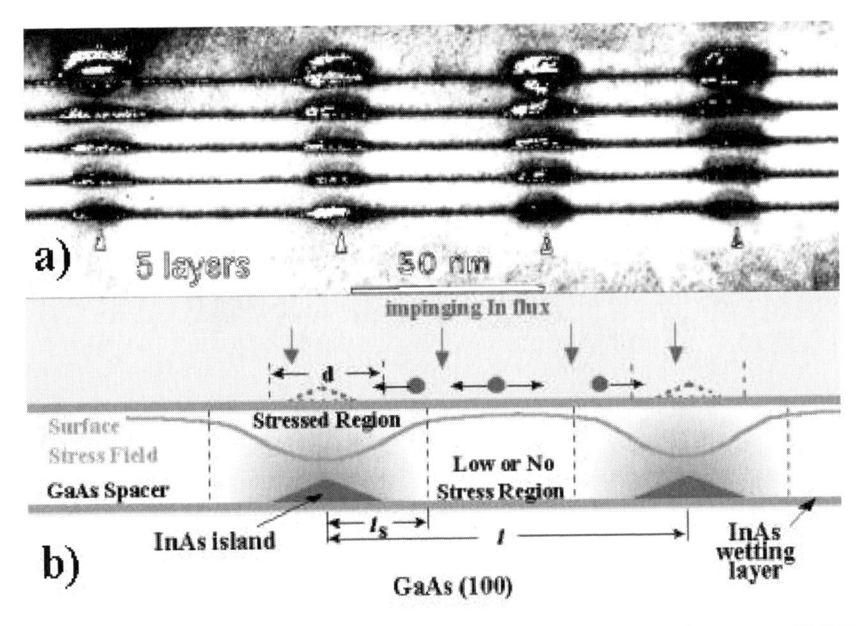

Figure 7.5. (a) Cross-sectional dark-field transmission electron microscopy (TEM) image of vertically aligned InAs quantum dots separated by 36 monolayers of GaAs. (b) Schematic of In adatom migration during epitaxial growth on a stressed surface [26].

Most efforts to improve the growth of quantum dots are focused on resolving the two difficulties of ordering dots and maintaining size uniformity within dot layers. Depending on the potential application, one of these may be more important than the other. While QD size uniformity can be influenced by epitaxial growth parameters, achieving total size uniformity by varying growth parameters has not yet been proved to be possible. Natural ordering of IV–IV QDs has been demonstrated [25], though the same effect has not been seen for dots grown on GaAs or other III–V semiconductor substrates.

One interesting feature of QD growth is that the dot layers align vertically of their own accord. While the island nucleation of the first epitaxial layer of dots may not be ordered within the growth plane, subsequent dot layers, if grown with a thin enough spacer layer between them and the lower layer, tend to align above the lower layer of dots, as can be seen in figure 7.5. Researchers believe this alignment is a result of the strain field associated with each QD propagating through the overgrown cap layer and providing preferential island nucleation sites for the next layer [26].

While the vertical self-alignment of QDs allows ordering in one dimension, spatial alignment along the growth plane remains a subject of

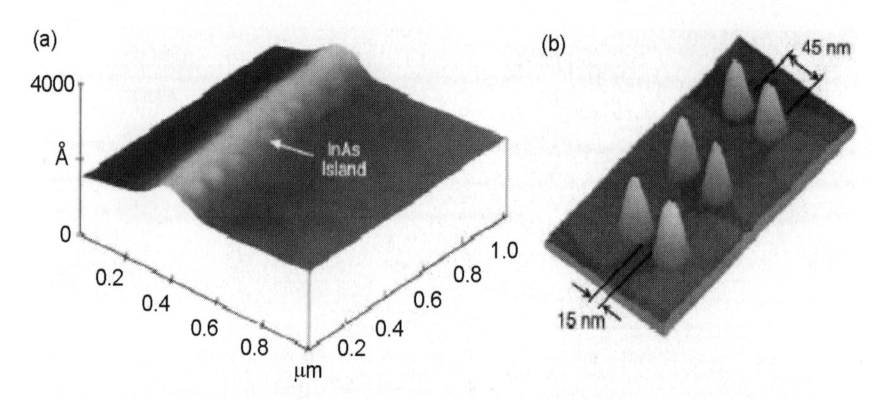

Figure 7.6. (a) One-dimensional array of InAs quantum dots formed on the edge of an etched GaAs ridge [28] [copyright (1995)]. (b) Two-dimensional array of InAs quantum dots formed by selective scanning tunneling assisted nanolithography [31].

substantial research. Numerous groups have used novel techniques in attempts to laterally order self-assembled dot growth. One such approach is epitaxial growth on patterned substrates. In this scheme, the growth substrate is patterned before growth of the dot layer. Various patterning schemes have been employed, including holographic lithography [27, 28], electron beam lithography [29, 30], scanning tunnelling probe assisted nanolithography [31], atomic force microscope direct patterning [32], and area-controlled growth using oxide masks defined by conventional photolithography [33, 34]. The above schemes all rely on a similar approach, which is to pattern the substrate in such a way that the QDs are forced to grow, or at least prefer to grow, in select nano-scale regions.

The patterning in figure 7.6(a) was achieved by etching a ridge on a (100) GaAs substrate, upon which InAs is deposited. The InAs accumulation rate depends on surface orientation, and thus the ridge edge reaches the critical InAs thickness for QD formation before other areas on the sample, creating a one-dimensional array of QDs along the side of the etched ridge. In figure 7.6(b), the GaAs growth surface is selectively oxidized by a scanning tunnelling probe. GaAs growth on top of the oxidized point is significantly slower than elsewhere, thus upon overgrowth of the substrate, an indent is left over the oxidized GaAs. When the InAs dot layer is deposited upon this surface, InAs preferentially nucleates on the indents in the GaAs.

While growth on patterned substrates has proven to be an effective means of ordering dot growth, patterned growth has significant drawbacks. Firstly, from an opto-electronic standpoint, in order to get substantial light emission from a patterned growth one must be able to write nanoscale patterns over a significant area, making the potential fabrication of such

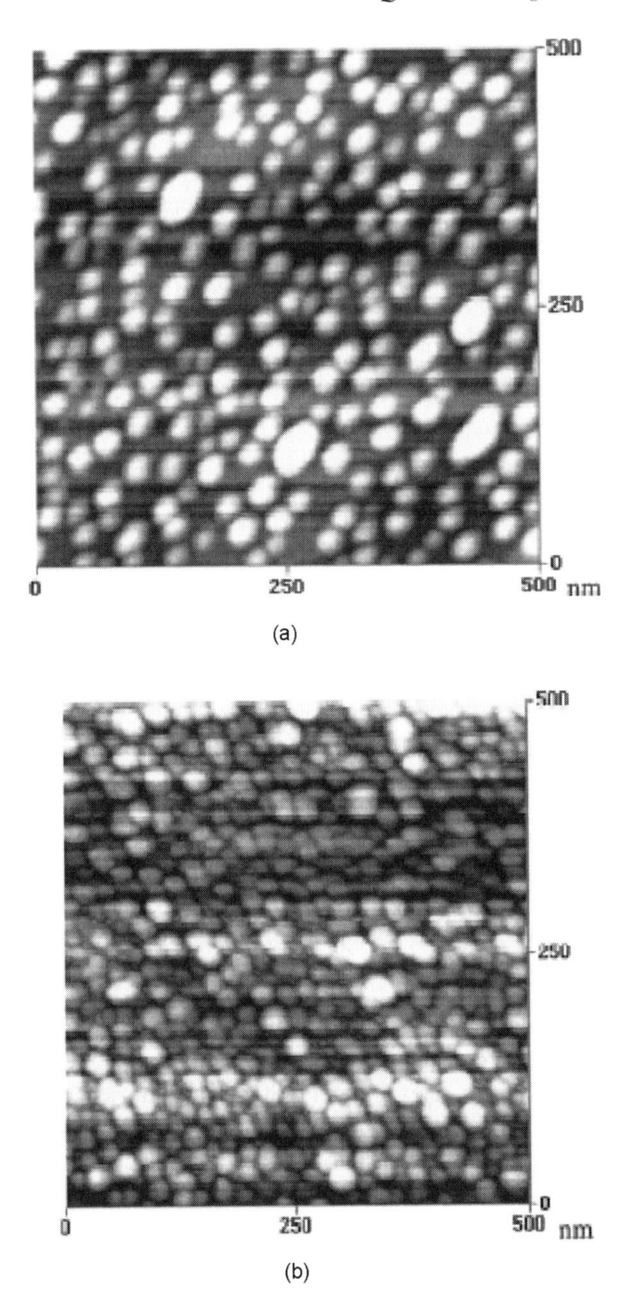

Figure 7.7. AFM images of quantum dot layer grown with (a) no stressor dots and (b) one layer of buried $In_{0.4}Al_{0.6}As$ under surface dots imaged. The PL linewidths from samples grown with one or more stressor layer show a decrease in luminescence linewidth by more than a factor of two, an indication of increased size uniformity [36].

devices rather labour-intensive. In addition, the patterning process very often degrades the electrical characteristics of the growth matrix surrounding the dots and the electrical and optical properties of the dots themselves. While much work has gone into ordering dot growth, there has been relatively little published work studying the optical properties of these dots. Nonetheless, ordered dots have been shown to luminesce at room temperature, though without the improvement in linewidth one might expect from the ordered, uniform dots in figure 7.6(b) [35].

While size uniformity tends to be a beneficial by-product of patterned growth, there have been attempts to improve size uniformity with methods which do not necessarily result in ordered dots. One such method involves the growth of a thin stressor layer below the self-assembled dot layer. An example of such a stressor layer would be 20 nm of $In_{0.2}Ga_{0.8}As$ grown approximately 10 nm below the first QD layer. Other stressor layers studied consist of one or more buried layer of InAlAs 'stressor dots'. These stressor layers have been shown to improve the size uniformity, and thus narrow the luminescence linewidth of the QD layers (see figure 7.7) [36]. Stressor layers have also been incorporated into patterned substrates, and seem to be able to exercise some spatial control over island nucleation, though no improvement in luminescence linewidth has been observed [27, 37].

7.3 Carriers in quantum dots

In order to design light-emitting devices using self-assembled quantum dots, it is important to have a solid understanding of both electron and hole energy levels within the dot. While it is not particularly difficult to qualitatively explain the manipulation of electron and hole energy states as a function of cladding material, dot composition, or dot size, developing a quantitative understanding of energy levels in quantum dots is a somewhat more rigorous task.

Before immersing ourselves too deeply into the theoretical aspects of QDs, it might be worthwhile to first determine which portions of QD theory will assist us in the understanding and fabrication of electroluminescent devices. Controlling electron and hole energy levels within QDs will be important, as the spacing between hole and electron states will determine the wavelength of emitted light. Once able to determine the energy states and wavefunctions of carriers in QDs, we will want to make use of this information to better understand optical transitions within QDs. In addition, we will want to have some understanding of electron and hole lifetimes in QD states when we begin to discuss QD lasers. We will attempt to give an overview of the theoretical studies of QDs bearing in mind the three issues of electronic structure, optical transitions, and carrier lifetimes in dots.

7.3.1 Electronic structure

7.3.1.1 Theory

At first glance, modelling the electronic structure of QDs might seem fairly straightforward. The band gaps of the semiconductors used in QD growth are well known, as are the band offsets for adjacent materials in heterostructure devices [38]. If the dot size and shape and the band offsets are known, one might argue the problem could be solved simply using the three-dimensional Schrödinger equation. In fact, modelling QDs is nowhere near this easy. This is because properties such as band gaps, band offsets and effective masses (to be discussed later) have been determined by studying bulk semiconductors. Unfortunately, the bulk properties of these materials may not be accurate for models on the nanometre scale. In addition, QDs are strained layers, and this strain will also affect the properties of these structures. Because of the strain introduced into the system with the deposition of the dot layer, any calculation must take into account the effect of the strain on the electronic structure of both the matrix and QD materials.

Strain can have a significant effect on the band structure of a semiconductor. We will assume that all QDs to be studied are pseudomorphic, which means that the strain in the QD layer is accommodated through elastic deformation, and not by the creation of defects or dislocations. Though this may not always be the case, defects and dislocations are extremely damaging to the electrical and optical properties of the systems we will be looking at, rendering them effectively unusable as electroluminescent devices. The most significant effect of strain in the QD layer is to separate the normally degenerate valence bands. For semiconductors, the biaxial compression of the strained layer causes not only an increase in the bandgap of the dot layer material, but also splits the valence bands of the dots [39]. This valence band splitting is actually advantageous for fabrication of QD lasers. Because the light hole and heavy hole bands are now separated by some distance in energy, one can essentially 'fit' more holes into the lowest energy of the valence band before one starts populating the next energy level, allowing for higher light output than degenerate heavy and light hole energy levels would. Strained layers have in fact been used to improve solid state laser performance for over two decades [40]. Another important effect of strain is the piezoelectric effect, where strain in the semiconductor shifts atoms from their unstrained equilibrium positions and thus produces an electric field which changes the potential (bands) of the semiconductor by some energy.

Early attempts to model the electronic structure of QDs used a single-band effective mass approximation, assumed all strain existed within the QDs and wetting layer, and altogether ignored the piezoelectric effect [41]. The problem was solved for a single electron (hole) in the conduction (valence) band, ignoring electron–electron (hole–hole) interactions. The QDs in this calculation were assumed to be conical in shape, with a base angle of 12°, sitting on a wetting layer of fixed width.

As mentioned earlier, we will assume we know the conduction and valence band offsets for the material system with which we are working. The effective mass approximation allows us to simplify the QD problem to a system which can be solved with the three-dimensional Schrödinger equation. The only adjustment which must be made in order to simplify the problem in this manner is to assign an 'effective mass' for the holes or electrons in each layer of the structure one is modelling. The effective mass for heterostructures is approximated theoretically by assuming the electron (hole) is sitting at the minimum (maximum) of the conduction (valence) band of the bulk material.

For a three-dimensional problem, the one-dimensional effective mass becomes an effective mass tensor. The effective masses of carriers in semiconductors have been studied extensively and are well known. Knowing the effective masses of the materials system in question, and the conduction and valence band offsets, the effective mass approximation has us solve the three-dimensional Schrödinger equation

$$\left[\frac{h^2}{2m(\mathbf{R})}\nabla^2 + V(\mathbf{R})\right]\Psi(\mathbf{R}) = E\Psi(\mathbf{R}) \tag{7.1}$$

where $m(\mathbf{R})$ is a function giving the effective mass of the carrier both inside and out of the dot layer and $V(\mathbf{R})$ describes the conduction (or valence) band profile in space. While this problem is still not easy to solve, it is a significant simplification of the initially complex problem.

A subsequent theoretical study of dots again used the single band, single particle effective mass approximation, but employed a more rigorous approach towards strain distribution in the system [42]. This study modelled the dots as pyramidal structures with square bases and a height equal to half the length of a base side. The strain distribution throughout the dot was then calculated numerically using elastic continuum theory, which treats the material system as a continuum with a continuous strain distribution and then minimizes the total energy of the system. Once the strain distribution is solved, it can be applied to determine the strain effect on the material band structure. Although strain will also affect the effective mass of carriers in the structure, this effect was not taken into consideration. The problem, with the new band structure, is then discretized into a three-dimensional grid, and solved numerically as in the previous example. The effective mass approximation calculations with and without strain effects both concluded that only one electron bound state existed in the dot while, because of the significantly heavier effective mass, numerous bound hole states could exist (figure 7.8).

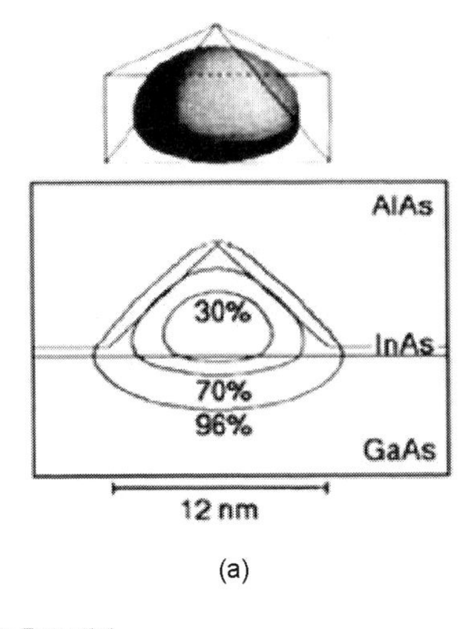

(a)

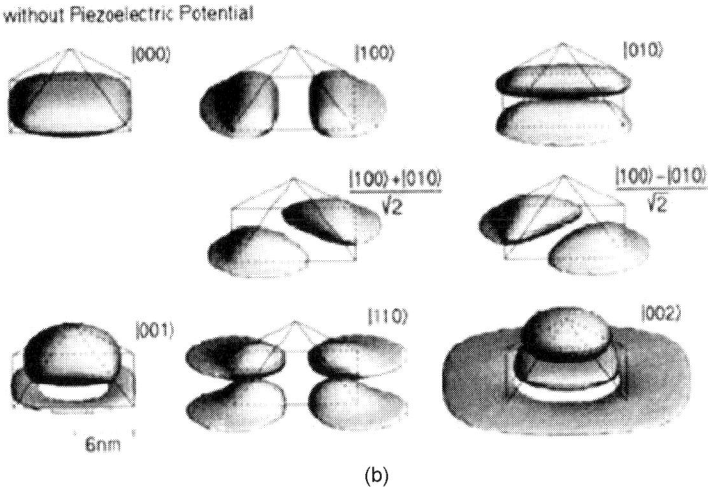

(b)

Figure 7.8. Calculated isoenergies for (a) single electron bound state in pyramidal QD and (b) first eight heavy hole bound states in QD of same size [42].

Each successive theoretical study of electronic structure in QDs built on the previous, adding further effects not yet considered in, or at least not yet incorporated into, other models. The next study incorporated the effect of strain on effective mass. The conduction band and valence band were then solved separately. The conduction band was solved as a single particle effective mass problem, while the valence band was solved using a four-band Schrödinger equation [43]. Using a multiple band approach for the valence band allowed for the consideration of valence band mixing between the heavy hole and light hole bands. This approach did predict the existence of an excited electron state separated by approximately 150 meV from the ground state, and only bound within the dot for the larger dot diameters considered.

The real problem with the effective mass approximation for QDs is that the effective mass is determined using the bulk band structure of the material in question. Because of the small size of QDs, approximating the properties of the dot layer by the bulk properties of the dot material is somewhat suspect. Despite this, perhaps the biggest motivation to move beyond the earliest models used to study QDs was the fact that experimental evidence [44, 45] began to appear suggesting that QDs held more than one bound electron state, which promptly necessitated a re-evaluation of previous theoretical QD models.

In the above theoretical models of QDs, holes in the valence band and electrons in the conduction band are treated separately. More sophisticated models attempt to derive the energy states of the QD by expanding them in a basis made up of the bulk material's valence band minima (heavy hole, light hole, and spin–orbit split-off, including spin) and conduction band minima (two states, including spin) [46–48]. When modelling electron and hole energies in most semiconductor heterostructures, it is often not necessary to consider further mixing between conduction and valence bands, as they are separated by large energies [49]. In the lower bandgap dots, however, the conduction–valence band separation is significantly smaller than that of, for example, an AlAs/GaAs/AlAs quantum well, and on the order of magnitude of inter-level energy spacings within the dot. Thus, rather than solving for the conduction and valence bands separately, a more accurate picture of dot electronic structure might be obtaining by considering mixing between these bands.

This eight-band $k \cdot p$ approximation, as it is known, solves for a Hamiltonian consisting of two 8×8 matrices, $H_{\mathrm{tot}} = H_0 + H_{\mathrm{str}}$, where H_0 represents the usual kinetic and potential terms and H_{str} is the addition to the normal Hamiltonian due to the strain in the QD. The eight-band $k \cdot p$ approximation proves able to account for multiple excited electron states (figure 7.9). However, the model is very much dependent on the modeller's choice of parameters, which can often be chosen to fit empirical data.

Other theoretical studies try to avoid the ambiguity associated with parameter-dependent methods. One such approach is to actually solve the QD problem atom by atom [50, 51]. The Hamiltonian for such a system is referred to as the pseudopotential Hamiltonian, and takes the form

$$H = -\tfrac{1}{2}\nabla^2 + \sum_{\alpha n} \nu_\alpha(\mathbf{r} - \mathbf{R}_{\alpha,n}). \tag{7.2}$$

For this expression, ν_α refers to the potential of an atom of type α (for InAs QDs on GaAs, the α will designate the atom as an In, Ga, or As atom). The

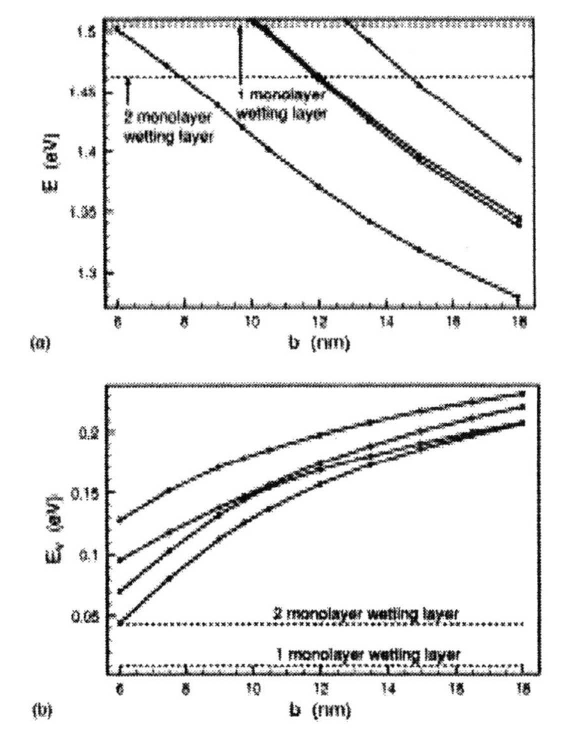

Figure 7.9. Calculated bound states (using eight-band $k \cdot p$ approximation) as a function of pyramidal base length for (a) electrons in the QD conduction band and (b) holes in the valence band. Wetting layer energies calculated for wetting layers of both 1 and 2 ML [47].

positions of individual atoms in the model are determined by first solving for strain distribution throughout the region being studied. Once the equilibrium strained crystal lattice is determined, the above Hamiltonian is expanded to include n atoms, sitting at the calculated lattice sites, $\mathbf{R}_{\alpha,n}$, with pseudopotentials determined by the atom type α.

The Schrödinger equation is then solved numerically using the expanded Hamiltonian. The limitations inherent in the pseudopotential approach are simply a result of the large number of atoms required to accurately model the QD and the surrounding material. Solving numerically for the Hamiltonian requires massive computational power. The dots examined by such an approach tend to be smaller in size than most of the dots pictured earlier in this chapter, as the larger dots include more atoms and are thus more difficult to model. Nonetheless, the results from this approach (figure 7.10) have matched experimental results quite nicely [52].

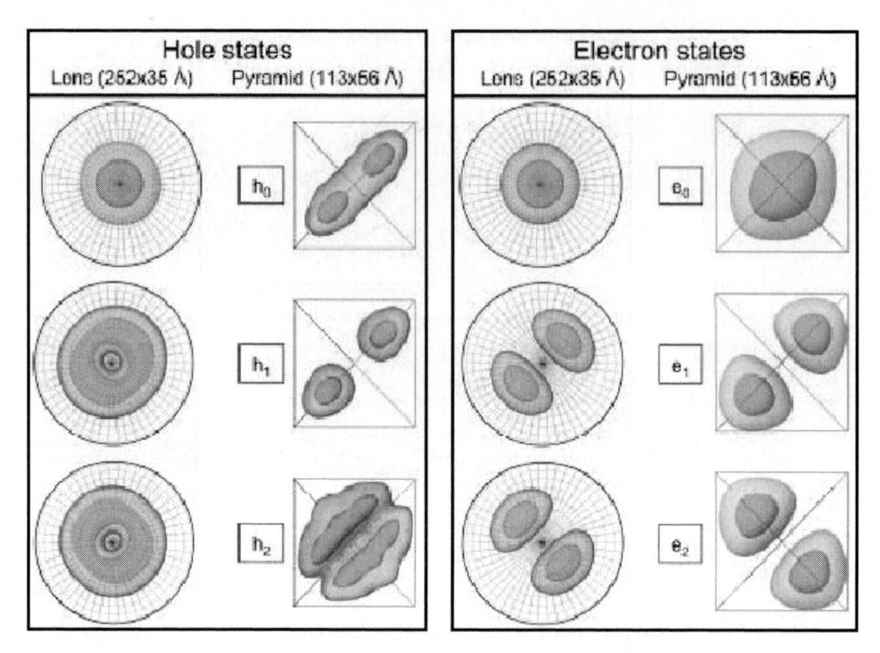

Figure 7.10. Lowest three energy states for electrons and holes in lens-shaped and pyramidal QDs. Darker and lighter surfaces represent 20% and 60% of the maximum charge density $|\psi|^2_{\rm max}$ [52].

7.3.1.2 Experimental determination of electronic structure

While theory continues to make advances in modelling quantum dots, it bears noting that energy states in QDs have been determined experimentally in a simple, but beautiful, magneto-tunnelling experiment [53]. Researchers placed a layer of InAs QDs in GaAs between $Al_{0.4}Ga_{0.6}As$ tunnelling barriers, which in turn are separated from an $n+$ GaAs emitter and collector by a layer of undoped GaAs on each side (figure 7.11(a)).

A bias is put on the sample and electrons are tunnel-injected from the emitter into the heterostructure. As the bias is increased, the electron energy levels in the QDs come into resonance with the energy of the injected electrons, and transmission through the structure increases. In this manner, as one scans voltage, one sees a series of current peaks, corresponding to electron transmission through the QD energy levels.

The same IV measurement is then performed, but with a magnetic field parallel to the growth plane. The magnetic field exerts a force on the electron giving it a momentum $\mathbf{k} = \mathbf{k}_z + \mathbf{k}_\beta$, with $\mathbf{k}_\beta$ being the direction parallel to the growth plane and perpendicular to the direction of the magnetic field. The strength of the magnetic field determines the magnitude of the electron $\mathbf{k}_\beta$ when it reaches the dot layer. By scanning the magnitude and direction of

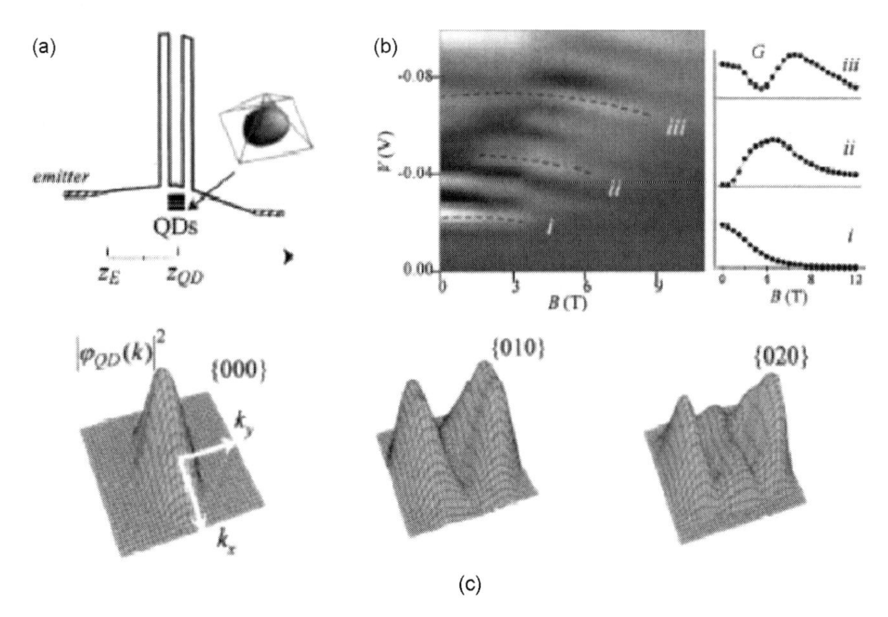

Figure 7.11. (a) Schematic of device. (b) Example of raw data from magneto-tunnelling experiment. Grey-scale plot shows conductance as a function of B field and voltage while three plots on the right show B-field dependence of conductance for the three conductance peaks labeled *i*, *ii* and *iii*. (c) Lowest three electron wavefunctions of quantum dot (in k-space) [54].

the magnetic field, one can scan through k-space (in the growth plane). The transmission through the dot at a given $\mathbf{k}$ corresponds to the magnitude of the electron wavefunction $\Psi(k_x, k_y)$ in the dot for that $\mathbf{k}$ ($\mathbf{k} = \mathbf{k}_z + \mathbf{k}_\beta = \mathbf{k}_x + \mathbf{k}_y + \mathbf{k}_z$). Figure 7.11(c) shows the resulting wavefunctions (in k-space) for the lowest three electron energy levels in the dot [54].

7.3.2 Optical transitions

Once the form of the carrier wavefunctions within the QDs has been determined, the strength of optical transitions between these states can be determined by finding the oscillator strength of each transition [48, 55]. For an optical transition between an initial state Ψ_i and a final state Ψ_f, the dimensionless oscillator strength is given by

$$f_{ij} = \frac{2}{m(E_\mathrm{f} - E_\mathrm{i})} |\langle \Psi_\mathrm{i} | \mathbf{e} \cdot \mathbf{p} | \Psi_\mathrm{f} \rangle|^2 \tag{7.3}$$

where $\mathbf{e}$ is the unit vector associated with the output photon's polarization, $\mathbf{p}$ is the momentum operator, and $(E_\mathrm{f} - E_\mathrm{i})$ is the energy separation between the initial and final states. While the above approach describes the strength of an optical transition in a dot assuming the presence of an electron and hole, determining the optical properties of an array of non-uniform dots all with different electron and hole occupation probabilities becomes significantly more difficult. Additionally, an electron and hole trapped in a QD often form an exciton, namely a bound electron–hole pair, which recombines at a lower energy with an altered oscillator strength, when compared with an unbound electron–hole pair. Although the above equation does not take into account excitonic, multi-exciton (multiple electrons bound to an equal number of holes), charged excitons (multi-excitons with an unequal number of electrons and holes), or the variation in occupation probabilities through the QD layer, it will suffice as good zeroth-order approximation for optical transition strength within dots.

7.3.3 Electron lifetimes

The behaviour of electrons in QDs is of great importance for opto-electronic applications of QD structures. Early studies of electron lifetimes in QDs suggested groundstate lifetimes in the $\sim$1 ns range, a factor of 2 greater than those found for quantum well structures [56]. This result is somewhat surprising, as the tight spatial confinement of carriers in QDs was expected to increase the oscillator strength of interband transitions due to expected spatial overlap of the electron and hole wavefunctions. However, as we have seen in the modelling of QD structures, it has been shown that overlap is somewhat decreased due to the spatial separation of the electron and hole wavefunctions (figure 7.12(b)). Another possible explanation for the longer than expected lifetime in QDs, the 'phonon bottleneck', has serious implications for the drive towards efficient QD light-emitters [57].

Very early in the study of QDs, concern was voiced regarding the potential efficiency of QD light-emitting devices due to this so-called phonon bottleneck [58]. In most solid-state light emitting devices, a carrier injected into the 'active', or light emitting, region will relax to the lowest available energy level by the emission of phonons, which can be thought of as vibrational modes of the crystal lattice. As the carrier emits phonons it descends, in steps equal to the energy of the emitted phonons, through the continuum of energy states. However, in a QD, there exists no continuum of energy states, since, as was earlier discussed, the energy levels in dots are discretized.

For a transition between energy states in the conduction (or valence) band of the dot, both energy and momentum must be conserved. However, because carriers are so tightly confined in QDs, they are not believed to have significant momentum. In the dispersion relation for phonons in GaAs shown in figure 7.12(a) [59], one can see that for low momentum (far left on the x axis), the optical phonons (upper branch) have energies of $\sim$30 meV while the acoustic phonons (lower branches)

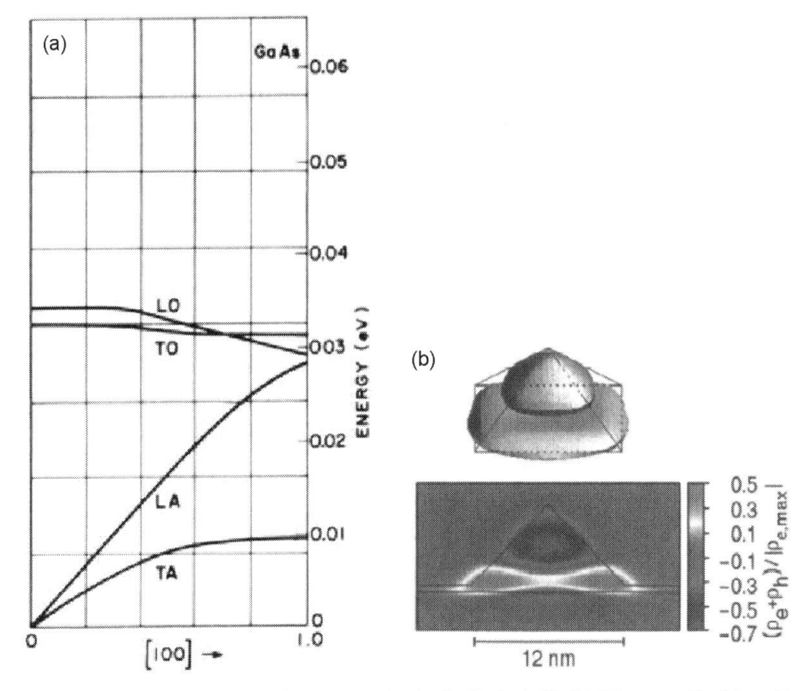

Figure 7.12. (a) Phonon dispersion curves for bulk GaAs [59]. (b) Electron (dark) and hole (light) ground state wavefunctions [42].

have very low energies. If the energy spacing between electron levels is $\gg 30\,\mathrm{meV}$, then, it was argued, carrier relaxation via a single phonon process is difficult, significantly increasing relaxation time in the QDs, especially at low temperature.

The issue of the phonon bottleneck in QDs remains one of some controversy. Because ground-state recombination seems to be the dominant source of luminescence for QDs, it is generally agreed that there exists some relaxation mechanism which overcomes the proposed bottleneck. What this mechanism is, exactly, remains a point of debate. Theoretical studies have accounted for (or predicted) these rapid relaxation times by Auger processes [60], phonon-assisted relaxation [61], or some combination of the two, where electrons are captured by the QDs from the wetting layer via phonon emission and relax within the dot by Auger processes [62]. Time-resolved photoluminescence spectroscopy experiments suggest that excited state lifetimes in QDs are on the order of tens of picoseconds [63–65], though different conclusions are reached concerning the mechanisms enabling these times. One explanation suggests that multiphonon processes allow rapid relaxation [63]. Others argue that Auger processes dominate as the number of carriers within each dot is increased [64, 65]. Yet another study argues for the existence of a continuum of states within the QDs, which facilitates rapid recombination by the emission of localized phonons [66].

If, in fact, the discretization of carrier energy states in QDs does produce a phonon bottleneck (which seems unlikely), the potential improvements expected from QD light emitting devices could be limited. However, it is important to remember these detrimental effects result from the very same phenomenon which accounts for the low-threshold currents in QD lasers up to temperatures of $\sim$150 K. The discrete density of states makes it difficult for carriers to thermalize out of the dots below 150 K, increasing carrier confinement in the active region [67]. In addition, the phonon bottleneck would make QDs ideal structures for mid-infrared emission, which we will discuss further in the coming section.

7.4 Electroluminescence from quantum dots

7.4.1 Near-infrared luminescence

Near-infrared emission from QDs necessarily involves the recombination of an electron with a hole, both of which must be generated electrically. The most obvious way to satisfy this requirement is to put the QDs in the depletion region of a p–n junction diode. Under forward bias, holes from the p-doped region diffuse into the dots, while electrons from the n-doped GaAs diffuse into the dots from the opposite side. Once in the dots, carriers will usually relax quickly to the ground state from where they will recombine and emit a near-infrared photon.

There is substantial interest in near-infrared light emitters from QDs and it is probably the most developed of all QD applications. The main motivation for this interest is the desire to make near-infrared lasers out of QD structures, in order to take advantage of their unique optical properties. Specifically, researchers are interested in fabricating QD lasers for telecommunications purposes, an application which will be discussed in the following section.

7.4.2 Quantum dot lasers

7.4.2.1 Fundamentals of semiconductor lasers

In order to fabricate a semiconductor laser, it is desirable to utilize an active region which is an efficient light emitter. Ideally, for a perfect near-infrared semiconductor laser, every electron and hole injected into the active region of the device results in an emitted photon. The first semiconductor lasers were made with a very heavily doped p^+n^+ junction diode [68–71] (figure 7.13(a)). In p–n junction diodes without QDs in the depletion region, electron–hole recombination occurs primarily in the n and p

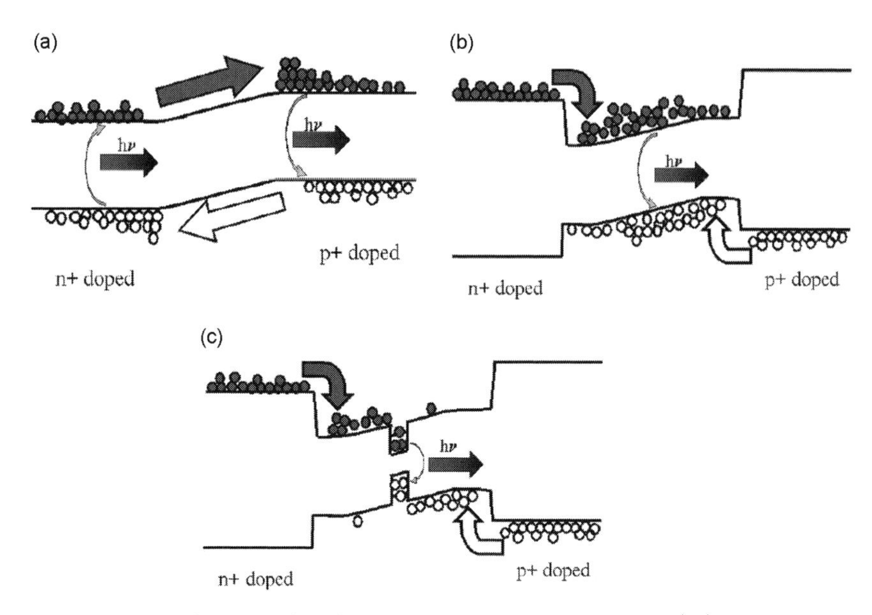

Figure 7.13. Development of semiconductor laser. Schematic of (a) p^+n^+ junction diode laser, (b) double heterostructure p–n junction diode laser, (c) quantum well laser.

regions, as minority carriers diffuse across the junction and recombine, on average, within a diffusion length of the junction. Since diffusion lengths for carriers in p–n junctions can be on the order of microns, light emission in these early devices occurs over a large volume. This means that a significant amount of current is necessary to achieve population inversion and thus lasing.

Improvements were made when it became possible to define a smaller active region using a double heterostructure design [72] (figure 7.13(b)). In these devices, the active region is bound on either side by a semiconductor of higher bandgap. Such a design has two main advantages. First, the potential barriers surrounding the active region prevent carrier diffusion and thus localizes electron–hole recombination in the lower bandgap active layer. Secondly, because the higher bandgap material has lower refractive index n, light is also confined to the active region. The result of this innovation was a dramatic increase in semiconductor laser efficiency leading to continuous-wave room temperature operation of these devices, and, incidentally, the 2000 Nobel Prize in physics.

As fabrication, and more specifically, growth, techniques improved, researchers became able to exploit the quantum mechanical properties of heterostructures, and lasers using quantum wells became state of the semiconductor laser art [73] (figure 7.13(c)).

As semiconductor laser devices utilized smaller and smaller active regions, device properties improved significantly. It is thus not surprising that QDs were expected to exhibit additional improvements in emission efficiency. As discussed earlier, QDs are extremely efficient light emitters as a result of their discretized energy states and because they are able to confine electrons and holes in close spatial proximity. The drawbacks of using QDs for the active region of a laser device are twofold. Firstly, the inhomogeneity of QD size distribution significantly broadens the active layer's gain spectrum and diminishes the predicted benefits of the discrete QD energy states. Secondly, QDs do not occupy, in volume, a large fraction of the active region. In a quantum well laser, a photon can stimulate an optical transition anywhere in the active region. In a QD laser, a photon can only efficiently stimulate an optical transition in a QD if the photon is near the small volume in space occupied by the dot, and furthermore, only if the QD has an allowed transition at the energy of the photon, which becomes less and less likely if size uniformity of the dots is not controlled. It is important, for these devices, that emitted photons are tightly confined in or near to the active (dot) layer. The ability to confine the optical modes of a QD laser, to a large extent, will determine the quality of the laser in question.

Optical mode confinement in semiconductor lasers can be achieved by a variety of means. The most common method for confinement is the edge-emitting laser. This device utilizes thick, low index layers below and above the active region. For a laser grown on a GaAs substrate, these layers most often consist of $Al_xGa_{1-x}As$, whose thickness determines the extent to which optical modes in the active region are vertically confined. Lateral confinement is usually achieved in these structures by etching a ridge through the active region, leaving only a stripe-shaped device, and thus confining the active region in an additional direction with an interface between the active region material and either atmosphere or a deposited layer of protective oxide (figure 7.14(a)). The mirrors of the optical resonator for edge-emitting lasers are the cleaved ends of the ridge structure, which are often covered with high-reflectivity coatings to adjust the transmission characteristics of the semiconductor/atmosphere interface.

Another method of mode confinement is used to make vertical cavity surface emitting lasers (VCSELs), shown in figure 7.14(b)–(d). A VCSEL, as the name suggests, emits light through the surface of the device (as opposed to the edge), and the device geometry more closely resembles a post or pillar (as opposed to a ridge). In a VCSEL, optical (and electrical) confinement is achieved by etching a post structure, implanting the surrounding area with protons degrading its ability to conduct electricity, or by selective oxidation of Al-containing layers above and below the active region. Distributed Bragg reflectors (DBRs) are used as mirrors for VCSELs. For an operation wavelength of λ, a DBR consists of a superlattice

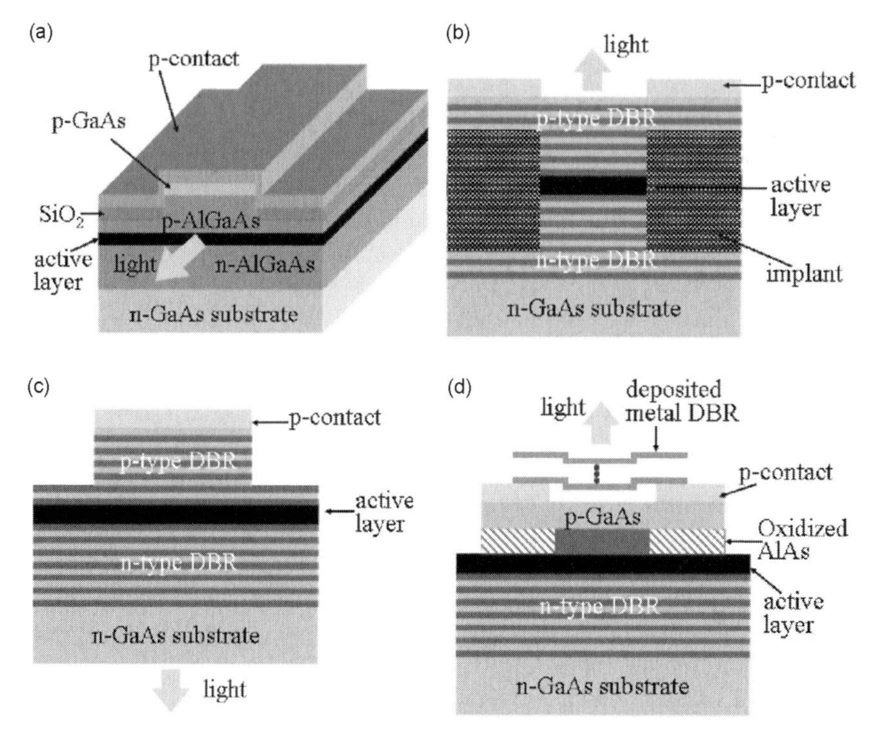

Figure 7.14. Examples of different laser structures. (a) Edge-emitting ridge laser, (b) VCSEL defined by ion-implantation, (c) VCSEL defined by etched post, (d) VCSEL with mode confinement defined by selective oxidation (oxide layer is created by oxidizing high x $Al_xGa_{1-x}As$ layers inward from side of structure).

made up of alternating, equal thickness, high-index and low-index layers. Each period of the superlattice has a thickness equal to $\lambda/4$, and though the reflection from a single layer is not large, when many layers are stacked, the reflectances add in phase and can give total reflection over 99%. If grown correctly, DBRs above and below the active region, separated by a distance λ, confine the optical mode with a peak at the midpoint between the DBRs.

Of significant concern for semiconductor lasers are temperature-related device problems. Most important among these is the device's ability to confine carriers at high operating temperatures (300 K). For a quantum well or QD laser, electrons or holes can more easily escape their confining potentials at higher temperatures. This can significantly increase the current (power) necessary to achieve lasing in a quantum-confined light emitter. Typically, the temperature relation for lasing threshold current density is

$$J_{\mathrm{th}} = J_{\mathrm{o}}\,\mathrm{e}^{-T/T_0} \tag{7.4}$$

where T_0 is the characteristic temperature for the device in question. A large T_0 indicates stable temperature characteristics for the laser. Often, lasers can exhibit different T_0 in different temperature ranges. This is the case for many lasers using quantum confined active regions.

Despite the decrease in carrier confinement at room temperature for quantum confined lasers, these devices are of interest because of their small size and because light emission from these devices can be obtained at wavelengths useful for communications applications. The motivation for the development of semiconductor lasers operating in the near-infrared, in large part, has been the desire to fabricate laser devices for telecommunications. For telecommunications purposes, light-emitters operating at 1.3 μm and 1.55 μm are ideal, as these are the wavelengths at which silica fibres exhibit attenuation minima.

Current communication lasers are grown on InP substrates, using quantum well active regions. Quantum well lasers suffer from relatively high threshold currents, as strong confinement of carriers to the active region is difficult at room temperature. The earliest QD lasers showed the potential for very strong carrier confinement, and thus low threshold currents for lasing. In addition, communication lasers fabricated on GaAs substrates would allow GaAs-based electronics and light-emitting devices to coexist on the same substrate.

7.4.2.2 Quantum dot lasers

The first quantum dot lasers were fabricated with a fairly basic structure, placing a GaAs–QD–GaAs heterostructure in the depletion region of an $Al_{0.4}Ga_{0.6}As$ p–n junction diode (figure 7.15) [74]. Initial studies of QD lasers showed almost no change in threshold current up to temperatures of 180 K. These early QD lasers achieved $T_0 = 430$ K for low temperature operation ($T < 180$ K), exceeding the $T_0 = 130$ K for single layer InGaAs quantum well lasers in the same temperature range [75]. It must be noted, however, that at room temperature, the QW laser T_0 of 130 K still outperformed early QD laser devices ($T_0 = 52$ K).

The first QD lasers emitted light in the 1 μm range. Once these devices were demonstrated, efforts to design QD lasers operating at communication wavelengths began. Raising the emission wavelength required changes in the typical methods used to grow QDs. One effort sought to decrease confinement of carriers in the dots and increase dot size (and thus lower QD ground state luminescence energy) by growing the InAs dots in an $In_xGa_{1-x}As$ quantum well, with $0.1 < x < 0.3$. Electroluminescence at 1.3 μm was seen from these structures [76]. It was determined that with slight modifications to the growth of these structures, significant improvements could be made in their device properties. Instead of growing the dots in the centre of a strained quantum well, InAs QDs were grown on

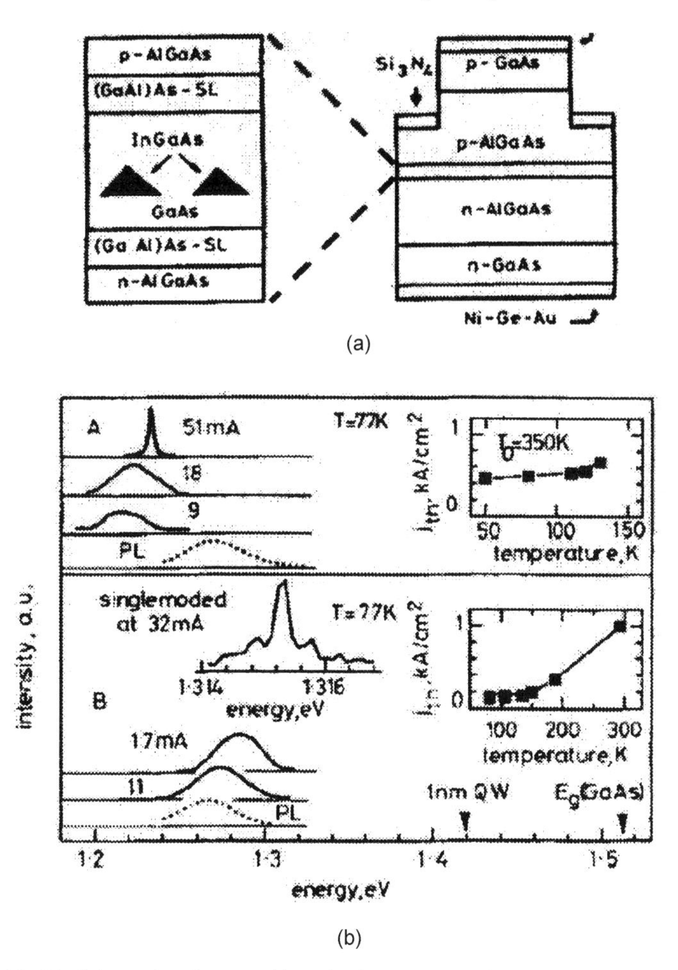

Figure 7.15. (a) Schematic of mesa ridge single layer InGaAs QD laser structure. Two samples were studied, and dots were grown with a substrate temperature of 460 °C (sample A) and 490 °C (sample B). (b) Luminescence and lasing characteristics of both samples. Dotted line indicates photoluminescence spectrum. Inserts show threshold current density versus temperature for both devices [74].

GaAs, and then overgrown with between 5 and 10 nm of $In_xGa_{1-x}As$ with $x \approx 0.13$. During this overgrowth, a process dubbed activated spinodal decomposition (ASD) was found to increase the QD size by solid phase segregation of the InGaAs cap layer (figure 7.16(a)–(c)) [77].

Utilizing ASD grown InGaAs QDs, a stripe laser operating at 1.3 μm was fabricated, with $T_0 = 160$ K at temperatures as high as 293 K [78]. The edge-emitting structure quickly led to the fabrication of an ASD-grown QD VCSEL emitting at 1.3 μm at room temperature [80].

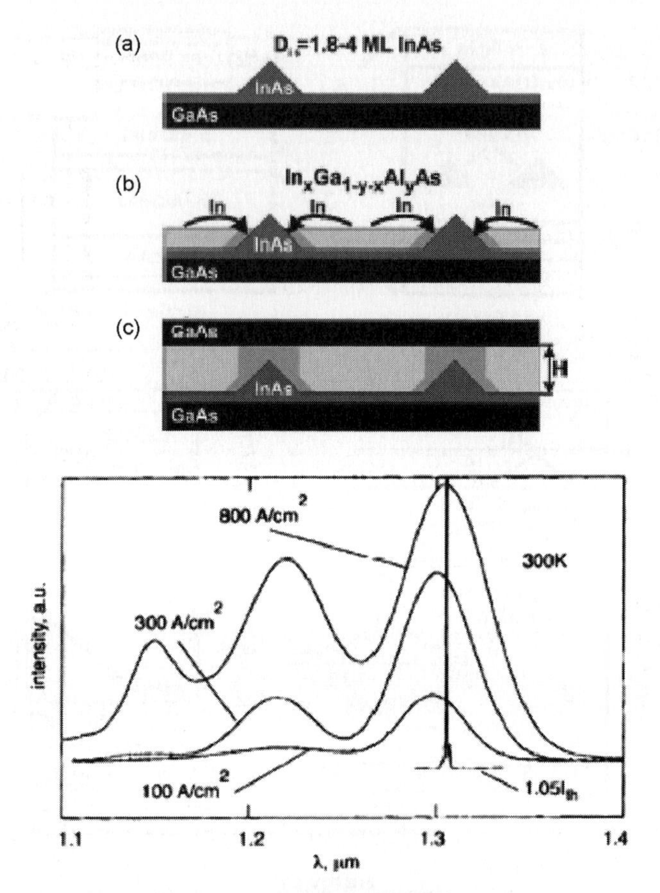

Figure 7.16. Schematic demonstrating ASD-grown QDs. (a) Initial InAs 'stressor' dots, (b) overgrowth of stressor dots with InGaAlAs showing In segregation near stressor dots, (c) structure after GaAs cap layer is added [79] [copyright (2000) by the American Physical Society]. (d) Spectrum at 300 K for VCSEL using ASD-grown dots as active region with current density $J = 100$, 300, 800 and 2800 A/cm^2 [80].

ASD-grown QDs were able to demonstrate lasing at communications wavelengths in both edge and surface-emitting structures. However, 1.3 μm QD emitters can be fabricated by means other than activated spinodal decomposition. One such method is MBE growth of QDs by atomic layer epitaxy (ALE) [81], where atomic layers of InAs are alternatingly grown with lattice mismatched GaAs. Instead of the expected strained superlattice formation, researchers saw the appearance of self-organized InGaAs QDs. These QDs, it was found, emit at significantly lower energy than typical Stranski–Krastanow (SK) QDs. As the ALE growth mode was studied further, it was noted that growths using sub-monolayer thicknesses for

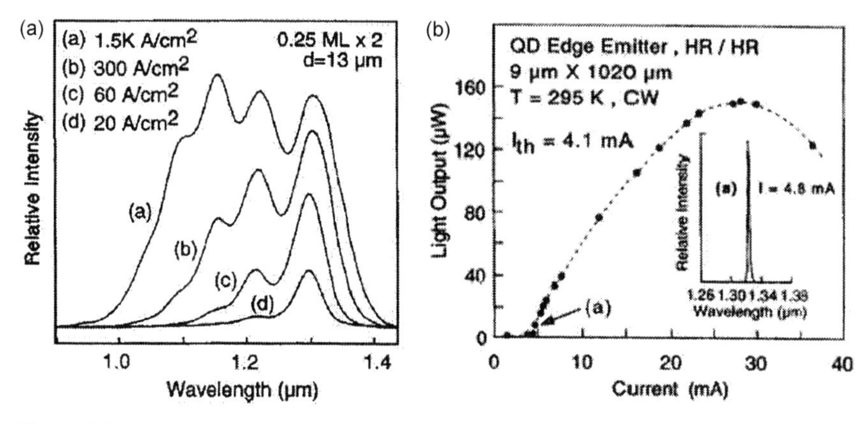

Figure 7.17. (a) Electroluminescence spectrum of submonolayer cycled growth QDs for varying current densities [82] [copyright (1995)]. (b) Lasing spectrum and light output versus current for QD laser fabricated from similar dots [83].

each InAs and GaAs layer gave growers better control of dot growth [82]. This variation of ALE growth is referred to as cycled submonolayer epitaxy.

Separate studies on submonolayer epitaxy QDs demonstrated ground state photoluminescence and electroluminescence at 1.3 μm [82, 83]. This was followed quickly by lasing at 1.32 μm from a similar structure (figure 7.17) [84]. These devices were shown to lase effectively under continuous wave operation at room temperature. In addition, it was found that by doping the QDs with holes, edge emitting lasers could be made with operating temperatures as high as 167 °C, with $T_0 = 161$ K at 80 °C [85].

Our discussion of QD emitters at 1.3 μm has concentrated primarily on InGaAs overgrown QDs and atomic layer or submonolayer cycled epitaxy QDs. Low energy emission from QDs, however, has also been achieved by MBE growth with extremely slow growth rates [86]. In fact, a combination of slow growth rate and InGaAs overgrowth was used to fabricate the first continuous-wave room temperature QD laser [87].

Future research on near-infrared QD lasers will most likely be centred on improving the laser characteristics of existing designs for 1.3 μm lasers and the development of QD lasers operating at 1.55 μm. Electroluminescence in the 1.55 μm range has been seen from so-called laterally associated dots (LAQDs) which are essentially clusters of SK dots which can form during low temperature (320 °C) MBE growth [88]. Other schemes involve InGaAsN or InNAs QD growth with very small percentages of nitrogen. These structures have demonstrated photoluminescence at both 1.3 and 1.55 μm [18]. However, both LAQDs and nitrogen-containing dots emit weakly at the desired wavelength of 1.55 μm and lasing has not yet been demonstrated from either.

7.4.3 Mid-infrared electroluminescence

There has been much interest in the fabrication of mid-infrared (5–20 μm) light emitting devices. Much of the motivation for development of these devices lies in remote gas sensing applications. Many gases have absorbance peaks in the mid-infrared and the efficient detection of the presence of these gases requires both mid-infrared detectors and mid-infrared emitters. Figure 7.18 depicts the absorbance spectrum for carbon dioxide and a schematic illustrating a possible remote monitoring system for gases with mid-infrared absorption peaks.

While near-infrared emitters fabricated with QDs emit light when an electron in the conduction band recombines with a valence band hole, mid-infrared QD emitters rely on intersubband transitions within the conduction or valence band. In a quantum well, although the energy levels in the well may be separated by energies as large as 100 meV, transitions between levels tend to be non-radiative. This is because there exists a continuum of states between levels, through which phonon-assisted relaxation can occur. Since the phonon-assisted transitions are significantly faster than radiative transitions, inter-sub-band transitions resulting in the emission of mid-infrared light are effectively suppressed. In QDs, there exists no continuum of states between energy levels. Thus, it has been suggested that significant light emission can be achieved by inter-sub-band transitions within the dot [90]. For mid-infrared luminescence, the supposed phonon bottleneck would become an advantage.

The first electroluminescence from QDs in the mid-infrared was seen from structures similar to the near-infrared emitters discussed earlier. Dots

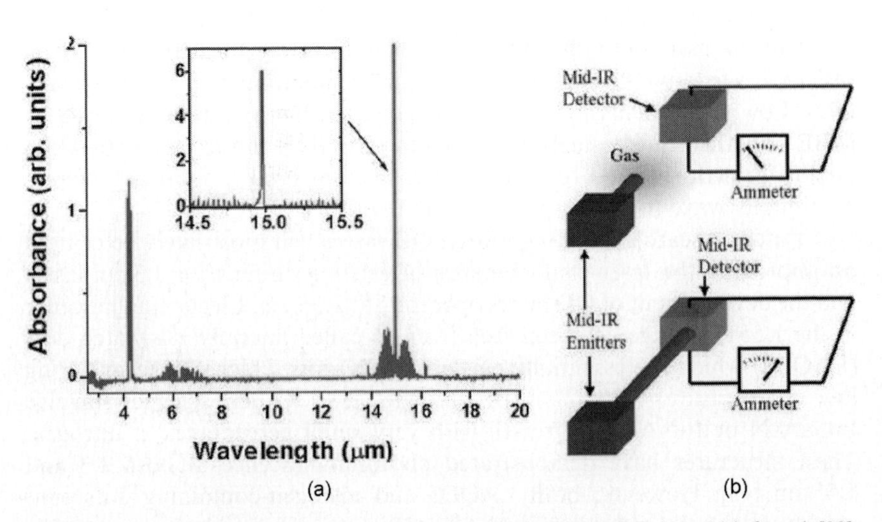

Figure 7.18. (a) Absorption spectrum for carbon dioxide (CO_2) in mid-infrared [89]. (b) Schematic of remote gas detection set-up (upper illustration depicts presence of gas).

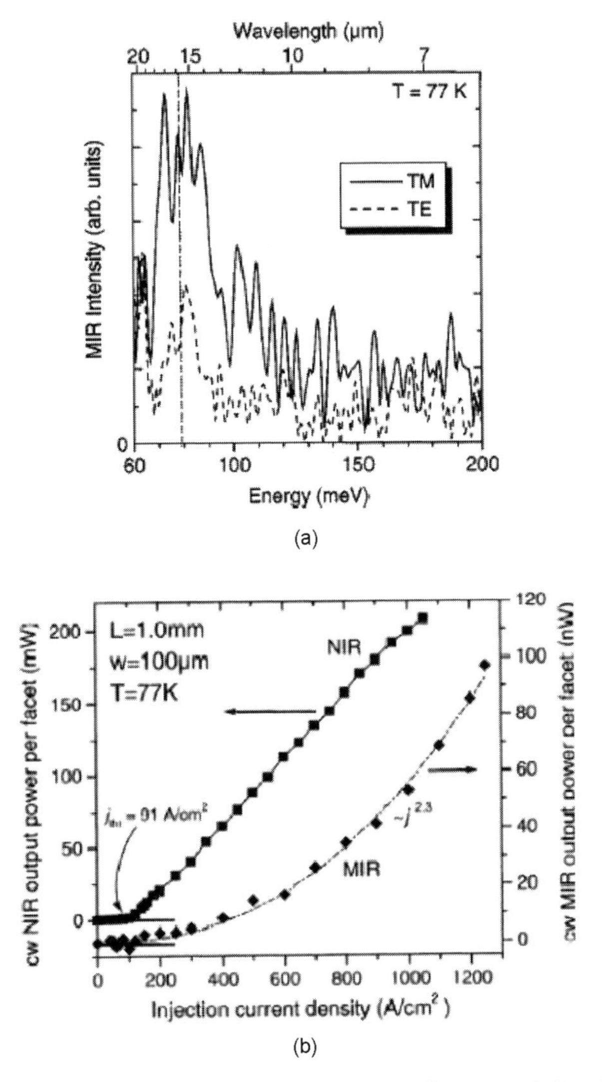

Figure 7.19. (a) Mid-infrared luminescence spectrum from near-infrared QD laser. (b) Light output versus injection current density for emission in both mid- and near-infrared [93].

were simply grown in the depletion region of a p–n junction diode and mid-infrared luminescence was reported, though initially no spectrum of the luminescence was shown [91]. Later results indicated emission was between 10 and 20 μm and was seen concurrently with near-infrared lasing [92, 93], as seen in figure 7.19. The goal of these devices was to make use of the rapid recombination time for electrons in the QD ground state. Ideally, this

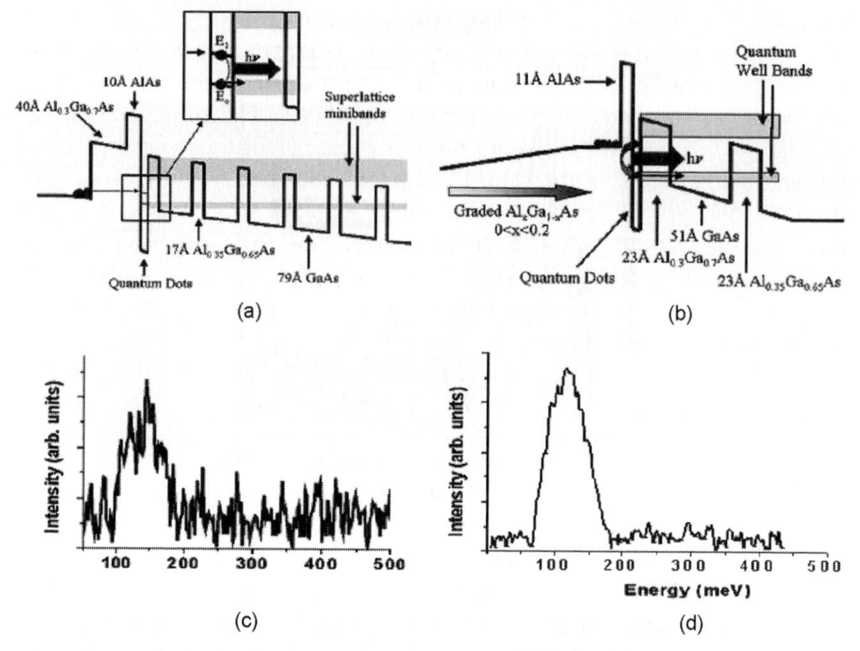

Figure 7.20. (a) Schematic of unipolar QD mid-infrared emitter with tunnel barrier injector and superlattice filter. (b) Mid-infrared emission from device in (a). (c) Unipolar mid-infrared emitter with graded AlGaAs injector and quantum well filter. (d) Mid-infrared emission from device in (c).

would empty out the ground state and create population inversion between excited states and the ground state, leading to mid-infrared lasing.

However, mid-infrared luminescence does not inherently require bipolar devices. Technically, long-wavelength light emission from QDs requires only one type of carrier. Since electron energy states in have greater energy spacing than hole states (due to the lighter effective mass of the electrons), the ideal mid-infrared light emitter is actually unipolar. There are two major difficulties in the design of unipolar mid-infrared light emitters: one must be able to first inject carriers into an excited state and second, remove them rapidly from the ground state in order to allow the next excited state electron to transition to the ground state. Devices were designed to achieve these effects, and are pictured in figure 7.20.

In figure 7.20(a) the electron is injected through an AlGaAs/AlAs barrier into the QD when a bias is applied to the device. If the electron reaches the dot it is expected to enter at an excited state. The electron is prevented from tunnelling directly through the dot by a superlattice filter, like those used for quantum cascade lasers [94]. The filter is designed to have a 'forbidden region' at the electron injection energy and a transmission band at the ground state of the QD. This forces the electron to transition to the lower state in the dot in order to escape through the transmission band. The inter-sub-band transition is hoped to be radiative, and thus result in the emission of a mid-infrared photon. Mid-infrared luminescence was seen from this device, with a peak at ~150 meV (8–9 μm) [95, 96]. Subsequent devices used graded AlGaAs injectors to more accurately control injection energy and filtered electron escape from the dots with a simple quantum well [97] (figure 7.20(c,d)). These devices demonstrated significantly larger emission intensity when compared with the superlattice filter devices, with emission now peaked between 9 and 10 μm. While mid-infrared lasers utilizing QDs as the active region have not yet been demonstrated, there remains significant interest in the field.

7.4.4 Single dot luminescence and single photon emitters using QDs

The first evidence of the discrete nature of QD energy states came with single-dot luminescence experiments, where carrier recombination in very small ensembles of dots was analysed. Researchers were able to see ultra-narrow spectral lines which corresponded to electron–hole recombination within a single dot [98–100]. These experiments, difficult in and of themselves, become significantly more complicated if one chooses to excite the dot electrically as opposed to optically. This is simply because studying such small ensembles of dots (grown with typical densities) requires a sub-micron length scale, and making electrical contact to structures this small is extremely difficult. However, electroluminescence experiments, of a sort, have shown single dot luminescence. Instead of making a permanent contact to the sample, one can use a scanning tunnelling microscope tip to selectively bias the top surface of the sample directly under the tip [101]. The bias drives electrons and holes into the few dots underneath the tip, where they recombine and emit near-infrared photons. Using this method, one can make out luminescence signals from individual dots, as is shown in figure 7.21.

While single dot luminescence might appear to be useful solely for understanding the physical nature of QDs (with few practical applications), this is not entirely true. There has been significant interest in building single-photon emitters. This would allow for 'photons on demand', desirable for many quantum communication and cryptography purposes. Such devices would require emission from isolated dots. Quantum dots are uniquely able to provide single photon emission. In a QD, the light emitted by the recombination of a single electron–hole pair occurs at a distinct wavelength. If the dot in question contains multiple electrons and holes, the energy of the emitted photons (now two or more at one time) will be shifted (as discussed previously). Thus, by using only photons emitted at the wavelength of single exciton recombination, one can ensure that the photons emitted are, in fact, single photons. This phenomenon is known as

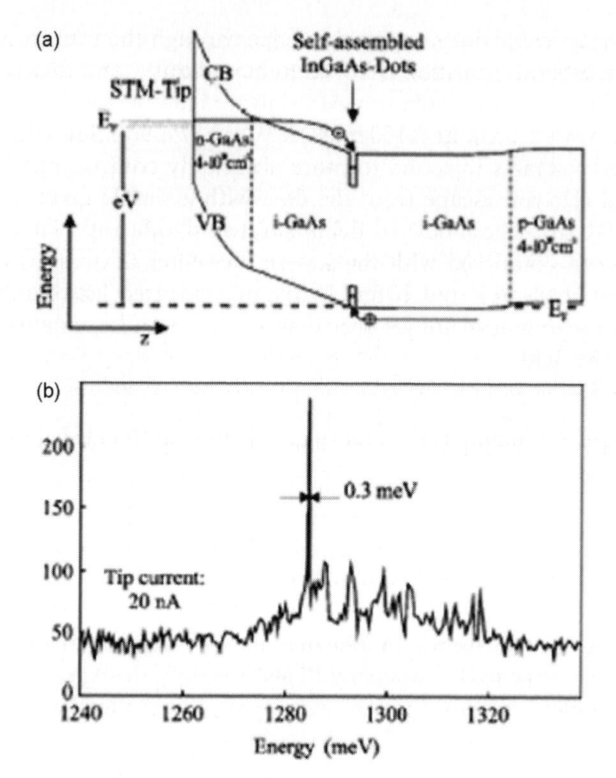

Figure 7.21. (a) Schematic of sample structure with presence of STM tip. (b) EL spectrum of single QD excited electrically by STM tip [101].

'photon anti-bunching', and has been experimentally demonstrated for QD photon emission [102].

Initial examples of these 'single-photon turnstiles' were fabricated with mesoscopic heterostructure p–n junctions defined by electron beam lithography [103], like the lithographically-defined dots briefly discussed earlier. However, because of the resolution limitations of electron beam lithography, these dots are 'large', and the spacing between energy levels in these devices is thus small. Because of the small spacing between energy levels in these structures, operating temperatures of less than 100 mK are required in order to see only photons at the ground state transition energy.

Other attempts to create single-photon emitters have used optically excited self-assembled QDs as the photon sources [104–106]. Recently, single photon electroluminescence from QDs has been demonstrated using very low density dot samples (5 QDs/μm^2) [107]. Square mesas (100 μm^2) containing dots in a p–n junction were etched, and contacts with apertures were placed over the mesas. Both single dot electroluminescence and photon anti-bunching were seen from these devices. These results are a large step forward towards the creation of an electrically-triggered single photon source, which remains a possible future application of QD electroluminescence.

7.5 Conclusion

This chapter has only scratched the surface of the study of QDs. The field is varied and vibrant, encompassing the categories of computer science, materials science, electrical engineering and physics, to name a few. More specifically, concerning electroluminescence from QDs, the areas of study range from the applied to the fundamental. Research into single dot electroluminescence aids in the study of electronic structure of these dots, while QD lasers in the near-infrared are already a viable alternative to present telecommunications technology. Other areas of interest, such as mid-infrared luminescence from QDs, hold the potential for efficient devices for a variety of applications. Less developed are proposals to utilize QDs as the single-photon sources required for many quantum cryptography and information devices. While it is difficult to predict what direction future research will take, the unique properties of QDs promise the possibility of a wealth of potential applications and the means to a better understanding of the physics of the solid state.

References

[1] Arakawa Y and Sakaki H 1982 *Appl. Phys. Lett.* **40** 939

[2] Stranski I N and Von Krastanow L 1939 *Akad. Wiss. Lit. Mainz Math.-Natur. Kl. IIb* **146** 797

[3] Eaglesham D J and Cerullo M 1990 *Phys. Rev. Lett.* **64** 1943

[4] Leonard D, Krishnamurthy M, Reaves C M, Denbaars S P and Petroff P M 1993 *Appl. Phys. Lett.* **63** 3203

[5] Huffaker D L, Baklenov O, Graham L A, Streetman B G and Deppe D G 1997 *Appl. Phys. Lett.* **70** 2356

[6] Leon R, Fafard S, Leonard D, Merz J L and Petroff P M 1995 *Appl. Phys. Lett.* **67** 521

[7] Bennett B R, Magno R and Shanabrook B V 1995 *Appl. Phys. Lett.* **68** 505

[8] Ahopelto J, Yamaguchi A A, Nishi K, Usui A and Sakaki H 1993 *Jpn. J. Appl. Phys. Part 2* **32** L32

[9] Ustinov V M, Kovsh A R, Zhukov A E, Egorov A Y, Ledentsov N N, Lunev A V, Shernyakov Y M, Maximov M V, Tsatsul'nikov A F, Volovik B V, Kop'ev P S and Alferov Z I 1998 *Tech. Phys. Lett.* **24** 22

[10] Zhukov A E, Egorov A Y, Kovsh A R, Ustinov V M, Ledentsov N N, Maximov M V, Tsatsul'nikov A F, Zaitsev S V, Gordeev N Y, Kop'ev P S and Alferov Z I 1997 *Semiconductors* **31** 411

[11] Carlsson N, Junno T, Montelius L, Pistol M-E, Samuelson L, Siefert W 1998 *J. Cryst. Growth* **191** 347

[12] Nishi K, Mirin R, Leonard D, Medeiros-Ribeiro G, Petroff P M and Gossard A C 1996 *J. Appl. Phys.* **80** 3466
[13] Leonard D, Pond K and Petroff P M 1994 *Phys. Rev. B* **50** 11687
[14] Moison J M, Houzay F, Barthe F, Leprince L, Andrè E and Vatel O 1994 *Appl. Phys. Lett.* **64** 196
[15] Xie Q, Chen P, Kalburge A, Ramachandran T R, Nayfonov A, Konkar A and Madhukar A 1995 *J. Crystal Growth* **150** 357
[16] Kim Y S, Lee U H, Lee D, Rhee S J, Leem Y A, Ko H S, Kim D H and Woo J C 2000 *J. Appl. Phys.* **87** 241
[17] Berryman K W, Lyon S A and Segev M 1997 *J. Vac. Sci. Technol. B* **15** 1045
[18] Sopanen M, Xin H P, Tu C W 2000 *Appl. Phys. Lett.* **76** 994
[19] Solomon G S, Trezza J A and Harris J S Jr. 1995 *Appl. Phys. Lett.* **66** 991
[20] Berryman K W, Lyon S A and Segev M, unpublished
[21] Solomon G S, Trezza J A and Harris J S Jr. 1995 *Appl. Phys. Lett.* **66** 3161
[22] Malik S, Roberts C, Murray R and Pate M 1997 *Appl. Phys. Lett.* **71** 1997
[23] Xu S J, Wang X C, Chua S J, Wang C H, Fan W J, Jiang J and Xie X G 1998 *Appl. Phys. Lett.* **72** 3335
[24] Hsu T M, Lan Y S, Chang W H, Yeh N T and Chyi J-I 1999 *Appl. Phys. Lett.* **76** 691
[25] Springholz G, Holy V, Pinczolits M and Bauer G 1998 *Science* **282** 734
[26] Xie Q, Madhukar A, Chen P and Kobayashi N P 1995 *Phys. Rev. Lett.* **75** 2542
[27] Lee H, Johnson J A, Speck J S and Petroff P M 2000 *J. Vac. Sci. Technol. B* **18** 2193
[28] Mui D S L, Leonard D, Coldren L A and Petroff P M 1995 *Appl. Phys. Lett.* **66** 1620
[29] Jeppesen S, Miller M S, Hessman D, Kowalsji B, Maximov I and Samuelson L 1996 *Appl. Phys. Lett.* **68** 2228
[30] Ishikawa T, Kohmoto S and Asakawa K 1998 *Appl. Phys. Lett.* **73** 1712
[31] Kohmoto S, Nakamura H, Ishikawa T and Asakawa K 1999 *Appl. Phys. Lett.* **75** 3488
[32] Hyon C K, Choi C, Song S-H, Hwang S W, Son M H, Ahn D, Park Y J and Kim E K 2000 *Appl. Phys. Lett.* **77** 2607
[33] Tatebayashi J, Nishioka M, Someya T and Arakawa Y 2000 *Appl. Phys. Lett.* **77** 3382
[34] Tsui R, Zhang R, Shiralagi K and Goronkin H 1997 *Appl. Phys. Lett.* **71** 3254
[35] Kohmoto S, Nakamura H, Nishikawa S and Asakawa K 2002 *J. Vac. Sci. Technol. B* **20** 763
[36] Krishna S, Sabarinathan J, Linder K, Bhattacharya P, Lita B and Goldman R S 2000 *J. Vac. Sci. Technol. B* **18** 1502
[37] Lee H, Johnson J A, He M Y, Speck J S and Petroff P M 2000 *Appl. Phys. Lett.* **78** 105
[38] Tersoff J, Margaritondo G, Perfetti P, Katnani A D, Grant R W, Kraut E A, Waldrop J R, Kowalczyk S P, Duggan G, Wolford D J, Keuch T F, Jaros M, Faurie J P, Guldner Y, Forrest S R and Lang D V 1987 *Heterojunction Band Discontinuities: Physics and Device Applications*, F Capasso and G Margaritondo (Amsterdam: North-Holland Physics Publishing) p 1-396
[39] O'Reilly E P 1989 *Semicond. Sci. Technol.* **4** 121
[40] Ludowise M J, Dietze W T, Lewis C R, Holonyak N Jr, Hess K, Camras M D and Nixon M A 1983 *Appl. Phys. Lett.* **42** 257
[41] Marzin J-Y and Bastard G 1994 *Solid State Commun.* **92** 437
[42] Grundmann M, Stier O and Bimberg D 1995 *Phys. Rev. B* **52** 11969

[43] Cusack M A, Briddon P R and Jaros M 1996 *Phys. Rev. B* **54** R2300
[44] Drexler H, Leonard D, Hansen W, Kotthaus J P and Petroff P M 1994 *Phys. Rev. Lett.* **73** 2252
[45] Schmidt K H, Medeiros-Ribeiro G, Oestreich M and Petroff P M 1996 *Phys. Rev. B* **54** 11346
[46] Jiang H and Singh J 1997 *Phys. Rev. B* **56** 4696
[47] Pryor C 1998 *Phys. Rev. B* **57** 7190
[48] Stier O, Grundmann M and Bimberg D 1999 *Phys. Rev. B* **59** 5688
[49] Bastard G, 1989 *Wave Mechanics Applied to Semiconductor Heterostructures* (New York: Halsted Press)
[50] Kim J, Wang L-W and Zunger A 1998 *Phys. Rev. B* **57** R9408
[51] Williamson A J and Zunger A 1999 *Phys. Rev. B* **59** 15819
[52] Williamson A J, Wang L W and Zunger A 2000 *Phys. Rev. B* **62** 12963
[53] Vdovin E E, Levin A, Patane A, Eaves L, Main P C, Khanin Y N, Dubrovskii V, Henini M and Hill G 2000 *Science* **290** 122
[54] Patané A, Hill R J A, Eaves L, Main P C, Henini M, Zambrano L, Levin A, Mori N, Hamaguchi C, Dubrovskii Y V, Vdovin E E, Austing D G, Tarucha S and Hill G 2002 *Phys. Rev. B* **65** 165308
[55] Sheng W and Leburton J-P 2002 *Appl. Phys. Lett.* **80** 2755
[56] Wang G, Farfard S, Leonard D, Bowers J E, Merz J L and Petroff P M 1994 *Appl. Phys. Lett.* **64** 2815
[57] Suguwara M, Mukai K and Shoji H 1997 *Appl. Phys. Lett.* **71** 2791
[58] Benisty H, Sotomayor-Torres C M and Weisbuch C 1991 *Phys. Rev. B* **44** 10945
[59] Waugh J L T and Dolling G 1963 *Phys. Rev.* **132** 2410
[60] Bockelmann U and Egeler T 1992 *Phys. Rev. B* **46** 15574
[61] Li X-Q, Nakayama H and Arakawa Y 1999 *Phys. Rev. B* **59** 5069
[62] Ferreira R, Bastard G 1999 *Appl. Phys. Lett.* **74** 2818
[63] Heitz R, Veit M, Ledentsov N N, Hoffman A, Bimberg D, Ustinov V M, Kop'ev P S and Alferov Z I 1997 *Phys. Rev. B* **56** 10435
[64] Ohnesorge B, Albrecht M, Oshinowo J, Forchel A and Arakawa Y 1996 *Phys. Rev. B* **54** 11532
[65] Morris D, Perret N and Fafard S 1999 *Appl. Phys. Lett.* **75** 3593
[66] Toda Y, Moriwaki O, Nishioka M and Arakawa Y 1999 *Phys. Rev. Lett.* **82** 4114
[67] Paillard M, Marie X, Vanelle E, Amand T, Kalevich V K, Kovsh A R, Zhukov A E and Ustinov V M 2000 *Appl. Phys. Lett.* **76** 76
[68] Hall R N, Fenner G E, Kingsley J D, Soltys T J and Carlson R O 1962 *Phys. Rev. Lett.* **9** 366
[69] Holonyak N Jr. and Bevacqua S F 1962 *Appl. Phys. Lett.* **1** 82
[70] Nathan M I, Dumke W P, Burns G, Dill F H Jr. and Lasher G 1962 *Appl. Phys. Lett.* **1** 62
[71] Quist T M, Rediker R H, Keyes R J, Krag W E, Lax B, McWhorter A L and Zeiger H J 1962 *Appl. Phys. Lett.* **1** 91
[72] Alferov Z I, Andreev V M, Korol'kov, Portnoi E L and Tret'yakov N, 1968 *Fiz. Tekh. Poluprovodn.* **2** 1545
[73] van der Ziel, Dingle R, Miller R C, Wiegmann and Nordland W A Jr. 1975 *Appl. Phys. Lett.* **26** 463
[74] Kirstaedter N, Ledentsov N N, Grundmann M, Bimberg D, Ustinov V M, Ruvimov S S, Maximov M V, Kop'ev P S, Alferov Zh I, Richter U, Werner P, Gosele U and

Heydenreich J 1994 *Electron. Lett.* **30** 1416

[75] Alferov Zh I, Bert N A, Egorov A Yu, Zhukov A E, Kop'ev P S, Kosogov A O, Krestnikov I L, Ledentsov N N, Lunev A V, Maximov M V, Sakharov A V, Ustinov V M, Tsapul'nikov A F and Shernyakov Y M 1996 *Semiconductors* **30** 194

[76] Ustinov V M, Maleev N A, Zhukov A E, Kovsh A R, Egorov A Y, Lunev A V, Volovik B V, Krestnikov I L, Musikhin Y G, Bert N A, Kop'ev P S, Alferov Zh I, Ledentsov N N and Bimberg D 1999 *Appl. Phys. Lett.* **74** 2815

[77] Zhukov A E, Kovsh A R, Egorov A Y, Maleev N A, Ustinov V M, Volovik B V, Maximov M V, Tsatsul'nikov A F, Ledentsov N N, Shernyakov Y M, Lunev A V, Musikhin Y G, Bert N A, Kop'ev P S and Alferov Zh I 1999 *Semiconductors* **33** 153

[78] Shernyakov Y M, Bedarev D A, Kondrat'eva E Y, Kop'ev P S, Kovsh A R, Maleev N A, Maximov M V, Mikhrin S S, Tsatsul'nikov A F, Ustinov V M, Volovik B V, Zhukov A E, Alferov Zh I, Ledentsov N N and Bimberg D 1999 *Electron. Lett.* **35** 898

[79] Maximov M V, Tsatul'nikov A F, Volovik B V, Sizov D S, Shernyakov Y M, Kaiander I N, Zhukov A E, Kovsh A R, Mikhrin S S, Ustinov V M, Alferov Zh I, Heitz R, Shchukin V A, Ledentsov N N, Bimberg D, Musikhin Y G and Neumann W 2000 *Phys. Rev. B* **62** 16671

[80] Lott J A, Ledentsov N N, Ustinov V M, Maleev N A, Zhukov A E, Kovsh A R, Maximov M V, Volovik B V, Alferov Zh I and Bimberg D 2000 *Electron. Lett.* **36** 1384

[81] Mukai K, Ohtsuka N, Suguwara M and Yamakazi S 1994 *Jpn. J. Appl. Phys., Part 2* **33** L1710

[82] Mirin R P, Ibbetson J P, Nishi K, Gossard A C and Bowers J E 1995 *Appl. Phys. Lett.* **67** 3795

[83] Huffaker D L and Deppe D G 1998 *Appl. Phys. Lett.* **73** 520

[84] Park G, Shchekin O B, Csutak S, Huffaker D L and Deppe D G 1999 *Appl. Phys. Lett.* **75** 3267

[85] Shchekin O B and Deppe D G 2002 *Appl. Phys. Lett.* **80** 3277

[86] Murray R, Childs D, Malik S, Siverns P, Hartmann J-M and Stavrinou P 1999 *Jpn. J. Appl. Phys.* **38** 528

[87] Mukai K, Nakata Y, Otsubo K, Sugawara M, Yokoyama N and Ishikawa H 1999 *IEEE Phot. Tech. Lett.* **11** 1205

[88] Zhukov A E, Volovik B V, Mikhrin S S, Maleev N A, Tsatsul'nikov A F, Nikitina E V, Kayander I N, Ustinov V M and Ledentsov N N 2000 *Tech. Phys. Lett.* **27** 734

[89] Data taken from Environmental Protection Agency website at www. epa. gov (http://www.epa.gov/ttn/emc/ftir/addcas.html)

[90] Hsu C F, O J-S, Zory P and Boetz D 2000 *IEEE J. Select. Topics Quantum Electron.* **60** 491

[91] Vorob'ev L E, Firsov D A, Shalygin V A, Tulupenko V N, Shernyakov Y M, Ledentsov N N, Ustinov V M and Alferov Zh I 1998 *JTEP Lett.* **67** 275

[92] Krishna S, Bhattacharya P, McCann P J and Nanjou K 2000 *Electron. Lett.* **36** 1550

[93] Grundmann M, Weber A, Goede K, Ustinov V M, Zhukov A E, Ledentsov N N, Kop'ev P S and Alferov Zh I 2000 *Appl. Phys. Lett.* **77** 4

[94] Faist J, Capasso F, Sivco D L, Hutchinson A L and Cho A Y 1994 *Science* **264** 533

[95] Wasserman D and Lyon S A 2001 *Phys. Stat. Sol. (b)* **224** 585

[96] Wasserman D and Lyon S A 2002 *Appl. Phys. Lett.* **81** 2848
[97] Wasserman D and Lyon S A (unpublished)
[98] Marzin J-Y, Gerard J-M, Izrael A, Barrier D and Bastard G 1994 *Phys. Rev. Lett.* **73** 716
[99] Grundmann M, Christen J, Ledentsov N N, Bohrer J, Bimberg D, Ruvimov S S, Werner P, Richter U, Gosele U, Heydenreich J, Ustinov V M, Egorov A Y, Zhukov A E, Kop'ev P S and Alferov Zh I 1995 *Phys. Rev. Lett.* **74** 4043
[100] Dekel E, Gershoni D, Ehrenfreund E, Spektor D, Garcia J M and Petroff P M 1998 *Phys. Rev. Lett.* **80** 4991
[101] Zrenner A, Findeis F, Beham E, Markmann M, Bohm G and Abstreiter G 2001 *Physica E* **9** 114
[102] Becher C, Kiraz A, Michler P, Imamŏglu A, Schoenfeld W V, Petroff P M, Zhang L and Hu E 2001 *Phys. Rev. B* **63** 121312
[103] Kim J, Benson O, Kan H and Yamamoto Y 1999 *Nature* **397** 500
[104] Santori C, Pelton M, Solomon G, Dale Y and Yamamoto Y 2001 *Phys. Rev. Lett.* **86** 1502
[105] Michler P, Kiraz A, Becher C, Schoenfeld W V, Petroff P M, Zhang L, Hu E and Imamŏglu A 2000 *Science* **290** 2282
[106] Pelton M, Santori C, Vučković J, Zhang B, Solomon G S, Plant J and Yamamoto Y 2002 *Phys. Rev. Lett.* **89** 233602
[107] Yuan Z, Kardynal B E, Stevenson R M, Shields A J, Lobo C J, Cooper K, Beattie N S, Ritchie D A and Pepper M 2002 *Science* **295** 102

Chapter 8

Gallium phosphide and its wide-band gap ternary and quaternary alloys

Alexander N Pikhtin[1] *and Olga L Lazarenkova*[2]
[1]St Petersburg State Electrotechnical University, St Petersburg, Russia
[2]Jet Propulsion Laboratory, California Institute of Technology, Pasadena, USA

8.1 Introduction

Gallium phosphide (GaP) is a well-studied typical wide band gap III–V semiconductor with a diamond-like crystal lattice. It was first created as a powder in 1928 by geochemist, crystallographer and chemist V M Goldschmidt [1], who also observed the analogy between the crystallographic structures of GaP and diamond. In 1951 Welker [2] began to study the semiconductor properties of GaP. The effect of electroluminescence in the optical range discovered in silicon carbide by Losev in 1923 [3] was later investigated in detail in crystals of GaP [4, 5]. GaP used to be the main material for red, yellow and green light-emitting diodes (LEDs) for a long time since it combines the benefits of both quite a wide energy band gap (2.27 eV at room temperature) and relatively simple technology of doping, and the creation of p–n junctions. The disadvantage of GaP as an active material for electroluminescent devices is its indirect band structure. It confined the GaP LED quantum efficiency to a value less than 1% [4–6]. This problem is now solved in ternary and quaternary alloys of GaP, where it is possible to create direct band gap materials with a band gap up to 2.3 eV. The quaternary alloy (Ga–In–Al)P is the current base material for the creation of red, orange and yellow LEDs. The internal quantum efficiency of these LEDs approaches 100%, while brightness is extremely high as it will be discussed in section 8.4 of this chapter. Since GaP is used as a transparent substrate for these LEDs, the investigation of its physical properties and technology of growing is of current interest.

Gallium phosphide and its wide band gap alloys are very promising as materials for short wavelength (down to ultraviolet) photodiodes, high temperature electronic devices etc., along with well-developed applications in different types of EL devices.

8.2 Gallium phosphide

8.2.1 Lattice, physical–chemical properties and technological data

Gallium phosphide has a cubical zincblende (sphalerite) type of crystallographic structure that corresponds to T_d^2–$^{3m}F_4$ space group of symmetry. GaP crystals are optically transparent and slightly tinted in spectrum from lemon-yellow to amber-orange depending on dopant type and concentration. They are highly chemically stable, do not dissolve in water, rigid to organic acids, alkali, sulphur and hydrochloric acids. GaP is stable in air when heated up to 750 °C and begins to dissociate in vacuum at about 1000 °C.

The fact that GaP crystals have only (110) cleavage surface confirms that the bonds in this compound are more ionic by nature than those of most III–V semiconductors. The absence of the symmetry centre leads to the formation of two polar axes along [111] and $[\bar{1}\bar{1}\bar{1}]$ crystallographic directions. The plane (111) differs from $(\bar{1}\bar{1}\bar{1})$ and consists of gallium atoms, while $(\bar{1}\bar{1}\bar{1})$ consists of those of phosphorus. Therefore these faces have drastically different properties. Thus the $(\bar{1}\bar{1}\bar{1})$ surface, formed by atoms of group V, may be destroyed easier than the (111) surface with etching or mechanical treatment.

The mixture of $HNO_3 + HCl$ (1 : 3) acids at 10–100 °C is mainly used for chemical polishing of the crystals. The etching speed at room temperature is about 8 mg/cm^2 h^{-1}.

An 85% solution of orthophosphorous acid (H_3PO_4) at 150–200 °C serves for a selective etching to obtain dislocations and other defects of the crystal lattice. The etching speed rises from 5 to 25 mkm/min when temperature increases from 150 to 200 °C and it is 20 times faster for the $(\bar{1}\bar{1}\bar{1})$ surface than for (111) and (100). Sometimes the mixture HNO_3 (3 parts)+HF (1 part)+H_2O (2 parts) at 50 °C etchant may be used for selective etching as well. Standard brom-methanol etchant may be used for polishing.

The main physical and chemical properties of GaP are presented in table 8.1 [5–10].

Temperature dependences of the lattice constant a [11, 12], linear thermal expansion coefficient α_ℓ [12], thermal conductivity κ [13], and Debye temperature Θ_D [14] are presented in figures 8.1 and 8.2. Note that these dependences may vary significantly for crystals or epitaxial layers of different quality. For example, the higher the concentration of impurities, the lower the thermal conductivity of the sample, especially at low temperature. The curves in figure 8.1–8.4 correspond to the GaP crystals of high quality.

Table 8.1. Main physical and chemical properties of gallium phosphide.

Parameter	Units	Value
Lattice constant, α	nm	0.54505 (290 K)
		0.5470 (900 K)
Mass density, d	g/cm^3	4.138 (300 K)
Linear thermal expansion coefficient, α_ℓ	10^{-6} K^{-1}	4.65 (300 K)
		5.97 (800 K)
Melting point, T_m	K	1730
Equilibrium vapour pressure at T_m	atm	32
	Pa	32×10^5
Debye temperature, Θ_D	K	468
Thermal conductivity, κ	W/cm K	1.1 (300 K)
Molar specific heat, C_p	J/mol K	43.903 (300 K)
Standard enthalpy of formation, ΔH_1^o	kJ/mol	−100.2 (298 K)
Standard entropy, S°	J/mol K	54.89 (298 K)
Heat of fusion, ΔH_m	kJ/mol	123.8
Hardness in Moos scale	—	5
Elastic moduli at 300 K:		
C_{11}	10^{10} N/m^2	14.05
C_{12}		6.20
C_{44}		7.03
Phonon energies at Γ-point:	meV	
LO(Γ)		49.96 (300 K)
		45.54 (300 K)
TO(Γ)		45.33 (4 K)
Sound velocities at 300 K:	m/s	
V_1^L [001]/[001]		5847
V_2^T [001]/[110]		4131
V_6^L [111]/[111]		6648
V_7^T [111]/[110]		3466
Dielectric constants:	—	
ε_0		11.11 (300 K)
		10.86 (75.7 K)
ε_∞		9.1 (300 K)
$d\varepsilon_0/dT$	K^{-1}	1.4×10^{-3}
$d\varepsilon_0/dP$	GPa^{-1}	−0.12

GaP mono-crystals are grown from stoichiometric or almost stoichiometric melt. Most commercially available crystals are grown by the modified Czochralski method along [111] or [100] crystallographic directions with nominal diameters from 25 to 75 mm in 5 mm steps. The typical dislocation density in such samples is about $(2-10) \times 10^4$ cm^{-2}. Extrinsic types of electrical conductivity may be achieved by doping GaP during its growth with sulphur or tellurium with concentrations varying from $N_D = 1 \times 10^{16}$

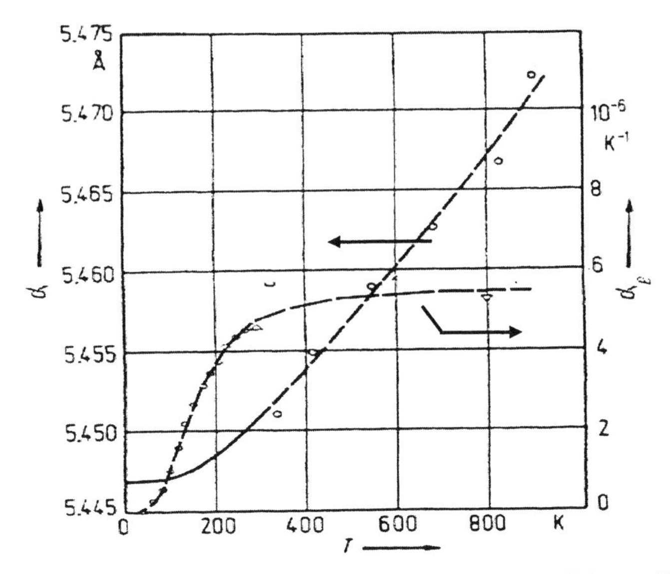

Figure 8.1. Lattice constant α [12] and linear thermal expansion coefficient α_ℓ [9, 12] as a function of temperature.

to $5 \times 10^{18}\,\mathrm{cm}^{-3}$ (n-type GaP), or with zinc with concentration from $N_A = 1 \times 10^{17}$ to $2 \times 10^{19}\,\mathrm{cm}^{-3}$ (p-type GaP). Undoped crystals usually have n-type electrical conductivity with a residual concentration of carriers $n \approx 10^{16}\,\mathrm{cm}^{-3}$. Semi-insulating GaP with electrical resistivity in a range

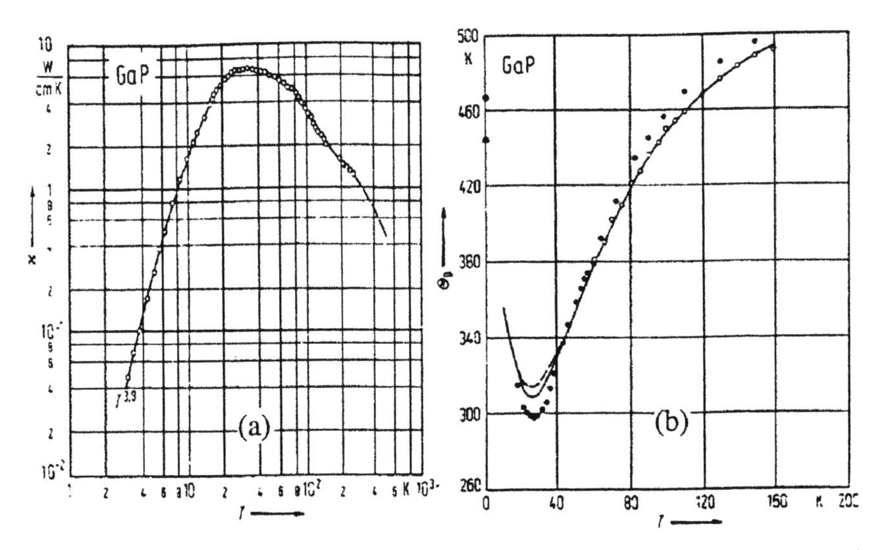

Figure 8.2. Thermal conductivity κ (a) [13] of p-GaP and Debye temperature Θ_D and (b) [14] as a function of temperature.

from 10^2 to 10^{11} Ω cm has to be doped with chromium (Cr), manganese (Mn) or iron (Fe).

The floating-zone method was used in some laboratories to grow high-quality monocrystals of GaP; however, the diameter of the samples was not larger than 1 cm.

Epitaxial layers of GaP grown by any common method such as liquid phase epitaxy (LPE), vapour phase epitaxy (VPE), metal-organic chemical vapour deposition (MOCVD) and molecular beam epitaxy (MBE) have the best quality. The thickness of the epitaxial layers may vary from few nm (sub-molecular layers are achievable in MBE and MOCVD) to several tens and hundreds of mkm (usually grown by LPE). These methods are also used for fabrication of different types of GaP-based structures, including quantum-sized heterostructures, attractive for nanoelectronic devices.

High pressure, about 215 kbar, causes the phase transition of GaP from its cubical GaP-I to the tetragonal phase GaP-II of D_{4h}^{19}–I_1^4/amd group symmetry. The corresponding volume change is about $\Delta V/V \approx 0.175$. The lattice parameters of GaP-II at $P = 240$ kbar and $T = 28\,^{\circ}$C are the following: $a = 0.4720$ nm and $c = 0.2468$ nm. If the pressure is higher than 296 kbar, only GaP-II may exist, while in the intermediate range both phases could coexist (there is about 75% GaP-II and 25% GaP-I at 250 kbar) [8, 9].

8.2.2 Electronic properties, electrical conductivity, impurities and defects

Gallium phosphide is an indirect band gap material with the electron energy band structure typical of III–V semiconductors of the zincblende symmetry (see figure 8.3). The main parameters of the band structure are presented in table 8.2.

The lowest minimum of the conduction band is situated near the Brillouin zone boundary in the [100] crystallographic direction. It is classified as X_1^c or as X_6^c in the double space group representation. This minimum has a complicated 'camel's back' structure as shown in detail in figure 8.4. Gallium phosphide was the very material where this specific band structure was discovered in 1978 and studied in detail [16–19]. The brief explanation of the appearance of this unique structure is the following. The absence of the inversion centre in the zincblende structure lifts the degeneration in the X point of the Brillouin zone that takes place in the diamond-like crystals. The $\mathbf{k} \cdot \mathbf{p}$ interaction between X_6^c and X_7^c (as shown in figure 8.4) states in the conduction band modifies the electron dispersion near the X_6^c minimum leading to the 'camel's back' energy dependence on the wave vector

$$E(k) = \hbar^2 k_{\parallel}^2/2m_\ell + \hbar^2 k_{\perp}^2/2m_{\perp} + \frac{\Delta}{2} - \left[\left(\frac{\Delta}{2}\right)^2 + \Delta_0 \frac{\hbar^2 k_{\parallel}^2}{2m_\ell}\right]^{1/2}, \qquad (8.1)$$

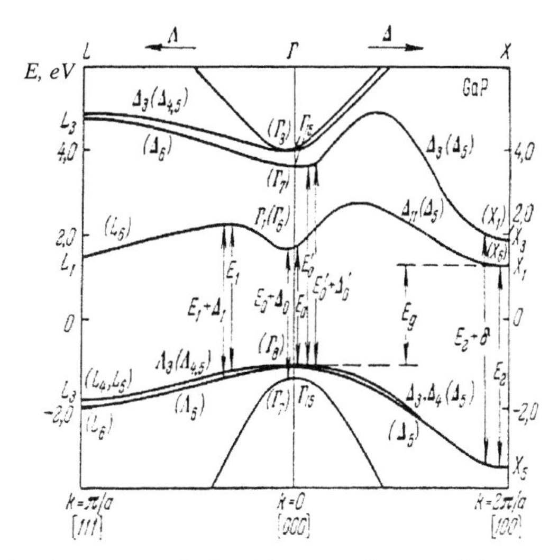

Figure 8.3. GaP-band structure calculated by a pseudo-potential method [15].

where $k_{\parallel}$ and $k_{\perp}$ are parallel and perpendicular to the [100] crystallographic direction wave vector components, $m_{\perp}$ is effective mass in the direction perpendicular to [100] direction; Δ_0 is non-parabolicity parameter determined by the matrix element of $(X_6^C\text{–}X_7^C)$ $\mathbf{k}\cdot\mathbf{p}$ interaction; Δ is the distance between X_6^C and X_7^C states on the Brillouin zone boundary; ΔE is the difference of the energy in the conduction band minimum and energy of the X_6^c on the Brillouin zone boundary

$$\Delta E = \frac{\Delta_0}{4}\left(1 - \frac{\Delta}{\Delta_0}\right)^2, \qquad k_m = \left(\frac{2m_\ell\,\Delta E}{\hbar^2}\right)^{1/2}, \qquad m_{\parallel} = \frac{m_\ell}{1 - (\Delta/\Delta_0)^2}.$$

Due to the camel's back structure both electrical conductivity and density of states effective masses increase with the temperature drop. This change in GaP is as large as 25% in the temperature range from 300 down to 80 K. It dramatically affects the electrical properties, cyclotron resonance, energy dispersion and density of states of the indirect excitons, shallow donors, electron–hole droplets etc. The camel's back structure in GaP is most evidently exhibited in exciton absorption and luminescence spectra as will be discussed in section 8.2.3.

The temperature dependence of the indirect band gap of GaP may be approximated as

$$E_g^{ind}(T) = E_g^X = E_g(\Gamma_8^V - \Delta_6^C) = 2.350 - (6.6 \times 10^{-4})\frac{T^2}{T + 460}, \qquad (8.2)$$

and it changes with the hydrostatic pressure as

$$\delta E_g^X(P) = E_g^X(P) - E_g^X(0) = 2.14 - (2.4 \times 10^{-2})P - (4.8 \times 10^{-4})P^2. \quad (8.3)$$

Table 8.2. Band structure parameters of GaP.

Parameter	Units	Value
Band gaps:		
$E_g^X = E_g(\Gamma_8^v) - E(\Delta_6^c)$	eV	2.350 (4.2 K) 2.272 (300 K)
$\Delta E = E(X_6^c) - E_{min}(\Delta_6^c)$	meV	3.5 (4.2 K)
$E_g^L = E(\Gamma_8^v) - E(\Gamma_6^c)$	eV	2.637 (78 K)
$E_0 = E_g^\Gamma = E(\Gamma_8^v) - E(\Gamma_6^c)$	eV	2.886 (4.2 K) 2.780 (300 K)
$E_1 = E(\Lambda_3^v) - E(\Lambda_1^c)$	eV	3.7 (300 K)
E_2 (along $\Sigma = [110]$ crystallographic direction)	eV	5.3 (300 K)
$\Delta = E(X_6^c) - E(X_7^c)$	eV	0.375 (4.2 K)
$\Delta_{SO} = E(\Gamma_7^v) - E(\Gamma_8^v)$	eV	0.080 (100 K)
dE_g^X/dT	meV K^{-1}	−0.46 (300 K)
dE_g^Γ/dT	meV K^{-1}	−0.65 (300 K)
dE_g^X/dP	eV/GPa	−0.024 (300 K)
dE_g^Γ/dP	eV/GPa	0.097 (300 K)
Exciton energies:		
$E_{gx}(1S, X_7)$	eV	2.3284 (4.2 K)
$E_{ex}(X_7)$	meV	21.5 (4.2 K)
$E_{ex}(X_6)$	meV	19.5 (4.2 K)
$\Delta_{ex}(X_7)$	meV	2.4 (4.2 K)
$\Delta_{ex}(X_6)$	meV	2.8 (4.2 K)
$E_{ex}^\Gamma = E_{ex}(\Gamma_6)$	meV	9.5 (4.2 K)
Effective masses:		
$m_n(\Gamma_6^c)$	m_0	0.0925
$m_\perp(X_6^x)$	m_0	0.25
$m_\ell(X_6^x)$	m_0	0.91
$m_\parallel(X_6^c)$	m_0	10.9
$m_{ph}\langle 111\rangle$	m_0	0.67
$m_{ph}\langle 100\rangle$	m_0	0.42
$m_{p\ell}\langle 111\rangle$	m_0	0.17
$m_{p\ell}\langle 100\rangle$	m_0	0.16
m_{pSO}	m_0	0.46

The analogous expressions for the direct band gap in the Γ-point are

$$E_g^{dir}(T) = E_g^\Gamma = E_g(\Gamma_8^V - \Delta_6^C) = 2.876 - 0.1081[\coth(164/T) - 1] \quad (8.4)$$

$$\delta E_g^\Gamma(P) = E_g^\Gamma(P) - E_g^\Gamma(0) = 2.76 + (9.7 \times 10^{-2})P - (3.5 \times 10^{-4})P^2. \quad (8.5)$$

The numeric coefficients in equations (8.2)–(8.5) assume that energy is measured in eV, temperature in kelvins, and pressure in GPa.

Impurities affect GaP properties significantly. They have been studied very well. The ionization energies of the main impurities in GaP are presented in table 8.3. Elements of group VI, such as S, Se and Te, substitute

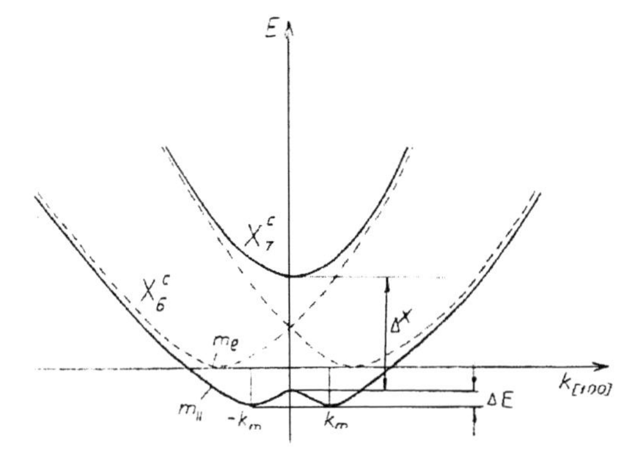

Figure 8.4. Camel's back structure of the lowest minimum in the conduction band of gallium phosphide.

phosphorus and become donors. Elements of group II, such as Be, Mg, Zn and Cd, substitute gallium and became acceptors. Elements of group IV may be either acceptors or donors, depending on the atoms they substitute. Sn atoms usually substitute Ga and are shallow donors. Si and Ge are amphoteric impurities, i.e. dependent on technology of growing and concentration. They may be situated either in nodes of Ga or P sub-lattices, becoming either donors or acceptors, respectively. Ge is an acceptor if its concentration is less than $10^{18}\,cm^{-3}$, while becoming a donor at higher doping. Atoms of elements with an unfilled 3*d* electron shell (i.e. Fe etc.) have a complicated system of energy levels and mostly have properties of

Table 8.3. Ionization energies of the impurities in GaP.

Donors	E_D (meV)	Acceptors	E_A (meV)
Theory EM	62 ± 2	Theory EM	45.3
Sn_{Ga}	72 ± 1	C_P	54.3 ± 1
Si_{Ga}	85 ± 1	Be_{Ga}	56.6 ± 1
Te_P	92.6 ± 1	Mg_{Ga}	60 ± 1
Se_P	105 ± 1	Zn_{Ga}	69.7 ± 0.5
S_P	107 ± 1	Cd_{Ga}	102.2 ± 0.5
Ge_{Ga}	204 ± 1	Si_P	210 ± 1
O_P	897 ± 1	Ge_P	265 ± 1
Li_{Ga}	61 ± 2	$Mn^{3+}(d^4)$	400
Li_P	91 ± 1	$Co^{3+}(d^6)$	410
		$Ni^{3+}(d^7)$	500
		$Fe^{3+}(d^5)$	700
		$Cu^{3+}(d^8)$	540
			620
		$Cr^{3+}(d^3)$	1125

deep acceptors. Oxygen is a deep donor. Some impurities form different kinds of molecular complexes. Also, isoelectronic traps on point defects (N_P or Bi_P), molecular complexes of two nitrogen atoms in the P sub-lattice (so-called NN_m couples, where m is the number of the coordination sphere), and (Zn–O) and (Cd–O) complexes are well-studied in GaP. They may easily bound excitons, as will be discussed in sections 8.2.3 and 8.2.4.

Silicon, oxygen, nitrogen and carbon are the typical residual technological impurities in GaP mono-crystals. The most unwanted impurities are Cu and Fe since they create deep centres of effective non-radiative recombination. Intrinsic defects, such as P and Ga vacancies and antiside defects, are analogous to those in GaAs.

As a result of the wide band gap, intrinsic electrical conductivity in GaP of the best quality may be achieved only at temperatures higher than 600 °C. Therefore, all transport properties are determined mostly by impurities and defects. Electron mobility may be up to 200 $cm^2/V\,s^{-1}$, while that of holes may be up to 135 $cm^2/V\,s^{-1}$ in high-quality epitaxial layers at room temperature.

Temperature dependences of the hole mobility for electrons and holes in LPE-grown GaP epitaxial layers are presented in figures 8.5(a) and (b),

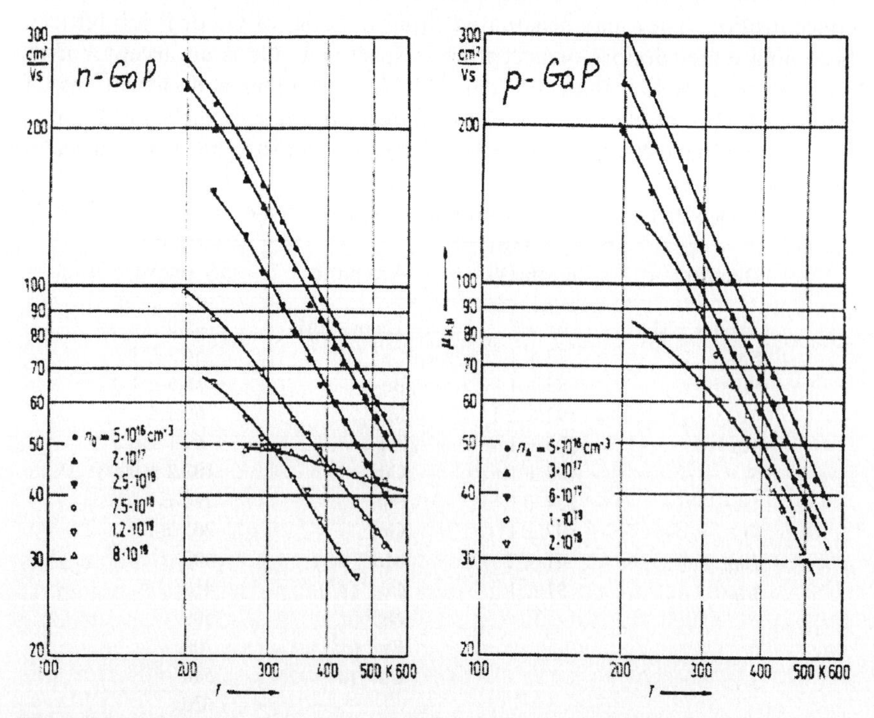

Figure 8.5. Temperature dependence of Hall mobility in (a) n-type and (b) p-type GaP epitaxial layers grown by LPE [20].

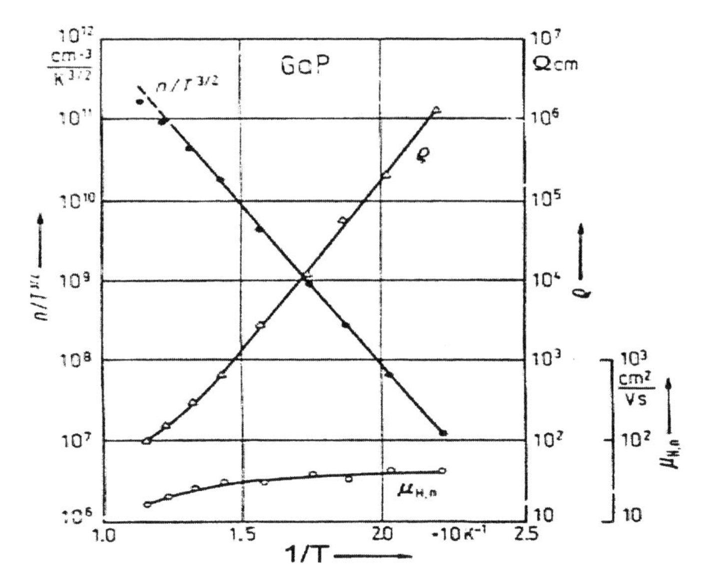

Figure 8.6. Temperature dependence of resistivity ρ, Hall mobility $\mu_{H,n}$ and electron concentration n for a semi-insulating GaP [20].

respectively. Figure 8.6 shows temperature dependences of the resistivity, Hall mobility, and free electron concentration in semi-insulating GaP. Note that resistivity of GaP may be as high as 10^{11} Ω cm at room temperature. The minority carrier lifetime varies in a wide range and is limited by existing recombination centres. In ideal intrinsic GaP crystal the lifetime would be as long as several tens of years, while in ideal GaP doped with donors or acceptors with concentration 10^{16} cm^{-3} it would be several msec. In reality, the relaxation time is much shorter and is determined by the material quality and the impurities concentration.

8.2.3 Optical properties

The spectra of the real ε_1 and imaginary ε_2 parts of the dielectric susceptibility of GaP obtained by ellipsometry measurements by Aspnes and Studna [21] are presented in figure 8.7. The arrows show the energies listed in table 8.2.

The refractive index spectrum is presented in figure 8.8. In the energy range corresponding to the low absorption it may be approximated as [22, 23]

$$n(\hbar\omega) = \left[1 + \frac{A}{\pi}\ln\frac{E_1^2 - (\hbar\omega)^2}{E_0^2 - (\hbar\omega)^2} + \frac{G_1}{E_1^2 - (\hbar\omega)^2} + \frac{G_2}{E_2^2 - (\hbar\omega)^2} + \frac{G_{TO}}{E_{TO}^2 - (\hbar\omega)^2}\right]^{1/2} \tag{8.6}$$

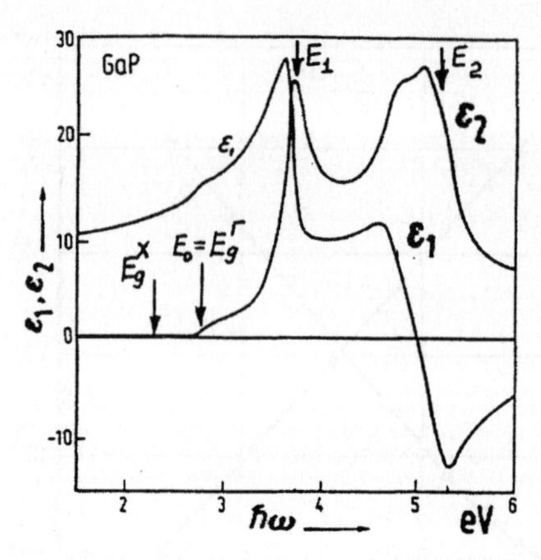

Figure 8.7. Spectra of the real and imaginary parts of dielectric susceptibility of GaP [21].

where parameters at 300 K are the following: $A = 0.420$, $G_1 = 31.4388\,\text{eV}^2$, $G_2 = 160.557\,\text{eV}^2$, $G_{\text{TO}} = 4.49 \times 10^{-3}\,\text{eV}^2$, and energies E_0, E_1, E_2 and $E_{\text{TO}} = \hbar\Omega_{\text{TO}}$ may be found in tables 8.1 and 8.2.

Sometimes it is more useful to express the refractive index as a function of wavelength. In the range $\lambda = 0.5$–$1000\,\mu\text{m}$ at $T = 297\,\text{K}$ it may

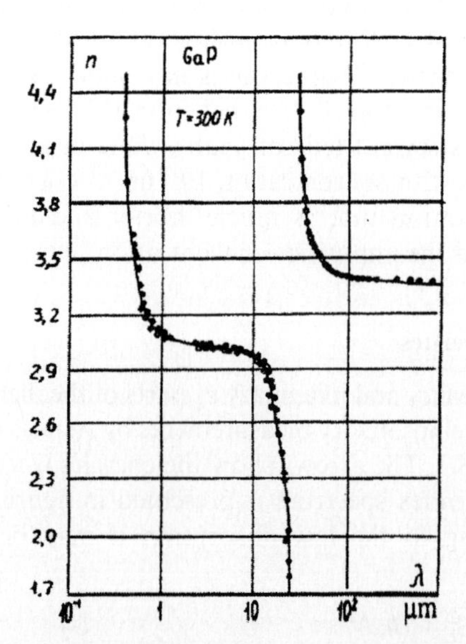

Figure 8.8. Refractive index of undoped GaP in the wavelength range from $\lambda = 0.5$–$1000\,\mu\text{m}$. Solid line shows the theoretical dependence (equation 8.6), while circles correspond to experimental data [22].

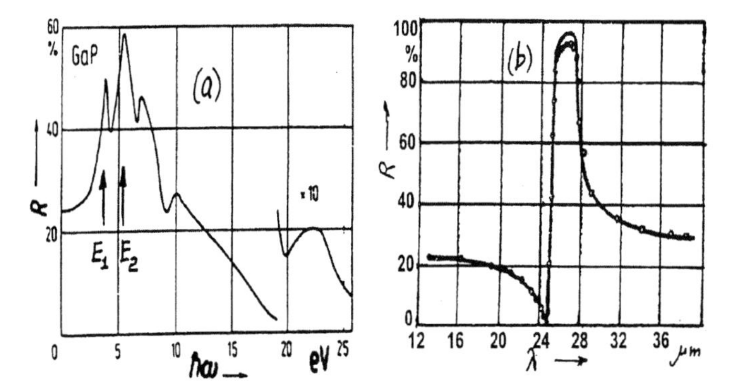

Figure 8.9. Reflectance of undoped GaP as a function of photon energy at room temperature for (a) short and (b) long wavelength spectral regions [24, 25].

be approximated as

$$n(\lambda) = \sqrt{1 + \frac{2.5493}{1 - (0.3351/\lambda)^2} + \frac{5.5255}{1 - (0.2339/\lambda)^2} + \frac{1.9407}{1 - (27.549/\lambda)^2}} \qquad (8.7)$$

where λ is measured in μm.

Refractive index dependences on temperature, hydrostatic pressure and impurity concentration are presented in [22, 23].

The short-wavelength part of the GaP reflectance spectrum (see figure 8.9(a)) is determined by the electronic band structure and the corresponding density of states, which is typical for semiconductors with tetrahedral coordination of the nearest-neighbour atoms. The specific structure appears in the long-wavelength part of the spectrum near the transverse optical (TO) phonon energy in the Γ point, as shown in figure 8.9(b). It is evidence of the strength of the ionic component in the chemical bond between atoms of gallium and phosphorus. The spectrum in this range is well described in the simple one-oscillator model

$$\varepsilon(\omega) = \varepsilon_\infty + [\varepsilon_0 - \varepsilon_\infty] \frac{\omega_{TO}^2}{\omega_{TO}^2 - \omega^2 + i\gamma\omega}, \qquad (8.8)$$

where ε_8 and ε_0 are high- and low-frequency dielectric constants, respectively, while ω_{TO} denotes the transverse optical phonon frequency in the centre of the Brillouin zone. The values of the parameters in equation (8.8) are listed in table 8.1 and the damping parameter in GaP is $\gamma = (0.003 \pm 0.0005)\ \mathrm{s}^{-1}$ [25]. The plasmon–phonon interaction causes a significant change of the reflectance coefficient if the concentration of the major carriers is higher than $n, p > 10^{18}\ \mathrm{cm}^{-3}$ [26].

The fundamental absorption edge of GaP presented in figure 8.10 is determined by the indirect exciton transitions $\Gamma_{15}^{V}(\Gamma_8^{V})$–$X_1^{C}(X_6^{C})$ with

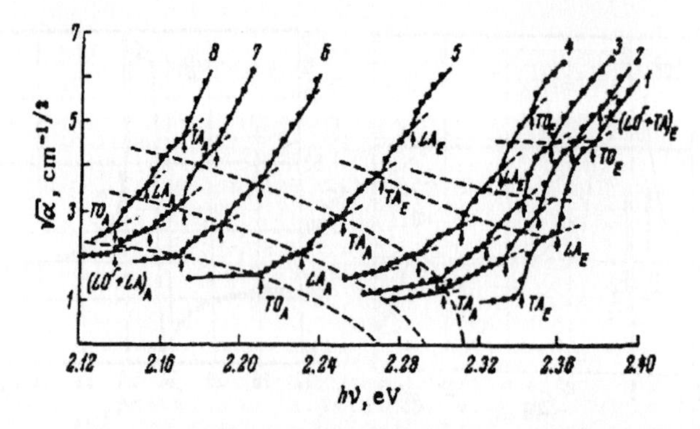

Figure 8.10. Fundamental absorption edge of undoped n-type GaP at different temperatures: (1) 4.2, (2) 85, (3) 114, (4) 160, (5) 295, (6) 381, (7) 445 and (8) 486 K. The arrows show the threshold energies of the components involving absorption (index A) and emission (index E) of the corresponding phonons [27].

phonon assistance. The excitonic effects have to be taken into account up to room temperature and even higher, depending on the quality of the crystal. The transitions with virtual participation of the lowest state of the conduction band in the centre of the Brillouin zone $\Gamma_1^C(\Gamma_6^C)$ are allowed with assistance of the longitudinal acoustic phonon LA(X) only. If the assisting phonons are TA(X) (transverse acoustical phonons with wave vector near the X-point of the Brillouin zone), TO(X) or LO(X), the optical transitions are allowed only trough higher intermediate states in the conduction $\Gamma_{15}^C(\Gamma_7^C)$ or valence Γ_5^V band. The square root of the absorption coefficient as a function of photon energy near the fundamental absorption edge of GaP in figure 8.10 is shown for different temperatures.

This 'unusual' form of phonon-assisted exitonic components was discovered in the absorbtion spectra of high-quality GaP crystals [28, 29]. It was found later that the excitonic absorption spectrum can be described quantitatively only by taking into account the 'camel's back' structure of the minimum in the conductance band [16–18] as illustrated in figure 8.11. The same fine structure appears in every phonon-assisted component and in the component related to the momentum scattering on point defects, such as impurities, vacancies and isoelectronic N_P centres [30].

The inset in figure 8.11(a) shows two dispersion branches of the lowest excitonic states corresponding to the electron-heavy (indicated by its spin-$\frac{3}{2}$) and electron-light (spin-$\frac{1}{2}$) holes. The energy is counted down in meV from the energy gap at the X-point: $E_g(\Gamma_8^V - X_6^C) = (E_g^{ind} + \Delta E)$. This dispersion determines excitonic density of states [16] (solid line in figure 8.11(a)) in perfect agreement with the experimental data [28] shown with circles in figure 8.11(a)), for the TA_E component of the absorption spectrum.

Comparison of the experimental spectra of the derivative of the absorption coefficient with respect to the photon wavelength in figures 8.11(b), (c)

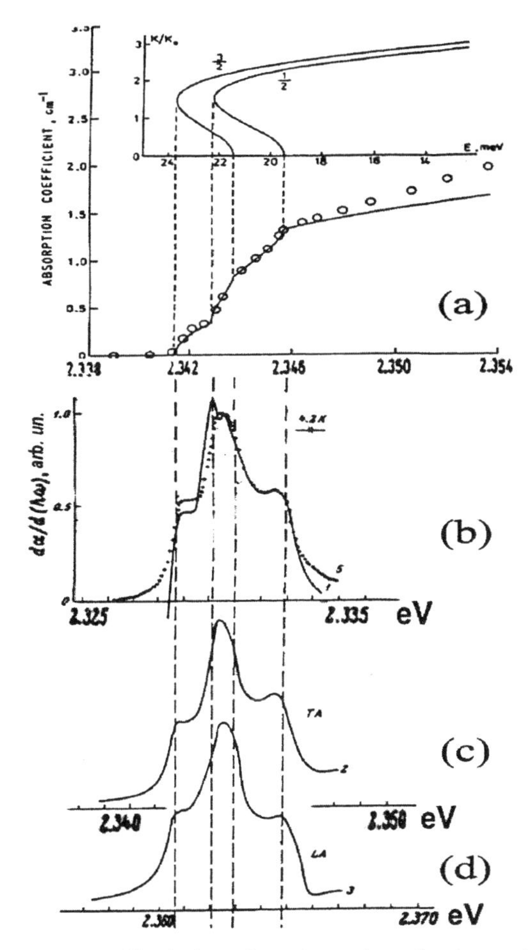

Figure 8.11. Fine structure of intrinsic exciton absorption edge due to the 'camel's back' structure of the conduction band minimum of GaP in the regions of TA_e (a, c), no-phonon (b) and LA_e components. Circles reproduce experimental data [28]. Solid line (a) is the theoretical exciton density of state [16]. Experimental differential spectra $d\alpha/d(\hbar\omega)$ for (b) no-phonon, (c) TE_e and (d) LA_e components are reproduced from [30]. Point in (b) is the calculated differential absorption spectrum with broadening parameter $\Gamma = 0.4$ meV. Corresponding exciton dispersion calculated for GaP is plotted in the insert.

and (d) proves that the same fine-structure appearance in all components of the spectrum is originated from the 'camel's back' excitonic dispersion [16], as traced with the dashed lines through the whole of figure 8.11. The theoretical spectrum plotted with the dotted line in figure 8.11(b) was calculated as a derivative of the convolution of the excitonic density of states and Lorentzian with the damping parameter $\Gamma = 0.4$ meV [30].

The doping of GaP with shallow donors and/or acceptors affects the fundamental absorption edge for several reasons.

First, such impurities create point defects in the crystal and effectively scatter quasi-momentum of the carriers. This process is responsible for the so-called 'no-phonon' component in the spectrum, as already discussed. This component would be absent in the ideal crystal.

Second, the variety of different scattering centres in GaP crystals smoothes the fine-structure in the absorption spectrum. This enables one to estimate the quality of the particular sample from its optical spectra. For example, the very pronounced fine-structure in figure 8.11 was measured in GaP crystals with a concentration of the residual impurities less than $2 \times 10^{16}\,\mathrm{cm}^{-3}$.

Third, the excitons and excitonic complexes may bind to the neutral impurities, resulting in the appearance of narrow lines in the absorption and luminescence spectra at low temperatures, as will be discussed in section 8.2.4. Group VI impurities (S, Te and Se) cause the strong resonance absorption (so-called C-lines) at 2.3098 eV at 4.2 K; and 2.295 eV at 100 K for C_S and C_{Te}. It corresponds to the exciton binding energy at about 18.7 meV with respect to the excitonic energy band gap E_{gx}^{x} of the free exciton. The exciton binding energy on the neutral Se centres is about 20.3 meV. The total absorption cross-sections are $\sigma_S = 9.8 \times 10^{-21}$ and $\sigma_{Te} = 8 \times 10^{-21}\,\mathrm{cm}^2/\mathrm{eV}$ [31]. This structure reduces to the small step at the edge of the absorption spectrum at room temperature. Optical absorption at C-lines is used for non-destroying measurement of the neutral impurities concentration larger than $10^{15}\,\mathrm{cm}^{-3}$.

The relation between the covalent radius and electronegativity of the isoelectronic impurities of group III (B, Al, In and Tl) and group V (N, As, Sb and Bi) and those of Ga and As, respectively determines the solutability of these elements in GaP. In agreement with the Hume-Rothery [32] rule saying that the atomic radii have to be less than 15% different to ensure the wide solutability range of one metal in another, Al, In and As in GaP form alloys in the whole range of concentration. The most useful for the applications are the following solid solutions: Ga(As,P), (Ga,In)P, (Ga,Al)P, (Ga,In,Al)P, (Ga,In)(As,P), (Ga,Al)(As,P) and (Ga,In,Al)(As,P). We will discuss their properties in sections 8.3 and 8.4. N, Bi, Sb and Tl deform the crystal lattice of GaP significantly and form so-called isoelectronic traps. They may bind electron or hole, depending on the form of the short-range potential. In contrast to regular donors and acceptors, isoelectronic impurity becomes charged by the trapped carrier and attracts a carrier of the opposite sign. The comparison of the absorption edge spectra in undoped GaP and GaP:N doped with nitrogen is presented in figure 8.12. Excitons bound on single nitrogen atoms are responsible for the sharp A-line in the GaP:N absorption spectrum. It was first observed by Gross and Nedzvetskii [33], and then it was associated with excitons bound on neutrogen neutral centres

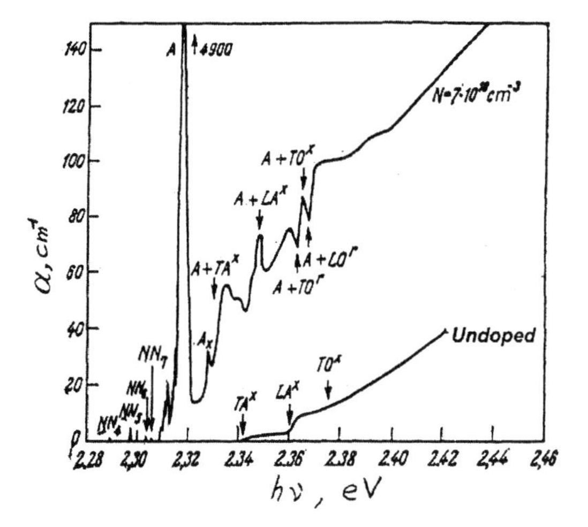

Figure 8.12. Absorption edge of undoped and nitrogen-doped GaP at 2 K [35].

by Thomas *et al* [34]. The absorption is so strong that GaP crystals doped up to $10^{19}\,\mathrm{cm}^{-3}$ become reddish. Nitrogen couples may form molecular complexes NN_n and bind excitons as well. Since the concentration of them is much less than that of single N atoms, the intensity of the corresponding lines is much smaller (see figure 8.12). However, they play a significant role in the luminescent processes. The complicated structure of the spectrum at higher than A-line energies is the joint effect of phonon replicas of the A-line, free excitons formed without phonon assistance, and interference between intermediate states [35]. The integrated absorption cross-section for the A-line is well known, $\sigma = 1.27 \times 10^{-15}\,\mathrm{cm}^2/\mathrm{eV}$, that allows determination of the nitrogen concentration from optical absorption measurements. The sensitivity of this method is $10^{15}\,\mathrm{cm}^{-3}$ for 30 μm epitaxial layers.

Optical properties of the isoelectronic traps are different in anion and cation sublattices [36]. In the anionic sublattice, trapped electrons have an s-type wave-function of X_1^C symmetry. It ensures strong binding with the conduction band minimum in the centre of the Brillouin zone Γ_1^C. The corresponding optical transitions are allowed. In the cationic sublattice trapped electrons have a p-type wave-function of X_3^C symmetry. This state does not mix with the Γ_1^C state and the corresponding optical transitions are forbidden. It explains the lack of the optical observation of excitons bound on B and Tl isoelectronic centres in GaP.

The direct optical transitions in the centre of the Brillouin zone exhibit themselves in the absorption spectrum through a well-pronounced exciton peak at about 2.87 eV and a relatively weak structure at about 2.96 eV corresponding to the direct transitions between spin–orbit splitted Γ_7^V sub-band and Γ_6^C minimum in the conductance band, as shown in figure 8.13 [9].

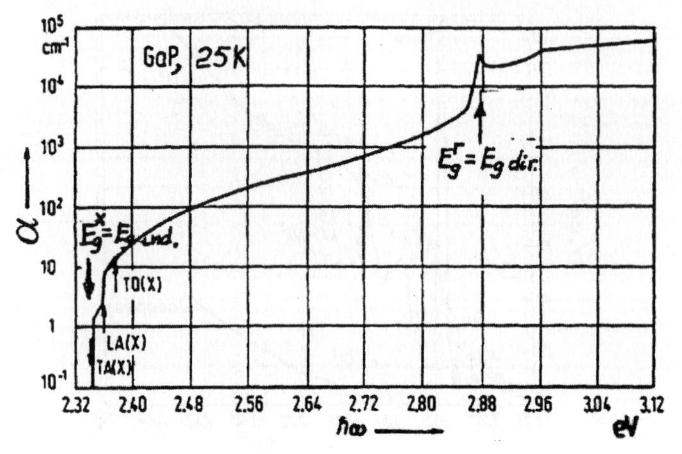

Figure 8.13. Absorption spectrum in the region of the indirect (E_g^X) and direct (E_g^Γ) energy gaps at 25 K [9].

Absorption in the infrared region is determined by the excitation of phonon vibrations. The most effective are transitions with the excitement of the transverse optical phonon in the centre of the Brillouin zone the TO(Γ). The series of the absorption peaks caused by the combination scattering are much weaker [25]. Free electrons in GaP take part in the direct optical transitions between $X_1^C(X_6^C)$ and $X_3^C(X_7^C)$ sub-bands of the conductance band and at high temperatures (>400 K) the corresponding absorption peaks near $\lambda = 3 \mu m$ may be described as [37, 38]

$$\alpha = BN \frac{(\hbar\omega - \Delta)^{3/2}}{\hbar\omega (kT)^{3/2}} \exp\left[-\frac{\hbar\omega - \Delta}{kT} \frac{m_2^*}{m_1^* - m_2^*} \right],$$

where N is free carrier concentration, $m_2^* = m(X_3^C)$ and $m_1^* = m(X_1^C)$ are effective masses in $X_3(X_7)$ and $X_1(X_6)$ minima of the conduction band, Δ is the energy gap between them. At high temperatures one can omit the fine-structure origin of the 'camel's back' and use the following parameters down to room temperature: $m_2^*/m_1^* = 0.2$, $\Delta = 0.30$ eV and $B = (1.4 \pm 0.3) \times 10^{17}$ eV cm^{-2} [38]. At lower temperatures the dominant process will be the photo-ionization and photoexcitation of the shallow donors S, Te and Se to the second $X_3^C(X_7^C)$ minimum of the conductance band [18, 38]. Both mechanisms exist at room temperature (figure 8.14), which means that not all of the impurities are ionized yet. Intensive absorption on the free carriers, which is proportional to $\sim\lambda^r$, also exists in the infrared region of the spectrum.

Doping of GaP with shallow acceptors C, Be, Mg, Zn and Cd increases absorption on the scattering on free holes and inter-band transitions V_1–V_2 and V_1–V_3 in the valence band at high temperature. Photo-excitation and

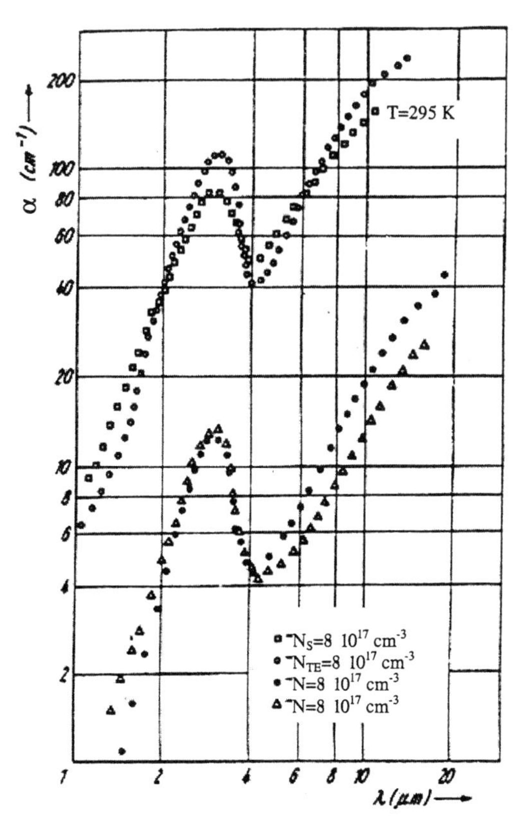

Figure 8.14. Infrared-part of the absorption spectra of n-GaP doped with S and Te at 295 K [38].

photo-ionization are dominant at low temperatures. All these processes coexist with the combination scattering on lattice vibrations.

In 1978 Kopylov and Pikhtin [18] showed that for the correct interpretation of the photo-excitation spectra of shallow donors in GaP one should account for the effect of the 'camel's back' structure of the conduction band minimum. Moreover, they measured the new values of ionization energies of Zn, Cd and C acceptors [39, 40]. The corresponding correction increased the ionization energy by 8 meV for acceptors and by 3 meV for donors [40]. Thus all the data reported in the literature before 1978 (including those in [1–4]) have to be corrected.

The Stark effect was observed on the shallow donors [41] and bound on single nitrogen atoms excitons [42]. The Zeeman effect on bound excitons was investigated in [43]. The A-line in the absorption spectrum also exhibits an 'abnormal' photo-effect [44] of a still undefined nature.

GaP has a relatively high electro-optic coefficient $r_{41} = 10^{-10}$ cm/V [45]. Spectra of piezo-optic coefficients in GaP are summarized in [46].

3*d* and 4*f* electron shells of the elements from the iron and lanthanum groups create deep discrete levels in the energy gap of GaP. The optical transitions between them affect spectra of optical absorption, luminescence, electron paramagnetic resonance and Raman scattering. The parameters of these centres may be found in [9, 47].

Excitation of the vibrations of single impurities and their molecular complexes results in the appearance of narrow resonances in infrared absorption spectra, exciton luminescence and Raman spectra. Isotope shift was observed for some of the impurities [9].

8.2.4 Luminescence

Gallium phosphide has been the model semiconductor for studying different mechanisms of luminescence for a long time. Some very important luminescence mechanisms (i.e. emitting recombination through donor–acceptor pairs and isoelectronic traps) were discovered specifically in GaP. This was caused by several reasons. First, the energy band gap of GaP corresponds to the very attractive 0.4–0.8 μm wavelength range, which is very practical for spectroscopic measurements with high resolution and sensitivity. Second, in contrast to most wide band gap semiconductors, such as group II–VI and III nitrides, one may easily tune its type of electrical conductivity by doping with accessible donors and acceptors and create p–n junctions. Third, the investigation of GaP as a base material for light emitting diodes (LEDs) has been very well financially supported. Almost every luminescence mechanism works in gallium phosphide.

Intrinsic luminescence is related to the fundamental absorption with the Van Roosbroeck–Shockley relation [48, 49]. Intrinsic luminescence in high-quality samples is mostly determined by the recombination of free excitons since they determine the fundamental absorption edge up to room temperature, as discussed in section 8.2.3 and illustrated in figure 8.10. This process in the indirect band gap GaP is accompanied by absorption or emission of phonons. At $T > 400$ K this mechanism of the luminescence is determined mostly by phonon-assisted band-to-band transitions. The exciton interaction increases the probability of intrinsic light-emitting recombination, which may be as high as $B \approx 2 \times 10^{-13}$ cm^3 s^{-1} at room temperature—that is, almost an order of magnitude larger than estimated for negligible excitonic interaction [49]. Nevertheless, B is still three orders of magnitude less than in such direct band gap semiconductors as GaAs and InP. However, a significant rise of the luminescence efficiency may be achieved in clean GaP crystals due to the wide band gap (2.27 eV at room temperature). In an ideal crystal the light-emitting lifetime would be as long as 10^{13} s (i.e. more than 1000 years!) since the intrinsic concentration would be as little as $n_i \approx 1$ cm^{-3}. At free carrier concentration n, $p \approx 10^{16}$ cm^{-3}, the lifetime reduces to 0.2 ms, still very high. Thus it is

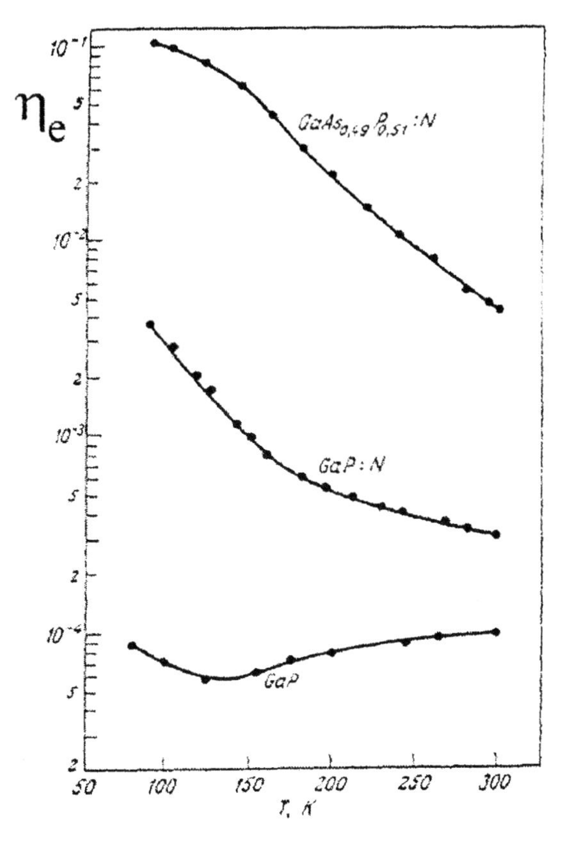

Figure 8.15. External quantum efficiency η_e of luminescence of pure GaP and of GaP and gallium arsenide phosphide doped with nitrogen as a function of temperature [50].

necessary to have GaP crystals with as low as possible concentration of deep non-radiative centres (especially O and Cu) to achieve high efficiency of fundamental luminescent recombination.

The relative intensity of the intrinsic luminescence rises with increasing temperature and pumping intensity, since deep impurity centres saturate, while carriers escape from shallow ones. The self-absorption in GaP is not as important as in direct band gap semiconductors. The lower curve in figure 8.15 traces the external quantum efficiency η_e of an undoped n-GaP as a function of temperature. The presented data were obtained in 1973 by Craford *et al* [50]. Significant progress has been made since then in the reduction of non-radiative relaxation in GaP. The non-radiative lifetime has been increased as a result of decreased residual deep trap concentration. It leads to quantum efficiency improved on an order of magnitude comparable with that presented in figure 8.15.

The intrinsic radiative luminescence is the main luminescence type in so-called 'pure green' GaP LEDs. Sometimes the active region of such LEDs is doped with isoelectronic centres like In etc. These impurities supply momentum scattering centres, which do not form local centres of additional recombination. Concentration of residual impurities may be lowered by additional doping with some rare-earth elements (e.g. yttrium). It leads to further improvement of the intrinsic luminescence near 555 nm at 300 K [51]. However, the intrinsic quantum efficiency of intrinsic photoluminescence in GaP is still no more than a few percent. High excitation density at temperatures lower than 45 K luminescence of the electron–hole liquid in GaP was obtained by Bimberg *et al* [52].

A donor–acceptor pair (DAP) radiative recombination always presents in the luminescence spectrum since residual donors and acceptors coexist in any semiconductor with main doping impurities. Usually, the order of compensation is even larger than 10%. Minority carriers created by excitation mechanisms are effectively trapped by these charged centres at low temperature, i.e. when E_A, $E_D \gg kT$. The energy of the emitted photon at the recombination process between electron trapped at donor with ionization energy E_D and hole trapped at acceptor with the ionization energy E_A separated by distance R from each other may be found as

$$\hbar\omega = E_g - E_A - E_D + \frac{e^2}{4\pi\varepsilon_0\varepsilon_r R} + \Delta E \tag{8.9}$$

where the Coulomb term reflects charging of donors and acceptors after recombination. The last term takes into account van der Waals interaction of the charged impurities

$$\Delta E = -\frac{e^2 b^5}{4\pi\varepsilon_0\varepsilon_r R^6}.$$

Here b is the constant of dipole–dipole interaction between radiative recombination centres. According to perturbation theory, it is related to the Bohr's radius $a_B$of the carrier trapped at a localization centre with large bound energy as

$$b \approx (6.5)^{1/5} a_B.$$

The last two terms in equation (8.9) may have only discrete values since the distance between trapping centres R is determined by their possible positions in the host crystal lattice. It results in the multi-line structure of the corresponding part of the luminescence spectrum (see figure 8.16) [53]. Numbers in brackets denote the number of the coordination sphere m, while numbers without brackets show degeneracy of the equivalent donor–acceptor pairs in the corresponding coordination sphere. DAP luminescence of type I takes place if both donors and acceptors substitute atoms of the

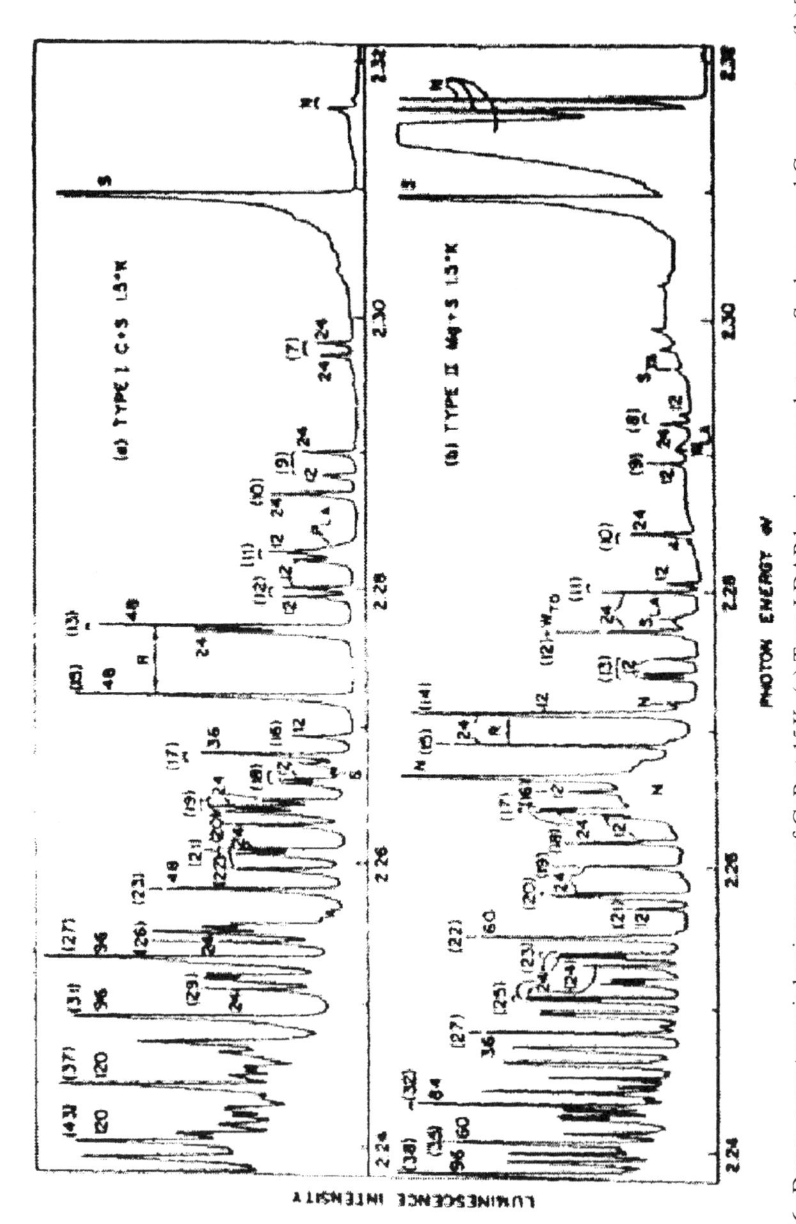

Figure 8.16. Donor–acceptor pair luminescence of GaP at 15 K. (a) Type I DAP luminescence between S_P donors and C_P acceptors. (b) Type II DAP luminescence between S_P donors and Mg_{Ga} acceptors. The luminescence lines at the higher energy are determined by recombination of bound on N and S excitons. After [53].

same sub-lattice, e.g. (C_P–S_P) DAPs in the P sub-lattice. DAP luminescence of type II takes place if donors and acceptors substitute atoms of different sub-lattices, e.g. Mg,Cu–S_P DAPs. Note that DAP luminescence is mostly determined by pairs of donors and acceptors separated quite far from each other ($m > 7$ as in figures 8.16 and 8.17). If the distance R between them becomes smaller, neutral molecular complexes are formed. The probability of the radiative transition of electron from donor to acceptor is proportional to the square of the dipole matrix element $|V_{\text{e-p}}|^2$. Since the overlap of electron and hole wave-functions exponentially decays at large R, this probability is approximately scaled with R as

$$W = W_0 \exp(-2R/a_B) \tag{8.10}$$

where a_B is the Bohr radius of a carrier with smaller bound energy and W_0 is the constant for every type of DAP.

The possibility of the DAP radiative recombination was first pointed out by Prener and Williams [55] and its first experimental observation was reported in [56] in GaP. Since then GaP has become the model object for study of this recombination mechanism, which plays an important role in all semiconductors. The DAP luminescence may be characterized by the following unique features:

(i) The low wavelength part of DAP luminescence at low temperature consists of multiple narrow lines (equation (8.9)). Their density per unit energy increases at lower energy. It corresponds to an increase of the distance R between donors and acceptors of the pair. It leads to the eventual vanishing of the fine structure of the spectrum due to overlapping of a number of lines and decreasing of the optical transition probability (equation (8.10)). The broad maximum in the emission spectrum appears along with phonon replicas corresponding to the emission of TO, LO, LA(X) and TA(X) phonons, as shown in figure 8.17 [54]. The structure of the phonon replicas depends on electron density distribution between donor and acceptor. Any perturbations in the crystal, such as internal strain in epitaxial layers, boundaries, built-in electric field etc., result in inhomogeneous broadening of all DAP multiple lines. It may wipe out the fine-structure of the luminescence, especially electroluminescence spectra.

(ii) The relaxation lifetime decreases at higher energies (i.e. at small R according to equations (8.9) and (8.10)). The integrated luminescence decays non-exponentially with time. Most DAPs are characterized by the low optical transition probability at low temperature resulting from small wave-function overlap. The long lifetime of the carriers means that the DAP luminescence may be damped by temperature due to the ionization of the impurities. It vanishes in GaP at $T > 150$ K for shallow DAPs.

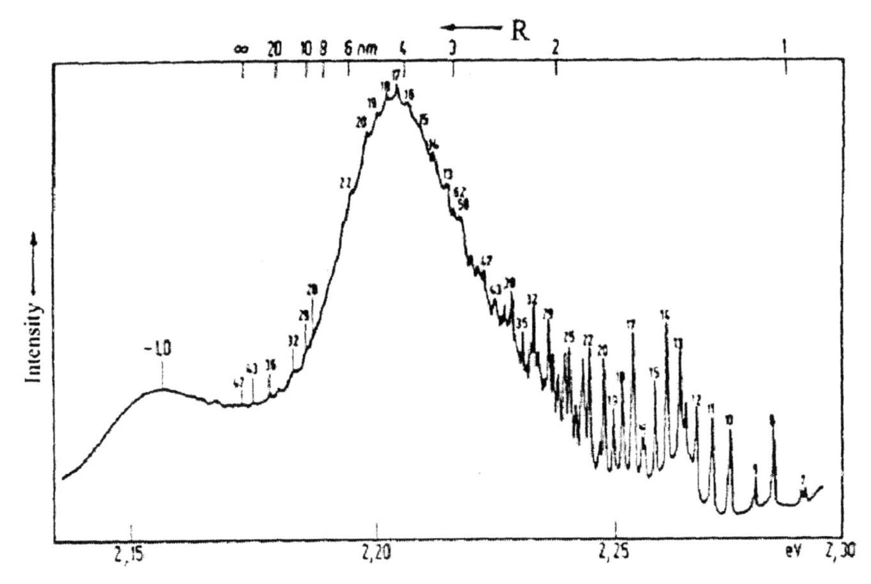

Figure 8.17. Photoluminescence spectrum of GaP doped with S and Zn. The distance R between donor and acceptor in DAP is shown on the top axis.

(iii) The maxima of the no-phonon DAP luminescence line and its phonon replicas shift up in energy with increasing excitation intensity. The emission spectra shift towards the larger wavelengths at longer time delay of the detector.

All these features of different DAPs in GaP were studied experimentally in detail [6, 9, 53, 57], including shift and splitting of the lines in magnetic field. For every DAP the symmetry of the impurity centres and the value of the first term $[E_g - (E_A + E_D)]$ in equation (8.9) were determined with high accuracy. (As already discussed in section 8.2.3, the ionization energy of all donors reported before 1978 have to be increased by 11 meV. However, it did not lead to the misinterpretation of the experimental spectra, since the indirect band-gap has to be increased by 11 meV as well.)

Table 8.4 summarizes the energy of the no-phonon maximum and lifetime in DAP luminescence for various DAPs.

Neutral-charged inter-impurity radiative recombination is analogous to DAP luminescence with the only, although extremely important, feature that one of the impurity centres is neutral and binds electron or hole with non-Coulomb potential. The absence of the Coulomb interaction between centres leads to the more narrow emission line and absence of the thin structure of the luminescence spectrum associated with discrete R. The position of the maximum of the no-phonon line may be found as

$$\hbar\omega = E_g - (E_D + E_h) \quad \text{or} \quad \hbar\omega = E_g - (E_A + E_e)$$

Table 8.4. Energy of no-phonon maximum (at 80 K) and lifetime (at 2 K) in DAP luminescence for different donor–acceptor pairs in GaP.

DAP	S_P–C_P	S_P–Si_P	S_P–Zn_{Ga}	S_P–Cd_{Ga}	Te_P–C_P	Te_P–Si_P
$\hbar\omega_{max}$ (eV)	2.21	2.05	2.19	2.16	2.22	2.06
t (μs)	2.0	2.5	1.7	2.0	11	—
DAP	Te_P–Zn_{Ga}	Te_P–Cd_{Ga}	Si_{Ga}–C_P	Si_{Ga}–Si_P	O_P–C_P	O_P–Zn_{Ga}
$\hbar\omega_{max}$ (eV)	2.21	2.17	2.22	2.08	1.42	1.41
t (μs)	—	—	250	250	0.66	0.66

for recombination between electron (hole) bounded to donor (acceptor) with the ionization energy E_D (E_A) and hole (electron) trapped at a neutral centre with the binding energy E_h (E_e). The effects of large relaxation time and non-exponential luminescence dumping discussed for DAP are also important for this recombination mechanism. The neutral-charged inter-impurity radiative recombination was experimentally obtained in GaP for different impurity pairs: (i) trapped on neutral donors holes recombined with electrons bound at shallow donors of group IV (S, Se, Te); (ii) trapped on neutral molecular complexes Zn–O and CdO electrons recombined with holes bound at shallow acceptors of group IV (Zn and Cd); (iii) trapped on iso-electronic Bi_P centre hole with electron bound to a shallow donor [6, 53, 57].

Band-to-impurity and impurity-to-band luminescence was observed in GaP for radiative recombination between free electrons and bound at acceptors Si_P and Cd_P holes and between free holes and bound at shallow (S, Te and Se) and deep (O) donors electrons. These transitions appears in luminescence spectra as a transformation of DAP recombination at temperature $T > 100$ K when one of the paired impurities had ionized. Usually, this would be a shallow acceptor in GaP. The mixing of Γ_1^C states of conduction band in those of a shallow donor (mostly formed by the lowest X_1^C states of the conduction band) increases the probability of optical transitions between shallow donor and Γ_1^V maximum of the valence band. Nevertheless, due to the indirect band gap, the probability of the radiative recombination in GaP is two to three orders of magnitude lower that that in direct band gap semiconductors. The spectral line of such transitions is determined by free carrier distribution in the valence band. Thus its maximum may be found as $\hbar\omega_{max} = E_g - E_D + kT/2$. If recombination takes place between free holes and bound at deep oxygen donor electrons, the form of very wide luminescence line in the infrared range is mostly determined by strong electron–phonon interaction [57–59].

Luminescence of excitons trapped on neutral donors or acceptors was obtained in GaP for many impurities [5, 6, 9, 10, 53, 57]. It exhibits itself at low temperature as a narrow line, often with a thin structure. The main

Table 8.5. Parameters of excitons bound to neutral donor in GaP.

Donor	Localization energies (meV)		Transition lifetimes (ns)
	Lowest components	Excited states splittings	
S	18.7	9.2 [1s(A), 1s(E)] 16.0; 16.6 (2s) 22.5; 22.4; 27.0; 27.5 (3s)	21
Se	20.3	16.7; 16.9 (2s) 23.0; 27.8; 30.2; 30.1 (3s)	21
Te	18.7	5.0 [1s(A), 1s(E)] 16.1; 21.6 (2s) 22.4; 26.9; 29.1 (3s)	21
Si_{Ga}	13.8	—	≤500
Ge_{Ga}	63.6	—	≤50
Sn_{Ga}	10.6	—	90
Li_P	16.3	—	—
Li_{Ga}	7.8	—	—

no-phonon line is accompanied by its phonon repetitions. More than 90% of the integral absorption cross-section for excitons trapped by shallow donors (S, Te and Se) corresponds to no-phonon absorption lines C_S, C_{Te} and C_{Se}, which are also replicated in luminescence spectra. They are usually denoted as C-lines with a subscript corresponding to the trapping atom, but on figure 8.16 they are denoted as S-line. The radiation lifetimes, excited states splitting, and energies of exciton localization on donors are summarized in table 8.5, while those on acceptors are listed in table 8.6 [5, 6, 9, 10, 53, 57, 60]. The energy origin corresponds to the energy of free exciton E_{gx}.

Table 8.6. Parameters of excitons bound to neutral acceptor.

Acceptor	Localization energies (meV)		Transition lifetimes (ns)
	Lowest components	Excited states splittings	
C_P	6.3	0.26	280
Be	6.4	0.33	—
Mg	6.5	0.20	190
Zn	7.15	0.20	110
Cd	8.9	0.88	14
X	5.6	0.99	300

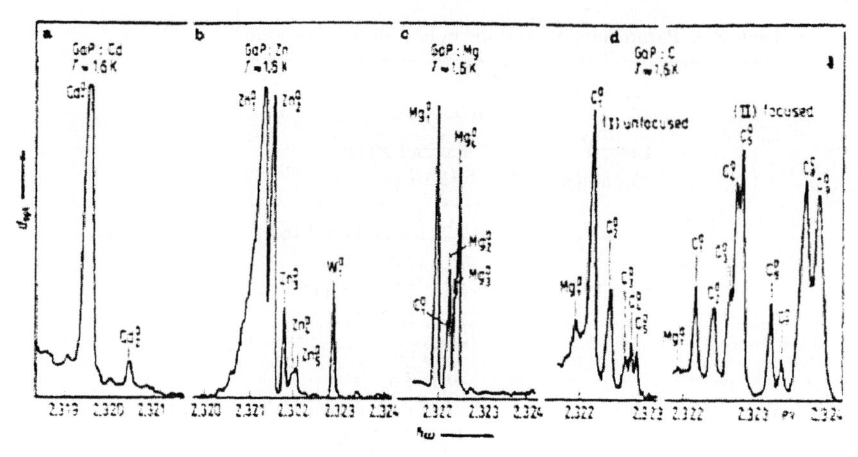

Figure 8.18. The no-phonon luminescence lines of excitons bound to the indicated acceptors in GaP [60].

Note that theoretical radiative lifetime $t \approx 10\,\mu s$, estimated from the absorption cross section presented in section 8.2.3, is three orders of magnitude larger than its experimental value. It means that non-radiative relaxation processes of excitons bound to neutral donors and acceptors are very intensive. Indeed, this is a very complex system consisting of a neutral donor (acceptor), two electrons (holes) and a hole (electron), thus the probability of Auger processes is very high. The corresponding low-temperature luminescence spectra [60] are presented in figure 8.18 for different types of acceptors. When temperature rises above 150 K the luminescence bound on shallow impurities excitons of quenches. This luminescence type is widely used for identification of impurities and determination of their concentration in low range [53]. Exciton localization energy monotonously increases with the corresponding donor or acceptor ionization energy, as traced in figure 8.19 [57]. Two electron transitions were also obtained in luminescence of excitons trapped at neutral donors [61]. They correspond to optical relaxation when part of the energy is spent to transfer enable transition of the donor to one of its excited states. Fieseler *et al* [62] reported on a thin structure consisting of 19 narrow lines of bound at molecular clusters with Te_2 exciton luminescence.

Luminescence of excitons bound to isoelectronic traps is the most effective and, therefore, useful for application in the LED luminescence mechanism in GaP. It significantly differs from radiative recombination of excitons bound on shallow impurities.

This system consists of isoelectronic centre, hole and single electron, i.e. there is no third carrier that could absorb energy in a non-radiative Auger process.

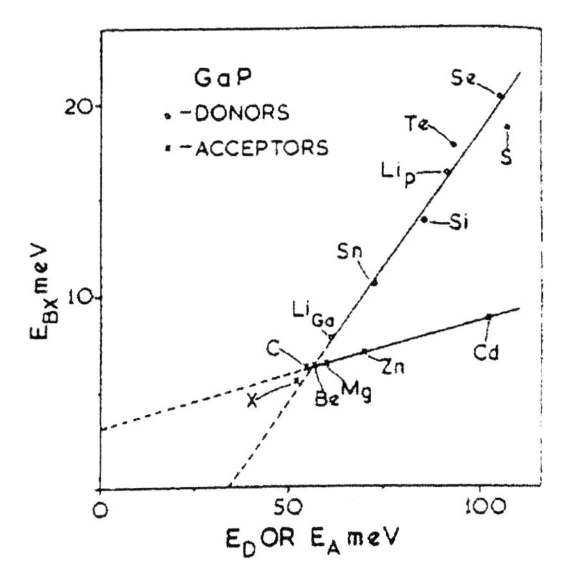

Figure 8.19. Dependence of the exciton localization energy E_{Bx} on corresponding donors E_{D} and acceptors E_{A} ionization energies [57].

In contrast to Coulomb potential of shallow impurities, the localization potential of an isoelectronic trap is short-range. It means that the wave-function of the localized exciton is delocalized in **k**-space, which leads to significant contribution of the conduction band states from the centre of the Brillouin zone in the excitonic state. It increases the probability of no-phonon radiative recombination.

As already pointed in section 8.2.3, nitrogen atoms are the most common isoelectronic traps in GaP. The splitting of its fine structure in magnetic field [43, 57], under hydrostatic [63] and uniaxial [64] deformation; shift and broadening of A-line in electric field [42] and its $N^{14} \rightarrow N^{15}$ isotope shift [65]; heat quenching, kinetics and excitation spectra [53, 57, 66] of the N-bound excitons luminescence have been studied in great details. Lupal and Pikhtin [67] experimentally observed acceptor-like excited states in differential absorption spectra at $T = 2\,\mathrm{K}$.

Unfortunately, an adequate quantitative theory of isoelectronic centres or deep impurities does not exist yet. Nevertheless, most experimental results validate applicability of the isoelectronic acceptor model to N-centres in GaP [57, 68, 69]. Figure 8.20 shows a typical luminescence spectrum of GaP with low concentration of nitrogen. From comparison of this spectrum with one presented in figure 8.12 we may see that the no-phonon A-line in the luminescence spectrum (figure 8.20) (in fact a doublet consisting of intensive allowed in the dipole approximation A-line and weak forbidden in dipole approximation B-lines with super-fine inner structure) resonates

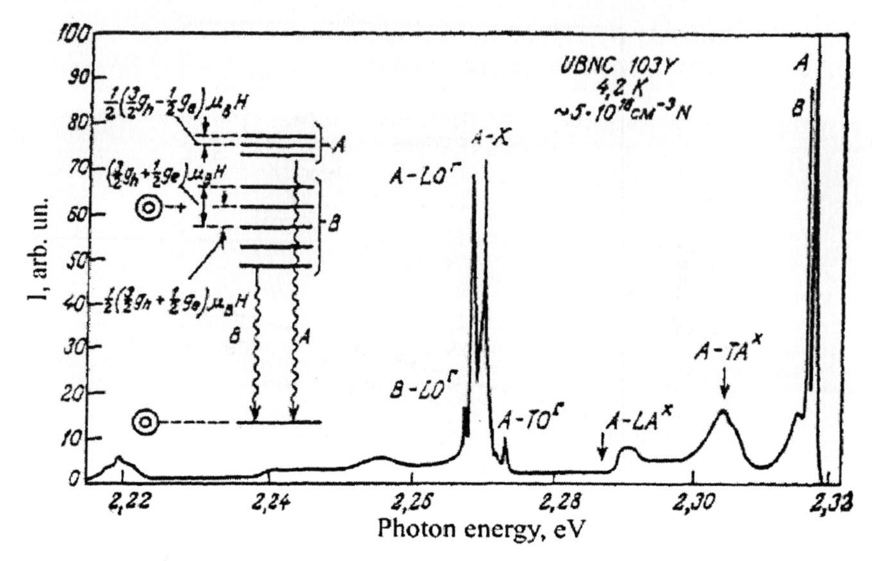

Figure 8.20. Photoluminescence spectrum of GaP with nitrogen concentration $N \approx 5 \times 10^{16}\,\mathrm{cm}^{-3}$. The inset schematically shows the splitting of the A- and B-lines of bound excitonic states in magnetic field [43, 68].

with A-line (in fact with a superposition of A- and B-lines) in the absorption spectrum (figure 8.12). The phonon replicas indicated in figures 8.12 and 8.20 also correspond with each other. The inset in figure 8.20 schematically shows the splitting of the excitonic states in magnetic field [68].

The radiative lifetime $t_N \approx 50\,\mathrm{ns}$ estimated from the absorption cross-section of the no-phonon A-line ($\sigma_A = 1.2 \times 10^{-15}\,\mathrm{cm}^2/\mathrm{meV}$), which corresponds to about one third of the total oscillator strength, is almost three orders of magnitude less than that of bound to donor exciton (see above). This value is in very good agreement with the experimentally measured lifetime $t \approx 40\,\mathrm{ns}$ [72] that means that radiative recombination is more intensive than the non-radiative one. Intrinsic quantum efficiency at low temperature is close to 100%.

NN_m molecular complexes are formed in GaP with high N concentration. Here m denotes the number of the coordination sphere of relative position of nitrogen atoms. Excitons bound to such complexes exhibit themselves as sequences of narrow lines in the luminescence spectrum. The NN_1 complex consisting of two N atoms substituting the neighbour P atoms has the largest binding energy. The binding energies of exciton E_b at molecular complexes counted from the energy of free exciton in GaP are listed in table 8.7. Numbers in parentheses correspond to the energy necessary to bind the electron or hole to the centre. The difference of the

Table 8.7. Exciton localization energies at point defects and at molecular isoelectronic traps.

Isoelectronic traps	Localization energies E_P (in meV)
N_P	11.4
NN_1 [110]	143.1 ($E_\ell = 124$)
NN_2 [200]	138.4
NN_3 [211]	64.1 ($E_\ell = 45.5$)
NN_4 [220]	39.1 ($E_\ell = 12$)
NN_5 [310]	31.4
NN_6 [222]	25.2 ($E_\ell = 12.8$)
NN_7 [321]	22.6
NN_8 [400]	20.9
NN_9 [330], [411]	18.4
NN_{10} [420]	17.5
Bi_P	97.3 ($E_h = 46.2$)
$(Sb_P–Sb_P)$ [111]	3.0
$(Zn_{Ga}–O_P)$ [111]	319 ($E_\ell = 296$)
$(Cd_{Ga}–O_P)$ [111]	420 ($E_\ell = 400$)
$(Mg_{Ga}–O_P)$ [111]	165
$(Li_{Ga}–Li_{Ga}–O_P)$	238.1
(Cu–Li)	84.5

two numbers gives the binding energy for the second particle forming the bound exciton.

In contrast to the luminescence of excitons bound to shallow impurities, the luminescence of excitons bound to isoelectronic traps in GaP is quite intensive up to room temperature. The larger the N concentration, the higher the luminescence intensity at room temperature, as illustrated in figure 8.21. In the same time the spectrum is shifted towards long wavelengths due to increased radiative recombination of bound on NN_m complexes excitons (see also figure 8.15).

Molecular complex $(Zn_{Ga}–O_P)$ formed by zinc and oxygen atoms substituting the Ga and P nearest to each other, respectively, is another important isoelectronic centre in GaP. Similar to N and NN_m, molecular complexes (Zn–O) and (Cd–O) have some properties of an isoelectronic acceptor. It meant that it first binds the electron by its short-range potential and then binds the hole by the acquired Coulomb potential. However, their short-range potential well is much deeper than in NN_m. Thus these isoelectronic centres have such properties of deep traps as high binding energies (see table 8.7), interaction with phonons, and their own local vibrations. It results in red shift and broadening of the emission spectrum. Also, the luminescence heat quenching takes place at higher temperature. This allows use of

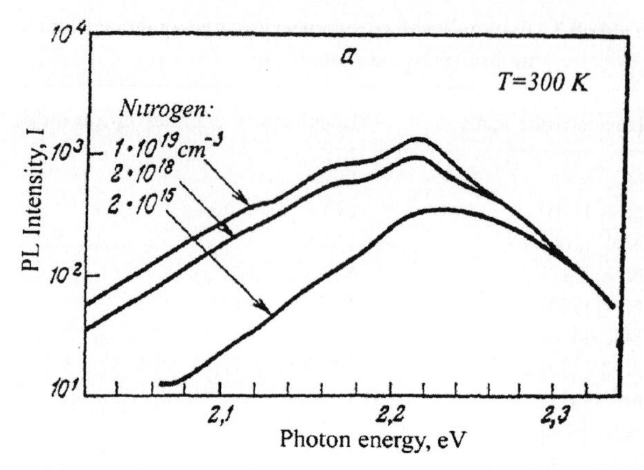

Figure 8.21. Photoluminescence spectrum of GaP at 300 K with different nitrogen concentrations [6].

isoelectronic complexes (Zn_{Ga}–O_P) as effective red light emission sources at room temperature.

The A-line of the bound on (Zn–O) and (Cd–O) complex excitons is well pronounced at low temperature and accompanied with series of equidistant phonon replicas. The distance between these replicas is determined by the energy of local vibration of the complex. The isotope shifts $O^{16} \rightarrow O^{18}$ and $Cd^{114} \rightarrow Cd^{110}$ confirm that the corresponding trapping centre consists of atoms of both types [73, 74]. The oscillator strength for recombination of bound on the (Zn–O) exciton is $f = 0.06$ [75] and the corresponding lifetime is $\tau = 0.1$ μs. Figure 8.22 shows the transformation of the luminescence spectrum of GaP:(Cd–O) with temperature. The technology of the samples growing ensured the minimal fraction of the closely situated Cd–O pairs. Thus the luminescence line caused by recombination of an electron trapped by the neutral (Cd–O) complex and a hole bound at spatially separated Cd acceptor superimposed with excitonic spectrum at low temperature. This neutral-charged inter-impurity radiative recombination line is shifted to the lower energies as indicated in figure 8.22 by arrows. Even though the concentration of neutral molecular complexes (Cd–O) is much smaller than the concentration of isolated impurities, the corresponding recombination ensures the inter-impurities' luminescence at room temperature. The same is valid for (Zn–O) molecular complexes [75]. Annealing at 400–600 °C enriches the sample with nearest-neighbour pairs Zn–O [76, 77] and increases the luminescence efficiency. According to these experimental data, about 50% of oxygen atoms in GaP with $N_O = 10^{17}\,cm^{-3}$ form (Zn_{Ga}–O_P) complexes after optimal annealing at 400 °C.

The detailed investigations of GaP luminescence kinetics [78] and lifetime temperature dependence confirms that its room-temperature effective

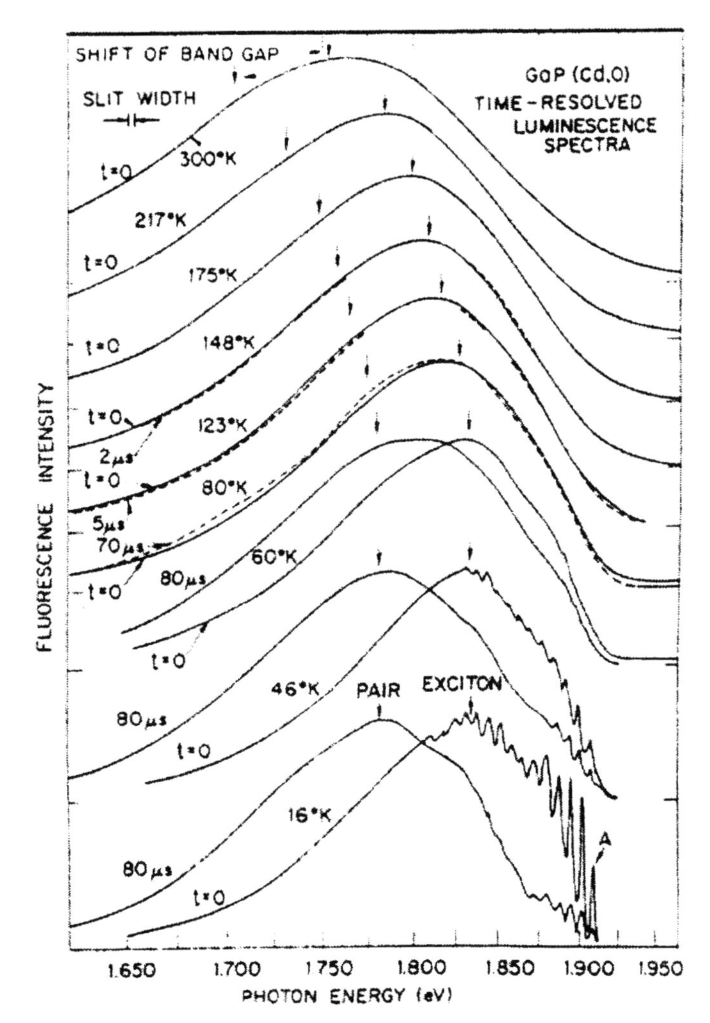

Figure 8.22. Temperature transformation of luminescence spectra of GaP:(Cd–O). The lower energy arrows indicate the maximum of the emission line corresponding to neutral (Cd–O)-charged Cd inter-impurity radiative recombination, while the upper one indicates the maximum of the no-phonon A-line of bound on neutral (Cd–O) exciton luminescence [75].

red luminescence associates with radiative recombination of bound on isoelectronic (Zn–O) molecular complex excitons.

As a consequence of relatively low binding energy (see table 8.7 and discussion above), luminescence of bound at isoelectronic donors Bi_P [80], (Cu–O) [81] and (Sb–O) [57] molecular complexes has low efficiency. Radiative recombination of bound at (Li–Li–O) [82], [$(B_{Ga}–N_P)–N_P$] and

[(B_{Ga}–N_P)–(B_{Ga}–N_P)] [83] molecular complexes has been obtained as well. However, it has relatively low intensity since the concentration of such complicated systems in GaP is quite low. At high excitation density N-bound excitons may condense and form electron–hole droplets in GaP:N [84].

Localized extrinsic and hot luminescence were also obtained in GaP. The oscillator strength of $d \leftrightarrow d$ and $f \leftrightarrow f$ forbidden in dipole approximation transitions in elements of iron and lanthanum groups is as low as 10^{-4}–10^{-8}. Thus the reported intensity of the corresponding emission was very small [5, 6, 9, 47, 53, 57]. Rather weak low-inertial ($\tau < 0.5$ ns) hot luminescence due to radiative inter-band transitions of hot electrons in broken-down GaP p–n junctions was reported in [85].

8.2.5 Device technology and applications

Gallium phosphide and $GaAs_{1-x}P_x$ alloy were the first materials used for commercial visible-range LED fabrication [6]. This followed after intensive investigations of radiative recombination in GaP in 1960–1970. Then green LEDs were created using luminescence through isoelectronic nitrogen traps in GaP (see figures 8.20 and 8.21) and red LEDs were created based on the luminescence of molecular isoelectronic complex Zn–O (see figure 8.22). The common diffusion technology of p–n junction creation was employed to fabricate these first LEDs. Metal Zn, ZnP_2, (Zn + P) and (70% Zn + 30% Ga) were used as diffusion sources. The diffusion temperature was in the range 650–1000 °C. Besides diffusion of Zn and Cd, ion implantation of Te, S and Si donors as well as Zn, Cd, Be and Mg acceptors followed by annealing are widely used to form highly conductive surface layers for low resistivity contacts in GaP-based devices. The first rapid improvement of red LED efficiency in the mid-1970s resulted from the epitaxial technology development as shown in figure 8.23 [86]. Later steps in the efficiency are related to single, double and quantum-sized heterostructure introduction in the device active region. Almost 100% internal luminescence efficiency is now achieved in red, orange and yellow LEDs based on ternary and quaternary wide band gap (Ga–In–Al)P/GaP nanostructures, leading to the extreme values of luminous efficiency.

The same GaP-based solid solutions are used for laser fabrication. The laser structures are grown by molecular beam epitaxy (MBE), including novel solid source molecular beam epitaxy (SSMBE), metal-organic chemical vapour deposition (MOCVD), and liquid phase epitaxy (LPE). In general, the technology of GaP-based devices has many common features with that of the analogous GaAs-based devices [87].

Since GaP is indirect band gap material, the efficiency of the corresponding red and green LEDs is less than that of up-to-date LEDs based on direct-band alloys such as (Ga–Al)As (red) and (Ga–In)N (green). Nevertheless,

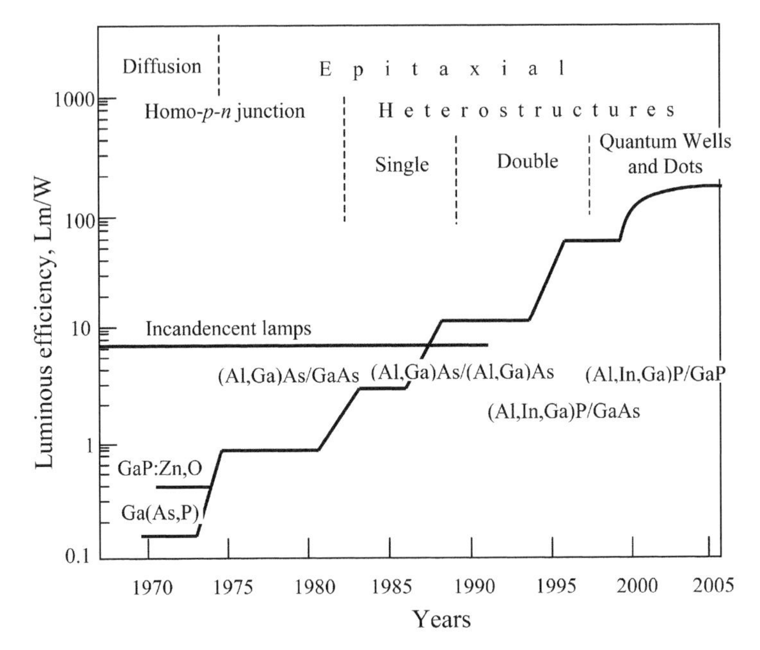

Figure 8.23. Evolution of red light emitting diodes efficiency.

GaP LEDs are less expensive and very reliable (more than 100 000 h of continuous work). Thus modifications of GaP LEDs are still produced and used in devices, where there is no need for super-brightness.

So-called 'pure green' LEDs with the emission spectrum maximum at $\lambda_{max} = 0.55\,\mu m$ are based on intrinsic radiative recombination of free excitons in high-quality epitaxial layers of GaP. Doping of GaP epitaxial layers in yellow-green LED structures with nitrogen increases the efficiency and shifts the spectrum towards the yellow wavelength range (see figure 8.21). Thus at Nitrogen concentration about $N \approx 10^{17}\text{cm}^{-3}$ electroluminescence maximum is at $0.55\,\mu m$, while at $N \approx 10^{19}\text{cm}^{-3}$ it shifts up to $0.57\,\mu m$. This shift results from the increased input of NN_m pairs to the luminescence spectrum.

'Pure red' GaP LEDs are based on radiative recombination of bound on (Zn–O) molecular complex excitons. Besides separate LEDs of various types, LED matrices, which may consist of more than 32 elements, and two-colour (bicolour) elements are currently commercially available.

Si–GaP and Si–GaP–InP based nanostructures are promising for future applications in nano-electronics and opto-electronics since, due to lattice mismatch between GaP and Si as little as 0.36%, these structures may be easily incorporated into silicon-based electronic devices.

The wide band gap of GaP allows using it in high-temperature electronics. Thus the working temperature interval for a common bipolar GaP transistor is from $-195\,^{\circ}C$ to $+550\,^{\circ}C$ [88].

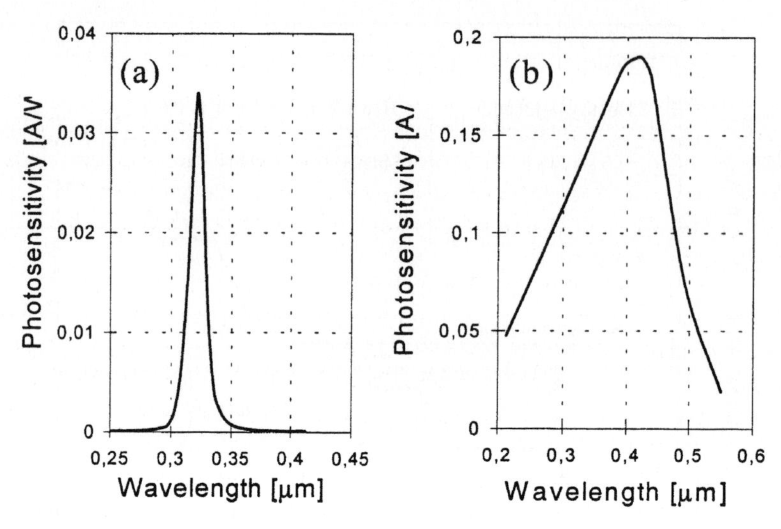

Figure 8.24. Photo-response spectrum of (a) selective, (b) broadband ultraviolet Ag–GaP photodiode.

Schottky diodes based on the contact Me–GaP (where Me is Ag, Pt etc.) are used in different types of ultraviolet photodiodes, including solar-blind selective one [89]. Figure 8.24(a), shows the spectrum of the photo-response of selective, while figure 8.24(b) shows that of the broadband ultraviolet photodetector. Some other parameters are listed in table 8.8 as examples.

Table 8.8. Parameters of ultraviolet Ag–GaP photodetectors.

Parameter	Unit	Symbol	Measurements conditions	Selective	Broad-band
Wavelength range	nm	λ_{min}–λ_{max}		300–350	200–560
	nm	$\Delta\lambda_{0.5}$		15	
	nm	$\Delta\lambda_{0.05}$		50	
Peak sensitivity wavelength	nm	λ_p		322	420
Photo-response	A/W	S_i	λ_p	0.034	0.19
Dark current	pA	I_T	$V = -1B$	0.01	0.02
			$V = -5B$	0.07	0.2
			$V = -10B$	0.5	1
Capacity	pF	C	$V = -1B$	250	400
			$V = -5B$	150	200
Active area	mm^2			0.8 (8.0)	1.8 (8.0)

8.3 Ternary alloys

Use of semiconductor alloys of different composition allows smooth variation of material properties in a wide range. Following Moor's law, the further progress of electronics is determined by quantum-sized structures, such as quantum wells (QW), nanowires (NW), quantum dots (QD) and more complicated structures of QW and NW superlattices as well as QD molecules and crystals [90].

According to the Hume-Rothery rule [32], continuous series of solid solutions exist only between compounds with less than 7–8% difference $\Delta a/a$ of the lattice constant. The less the ratio $\Delta a/a$ the closer the alloy to an ideal one, where all different kinds of atoms are statistically independently distributed in nodes of the crystal lattice linearly interpolated lattice constant according to Vegard's law. Internal strain in epitaxial structure grown on lattice-mismatched substrate and spontaneous long-range ordering (see below) may cause pseudo-deviation from Vegard's rule. Mass density and linear thermal expansion coefficient are supposed to change linearly with the alloy composition as well.

Even though we assume an ideal periodicity of the crystal lattice, the difference of the chemical nature of the atoms randomly distributed in lattice bonds breaks the translation symmetry of the Hamiltonian. The electron wave vector $\mathbf{k}$ is no longer a 'good quantum number'. However, in isovalent A^3B^5 alloys this perturbation is small, that allows considering the potential energy as

$$V(\mathbf{r}) = V_0(\mathbf{r}) + V_1(\mathbf{r}), \tag{8.11}$$

where $V_0(\mathbf{r})$ is a periodic potential with translation symmetry and $V_1(\mathbf{r})$ is non-periodic perturbation.

In virtual crystal approximation $V_1(\mathbf{r})$ is neglected and an A_xB_{1-x} alloy is considered as a virtual ideal crystal with $a(x)$ lattice constant determined by Vegard's law and analogously averaged pseudo-potential $V_0(x)$ [91]. All conceptions introduced for an ideal crystal are valid in this approximation. Thus one may use the pseudo-potential method [92] or a dielectric model [93] to compute the band structure of an alloy. They lead to smooth change of the energy structure parameters and energy gaps E_g^Γ, E_g^X, E_g^L, E_1, E_2 may be approximated by the parabolic function of x:

$$E_g(x) = xE_g(1) + (1-x)E_g(0) - cx(1-x). \tag{8.12}$$

The bowing coefficient, c, is usually positive and increases with atomic difference of the substituting elements.

The energy gaps monotonously decrease with increasing of the atom number sum $\sum Z = Z^{\mathrm{III}} + Z^{\mathrm{V}}$ as shown in figure 8.25. This monotonic reduction in the corresponding energy gaps probably represents a general feature of pseudopotentials in the periodic system of elements. The rate of

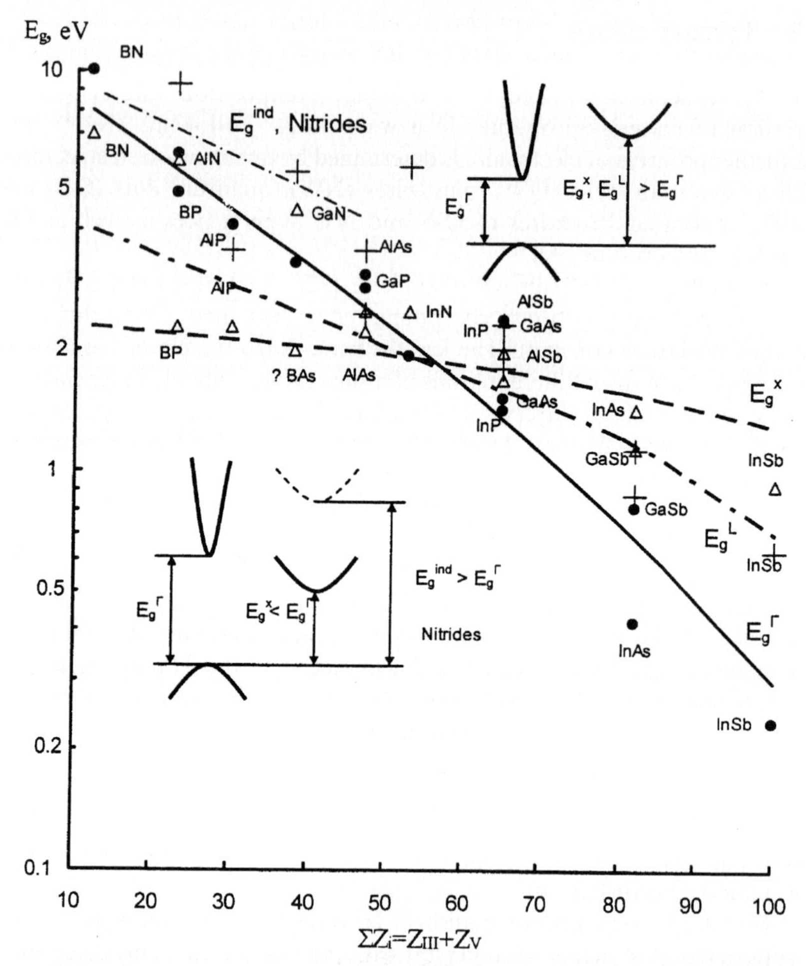

Figure 8.25. Change of energy gaps with the sum of atomic numbers of the constituting elements in binary compounds A^3B^5. Solid line traces the direct energy gap E_g^Γ, dashed line traces the energy gap E_g^X and dot-dashed lines trace the energy gap E_g^L. Circles, triangles, and crosses correspond to E_g^Γ, E_g^X and E_g^L experimental values, respectively.

change in the direct energy gap E_g^Γ is considerably greater than that of the indirect gaps E_g^X and E_g^L. It is quite interesting to note that the energy gaps at the centre of the Brillouin zone of nitrides with the wurtzite structure fit well the general relationship of other compounds with zincblende structure. However, the indirect energy gaps of nitrides are considerably greater than those for other A^3–B^5 compounds.

Boron phosphide (BP) is an indirect gap material with an energy gap close to that of GaP. The technology of BP crystal growing is almost undeveloped since it is almost never used in electronics. The covalent radius of boron (0.086 nm) is much less than that of gallium (0.126 nm) and thus the alloy (GaP–BP) does not exist.

Aluminium phosphide (AlP) forms a continuous series of close-to-ideal alloys with GaP since covalent radiuses of Ga and Al differ very little. In spite of the hygroscopicity of AlP and AlAs, which are unstable in air, $Al_xGa_{1-x}P$ alloys are stable in the $x \leq 0.3$–0.4 composition range.

Indium phosphide (InP) along with GaAs is one of the most important materials of modern semiconductor electronics with well-developed technology [87, 90]. The covalent radiuses difference of Ga and In is close to the critical value ($\Delta R/R = 13.3\%$), while that of P (0.110 nm) and As (0.118 nm) is about half. Both materials form continuous series of alloys with GaP.

Thallium phosphide (TlP) is toxic and does not exhibit other useful properties.

The main properties of the binary compounds of interest and silicon are summarized in table 8.9 [9, 10, 95, 97].

Note that the value of direct energy gap $E_g^\Gamma = 4.1$ eV of AlP recommended in table 8.9 differs from the commonly used value (3.63 eV at 4.2 K) taken from

Table 8.9. Main properties of Si and binary A^3B^5 compounds that form continuous series of wide-band gap ternary alloys with GaP.

Parameter	Units	GaAs	GaP	InP	AlP	Si
Lattice constant, a	nm	$0.565325 + 3.88 \cdot 10^{-6} \times (T - 300)$	$0.54505 + 2.92 \cdot 10^{-6} \times (T - 300)$	$0.58697 + 2.79 \cdot 10^{-6} \times (T - 300)$	$0.54672 + 2.92 \cdot 10^{-6} \times (T - 300)$	$0.543107 + 1.8 \cdot 10^{-6} \times (T - 300)$
Linear thermal expansion coef. (300 K), α_ℓ	10^{-6} K^{-1}	5.98	4.65	4.75	4.2	2.59
Lattice thermal conductivity (300 K), κ	W/cm K	0.45	1.1	0.9	0.9	1.56
Direct en. gap E_g^Γ	eV					
300 K		1.424	2.780	1.344	4.0	4.1
4.2 K		1.519	2.866	1.4235	4.1	4.185
Indirect en. gap E_g^X	eV					
300 K		1.88	2.272	2.27	2.45	1.1242
4.2 K		1.981	2.350	2.384	2.52	1.1695
Indirect en. gap E_g^L	eV					
300 K		1.72	2.64	1.82	3.5	—
4.2 K		1.815	2.72	2.014	3.57	—
Dielectric constants, 300 K						
ε_0		12.9	11.11	12.56	9.8	11.9
ε_∞		10.89	9.1	9.6	7.5	11.9

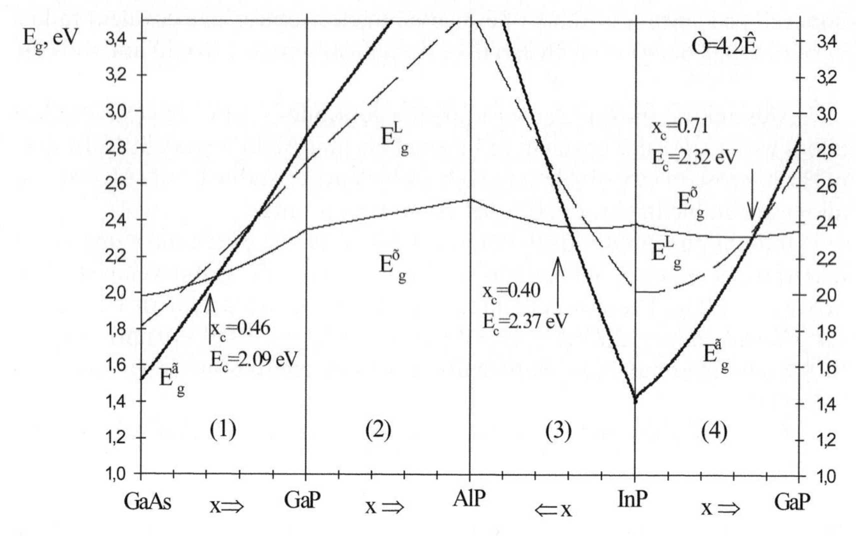

Figure 8.26. Compositional dependences of the direct E_g^Γ and indirect E_g^X and E_g^L energy gaps at 300 in quasi-binary GaP wide-band gap solid solutions.

[98]. That widely used little band gap contradicts experimental data reported for $Al_xIn_{1-x}P$ [99], $Al_xGa_{1-x}P$ [100], and lattice-matched to GaAs $(Al_xGa_{1-x})_{0.52}In_{0.48}P$ [101] alloys. Also it does not fit the general tendencies traced in figure 8.25. These contradictions forced the authors of the last review [97] to recommend the negative bowing parameter $c = -0.48$ eV to be used in equation (8.12), that is hardly possible. Thus we recommend a different value (see table 8.9) found from extrapolation of the quite reliable data of [103].

The composition dependence (equation (12)) of the energy gaps of GaP wide band gap ternary alloys are presented in figure 8.26. All necessary parameters are taken from tables 8.9 and 8.10.

The dependences for $GaAs_{1-x}P_x$ and $Ga_xIn_{1-x}P$ are analogous to those reviewed in [97]. For ideal $Al_xGa_{1-x}P$ alloy that has not been well studied yet, we assume that there is little or no bowing of the compositional variation of the direct and indirect energy gaps (see table 8.10). This result can be compared with recent studies of the compositional dependence of band gaps of $Al_xGa_{1-x}As$. An experimental study found a very small bowing parameter [102]; and a theoretical study, using empirical pseudopotential methods, predicted a linear compositional dependence of the electronic band structure [103]. This behaviour was explained by very similar lattice constants of AlAs and GaAs and by the fact that the alloy potential form factors are a linear function of composition [103]. It is probable that similar arguments hold for the AlP–GaP system. Moreover,

Table 8.10. Bowing parameters c in equation (8.12) and crossover points x_c, $E_g^{\Gamma\text{-X}}$ for direct E_g^{Γ} and indirect E_g^X and E_g^L energy gaps of GaP wide band gap ternary alloys.

Parameters	Units	$GaAs_{1-x}P_x$	$Ga_xIn_{1-x}P$	$Al_xIn_{1-x}P$	$Al_xGa_{1-x}P$
E_g^{Γ}; c^{Γ}	eV	0.19	0.65	0.5	0
E_g^X; c^X	eV	0.24	0.20	0.3	0
E_g^L; c^L	eV	0.16	0.80	0.4	0
300 K					
x_c	—	0.46	0.71	0.39	—
$E_g^{\Gamma\text{-X}}$	eV	2.0	2.23	2.27	—
4.2 K					
x_c	—	0.46	0.71	0.40	—
$E_g^{\Gamma\text{-X}}$	eV	2.09	2.32	2.37	—

it may be proved by recent experimental observation of linear dependencies of $E_g^{\Gamma}(x)$ and $E_g^X(x)$ in $(Al_xGa_{1-x})_{0.52}In_{0.48}P$ [101].

As recommended in table 8.10, $Al_xIn_{1-x}P$ bowing parameters are very different from those reported in [87, 97, 99] since we use another value of E_g^{Γ} (AlP). We believe that negative bowing parameter ($c^{\Gamma} = -0.48$ eV) [97] would contradict the basic assumptions of the virtual crystal approximation [93, 95, 103] and most experimental data. We found recommended c^{Γ} and c^X parameters from extrapolation of those reported in [101] experimental data for $Al_{0.52}In_{0.48}P$ lattice matched to GaAs.

Bowing parameters may differ for those grown by different techniques, epitaxial layers depending on intrinsic strain and substrate orientation. Moreover, the spontaneously ordered CuPt-type structures were found in $Ga_{0.5}In_{0.5}P$ [104], $Ga_{0.7}In_{0.3}P$ [105], (Al,In)P and other systems with close to critical difference of lattice constants of constituting binary compounds [106]. An unexpected feature of A^3B^5 alloys with relatively large $\Delta a/a$ is spontaneous CuPt-like long-range ordering of the group III elements in the cationic sublattice. This ordered structure has a trigonal symmetry C_{3v} instead of initial cubical T_d one with period in the (100) plane twice as large as lattice constant. It leads to the valence-band splitting and band gap reduction that affects optical properties dramatically.

All values listed in table 8.10 are valid for structures with infinitesimal strain and free from long-range ordering.

Note that near the critical composition x_c of solid solution all three minima in conduction bands E_g^{Γ}, E_g^X and E_g^L have very close energies. It increases their sensitivity to strain and ordering effects. Especially this is exhibited for $Ga_xIn_{1-x}P$ (see figure 8.26).

Most non-nitride wide direct band gap A^3B^5 semiconductors are based on $Ga_xIn_{1-x}P$ and $Al_xIn_{1-x}P$ alloys (see figures 8.25 and 8.26 and tables 8.9

and 8.10). The direct band gap of $Al_{0.39}In_{0.61}P$ at critical composition corresponds to green light.

If the electron mean free path $L \gg 1/k$, where k is the electron wave vector, and electric properties of the solid solutions may be described in virtual crystal approximation. Effective mass in all extrema changes almost linearly with composition. Additional alloy scattering reduces carrier mobility. This is especially evident in low-doped materials. At temperature lower than 80 K carriers may localize on alloy fluctuations in (Ga,In)P. Electrical properties change abruptly at critical alloy composition when the lowest minimum is switched, and it is necessary to take into account carrier redistribution between all of them.

Optical effects resulting from direct vertical electron transitions may be described within virtual crystal approximation [95, 96]. It means that all optical constants of an alloy may be found from those of binary compounds. All maxima in imaginary and real parts of dielectric susceptibility (see figure 8.7) and fundamental absorption and reflection spectra smoothly shift while composition is changed. The refractive index spectrum depends on composition, temperature and pressure according to equation (8.6) in the $E_{TO} \ll \hbar\omega \le E_g^\Gamma$ range. Here one should substitute in equation (8.6) linear extrapolation of $A(x)$, $G_1(x)$ and $G_2(x)$ and quadratic extrapolation (equation (8.12)) for $E_0 = E_g^\Gamma$, E_1 and E_2 [23, 107]. However, one should remember that this extrapolation is valid in the lattice absorption range ($\hbar\omega \approx E_{TO} = \hbar\Omega_{TO}$) for one-mode alloys only, since VCA is invalid for vibration properties [108]. Figure 8.27 presents the comparison of the experimental data and computed in the VCA dependencies for the refractive index of $Al_xGa_{1-x}P$ [107].

Alloy thermal conductivity $\kappa(x)$, mostly determined by crystal lattice vibrations, cannot be described in virtual crystal approximation. In fact, there is a deep minimum in its dependence on the alloy composition resulting from decreased phonon lifetime and mean free pass.

Determined by vertical optical transitions, fundamental absorption edge and edge luminescence smoothly shift with composition of direct band gap alloys. This shift may be adequately described in VCA. Exciton effects may exhibit themselves up to room temperature in samples of high quality, since broadening of the spectra due to local fluctuations of alloy composition is as small as 1–2 meV for Wannier excitons with excitonic Bohr radius as large as several tenths of lattice constants [95, 110]. Besides optical transitions to the bound excitonic state ($n = 1$), those to the first excited ($n = 2$) state were experimentally observed in high quality samples of $GaAs_{1-x}P_x$ [111]. The width of the edge luminescence bands and their intensities are governed largely by the quality of the crystals. For example, in the case of $Ga_xIn_{1-x}P$ epitaxial films grown on GaP substrates, a relatively small difference between the lattice parameters of the substrate and film results in a strong broadening of the short-wavelength bands in the photoluminescence spectra [112].

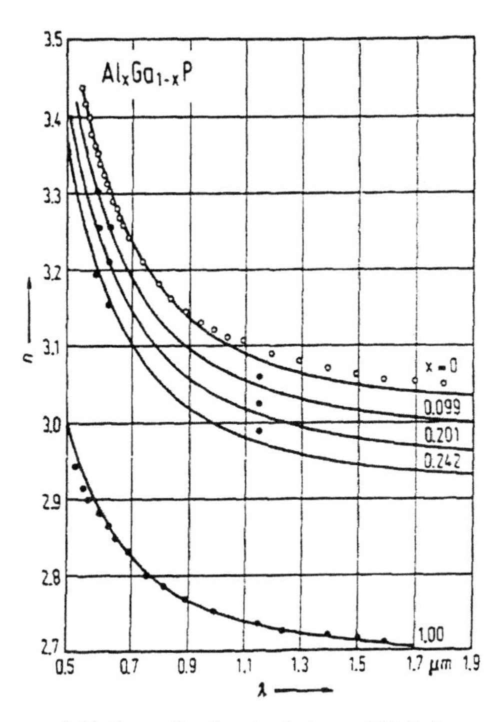

Figure 8.27. Spectrum of $Al_xGa_{1-x}P$ refractive index at 298 K for samples with different alloy composition x. Circles correspond to experimental data, while data calculated in virtual crystal approximation spectra are shown by solid lines [107].

Indirect optical transitions in solid solutions are strongly affected by breaking of the translation symmetry of the Hamiltonian [95]. It results in breaking of the quasi-momentum conservation law and appearance of so-called 'indirect no-phonon' or 'quasi-direct' transitions between extrema in different points of the Brillouin zone. The intensity of such transitions depends on the alloy disorder $x(1-x)$ and energy difference $\Delta(x)$ between virtual and final states as

$$A(x) = A_0 x(1-x)[1/\Delta(x)]. \tag{8.13}$$

The lowest $\Gamma_1(\Gamma_6)$ minimum in the conduction band often serves as a virtual state for this transition. Thus near the critical alloy composition x_c when energy difference $\Delta(x \approx x_c)$ tends to zero, the intensity of the indirect no-phonon transitions is resonantly enhanced as illustrated in figures 8.28 and 8.29(a). Its intensity becomes comparable with the intensity of vertical direct transitions in the centre of the Brillouin zone. Parameter A_0 depends on the perturbation V_1 (see equation (8.11)) introduced by disorder in a particular solid solution. Note that its value is about 1.4 times larger in $Ga_xIn_{1-x}P$ than in $GaAs_{1-x}P_x$ [95, 113].

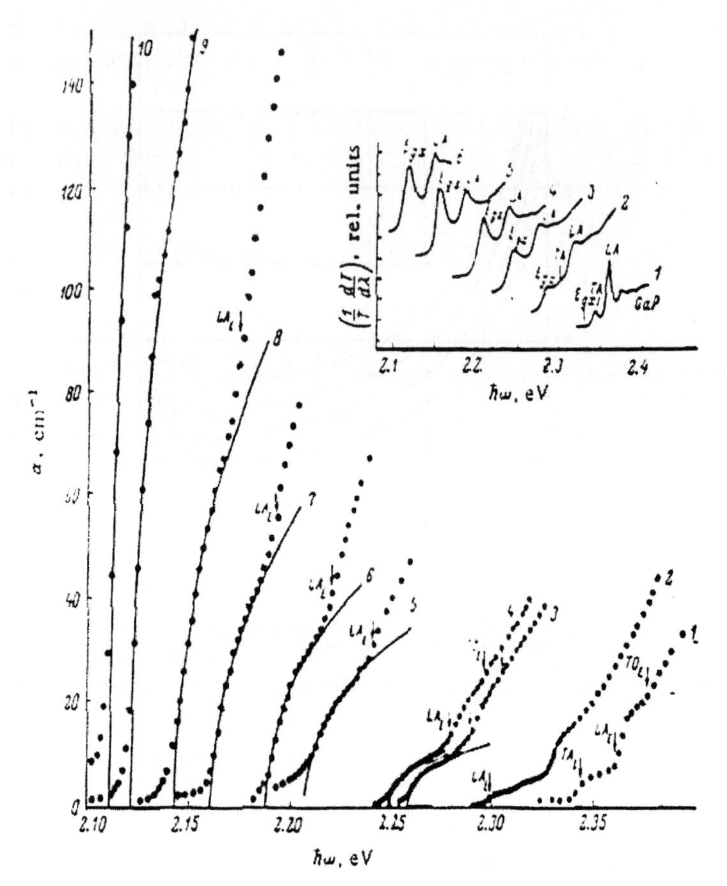

Figure 8.28. Fundamental absorption edge of $GaAs_{1-x}P_x$ crystals at 4.2 K in the range of alloy compositions corresponding to indirect energy band structure [95, 113]. Numbers correspond to the following phosphorus content x: (1) 1, (2) 0.931, (3) 0.86, (4) 0.845, (5) 0.764, (6) 0.71, (7) 0.65, (8) 0.61, (9) 0.56, (10) 0.535. Theoretical curves are plotted with solid lines, while circles correspond to experimental data. The insert gives the differential absorption spectra plotted with logarithmic scale for samples with the following values of x: (1) 1, (2) 0.924, (3) 0.856, (4) 0.785, (5) 0.653, (6) 0.552.

The absorption spectrum taking into account excitonic effects at parabolic dispersion near no-phonon indirect component edge may be approximated with a square root dependence

$$\alpha(\hbar\omega) = A(x)\sqrt{\hbar\omega - E_{gx})} \tag{8.14}$$

where $E_{gx} = (E_g^x - E_{ex}^x)$ is the indirect excitonic energy gap and $A(x)$ is the same as in equation (8.13). However, experimentally observed no-phonon excitonic absorption edge in $GaAs_{1-x}P_x$ and $Ga_xIn_{1-x}P$ may only roughly be approximated with equation (8.14) [114]. It is explained by the 'camel's back' structure of the conduction band minimum discussed in sections 8.2.2 and 8.2.3 and shown in figures 8.4 and 8.11.

There are phonon-assisted exciton transitions besides no-phonon component of the absorption spectrum of solid solutions with indirect energy band structure. Their intensity $B(x)$ may be describes in VCA as

$$B(x) = B_0\left(\frac{1}{\Delta \mp \hbar\Omega}\right)^2 \tag{8.15}$$

where Δ is the energy difference between virtual and final state and $\hbar\Omega$ is the energy of emitted (−) or absorbed (+) phonon. The B_0 value is mostly determined by the exciton–phonon matrix element that remains constant for the given phonon type. The corresponding absorption spectrum may only roughly be approximated with square-root dependence

$$\alpha(\hbar\omega) = B(x)\sqrt{\hbar\omega - E_{gx} \mp \hbar\Omega}$$

because of the 'camel's back' structure of the conductance band minimum (figure 8.11). The strong resonance enhancement at almost critical alloy composition was observed for LA-phonon-assisted transitions only [95, 113] as shown in figure 8.29(b). This is determined by the selection

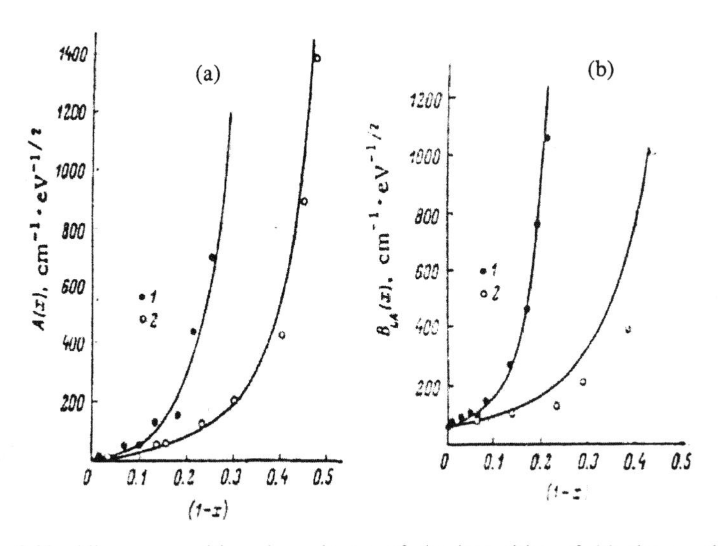

Figure 8.29. Alloy composition dependences of the intensities of (a) the no-phonon component $A(x)$ and (b) the longitudinal acoustical phonon-assisted component $B_{LA}(x)$ in the exciton absorption spectra for 1-$Ga_xIn_{1-x}P$ and 2-$GaAs_{1-x}P_x$ alloys. Solid lines correspond to analytical expressions (8.13) and (8.15), while experimental data are represented by solid and open circles for $Ga_xIn_{1-x}P$ and 2-$GaAs_{1-x}P_x$, respectively [95].

rules for zincblende crystal structures that allow Γ_8^V–X_6^Γ transitions through Γ_6^c virtual state with LA phonon assistance only and through deep Γ_7^c and Γ_3^V virtual states with TA, TO and LO phonon assistance. The transitions through deep states superimpose with more intensive continuum of direct band-to-band transitions and thus are not revealed in experimental spectra for $GaAs_{1-x}P_x$ and $Ga_xIn_{1-x}P$.

Note that the considered alloys represent a unique possibility for fundamental investigations of the effect of band structure and disorder on main physical properties of matter [95].

Both no-phonon and phonon-assisted radiative recombination of free excitons were also experimentally observed in luminescence spectra of indirect-band $GaAs_{1-x}P_x$, $Ga_xIn_{1-x}P$ and $Ga_{1-x}Al_xP$ [95, 96]. The intensity of no-phonon and LA-phonon assisted components enhanced at $\Delta(x) \to 0$ and $\Delta(x) \to \hbar\Omega_{LA}$ in accordance with equations (8.13) and (8.15).

Modulation spectroscopy techniques allowed experimental observation of transitions to the second minimum of the conductance band $X_3^c(X_7^c)$ [115]. The intermediate virtual state for these transitions is $\Gamma_1^c(\Gamma_6^c)$ minimum and in the $GaAs_{0.158}P_{0.842}$ alloy of nearly critical composition the intensity of indirect no-phonon $\Gamma_{15}^V(\Gamma_8^V) \longrightarrow X_3^c(X_7^c)$ transition was comparable with that of the direct $\Gamma_{15}^V(\Gamma_8^V) \longrightarrow \Gamma_1^c(\Gamma_6^c)$ one.

The degree of perturbation of an impurity energy spectrum by the local alloy composition fluctuations depends on the state localization radius. If it is several tenths of lattice constants, as for shallow donors bound to the conduction band minimum in Γ point and excited states of deep impurities, all parameters are averaged in virtual crystal approximation and corresponding line broadening may be neglected. VCA is invalid for ground states of deep impurities, isoelectronic traps, shallow acceptors, and shallow donors mostly bounded to X and L minima of the conduction band, since their localization radius is less than ten lattice constants and fluctuation of the local environment strongly affects the energy states. The corresponding splitting and broadening of the ground state of Si, S and Te donors bound to $X_1^c(X_6^c)$ minimum in indirect band gap $GaAs_{1-x}P_x$ and $Ga_xIn_{1-x}P$ were experimentally observed in photo-excitation and photo-ionization spectra [116, 117]. Besides pseudo-potential fluctuations, local deformation of crystal lattice affects impurity spectrum by changing its central cell potential. It explains very different broadening of silicon and telluride donor states observed in $GaAs_{1-x}P_x$ and $Ga_xIn_{1-x}P$ [117].

Splitting and broadening of impurity states change the meaning and measurement techniques of 'ionization energy' of deep centres in alloys. If there is no detail information, it may be characterized by average ionization energy $E_{D,A} = \langle E \rangle$ and statistical Gaussian distribution dispersion ε

$$g(E) = \frac{N}{\sqrt{2\pi}\varepsilon} \exp - \frac{(E - \langle E \rangle)^2}{2\varepsilon^2} \tag{8.16}$$

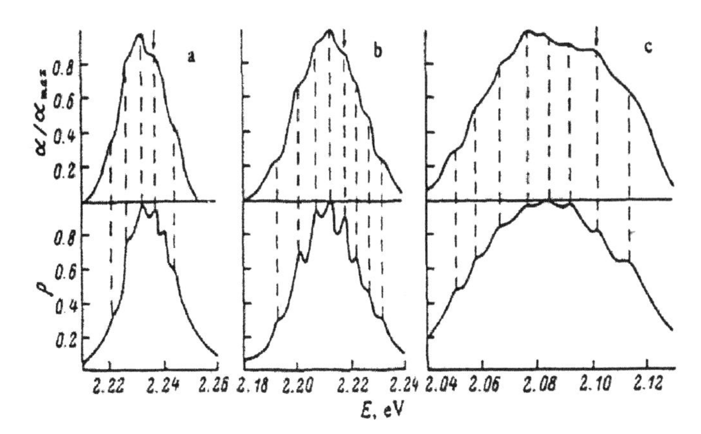

Figure 8.30. Experimental absorption spectra and theoretical density of states of an exciton bound to a nitrogen atom in $GaAs_xP_{1-x}$ solid solutions with the following compositions of x: (a) 0.099, (b) 0.15, (c) 0.33 [119]. The arrows identify the line position in the virtual crystal approximation.

where dispersion ε is

$$\varepsilon = \sqrt{U_0 \alpha x(1-x)}. \tag{8.17}$$

Potential U_0 determines shift of the averaged ionization energy in alloy in the first order of perturbation theory. Parameter $1/\alpha$ is the number of primitive cells in the volume of the bound electron or hole localization [117]. All electrical properties of wide band gap alloys also have to be analysed taking into account these specifics. The dispersion parameter ε for donors in $GaAs_{0.3}P_{0.7}$ and $Ga_{0.7}In_{0.3}P$ is about 1–2 meV for Te [117], about 8–10 meV for Si [116], and about 30 meV for deep Y-state in $GaAs_{0.3}P_{0.7}$ [118].

Analogous to GaP, nitrogen isoelectronic trap in GaP solid solutions is of utmost interest. Figures 8.30 and 8.31 show experimental and theoretical values of splitting and shift of A_N lines of an exciton bound to a nitrogen trap in $GaAs_{1-x}P_x$:N [119]. The well-pronounced fine structure of the absorption spectra corresponds to the series of energy levels at different nitrogen surrounding in the solid solution. The same structure appeared in phototransmittance spectra and involves bound excitons in $GaAs_{1-x}P_x$:N [120]. A shift of the luminescence band relative to the absorption (luminescence-excitation) band was also observed in $GaAs_{1-x}P_x$:N [121]. The difference between the positions of the luminescence and absorption bands is explained by the presence of a system of N energy levels. The position of a band in the absorption spectrum is governed by the density of states and, in the case of luminescence spectra, is moreover related to the occupancy of these states [119]. At low excitation rates the radiative recombination process involves mainly the lower energy levels, which is the reason for the observed energy

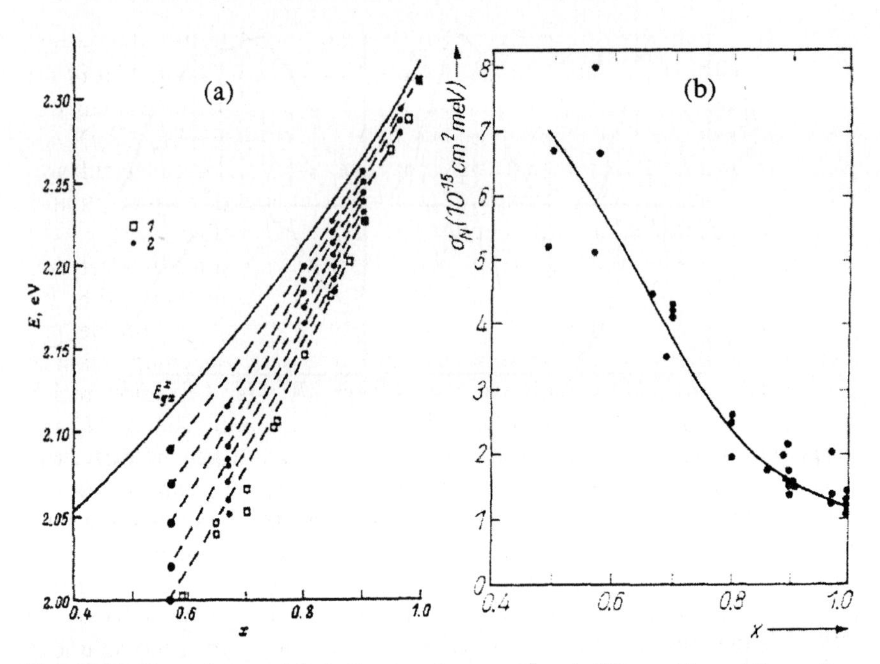

Figure 8.31. Dependences of the indirect exciton gap E^x_g and of the positions of the exciton levels bound to nitrogen atoms on the composition of $GaAs_{1-x}P_x$:N at $T = 77K$: (a) (1) luminescence [96] and (2) optical absorption [119] and (b) the integral absorption cross-section A_N line as a function of $GaAs_{1-x}P_x$ alloy composition [125]. The dashed lines represent interpolation of the experimental data on the basis of the absorption spectra.

shift. Moreover, this shift is evidence of the interaction and energy transfer between the spatially separated isoelectronic traps [122].

As x is decreased, the Γ and X minima of the conductance band in indirect gap $GaAs_{1-x}P_x$ and $Ga_xIn_{1-x}P$ become closer, which increases the weight of the $\Gamma^c_1(\Gamma^c_6)$ minimum in a wave-function of an electron bound to a shallow donor. It increases the probability of no-phonon radiative recombination through such centres. This effect was experimentally observed for shallow donor–acceptor pairs in $GaAs_{1-x}P_x$ [95].

The experimental absorption spectra of GaAsP doped with nitrogen ($GaAs_{1-x}P_x$:N) exhibit an increase in the oscillation strength of the no-phonon A_N-line of excitons bound to nitrogen [123]. When the composition is altered from $x = 1.0$ to $x = 0.53$, the oscillator strength increases by a factor exceeding 20, which is due to a change in the band structure parameters (reduction in $\Delta = \Gamma^c_1 - \Gamma^c_6$) as a result of a change in the composition. The Koster–Slater approximation was used in [6, 57, 124] to calculate the influence of changes in the band structure parameters of $GaAs_{1-x}P_x$ and $Ga_xIn_{1-x}P$ on the wave function of an electron bound to an N-isoelectronic centre. An exciton was regarded as an electron bound by the short-range potential of the centre and a hole bound by the Coulomb potential of the captured electron. The experimental data and results of numerical calculations in virtual crystal approximation carried out for $GaAs_{1-x}P_x$:N are plotted in figure 8.31(b) [124, 125]. Even though these analytical dependences do not take into account either splitting or broadening of the impurity states, they qualitatively explain experimental data presented in figure 8.15.

Radiative recombination with and without participation of isoelectronic nitrogen traps in GaP wide band gap solid solutions was investigated by the groups led by Holonyak [123, 126] and Alferov [112, 127]. A considerable increase was found in the luminescence efficiency when the composition of alloys with the indirect band structure approached the transition region corresponding to $\Delta(x) \to 0$, so that even stimulated emission of N- and NN_m-bound exciton lines was observed for crystals with indirect band structure [128]. These studies and progress in $GaAs_{1-x}P_x$ and $Ga_xIn_{1-x}P$ epitaxial growing technology resulted in enhancement of external quantum efficiency of radiative recombination almost an order of magnitude compared with the data in figure 8.15. However, the efficiency did not reach the values obtained later on lattice-matched heterostructures based on alloys in the system $Ga_{1-x}Al_xAs$/GaAs or on quaternary solid solutions (Ga–In–As–P) and (Ga–In–Al–P).

8.4 Quaternary alloys

There are two types of A^3B^5 quaternary alloys: $A_x^{III}B_{1-x}^{III}C_y^{V}D_{1-y}^{V}$ and $A_x^{III}B_y^{III}C_{1-x-y}^{III}D$ or $AB_x^{V}C_y^{V}D_{1-x-y}^{V}$. In contrast to ternary alloys, they have two independent degrees of freedom allowed to tune two parameters of the material independently. The possibility of changing energy band gap and lattice constant are particularly useful since they allow one to fabricate lattice matched heterostructures.

Most features of electronic spectrum, impurity states, electrical, optical and luminescent properties are similar to those discussed in section 8.3 for ternary alloys. Vegard's law is usually a good approximation for the lattice constant in quaternary alloys. Perturbation potential V_1 in equation (8.11) is small and virtual crystal approximation is valid for states with relatively large localization.

Alloys of the first type $A_xB_{1-x}C_yD_{1-y}$ may be treated as a combination of four binary compounds AC(1); BC(2); AD(3) and BD(4). Then Vegard's law may be written as

$$a(x, y) = xya_1 + (1-x)ya_2 + x(1-y)a_3 + (1-x)(1-y)a_4 \qquad (8.18)$$

where x, $(1-x)$, y and $(1-y)$ are molar fractions, while $a_1 \cdots a_4$ are corresponding lattice constants of the initial binary compounds. Energy gap $E_g(x, y)$ in any point of the Brillouin zone may be approximated as

$$E(xy) = xyE_1 + (1-x)yE_2 + x(1-y)E_3 + (1-x)(1-y)E_4 \\ -[c_{24} + (c_{13} - c_{24})x]y(1-y) - [c_{34} + (c_{12} - c_{34})y]x(1-x). \tag{8.19}$$

Here E_n is energy gap in the corresponding point of the Brillouin zone for the nth binary compound and c_{mn} are bowing parameters for a ternary alloy formed by mth and nth binary compounds. This equation follows from interpolation of equation (8.12). The corresponding charts may be represented in three dimensions using coordinates analogous to those in figure 8.26. However, more often they plotted as series of iso-energetic lines projections on the (x, y) surface to make them easier to use.

$Ga_xIn_{1-x}P_yAs_{1-y}$ alloy is of particular practical interest (see reviews [90, 95, 97, 129]). Figure 8.32(a) shows its energy gap dependence on the alloy composition. The hashed area corresponds to indirect-band-gap alloys. Dot-dashed lines trace alloy compositions that are lattice matched to InP and GaAs. The corresponding relation between x and y obtained from equations (8.18) and (8.19) with parameters a and C_{mn} presented in tables 8.9 and 8.10 at room temperature is

$$y = \frac{0.1886 - 0.40505x}{0.1886 + 0.01415x} \tag{8.20}$$

for lattice matched to InP alloys and

$$y = \frac{0.40505(1-x)}{0.1886 + 0.0145x} \tag{8.21}$$

for those matched to GaAs.

The first group corresponds to direct-band-gap alloys with E_g^Γ from 0.74 to 1.35 eV. They are widely used for fabrication of heterolasers and photodiodes working in the 1.3–1.55 μm spectral range that corresponds to minimal losses and zero dispersion of optical waveguides. Also lattice matched to InP alloys are used in heterolasers for the pumping of optical amplifiers at 0.98 and 1.48 μm.

Alloys lattice-matched to InP have direct band gap E_g^Γ varying from 1.42 to 1.92 eV and may be used for fabrication of reliable high-power injection lasers working from near infrared to 0.65 μm.

Since $Ga_xIn_{1-x}P_yAs_{1-y}$ alloys are widely used in opto-electronics, especially in fibre optical communication lines, their technology of growing (mostly MBE and MOCVD) is well-developed [9, 10, 23, 86, 87, 97, 129] and internal quantum efficiency in quantum dot and quantum well structures is approaching 100% at room temperature. Injection lasers using $Ga_xIn_{1-x}P_yAs_{1-y}$ in an active region are commercially produced with lifetimes of more than 100 000 h. The external differential quantum efficiency of

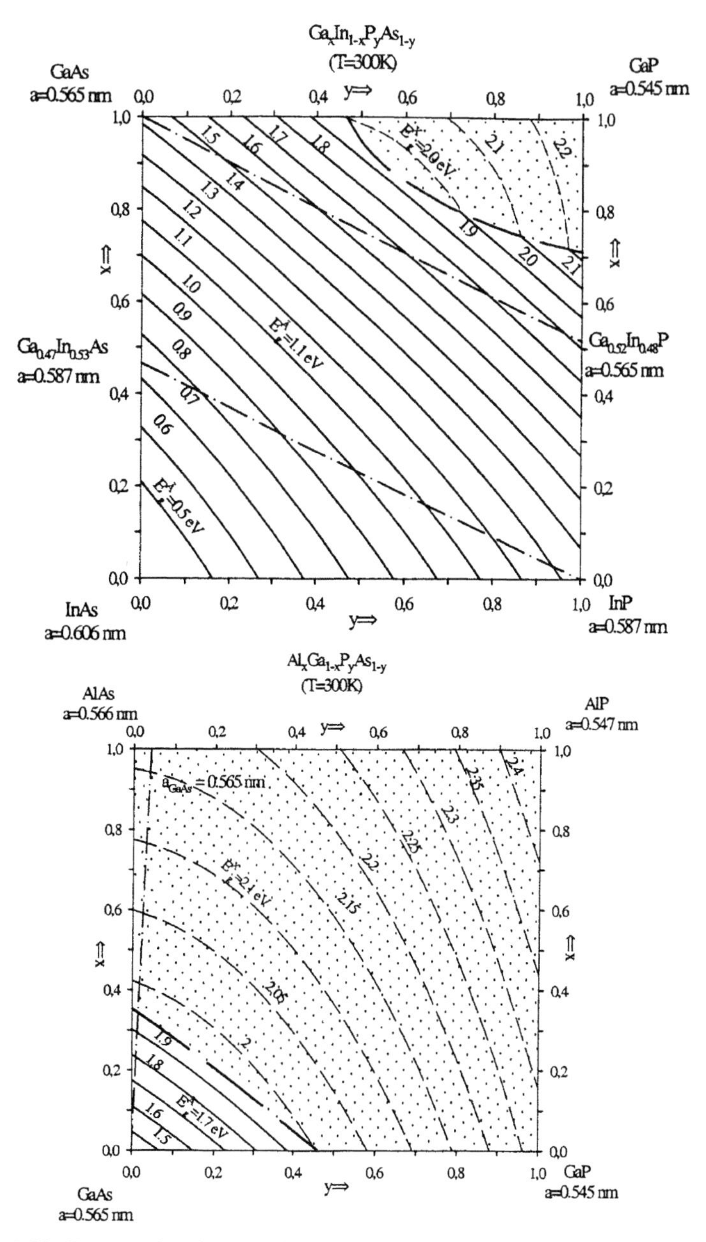

Figure 8.32. Energy band gap dependence on the alloy composition for (a) $Ga_xIn_{1-x}P_yAs_{1-y}$ and (b) $Al_xGa_{1-x}P_yAs_{1-y}$ alloys at $T = 300$ K. Solid lines correspond to direct band gap iso-energetic compositions, while dashed lines denote indirect compositions. Dot-dashed lines shows the compositions that are lattice-matched to GaAs (in both charts) and to InP (for $Ga_xIn_{1-x}P_yAs_{1-y}$ only). The indirect-band gap alloy composition areas are hashed.

high-power laser diodes with separate optical and electronic confinement in double quantum well InGaAs–InGaAsP–InGaP grown by molecular beam epitaxy reached 85–86%. They generate at $\lambda = 0.97\,\mu m$ an optical power 10.6 W in continuous [131] and 14.3 W in quasi-continuous regimes [132]. The highest optical power, 12 W, was continuously generated at 0.98–1.03 μm with 66% absolute efficiency and 93% differential efficiency [133]. Absolute efficiency was higher than 60% in the 2–6 W emission power range. The emission power 5.2 W in the continuous regime was achieved in InGaAsP/InP heterolasers at the (very important for fibre optics) wavelength of 1.55 μm [134]. In the 0.715–0.735 μm range optical power achieved 7 W in the continuous regime with 50% efficiency [135]. Note that the achieved power density of catastrophic optical degradation of $40\,MW/cm^2$ [133] is close to the limit determined by optical break-down of the material [86].

The $Ga_xIn_{1-x}P_yAs_{1-y}$ lattice matched to GaAs has the largest energy band gap at $y = 1$, i.e. in ternary alloy $Ga_xIn_{1-x}P$ at $x = 0.52$ and $E_g = 1.92\,eV$. This limits the spectral range of luminescent devices within the red part of the spectrum, $\lambda \geq 0.65\,\mu m$. Properties of direct-band-gap alloys $Ga_xIn_{1-x}P$ with E_g^Γ up to 2.23 eV were discussed in section 8.3.

$Al_xGa_{1-x}P_yAs_{1-y}$ is another important example of the quaternary alloy of the first type. Figure 8.32(b) shows energy band gap composition dependence calculated with equation (8.19). The direct-band-gap area corresponds to the energy range from 1.42 eV (GaAs) to 2.0 eV ($GaAs_{0.46}P_{0.54}$). This alloy has not been explored enough since it has characteristics comparable with GaInPAs, and AlAs and AlP are unstable.

Substitution of the components of a quaternary alloy of the second type is within only one sublattice, i.e. either cationic $A_x^{III}B_y^{III}C_{1-x-y}^{III}D$ or anionic $AB_x^VC_y^VD_{1-x-y}^V$. These alloys may be considered as consisting of three pseudo-binary compounds: AD(1), BD(2) and CD(3) for $A_xB_yC_{1-x-y}D$, and AB(1), AC(2) and AD(3) for $AB_xC_yD_{1-x-y}$, those corresponding to the vertices of the triangle chart of $E_g(x, y)$. Equations for calculation of the lattice constant and energy gap for a given composition (x, y) are analogous to equations (8.18) and (8.19):

$$a(x, y) = xa_1 + ya_2 + (1 - x - y)a_3, \tag{8.22}$$

$$E(x, y) = xE_1 + yE_2 + (1 - x - y)E_3 - c_{12}xy - c_{13}x(1 - x - y) - c_{23}y(1 - x - y). \tag{8.23}$$

The physical meaning of bowing coefficients c_{mn} and energy band gaps E_n are the same as in equations (8.18) and (8.19). From figure 8.25 one may conclude that $Al_xGa_yIn_{1-x-y}P$ has the largest direct energy band gap, excluding nitrides. The corresponding dependences calculated using equations (8.22) and (8.23) with the parameters from tables 8.9 and 8.10 are presented in figure 8.33. Indeed, $Al_xGa_yIn_{1-x-y}P$ may have as large a direct energy band gap as $E_g^\Gamma = 2.27\,eV$ at 300 K and 2.37 at 4.2 K.

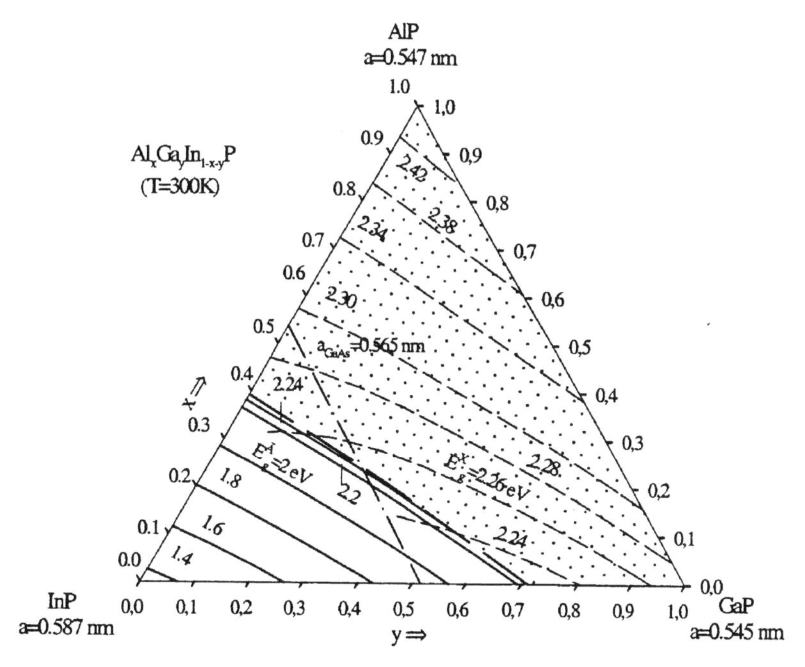

Figure 8.33. Energy band gap dependence on the alloy composition for $Al_xGa_yIn_{1-x-y}P$ alloy at $T = 300$ K. Solid lines correspond to direct band gap iso-energetic compositions, while dashed lines show indirect ones. Dot-dashed lines show the compositions that are lattice-matched to GaAs. The indirect-band gap alloy composition area is hashed.

The first time the quaternary alloy AlGaInP was mentioned in was in 1975 [136] in relation to the created heterojunctions based on broadband isolattice solid solutions of GaInP in order to exhibit effective injection luminescence in the red region of the spectrum. It was the first step towards the creation of high quality heterojunctions in GaInAsP and AlGaInP systems for use in injection light sources. Note that $Al_xGa_yIn_{1-x-y}P$ alloys are lattice matched to GaAs$(x + y) = 0.52$, i.e. $(Al_zGa_{1-z})_{0.55}In_{0.48}P$ alloy is lattice-matched to GaAs. After the substitution of the parameters in equation (8.23) with values taken from tables 8.9 and 8.10 one can get the following dependences on the Al concentration of the energy gaps of $(Al_zGa_{1-z})_{0.52}In_{0.48}P$:

$$\begin{aligned} E_g^\Gamma(z) &= 1.928 + 0.672z \quad (300\,\text{K}) \\ E_g^\Gamma(z) &= 2.011 + 0.679z \quad (4.2\,\text{K}) \\ E_g^X(z) &= 2.221 + 0.063z \quad (300\,\text{K}) \\ E_g^X(z) &= 2.316 + 0.063z \quad (4.2\,\text{K}). \end{aligned} \tag{8.24}$$

Composition dependences $E_g^\Gamma(z)$ and $E_g^X(z)$ calculated using equation (8.23), are close to linear and are in agreement with experimental results reported in

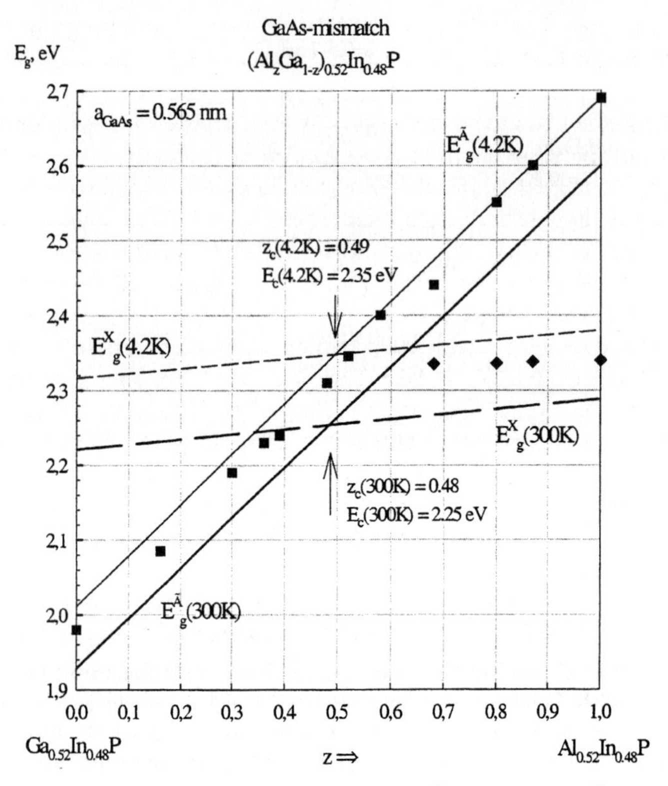

Figure 8.34. Composition dependence of the direct E_g^Γ and indirect E_g^X energy gap of $(Al_zGa_{1-z})_{0.52}In_{0.48}P$ alloy. Experimental data from [101] for 2 K; solid and dashed lines from equation (8.24) for E_g^Γ and E_g^X respectively.

[99], as shown in figure 8.34. At close to the critical composition $z_c = 0.48$, when direct and indirect band gaps are almost equal, $E_g^\Gamma = 2.25\,\text{eV}$ (300 K) and $E_g^\Gamma = 2.35\,\text{eV}$ (4.2 K).

As may be seen from the figures 8.25 and 8.26, *L*-minimum affects the properties of the (Al,Ga,In,P) alloy at the composition close to the transition from direct to indirect band structure. Especially importantly, this effect may be exhibited in strained layers, where *L*-minimum at certain conditions may have the lowest energy in the conduction band.

The virtual crystal approximation in quaternary alloys has the same limitations of applicability as in ternary alloys. Thus as soon as one knows the parameters of the energy band of the quaternary alloy, one may predict its optical properties with good accuracy. The properties determined by vertical optical transitions may be calculated the same way as for ternary alloys. Near the crossover point the virtual and the final states in the conductance band become closer in energy, resulting in resonant enhancement of both no-phonon and LA-phonon assisted indirect optical transitions. This may be traced in figure 8.29 and equations (8.13) and (8.15). This effect has to be taken into account while analysing luminescent properties of the alloy near the crossover point.

Impurity properties and electrical conductivity are analogous to those in the corresponding ternary alloys (see section 8.3). The effective mass of carriers almost linearly changes with the alloy composition. This was proved by cyclotron resonance experiments in GaInAsP (see for example [97] and references therein).

Spontaneous long-range ordering may occur in quaternary alloys grown under certain conditions. A clear dependence of the optical properties of AlGaInP on long-range ordering of the group III sublattice was observed in epitaxial layers grown by metal-organic vapour phase epitaxy [137]. The order-induced band-gap reduction in macroscopic areas may significantly affect transport properties changing their electrical properties. The long-range ordering may be very useful for creation of heterostructures with self-assembled quantum dots [138].

The main advantage of the quaternary alloys in comparison with ternary is the possibility of independent change of two parameters (i.e. energy gap and lattice constant). This makes it possible to fabricate lattice-matched heterostructures and quantum wells of high quality. The values of the band offset in valence ΔE_V and conductance ΔE_C band are crucial for this purpose. They are determined by both properties of the materials forming the hetero-junction and internal strain distribution. Fortunately, the experimental values of the energy band offsets in the application important for the electroluminescent device (Ga,In)P–$(Al_zGa_{1-z})_{0.5}In_{0.5}P$ heterostructures may be extrapolated by parabolic function of z [139]. The band offsets in the nominally unstrained $Ga_{0.5}In_{0.5}P$–$(Al_zGa_{1-z})_{0.5}In_{0.5}P$ heterostructure are

$$\begin{aligned} \Delta E_V &= 63z + 157z^2 \quad [\text{meV}] \\ \Delta E_\Gamma &= \Delta E_C = 547z - 157z^2 \quad [\text{meV}]. \end{aligned} \tag{8.25}$$

The band offsets for 1% compressively strained $Ga_{0.38}In_{0.62}P$–$(Al_zGa_{1-z})_{0.5}In_{0.5}P$ are

$$\begin{aligned} \Delta E_V &= 72 + 63z + 157z^2 \quad [\text{meV}] \\ \Delta E_\Gamma &= \Delta E_C = 72 + 547z - 157z^2 \quad [\text{meV}]. \end{aligned} \tag{8.26}$$

These dependences are shown in figure 8.35 together with the offset ΔE_X between the quantum well conduction band (Γ_C) and the barrier X conduction band (X_C). The pressure coefficients of the Γ_C–Γ_V and X_C–Γ_V band gaps

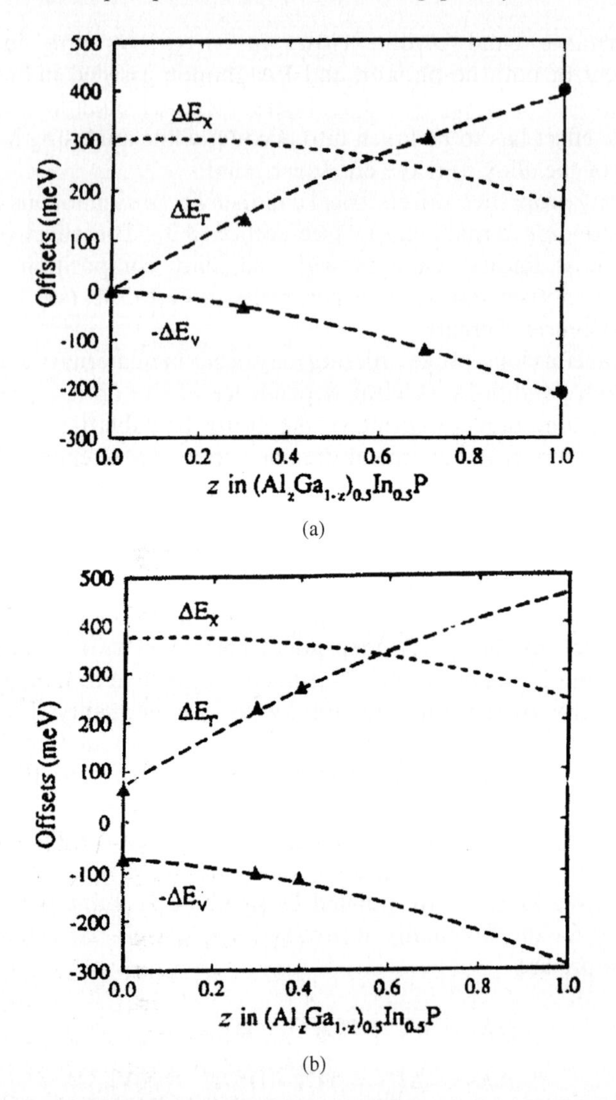

Figure 8.35. Composition dependence of the band offsets (a) for unstrained $Ga_{0.5}In_{0.5}P$–$(Al_zGa_{1-z})_{0.5}In_{0.5}P$ and (b) for 1% compressively strained $Ga_{0.38}In_{0.6}2P$–$(Al_zGa_{1-z})_{0.5}In_{0.5}P$ [139].

across the alloy range giving [139]

$$\begin{aligned}\frac{\mathrm{d}E_{\mathrm{g}}^{\Gamma}}{\mathrm{d}P} &= 8.0 + 0.2z \quad [\mathrm{meV/kbar}]\\ \frac{\mathrm{d}E_{\mathrm{g}}^{\mathrm{X}}}{\mathrm{d}P} &= -1.8 + 0.1z \quad [\mathrm{meV/kbar}].\end{aligned} \tag{8.27}$$

As the result of all the advantages discussed, In–Ga–Al–P quaternary alloys are very suitable for red, orange, yellow and yellowish-green LEDs and for injection heterolaser creation. Contemporary technology (MOVPE, MBE and SSMBE) allows one to create epitaxial layers and heterostructures on high-quality AlGaInP. Continued progress in epitaxial growth, processing technology and the design of structures for effective light extraction enabled the fabrication of AlGaInP light-emitting diodes with external efficiencies above 50% [140, 141].

At present the development of very bright electroluminescent devices has been enabled by the two material systems: AlGaInP for long-range and AlGaInN for short-range of visible spectra.

Internal electroluminescence efficiency approaching 100% was achieved in AlGaAs heterostructures about 20 years ago (see for example [127]). The external luminescence efficiency of (AlGaInP) LEDs is limited by the extraction of light from the active structure of LED [86]. Only 2–3% of the light emitted with almost 100% efficiency is getting out so far in standard constructions of LEDs [6]. This is caused by high refractive index and the effect of complete internal reflection.

A typical high-efficiency (Al,Ga,In)P LED structure consists of an elaborate sequence of epitaxial layers. For the active material, a double heterostructure design or a multiple-quantum-well structure, embedded between larger band gap layers, is used to optimize the carrier confinement. The barriers are maximized by doped AlInP for the confinement layers. A tensile strain barrier cladding layer can be used for AlGaInP yellow-green LEDs [142]. The multiple quantum barrier structures, a combination of barrier layer and short-period strained superlattice, is used such that the lowest miniband accessible for electrons tunnelling out of the active layer is above the largest conduction band energy of any of the constituting materials. Additional setback layers can be used to control the doping profiles, since the p-dopants (Mg, Zn) and source n-dopants (S, Te) tend to diffuse during material growth [141]. Usually, the structures are completed by a top window/current-spreading layer and an optional bottom distributed Bragg reflector (DBR) when grown on GaAs substrate (figure 8.36).

The common LED structure is a square die cut from a GaAs–AsGaInP epitaxial structure (figure 8.36) with electrodes on the bottom and top sides. The GaAs substrate is not transparent in the visible spectral range and only the emission into the top escape cone contributes 3–4% to the extraction efficiency. To reduce the effect of substrate absorption the LED structure is grown on AlGaAs- or AlGaInP-based DBRs formed as high and low index quarter wavelength layers. The integral reflectivity of the DBR is increased when the stop band centre of the DBR is detuned away from the emission maximum to longer wavelengths [141]. A DBR with 20 pairs of $Al_{0.5}Ga_{0.5}As$–AlAs has integral reflectivity from 50 to 80% for red and yellow LEDs. A transparent window layer of some microns thickness is

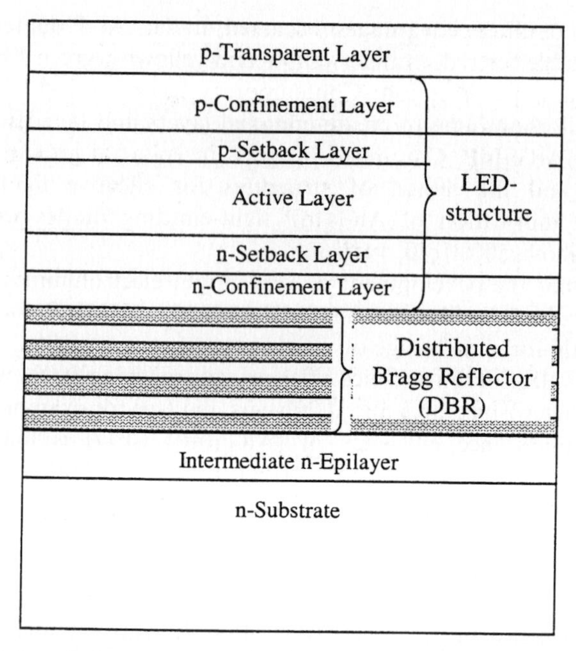

Figure 8.36. Schematic drawing of the layer structure of a typical high-brightness AlGaInP LED.

grown on top of the active layers (figure 8.36). This results in improved current spreading and additionally increases the extraction of light through the side facets [143].

One of the biggest challenges in the fabrication of all light-emitting diodes is the extraction of light from the semiconductor material into the surrounding media [6, 86]. Considering a flat interface, the total internal reflection limits the light extraction per out-coupling surface to a few percent. A positive contribution to the light extraction is made by the so-called photon-recycling effect (self-absorption and re-emission of photons into an active region) [140, 144], when the internal quantum efficiency is high. Much progress has been made in AlGaInP technology resulting in LEDs with conversion efficiencies comparable with semiconductor laser diodes [140, 141]. The simplest methods to increase the extraction efficiency are to texture surfaces of the chip [145] or to modify the refractive properties of the chip surface by using diffractive structures on top [146]. Geometrical structures with the shape of truncated tetrahedrons are etched into a GaP window layer (transparent layer on figure 8.36) [145]. Luminous efficiencies of more than 30 lm/W have been achieved in a wavelength range from 585 to 625 nm for GaAs-absorbing substrate LEDs.

An efficient technique to remove the absorbing GaAs substrate and to transfer the AlGaInP LED layers to a transparent GaP wafer was developed

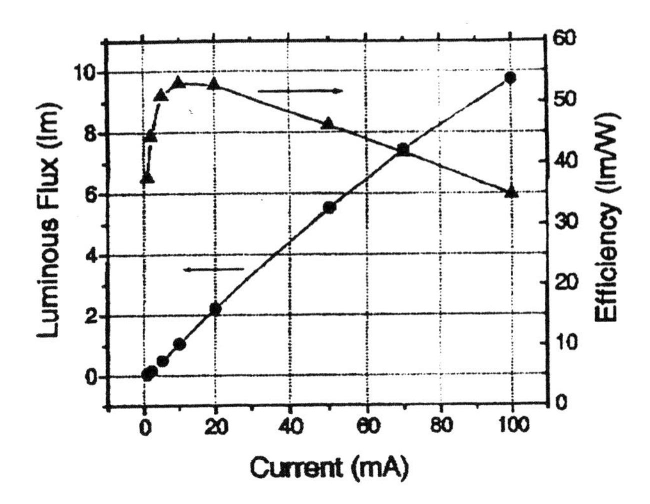

Figure 8.37. Optical performance of a 615 nm thin-film LED operated under d.c. conditions [141].

by Kish and co-workers [147, 148]. With the very thick GaP current spreading layer on top of the AlGaInP active region and the transparent GaP wafer below, the transparent-substrate LEDs achieve record high levels of light extraction. High-power truncated-inverted-pyramid $(Al_xGa_{1-x})_{0.5}In_{0.5}P/GaP$ light-emitting diodes exhibiting >50% external quantum efficiency [149] were fabricated by this technique. A luminous efficiency of 102 lm/W at 610 nm and a peak external efficiency of 55% at 650 nm were measured.

So-called high efficiency thin-film LEDs have in common that the AlGaInP structure is at first deposited on GaAs wafers, but the absorbing GaAs is subsequently removed in the fabrication process. The final result is a device in which the thin film of AlGaInP active material (structure) has been transferred from one wafer to another [141]. Transfer to a new carrier can be facilitated either on chip- or wafer-level and usually involves an intermediate metal layer for soldering. Geometrical shapes such as cones, prisms or spheres (or the carrier surface) might be etched before they are covered with metal and embedded inside the LED structure. Figure 8.37 depicts the optical characteristics of LED fabricated by these techniques. The LED emits up to 9.7 lm at 100 mA d.c. In the range 10–20 mA of operation current, the luminous performance is above 50 lm/W. One of the advantages of the thin-film technology is the low forward voltage that can be achieved. Operated at a d.c. current of 10 mA, the forward voltage of the 615 nm LEDs is still low, 2.0 V [141]. A novel method was proposed to glue a highly reliable AlGaInP LED on to a transparent sapphire substrate [150].

A completely different approach to handle the problem of light extraction is that of a resonant-cavity light-emitting diode (RCLED) (see overview [151]). In these devices, the active layer is embedded in a cavity with at least one dimension of the order of the wavelength of the emitted light. Under those circumstances, the spontaneous emission process itself is modified, such that the internal emission is no longer isotropic. The external efficiency of 12% for red RCLED was achieved in [141] by fine tuning the structure. Internal quantum efficiencies of 90% and 40% were determined for AlGaInP LED and microcavity LED, respectively [152]. High-speed RCLEDs at 650 nm optimized for short-haul communication systems on polymethyl methacrylate plastic optical fibre are presented in [153].

Quaternary (Al,Ga,In)P alloy is the basic material semiconductor for lasing in red, orange, yellow and yellow-green spectral regions. Strained and unstrained heterostructures, single and multiple quantum wells, superlattices and quantum dots are used to form the active region. The properties of $(Al,Ga)_{0.5}In_{0.5}P$, strained $Ga_xIn_{1-x}P/(Al,Ga)_{0.5}P$ heterostructures and single quantum well laser diodes with $Al_{0.5}In_{0.5}P$ cladding layers, prepared by low pressure organometallic vapour phase epitaxy, are described in [154].

The high-power 600 nm spectral band $(Al_xGa_{1-x})_yIn_{1-y}P/Ga_zIn_{1-z}P/GaAs$ quantum well laser was prepared using all-solid-source molecular beam epitaxy for layer growth [155]. An output power up to 3 W in continuous wave mode has been demonstrated at the wavelength of 670 nm, 2 W at 650 nm and 1 W at 630 nm. Preliminary life tests carried out for the 650 and 670 nm lasers suggest that these lasers will be quite reliable in operation. The high-performance vertical-cavity surface-emitting lasers with emission wavelength between 650 and 670 nm were fabricated on the basis of AlGaInP/AlGaAs heterostructures grown by MOVPE [156]. The continuous-wave room temperature output power of 4.1 mW at 650 nm and 10 mW at 670 nm have been achieved from oxide-confined VCSE lasers.

One of the main advantages of AlGaInP electroluminescence devices besides record high luminous efficiency in the red-yellow spectral range is the considerably improved reliability. It was demonstrated for LED1 [140, 141, 157, 158] and laser diodes (LD) [159–162]. By extrapolating long-term reliability data, lifetimes of more than 100 000 h have been calculated for LEDs. The excellent degradation behaviour of AlGaInP lasers has been attributed to a decreased sensitivity of the devices to oxidation due to reduced Al content in the active zone, compared with dark-line defects which decreased owing to incorporation of In in the compound [162].

High reliability requires an optimized epitaxial growth. The effect of long-range ordering may appear both in quaternary AlGaInP and ternary InGaP and AlGaP alloys. This is necessary to take into account while choosing the optimal technological conditions for the formation of active region of devices.

8.5 Conclusion

GaP and AlGaP alloy were the first materials for the commercial production of LEDs and injection lasers emitting in the visible range. Currently, (Ga–In–Al)P quaternary alloys are the main materials for most effective electroluminescent sources of coherent and incoherent light emission in red, orange, yellow and yellowish-green range. The internal quantum efficiency of the luminescence in the corresponding heterostructures achieves 100%. The external efficiency exceeded 50% due to development of special structures to improve light-extraction mechanisms. This is comparable with the efficiency of modern laser diodes. The light-emitting structures based on (Ga–In–Al)P quaternary alloys are very reliable (the warranted lifetime is 10 000 h and life expectance up to 100 000 h) and inexpensive. It makes them very profitable for use in lighting, light signalling and high-speed resonant cavity light-emitting diodes for low-cost short-haul communication system using plastic optical fibres.

The technology of growing of GaP and its ternary and quaternary alloys is well developed. Intrinsic and extrinsic physical, chemical, electrical, optical properties and luminescence mechanisms have been investigated in detail and are reliably identified.

Contemporary epitaxial techniques make it possible to fabricate both relaxed and strained (Ga–In–Al)P-based heterostructures, including DHSs, SQWs, MQWs, SLs and QDs. These materials and structures may be also used to design selective and broad-band photodiodes with a desired photosensitivity spectrum for visible and ultraviolet ranges, devices of high-temperature electronics etc. The so-called 'camel's back' structure of the absolute minimum in the conduction band of GaP probably may be interesting for some novel devices of UHF electronics.

Acknowledgments

The authors would like to thank Dr Sergey A Tarasov and Oleg Komkov (Laboratory of Optical Characterization of Electronic Material and Structures, St. Petersburg Electrotechnical University) and Dr Nikita A Pikhtin (A F Ioffe Physico-Technical Institute) for useful discussions and their help with the manuscript preparation. This work was performed when one of the authors (O.L.L.) held a National Research Council Research Associateship Award at Jet Propulsion Laboratory.

References

[1] Goldshmidt V M 1929 *Trans. Faraday Soc.* **25** 253; 1926 *Naturw.* **14** 477; 1931 *Structurber.* **136** 132

[2] Welker H 1952 *Z. Naturforsch.* **11** 744
[3] Losev O W 1923 *Telegr. Telef. Provod.* **18** 61
[4] Gershenzon M 1966 *Luminescence in Inorganic Solids* ed. P Goldberg (New York: Academic Press) ch. 11
[5] Yunovich A E 1972 Radiative recombination and optical properties of gallium phosphide. In *Radiation Recombination in Semiconductors* ed. Ya E Pokrovsky (Moscow: Nauka) ch. 3 (in Russian)
[6] Bergh A A and Dean P J 1976 *Light-emitting Diodes* (Oxford: Clarendon)
[7] Hilsum C and Rose-Innes A C 1961 *Semiconducting III–V Compounds* (Pergamon Press)
[8] Madelung O 1964 *Physics of III–V Compounds* (New York: Wiley)
[9] Landolt-Börnstein 1982 *Numerical Data and Functional Relationships in Science and Technology*, New Series, vol. 17a; 1982 vol. 22a ed. O Madelung (Berlin: Springer)
[10] Adachi S 1992 *Physical Properties of III–V Semiconductor Compounds: InP, InAs, GaAs, GaP, InGaAs and InGaAsP* (New York: Wiley)
[11] Kudman I and Pfaff R J 1972 *J. Appl. Phys.* **43** 3760
[12] Deus P, Voland U and Schneider H A 1983, *Phys. Status Solidi (a)* **80** K29
[13] Muzhdaba V M, Nashel'sky A Ya, Tamarin P V and Shalyt S S 1968 *Fiz. Tverd. Tela* **10** 2866 (English translation: *Sov. Phys. Solid State* **10** 2265)
[14] Irwin J C and La Combo J 1974 *J. Appl. Phys.* **45** 569
[15] Pollak F H, Higginbottam C W and Cardona M 1966 *J. Phys. Soc. Japan, Suppl.* **21** 20
[16] Glinskii G F, Kopylov A A and Pikhtin A N 1979 *Solid State Commun.* **30** 631
[17] Altarelli M, Sabatini R A and Lipari N O 1978 *Solid State Commun.* **25** 1101
[18] Kopylov A A and Pikhtin A N 1977 *Phys. Tekh. Polupr.* **11** 867 (English translation: 1977 *Semiconductors* **11** 510)
[19] Kopylov A A and Pikhtin A N 1978 *Solid State Commun.* **24** 801
[20] Kao Y C and Eknoyan O 1983 *J. Appl. Phys.* **54** 2468
[21] Aspnes D E and Studna A A 1983 *Phys. Rev. B* **27** 985
[22] Pikhtin A N, Prokopenko V T and Yaskov A D 1976 *Fiz. Tekh. Polupr.* **10** 2053 (English translation: 1976 *Semiconductors* **10** 1224)
[23] Pikhtin A N and Yaskov A D 1988 *Fiz. Tekh. Polupr.* **22** 969 (English translation: 1988 *Semiconductors* **22** 613)
[24] Philipp H R and Ehrenreich H 1963 *Phys. Rev.* **129** 1550
[25] Kleinman D A and Spitzer W G 1960 *Phys. Rev.* **118** 110
[26] Yas'kov D A and Pikhtin A N 1969 *Mat. Res. Bull.* **4** 781
[27] Pikhtin A N and Yas'kov D A 1969 *Fiz. Tverd. Tela* **11** 561
[28] Dean P J and Thomas D G 1966 *Phys. Rev.* **150** 690
[29] Capizzi M, Evangelisti F, Fiorini P, Frova A and Patella F 1978 *Solid State Commun.* **24** 801
[30] Lupal M V and Pikhtin A N 1981 *Fiz. Tekh. Polupr.* **15** 822 (English translation: 1981 *Semiconductors* **15** 471
[31] Pikhtin A N and Yaskov D A 1969 *Fiz. Tverd. Tela* **11** 2213
[32] Hume-Rothery W 1936 *The Structure of Metals and Alloys* (London: Institute of Metals)
[33] Gross E F and Nedzvetskii D S 1963 *Doklady Acad. Sci. USSR* **152** 309
[34] Thomas D G, Hopfield J J and Frosch C J 1965 *Phys. Rev. Lett.* **15** 857; 1966 *Phys. Rev.* **150** 680

[35] Hopfield J J, Dean P J and Thomas D G 1967 *Phys. Rev.* **158** 748
[36] Morgan T N 1968 *Phys. Rev. Lett.* **21** 819
[37] Spitzer W G, Gershenson M, Frosch C J and Gibbs D F 1959 *J. Phys. Chem. Solids* **11** 339; Allen J W and Hodby J W 1963 *Proc. Phys. Soc.* **82** 315
[38] Pikhtin A N and Yaskov D A 1969 *Phys. Status Solidi* **34** 815
[39] Berndt V, Kopylov A A and Pikhtin A N 1974 *Pisma Zh. Eks. Teor. Fiz.* **22** 578; 1977 *Fiz. Tekh. Poluprovodn.* **11** 1782 (English translation: 1977 *Semiconductors* **11** 1044)
[40] Kopylov A A and Pikhtin A N 1978 *Solid State Commun.* **26** 735
[41] Kopylov A A, Medvedev S P and Pikhtin A N 1979 *Pisma Zh. Eks. Teor. Fiz.* **30** 506; *Fiz. Tehn. Poluprovodn.* **13** 1586 (English translation: 1979 *Semiconductors* **13** 924)
[42] Glinskii G F and Pikhtin A N 1975 *Fiz. Tehn. Poluprovodn.* **9** 2139 (English translation: 1975 *Semiconductors* **9** 1393)
[43] Thomas D G, Hopfield J J and Frosch C J 1966 *Phys. Rev.* **150** 680
[44] Pikhtin A N and Popov V A 1980 *Pisma Zh. Eks. Teor. Fiz.* **31** 723
[45] Nelson D F and Turner E H 1968 *J. Appl. Phys.* **39** 3337; Glinskii G F, Pikhtin A N and Yaskov D A 1972 *Fiz. Tverd. Tela* **14** 350
[46] Cardona M, Grimsditch M and Olego D 1979 in *Light Scattering in Solids* ed. J Birman (New York: Plenum) p. 249
[47] Singh V A and Zunger A 1985 *Phys. Rev. B* **31** 3729; Masterov V F 1978 *Fiz. Tekh. Poluprovodn.* **12** 625 (English translation: 1975 *Semiconductors* **9** 363); *Fiz. Tekh. Poluprovodn.* **18** 3 (English translation: 1975 *Semiconductors* **18** 1)
[48] Van Roosbroeck W and Shockley W 1954 *Phys. Rev.* **20** 9
[49] Varshni Y P 1967 *Phys. Status Solidi* **20** 9
[50] Craford M G, Keune D L, Groves W O and Herzog A N 1973 *J. Electron. Mater.* **2** 137
[51] Gorelenok A T and Shpakov M V 1996. *Fiz. Tekh. Poluprovodn.* **30** 488 (English translation: 1996 *Semiconductors* **30** 269)
[52] Bimberg D, Skolnick M S and Sander L M 1979 *Phys. Rev.* **19** 2231; Shah J, Leheny R F, Harding W R and Wight D R 1977 *Phys. Rev. Lett.* **38** 1164
[53] Dean P J 1982 *Prog. Crystal Growth Charact.* **5** 89
[54] Dean P J 1978 private unpublished data
[55] Prener J S and Williams F E 1956 *J. Electrochem. Soc.* **103** 342
[56] Hopfield J J, Thomas D G and Gershenzon M 1963 *Phys. Rev. Lett.* **10** 162
[57] Dean P J 1970 *J. Luminescence* **1** 398; 1973 **7** 51; 1973 *Luminescence of Crystals, Molecules and Solutions* ed. F E Williams (New York: Plenum) p. 538
[58] Kopylov A A and Pikhtin A N 1974 *Fiz. Tehn. Poluprovodn.* **8** 2398 (English translation: *Semiconductors* **8** N12)
[59] Bhargava R N 1970 *Phys. Rev. B* **2** 387; Dishman J M 1971 *Phys. Rev. B* **3** 2588
[60] Dean P J, Faulkner R A, Kimura S and Illegems M 1971 *Phys. Rev. B* **4** 1926
[61] Dean P J, Cuthbert J D, Thomas D G and Lynch R T 1967 *Phys. Rev. Lett.* **18** 122
[62] Fieseler H, Haufe A, Schwabe R and Streit I 1985 *J. Phys.* **18** 3705
[63] Wolfe M I, Kressel H, Halpern T and Raccah P M 1970 *Appl. Phys. Lett.* **24** 279
[64] Onton A and Morgan T N 1970 *Phys. Rev. B* **1** 2592
[65] Thomas D G and Hopfield J J 1966 *Phys. Rev.* **150** 580
[66] Zhang X, Hong Q and Dou K 1990 *Phys. Rev. B* **41** 1376
[67] Lupal M V and Pikhtin A N 1985 *Pisma Zh. Eks. Teor. Fiz.* **42** 201
[68] Thomas D G, Hopfield J J and Frosch C J 1965 *Phys. Rev. Lett.* **15** 857

[69] Faulkner R A 1968 *Phys. Rev.* **175** 991
[70] Lightowlers E C, North J C and Lorimor O G 1974 *J. Appl. Phys.* **45** 2191
[71] Kloth B, Lupal M V, Pikhtin A N, Richter C E, Ries R, Stegmann R and Trapp M 1987 *Phys. Status Solidi (a)* **154** 545
[72] Cuthberg J D and Thomas D G 1967 *Phys. Rev.* **154** 763
[73] Morgan T N, Welber B and Bhargava R N 1968 *Phys. Rev.* **166** 751
[74] Henry C H, Dean P J, Thomas D G and Hopfield J J 1968 *Proc. Conf. Localized Excitations* ed. R F Wallis (New York: Plenum) p. 267
[75] Cuthbert J D, Henry C H and Dean P J 1968 *Phys. Rev.* **170** 739
[76] Onton A and Lorenz M 1968 *Appl. Phys. Lett.* **12** 115
[77] Toyama M and Kasami A 1972 *Jap. J. Appl. Phys.* **11** 860
[78] Dishman J M and DiDomenico M Jr 1970 *Phys. Rev. B* **1** 3381; *Phys. Rev. B* **2** 1988
[79] Jayson J S, Bhargava R N and Dixon R W 1970 *J. Appl. Phys.* **451** 4972
[80] Dean P J and Faulkner R N 1969 *Phys. Rev.* **185** 1064
[81] Dean P J, Faulkner R N and Kimura S 1970 *Phys. Rev. B* **2**(10) 4062; Gislason H P, Monemar B, Pistol M E, Kanaa H K and Cavanett B C 1986 *Phys. Rev. B* **33** 1233
[82] Dean P J 1971 *Phys. Rev. B* **4** 2596
[83] Dean P J, Thomas P J and Frosch C J 1984 *J. Phys. C: Solid State Phys.* **17** 747
[84] Kardontchik J E and Cohen E 1979 *Phys. Rev. B* **19** 3181
[85] Pilkuhn M H 1970 *J. Appl. Phys.* **40** 3162
[86] Pikhtin A N 2001 *Optical and Quantum Electronics* (Moscow: High School Publishers) (in Russian) p. 573
[87] Landolt-Börnstein 1984 *Numerical Data and Functional Relationships in Science and Technology*, New Series, vol. 17d, *Technology of III–V, II–VI and Non-Tetrahedrally Bounded Compounds* ed. M Schulz and H Weiss (Berlin: Springer)
[88] Zipperian T E and Dawson L R 1983 *J. Appl. Phys.* **54** 6019
[89] Pikhtin A N, Tarasov S A and Bernd K 2003 *IEEE Trans. Electron Devices* **50** 215
[90] Alferov Zh I 1998 *Fiz. Tekh. Poluprovodn.* **32** 3 (English translation: *Semiconductors* **32** N1 1)
[91] Nordheim L 1931 *Ann. Phys.* **9** 607
[92] Heine V and Weaire D 1970 *Solid State Phys.* **24** 249
[93] Van Vechten J A and Bergstresser T K 1970 *Phys. Rev. B* **1** 3351
[94] Bolero O, Wooley J C and Van Vechten J A 1973 *Phys. Rev. B* **8** 3794
[95] Pikhtin A N 1977 Optical transitions in semiconductor solid solutions review *Semiconductors* **11** 245
[96] Nelson R J 1982 in *Excitons* ed. E I Rashba and M D Sturge (Amsterdam: North-Holland) p. 319
[97] Vurgaftman I, Meyer J R and Ram-Mohan L R 2001 *J. Appl. Phys.* **89** N11 5815
[98] Monemar B 1973 *Phys. Rev. B* **8** 5711
[99] Dawson M D, Najda S P, Kean A H, Duggan G, Mowbray D J, Kowalski O P, Skolnik M S and Hopkinson M 1994 *Phys. Rev. B* **50** 11190
[100] Bour D P, Shealy J R, Ksendzov A and Pollak F 1988 *J. Appl. Phys.* **64** 6456
[101] Mowbray D J, Kowalski O P, Hopkinson M, Skolnik M S and David J P R 1994 *Appl. Phys. Lett.* **65** 213
[102] Bosio C, Staehli J L, Guzzi M, Burri G and Logan R A 1988 *J. Phys. C* **10** 4709
[103] Baldareschi A, Hess E, Maschke K, Neuman H, Schulze K-R and Unger K 1977 *J. Phys. C* **10** 4709
[104] Mbaye A A, Fereira L G and Zunger A 1987 *Phys. Rev. Lett.* **58** 49

[105] Kondow M, Kakibayashi H, Tanaka T and Minagawa S 1989 *Phys. Rev. Lett.* **63** 884

[106] Srivastava G P, Martines J I and Zunger A 1988 *Phys. Rev. B* **38** 12694

[107] Pikhtin A N and Yaskov A D 1989 *Fiz. Tekh. Poluprovodn.* **14** N4 661 (English translation: 1989 *Semiconductors* **14** 389)

[108] Barker A S and Sivers A J 1975 *Rev. Mod. Phys.* **47** S175; Genzel L and Bauhofer W 1976 *Z. Phys. B* **25** 13

[109] Adachi S 1983 *J. Appl. Phys.* **54** 1844

[110] Alferov Zh I, Portnoj E L and Rogachev A A 1968 *Fiz. Tekh. Poluprovodn.* **2** 1194; Baranovsky S D and Efros A L 1978 *Fiz. Tekh. Poluprovodn.* **12** 2233 (English translation: *Semiconductors* **12** N11 1328)

[111] Nelson R J, Holonyak N and Groves W O 1976 *Phys. Rev. B* **13** 5415

[112] Alferov Zh I, Garbusov D Z, Mishurnyi V A, Rumyantsev V D and Tret'yakov D N 1973 *Fiz. Tekh. Poluprovodn.* **7** 2305 (English translation: 1973 *Semiconductors* **7** 337)

[113] Pikhtin A N, Razbegaev V D and Yaskov D A 1972 *Phys. Status Solidi B* **50** 717; 1973 *Fiz. Tekh. Poluprovodn.***7** 471 (English translation: 1973 *Semiconductors* **7** 1534

[114] Dean P J, Kaminsky G and Zetterstrom R B 1969 *Phys. Rev.* **181** 1149

[115] Lupal M V and Pikhtin A N 1980 *Fiz. Tekh. Poluprovodn.* **14** 2178 (English translation: *Semiconductors* **14** 1291)

[116] Berndt V, Kopylov A A and Pikhtin A N 1978 *Fiz. Tekh. Poluprovodn.* **12** 1628 (English translation: 1978 *Semiconductors* **12** 964)

[117] Kopylov A A and Pikhtin A N 1981 *Fiz. Tekh. Poluprovodn.* **15** 2164 (English translation: 1981 *Semiconductors* **15** 1257)

[118] Zubkov V I, Pikhtin A N and Solomonov A V 1989 *Fiz. Tekh. Poluprovodn.* **23** 64 (English translation: *Semiconductors* **23** N1 39)

[119] Glinskii G F, Lupal M V, Parfenova I I and Pikhtin A N 1991 *Fiz. Tekh. Poluprovodn.* **26** 641 (English translation: *Semiconductors* **26** N4 364)

[120] Ivkin A N and Pikhtin A N 1998 *Pis'ma Zh. Tekh. Fiz.* **24** 18 (English translation: *Technical Physics Lett.* **24** N6 419)

[121] Nelson R J, Holonyak N, Coleman J J *et al* 1976 *Phys. Rev. B* **14** 685

[122] Pikhtin A N, Popov V A and Unis M 1988 *Fiz. Tekh. Poluprovodn.* **22** 1107 (English translation: 1988 *Semiconductors* **22** 698)

[123] Campbell J C, Holonyak N, Craford M G and Keune D L 1974 *J. Appl. Phys.* **45** 4543

[124] Glinskii G F, Loginova T N, Lupal M V and Pikhtin A N 1986 *Fiz. Tekh. Poluprovodn.* **20** 676 (English translation: 1986 *Semiconductors* **20** N4)

[125] Kloth B, Lupal M V, Pikhtin A N, Richter C E, Ries R, Stegmann R and Trapp M 1987 *Phys. Status Solidi* (*a*) **100** 545

[126] Holonyak N Jr *et al* 1972 *J. Appl. Phys.* **43** 4148; 1973 **44**, 5517; 1972 *Phys. Rev. Lett.* **28** 230; 1972 *Appl. Phys. Lett.* **20** 11

[127] Alferov Zh I *et al* 1971 *Semiconductors* **5** 982; 1972 **2** 589; 1972 **6** 1620; 1973 **7** 1449; 1973 **7** 435; 1972 **6** 1930; 1975 **9** 435; 1972 **6** 1930; 1974 **7** 1534; 1975 **9** 305

[128] Coleman J J, Holonyak N, Ludwise M J, Nelson R J, Wright P D, Groves W O, Keune D L and Craford M G 1975 *J. Appl. Phys.* **46** 4835

[129] Adachi S 1982 *J. Appl. Phys.* **53** 8775

[130] Beister G, Erbert G, Knauer A, Maege J, Ressel P, Sebastian J, Staske R and Wenzel H 1999 *Electronics Lett.* **35** 1641

[131] Al-Muhanna A, Mawst L J, Botez D, Garbusov D Z, Martinelli R U and Connoly J C 1998 *Appl. Phys. Lett.* **73** 1182

[132] Al-Muhanna A, Mawst L J, Botez D, Garbusov D Z, Martinelli R U and Connoly J C 1997 *Appl. Phys. Lett.* **71** 1142

[133] Livshits D A, Kochnev I V, Lantratov V M, Ledentsov N N, Nalyot T A, Tarasov I S and Alferov Zh I 2000 *Electronics Lett.* **36** 1878

[134] Golikova E G, Kureshov V A, Leshko A Ju, Livshits D A, Lutetskij A V, Nikolaev D N, Pikhtin N A and Tarasov I S 2000 *Tech. Phys. Lett.* **26** 913

[135] Knauer A, Erbert G, Wenzel H, Bhatacharya A, Bugge F, Maege J, Pittroff and Sebastian J 1999 *Electronics Lett.* **35** N8 638

[136] Alferov Zh I, Arsent'ev I N, Garbuzov D Z, Konnikov S G and Rumyantsev V D 1975 *Pis'ma Zh Tekh. Fiz.* **1** 305 (English translation: 1975 *Technical Physics Lett.* **1** N7)

[137] Kowalski O P, Wegerer R M, Mowbray J, Skolnick M S, Button C C, Roberts J S, Hopkinson F, David J P R and Hill G 1996 *Appl. Phys Lett.* **68** 3266

[138] Ledentsov N N, Ustinov V M, Shchukin V A, Kop'ev P S, Alferov Zh I and Bimberg D 1998 *Semiconductors* **32** 343; Bimberg D, Grundman M and Ledentsov N N 1999 *Quantum Dot Heterostructures* (Chichester: Wiley) p. 328

[139] Meney A T, Prins A D, Phillips A F, Sly J L, O'Reilly E P, Dunstan D J, Adams A R and Valstor A 1995 *IEEE J. Select Topics Quantum Electron.* **1** 697

[140] Kish F and Fletcher R 1997 *AlGaInP light-emitting diodes* in *Semiconductors and Semimetals, High Brightness Light Emitting Diodes* ed. G B Stringfellow and M G Craford **48** 149

[141] Streubel K, Linder N, Wirth R and Jaeger A 2002 *IEEE J. Select Topics Quantum Electron.* **8** 321

[142] Chang S J, Chang C S, Su Y K, Chang P T, Wu Y R, Huang K H and Chen T P 1997 *IEEE Photon Technol. Lett.* **9** 1199

[143] Vanderwater D A, Tan L-H, Höfler G E, Defevere D C and Kish F A 1997 *Proc. IEEE* **85** 1752

[144] Garbuzov D Z 1976 Proc. *Summer School on Optoelectronics and Integrated Optics* (Czechoslovakia: Marianske Lazne) p. 327

[145] Linder N, Kugler S, Stauss P, Streubel K P, Wirth R and Zull H 2001 *Proc. SPIE* **4278** 19

[146] Windish R, Rooman C, Meinlschmidt S, Kiesel P, Zipperer D, Döhler H, Dutta B, Kuijk M, Borghs G and Heremans P 2001 *Appl. Phys. Lett.* **79** 1

[147] Kish F A *et al* 1994 *Appl. Phys. Lett.* **64** 2839; 1995 **67** 2060

[148] Höfler G E, Vanderwater D, Defere D C, Kish F A, Carnras M, Steranka F and Tan L-H 1996 *Appl. Phys. Lett.* **69** 803

[149] Krames M R, Ochinai-Holocomb M, Höfler G E, Carter-Coman C, Chen E I, Tan L-H, Grillot P, Gardner N F, Chui H C, Huang J-W, Stockman S A, Kish F A and Craford M G 1999 *Appl. Phys. Lett.* **75** 2365

[150] Chang S-J, Su Y-K, Yang T, Chang C-S, Chen T-P and Huang K-H 2002 *IEEE J. Quantum Electron.* **38** 1390

[151] Delbeke D, Bockstaele R, Bienstman P, Baets R and Benisty H 2002 *IEEE J. Select. Topics Quantum Electron* **8** 189

[152] Royo P, Stanley R P, Ilegems M, Streubel K and Gulden K H 2002 *J. Appl. Phys.* **91** 2563

[153] Dumitrescu M M, Saarinen M J, Guina M D and Pessa M V 2002 *IEEE J. Select. Topics Quantum Electron.* **8** 219

[154] Bour D P, Geels R S, Treat D W, Paoli T L, Ponce F, Thornton R L, Krusor B S, Bringans R D and Welch D F 1995 *IEEE J. Quantum Electron.* 30 **593**

[155] Savolainen P, Toivonen M, Pessa M, Corvini P, Jansen M and Nabiev R F 1999 *Semicond. Sci. Technol.* **14** 425

[156] Knigge A, Zorn M and Tränkle G 2002 *Electronics Lett.* **38** 882

[157] Kish F A, Vanderwater D A, DeFevere D C, Steigerwald D, Höfler G E, Park K and Steranka F 1996 *Electronics Lett.* **32** 132

[158] Lacey J, Morgan D, Aliyu Y and Thomas H 2000 *Qual. Reliab. Engng. Int.* **16** 45

[159] Watanabe M 1999 *IEEE J. Select. Topics Quantum Electron.* **5** 750

[160] Herrick R W and Petroff P M 1998 *Appl. Phys Lett.* **72** 1799

[161] Choy W J, Chang J H, Choi W T, Kim S H, Kim J S, Leem S J and Yoo T K 1995 *IEEE J. Select. Topics Quantum Electron.* **1** 717

[162] Endo K, Kobayashi K, Fujii H and Ueno Y 1994 *Appl. Phys Lett.* **64** 146

Chapter 9

Gallium nitride and related materials

M Godlewski[1,2] *and A Kozanecki*[1]
[1]Institute of Physics PAS, Warsaw, Poland
[2]College of Science, Cardinal S. Wyszyński University, Warsaw, Poland

9.1 Introduction

Gallium nitride (GaN)-based materials found applications in short wavelength light-emitting devices and also in white light sources. Properties of these materials and also light emission properties are discussed in numerous review articles, conference proceedings and special issues of regular journals (see e.g. [1–5]), thus will not be repeated here.

Most of the authors agree that the most important problems to be solved is selection of appropriate substrates materials for GaN epitaxy, selection of buffer layers and elucidation of their properties and finally mastering laser emission in the green–violet spectral region. These topics will be reviewed in this chapter. We will also discuss new possible applications of GaN light-emitting devices based on GaN activated with rare earth ions.

9.2 Properties of buffer layers for GaN epitaxy

Growth of GaN-based devices on lattice mismatched sapphire requires special technological steps. These steps are necessary for strain relaxation and reduction of dislocation density, since lattice mismatch is exceptionally large (about 14%). Both strain relaxation and reduction of dislocation density are crucial. Dislocations affect rate of radiative recombination, whereas appearance of the strain leads to strong piezoelectric effects.

Not surprisingly most of the researches agree that the first major breakthrough in GaN technology came with the introduction of so-called buffer layers. A two-step growth mode was introduced. The first step is deposition of a low-temperature (LT) AlN [6–8], GaN [9], AlGaN or multiple-pair buffer layers [10–12]. This was followed by GaN epitaxy performed at 'normal' conditions. The so-obtained GaN films show better structural quality, compared with those deposited on sapphire without a buffer layer.

Several other attempts gave similar improvement of GaN morphology. For example, growth on sapphire can also be promoted by initial nitridation of sapphire in an ammonia stream [13–15]. Mechanism of the growth is not clear in this case. There are conflicting models explaining improved morphology of GaN films. Nitridation results in formation of either a thin AlN [16], BN [17], or AlNO [18] amorphous layers. These layers form elastic buffers for GaN epitaxy and thus can promote growth of structures of a good quality.

These inventions helped to improve considerably not only the morphology of GaN epilayers, but noticeable improvements of electrical and optical properties of GaN epilayers were also reported [19].

It was found that AlN and GaN buffer layers should be fairly thin. Optimal width is 50–100 nm in the case of AlN and less than 30 nm for GaN. Even at this width buffer layers reduce detrimental effects related to lattice mismatch. The latter occurs since buffer layers contain a very large density of dislocations. This helps to relax strain at the interface, which is very crucial considering nearly 14% misfit in the GaN/sapphire system. It was found that from initial 13.8% only 1.1% misfit remains as the residual strain [16]. The rest is relieved by misfit dislocations at the interface region [16].

Reduction of strain is not the only advantageous property of the system. Buffer layers also supply nucleation centres for growth of an improved quality GaN [19]. Not surprisingly, the problem of optimized AlN, AlGaN or GaN buffer layers is one of the most studied in GaN technology, as discussed in numerous recent publications on this topic [5–20].

9.2.1 LT GaN buffer layer

The LT AlN buffer was initially considered as a better choice. AlN shows a smaller lattice mismatch to sapphire than GaN, i.e. a more efficient release of a stress near the interface is expected in the AlN/sapphire system. However, LT GaN buffer is easier to be introduced. It can be introduced in the same growth process, i.e. when using the same growth procedure, only by lowering the growth temperature. Thus, most of the recent studies concentrate on the properties of this buffer.

Structural properties of LT GaN buffer were recently studied by several groups [16, 20–23]. All these studies indicate that buffer layer is deposited as a fairly uniform and polycrystalline GaN film. The presence of mis-oriented three-dimensional grains was observed. These grains partially coalesce under annealing [16]. These studies also confirmed that LT GaN buffer should be very thin, of less than 30 nm [24, 25], i.e. much thinner than LT AlN buffers.

High efficiency of strain relaxation at GaN/sapphire interface by such thin LT GaN buffers was rather an unexpected result. Two mechanisms were proposed to explain this puzzling property of the LT GaN buffer [20, 21, 23]. The first mechanism assumes a mixed cubic-hexagonal structure of the GaN buffer. Islands of a cubic-phase GaN surrounded by hexagonal-phase GaN were detected in the region of the LT GaN buffer [23]. Such a mixture of GaN phases can probably help in stress relaxation at the GaN/sapphire interface [23].

The second model proposed relates an enhanced stress relaxation to Al inter-diffusion from the sapphire to the LT GaN buffer [20, 21]. Al inter-diffusion results in a formation of either AlGaN or even of pure AlN micro-crystals. These micro-crystals, when present in the interface region, i.e. close to the GaN/sapphire interface, can help in strain relaxation.

We have recently confirmed the latter model by studying properties of GaN/sapphire interface in InGaN/GaN quantum well structures. The structures were grown by metal-organic chemical vapour deposition (MOCVD) on (0001) sapphire with a LT GaN buffer layer, using a horizontal cell MOCVD system [20]. Thin LT GaN buffer layer was first grown at 500 °C. Then, the growth was stopped and the temperature was increased to 1075 °C. Such increase in the temperature led to recrystallization of the buffer, but also in smoothing of the buffer surface and in coalescence of the grains present in the interface region [16]. The recrystallized buffer forms a nucleation layer helping in the growth of an improved quality GaN film.

In figures 9.1 and 9.2 we show the topography of the samples studied. A flat surface was seen by scanning electron microscopy (SEM) and by atomic force microscopy (AFM). Relatively few micro-holes and large-size holes, penetrating to the region of the LT GaN buffer, were observed. The presence of such micro-holes and large holes helped to observe the depth-dependent properties of GaN films and compare the growth modes in two regions of the film, i.e. directly over the sapphire substrate (in the middle of the large hole) and at the surface of the films [20]. Figure 9.1 shows the edge of one of the micro-holes. In this area the LT GaN buffer is exposed and its properties can be studied directly, as shown in figure 9.2. This figure shows buffer features observed in a high-resolution SEM image taken in the middle of the hole.

A granular-like micro-structure of the LT GaN buffer is resolved in the SEM image. After the recrystallization the LT GaN buffer is relatively smooth, with coalescent grains of 50–100 nm size. Buffer initializes growth of ordered GaN in form in hexagons. One of such hexagons is shown in figure 9.3.

These hexagons show the initial stages of the GaN growth. Large and hexagonal islands are deposited first, which then merge and form flat and regular GaN films.

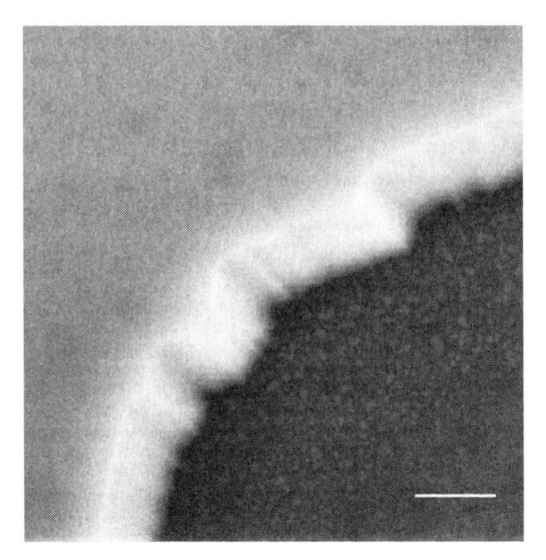

Figure 9.1. A SEM image taken at edge of a large hole at 8000× magnification, i.e. from a 12.5 μm × 12.5 μm region of the structure [20]. Horizontal bar indicates 2 μm distance in the plane of the picture.

Detailed cathodoluminescence (CL) and the depth-profiling CL investigations were carried out [20]. These investigations were performed to determine buffer properties. The idea behind the depth-profiling CL is described elsewhere [26]. For further discussion it is enough to know that

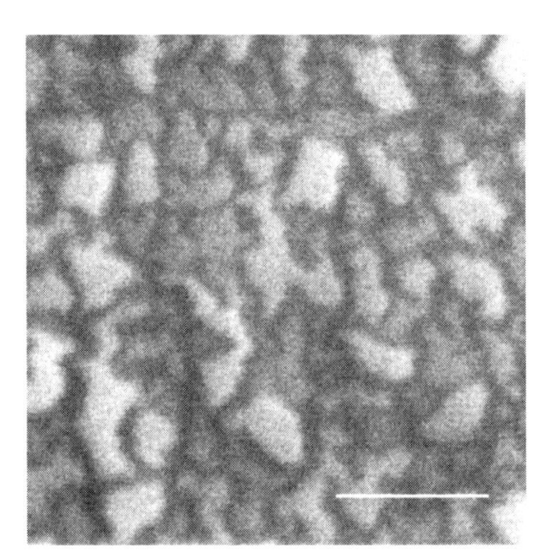

Figure 9.2. SEM image of granular microstructure of LT GaN buffer directly overgrowing the sapphire substrate taken at 60 000× magnification, i.e. from 1.65 μm × 1.65 μm region of the structure [20]. Horizontal bar indicates 0.5 μm distance in the plane of the picture.

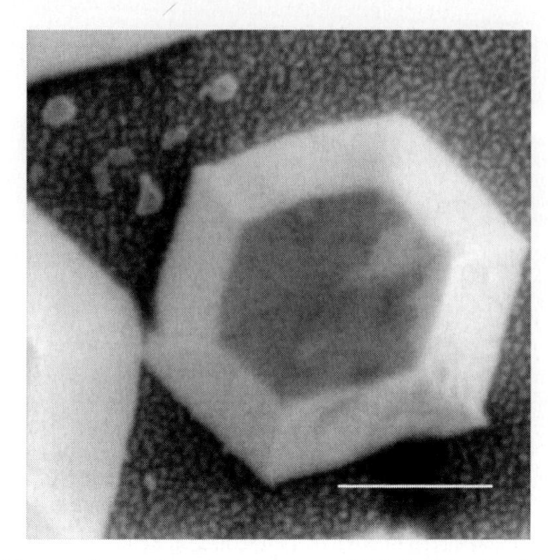

Figure 9.3. SEM image of granular microstructure of LT GaN buffer directly overgrowing the sapphire substrate taken at 60 000× magnification, i.e. from 1.65 μm × 1.65 μm region of the structure [20]. Horizontal bar indicates 0.5 μm distance in the plane of the picture.

depth of electron penetration depends on accelerating voltage in the electron microscope. For example, at 30–35 kV accelerating voltage primary electrons have a sufficient energy to penetrate through whole the structure, i.e. they excite CL from the interface region of the structure. This could be confirmed experimentally by observing the appearance of a weak sapphire-related emission, not shown in figure 9.4. Depth profiling CL thus allows study of the interface region of the structure. In turn, at much lower accelerating voltages one can study CL emission coming from the upper regions of GaN epilayers.

The origin of the observed CL emissions could be identified by studying the CL excited at different regions of the film, i.e. from the spot mode CL. Such an approach enables one to compare the CL spectra excited from the regions of a flat surface of the film and at a micro-hole. Three bright CL emissions were observed at the large hole, as shown in figure 9.4.

Of the three edge CL emissions originating from the GaN buffer region, the first, with the maximum at about 3.7 eV, has energy larger than the GaN band-gap energy. This line was related to CL emission of the AlGaN layer, with about 10% Al fraction [20]. The origin of the remaining two CL emissions could also be explained. The second CL emissions was attributed to an excitonic 'edge' emission of GaN, but red-shifted and much broader than that observed in a strain relaxed film. The third CL is relatively broad and strong. This is the so-called blue emission (BL) band of GaN, with the maximum at about 2.95–3.0 eV [20]. The donor–acceptor

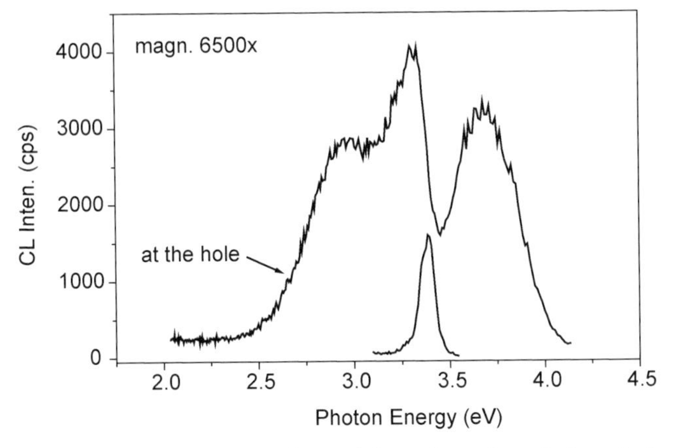

Figure 9.4. The spot-mode CL spectra detected from the region of a large hole. For comparison we also show the 'edge' GaN CL (lower curve) observed from the film excited outside of the large hole [20].

pair recombination origin of the BL band was demonstrated previously [19].

Such an origin of the CL emissions was deduced from their in-plane properties, as shown in figure 9.5. This region of a double hole was selected for this study. Then scanning CL investigations were performed by setting the CL detection at different wavelengths corresponding to the bands observed in the CL spectrum shown in figure 9.4.

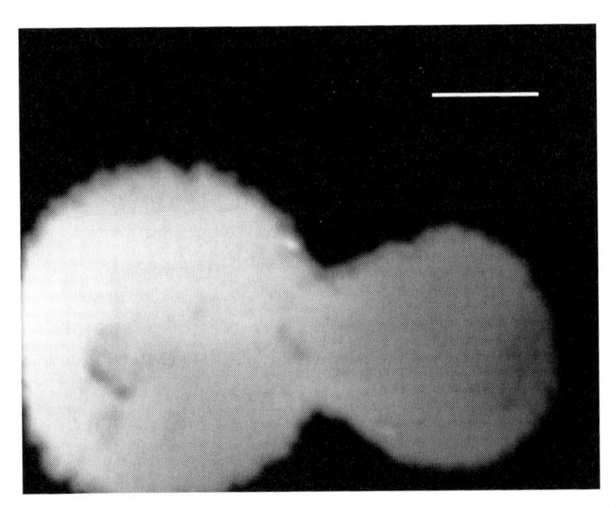

Figure 9.5. Scanning CL image of in-plane dependence of the AlGaN-related CL emission taken in the double hole region at 360× magnification [20]. Horizontal bar indicates 50 μm distance in the plane of the picture.

Figure 9.5 shows that AlGaN-related emission comes from the double hole region only. Also the BL comes from the interface region, which indicates donor and acceptor accumulation at the interface. The spot-mode CL data indicate also that the interface region is strained and contains a large density of defects. For example, the 'edge' GaN emission is strongly broadened and red-shifted as compared with its spectral position in high-quality thick GaN films (see figure 9.4).

The CL study, briefly outlined above, clearly supports the second mechanism of stress relaxation at the LT GaN buffer. The CL emission from the interface region contains contribution of both GaN- and AlGaN-related emission bands.

Moreover, the depth-profiling CL investigations showed a gradual improvement of the sample quality with increasing distance from the interface. A granular microstructure of the LT GaN buffer results first in a strong fluctuation of the CL intensity at the GaN 'edge' CL. With increasing distance from the interface the CL intensity fluctuations become less pronounced and finally a homogeneous emission is observed [20]. A very pronounced acceptor-related BL, observed from the interface region of the structure, indicates an enhanced doping at the interface region. Oxygen is the most likely donor species in this region of the sample [21]. The origin of acceptor centres is less clear. Carbon is certainly one of the potential candidates for acceptor contamination at the interface region. Intrinsic acceptor-like centres (vacancies) should also be abundant in the highly-defected interface region of the structures.

9.3 Freestanding GaN layers for GaN epitaxy

Whereas GaAs-based electronic and opto-electronic devices are grown on high-quality GaAs substrate, most GaN-based devices are grown on sapphire, with surprisingly large lattice mismatch to wide band gap nitrides. Not surprisingly the major obstacle for further improvement of GaN-based devices is lack of lattice-matched substrate materials.

The most suitable substrate is of course bulk GaN. Progress in developing bulk GaN substrates is very encouraging, but rather slow [27]. In consequence of these difficulties three approaches are at present intensively tested. The first one is based on development of the technology of free standing GaN layers [28]. One such approach is described below. The second one is based on introduction of buffer layers on top of sapphire, which help to initialize GaN growth. This approach was already described in this chapter. The third approach is based on introduction of so-called universal substrates and the following lateral growth of GaN with a reduced dislocation density [29]. This approach is extensively discussed in several reviews and will not be described here.

9.3.1 Freestanding GaN layers—an alternative approach

ZnO films of a relatively good quality can be grown by atomic layer deposition using sapphire or quartz glass as substrate materials [30]. Figure 9.6 shows x-rays diffraction patterns in a broad angular range for the layers grown on a glass. X-ray examination indicated a strong preferred orientation of the grains, with the tendency for growth in the 0001 direction. This manifested as a relatively high intensity of 00*l* reflections.

Moreover, AFM shows very flat surfaces of the ZnO films, with mean roughness fluctuations (root mean square, RMS) of 0.34–0.38 nm from up to 100 μm^2 area for films grown on a glass substrate. ZnO films show a granular microstructure, with grains of 10–50 nm size for ZnO/glass. Larger grains of 300–500 nm size are seen for films grown on sapphire. In figure 9.7 we show an AFM image obtained for a ZnO/glass layer grown using extremely simple precursors [30].

High flatness of ZnO/glass films allows them to be used as substrates for GaN epitaxy [31]. The project requires an application of a low growth temperature technique, since ZnO is not resistant to ammonia treatment when growth temperatures are higher than 650 °C. Remote plasma-enhanced laser-induced chemical vapour deposition (LCVD) was used with high-power excimer laser pulses, applied parallel to the film surfaces. This allowed reduction of a growth temperature below a required value, and growth of relatively good quality GaN epilayers.

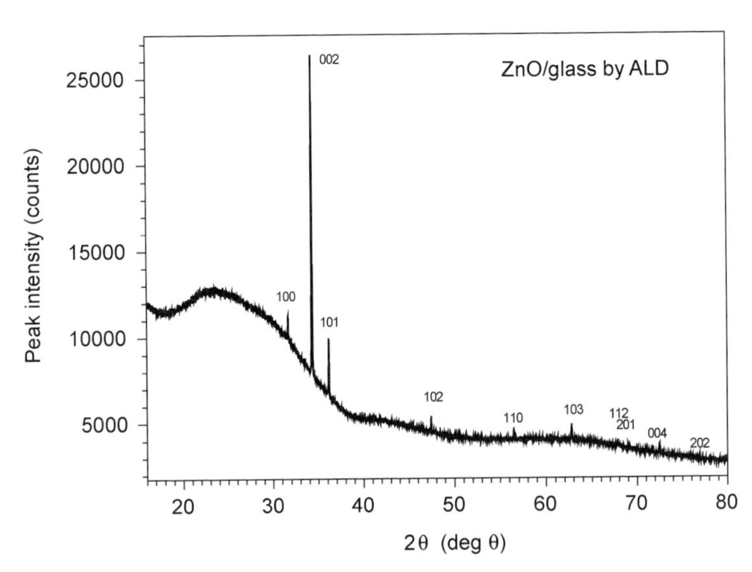

Figure 9.6. X-ray diffraction pattern of ZnO/glass layer grown by atomic layer deposition (after K Kopalko *et al*, to be published).

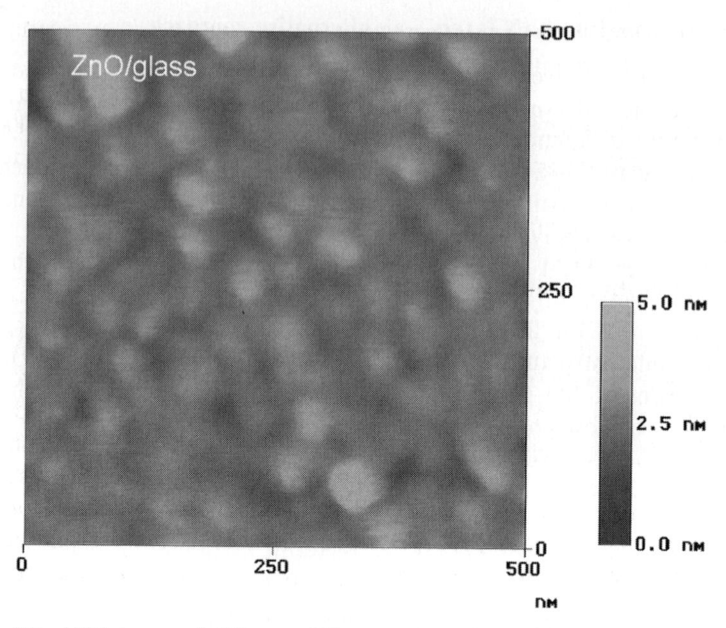

Figure 9.7. AFM image of 500 nm × 500 nm region of the ZnO/glass layer grown by atomic layer deposition (after K Kopalko *et al*, unpublished).

Preferential c-axis orientation of polycrystalline ZnO/glass films helps to initialize correct nucleation of GaN layer. The so-obtained GaN films were polycrystalline, but have relatively good structural and electrical properties [31].

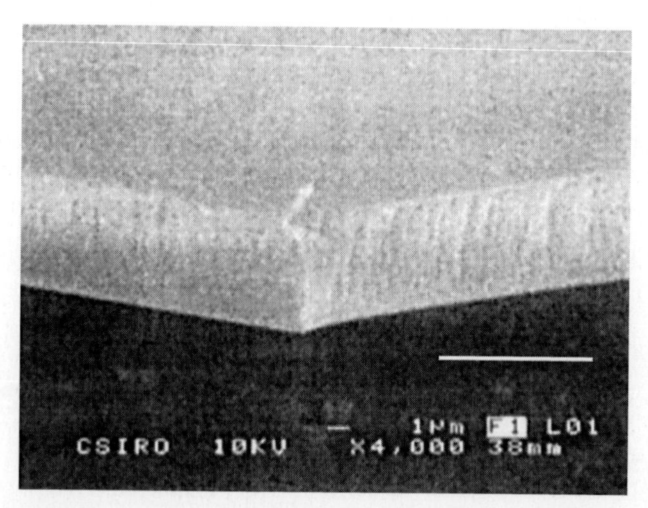

Figure 9.8. SEM image of a free-standing GaN epilayer grown by LCVD method on ZnO/glass substrate [31]. Horizontal bar indicates 10 μm distance in the plane of the picture.

Glass and ZnO layer can easily be etched away [31]. Then, freestanding layers of GaN can be obtained, as is shown in figure 9.8.

The so-obtained freestanding GaN layer is polycrystalline, but can easily be recrystallized by low-temperature (below 570 °C) annealing. A crystalline structure of these films was verified with x-ray diffraction technique [31].

9.4 GaN-based laser diodes

The first commercial GaN-based laser diodes (LDs) (with InGaN/GaN active regions—InGaN quantum well and GaN barrier) were introduced about 4–5 years ago [32, 33]. These were the first commercialized semiconductor-based LDs emitting in a blue-violet spectral region. The market for such laser diodes is huge and there are predictions of their widespread applications in memory devices, DVD players, multi-colour projectors etc.

Unfortunately, the progress in this field is very slow. GaN-based LDs are still very expensive, have low output power, and unfortunately have relatively short lifetimes.

Most of the stability problems, high device costs, and also relatively low LD performance relate to a lack of lattice-matched substrates. Available LDs are grown on a lattice-mismatched sapphire, which results in a high dislocation density in these structures, despite all growth related tricks. This is a major obstacle, since dislocations act as efficient centres of non-radiative recombination in most of the semiconductors, including nitrides [2]. Thus, formally GaN should not be a suitable material for opto-electronic applications, which of course is not the case.

Dislocation-related detrimental effects are partly reduced by the large magnitude of potential fluctuations present in InGaN quantum wells. The resulting short diffusion lengths of carriers and of excitons limit the role of dislocations as centres of non-radiative recombination [1, 2], since most free carriers and excitons will be trapped and will not approach dislocations to decay there non-radiatively. This is why nitrides are so successfully used in opto-electronics for light emitting diodes.

Unfortunately, at high excitation density, which is required to obtain a stimulated emission, potential fluctuations present in QW planes can be at least partly screened by free carriers. Then, diffusion lengths of carriers and excitons should increase, due to a partial screening of potential fluctuations. Dislocations become then more effective as centres of a non-radiative recombination. This explains difficulties in achieving laser emission, despite massive production of GaN-based light emitting diodes (LEDs).

The density of dislocations can be reduced either by introducing special growth tricks or by introducing lattice-matched substrates. The first approach is now very well documented. It was demonstrated that the density of dislocations can be lowered by two to three orders in magnitude when using the epitaxial lateral overgrowth (ELO) technique (see [34] for the explanation). Use of the ELO and related growth methods was essential for achieving a stimulated emission [32, 33].

Unfortunately, use of the so-called universal substrates is not the best solution. The so-obtained GaN-based LDs still contain relatively large dislocation densities of about 10^6–$10^7\,cm^{-2}$. Moreover, the ELO process considerably increases costs of the LD devices and limits width of a laser cavity, since width of overgrowth regions with an improved morphology is limited. In the following part we will concentrate on possibilities of homo-epitaxial LDs, i.e. on advantages of using GaN substrates for growth of LDs.

9.4.1 Homo-epitaxial laser diodes

Use of lattice-matched substrate, in particular of bulk GaN, allows one to reduce density of dislocations. In fact, homo-epitaxial LDs contain about $10^2\,cm^{-2}$ or even fewer dislocations, i.e. about 10^4 to 10^5 times less, as compared with hetero-epitaxial LD structures [34, 35]. At such low density of dislocations, considering the small size of active parts of the laser cavities, most of the so-obtained homo-epitaxial LD structures are statistically free of dislocations.

This fact attracts a lot of interest in the developing homo-epitaxial LD structures. Already the first attempts were successful. Use of bulk GaN templates enabled to achieve laser action of homo-epitaxial LDs under carrier injection conditions [35]. The homo-epitaxial LD structures show superior properties. For example, laser action at record low threshold powers under optical pumping was reported [36].

Homo-epitaxial structures studied under optical pumping were separate confinement heterostructure device (SCHD) LDs. These structures were undoped. Their GaN cap layer was made thinner, to reduce unwanted light absorption. The active region of the structure consisted of five $In_{0.09}Ga_{0.91}N/In_{0.01}Ga_{0.99}N$ QWs. This active part of the LD was embedded between two 0.1 μm thick GaN wave-guiding layers and two cladding layers. The lower cladding layer consisted of a $GaN/Al_{0.15}Ga_{0.85}N$ superlattice (2.5 nm/2.5 nm) with 120 repetitions. The upper cladding layer was in the form of 0.36 μm thick $Al_{0.08}Ga_{0.92}N$, covered with a thin GaN cap layer. LD cavity length was $L = 300$, 500, 800 or 1000 μm [36].

In figure 9.9 we show the threshold dependence of a stimulated emission under optical pumping achieved for the homo-epitaxial LDs. Room temperature results are shown taken for the LD structure with a 300 μm long cavity. Sharp laser modes (with 0.25 nm width) are observed at pumping densities above the threshold power. At these conditions laser emission is about 100% polarized (figure 9.10), with the polarization plane perpendicular to the active layer plane, i.e. the TE cavity modes are excited.

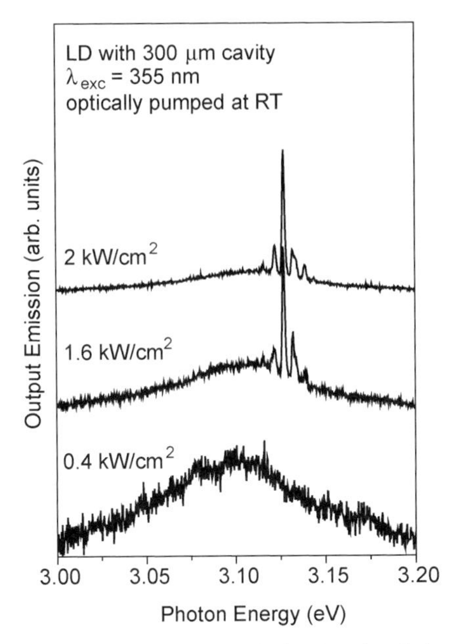

Figure 9.9. Room temperature spontaneous and stimulated emission from homo-epitaxial LD structure with 300 μm laser cavity [36].

The threshold powers under optical pumping and at room temperature are record low. These threshold powers are 5.8 kW/cm^2 for the LD with 300 μm cavity, 5.6 kW/cm^2 for the LD with 500 μm cavity, 2.5 kW/cm^2 for the LD with 800 μm cavity, and 2.4 kW/cm^2 for the LD with 1000 μm cavity [36]. They are by factor of 5–6 lower than the lowest threshold

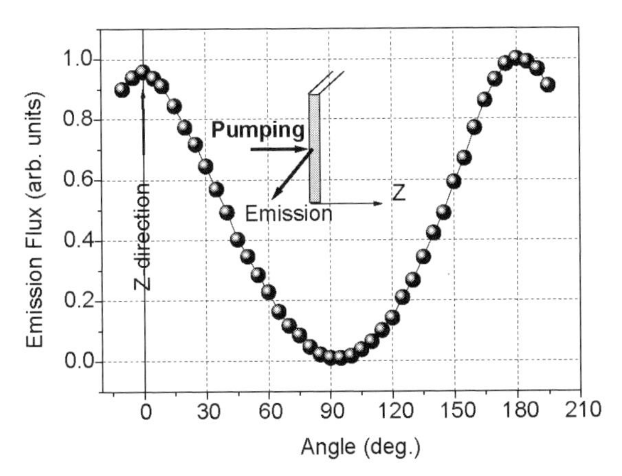

Figure 9.10. Room temperature linear polarization of a stimulated emission from LD structure with 300 μm laser cavity (V Yu Ivanov *et al*, unpublished results).

powers reported in the literature for LDs grown on lattice-mismatch substrates (see references given in [36]).

Typical threshold powers for optically pumped stimulated emission of hetero-epitaxial LD structures are in the order of MW/cm^2 [37], i.e. these threshold powers are one to two orders of magnitude larger than those observed for homo-epitaxial LDs. Such a difference in threshold powers indicates that elimination of most (all) dislocations is essential.

Moreover, use of bulk GaN as a substrate means that the laser mirrors can be simply cleaved, which considerably improves their quality. Also the internal wave-guide losses are low in the case of dislocation-free LD structures.

9.4.2 Electron beam pumping

As already mentioned, potential fluctuations, which lead to strong localization effects, are responsible for efficient light emission from GaN-based LED structures. Unfortunately, these potential fluctuations are assumed to be screened in laser diodes. This we tested by studying light emission properties of GaN epilayers and structures under electron beam pumping [38].

Basov *et al* were the first to demonstrate (for liquid xenon) the possibility of generating laser radiation by excitation with an electron beam [39]. The method was then applied to several semiconductor-based heterostructures. Blue-colour stimulated emission was induced by e-beam for ZnCdSe/ZnSe heterostructures [40]. It was demonstrated that the method has important advantages. The most important is that e-beam pumped lasers can avoid limitations of p-type doping of ZnSe [40]. Even though doping of the active layer of laser reduces the threshold current, doping is essential only in the case of short non-radiative lifetimes in the device [41]. Otherwise, undoped structures can be used in e-beam pumped structures.

Record low threshold currents of laser action under electron beam pumping indicate also that limitations related to contact technology can be avoided [42]. Threshold currents as low as $5\,A/cm^2$ (for ZnSe), $12\,A/cm^2$ (for ZnSe/ZnSSe superlattice) [43] and $3\,A/cm^2$ [44] were reported at e-beam excitation conditions. Achievement of such low currents is important, since high excitation density often results in shortening of the device lifetime. Lowering of the threshold current can hopefully help to eliminate some of degradation processes encountered in ZnSe-based laser devices [45]. A better insight to the origin of the degradation is also expected. In fact, the method turned out to be very useful in studying degradation mechanisms in LD structures [46, 47].

Importantly, laser action under electron beam pumping can be achieved not only for advanced laser structures, such as VCSEL structures [48], but also for simple structures covered with metal mirrors [49], from large area single crystals [50]. Laser modes were reported only in the former case [50].

9.4.2.1 Electron beam pumping of homo-epitaxial GaN structures

The important advantage of experiments under electron beam pumping is the possibility of in-depth and in-plane characterization of light emission properties. In-depth properties can be studying by measuring light emission properties at different e-beam energies. In turn in-plane instabilities of the emission can be determined by scanning excitation through the surface of the structure.

These approaches enable testing of the present models of light emission properties of GaN-based LDs. We hope to estimate the role of strong potential fluctuations present in the active region of the devices, evaluate threshold powers upon electron beam pumping, and also to visualize inter-links between different regions of a complicated, multi-layer LD structure. The results of relevant CL experiments are discussed below.

Even though homo-epitaxial GaN samples have the highest structural quality achievable at present, they still contain dislocations. Their concentration is luckily low and is typically below $10^4\,cm^{-2}$. The results discussed in [38] were taken for two series (numbered 511 and 534) of homo-epitaxial quantum well (QW) structures. Each of the structures contained an active region consisting of 20 InGaN QWs embedded between GaN barriers. The indium fraction in QWs was about 2% in the 511 series and 2.8–3% in the 534 series. The structures had a different width of QWs and of barrier regions—2.8 nm (QW)/5.3 nm (barrier) for the 511 series and 4 nm (QW)/ 8 nm (barrier) in the 534 series. Growth temperature was 790 °C for the 511 series and 785 °C for the 534 series. Structures were grown on GaN bulk substrate covered with GaN buffer layer (1 μm thick in the 534 series and 0.55 μm thick in the 511 series). Samples varying in structural quality were selected to get better insight into the origin of the observed emission instabilities.

CL was excited by an electron beam set normal to the surface of the films, i.e. along the growth direction, and was collected by a cylindrical mirror mounted over the structure. Most of the CL experiments discussed below were performed at relatively low magnification of about 2000 times.

First the optimal conditions for the detection of the edge CL emission were determined, by selecting an accelerating voltage for which primary electrons excite the most intensive CL emission from the active region of the structures. Then we studied the dependence of this CL emission on excitation density by varying current density of primary electrons.

For three of the five samples studied we observed threshold-like dependence on the e-beam density (figures 9.11–9.13), the one commonly observed for the stimulated emission in laser structures.

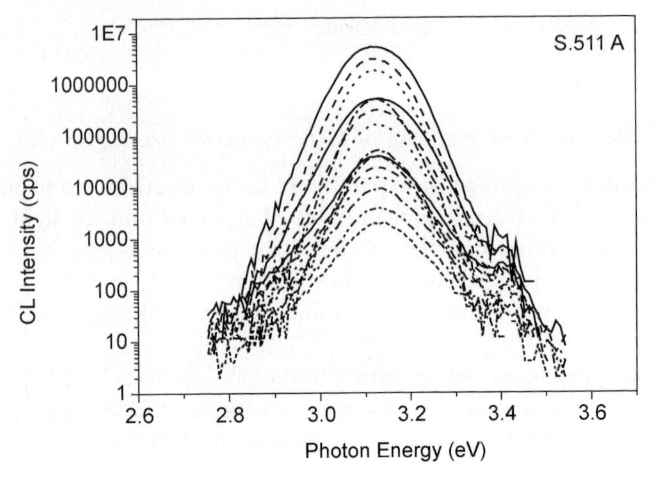

Figure 9.11. Dependence of the edge CL intensity on density of excitation current for the homo-epitaxial structure labelled 511 A [38].

As already mentioned, three of the five samples studied show threshold-like dependences on the e-beam current density. For these structures, after a slow initial rise, the CL intensity increases very fast at larger e-beam densities. This property of the CL emission we relate to the observation of the stimulated emission from the homo-epitaxial structures studied. The relevant results of CL experiments are shown in figures 9.11–9.13.

Laser modes and their strong polarization should be observed to prove that laser action is induced. Unfortunately, we could not resolve laser modes

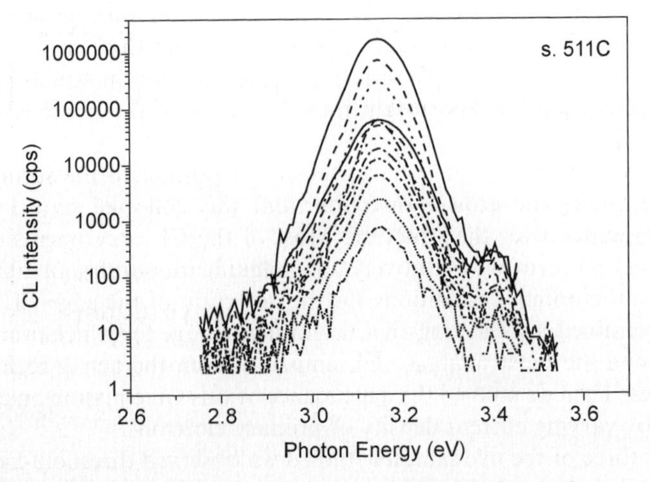

Figure 9.12. Dependence of the edge CL intensity on density of excitation current for the homo-epitaxial structure labelled 511 C [38].

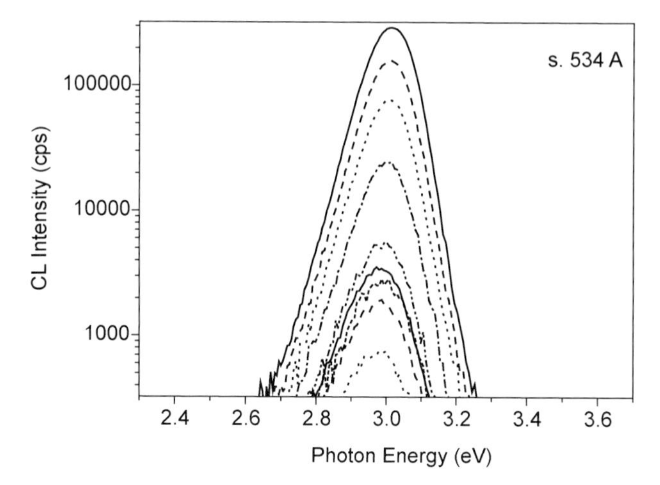

Figure 9.13. Dependence of the edge CL intensity on density of excitation current for the homo-epitaxial structure labelled 534 A [38].

in the CL spectra, for the CL excited with e-beam current above the threshold current density. This was due to large sizes of the samples used, lack of mirrors in the emission direction, the irregular shape of the samples, but also due to a limited resolution of the CL set up.

The most intensive CL emission, with the smallest threshold density, was observed for samples 511 A and 534 A, as shown in figures 9.14 and 9.15. In the latter figure we compare threshold densities observed for the different structures showing a stimulated emission.

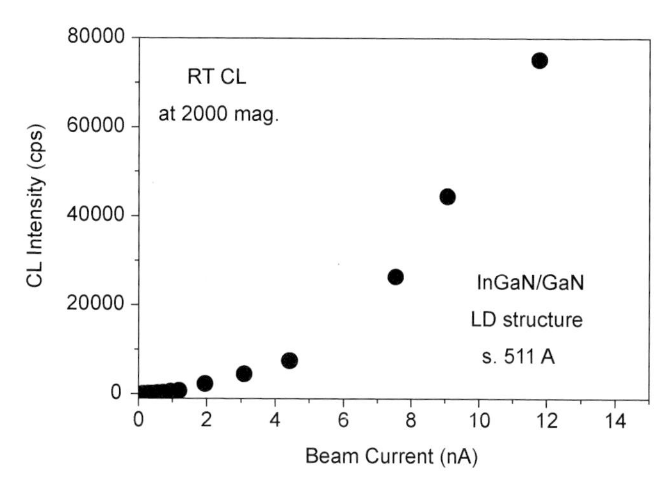

Figure 9.14. Dependence of the edge CL intensity on density of excitation current for structures 511 A [38].

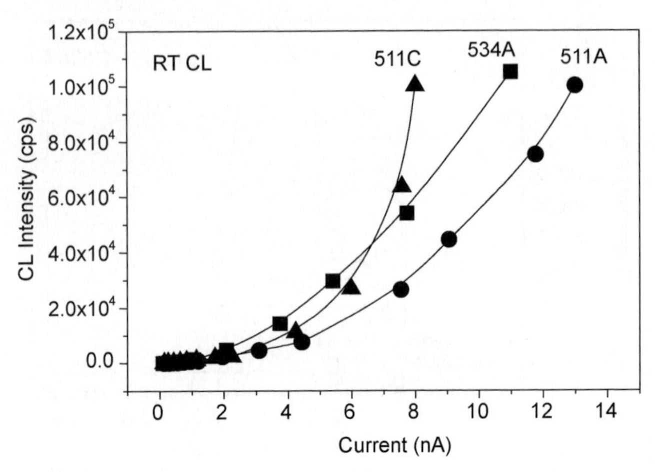

Figure 9.15. Summary of dependence of CL intensity on density of excitation current [38].

We can evaluate threshold density of the e-beam current taking, as the e-beam radius, the radius of a cloud of primary and secondary electrons. This radius is fairly small since diffusion length of carriers and excitons is limited by strong localization effects. We estimate this radius to be in the range of 50 nm.

The estimated threshold current densities of the e-beam are in the range of 100 A/cm^2, i.e. these threshold densities are fairly low, as compared with those necessary to achieved laser action under carrier injection conditions.

For the 534 A sample we observed fairly weak, but large scale, variations of the threshold density. Due to a large homogeneity of the CL emission, dependence of the CL on an electron beam density is similar, when measured at different areas of the film, as is shown in figure 9.16. The threshold was measured from two different areas on the sample at fairly low (2000) magnification, confirming good large-scale homogeneity of the CL in the structures studied.

It was reported that efficiency of light emission in GaN and in InGaN epilayers and their QW structures can be significantly enhanced upon n-type doping [1, 51, 52]. The effect was tentatively related to screening of piezoelectric fields, improvement of morphology (smoother interfaces) and to saturation of deep dislocation-related centres, expected in heavily doped samples. In consequence, at high doping levels or at high excitation densities PL or CL should be fairly in-plane homogeneous and emission fluctuations should be screened.

The above model could be tested with CL by studying spot dependence of threshold for a stimulated emission (figure 9.16), sample morphology

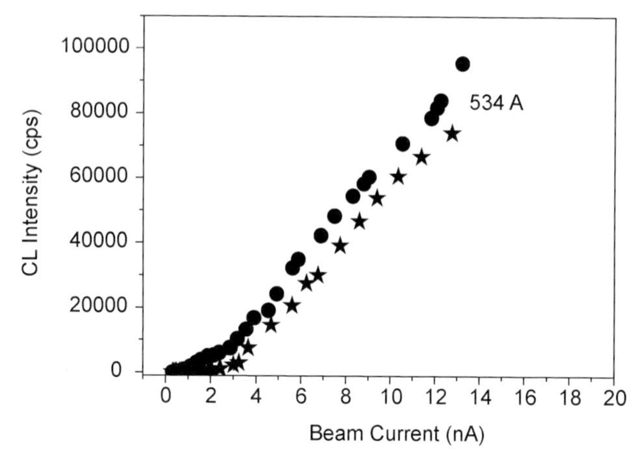

Figure 9.16. Spot dependence of a threshold for the stimulated emission for QW structure 534 A [38].

(figure 9.17) and in-plane fluctuations of the emission with their correlation to details of a micro-structure.

After selecting structures showing a stimulated emission, their morphology was studied with SEM and scanning CL, taking images from the same regions of the sample. This allowed us to estimate relations between morphology of the structures and in-plane fluctuations of the QW emission. The relevant results of the scanning CL experiments are shown in figures 9.18–9.21.

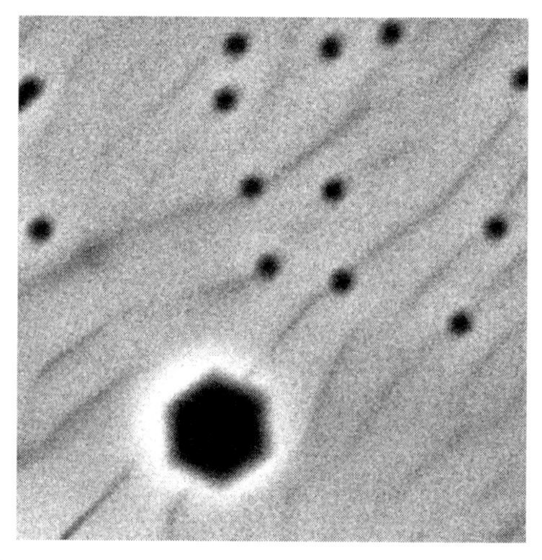

Figure 9.17. Morphology of 534 A QW structure showing the lowest threshold for the stimulated emission under electron beam pumping [38].

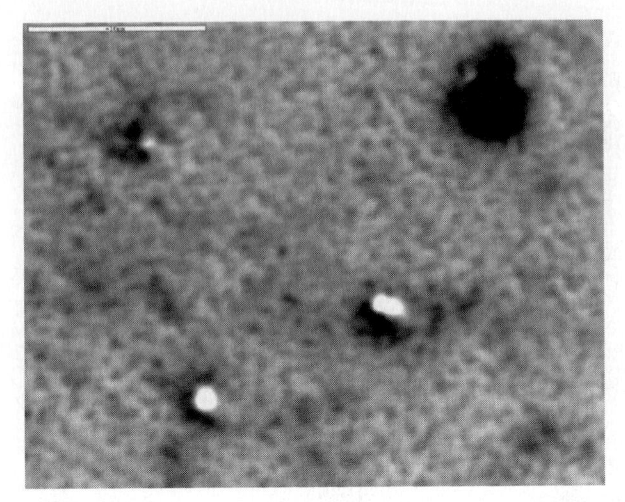

Figure 9.18. SEM images of structure 511 A measured at 10 kV and at 10 000 magnification [38].

A clear correlation between CL properties and structural quality of the structures could be observed. The samples showing the lowest threshold densities showed the best structural quality. Density of pin-holes and of dislocations was low. Atomic size growth step were observed as bright regions in CL images, likely due to decoration of growth steps with impurities.

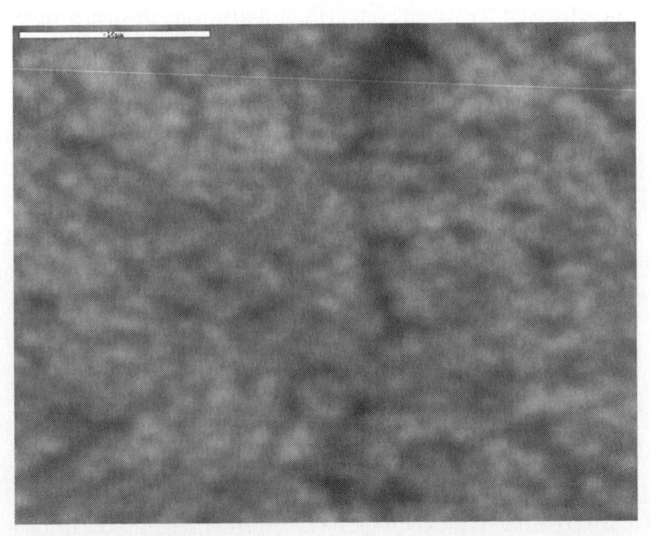

Figure 9.19. CL image of structure 511 A measured at 10 kV and at 10 000 magnification. For CL scan detection was set at the edge CL, 2.66 nA (below edge for lasing) [38].

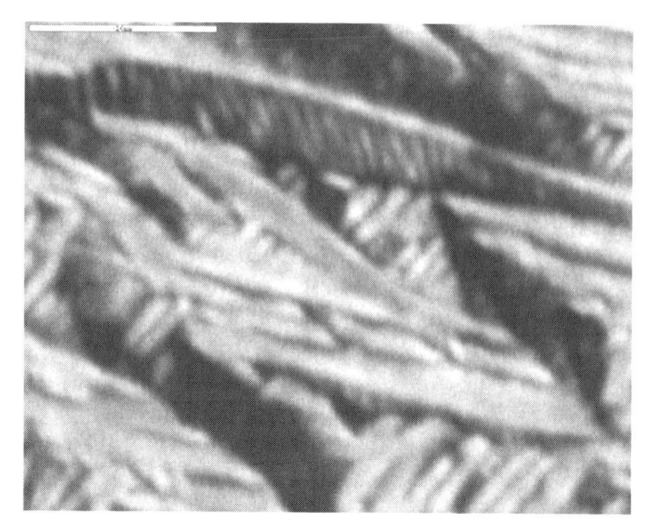

Figure 9.20. CL image of structure 511 C measured at 10 kV and at 6000 magnification. For CL scan detection was set at the edge CL either below or above edge for lasing [38].

These CL instabilities were observed even at excitation densities larger than those required for the stimulated emission, indicating that even at high excitation densities potential fluctuations are not fully screened in InGaN QW planes.

Figure 9.20 shows in-plane fluctuations of the QW stimulated emission on the example of the 511 C structure. Strong CL intensity fluctuations are observed, which directly reflect growth details. Atomic size growth steps,

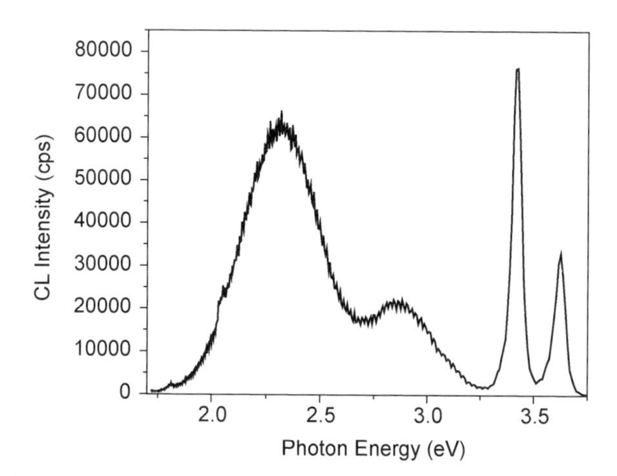

Figure 9.21. CL spectrum of homo-epitaxial LD structure measured at 10 kV accelerating voltage, 17.6 nA primary current density and at 2000 magnification (after M Godlewski *et al*, to be published).

also observed in the SEM images, are even more pronounced in the scanning CL study. CL fluctuates in the intensity for the e-beam excitation below but also above the threshold current density. The observed fluctuations reflect microstructure of the samples and not indium fraction fluctuations, which typically are of a much smaller scale.

Further studied were performed for a series of homo-epitaxial LD samples with identical layer structure, but cleaved with the different LD cavity lengths, of $L = 300$, 500, 800 and 1000 μm. The cavities were formed by cleaving, which is a simple procedure in homo-epitaxial structures and results in atomically flat surfaces. The so-obtained high reflection and output mirrors had a reflection coefficient of $R = 16\%$. These were LD structures used for optical pumping, discussed above. The results will be presented for the structure with 300 μm laser cavity.

Relatively bright emission is observed under the e-beam pumping, as is shown in figure 9.21. At 10 kV accelerating voltage only the upper part of the device is excited, as is explained in [53], in which basics of the depth-profiling CL are outlined. Bright emission from the upper GaN cap layer (likely with some contribution of the underlying GaN wave guiding layer) is observed together with edge emission from the AlGaAs upper cladding layer. Surprisingly strong yellow CL emission is detected, together with another DAP emission in a blue spectral region. The latter we attribute to the BL emission, already discussed in this chapter.

A better insight into the origin of the CL emissions can be obtained from depth-profiling CL measurements, as shown in figures 9.22(a–d). These data were collected at constant current conditions, at different accelerating voltages, indicated in figures 9.22(a–d).

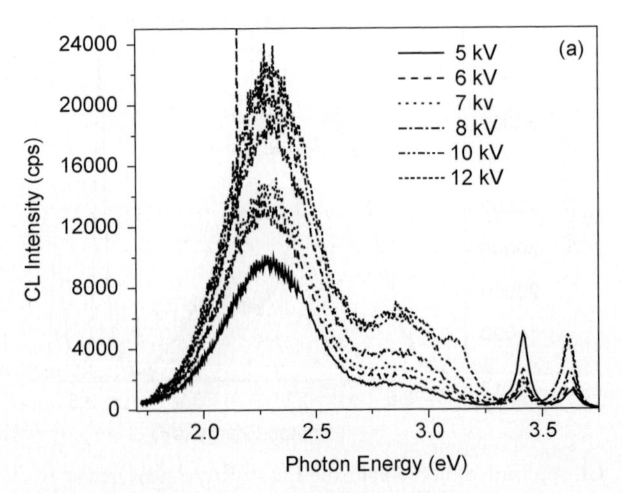

Figure 9.22. (a) Depth profiling CL data taken at low accelerating voltages (after M Godlewski *et al*, to be published).

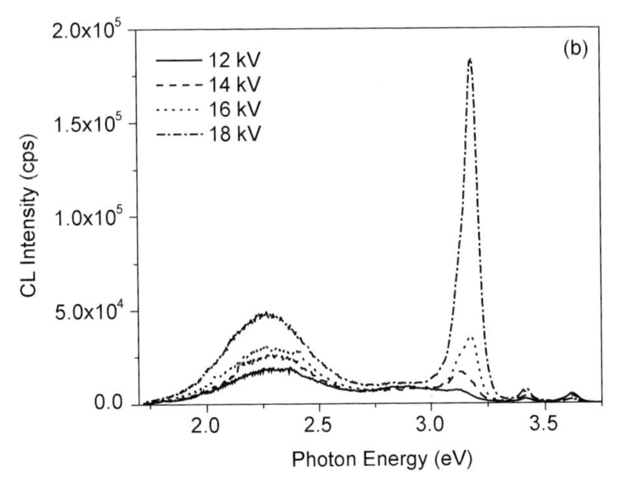

Figure 9.22. (b) Depth-profiling CL spectra taken at larger accelerating voltages (after M Godlewski *et al*, to be published).

Depth-profiling experiments indicate that GaN excitonic emission mostly comes from the upper cap layer. It decreases in intensity with increasing accelerating voltage, i.e. when primary electrons penetrate deeper into the structure. Similar emission is also excited from the GaN wave guiding layer, but is weaker if excited there.

Two GaN-related DAP emissions also come from the wave guiding region of the device. Emission from the QW active region of the LD structure

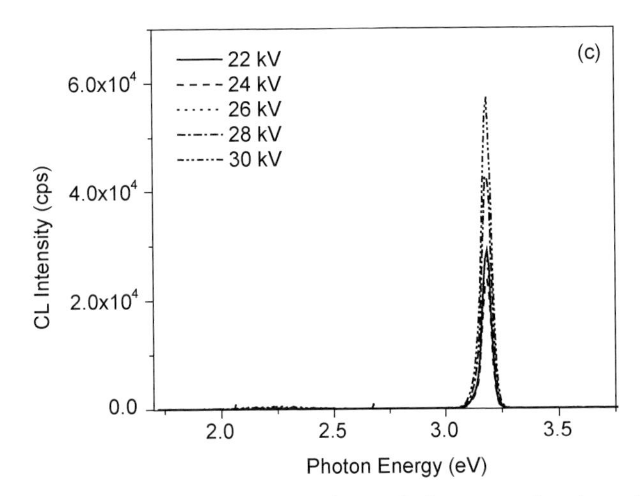

Figure 9.22. (c) Depth-profiling CL spectra taken at the largest accelerating voltages (after M Godlewski *et al*, to be published).

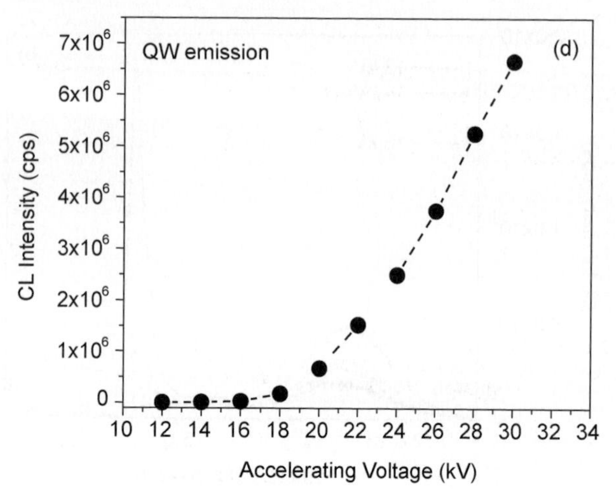

Figure 9.22. (d) Summary of the depth-profiling CL experiments (after M Godlewski *et al*, to be published).

is not seen, when the upper layers of the device are excited, which means that energy transfer, photon recycling or carrier diffusion from the upper layers to the QW region of the device are inefficient.

Experiments at further increased accelerating voltages confirm that the GaN edge emission comes from both the upper cap layer and the GaN wave guiding layer. Parasitic yellow GaN emission is also observed from both these GaN layers and is the brightest from the wave guiding layer and at the GaN/AlGaN interface.

A dramatic increase of the laser emission is observed for accelerating voltage exceeding 18 kV, as shown in figures 9.22(c,d). This emission dominates and is so strong that other CL emissions are no longer detected at the same experimental settings.

Once conditions (accelerating voltage) for the excitation were optimized, threshold power for a laser emission could be determined. A clear threshold dependence of the emission is observed. First, the QW emission increases nonlinearly, but then, for currents above about 10 nA, this increase is extremely rapid and is described by a power low dependence (figures 9.23 and 9.24).

For excitation below a threshold current density CL consists of excitonic emission and a weak yellow emission (figure 9.25(a)). GaN and AlGaN emission are not seen. They are much weaker than those coming from the active part of the device. InGaN CL, at such excitation conditions, shows in-plane fluctuations, with some dark areas and indications of growth steps (figure 9.26). YL is fairly in-plane homogeneous and is only slightly enhanced in the region of a dark spot.

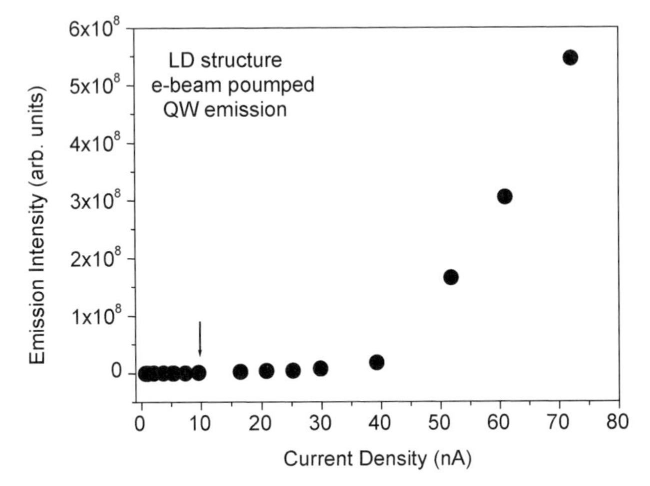

Figure 9.23. Threshold current dependence for a stimulated emission (after M Godlewski *et al*, to be published).

For the e-beam excitation above the threshold current YL is no longer observed (figure 9.25(b)). CL is dominated by a stimulated emission. In-plane contrasts of the CL are still observed, or even are enhanced. Growth steps are clearly resolved, together with totally dark area, as shown in figures 9.27–9.30 for the scanning CL spectra taken at four different regions.

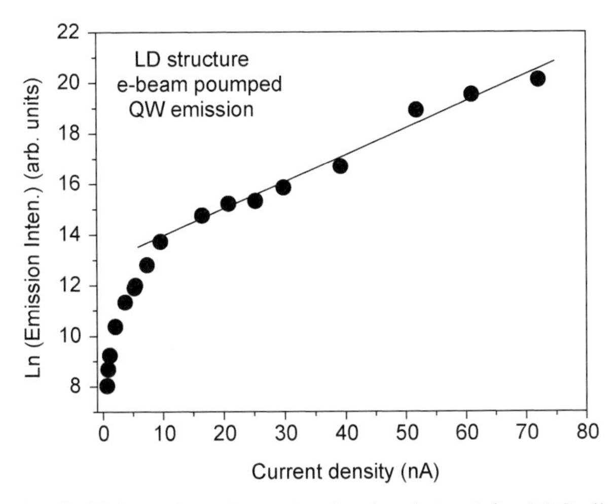

Figure 9.24. Threshold dependence for a stimulated emission (after M Godlewski *et al*, to be published).

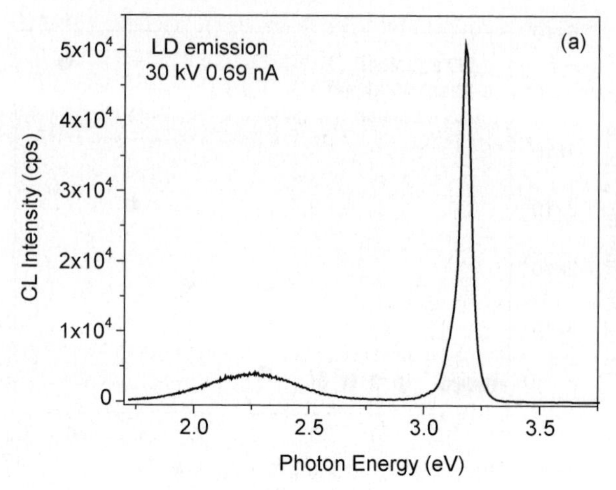

Figure 9.25. (a) Emission from the active region of the LD structure observed for excitation below the threshold for a stimulated emission (after M Godlewski, *et al*, to be published).

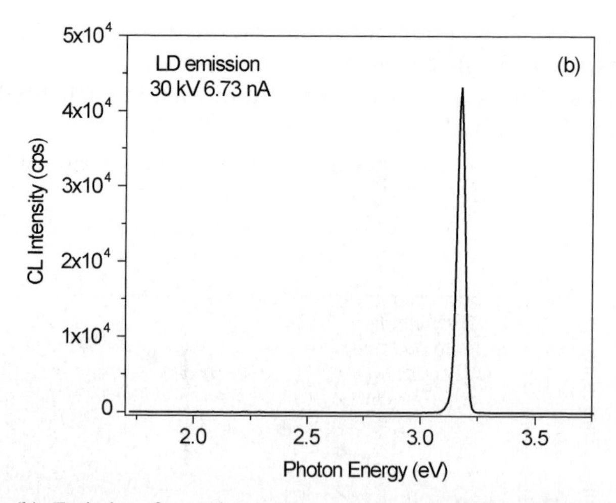

Figure 9.25. (b) Emission from the active region of the LD structure observed for excitation above the threshold for a stimulated emission (after M Godlewski, *et al*, to be published).

The magnitude of the observed CL intensity fluctuations can be determined from spot-mode CL investigations, as shown in figure 9.32. These measurements indicate that nearly no edge emission come from the darkest regions in figures 9.27–9.30. Spot mode CL measurements (figure 9.31) indicate not only that large intensity fluctuations are present in the

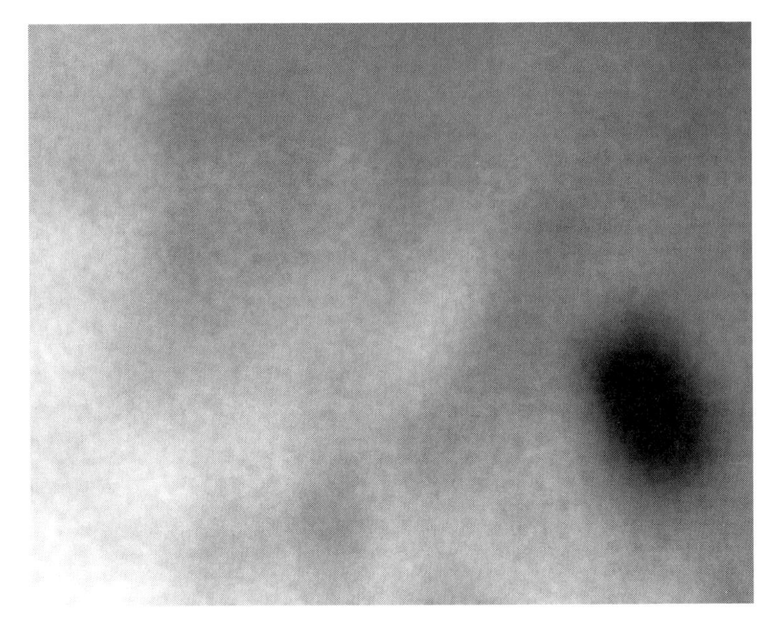

Figure 9.26. In-plane fluctuations of the InGaN QW emission (below threshold excitation) (after M Godlewski *et al*, to be published).

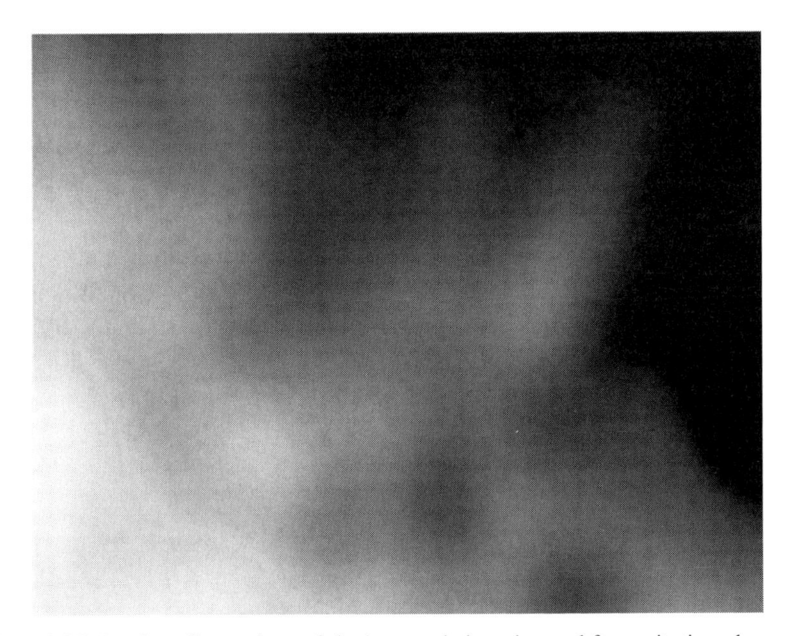

Figure 9.27. In-plane fluctuations of the laser emission observed for excitation above the threshold value (after M Godlewski *et al*, to be published).

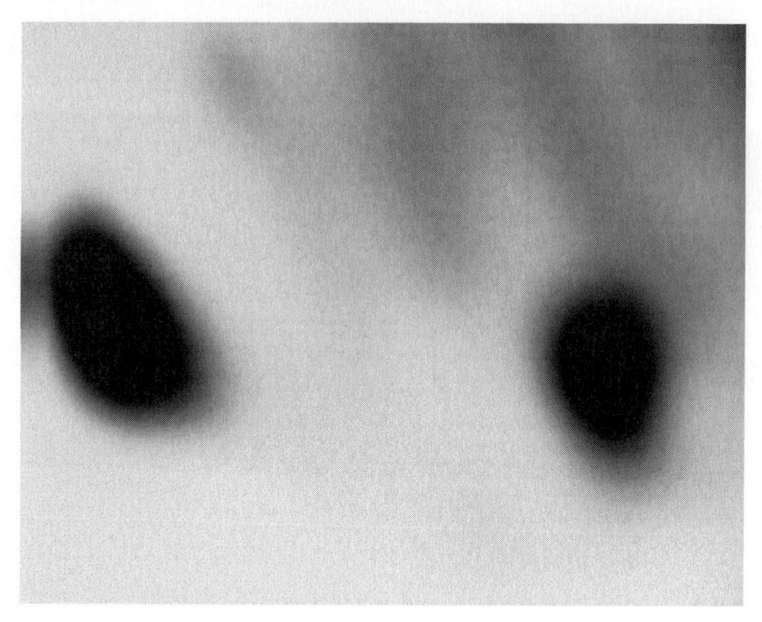

Figure 9.28. In-plane fluctuations of the laser emission observed for excitation above the threshold value (after M Godlewski *et al*, to be published).

Figure 9.29. In-plane fluctuations of the laser emission observed for excitation above the threshold value (after M Godlewski *et al*, to be published).

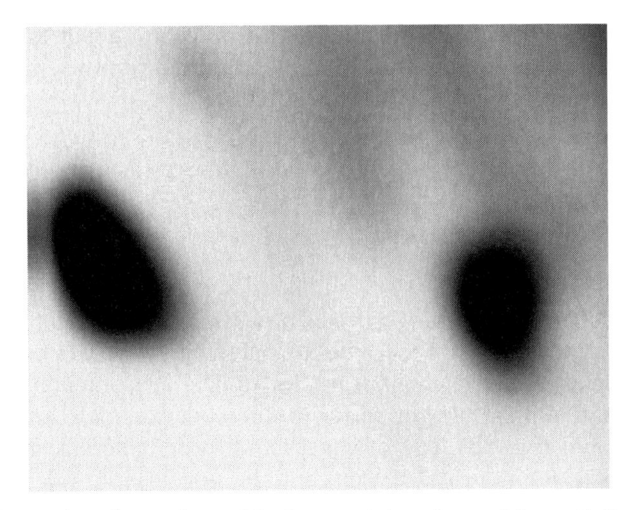

Figure 9.28. In-plane fluctuations of the laser emission observed for excitation above the threshold value (after M Godlewski *et al*, to be published).

LD emission, but also that noticeable spectral position shifts of the emission occur, which may originate from indium fluctuations, various strain conditions, etc.

Summarizing, detailed CL investigations indicate that potential fluctuations, so crucial in LED structures, are still present in LD structures, despite large excitation densities. These measurements confirm advantages of using

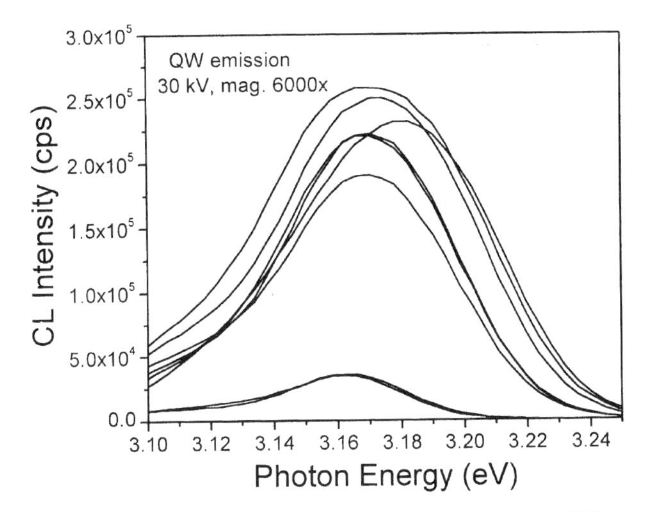

Figure 9.31. Spot mode CL of the LD structure for excitation of the QW emission above the threshold value (after M Godlewski *et al*, to be published).

lattice-matched substrates. The latter results in record low threshold powers for a laser emission upon optical pumping and led to introduction homo-epitaxial LD working under electrical injection.

9.5 GaN doped with rare earth ions

9.5.1 Introduction

The characteristic optical properties of rare earth (RE) elements have led to many important photonic applications, including solid state lasers, optical storage devices, displays, components for optical telecommunication (fibre lasers, optical amplifiers). In most of these devices RE impurities are introduced into various forms of glasses and oxide materials. The range of applications of RE-doped materials could be significantly widened to opto-electronic devices and integrated opto-electronic circuits by introducing REs into semiconductors. In practical semiconductor devices the intra-4*f*-shell emission of REs will be excited by electrons and holes generated by electric field. However, excitation of the emission of REs is a difficult problem to be solved, as energy transfer from primarily excited *e–h* pairs to the 4*f*-shell can be inherently inefficient, as the 4*f*-shell is screened from charge carriers by the outer 5*s* and 5*p* electron shells of lanthanide ions. Competing processes of radiative and non-radiative recombination on other defects and impurities may reduce luminescence efficiency of REs even further.

Erbium-doped semiconductors are particularly interesting as materials for temperature stable optical sources operating at 1.54 μm, i.e. in the minimum loss window of optical fibre communication. Si:Er is most important in this respect, because of the possibility of integration of micro- and opto-electronic circuits using processing technology compatible with the technology of VLSI circuits [54]. In spite, however, of progress in enhancing photoluminescence (PL) of Er^{3+} in silicon, the PL intensity still experiences significant loss at room temperature [55]. Oxygen co-doping is one of the ways for improving the efficiency of PL and its thermal stability [55], but even in O-doped Si:Er the PL efficiency is not high enough for practical utilization in devices.

Favennec *et al* [56] showed that thermal quenching of the Er PL can be reduced in comparison with silicon, in wide bandgap semiconductors. In fact, in erbium-doped GaN the reduced temperature quenching of Er photoluminescence at 1.54 μm was confirmed by a few groups of authors [57–61]. The reasons for the dependence of quenching on the band gap width can be easily understood from figure 9.32, which presents a simple scheme of energy flow in Er-doped silicon. This scheme is also valid for other RE-doped semiconductors.

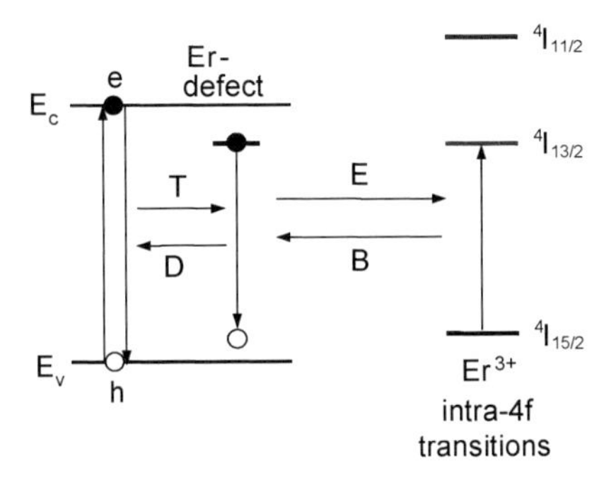

Figure 9.32. Excitation and de-excitation of Er^{3+} PL in silicon.

In the first stage photo-excited electrons and holes form excitons. These excitons can then be trapped at Er-related levels. If trapped, they can non-radiatively transfer their energy to the 4*f*-shell of Er ions in an Auger-like process. At higher temperatures the Er-related traps can be thermally depopulated or the excitation energy can be back transferred from the excited Er^{3+} ion to charge carriers or excitons. Therefore, depth in the forbidden gap of electronic levels mediating transfer of excitation to Er ion is critical for the efficiency of quenching of the Er^{3+} PL.

The recent demonstration of visible and infrared PL as well as electroluminescence (EL) from RE-doped GaN has spurred significant interest in this class of materials for possible applications in opto-electronic devices. Steckl *et al* [62] obtained blue electroluminescence concentrated at 477 nm from Tm-doped GaN layers grown by molecular beam epitaxy (MBE). Erbium impurity was used for generation of green PL as well as EL at 537 and 558 nm in MBE grown GaN [63]. Eu ions were used for obtaining red emission [64]. The first integration of primary colours in GaN:RE EL devices utilized a structure with two stacked layers of GaN:Eu and GaN:Er [65]. Finally, Lee and Steckl [66] showed lateral colour integration in GaN EL films doped with Er and Eu, thus showing potential for practical realization of full colour displays [67]. Recently the possibility of application of GaN doped with rare earth in optical storage devices was also presented [68].

9.5.2 Lattice location studies of RE atoms in GaN

The determination of the lattice location of RE impurities in crystalline lattice is an important step in studies of the local atomic environment and local symmetry of the emitting RE centres. In III–V compounds (arsenides and phosphides) irregular non-substitutional location of RE ions seems to be a rule [69]. Apparently, this is due to the very low solid solubility of REs in these compounds. The only exception from this rule is Yb impurity, which is highly soluble in InP and InP-based alloys. Kozanecki [70] found, using Rutherford backscattering (RBS) in combination with channelling, that Yb atoms locate substitutionally at In sites and solid solubility exceeds $10^{19}\,cm^{-3}$. According to the model of [71] the high solubility is a result of low strain energy due to Yb–In replacement. The ionic radii of Yb^{3+} and In^{3+} are comparable (0.858 and 0.92 Å, respectively) with that of Yb^{3+} a little smaller and therefore strain energy due to substitution is low and does not block that process.

Surprisingly, experiments indicate that for hexagonal GaN the substitutional location of RE ions is a rule, in spite of a big difference of ionic radii of RE ions (>0.9 Å) and Ga^{3+} (0.858 Å). Alves *et al* [72] showed, using RBS in combination with channelling of helium ions, that Er atoms locate substitutionally already in as-implanted wurtzite GaN for Er doses lower than $10^{15}\,cm^{-2}$. Substitutional fraction of Er atoms increases as a result of thermal annealing, but in the annealed layers the location of Er atoms seems to depend on the presence of co-implanted oxygen. Annealing at 900 °C either reduces slightly the substitutional fraction of Er, or causes small displacement of Er ions from the lattice sites. The best fit of the experimental results was obtained assuming 70% of the Er atoms in regular sites and 30% located randomly. In Er + O implanted GaN annealed at 900 °C Er atoms always occupy substitutional Ga sites. Authors suggested that the presence of O stabilizes the Er in Ga sites most probably through the formation of Er–O complexes [72].

Similar results were obtained for Pr-doped GaN. Monteiro *et al* [73] showed that Pr atoms locate at lattice sites. Thermal stability of Pr atoms at Ga sites was also confirmed by Wahl *et al* [74] with use of channelling of electrons emitted in β decay of the radioactive ^{137}Cs ions. Other RE impurities, such as Eu, were found to locate at sites close to substitutional [75]. Analysis of angular dependence of backscattering yield showed that they are displaced from the lattice positions by 0.2 Å along the $\langle 0001 \rangle$ axis. This result was then confirmed by de Vries *et al* [76] using emission channelling. They found that 65% of Eu atoms sit at Ga sites already in as-implanted samples. Damage removal as a result of thermal annealing increases the substitutional fraction by a few percent, but some displacement ~0.15 Å from the lattice position is still observed, in agreement with the RBS/channelling data of Alves *et al* [72].

De Vries *et al* [77] also measured lattice location of Nd in GaN and the results are quite similar to those for Eu: ~70% of the implanted atoms are located at the Ga sites, and in annealed samples a small displacement perpendicular to the $\langle 0001 \rangle$ crystallographic axis is slightly reduced. However, in all these experiments some uncertainty in the determination of the lattice positions of RE atoms exists due to incomplete recovery of the implanted layers even after high-temperature annealing, particularly for the Er doses exceeding $10^{15}\,cm^{-2}$.

Recently, Citrin *et al* [78] presented evidence, using EXAFS, that concentration of substitutional Er may exceed 10^{21} cm^{-3} with and without comparable concentration of co-doped oxygen. This is in some discrepancy with the results of Alves *et al* [72], who showed that without oxygen the substitutional fraction is lower than in O-co-doped GaN. The reason can be a different way of doping with Er: implantation [72] and during growth [78].

Citrin *et al* [78] have also found that the Er–N bond length is unusually short (2.17 ± 0.2 Å). They also showed using density functional theory calculations that high ionicity of this bonding prevents the formation of any Er–N rich phases the result opposite to Si:Er, where isolated point Er defects are unstable against the formation of Er_3Si_5 metallic phase.

9.5.3 Emission of Er^{3+} ions in GaN

Luminescence of Er^{3+} ions was studied by several groups of authors [57–67]. Two aspects were of interest: (i) emission at 1.54 μm for possible applications of GaN:Er devices in optical communications, and (ii) visible emission for applications in colour displays. In this part of the paper we will focus predominantly on studies of infrared emission.

In general, the PL spectra near 1.54 μm are complex thus suggesting the existence of several different emitting centres. It was shown that the integrated intensity of Er PL changes only slightly between 4 and 300 K, particularly for above band gap excitation energy [59]. Some redistribution of the intensity between individual lines was observed at elevated temperatures, which suggests energy migration between different centres. In recent site-selective PL and PLE experiments, four different Er^{3+} spectra having four different optical excitation bands were detected [79].

It was also found that oxygen co-doping, essential for the enhancement of the Er^{3+} PL intensity in silicon, also influences the PL intensity of Er^{3+} in GaN by orders of magnitude. It was shown that for relatively low Er contents (doses of the order of 10^{14} cm^{-2}) the maximum PL intensity is obtained for oxygen doses an order of magnitude higher. For high Er contents the PL intensity increases by a factor of 2–3 as a result of co-doping with oxygen at concentrations comparable with those of Er.

It seems that for above band gap photo-excitation trap mediated processes are responsible for excitation of Er ions in GaN [79] and also in AlN [80]. This is promising for EL devices in which impact excitation of Er by hot carriers is the main mechanism of generation of luminescence. In fact, it was found that the cross section for impact excitation is on the order of 5×10^{-16} cm^{-2} [59], five orders of magnitude higher than for resonant excitation of Er^{3+} ions in glasses.

The number of the emitting Er centres observed in photoluminescence of Er implanted GaN puts questions about their relationship to the observed location of Er atoms at Ga sites. Perhaps the problem is due to highly differing concentrations of Er ions detected with particular experimental techniques. For lattice location studies of atoms using RBS/channelling large concentrations of Er exceeding $10^{18}\,cm^{-3}$ are required. Similarly, for experiments with channelling of electrons emitted in a process of beta decay of radioactive ions implanted into semiconductors the implant doses are of the order of $10^{13}\,cm^{-2}$ and, consequently, the resulting concentrations exceed $10^{18}\,cm^{-3}$. On the other hand, PL can be easily observed for concentrations of the emitting centres several orders of magnitude lower. Incomplete recrystallization of the implanted layers can be responsible for the existence of a variety of slightly different centres. Therefore, to answer the question about the local structure of the emitting Er centres it is necessary to reduce the number of centres, minimizing the ion dose (less damage and better recrystallization) and avoiding non-uniform broadening of the PL lines.

In wide band gap materials optical resonant excitation of Er ions is possible, as several excited states of RE have energies lower than the band gap energy. The absorption cross section of Er for resonant excitation is in general low—of the order of $10^{-21}\,cm^{-2}$. This is because dipole–dipole transitions between the 4*f* states are parity forbidden (Laporte's rule). Splitting of higher Er energy levels ($^4I_{11/2}$, $^4I_{9/2}$) can be investigated using a tuned laser as the excitation source. It will give information about symmetry of the local crystal field. At resonance only selected centres are excited, thus helping in the identification of Er centres. It also seems that some competing channels of recombination can be switched off, making analysis of the symmetry of Er centres easier.

9.5.4 PL excitation spectroscopy of Er^{3+} in GaN

Samples studied were both heavily and weakly doped GaN layers grown on sapphire. The typical low temperature PL spectra of Er^{3+} are shown in figure 9.33. The spectra were measured at 7 K under different excitation wavelengths: above band-gap (351 nm) and at two below band-gap energies—514 nm corresponding to donor–acceptor transitions and in the near infrared, 984.6 nm corresponding to the direct $^4I_{15/2}$–$^4I_{11/2}$ absorption transition. It seems that in each case the dominant emission comes from different Er centres. The broad PL line at 1539 nm shows no structure for excitation at 514 nm nor for other lines of an Ar laser, indicating that the excitation does not proceed via intra-4*f*-shell transitions of Er^{3+} but via a broad, trap-related absorption band, most probably donor–acceptor pairs.

The temperature dependence of the integrated PL intensity for three different excitation regimes is shown in figure 9.34. It is seen that temperature evolution depends on pump wavelength. Resonant pumping at 984.6 nm to the $^4I_{11/2}$ excited state was found to be an order of magnitude less efficient than with both the above and below band gap light. The temperature

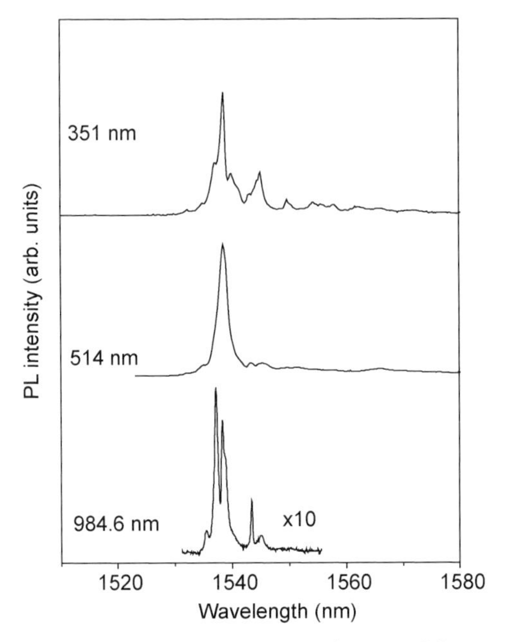

Figure 9.33. PL spectra of Er-implanted GaN at 7 K for three different excitation regimes: above band-gap (top), below band-gap (middle) and intra-4*f* (bottom) [82].

dependence of the PL intensity in all three excitation regimes is different. For the above band gap excitation a slight decrease of the PL intensity is observed already at low temperatures, with a deactivation energy of 3.1 ± 0.2 meV, which is a typical value for excitons bound on donors in GaN. It suggests that excitation of Er ions is mediated by excitons bound on shallow impurity

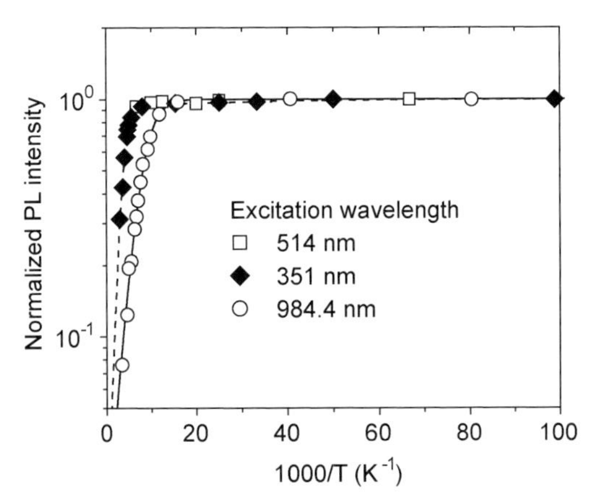

Figure 9.34. Temperature dependence of the integrated PL intensity for different excitation conditions [82].

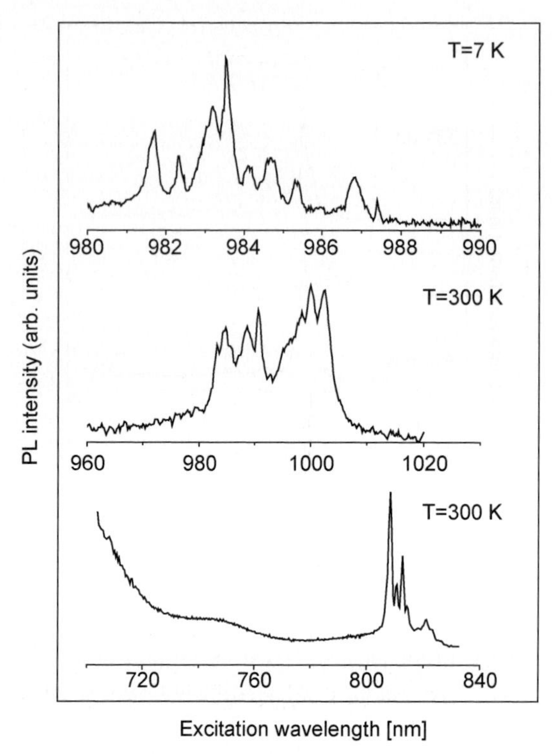

Figure 9.35. PLE spectra of Er^{3+} ions in GaN implanted with a dose of $2 \times 10^{15}\,cm^{-2}$ of 800 KeV Er^{+} ions and annealed at 1000 °C [82].

centres, probably in close vicinity of Er. Above 100 K another quenching process sets in, with a characteristic energy of 95 ± 10 meV. This process is responsible for a 70% loss of the PL intensity at room temperature. The dominant excitation mechanism seems to be energy transfer from recombining D–A pairs, and the deactivation energy is most probably related to the ionization energy of the shallower impurity involved. In contrast, the PL excited in the near band gap range of energies (457–514 nm) is constant up to room temperature suggesting that deeper traps are involved in excitation process. Surprisingly, the centres excited resonantly into the 4*f*-shell at 984.6 nm show the strongest temperature quenching of PL with a deactivation energy of 43 meV.

The reasons for this unusual result seem to be explained in figure 9.35, where photoluminescence excitation (PLE) spectra are presented. Intra-4*f*-shell PL excited selectively via $^4I_{11/2}$ state at 7 K is quite complex and the number of the dominant lines is higher than six indicating that they belong to more than one centre. At 300 K additional PLE lines appear at longer wavelength (middle curve). It may suggest that there is migration of excitation

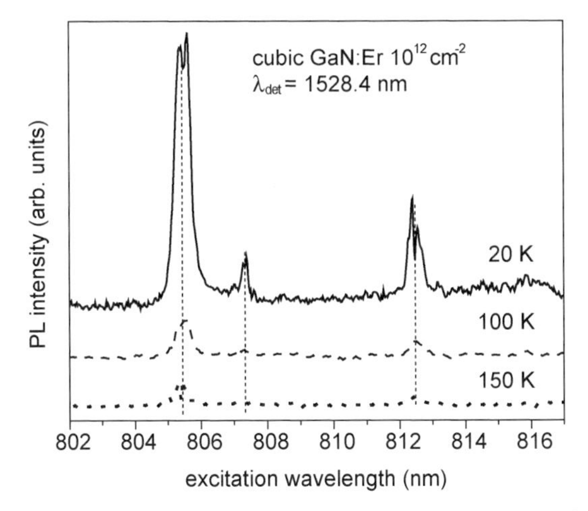

Figure 9.36. Temperature dependence of the PLE spectra for cubic GaN implanted with a low dose of Er ions (after V Glukhanyuk *et al*, to be published).

energy among Er centres at the applied implant dose. As mentioned above the absorption cross section for resonant excitation of Er is low, so at such excitation conditions the centres with the highest concentration will be predominantly excited. The migration of excitation energy among different centres will dominate as the inter-centre distances are the shortest. Migration of energy may also lead to an increased probability of non-radiative de-excitation at other defects, often not related to Er. It may explain the stronger PL quenching observed under direct excitation conditions.

The lowest curve in figure 9.35 shows PLE spectra excited in the red and near-infrared range of wavelength up to 820 nm reveal two different excitation mechanisms. For shorter wavelengths excitation proceeds via a broad absorption band due to deep traps, and near 800 nm absorption lines due to the $^4I_{15/2}$–$^4I_{9/2}$ resonant transitions dominate. The number of lines is more than five, so at least two centres exist for this implant conditions.

9.5.5 Analysis of Er site symmetry in GaN

Photoluminescence excitation spectroscopy makes it possible to distinguish between the numerous, optically active Er^{3+} centres observed in the GaN host and to gain insight into their excitation mechanisms [60, 79, 81, 82]. Due to the large variety of Er centres, however, it was so far impossible to determine the site symmetry and lattice location of Er with use of optical methods.

Identification of the site symmetry of Er implanted into GaN based on the analysis of the Stark splitting of the $^4I_{9/2}$ excited multiplet of Er^{3+} is possible to achieve using high purity material and very low Er implant

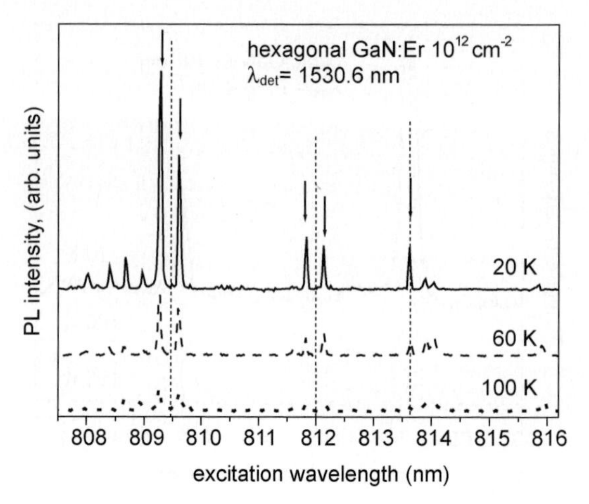

Figure 9.37. Temperature dependence of the PLE spectra for hexagonal GaN implanted with a low dose of Er ions (after V Glukhanyuk *et al*, to be published).

doses (10^{12} cm^{-2}, 600 keV), which ensured that only one dominant Er luminescent centre was produced. The analysis of the crystal field parameters allows one to determine the lattice location of Er in cubic and hexagonal GaN.

In all samples studied one dominant centre is observed in photoluminescence, both under above band-gap as well as under intra-4*f*-shell excitation. The PLE spectra to the $^4I_{9/2}$ manifold of Er^{3+} in cubic (figure 9.36) and hexagonal (figure 9.37) GaN, as measured in samples implanted with the smallest Er dose (10^{12} cm^{-2}) and annealed at 900 °C, are shown in figures 9.36 and 9.37. The measurement temperatures are indicated in the figures.

In cubic symmetry the $J = \frac{9}{2}$ manifold splits into three states, two quartets (Γ_8) and one doublet (Γ_7). In sites of lower symmetry the quartets will be further split and altogether five PLE lines can be expected. As seen in figure 9.38, in cubic GaN a small splitting of the Γ_8 states is observed, indicating a distortion from purely cubic symmetry. In hexagonal GaN:Er the splitting of the quartet states is slightly larger and the Er absorption lines extremely narrow (of the order of 0.05 nm). The PLE lines of the dominant centre are indicated in the figure with arrows. Dashed lines in figure 9.38 mark the expected positions for transitions in purely cubic symmetry, assuming that the distortion is small, i.e. the centre of gravity is taken for the position of the quartet states.

In the first approximation the cubic crystal field potential was taken into account. Assuming that distortions from cubic symmetry are small the energy positions of the Γ_8 states can be taken as centres of gravity of the split lines, as shown in figure 9.38.

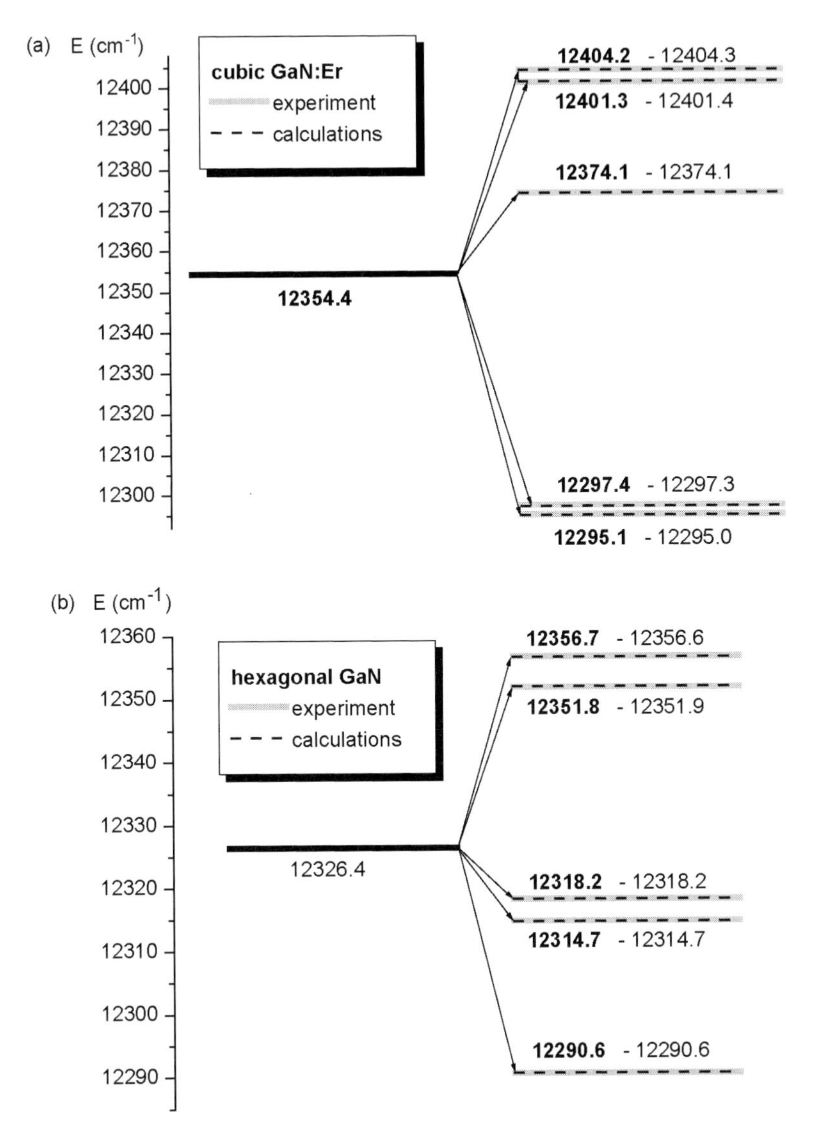

Figure 9.38. Energies of the $^4I_{9/2}$ crystal-field multiplet relative to the ground state (cm^{-1}) in (a) cubic and (b) hexagonal GaN.

The cubic symmetry crystal field Hamiltonian, for the quantization axis along one of the fourfold cube axes, can be expressed as

$$H_{\text{cub}} = B_4 O_4 + B_6 O_6 = B_4(O_4^0 + 5O_4^4) + B_6(O_6^0 - 21O_6^4)$$

where O_l^m are equivalent operators in the notation of Stevens [83], B_4 and B_6 are 4th- and 6th-order cubic crystal field parameters.

If the quantization axis is taken along the threefold rotation axis the cubic Hamiltonian takes the form

$$H^{(2)}_{\text{cub}} = B_4\left(-\frac{2}{3}O_4^0 + \frac{40\sqrt{2}}{3}O_4^3\right) + B_6\left(\frac{16}{9}O_6^0 + \frac{140\sqrt{2}}{9}O_6^3 + \frac{154}{9}O_6^6\right).$$

The parameters B_4 and B_6 are related to crystal field coefficients A_4 and A_6 by [84] as

$$B_4 = A_4\langle r^4\rangle\beta, \qquad B_6 = A_6\langle r^6\rangle\gamma$$

where β and γ are multiplicative factors dependent on L, S and J of the free ion state, and for the $J = \frac{9}{2}$ state of Er^{3+} are both positive.

The splitting of the $J = \frac{9}{2}$ level was then fitted with a crystal field Hamiltonian of tetragonal (D_{2d}):

$$H_{D_{2d}} = B_2^0O_2^0 + B_4^0O_4^0 + B_4^4O_4^4 + B_6^0O_6^0 + B_6^4O_6^4$$

and trigonal (C_{3v}) symmetry

$$H_{C_{3v}} = B_2^0O_2^0 + B_4^0O_4^0 + B_4^3O_4^3 + B_6^0O_6^0 + B_6^3O_6^3 + B_6^6O_6^6$$

in cubic and hexagonal GaN, respectively.

The crystal field parameters obtained from the fit are summarized in table 9.1. The calculated energies agree with the experimental ones within the accuracy of 0.1 cm^{-1}, as shown in figure 9.38.

As is seen, 6th-order parameters are small, suggesting that perturbation of regular symmetry of the centres is low. As a result it can be concluded that Er implanted into hexagonal GaN occupies a position with C_{3v} point symmetry, while in cubic GaN the site occupied has tetragonal point symmetry. The results for hexagonal GaN are fully consistent with EXAFS [78] and RBS [72] experiments which show substitutional location of Er. In the case of cubic GaN the distortion from cubic symmetry is relatively small. Analysis of cubic crystal field parameters suggests that Er atoms are located at interstitial sites, close to a regular tetrahedral. The small tetragonal distortion might be related to strain in the layer ($\langle 100\rangle$ is the growth direction),

Table 9.1. Crystal field parameters (in the units of cm^{-1}) for the Stark splitting of the $J = \frac{9}{2}$ state.

	Cubic GaN		Hexagonal GaN
T_d:	$B_4 = 0.0227$; $B_6 = 2.08 \times 10^{-4}$	T_d:	$B_4 = -6.8 \times 10^{-4}$; $B_6 = 2.12 \times 10^{-4}$
D_{2d}:	$B_2^0 = 0.11$; $B_4^0 = 0.0226$; $B_4^4 = 0.114$; $B_6^0 = 2.1 \times 10^{-4}$; $B_6^4 = 4.36 \times 10^{-3}$	C_{3v}:	$B_2^0 = 0.27$; $B_4^0 = 4.3 \times 10^{-4}$; $B_4^3 = -0.012$; $B_6^0 = 3.6 \times 10^{-4}$; $B_6^3 = 4.75 \times 10^{-3}$; $B_6^6 = 3.48 \times 10^{-3}$

though charge compensation by an extra defect cannot be excluded at this stage.

Summarizing, lattice location experiments reveal that RE ions in GaN are predominantly located at Ga substitutional sites. In the case of Er in GaN it was shown that most probably this is due to the very short Er–N bond lengths and resulting stability of Er-related point defects against precipitation of Er-rich phases. It can be assumed that the same rule applies to other REs in GaN, because of close similarity of chemical properties of RE family. Oxygen co-doping is beneficial for luminescence intensity and stabilization of Er at the Ga site. Local symmetry of Er in hexagonal GaN is C_{3v} for diluted concentrations of Er. The dominant excitation mechanism is defect mediated and relatively low deactivation energies indicate that other, shallow impurities take part in the excitation of Er. PLE measurements performed at $^4I_{11/2}$ excited state for high Er concentrations reveal energy migration among different Er centres. PLE spectroscopy and analysis of crystal field parameters from optical spectra confirmed that Er atoms occupy regular lattice sites.

References

[1] Nakamura S and Fasol G 1997 in *The Blue Laser Diode* (Berlin, Heidelberg, New York: Springer)

[2] Godlewski M and Goldys E M 2002 in chapter in *III—Nitride Semiconductors: Optical Properties* vol II *Optoelectronic Properties of Semiconductors and Superlattices* eds. Hongxing Jiang and M Omar Manasreh (New York: Taylor & Francis) p. 259

[3] Stepniewski R, Wysmolek A, Korona K P and Baranowski J M 2002 in chapter in *III—Nitride Semiconductors: Optical Properties* vol I *Optoelectronic Properties of Semiconductors and Superlattices* eds. Hongxing Jiang and M Omar Manasreh (New York: Taylor & Francis) p. 197

[4] In *Group III—Nitrides and Their Heterostructures* 2003, special issue of Physica Status Solidi (c) 0/6 1569–1949

[5] 2002 special issue of *Physica Status Solidi (c)* **0/1** 19

[6] Amano H, Sawaki N, Akasaki I and Toyoda Y 1986 *Appl. Phys. Lett.* **48** 353

[7] Amano H, Asahi T and Akasaki I 1990 *Jpn. J. Appl. Phys.* **29** L205

[8] Sasaki T and Matsuoka T 1995 *J. Appl. Phys.* **77** 192

[9] Nakamura S 1991 *Jpn. J. Appl. Phys.* **30** L1705

[10] Smart J A, Schremer A T, Weimann N G, Ambacher O, Eastman L F and Shealy J R 1999 *Appl. Phys. Lett.* **75** 388

[11] Yang C-C, Wu M C and Chi G-C 1999 *J. Appl. Phys.* **85** 8427

[12] Uchida K, Nishida K, Kondo M and Munekata H 1998 *J. Cryst. Growth* **189/190** 270

[13] Kawakami H, Sakurai K, Tsubougchi K and Mikoshiba N 1988 *Jpn. J. Appl. Phys.* **27** L161

[14] Yamamoto A, Tsujino M, Ohkubo M and Hashimoto A 1994 *J. Cryst. Growth* **137** 415

[15] Uchida K, Watanabe A, Yano F, Kouguchi M, Tanaka T and Minagawa S 1996 *J. Appl. Phys.* **79** 3487
[16] Pecz B, di Forte-Poisson M A, Huet F, Radnoczi G, Toth L, Papaioannou V and Stoemenos J 1999 *J. Appl. Phys.* **86** 6059 and references therein
[17] Ptak, A J, Ziemer K S, Millecchia M R, Stinespring C D and Myers T H 1999 *MRS Internet J. Nitride Semicond. Res.* **4S1** G3. 10
[18] Ohshima N, Yonezu H, Yamahira S and Pak K 1998 *J. Cryst. Growth* **189/190** 275
[19] Tansley T L, Goldys E M, Godlewski M, Zhou B and Zuo H Y 1997 in *Optoelectronic Properties of Semiconductors and Superlattices* vol 2 *GaN and Related Materials* ed. S J Pearton (Gordon and Breach Publishers) pp. 233–293 and references therein
[20] Godlewski M, Goldys E M, Phillips M R, Pakula K and Baranowski J M 2001 *Appl. Surf. Sci.* **177** 22
[21] Li S-Y and Zhu J 1999 *J. Cryst. Growth* **203** 473
[22] Kim K S, Oh C S, Lee K J, Yang G M, Hong C-H, Lim K Y and Lee H J 1999 *J. Appl. Phys.* **85** 8441
[23] Onitsuka T, Maruyama T, Akimoto K and Bando Y 1998 *J. Cryst. Growth* **189/190** 295
[24] Liu X, Wang L, Lu D-C, Wang D, Wang X and Lin L 1998 *J. Cryst. Growth* **189/190**, 287
[25] Yang C-C, Wu M C and Chi G-C 1999 *J. Appl. Phys.* **86** 6120
[26] Godlewski M, Goldys E M, Phillips M R, Langer R and Barski A 2000 *J. Mater. Res.* **15** 495
[27] Grzegory I 2002 *J. Phys.—Condens. Mat.* **14** 11055
[28] Miskys C R, Kelly M K, Ambacher O and Stutzmann M 2003 *Physica Status Solidi* (*c*) **0/6** 1627
[29] Zylkiewicz Z R 2002 *Thin Solid Films* **412** 64
[30] Godlewski M, Kopalko K, Szczerbakow A, Łusakowska E, Godlewski M M, Goldys E M, Butcher K S A and Phillips M R 2003 *Proc. Estonian Acad. Sci. Phys. Math.* **52** 277
[31] Butcher K S A, Afifuddin, Chen Patrick P-T, Godlewski M, Szczerbakow A, Goldys E M and Tansley T L 2002 *J. Crystal Growth* **246** 237
[32] Nakamura S, Senoh M, Nagahama S, Matsishita T, Kiyoku H, Sugimoto Y, Kozaki T, Umemoto H, Sano M and Mukai T 1999 *Jpn. J. Appl. Phys. Part 2* **38** L226
[33] Nakamura S, Senoh M, Nagahama S, Iwasa N, Matsushita T and Mukai T 2000 *Appl. Phys. Lett.* **76** 22
[34] Zytkiewicz Z R 2002 *Thin Solid Films* **412** 64
[35] Grzegory I, Boćkowski M, Krukowski S, Łucznik B, Wróblewski M, Weyher J L, Leszczyński M, Prystawko P, Czernecki R, Lehnert J, Nowak G, Perlin P, Teisseyre H, Purgał W, Krupczyński W, Suski T, Dmowski L H, Litwin-Staszewska E, Skierbiszewski C, Łepkowski S and Porowski S 2001 *Acta Phys. Polon. A* **100** Supplement 229
[36] Ivanov V Yu, Godlewski M, Teisseyre H, Perlin P, Czernecki R, Prystawko P, Leszczynski M, Grzegory I, Suski T and Porowski S 2002 *Appl. Phys. Lett.* **81** 3735
[37] Stocker D A, Schubert E F and Redwing J M 2000 *Appl. Phys. Lett.* **77** 4253
[38] Godlewski M, Ivanov V Yu, Goldys E M, Phillips M, Böttcher T, Figge S, Hommel D, Czernecki R, Prystawko P, Leszczynski M, Perlin P, Grzegory I and Porowski S 2003 *Acta Physica Polonica A* **103** 689
[39] Basov N G, Danilychev V A and Popov Yu M 1971 *Kvantovaya Elektronika* **1** 29

[40] Herve D, Accomo R, Molva E, Vanzetti L, Pagel J J, Sorba L and Franciosi A 1995 *Appl. Phys. Lett.* **67** 2144

[41] Lavrushin B M, Nabiev R F and Popov Yu M 1988 *Kvantovaya Elektronika* **15** 78

[42] Fitzpatric B, Khurgin J, Harnack P and de Leeuw D 1986 *International Electron Devices Meeting 1986*. Technical Digest Cat. No. 86CH2381-2 (New York, IEEE) p. 630

[43] Cammack D A, Dalby R J, Cornelissen H J and Khurgin J 1987 *J. Appl. Phys.* **62** 3071

[44] Fitzpatrick B J, Harnack P M and Cherin S 1986 *Philips J. Research* **41** 452

[45] Katsap V N, Kozlovskii V I, Kruchnov V Yu, Namm A V, Nasibov A S, Nosikov V B, Reznikov P V and Ulasyuk V N 1987 *Kvantovaya Elektronika* **14** 1994

[46] Bonard J-M, Ganiere J-D, Vanzetti La, Paggel J J, Sorba L, Franciosi A, Herve D and Molva E 1998 *J. Appl. Phys.* **84** 1263

[47] Borkovskaya L V, Dzhumaev B R, Korsunskaya N E, Papusha V P, Pekar G S and Singaevsky A F 1999 *Int. Soc. Opt. Eng. Proceedings of SPIE*—the International Society for Optical Engineering, USA, vol. **3724** 244

[48] Onischenko A and Surma J 1999 *Int. Soc. Opt. Eng. Proceedings of SPIE*—the International Society for Optical Engineering, USA, vol. **3211** 126

[49] Gribkovskii V P, Gurskii A L, Davydov S V, Lutsenko E V, Kulak I I, Mitkovets A I, Yablonskii G P and Gremerik V F 1993 *Jap. J. Appl. Phys.* **32** Supplement 3, 521

[50] Fitzpatrick B J, McGee T F III and Harnack P M 1986 *J. Crystal Growth* **78** 242

[51] Nakamura S, Mukai T and Senoh M 1993 *Jpn. J. Appl. Phys. Part 2* **32** 16

[52] Nakamura S, Senoh M, Nagahama S, Iwasa N, Yamada T, Matsushita T, Sugimito Y and Kiyoku H 1997 *Appl. Phys. Lett.* **70** 1417

[53] Godlewski M, Goldys E M, Phillips M R, Langer R and Barski A 2000 *J. Mater. Res.* **15** 495

[54] Coffa S, Franzo G and Priolo F 1998 *Mater. Res. Bull.* **23** 2325

[55] Polman A 1997 *J. Appl. Phys.* **82** 1

[56] Favennec P N, Haridon H L', Salvi M, Mountonnet D and LeGuillou Y 1989 *Electron. Lett.* **25** 718

[57] Wilson R G, Schwartz R N, Abernathy C R, Pearton S J, Newman N, Rubin M, Fu T and Zavada J M 1994 *Appl. Phys. Lett.* **65** 992

[58] Torvik J T, Feuerstein R J, Pankove J I, Qiu C H and Namavar F 1996 *Appl. Phys. Lett.* **69** 2098

[59] Torvik J T, Qiu C H, Feuerstein R J, Pankove J I and Namavar F 1997 *J. Appl. Phys.* **81** 6343

[60] Thaik M, Hömmerich U, Schwartz R N, Wilson R G and Zavada J M 1997 *Appl. Phys. Lett.* **71** 2641

[61] Kim S, Rhee S J, Turnbul D A, Reuter, E E, Li X, Coleman J J and Bishop S G 1997 *Appl. Phys. Lett.* **71** 231

[62] Steckl A J, Garter M, Lee D S, Heinkenfeld J and Birkhahn R 1999 *Appl. Phys. Lett.* **75** 2184

[63] Steckl A J, Garter M, Birkhahn R and Scofield J 1998 *Appl. Phys. Lett.* **73** 2450

[64] Heinkenfeld J, Garter M, Lee D S, Birkhahn R and Steckl A J 1999 *Appl. Phys. Lett.* **75** 1189

[65] Steckl A J, Heinkenfeld J, Lee D S and Garter M 2001 *Mater. Sci. Eng. B* **81** 97

[66] Lee D S and Steckl A J 2002 *Appl. Phys. Lett.* **80** 1888

[67] Wang Y Q and Steckl A J 2003 *Appl. Phys. Lett.* **82** 502

[68] Lee B K, Chi C J, Chyr I, Lee D S, Beyette F R and Steckl A J 2002 *Opt. Eng.* **41** 742

[69] Kozanecki A 1996 *Acta Physica Polonica A* **90** 73
[70] Kozanecki A 1989 *Phys. Status Solidi* (*a*) **112** 777
[71] Casey H C Jr and Pearson G I 1964 *J. Appl. Phys.* **35** 3401
[72] Alves E, DaSilva M F, Soares J C, Vianden R, Bartels J and Kozanecki A 1999 *Nucl. Instr. Meth. in Phys. Res. B* **147** 383
[73] Monteiro T, Boemare C, Soares M J, Sa Ferreira R A, Carlos L D, Lorenz K, Vianden R and Alves E 2001 *Physica B* **308–310** 22
[74] Wahl U, Vantomme A, Langouche G, Araujo J P, Peralta L, Correia J G and the ISOLDE Collaboration 2000 *J. Appl. Phys.* **88** 1319
[75] Lorenz K, Alves E, Wahl U, Monteiro T, Dalmasso S, Martin R W, O'Donnel K P, Vianden R, accepted for publication in *Mater. Sci. Eng. B*
[76] De Vries B, Wahl U, Vantomme A, Correia J G and the ISOLDE Collaboration, accepted for publication in *Mater. Sci. Eng. B*
[77] De Vries B, Wahl U, Vantomme A, Correia J G and the ISOLDE Collaboration 2002 *Phys. Status Solidi* (*c*) **1** 453
[78] Citrin P H, Northrup P A, Birkhahn R and Steckl A J 2000 *Appl. Phys. Lett.* **76** 2865
[79] Kim S, Rhee S J, Turnbul D A, Li X, Coleman J J, Bishop S G and Klein P B 1997 *Appl. Phys. Lett.* **71** 2662
[80] Wu X, Hömmerich U, Mackenzie J D, Abernathy C R, Pearton S, Schwartz R N, Wilson R G and Zavada J M 1997 *Appl. Phys. Lett.* **70** 2126
[81] Kim S, Rhee S J, Turnbul D A, Li X, Coleman J J and Bishop S G 2000 *Appl. Phys. Lett.* **76** 2403
[82] Przybylińska H, Jantsch W and Kozanecki A 2001 *Mater. Sci. Eng. B* **81** 147
[83] Stevens K W H 1952 *Proc. Phys. Soc. Lond.* **A65** 209
[84] Hutchings M T 1964 *Solid State Phys.* **16** 227

PART 3

IV GROUP MATERIALS

Chapter 10

Silicon and porous silicon

Bernard Gelloz and Nobuyoshi Koshida
Department of Electrical and Electronic Engineering, Tokyo University of Agriculture and Technology, Tokyo, Japan

10.1 Introduction

Silicon technology overwhelmingly dominates microelectronics. However, silicon is a very poor light-emitting material because of its indirect band gap. Furthermore, its gap being at 1.1 eV, silicon cannot emit light in the visible in its bulk crystalline form. As a result, fields such as optical communications and displays cannot be considered with 100% silicon technology. Other semiconductors, such as III–V and II–VI semiconductors, have been selected as the elementary brick for light-emitting devices. The use of semiconductors other than silicon makes problems arising such as higher cost, difficulties due to a technology less mature than silicon technology and compatibility drawbacks with existing silicon technology.

The so-called electrical interconnect bottleneck problem should emerge in the next decade. Today, the overall performance of a multichip computing system is limited by the interconnections between chips. It is predicted that this problem will migrate to the single chip level after one decade. The limitations of interconnects are speed and signal integrity, power use and heat dissipation. Without a solution to this problem, the growth of the semiconductor industry should experience very severe situations when the feature size approaches 50 nm. Optical interconnections would be very attractive for several reasons. For example, there is almost no distance-dependent loss or distortion of the signal, no deleterious fringing-field effects, no heat dissipation in the interconnections, and the interconnect does not have to be redesigned as the clock speed increases. One of the challenges in this field is the availability of suitable Si-based LEDs and lasers.

The hope that silicon could be used as a light-emitting material in the visible arose in 1990 when efficient visible light from porous silicon (PS) was demonstrated at room temperature [1]. In the next decade, a large amount of work has been devoted to the fabrication of PS-based LEDs. Efficient electroluminescence (EL) has first been obtained using an electrolyte as contact [2]. The first demonstration of injection mode solid state EL based on PS was made in 1992 by Koshida and Koyama [3]. Despite the large amount of work and motivation, over 12 years after the discovery of light emission from PS, a practical PS-based LED has yet to be successfully fabricated. Several problems must be overcome: poor efficiency, poor stability, the difficulty of obtaining green and blue EL, and efficient EL being usually obtained at too high operating voltages. However, recent significant progresses in PS EL efficiency and stability should renew motivation in this field.

After the effervescence created by light-emitting PS has decreased, other Si-based materials have been considered for visible light emission. These materials mainly involve the synthesis of Si nanocrystals (nc-Si). These nc-Si have been fabricated using various techniques, such as plasma-enhanced (PE) or low-pressure (LP) chemical vapour deposition (CVD), laser ablation and sputtering in various conditions. The host materials for the nc-Si are SiO_2, Si_3N_4, amorhous Si (a-Si), Si/SiO_2 or Si/CaF_2 superlattices. Recently, the report from Pavesi *et al* [4] about the observation of positive optical gain in an Si-based light-emitting material has created great motivation and hope in the research community for the quest of the first Si-based laser. So far, optical gain has been confirmed by a few other groups [5–7]. Valenta *et al* [8] have warned about a possible artefact in the gain measurement, manifesting itself as an apparent but false gain, and proposed some experimental procedures as complementary checks.

Recently, relatively high efficiency Si LEDs emitting at about 1150 nm have been successfully fabricated [9, 10]. This should motivate researchers to investigate the possibility of practical Si optoelectronics in the infrared. Most interesting for optical interconnects and communications is the infrared optical windows at around 1500 nm. A fruitful pathway towards Si-based infrared EL for this optical window is the use of erbium-doped Si-based materials. Indeed, Er is a rare-earth impurity with an internal transition at 1.54 μm. Er has been implanted or incorporated into various Si-based materials, containing nc-Si or not, and EL has been observed.

The purpose of this review is to survey the progress in the field of Si- and PS- based EL. Somewhat shorter, and more focused on a particular aspect, recent reviews have been published concerning Si-based LEDs for integrated circuits [11], and LEDs based on PS [12]. In section 10.2, PS preparation methods and main characteristics are presented. Definitions, and requirements concerning Si-based EL, are introduced. Then, section 10.3 deals with EL in the infrared region ($\lambda > 1$ μm). Light emission from bulk Si and Er-doped materials are discussed. Section 10.4 is devoted to EL originating from Si nanoclusters and Si-rich SiO_2 materials. Si-based superlattices are also discussed in this section. Most devices in this section emit light in the visible or/and in the 0.4–0.9 μm wavelength region. Due to an extensive amount of work having been undertaken in the field of PS-based EL, a special section (section 10.5) is devoted to PS EL. Section 10.6 discusses another application of PS directly connected to the field of Si-based EL: ballistic EL using porous Si as a cold electron emitter. In section 10.7, a few EL-related properties are presented. It includes PS-based filters, waveguides, optical switching and electro-optical memories. Finally this review is concluded in section 10.8.

10.2 Background

Especially for the reader unfamiliar with PS, this section first gives an overview of PS preparation methods and main characteristics. This short overview is not intended to cover everything about PS but still covers the main aspects that one should know in the field of applied EL from PS. For further reading about PS properties, such as PS formation mechanisms, luminescence mechanisms, structural properties and so on, the reader may refer to some reviews [13–17] available in the literature.

10.2.1 Porous Si formation

PS is usually formed by anodization of Si in dilute aqueous or ethanoic HF [18]. Ethanol is used in order to facilitate the evacuation of H_2 gas (produced by the Si dissolution) from PS pores. PS is readily obtained by anodization of p-type Si in the dark. Very well controlled thick PS layers can be obtained from p-type substrates. Anodization of n-type Si requires illumination in order to photogenerate holes in the substrate. Front-side illumination leads to PS layers with large in-depth porosity (percentage of vacuum in PS) gradients, whereas back-side illumination leads to uniform PS layers.

PS formation and characteristics depend on several parameters, the most important ones being HF concentration, anodization current density, anodization time, substrate doping density and type, and temperature. PS can be formed for current densities from 0 to a critical current density, J_{PS}. For current densities greater than J_{PS}, electro-polishing occurs. J_{PS} depends on the HF concentration and temperature. The general trend is that the porosity (percentage of vacuum in PS) increases with the current density and when the HF concentration decreases. Porosity increases when temperature diminishes [19].

Figure 10.1 shows a possible cell for PS formation. An ideal cell should be designed to get uniform current density over all the Si substrate area in contact with the electrolyte in order to generate PS with good lateral uniformity. More simply, Teflon tape (HF resistant) could be applied on the Si substrate, letting only the surface to be anodized in contact with the electrolyte.

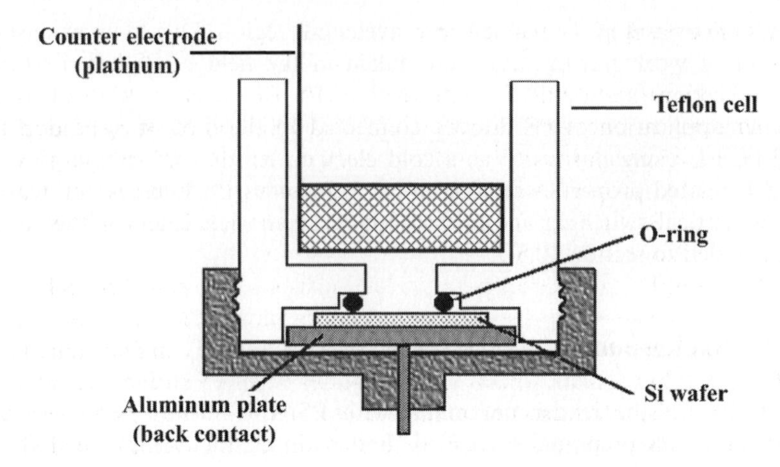

Figure 10.1. Cross-sectional view of a possible simple anodization cell for porous silicon formation.

10.2.2 Porous Si main characteristics

After anodization in aqueous HF, PS is rinsed (usually in water or ethanol) and dried. High porosity (>70%) PS may self-destroy upon drying [20–22]. This phenomenon is due to capillary stresses associated with nanometric pore size. Cracking of PS occurs if the PS thickness is greater than a critical value, which depends on the porosity and the surface tension of the liquid in the pores. Some techniques have been proposed in order to dry high-porosity PS. The use of pentane as the drying liquid has been shown to increase the PS thickness obtainable without cracking in the case of p^+ Si because its surface tension is significantly less than for water or ethanol [21, 22]. Another technique called freeze drying has been suggested. The liquid in the pores is first frozen and then sublimed under vacuum [23–25]. Finally supercritical drying is the most efficient drying technique though requiring sophisticated equipment [20, 26, 27]. The removal of the liquid inside PS (generally CO_2) is done above the critical point, thus avoiding interfacial tensions. Thick PS layers emitting bright luminescence can be dried without any special technique when a large porosity gradient is present in the PS depth. In this case, the upper part of PS is highly porous and luminescent, and is sustained by the lower part of PS whose porosity is low.

PS is basically described by its thickness, porosity and morphology. PS morphology depends on the doping level. PS made from p^- Si is highly isotropic whereas PS made from p^+ Si exhibits a columnar structure, the direction of the columns depending on the crystalline orientation of the substrate. The case of PS made from n-type Si is identical, but porosity gradients and structural changes in the PS depth should also be taken into account. More details on PS pore morphology can be found in [28].

10.2.3 Porous Si photoluminescence

PS can be luminescent only for sufficiently high porosities, in excess of 70%. The efficiency of photoluminescence (PL) as a function of type and doping level generally decreases in the following order: n^-, p^-, n^+, p^+. However, PS fabricated from n^+-type Si can include a highly luminescent nanoporous top PS sub-layer, providing luminescence efficiency of the same order as PS from n^--type Si.

The useful luminescence band of PS consists of the so-called 'S-band'. This luminescence originates from exciton recombination into Si nanocrystals. The luminescence mechanism of this band is mainly quantum confinement, but can also involve surface states when PS surface is contaminated [29]. The luminescence efficiency can be improved using various anodization post-treatments such as thermal oxidation, anodic oxidation, chemical oxidation, chemical or photo-assisted dissolution in HF. The last process, as well as anodic oxidation, is particularly efficient in increasing the PL efficiency.

Achievement of red and orange PL is rather easy. However, green and blue is rendered difficult by the presence of surface states introduced by contaminants on PS surface. More recently, Mizuno *et al* [30] showed that the PL of PS could be tuned from red to blue, depending on the formation conditions. They obtained a continuous tuning from red to blue by using a starting PS layer made from p-type Si, and subsequent photo-dissolution [30, 31], as shown in figure 10.2. The as-prepared PS emits in the red. After 5 min of photo-dissolution under a tungsten lamp, the PL is green and its intensity has dropped by half. After 15 min of photo-dissolution, the PL is blue and its intensity is 200 times less than the initial red one. The blue PL originates from oxide-free Si nanocrystals in PS. The blue emission shifts to red if PS gets oxidized. The exposure to air for a few seconds is enough to induce such a shift. The reason is that oxide at the surface of PS introduces surface states in the gap of blue-emitting PS, preventing the blue light to be efficiently emitted [29]. Increase of hydrogen terminations by exposure of PS to a H_2 gas at room temperature for 12 h has been shown to enhance the stability of blue-emission.

The PL lifetimes span from a few nanoseconds in the blue region [32] to several tens of microseconds in the red region. Details could be found, for example, in the review from Cullis *et al* [15]. These long lifetimes suggest indirect transition as seen in bulk Si. It has been confirmed by band structure calculation and spectroscopic characterization [13–17]. The rather long PL lifetimes make difficult high speed luminescence switching.

Luminescent PS is usually highly resistive due to depletion of free charge carriers as a result of quantum confinement. This makes the charge carrier injection into Si nanocrystals rather difficult.

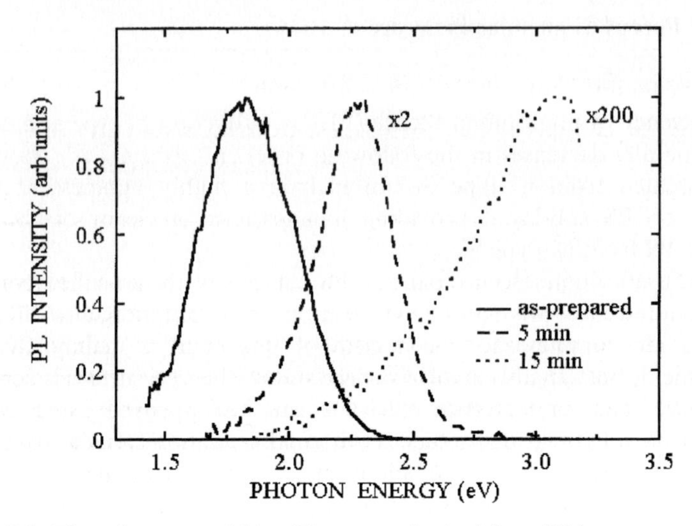

Figure 10.2. The red, green and blue PL spectra obtained from PS by post-anodization illumination with a tungsten lamp. Respective illumination times are also shown [30].

10.2.4 Definitions and requirements

To begin with, a few definitions should be given in order to avoid any confusion.

First, in this review, the term LED is used for any kind of light-emitting device. This includes light generation related to unipolar injection, and also various configurations other than pn junctions.

In order to evaluate light sources, several parameters can be used and are usually evaluated and reported by researchers. These include the EL onset voltage and current, the external quantum efficiency (EQE), the external power efficiency (EPE), the EL response time, and the EL stability. However, all these parameters are not always stated in reports. Sometimes, the internal quantum efficiency is given, but not the EQE. The EPE is rarely reported. The EL onset voltage and current sometimes refer to the apparatus detection threshold and sometimes to the naked eye detection threshold. The picture is most blurred concerning the stability. No clear definition has been used, so that when a report states a number of days of stability, it is not clear what the situation of the LED is after this number of days (initial EL intensity divided by 10, EL cannot be seen by naked eye anymore, etc), unless you read the paper and it is clearly defined in it. In this review, the data given will be as accurate as the original paper is.

The EQE is the number of photons emitted outside the device versus the number of charge injected. It can be reformulated as the product of the intrinsic quantum efficiency by the carrier injection efficiency by the light

Table 10.1. Display and interconnects requirements.

Application	Efficiency (%)	Modulation speed (MHz)	Peak emission	Spectral width	Stability
Displays	>1	>0.001	Red, green, blue	100 meV	10^5 h
Interconnects	>10	>100	Any	—	>10^5 h

extraction efficiency. The intrinsic quantum efficiency reflects the number of photons emitted versus the number of electron–hole pairs generated in the material. The carrier injection efficiency is the ability to efficiently inject carriers in the luminescent centres in the material (number of carriers injected into luminescent centres versus the total number of carriers injected into the material). Finally, the light extraction efficiency reflects the fraction of photons that actually exit the device versus the total number of photons generated. All these individual efficiencies can be acted on in order to improve the EQE.

The power efficiency is the power of light emitted outside the device versus the input electrical power.

The main applications for Si-based EL are displays and optical interconnects. The requirements are listed in table 10.1 for both applications.

10.3 EL from bulk Si and erbium

In bulk c-Si, radiative recombinations are very poorly efficient. This is due to the indirect nature of the c-Si bandgap [33]. Indeed, photon emission requires crystal momentum conservation, mostly accomplished via phonon emission (or absorption at temperature above about 100 K) [34–36]. The consequence of this three-body process is a relatively low radiative recombination rate compared with what is found in semiconductors with direct bandgap. The associated lifetime is long, typically ranging from 1 ms to 4.6 h [9, 37]. On the contrary, non-radiative transitions (i.e. Auger [38–41], Shockley–Read–Hall [42, 43], and surface recombination [44]) exhibit much faster recombination rates, typically spanning the range from the nanosecond to the millisecond. The result of the radiative recombination rates being much lower than non-radiative recombination rates is very low optical emission efficiency.

10.3.1 Bulk crystalline Si pn junction LEDs

The first device showing EL at room temperature from c-Si was reported in 1952 [45] using a forward biased simple pn junction. EL at room temperature as well as 77 K was then achieved and explained [46]. At 77 K, the EL spectrum shows mainly a peak at 1.100 eV. At this temperature, the EL emission is due to free carriers and exciton recombination. At room temperature, however, the situation is somewhat different. First, the radiative recombination rate is lower (drop of about 10 between 100 K and 300 K [43]) due to higher non-radiative transition rates and exciton ionization. Second, the EL peak is at a lower energy due to bandgap shrinkage [9, 46, 48]. Third, the luminescence spectrum is broader, due to the rise in the free carrier kinetic energy [36]. In addition, various spectral features are still not well understood yet. The dopants also affects the luminescence band (see [49] for a complete classification).

A pn junction exhibits a power efficiency of about 10^{-2}% at 1.1 μm in forward bias and of about 10^{-6}% under reverse bias (emission in the visible, in the avalanche breakdown regime) [50]. In order to improve the efficiency of EL from c-Si pn junctions, several strategies have been implemented. They include enhancement of internal efficiency and optimization of light extraction.

Nag *et al* [10] fabricated a device in which the non-radiative recombination rate is reduced as a result of carrier localization. Carrier localization is achieved by a three-dimensional local strain field induced by appropriately produced dislocation loops in the pn junction (see figure 10.3). These dislocation loops as well as the p part of the junction are produced in two simple steps: boron implantation at a dose of 1×10^{15} cm^{-2} at an energy of 30 keV, and subsequent annealing in nitrogen atmosphere for 20 min at 1000 °C. The dislocation loops form an array situated in a planar region parallel to, and around 100 nm from, the pn junction. They are about

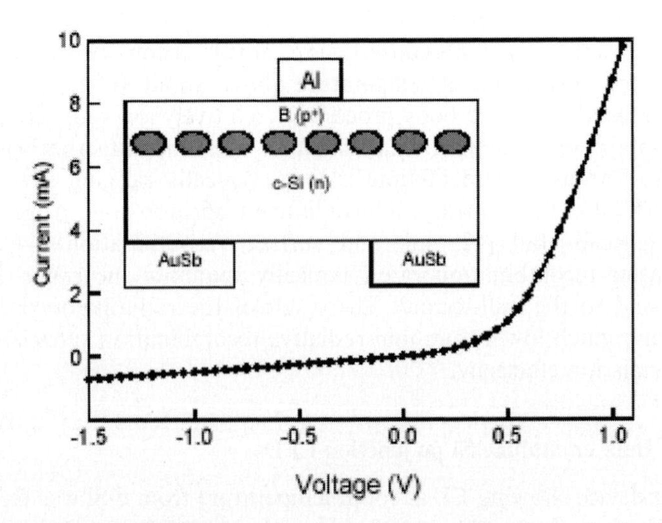

Figure 10.3. Current as a function of voltage for the device shown in the inset. The ellipses show the dislocation loops [10].

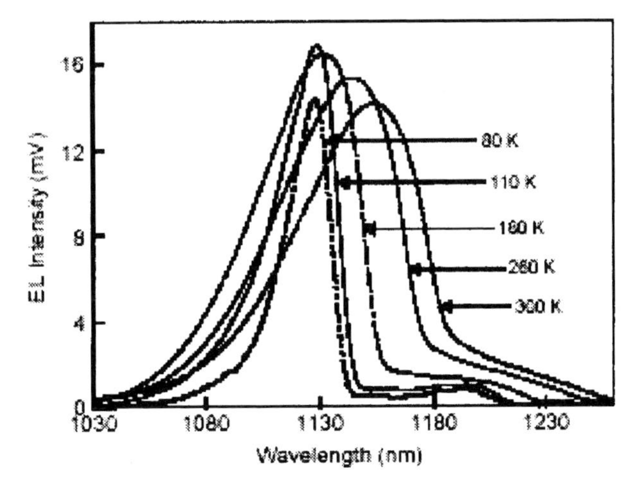

Figure 10.4. Spectra of electroluminescence intensity at various temperature for the device shown in the inset of figure 10.3. The device was operated at a forward current of 50 mA in all cases [10].

80–100 nm in diameter and are spaced around 20 nm apart. These loops are said to induce a blocking potential that confines carriers close to the junction region. At 100 mA forward current, the emitted light was 19.8 μW and the EQE was about $2 \times 10^{-2}\%$ at room temperature. The device response at room temperature was about 18 μs. At room temperature, the emission spectrum shows a peak at around 1150 nm. Finally, the EL and PL efficiencies of the device increase with the temperature, as shown in figure 10.4. This unusual property has not been explained yet.

Green *et al* [9] have used a combination of techniques aiming at enhancing both the internal quantum efficiency and light extraction efficiency (the device scheme is shown in figure 10.5). The internal efficiency can be improved by reducing the rate of non-radiative recombination. They have achieved this by using high quality Si float-zone wafers instead of CZ wafers. The surface recombination was reduced by using high quality thermal oxide for surface passivation. Shokley–Read–Hall recombination in the junction region was reduced by limiting metal areas and high doping regions to contact areas. They have also optimized light emission from their device by reducing the re-absorption of emitted photons in Si. For this purpose, low doping levels have been used [9]. Finally, light extraction was enhanced using a suitably textured Si surface. Textured surfaces are typically used in Si photovoltaic cells to increase the absorptivity. The spectral emissivity is equal to the absorptivity. In this way, light extraction can be enhanced by over an order of magnitude, compared with a planar surface. Green *et al* have thus obtained the highest EPE to date of about 1%. Figure 10.6 shows the wavelength dependence of perpendicular emission

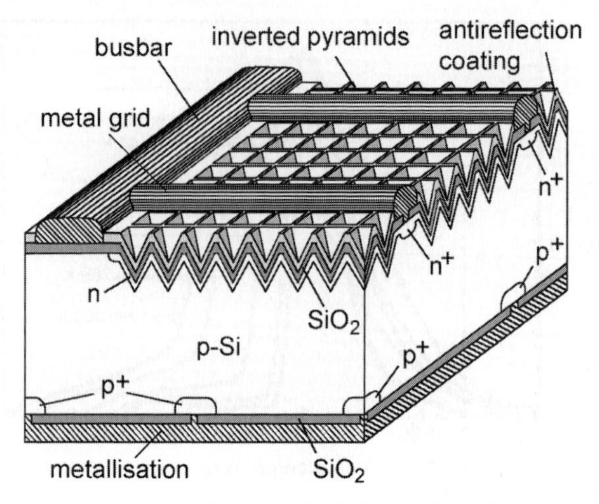

Figure 10.5. Schematic representation of high-efficiency silicon light-emitting diode [9]

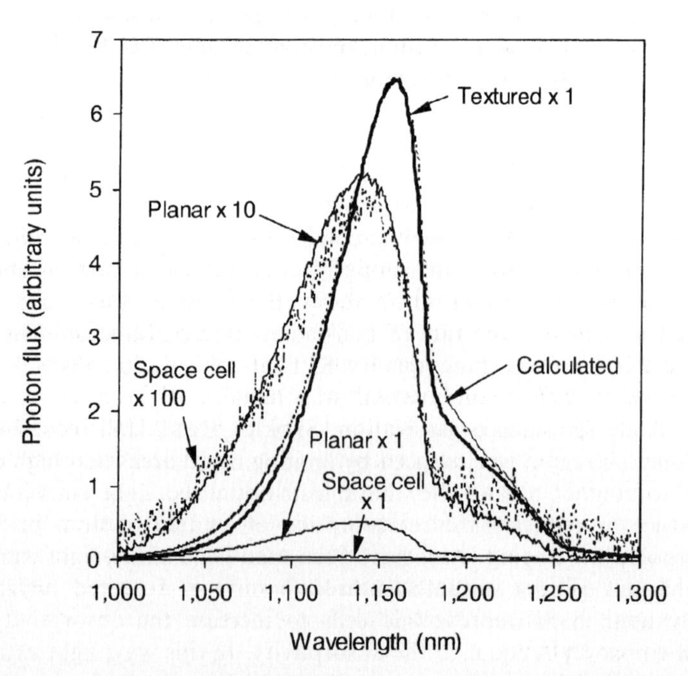

Figure 10.6. Electroluminescence spectra for textured, planar and baseline space solar cell diodes under 130 mA bias current at 298 K (diode area 4 cm^2). Calculated value assumes a rear reflectance of 96% [9].

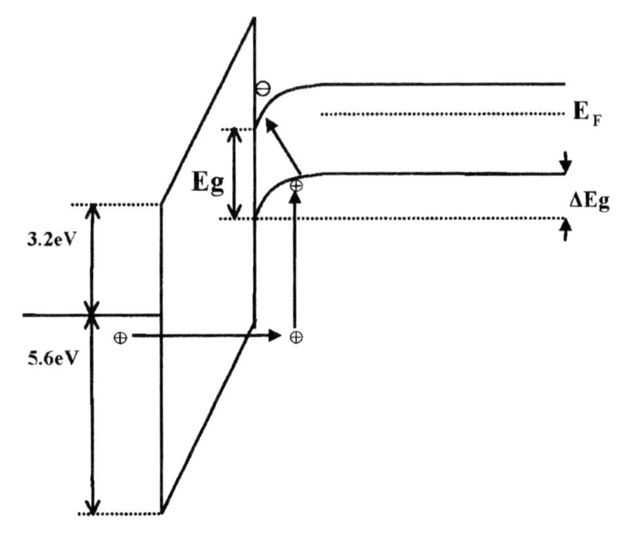

Figure 10.7. Schematic band diagram in the p-type MOS diode showing the band gap reduction, obtained from an electron–hole plasma model [51].

at 298 K for various devices. At room temperature, the emission spectrum shows a peak at around 1150 nm. The high radiative lifetimes in Si makes the device relatively slow to respond.

10.3.2 Other EL devices based on bulk crystalline Si

Room temperature EL at around 1050 meV has been achieved using a metal-oxide–silicon tunnelling diode [51]. Both n- and p-type Si has been used. The oxide layer was thin enough (2.7 nm) to allow carrier tunnelling under sufficient bias. Figure 10.7 shows the band diagram of a p-type diode under forward operation [51]. Electrons tunnel from the Si to the Al gate of the MOS structure. Holes tunnel from Al to Si and relax to the Si valence band edge. They then recombine with electrons in the accumulation region. The EL spectrum is very similar to that of Si pn junctions, but slightly shifted (100 meV) towards lower energies due to band bending effect. The EL spectrum is slightly voltage-tunable due the band bending effect. The EL EQE has not been measured. Significant loss of light occurs because of absorption by the Al contact. The EL is found much less temperature dependent than PL of Si. It was attributed to the field-induced carrier confinement in the emissive region, which contains much less impurity states than the unconfined region [52].

Room temperature EL (peak at about 1075 meV) has been achieved using a c-Si p-i-n structure under forward biased current pulses [53]. The concept of this EL is the following. Quantum efficiencies of several percent are possible for the PL of well-passivated c-Si at room temperature under pulsed laser irradiation [54]. The luminescence of the interband radiative transition at 1.1 eV increases with the $n \times p$ product due to the bimolecular recombination mechanism. The Auger recombination limits the efficiency at excess carrier concentrations of the order of 10^{18} cm^{-3}. Such high values of n and p can also be reached during short injection current pulses into the i region of the p-i-n structure. The advantages of the p-i-n structure are bipolar injection and the fact that the amount of injected electrons and holes are on the same order of magnitude. The dependence of the EL intensity on the current density is found parabolic for pulse duration below 3 μs due to the bimolecular recombination mechanism. A steady state is obtained for longer times (above 10 μs). The EL can be modulated up to 10 MHz. The maximum EL quantum efficiency is about 0.01%. Calculation shows that efficiencies of about 2 to 5% are achievable paying the price of a lower modulation speed (kHz region). Other different ways to increase the efficiency are discussed.

10.3.3 Er-doped bulk Si and SiO_x matrices

Er is a rare earth atom which in its 3+ ion state exhibits a radiative intra-4*f*-shell transition at 1.54 μm. This wavelength fits one of the optical communication windows. The transition is independent of host material and temperature, and exhibits a very sharp linewidth (0.01 nm). It can be electrically excited when Er is incorporated into a semiconducting host. Optical and electrical excitation of Er-doped Si has received a lot of attention [55, 56].

The first demonstration of room temperature sharp line EL from an Er-doped Si LED may be shared between two groups [57, 58] who published at the same time. The device of Zheng *et al* [57] consisted of a p/n^+ junction implanted with Er. The light intensity saturates at drive current density of 5 A/cm^2 due to the long excited state lifetime. The EL yield is evaluated to be four times greater than the PL one. Franzo *et al* [58] observed EL from Er both under forward (1.4 V) and reverse (−5.3 V) conditions with a p^+/n^+ Si diode implanted with Er. The room temperature EL intensity is greater under reverse bias. Also, the EQE is one order of magnitude greater under reverse bias than under forward bias. The temperature dependence of the EL intensity showed that, under forward bias, Er is excited through an energy transfer process involving recombination. In contrast, under reverse bias, Er excitation likely occurs via impact excitation. No degradation of the diodes was ever observed, even after several hours of operation at maximum power.

Several instances of highly doped pn junctions with Er in the active region of the device have been fabricated [58–60]. EL at 1.54 μm is achieved but with rather low efficiencies at room temperature. The efficiency is limited by the small number of excitable sites and by non-radiative processes.

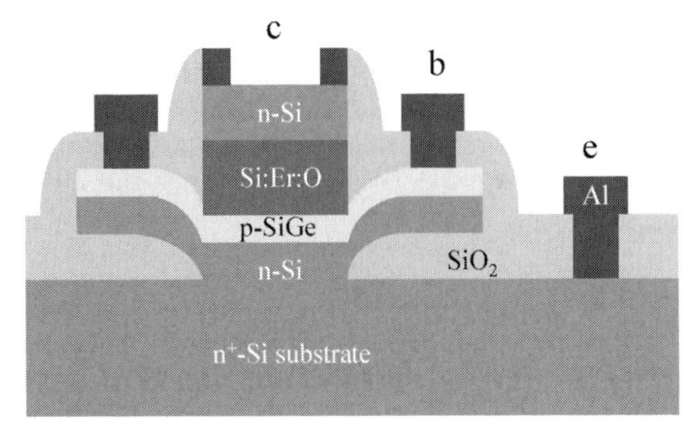

Figure 10.8. Schematic cross-sectional view of a SiGe/Si:Er:O HBT-type light-emitting device prepared by MBE [64].

However, under reverse condition, these devices are much more efficient [58–60]. Coffa *et al* [59] has obtained internal quantum efficiency of 10^{-2}%. The Er excitation cross-section was $6 \times 10^{-17}\ \text{cm}^2$. It has been shown that the excitation process under the reverse condition is impact ionization within the thin depletion region [60]. EL can be modulated faster than 1 MHz.

By additional co-dopants, such as oxygen, the excitation probability of Er transition at 1.54 μm can be increased [61–63]. Devices including Si:Er:O layers have been fabricated [64, 65]. The device from Du *et al* [64] includes an Si/SiGe/Si:Er:O npn heterojunction bipolar transistor in which Er is embedded in the collector–base depletion region (see figure 10.8). Electrons injected from the emitter are accelerated by the high electric field in the reverse biased collector–base junction and then excite Er by impact ionization. The current density is controlled by the emitter–base voltage and carrier energy by the collector–base voltage. Another advantage of this device compared with the conventional one is the absence of a dark region. Population inversion may be achievable in such a structure. Intense light emission is achieved at a driving current of $100\ \text{mA cm}^{-2}$ and at 3 V across the collector and emitter. The EQE is five times better than a control diode, but is still rather low: 10^{-3}%. Device optimization is expected to increase this value.

Markmann *et al* [65] reported efficient light emission at 1.54 μm from Er in Si excited by hot electron injection through thin suboxide layers. They fabricated both unipolar and bipolar devices in which electrons can be injected hot through suboxide barriers into Si:Er:O layers. The suboxide layer was embedded in an unipolar n-Si/Si:Er:O/n-Si or a bipolar p-Si/Si:Er:O/n-Si diode structure. At room temperature, these structures yield a two times higher EL output at 1.54 μm than an optimized reverse biased pn diode without a suboxide layer. Absolute output power of 250 nW and an EQE of 1.3×10^{-2}% are reported.

EL from Er has also been obtained from MOS structures, the Er being implanted into the oxide [66]. EL at 1.5 μm is reported. The Er^{3+} ions are excited by impact of hot electrons, tunnelling through the oxide. The excitation cross-section is found to be two orders of magnitude larger than that of an Er-doped Si pn junction (measured value: about 6×10^{-15} cm^2). The EL time response was about 1.5 ms.

10.3.4 Er-doped Si nanoclusters

The Er luminescence is greatly enhanced when incorporated into an SiO_2 matrix containing Si nanocrystals [67, 68]. Non-radiative de-excitation may be reduced due to enhanced localization. The emitting Er ions are located in the SiO_2 matrix or at the Si nanocrystal–SiO_2 interface [68]. The coupling between Si nanocrystals and Er is very strong. Transfer of excitation from Si nanocrystals to neighbouring Er ions is very effective.

Tsybescov *et al* [69] have obtained room temperature PL and EL from so-called Er-doped silicon-rich silicon oxide. PS made from p^+-type Si was the starting material. PS was first doped by Er ions by electroplating and later converted to silicon-rich silicon oxide by partial thermal oxidation at 900 °C. The LED structure is a n-i-p bipolar device with a 350 nm thick n-type poly-Si top contact, a 0.5 μm Er-doped silicon-rich silicon oxide, and a p-type c-Si substrate. Figure 10.9 shows the PL and EL spectra as well as the device structure (inset). EL was detected under forward bias above 10 V and current density near 1 A/cm^2. The low efficiency ($<10^{-4}$%) was attributed to the excitation of a small fraction of the Er ions only (volume concentration of Er ions was about 5×10^{18} cm^{-3}). The excitation mechanism was believed to be impact of hot electrons.

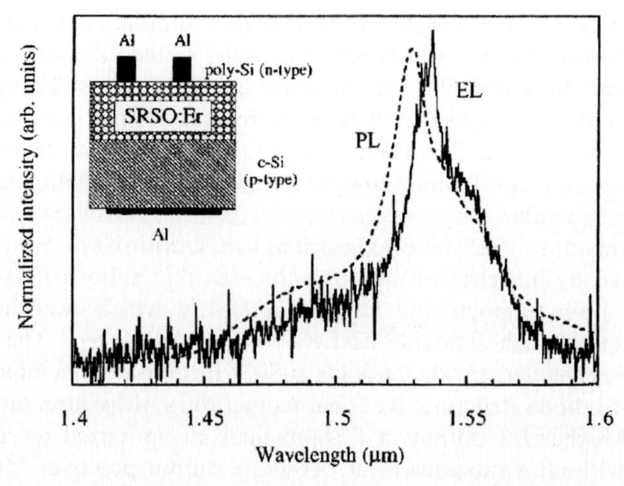

Figure 10.9. Room-temperature PL and EL spectra from the device shown in the inset [69].

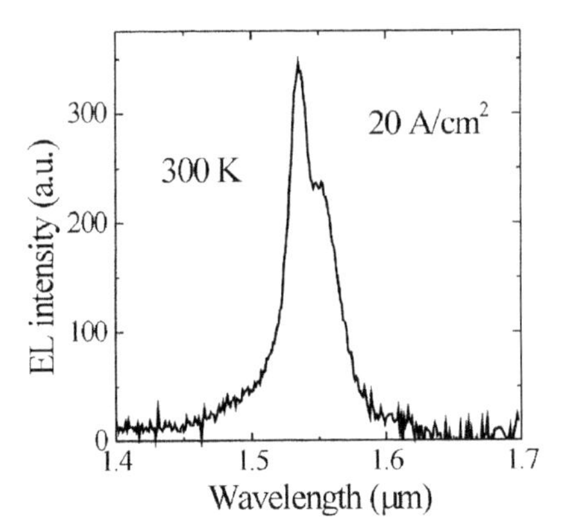

Figure 10.10. Room-temperature electroluminescence spectrum of an Er-doped Si nano-Si rich MOS device biased under 20 A/cm^2 [71].

A similar LED based on Er-doped PS was later reported [70]. The volume concentration of Er ions was about 10^{19} cm^{-3}. This time, stable EL was observed both under forward and reverse conditions. An EQE of 0.01% was achieved for highest input powers. Temperature dependence of the EL led the authors to conclude that, in reverse bias, the EL is a result of excitation by hot electrons, whereas excitation in forward bias occurs through an electron–hole-mediated recombination process as in PL.

Very recently, Iacona *et al* [71] have developed an LED based on Er-doped Si nanoclusters embedded in MOS structures. A 70 nm thick SiO_x ($x < 2$) was deposited on a low resistivity p-type Si substrate. Er is then implanted to a dose of 7×10^{14} cm^{-3}. Samples were annealed at 900 °C for 1 h in order to activate Er ions and to induce the separation of the Si and SiO_2 phases. The presence of the Si nanoclusters dispersed in the SiO_2 matrix makes Er electrical excitation efficient. The excitation cross-section for Er under electrical pumping (10^{-14} cm^2) is two orders of magnitude higher than the effective excitation cross-section of Er through Si nanoclusters under optical pumping. Stable EL (see spectrum in figure 10.10) is obtained at room temperature with an estimated quantum efficiency of 1%. EL rise and decay times were about 77 μs and 660 μs, respectively.

10.3.5 Other Er-doped Si-based devices

Gusev *et al* [72] reported EL at 1.54 μm from Er-doped amorphous hydrogenated Si at room temperature. The device consisted in an Al/a-Si:H(Er)/n-c-Si/Al structure. EL peaks at 0.8 eV and 1.075 eV were observed under reverse bias and forward bias, respectively. The former was attributed to Er and defect-oriented luminescence excited by electron injection from the

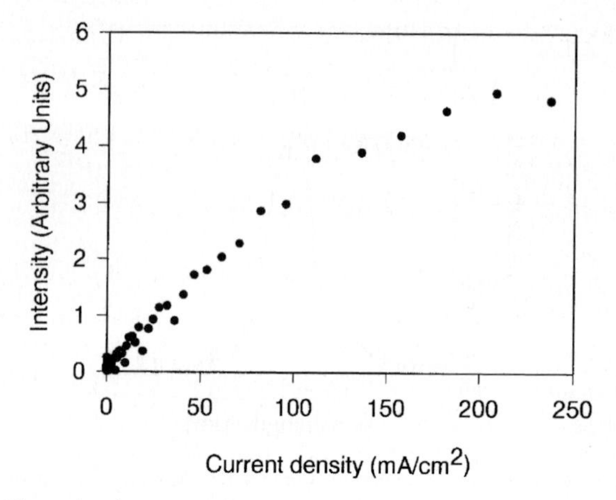

Figure 10.11. Electroluminescence intensity at 1532 nm as a function of the current density. Active diode area was 4 mm^2 [73].

AL contact, whereas the latter was attributed to electron–hole pair recombination in the n-type substrate (holes injected from Al travel through a-Si and recombine with electrons available in c-Si).

Curry *et al* [73] have fabricated a silicon-based organic light-emitting diode operating at a wavelength of 1.5 μm at room temperature. Er was present in the organic film. The LED was fabricated using a 40 nm layer of hole transport organic film deposited on to a p^+-type Si substrate. Then a 50 nm layer of Er tris(8-hydroxyquinoline) was deposited, and finally a 200 nm Al top contact was evapourated. The devices start to exhibit luminescence at current densities of about 3 mA/cm^2 (17 V). The EL intensity shows an unexplained sub-linear increase with current density, as shown in figure 10.11. The internal efficiency of this device at 33 V is of the order of 0.01%.

Finally, the semiconductor manufacturer STMicroelectronics has recently announced, in a press release (consultable in their web site), the fabrication of a silicon-based light emitter that matches the efficiency of traditional light-emitting compound semiconductor materials such as gallium arsenide (GaAs). No details were given.

10.4 EL from low-dimensional Si structures

Visible EL from Si nanocrystals embedded in an SiO_2 matrix was first demonstrated by DiMaria *et al* [74] in 1984. In 1990, visible light emission from PS was demonstrated [1]. Mostly EL from PS has been studied in the early 1990s, starting from the first demonstration of injection mode EL from PS in 1992 [2]. However, drastic problems such as low carrier injection efficiency and low stability have pushed researchers towards other systems such as arrays of Si nanocrystals or nanowires, either closely packed or embedded in a matrix (i.e. a-Si, SiN, but mostly SiO_2). These two approaches are briefly discussed below.

Si nanocrystals are usually directly deposited on to a substrate using diverse techniques such as CVD, laser ablation, sputterering etc. Recently Si nanocrystals have even been fabricated in the form of colloid suspension [75], but this approach has not been considered yet for EL applications. In this approach, the Si nanocrystal size distribution is usually quite narrow but the surface passivation and the layer mechanical strength are usually rather poor.

Concerning Si nanocrystals in SiO_2 matrix (see for example [76] and references therein), the excess of Si in SiO_2 can be generated either during SiO_x deposition (usually by CVD or sputtering) and further separation of the Si and SiO_2 phases by annealing, or by Si implantation in SiO_2 and subsequent annealing in order to fix damages induced by the implantation process, and agglomerate Si ions in clusters. The annealing process agglomerates the ions in clusters with different effectiveness depending on the temperature and time. Although low-temperature annealing eliminates most structural defects, the oxygen deficiency defects (believed to be at the origin of the ultraviolet and blue emission) are stable up to about 1100 °C. Higher-temperature annealing favours the formation of clusters emitting red light originating from quantum confinement. In general, this approach leads to large Si nanocrystal size distribution but rather good surface passivation and layer mechanical strength. The main problem of these structures is difficult carrier injection since the host matrix is usually an insulator. Another problem is the possibility of light emission from oxygen-related defects in SiO_2 (around 1.9 eV), as stated above.

Si-based superlattices have been considered in order to generate visible light from Si. Work in this field was triggered by the discovery of room temperature PL in Si/SiO_2 superlattices, with a size tunable emission energy [77].

In this section, we first address the issue of EL from Si nanowires, EL from Si nanocrystals deposited on a substrate, and devices using Si nanocrystals in an SiO_2 or another host matrix. Then, Si-based superlattices are presented.

10.4.1 Arrays of Si nanocrystals and nanowires

EL from Si nanopillars has been reported by Nassiopoulos *et al* [78, 79]. Si nanopillars were fabricated by using deep ultraviolet lithography, highly anisotropic Si reactive ion etching and high-temperature thermal oxidation for further thinning. The pillars (lying on, and perpendicular to, the original Si substrate) were below 10 nm in diameter and 0.4–0.6 μm high. The space

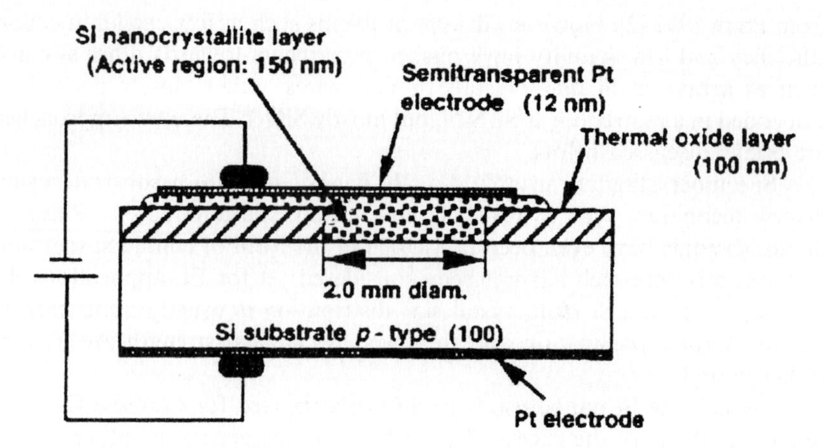

Figure 10.12. Schematic cross-sectional view of the LED of Yoshida *et al* [81].

between pillars was filled with an isolating transparent polymer. The top contact was either gold or ITO. Weak and low efficiency EL peaking at about 675 nm was observed at voltages above 12–14 V. EL was stable for several hours [79].

Toyama *et al* [80] have observed EL from a glass/SnO_2/p-type Si nanocrystals/Al structure by applying positive bias voltage to the SnO_2 electrode. P-type Si nanocrystals are first deposited on to SnO_2-cated glass by PE-CVD. This Si layer is then anodized in HF. Although the authors do not mention the word porous to characterize their Si film, this anodization is likely to lead to a porous amorphous Si layer. Si nanocrystals sizes were estimated to be in the range 3–5 nm. The EL showed a peak at 1.57 eV. The EL onset voltage was about 5 V. The EQE was below 10^{-2}%. The EL was attributed to radiative recombination of electrons injected from the Al electrode and holes in the p-type Si nanocrystal film. The authors suggest that electron injection is a result of a tunnelling process at the interface of Si nanocrystals/Al.

An example of EL from Si nanocrystals prepared by laser ablation is the device fabricated by Yoshida *et al* [81]. The diode structure was Pt/closely packed Si nanocrystals/p-type Si/Pt, as shown in figure 10.12. EL at 1.66 eV at room temperature was obtained. The EL energy peak was lower than the PL one (2.07 eV). The EL current and voltage threshold (photomultiplier detection) were 1 mA and 7 V, respectively. Detection thresholds of the naked eye in the dark were 15 mA and 25 V for the current and voltage, respectively. The active area was 0.031 cm^2. The estimated EQE was 10^{-4}%. A strong superlinear dependence of the EL intensity on the current density (see figure 10.13) was attributed to the EL generation mechanism, namely impact ionization by minority hot carriers.

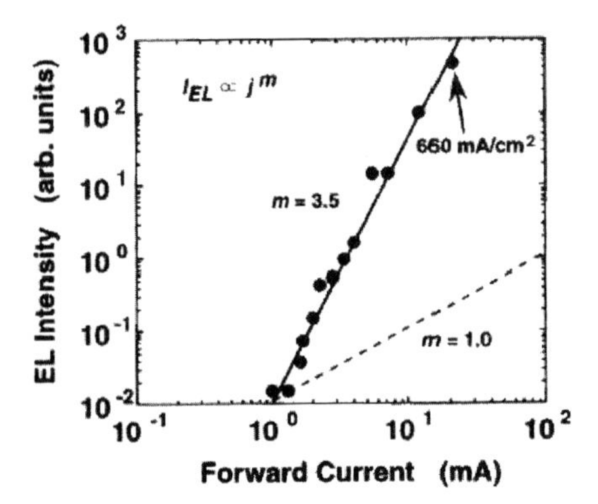

Figure 10.13. Nonlinear dependence of the electroluminescence intensity on the forward current [81].

Visible EL has been reported from Si nanocrystals embedded in a-Si:H films prepared by plasma enhanced chemical vapour deposition [82]. The EL spectrum shows two peaks, at about 630–680 nm and 730 nm. EL detection threshold is about 6 V. This value increases when the film conductivity decreases. In addition, the EL intensity increases with the film conductivity. The EL mechanism is not clear.

Park *et al* [83] have used a silicon nitride film as the host matrix for Si nanodots. Amorphous Si quantum dots were formed in SiN_x by PE-CVD. Control of dot size led to red, green and blue PL. An orange LED was fabricated using Si dots with a mean size of 2.0 nm. The turn-on voltage was below 5 V. The EQE was about 2×10^{-3}%.

10.4.2 Si-rich SiO_2 simple systems

The approach consisting of Si implantation in a matrix such as SiO_2 has led to several reports. Among these, Song *et al* [84] have studied the EL of Si^+-implanted thermally formed SiO_2. The oxide layer was 34 nm thick. Si was implanted (25 keV, 10^{16} cm^{-2}) mainly in the bottom half of the oxide layer. Both PL and EL show three bands, around 470, 600 and 730 nm. However, the 600 nm band is the strongest in EL whereas it is weaker in PL. The relative contributions from different luminescence bands to EL depend on the annealing conditions. From the study of different annealing conditions, the authors conclude that the 470 nm band needs too high excitation energy to be electrically excitable. The 730 nm band depends on the existence of Si nanocrystals, which are not numerously produced until high temperature annealing. The 600 nm band is easy to excite electrically due to its low energy. Moreover, the presence of Si nanocrystals improves the EL from oxygen-related defects.

Luterova *et al* [85] have observed red EL from Si^+-implanted sol-gel-derived SiO_2 films on n-type Si. The oxide layer was 250 nm thick. Four different energies and ion doses have been used for the Si implantation in order to get a flat ion profile across the oxide layer. Annealing was performed at 1100 °C for 1 h. EL is observed from 5 V and 1 A/cm^2. The device shows no rectifying behaviour but the EL can be obtained only at one bias polarity. EL at 295 K is emitted only from a small number of bright spots. The EL spectrum shows only a peak at 750 nm (attributed to Si nanocrystals) whereas the PL shows the same peak plus another one in the blue range attributed to the presence of defects. The EL is attributed to electron–hole recombination in Si nanocrystals as a result of carrier injection. Si nanocrystals are believed to create several conductive percolation paths across the oxide layer. A low efficiency (10^{-5}%) is attributed to shunting current paths due to defects. The EL decay is 8 μs.

Lalic and Linnros [86] have studied the fabrication and the electroluminescent properties of Si nanocrystals in SiO_2. The Si nanocrystals were generated by ion implantation and further annealing at 1100 °C for 1 h. Both the Si dose and SiO_2 thickness have been varied. The oxide thickness has been varied from 12 to 100 nm. The thermal oxide was grown on a p-type Si substrate. Then a 210 nm thick a-Si layer was deposited for use as a protective and contacting layer as well as to adjust the position of the implantation profile peak. A 160 nm thick highly phosphorus-doped poly-Si layer was then deposited and used to enhance the carrier injection into the Si nanocrystals. EL was observed from LEDs with oxide thickness less than 18 nm for bias voltages above 8 V. The EL spectrum shows a peak at about 1.55 eV. The EL efficiency is found more than one order of magnitude lower than that of their previously reported LED based on PS [87–89], which should give less than 2×10^{-2}%. However, the EL of the new device shows no degradation (due to Si nanocrystal passivation by SiO_2). This is very much better than the results they obtained with PS (only seconds of stable EL). Higher implantation doses leads to a higher level of interconnection between Si nanocrystals. More interconnected structures induces shorter EL time constants, larger leakage currents, and lower efficiency.

Qin *et al* [90] have studied visible EL from an Au/extra-thin Si-rich SiO_2/p-type Si structure. The oxide thickness was 4 nm. It was grown by magnetron sputtering. The EL was observed at voltage above 4 V. The EL showed a peak at 1.9 eV with a FWHM of 0.5 eV when the Si-rich oxide film was not annealed. Annealing at 800 °C induces a widening of the EL spectrum and the appearance of several shoulders at about 1.5, 2.2 and 2.4 eV. Furthermore, The EL peak energy blue-shifts with increasing forward bias. The luminescence is believed to originate from several types of luminescent centre in the Si-rich Si oxide film.

Irrera *et al* [91] have studied EL from Si quantum dots embedded within a MOS structure. A 25 nm thick substoichiometric SiO_x ($x < 2$) film was deposited on a p^+-type Si substrate by PE-CVD and annealed at 1100 °C for 1 h in N_2 in order to induce separation of the Si and SiO_2 phases. The Si nanocrystals thus formed had a mean radius of about 1.0 nm. The excitation cross-section of Si nanocrystals under electrical pumping (4.7×10^{-14} cm^2) is found to be two orders of magnitude higher than under optical pumping. This was also the case for Er excitation in their similar device including implanted Er ions [71]. The mechanism responsible for the light emission is the same under both optical and electrical pumping (self-trapped exciton recombination at a Si=O interfacial state). Efficient (no value of efficiency given) EL is obtained at 4 V, and 0.2 mA/cm^2. The EL spectrum exhibits a peak at about 900 nm. EL response time can be lowered down to about 25 μs.

Devices including a few nanometres thick amorphous Si layer sandwiched between two SiO_2 layers have been investigated [92, 93]. A forward biased p-type Si/SiO_2/a-Si/SiO_2/Au [92] and a reverse biased n^+-type Si/SiO_2/a-Si/SiO_2/Au [93] have been studied. The thickness of the a-Si layer has been varied between 0 and 4 nm. The upper and lower SiO_2 thicknesses were 3.0 and 1.5 nm, respectively. Room temperature EL is observed at about 5–7 V in the wavelength range from 600 to 700 nm depending on a-Si thickness. In both devices, EL peak intensity and peak wavelength synchronously swing with increasing the a-Si layer thickness, as shown in figure 10.14 [92]. The swing period is consistent with half the de Broglie wavelength of the carriers [93]. The EL is mainly attributed to luminescent centres in the oxide (excited by electrical breakdown), with a minor contribution of the a-Si layer as a result of quantum confinement.

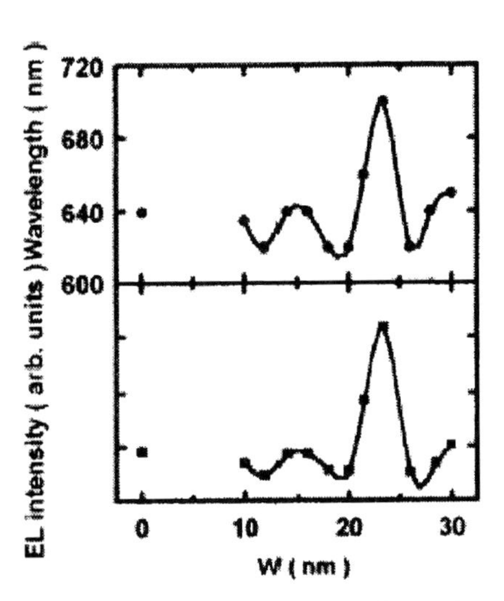

Figure 10.14. Electroluminescence peak intensities and wavelengths as a function of Si layer thicknesses for a Au/SiO_2 (4.5 nm)/p-type Si structure and Au/SiO_2/nanometre Si/SiO_2/p-type Si structures [92].

Photopoulos and Nassiopoulou [94] studied room- and low-temperature EL in the visible from a single layer of Si nanocrystals in between two thin (5 nm) SiO_2 layers. The EL was attributed to quantum confinement. The effects of size and size distribution were demonstrated. The EL peak wavelength exhibited a voltage-tunability from the red (800 nm) to the yellow (600 nm). The authors suggest different possible mechanism to explain the voltage-tunability of the peak wavelength: selective carrier injection in Si nanocrystals of different sizes, an enhanced Auger recombination rate at high voltages, Coulomb charging, and quantum-confined Stark effect.

Very recently, Osaka *et al* [95] reported EL at 1400 nm from a p-type Si/SiO_2-doped Si/metal diode. SiO_2-doped Si was deposited by an r.f. magnetron sputtering technique on p-type Si at a substrate temperature of 400 °C. EL is attributed to hole injection from the substrate followed by optical transition between electron-bound states in the film and injected holes. The EL spectrum is much narrower (FWHM 125 nm) than the PL one (FWHM 400 nm). This behaviour was attributed to the hole density of states being much broader than the electron-bound state distribution. The films show optical absorption in the range 1.4–2.0 eV, comparable with that of undoped a-Si:H films. The SiO_2-doped Si films may be useful for photovoltaic applications.

10.4.3 Superlattices

The first LEDs based on Si–SiO_2 superlattices were fabricated by MBE [96, 97]. The basic period of the superlattice was formed by a monolayer of adsorbed oxygen and a thin MBE-deposited single crystal Si layer. Both EL and PL were observed. The luminescence was attributed to quantum confinement as well as Si/O binding regions. The EL was stable for more than a year.

Another possibility for superlattice growth is PE-CVD. Poly-Si/SiO_2 superlattices were thus formed [98, 99]. The thickness of the Si layer was varied either from 1 to 3 nm (up to 60 periods) or from 75 to 150 nm (up to 4 periods) [98]. The oxide layer was around 1 or 2 nm. All structures contained Si nanodots. At low current injection, the emission was red and the quantum efficiency low (about 10^{-3}%). The authors later suggested that this EL originates from defect levels on the oxide near the interface [99]. At larger injection current, the EL intensity was superlinear with respect to the current density and a blue-shift occurred together with a narrowing of the FWHM down to about 5 nm. These phenomena are consistent with plasma emission in the intergrain regions.

Nanocrystalline Si/SiO_2 superlattices have been fabricated by alternate sequences of LP-CVD of thin Si layers and high-temperature thermal oxidation [100]. Initially amorphous Si films were crystallized during the oxidation step used to make the top oxide layer. Multi-layers with five or ten periods were fabricated, with nc-Si thickness between 1.5 and 5 nm, and SiO_2 thickness between 5 and 101 nm. The EL spectrum shows several peaks at 550, 650 and 750 nm.

Gaburro *et al* [101] have also fabricated light-emitting Si/SiO_2 superlattices by LP-CVD. The Si and SiO_2 thicknesses were 2 and 5 nm, respectively. The fabrication process is inexpensive and CMOS compatible. The PL band shows a peak at 800 nm due to quantum confinement, whereas EL shows two bands. One band in the infrared region is attributed to blackbody radiation. The other band is visible and is attributed to coupling of energetic electrons with surface plasmon-polaritons, or to hot-electron relaxation. EL is observed from 2 mA for a contact area of 5×10^{-5} cm^2. The EQE is 2×10^{-4}%.

Heng *et al* [102] fabricated a LED based on amorphous Si/SiO_2 superlattices deposited on a p-type Si substrate by sputtering. The thickness of the a-Si layers was varied in a range of 1.0 to 3.2 nm. The thickness of the SiO_2 layer was 1.5 nm in all devices. EL was observed above about 5 V. Every EL spectrum could be decomposed into two Gaussian bands with peak energies of 1.82 and 2.22 eV. The EL intensity and current swings synchronously with increasing the Si layer thickness with a period being consistent with half the De Broglie wavelength of the carriers. The authors had already observed the same behaviour with using only a single period instead of a superlattice [92, 93]. They attribute the EL emission to defect centres rather than levels in quantum dots.

In addition to Si–SiO_2 superlattices, a-Si:H/a-SiN_x:H multiquantum well structures have led to EL [103]. Layers are deposited by PE-CVD, and subsequent annealing is performed with a KrF excimer laser. The a-SiN_x:H layer was either 3 or 10 nm and the a-Si:H layer was varied from 2 to 4 nm. The LED structure was semitransparent electrode/crystallized a-Si:H/a-SiN_x:H multiquantum wells/crystallized n^+-a-Si:H/SiO_2/Si or quartz substrates. The EL spectrum showed multiple peaks around 600 and 700 nm. The onset of visible EL was 7 V in the best case. The authors attribute the EL mechanism to carrier injection into nanosized Si crystallites in the multiquantum wells, and radiative recombination via Si quantum well states.

Yet another configuration, including CaF_2/Si superlattices, has been investigated. PL [104–106] and EL [107–112] from these structures have been reported. A theoretical report attributes the EL emission from these structures to electron and hole tunnelling through the CaF_2 barriers, occurring via the Wentzel–Kramers–Brillouin mechanism [108]. Maruyama *et al* [109] studied the EL of such superlattices on p-type Si. The superlattice was grown by MBE and was followed by a rapid thermal annealing treatment. The EL is white. The El spectrum is such that the emission increases from below 400 nm to more than 700 nm, whereas the PL has a peak at about 575 nm. The authors published some improvements one year later [110]. The electrical transport in these superlattices has been studied [111]. One hundred periods were used. Si thickness ranged from 1.2 to 1.6 nm,

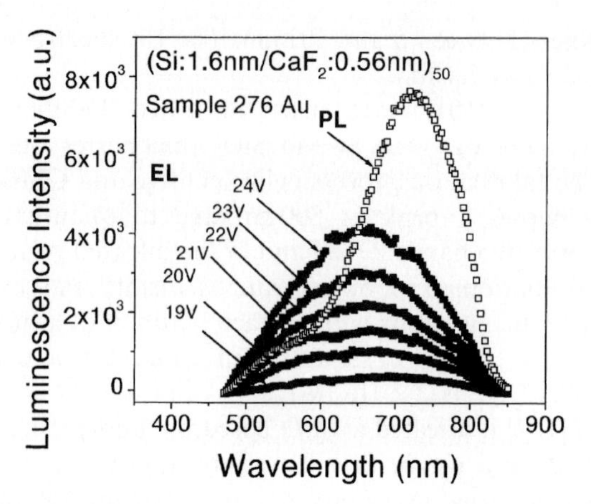

Figure 10.15. Electroluminescence spectra at various gate voltages and photoluminescence spectrum from a device including 50 bilayers of Si(1.6 nm)/CaF_2(0.56 nm) [112].

and CaF_2 thickness was below 1 nm. At voltages higher than 4 V (both in forward and reverse conditions), a Poole–Frenkel-type mechanism accounts for the observed electric-field-assisted conduction through the layers. The same authors found that the EL spectra are slightly current-tunable [112]. This was attributed to Auger quenching of PL and size-dependent carrier injection. Figure 10.15 shows EL spectra and PL of a typical device.

10.5 Visible electroluminescence from PS

PS-based EL has been extensively studied. As a result of the large amount of contributions, several approaches could be considered for the way to present this field. In a previous review [12] focusing exclusively on PS-based EL, we chose to sort the contributions by techniques used to (i) increase the efficiency, (ii) enhance the stability and (iii) tune the emission band. This present section is organized differently.

In the first paragraph of this section, we first briefly overview the three techniques stated above. In addition, we present an overview of the techniques used for PS EL integration and present the state of the art concerning the EL response time. This paragraph provides a short though fairly complete picture of the field. Also the advantage of this section is the direct and fast access to the techniques used for a particular target (enhancement of efficiency or stability, tuning or moulding the emission band, speed and integration aspects).

The other paragraphs in this section present in more details the contributions to PS-based EL sorted by device fabrication process. The first paragraph deals with wet EL. Then, a comparison of wet and solid-state

EL is given. The first and most simple LEDs (made from mere as-formed PS) are then introduced. The next paragraph focuses on devices including PS made from pn junctions. Then, devices with intentionally partially oxidized (by thermal, chemical and electrochemical oxidation) PS are presented. The next paragraph is devoted to device in which PS has been impregnated with another material, mostly to provide a better electrical contact to PS and with a view to imitating the efficient wet EL. In another paragraph, the influence of various top contact configurations are reviewed. Then, devices which include an optically active PS layer in a microcavity are discussed. The next paragraph deals with PS whose surface has been chemically modified, but not oxidized, and also PS capping techniques. Substitution of hydrogen by organic ligands is presented. The next paragraph discusses the modulation speed of PS based devices. Finally, the issue of integration of PS-based LEDs is discussed.

10.5.1 Overview

An overview of the field of EL from PS is now presented. This section focuses on the methods and the contributions which have led to significant advances in (i) enhancing the efficiency, (ii) enhancing the stability, (iii) tuning and moulding the emission band, and (iv) device speed and integration.

10.5.1.1 Enhancing the efficiency

The EL EQE and EPE are proportional to the density of Si nanocrystals in PS, to the intrinsic efficiency, to the carrier injection efficiency, and to the extraction efficiency. Furthermore, if the quantum efficiency remains constant in a given operating voltage interval, a voltage as low as possible is preferable for high power efficiency. All the above aspects have been considered in order to enhance the EQE and the EPE.

Figure 10.16 shows the evolution of the EQE during the past years since the discovery of PS luminescence based on selected contributions. Table 10.2 lists the devices selected in figure 10.16.

The earliest devices consist of a single layer of as-formed PS and a top contact. The efficiency of this type of device, without any further treatment, was usually low, i.e. $<10^{-3}\%$ [3, 113–116]. Moreover, the operating voltages for EL visible in daylight are rather high, usually greater than 10 V.

The efficiency has been improved by enhancing the carrier injection into luminescent nanocrystals. For this purpose, pn junctions have been considered. P^+n^- [87–89, 117–120], n^+p^- [121], n^+p^+ [122] and p^+n^+ [123] junctions have been used. EQE of about 0.2% have been reported under both CW [118] and pulsed operation [87–89, 124, 125]. 1.1% has been achieved under CW operation [123]. Some devices in this category have rather low operating voltages, one below 6 V [118].

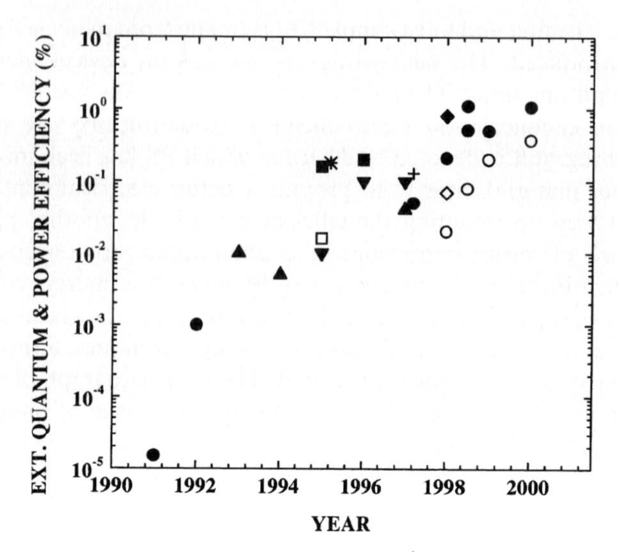

Figure 10.16. Selected values of external quantum and power efficiencies. Same symbol shape refers to same research group. Solid symbols refer to quantum efficiencies. Hollow symbols refer to power efficiencies. ■ □ [3, 123, 137, 138, 159, 196], ▼ [116, 117, 119, 126, 128], ∗ [118], ● ○ [87, 88], ◆ ◇ [120], ▲ [121, 135, 136], + [160]. Most devices appearing in this figure 10.6 are also listed in table 10.2.

In order to imitate the liquid contact (which provides very efficient EL), devices in which a PS layer is impregnated with another material have been fabricated. The material can be either a metal [124, 126–128] or a polymer [129–133]. Efficiency and stability is usually enhanced by polymer impregnation, but the attempt to imitate completely the liquid contact is not achieved, the efficiency and stability remaining very low in any case.

Devices in which PS has been partially oxidized are among the most efficient to date. Especially, thermal oxidation at around 850 °C [134–136], and ECO [123, 137, 138], are very effective. Efficiency of about 0.1% has been obtained [134–136]. Anodic oxidation has been shown to dramatically improve efficiency by reduction of the leakage current [123, 137, 138]. Anodically oxidized devices are the most efficient to date. EQE of 1.1% [123, 138] and power efficiency of about 0.4% [138] have been achieved, with bright EL below 5 V [138].

The reduction of the operating voltage is a way to reduce the power efficiency. As the series resistance due to PS increases with PS thickness, attempts have been made in order to fabricate thin PS layers. Highly homogeneous PS layers are required for high efficiency. This is challenging when considering sub-micrometer thick PS. Extreme care should be taken in the fabrication process. Simons *et al* [139] have fabricated a device including a 400 nm thick PS layer. The temperature is found especially important for

Table 10.2. Some characteristics of the devices of which EL efficiencies are plotted in figure 10.16.

Contact	Structure	Post-treatment	EL threshold ($V\,mA/cm^2$)	Stability of EL intensity	Emission peak (nm)	Highest efficiency EQE-EPE (%)	Year	Ref.
Au	p(D)		4–50		680	10^{-3}–	1992	3
Au	P^+n(L)		1.7–0.1	80 h	700	10^{-2}–	1993	117
Au	n(L:UV)	In electroplating	−0.1		480	0.005–	1994	126
ITO	p^+n(L)	1 min L	$2.3–10^{-3}$	Hours	600	0.18–	1995	118
Au	p^+n(L)			Seconds	630	0.16–0.016	1995	87
Au	p and p^+n(L)	2 min exposure in 10% HNO_3	5–		650	0.01–	1995	121
Au	*n*-poly-Si(L)				860–930	0.04–	1996	128
Au	p^+n(L)		3–1	Seconds	670–780	0.2–	1996	88
Al/poly-Si	p^+p(D)	Anneal in N_2 Or in 10% O_2 in N_2	1.5–2	1 month	620–770	0.1–	1996 1997	135, 136
Au	n(L:UV)	In electroplating	−0.1		480	0.01–	1997	119
Au	p^+n(L)		$>10–10^{-3}$		690	0.8–0.07	1997	120
Au/Al/pin (a-Si:H)	np				3 peaks: 455, 590, 670	0.13–	1997	160
Au	n^+(L)		−0.1		680	0.05–	1997	196
ITO	n^+(L)	ECO	$3.5–4 \times 10^{-4}$		640	0.51–0.05	1998	123
ITO	p^+n^+(L)	ECO	$5–1.5 \times 10^{-4}$		650	1.1–0.08	1998	123
ITO	n^+(L)	ECO	$3–10^{-4}$	Hours	640	0.21–0.02	1998	137
ITO	$n^+(D)n^+$(L)	ECO	$2 - 1.8 \times 10^{-3}$	Hours	680	0.5–0.2	1999	159
ITO	$n^+(D)n^+$(L)	ECO	$2.2 - 7 \times 10^{-4}$	Days, EQE is stable for months	680	1.07–0.37	2000	138

D and L mean that anodization has been conducted in the dark and under illumination, respectively. ECO refers to electrochemical oxidation, also referred to as anodic oxidation. The EL threshold voltage and current density are the lowest voltage and current density at which EL detection is possible with the author's detector. EQE and EPE are external quantum efficiency and external power efficiency, respectively. Stability data usually means that EL has dropped more than an order of magnitude after the given time period. When an exact value cannot be given, what is stated in the report is given.

good reproducibility. Their best device exhibits 0.18% of EQE at 4–6 V and they obtain EQE greater than 0.1% on a reproducible basis. Gelloz and Koshida [138] have achieved a record EPE of 0.37% below 6 V, with an optimized device including an ECO-treated thin PS layer ($<1\,\mu m$) formed at 0 °C under strictly controlled conditions. The incorporation of polymers has also shown the potential to reduce the operating voltage [129–133].

10.5.1.2 Enhancing the stability

EL and efficiency stability is still far below the 10 000 h minimum requirement for display purposes. One reason for the poor stability of PS EL is gradual oxidation of PS occurring during operation and storage in air. As a result of this oxidation, EL intensity and efficiency continually decrease due to driving current reduction at constant voltage and an increase in non-radiative defects.

The surface of as-anodized PS is mainly terminated with hydrogen atoms. Si–H_x ($x = 1, 2, 3$) groups do not protect PS against oxidation and contamination by numerous chemical species present in their storage environment, even at room temperature [140–149]. Furthermore, these groups can desorb at relatively low temperature [150] (from about 300 °C; the threshold temperature decreases when PS porosity increases [151]). Whether desorption of SiH_x groups or oxidation takes place, the intrinsic quantum efficiency decreases because of a rise in dangling bonds at the PS surface.

In order to enhance the stability, unwanted natural oxidation should be avoided. One approach is to change the chemical nature of the PS surface. Another approach is to substitute hydrogen by more stable organic groups. Another promising suggestion is PS capping, the idea being to prevent water molecules from entering into PS pores. The mechanical and electrical stability can also be enhanced.

Si nanocrystal passivation by SiO_2 is very efficient in enhancing the stability though it is not enough for the long run. Devices in which PS has been partially oxidized are among the most stable to date (and also the most efficient). Oxidation has been achieved by different methods: low temperature chemical oxidation [152], thermal oxidation at around 850 °C [134–136] and ECO [123, 137, 138]. Efficiency of about 0.1% with weeks of stability has thus been obtained [134–136]. ECO has been shown to improve the stability in addition to the efficiency [123, 137, 138]. No degradation of EQE has been detected even after one month, though low-rate oxidation still occurs upon storage [138]. The device from Pavesi *et al* [153] using ECO showed no degradation at all.

The substitution of hydrogen terminating as-formed PS by stronger ligands has been studied with a view to enhancing the stability. Devices including deuterium-terminated PS [154] and alkyl-terminated PS [155] have been realized. The latter gives very good stability. The alkyl groups protect the PS surface by a steric effect and by the high chemical stability of the Si–C bonds.

Layer capping and encapsulation techniques have been proposed in order to prevent the contamination of PS by all sorts of molecules present in ambient atmosphere. Especially, PS should be protected from water with a view to avoiding the ineluctable oxidation that occurs when it is left in air. The top contact (ITO) itself could act as an incomplete capping layer [156]. More complete capping of PS has been achieved by Lazarouk *et al* [157] who fabricated a device with an Al contact with transparent Al_2O_3 windows on to PS. The stability is reported to be more than a month. However, the capping process decreases the EQE since EL originating from under the non-oxidized Al cannot get out of the device. Recently, Koshida *et al* [158] have achieved effective capping of PS by ECR (electron cyclotron resonance)-sputtered SiO_2 films. The EQE is more effectively stabilized with increasing the capping layer thickness. The deposition in metal mode leads to much better stabilization than deposition in the oxide mode, because it presents lower permeability to water molecules.

The mechanical and electrical stability is now considered. The Si/PS interface is clean and allows good carrier injection. It is usually not the case of the PS/top contact. In order to enhance the electrical contact to PS, some authors [123, 134–136, 138, 159] have included a superficial compact porous layer between the optically active porous layer and the top contact. In the case of Gelloz and Koshida [138, 159], the superficial layer consists of a PS layer more compact than the underlying active porous layer. In some other cases, a p^+ PS superficial layer [123, 134–136] is used. A thin poly-Si layer has also been used between the top contact and PS [136]. The incorporation of the poly-Si layer reduces the surface-state concentration at the interface between PS and the top contact [136]. Filling of the pores with materials that are inert from the electrical and optical point of view has been proposed to enhance the mechanical stability of the PS skeleton [136]. The maximum increase in hardness achieved is 50%.

10.5.1.3 The EL emission and its tuning

The useful luminescence band of PS consists of the so-called 'S-band'. This luminescence originates from exciton recombination into Si nanocrystals. Red and orange emission is rather easy to obtain. Though possible generation of green and blue emission from PS nanocrystals has been demonstrated [30, 31], it is still challenging to reach efficiencies as high as for red-orange luminescence. Another problem is that the typical FWHM of the PL is usually relatively large, of about 300 meV.

Basically, tuning the emission band can be achieved by changing the size of the Si nanocrystals in PS. There are some general trends, such as the higher the PS porosity, the higher the emission energy. Tuning of EL can be achieved by varying the PS formation conditions. The electrochemical oxidation technique (ECO) can be used to blue-shift the EL emission peak. Succeeding in producing nanocrystals small enough that they can emit blue colour is unfortunately not enough. It has been shown that PS contaminated by oxygen cannot emit blue colour because of oxygen-related surface states that lie in the gap [29]. Therefore, for blue emission to be achievable, PS should be oxide-free [30, 31]. In addition different configurations of the top contact can lead to voltage-tunable EL. Finally, the tuning of the emission within a given band is achievable by considerably reducing the spectrum FWHM using microcavities.

It is relatively easy to form PS emitting red or orange colours. The EL can be tuned by changing the time of anodization [88], or the anodization conditions, or by using the ECO treatment [137]. These emission peaks are still in the red-orange colours and did not address the green and blue emission issue.

Low efficiency EL in the range 450–550 nm has been reported [124, 126–128], using PS fabricated under ultraviolet illumination. Blue EL (believed to originate from oxide-related centres) efficiency was improved by depositing indium into PS pores [126]. EL colours from red to blue were obtained by variation of the contact metal [124, 126–128]. In/Au, Al/Au, Ga/Au, Sn/Au and Sb/Au contacts lead to EL emission at 455, 455, 520, 555 and 700 nm, respectively [124]. Once again, the oxide in the PS layers is believed to be responsible for the short wavelength EL emission. Chen *et al* [160] have reported voltage-tunable EL with a device including a p-i-n a-Si:H multi-layer structure deposited on to PS. Three peaks are present in the EL spectrum. The relative intensities of these peaks change as a function of the voltage, resulting in voltage-tunable EL.

Green and blue colours are difficult to obtain, especially with high efficiency. The PS surface should be oxygen-free in order to emit blue light because oxygen-related defects in the PS bandgap prevent blue emission [29]. Blue EL has been demonstrated [32], but intensity and stability are very low at present.

Some EL devices based on PS, which include partially oxidized luminescent PS layers [119, 126–128, 157, 161–163] are believed to exhibit EL from oxide defects or contamination in the oxide, instead of recombination of excitons in Si nanocrystals. PL and EL spectra of these devices are quite different, suggesting that the two phenomena have different origins. Some of these devices emit white [157, 162] or green-blue [119, 126–128, 161, 163] colours.

The tunability of PL [164] and EL [165–167] colours has been achieved mostly in the red-orange part of the spectrum by inserting the emitting PS layer in a microcavity. FWHM for EL is reduced by about a factor of 3 [167].

The angular dependence of the emitted EL light has been found to be as a Lambert source by Kozlowski *et al* [67], whereas Linnros and Lalic [87] have found uniform emission up to an angle of 70°. High directionality has been achieved by using PS in microcavities [165, 166].

10.5.1.4 Speed and integration

The decay and rise times of all devices are below a millisecond and do not represent any problem for display applications. However, it is not enough for interconnects. There are some variations among devices. The modulation speed is usually influenced by the carrier mobility in PS, the radiative recombination processes and charge trapping. The EL modulation speed is typically found to be of the order of about ten of microseconds [87–89, 121, 138, 128, 169]. This would limit the frequency to about 100 kHz. However, one group has reported an efficient (0.1% of EQE) device based on partially oxidized PS that can be modulated at a frequency greater than 1 MHz [170]. A modulation frequency of 200 MHz has been reported [170] for a device showing lower efficiency and in which EL probably does not originate from recombination of excitons in Si nanocrystals. A reduction of the EL response time is potentially possible by using PS in a microcavity [165] since the PL lifetime may be reduced in this way. Another promising route for faster devices is the use of the fast (nanometre range) blue luminescence [32].

The use of PS EL in LSI technology has been demonstrated [171]. An alphanumeric display has been integrated with pnp bipolar transistors. PS-based EL was driven by transitors. EL from polycrystalline Si and its possible integration in large-area applications have also been demonstrated [172]. Porous polycrystalline Si diodes were shown to operate with efficiencies comparable with conventional PS diodes. It has also been confirmed that porous polycrystalline LEDs can be driven by a poly-Si-based switching TFT [172]. The fabrication of PS using silicon on insulator substrates has been demonstrated [173] using an alternating current electrochemical process. Furthermore, a PS LED has been fabricated using a process compatible with an industrial bipolar + complementary MOS + diffusion MOS technology [174]. Finally, the monolithic integration of a PS-based light emitter and a PS-based waveguide, using a simple anodization and standard silicon integrated circuit process techniques, has been demonstrated [175]. It seems that if the individual components of an Si-based opto-electronic circuit can be made practical, solutions will be found to integrate them together on the same chip.

10.5.2 Porous Si in contact with a liquid

The first demonstration of EL from PS has involved a liquid contact [2]. Anodically polarized p-type PS of high porosity in contact with an indifferent

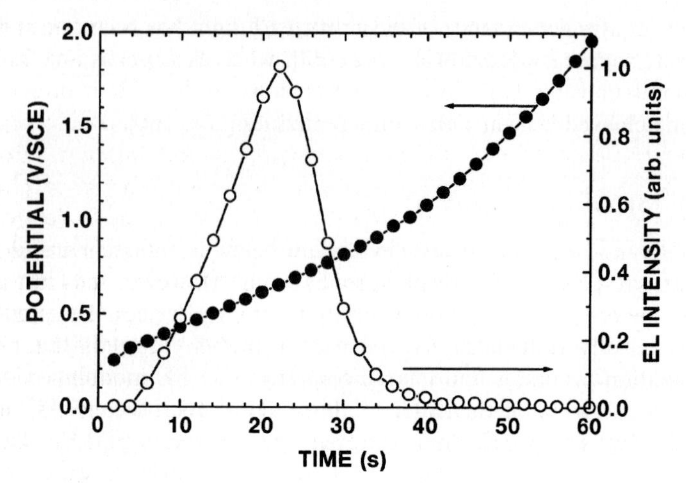

Figure 10.17. Potential and electroluminescence intensity as a function of time for an 80% porosity, 1 μm thick, p^+-type porous layer during anodic oxidation under 5 mA/cm^2 in H_2SO_4 (1 M). SCE refers to saturated calomel electrode.

conducting liquid phase exhibits efficient (~1%) and bright EL (visible with naked eye in daylight) at low voltages (~1–2 V). The EL has been attributed to radiative recombination of holes from the Si substrate with electrons resulting from the oxidation of Si atoms at the inner surface of PS [176]. Figure 10.17 shows the potential and the EL intensity as a function of time during anodic oxidation of a 1 μm thick, 80% porosity, p^+-type porous layer under constant current density. Holes supplied by bulk silicon are injected into silicon nanocrystals as a result of the polarization and trigger Si oxidation. This last process involves several charge exchange steps between the Si nanocrystal and its surface. In one of these steps, an electron is injected from the Si surface into the Si nanocrystal. EL becomes possible when a new hole is injected into the nanocrystal from the bulk [176]. The EL is initiated only after a delay time corresponding to oxidation of the coarser Si regions in PS, regions that cannot emit light due to too weak confinement. During the experiment, since oxidation tends to hinder the carrier flow, the potential increases in order to sustain a constant current. Later on, the potential becomes high enough so that holes can be injected into light-emitting nanocrystals, and EL begins. This EL is limited in time because Si nanocrystals are irreversibly oxidized during the process. The EL intensity eventually decreases and vanishes [2, 177–181]. Furthermore, the electrical contact between PS and the Si substrate will be broken at some point, resulting in a sharp rise of the potential.

Much more stable wet EL has been obtained under cathodic polarization, using the persulphate ($S_2O_8^{2-}$) ions [182, 183]. Electrons from the substrate are injected into Si nanocrystals under cathodic bias and reduction of persulphate ions occurs. The reduction of the $S_2O_8^{2-}$ ion is a two steps process. First, an electron from the Si conduction band is captured by an $S_2O_8^{2-}$ ion. This reaction leads to the unstable very reactive intermediate SO_4^- ion. The reduction of this ion takes place via hole injection into the Si valence band, to give the stable SO_4^{2-} ion [184]. Because the process includes a step in which a hole is injected into the valence band of the Si electrode, the two types of charge carriers can be injected in Si nanocrystals and therefore EL can be observed. Since the Si electrode is not consumed during the process, this EL is more stable than that resulting from anodic oxidation.

Another attractive feature of this system using persulphate ions is the voltage-induced spectral shift of the EL [185]. The tunability does not occur only by activation of higher and higher emission energies, but also by simultaneous quenching of the lowest emission energies. For example, at −1400 mV/ECS, emission at 2 eV is maximum, but emission at 1.6 eV has already completely vanished, after having peaked at −1150 mV/ECS. As the potential becomes more and more negative, electron injection in higher and higher energy levels is achieved, leading to EL at higher and higher energies. The quenching part has been explained by the Auger effect [186]. Injection of one electron into a nanocrystal can trigger EL. However, if two or more electrons are injected into one nanocrystal (when the applied potential is high enough), the non-radiative Auger recombination, which is much more probable than the radiative recombination, dominates thus quenching the EL. When one electron is present in a nanocrystal, a second electron may have to overcome the coulomb repulsion in order to penetrate into the same nanocrystal [186].

Finally, another type of wet EL has been observed, following the idea of one charge carrier injection from the substrate and one charge carrier injection from a species in the electrolyte. In the case of persulphate ions, the reduction of the ions leads to hole injection into PS. Methylviologen [187] and formic acid [188] oxidation lead to electron injection into PS, after capture of holes provided by the substrate. However, this type of EL is not stable due to simultaneous Si oxidation taking place under anodic bias.

10.5.3 Differences between wet and solid-state EL

The first demonstration of injection-mode solid-state EL based on PS was reported in 1992 [3]. The device structure was typical of most PS-based electroluminescent devices. First, a back contact is deposited on to the back-side of the Si substrate. After PS formation on the polished side of the substrate, a top contact is deposited on top of the PS layer. The voltage is applied between the back contact and the top contact. The EL is observed through the semi-transparent gold top contact. The device showed very poor external quantum efficiency (EQE) ($<10^{-4}$%) and had a high operating voltage (>10 V). The stability was only a few minutes. The EL emission was red band.

These characteristics show the main problems of solid state EL from PS: low efficiency, high operating voltage, poor stability and the difficulty of obtaining green and blue colours.

The characteristics of solid-state EL are very different from the results of wet PS EL, in which high efficiency, low operating voltages and tunable EL are easily achievable. The different conduction mechanisms involved in the two systems can explain most of these differences.

In the wet configuration, the conductive liquid impregnates completely the PS skeleton and short circuits the highly resistive PS layer. The current can be limited by one of three mechanisms: the supply of carriers by the substrate (regime 1), the supply of ions by diffusion in the electrolyte (regime 2) or the kinetic of the electrochemical reaction (regime 3) [189–191] (figure 10.18). In the first regime, charge carriers are consumed as soon as they

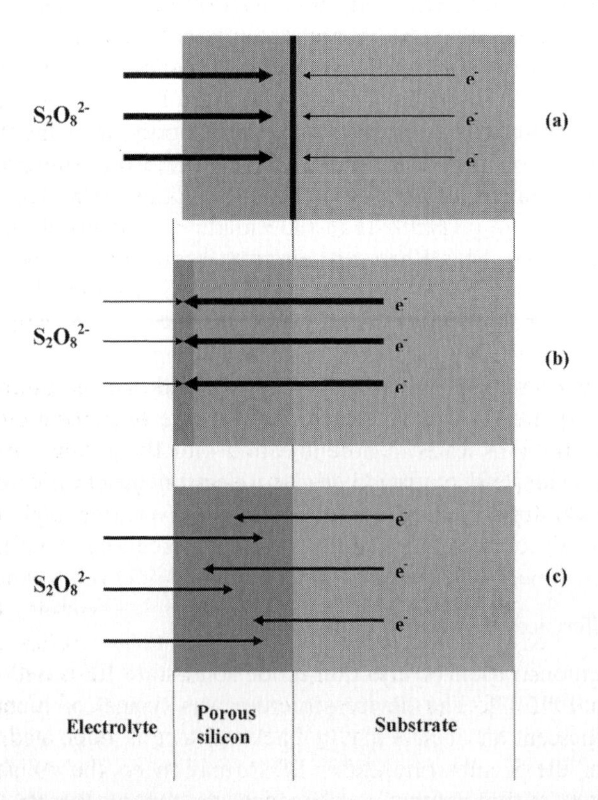

Figure 10.18. Schematic diagram of the three possible conduction regimes of the porous silicon/electrolyte system illustrated for the case of persulphate ions. Electrons or holes limited conduction (a), limitation by the mass transport in the electrolyte (b), and by the electrochemical reaction rate (c) [190].

are generated at the substrate–PS interface by the reactive ions. They do not penetrate into PS and no EL is observed. An example is during PS formation. The charge exchange between the electrolyte and Si takes place at the bottom of the pores, leading to the propagation of those pores in the depth of PS, further electrochemical etching of PS being avoided since no carrier penetrates into it. In the second regime, PS is accumulated with free charge carriers and the charge exchange with the electrolyte takes place at the top of the PS layer only. Ions supplied by the electrolyte bulk are consumed as soon as they reach the top of PS and cannot penetrate deeply into PS pores. Efficient EL is observed even though a very thin sub-layer of PS actually emits light. An example of this situation is the EL obtained with persulphate ions discussed earlier in this section [182, 183, 185, 189, 190, 192]. Due to a high reduction reaction kinetic, the persulphate ions can be consumed very quickly if a large amount of electrons is available at the electrode, as in the case of cathodically polarized n-type PS. In the third regime, PS is accumulated with free charge carriers and electro-active ions are present in the whole pores of PS. The charge exchange (and hence the EL) can take place in the whole thickness of PS. An example is anodic oxidation (discussed earlier in this section as the first instance of EL of PS). The kinetic of PS oxidation is low enough for the holes to accumulate completely the PS skeleton.

When charge carriers can penetrate into PS, their transport is a diffusion process since no significant electric field is set across PS (the electrolyte short circuit PS) [192–194]. Therefore the key features responsible for the high efficiency and low operating voltages are the absence of an important voltage drop in the liquid-impregnated porous skeleton and the fact that both electrons and holes can be efficiently injected into Si nanocrystals in part, or the whole thickness of PS in parallel.

Concerning solid-state devices, high voltages are necessary to get a significant current through PS because of the very high resistivity of nano-porous Si. The resistivity of nano-porous Si is usually greater than $10^5\,\Omega\,\mathrm{cm}$ at room temperature [195]. As a result, high electric fields are set across the PS layer. Locally, the electric field can be high enough to separate the electron–hole pairs, thus seriously limiting the EL efficiency [196]. This is a major difference with the wet configuration. Another great difference is the charge carrier injection mechanism. Indeed, in dry PS, both carrier types are injected into Si nanocrystals in series and not in parallel. Furthermore, there are some preferred low resistivity (but non-radiative) paths for the carrier flow. This reduces considerably the efficiency by shunting luminescent nanocrystals and inducing a large leakage. In wet PS, however, carriers accumulate the whole PS layer and all Si nanocrystals are potentially available for EL in suitable conditions of conduction.

The electron–hole pair generation mechanism in most PS-based EL devices can be described as follows [196]. Figure 10.19(a) shows a very simplified band diagram of Si quantum wire in PS at thermal equilibrium.

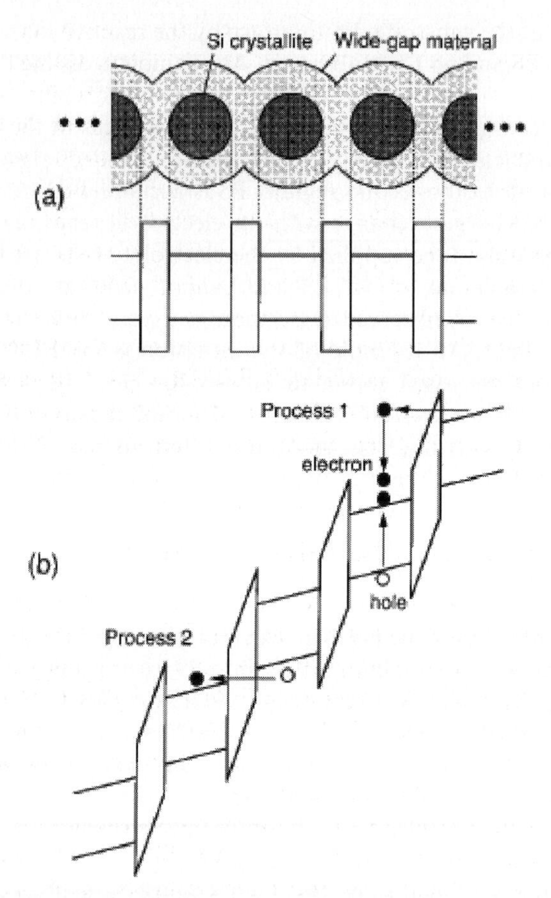

Figure 10.19. (a) Simplified schematic band diagram of a quantum wire in porous silicon under thermal equilibrium. (b) Band structure of the Si nanostructure system under a strong electric field. Two possible high-field effects are illustrated; process 1 (electron–hole pair generation by energetic hot electrons) and process 2 (tunnelling injection of electrons from the valence band) [196].

Si nanocrystals are connected by wider-bandgap regions. Under sufficiently high bias voltage, the band structure eventually reaches the state illustrated in figure 10.19(b). In this condition, two phenomena are likely to occur. First, hot electrons injected from the conduction band of a crystallite into the conduction band of the neighbouring one can create electron–hole pairs by impact ionization. Second, electron–hole pairs can be created by the tunnelling of electrons from the valence band of a crystallite into the conduction band of the next one. Therefore, although only one type of charge carrier is actually injected into PS, both electrons and holes can be generated in PS provided that a high electric field exists across PS [196].

Devices including a PS layer formed from a pn junction may allow injection of holes and electrons from the two different electrodes. However, it is not certain that true pn junctions exist in PS. Indeed, free carriers are mostly absent from PS, even when the bulk Si initial doping level is high, and should be trapped on some kind of surface states [197]. As for p-type Si, it seems that dopant atoms are still present in PS at concentration similar to bulk Si, but are passivated [197].

Recently, studies [198, 199] have shown that exposure of PS by NO_2 at low partial pressures ($<10^{-2}$ torr) induces a steep increase of over five orders of magnitude of conductivity. The mobility has been increased up to a value close to the bulk Si one. A chemisorption phenomenon has been suggested to explain the generation of free carriers in PS. This study may open a new route to Si nano-electronics and may allow better carrier injection for EL devices, provided that there is no significant quenching of the luminescence as a secondary effect.

10.5.4 Devices including an as-formed PS layer

Most devices in this section are listed in table 10.3. The very first instance of EL from PS was obtained from 75 μm thick PS formed from n^--type Si [200]. Very high voltages, greater than 100 V, were necessary in order to obtain very low efficiency red EL in d.c. regime, both in forward and reverse mode. The EL stability did not exceed 1 h. Charge carriers were not injected from the electrodes, but rather generated inside PS by electric field breakdown, as suggested by the high voltages required.

The first true demonstration of injection-mode solid-state EL based on PS was reported by Koshida and Koyama in 1992 [3]. PS was formed from p^--type Si. PS was then thinned in HF or in KOH before Au deposition. The device showed very poor external quantum efficiency (EQE) ($<10^{-4}$%) and had a high operating voltage (>10 V). The stability was only a few minutes. The EL emission was red band.

Following these two reports, a tremendous amount of work has been devoted to EL from PS, leading to many reports in the early 1990s. The earliest devices consisted of a single layer of as-formed PS and a top contact made of gold, ITO, or another conductive material. The best results are usually obtained with ITO, since it offers good transparency and fair conductivity. The efficiency of this type of devices, without any further treatment, is usually low, i.e. $<10^{-3}$% [3, 113–116], except in one case where 0.05% was obtained [196]. Operating voltages for EL visible in daylight are rather high, usually greater than 10 V.

Blue EL has been demonstrated [32] using a p-type PS layer with a gold contact. This shows the possibility of a red/green/blue LED based on PS. However, the efficiency of this device was very low and much work is needed to reach the same level of efficiency as for red-orange emitting

Table 10.3. Some characteristics of most devices discussed in section 10.5.4.

Contact	Structure	Post-treatment	EL threshold (V mA/cm^2)	Stability of EL intensity	Emission peak (nm)	Highest efficiency EQE-EPE (%)	Year	Ref.
Au	n(L)			45 min	650		1991	200
Au	p(D)		4–50		680	10^{-3}–	1992	3
ITO	P(D)		<2–<10	>5 h	580		1992	113–115
Au or ITO	n(L)		–1	Minutes	700		1992	116
Au	N^+(L)		–0.1		680	0.05–	1997	196
Au	Oxide-free PS from p(D)	L + H_2 exposure for 12 h			430		1999	32

D and L mean that anodization has been conducted in the dark and under illumination, respectively. ECO refers to electrochemical oxidation, also referred to as anodic oxidation. The EL threshold voltage and current density are the lowest voltage and current density at which EL detection is possible with the author's detector. EQE and EPE are external quantum efficiency and external power efficiency, respectively.

devices. Blue emission is difficult for several reasons: fabrication of small nanocrystals is difficult, keeping the nanocrystals oxide-free (to avoid the luminescence red-shift due to introduction of surface states by oxygen) is not easy and charge carrier injection into high energy levels is also not easy.

10.5.5 Porous Si formed from pn junctions

Most devices in this section are listed in table 10.4. The use of p^+n junctions, which are made porous by anodization, has been very promising. In this configuration, boron atoms are incorporated into an n-type substrate either by implantation or by diffusion. After PS is formed, the top contact is deposited on to the p^+ side. Devices in which a pn junction is porosified have shown better efficiencies than devices including one type only and without any post-anodization treatment. p^+n^- [87–89, 117–120], n^+p^- [121], n^+p^+ [122] and p^+n^+ [123] junctions have been used.

For a while, these devices were the most efficient reported, with efficiency of about 0.2% [87–89, 118], whereas devices including one type only and without any post-anodization treatment were showing efficiencies in the range 10^{-5}–10^{-3}% [3, 113–116], with one exception, 0.05% being reported by Oguro *et al* [196].

The advantage of the pn junction configuration is supposed to be that both electrons and holes could be injected into PS, without the need of impact ionization or the tunnelling generation described in figure 10.19. However, it is not certain that a pn junction is really present in PS in all the reported cases. The carrier injection mechanism might well be the one described in figure 10.19 for some devices in this category. The role of the p^+ superficial layer may be limited to an interface layer between n-type PS and the top contact in some cases.

Linnros and Lalic have used a porosified np^+ junction [87–89]. The junction was at a depth of 0.25 μm. Anodization is performed until the total PS thickness is in the range 20–60 μm. PS from the n-type part is composed of a nanoporous layer at the top and a macroporous underlying layer. EL is reported to originate from the n-type part. The device shows high series resistance (due to high PS thickness), inducing high operating voltages (>10 V) and perhaps shunting the pn behaviour. The device is pulse-operated. Their best device shows an EQE of about 0.2%. Nishimura *et al* [120] have also used a porosified pn junction. They obtained an EQE as high as 0.8% under pulsed operation, but for high voltages and low EL intensities.

Gelloz *et al* [123] reported a record EQE of 1.1% (external power efficiency (EPE) of 0.08%) using a porosified p^+n^+ junction. The PS layer was about 20 μm. This device also suffered from high voltage operation. In this case, the high EQE was attributed mainly to the post-anodization anodic oxidation of PS (described in section 10.3.3).

Table 10.4. Some characteristics of most devices discussed in section 10.5.5, about LEDs including a porous Si layer formed from a pn junction.

Contact	Structure	Post-treatment	EL threshold ($V\,mA/cm^2$)	Stability of EL intensity	Emission peak (nm)	Highest efficiency EQE-EPE (%)	Year	Ref.
Au	N^+p^+p(D)		– <600	>6 h	640	0.18–	1993	122
Au	p^+n(L)		1.7–0.1	80 h	700	10^{-2}–	1993	117
ITO	p^+n(L)	1 min L	2.3–10^{-3}	Hours	600	0.18–	1995	118
Au	p^+n(L)			Seconds	630	0.16–0.016	1995	87
Au	p and p^+n(L)	2 min exposure in 10% HNO_3	5–		650	0.01–	1995	121
Au	p^+n(L)		3–1	Seconds	670–780	0.2–	1996	88, 89
Au	p^+n(L)		>10–10^{-3}		690	0.8–0.07	1997	120
ITO	p^+n^+(L)	ECO	5–1.5×10^{-4}	Hours	650	1.1–0.08	1998	123

D and L mean that anodization has been conducted in the dark and under illumination, respectively. ECO refers to electrochemical oxidation, also referred to as anodic oxidation. EQE and EPE are external quantum efficiency and external power efficiency, respectively. Stability data usually means that EL has dropped more than an order of magnitude after the given time period. When an exact value cannot be given, what is stated in the report is given.

Simons *et al* [139] have used a p^+n porosified junction. They have made calculations supporting the fact that a pn junction exists in the PS layer included in their device [118], even though they do not anneal the boron-implanted atoms for electrical activation. EL is believed to take place in the p^+n junction. Due to the thin PS layer used (400 nm), their fresh device is CW operated at voltages below 6 V. The best EQE is 0.18% [118, 139, 169,] at 4–6 V and an EQE of about 0.1% is achieved on a reproducible basis. The EQE is highly dependent upon the boron dose. Best performance is found for a dose of about $10^{16}\,cm^{-2}$. The temperature is found especially important for good reproducibility.

Chen *et al* [122] have fabricated a device by anodizing an n^+p^+p substrate. PS is formed from the n^+ side and extends up to the p region. The top contact is Au. EL could be seen within a forward voltage of 5–10 V and current density of about $600\,mA/cm^2$. Rectifying properties were very good. A non-ideality factor of 3 has been measured over three decades of current, and a series resistance of only 250 Ω. The PL and EL peak is at about 650 nm. It was not possible to draw any conclusion about the doping dependence of light emission from the experimental data. However, light was observed to originate from a PS region a few micrometres below the top contact–PS interface. The authors then conclude that the EL has to be the result of electron–hole recombination at the porous n^+p^+ junction.

10.5.6 Partially-oxidized porous Si

Most devices in this section are listed in table 10.5. In order to replace the fragile $Si–H_x$ bonds terminating the surface of freshly prepared PS by SiO_2, PS oxidation has been conducted under various conditions: chemically [152], thermally [134–136] and electrochemically [123, 137, 138, 153, 159]. Kozlowski *et al* [152] have oxidized n-type PS in H_2O_2/water/ethanol mixture. This way, EL can be stabilized at about 50% of the initial value for several hours. EQE is found to increase during operation.

One group has oxidized PS using thermal oxidation in the range 800–900 °C, either in N_2 (nominally) or in N_2 with 10% O_2 [134–136]. The device consists of a porosified p^+p^--type junction. The superficial p^+-type PS layer is 500 nm thick whereas the p^--type underlying active PS layer is 1 μm thick. A 300 nm thick n^+-type poly-Si was deposited on to the p^+ side to act as the top contact. The resulting device has several advantages. First, the stability is rather good, with weeks of stable EL [134]. Second, the EQE of 0.1% is relatively high [134–136]. Third, the EL can be modulated by a square wave current pulse with frequencies greater than 1 MHz [135]. Finally, a bipolar device fully compatible with conventional Si microelectronic processing has been demonstrated [135, 171]. The high stability is due to replacement of the fragile hydrogen passivation by a thin Si oxide coverage.

Table 10.5. Some characteristics of most devices discussed in section 10.5.6, about LEDs including an oxidized porous Si layer.

Contact	Structure	Post-treatment	EL threshold (V mA/cm^2)	Stability of EL intensity	Emission peak (nm)	Highest efficiency EQE-EPE (%)	Year	Ref.
Au	n(L)	H_2O_2 oxidation			650–750			152
Au	n(UV)	H_2O_2 oxidation		>7 h	460–550			152
Al/poly-Si	p^+p(D)	Anneal in N_2 Or in 10% O_2 in N_2	1.5–2	1 month	620–770	0.1–	1996 1997	135, 136, 171
ITO	n^+(L)	ECO	$3.5–4 \times 10^{-4}$		640	0.51–0.05	1998	123
ITO	p^+n^+(L)	ECO	$5–1.5 \times 10^{-4}$		650	1.1–0.08	1998	123
ITO	n^+(L)	ECO	$3–10^{-4}$	Hours	640	0.21–0.02	1998	137
ITO	n^+(D)n^+(L)	ECO	$2–1.8 \times 10^{-3}$	Hours	680	0.5–0.2	1999	159
Al/n^+	p(L)	ECO		>1 week			1999	153
ITO	n^+(D)n^+(L)	ECO	$2.2–7 \times 10^{-4}$	Days, EQE is stable for months	680	1.07–0.37	2000	138

D and L mean that anodization has been conducted in the dark and under illumination, respectively. ECO refers to electrochemical oxidation, also referred to as anodic oxidation. EQE and EPE mean external quantum efficiency and external power efficiency, respectively. Stability data usually means that EL has dropped more than an order of magnitude after the given time period. When an exact value cannot be given, what is stated in the report is given.

The increase in response speed may be related to a different recombination mechanism involving the oxide rather than the interior of the Si nanocrystals.

Blue and green EL has been reported from partially oxidized PS. Mimura *et al* [161] obtained blue EL from porous SiC layers and green EL from PS formed under ULTRAVIOLET illumination. In this latter case, the PL and EL were different, and thought to have different mechanisms. The efficiency was very low ($<10^{-5}\%$).

Post-anodization partial anodic oxidation of PS has been studied with a view to increasing the efficiency and stability [137]. PS includes luminescent Si nanocrystals, but also regions of non-confined or weakly confined Si that cannot emit any light. A large current flows through weakly confined Si (energetically easiest paths), inducing a large leakage current. This is one reason for the low EQE of PS-based EL. The objective of anodic oxidation (ECO) is to selectively decrease the size of the non-confined Si skeleton without much affecting the confined Si nanocrystals, in order to reduce the leakage without damaging the luminescence. In addition ECO performed in given conditions optimizes charge carrier injection into luminescent crystallites.

The detailed mechanisms of ECO of luminescent p^+ PS can be found in [181]. It has also been discussed briefly in section 10.5.2 in the case of p^- PS. Figure 10.20 illustrates the ECO process. Oxidation occurs at the internal surface, where holes supplied by the substrate are injected. During the first stages of the process, hole injection, i.e. oxidation, occurs only in non-confined silicon regions. Hole injection in more energetic levels is later achieved, and injection eventually occurs in low-dimensional luminescent crystallites. EL can be observed at this stage. EL reaches a maximum when carrier injection in confined crystallites is optimal.

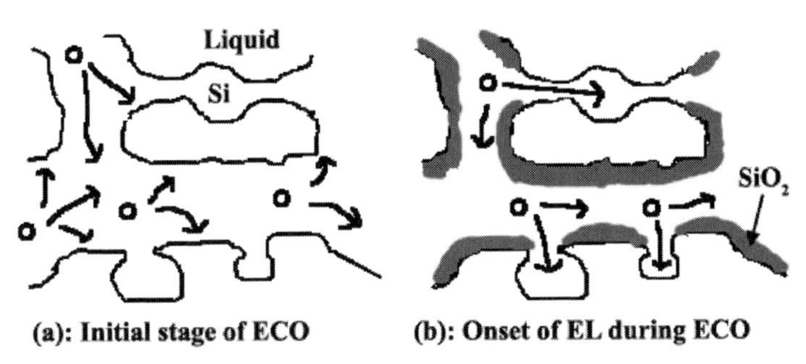

Figure 10.20. Schematic representation of anodic oxidation (also referred to as ECO in the text) in PS under galvanostatic condition. First, holes supplied by the substrate flow through non-confined silicon where energy levels are easily accessed (a). Later on, as a result of oxidation of the surface of non-confined silicon, the potential is increased in order to maintain the current constant. Then, carrier injection into luminescent crystallites becomes possible, leading to electroluminescence (b).

When oxidation is performed up to the maximum of EL, the size of non-emissive silicon is much more oxidized than luminescent crystallites. ECO decreases the leakage by several orders of magnitude and optimize carrier injection into luminescent crystallites. Thermal or chemical oxidation could not lead to such an enhancement since it occurs on the whole internal surface of PS (without selecting non-confined silicon from confined silicon). Moreover, contrary to thermal oxidation, ECO does not affect the PS structure and passivation with hydrogen [179].

ECO also enhances the luminescence homogeneity. If PS is not uniform, ECO acts as a kind of healing treatment in so far as it tends to homogenize the size distribution of the conductive paths. The EL during ECO is also a unique probe of PS homogeneity. Indeed, if it is seen by naked eye that EL is not uniform during ECO, then PS is not homogeneous on the millimetre scale. Devices showing highly uniform EL during ECO are found the most efficient ones.

The strong effect of ECO on the EL efficiency was first demonstrated by Gelloz *et al* [137], who obtained an EQE of 0.21% (EPE of 0.02%) with a single PS layer made from n^+ (100) Si. The treatment enhances the EL efficiency by several orders of magnitude, mainly by reducing the current density in the device. Figure 10.21 shows the PL before and after oxidation as well as the EL spectrum of the oxidized sample at -25 V. As expected [181], ECO treatment induces a blue-shift of PL. The EL spectrum fits very well the PL one. This indicates that the same carrier-recombination mechanisms is involved in the two phenomena (recombination of excitons localized within silicon nanocrystallites). EQE of 0.51% and 1.1% (EPE of 0.08%) was then obtained with n^+ (111) and p^+n^+ (111) (p^+ at the surface) PS layers, respectively [123]. However, the operating voltages of these devices were still high, exceeding 10 V. An optimized device, operating below 5 V, with an EQE of 1.07% and an EPE of 0.37% has later been fabricated [138]. The modulation speed was about 33 kHz. Strictly controlled conditions are important for good reproducibility. Especially, the temperature is maintained at 0 °C during PS formation of the active PS layer. This device exhibits the best characteristics to date in terms of quantum and power efficiencies and low operating voltage.

Figure 10.22 [138] shows the EL intensity and the current density as a function of voltage, for two devices. They have been prepared in the same conditions, except that one has been ECO-treated and the other one is not oxidized. EL of oxidized device can be seen with the unaided eye in room lighting at operating voltages below 5 V. ECO decreases the current density of about three orders of magnitude in this case. The EL intensity is also improved by ECO. The EQE of the device increases by more than four orders of magnitude due to ECO. The dramatic enhancement of EQE of PS EL due to ECO has been confirmed by Pavesi *et al* [153].

The ECO treatment also improves the EL stability [123, 137, 138]. The device of Gelloz and Koshida [138] shows no loss of efficiency during

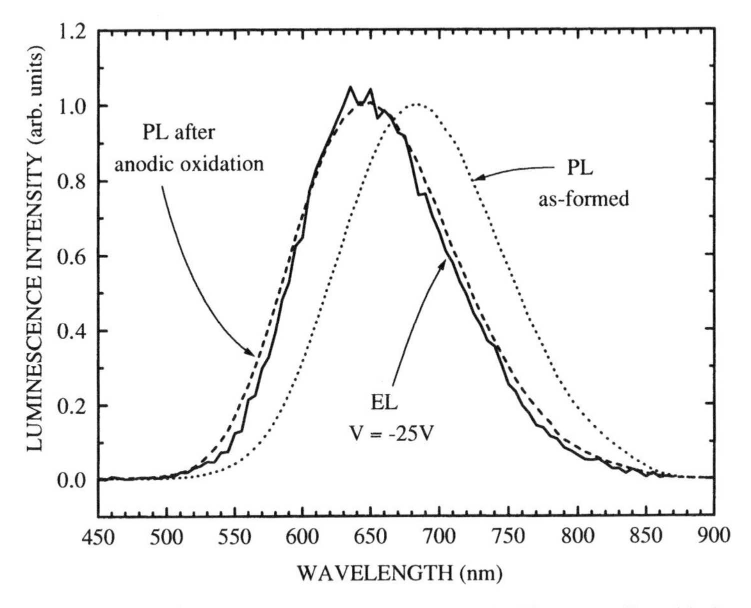

Figure 10.21. Normalized PL of porous silicon with and without anodic oxidation post-treatment. Excitation is 325 nm. Also represented is the EL spectrum obtained at −25 V with the oxidized porous layer [137].

operation, as shown in figure 10.23, and after one month's storage in air. However, the PS layer still gets slowly oxidized during operation and storage in air, hence decreasing the driving current and thus the EL intensity at a given voltage. ECO as carried out by Gelloz and co-workers [123, 137, 138] does not replace the hydrogen coverage of luminescent nanocrystals by silicon oxide [179]. It does not replace the hydrogen passivation by SiO_2 passivation. Pavesi *et al* [153] have also applied the ECO on their device and confirmed the enhancement of the stability. No degradation of EL intensity during several days has been obtained.

10.5.7 Porous Si impregnated with other materials

Most devices in this section are listed in table 10.6. In order to imitate the efficient liquid contact, impregnation of PS with conductive materials has been considered. The material has been either a metal or a polymer. One group has studied the incorporation of In [119, 126, 127], Al [119], Sn [119, 127] and Sb [119, 127] into PS pores by electrochemical techniques. The best device was obtained with In, with an EQE of 0.01% in a.c. conditions [119]. The In plating increases the EQE by a factor of 150. PS is also partially oxidized. The mechanism by which the EQE is increased by

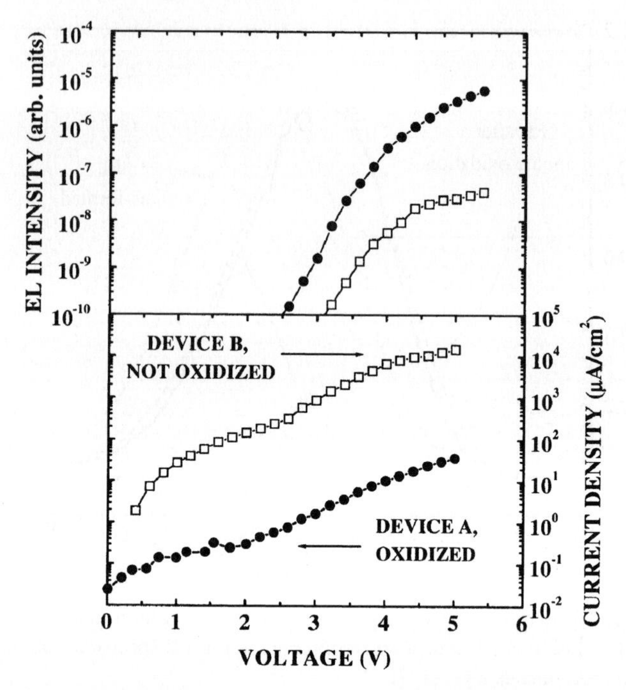

Figure 10.22. Electroluminescence intensity and current density as a function of applied voltage for an anodically oxidized (device A) and a non-oxidized device (device B) [138].

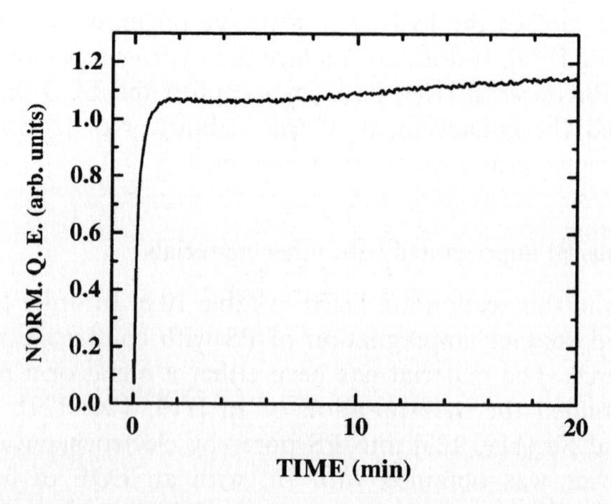

Figure 10.23. Normalized quantum efficiency at 10 V as a function of time for an efficient anodically oxidized device [138].

Table 10.6. Some characteristics of most devices discussed in section 10.5.7, about LEDs including a porous Si layer impregnated by another material.

Contact	Structure	Post-treatment	EL threshold (V mA/cm^2)	Stability of EL intensity	Emission peak (nm)	Highest efficiency EQE-EPE (%)	Year	Ref.
Au	p(D)	Polypyrrole electro-deposition	2–<0.01		590		1993, 1995	129, 130
Au	n^+(D)	PANI chemical deposition	3–400		800		1995	131
Au	n(L)	Polyaniline chemical deposition	−500		790		1996	132
Au	n(L:UV)	In electroplating	−0.1	Hours	455–480	0.01–	1994–1997	119, 126, 127
Au	n(L:UV)	Ga electroplating		Hours	520		1996	127
Au	n(L:UV)	Sn electroplating		Hours	550	0.0005–	1996, 1997	119, 127
Au	n(L:UV)	Sb electroplating		Hours	700–750	0.0001–	1996, 1997	119, 127
Au	n(L:UV)	Al electroplating	−0.1	Hours	480	0.005–	1997	119

D and L mean that anodization has been conducted in the dark and under illumination, respectively. EQE and EPE mean external quantum efficiency and external power efficiency, respectively. Stability data usually means that EL has dropped more than an order of magnitude after the given time period. When an exact value cannot be given, what is stated in the report is given.

In electroplating is not fully understood as the oxide may be also be involved. This device emits in the blue (480 nm). This EL may well be oxide-related rather than the result of exciton recombination in Si nanocrystals.

As for polymers, polypyrrole [129, 130] and polyaniline [131, 132] have been used. Efficiency is usually enhanced by the polymer impregnation, but the attempt to imitate completely the liquid contact could not be achieved in all cases, the efficiency remaining low. Koshida *et al* [129, 130] have studied a device in which polypyrrole had been electrochemically deposited into p-type PS pores and a gold contact deposited on to PS. The current–voltage characteristics and the voltage and current dependence of EL are significantly improved in comparison with a control device. A rectifying ratio at 15 V of about 1000 was obtained with electropolymerized contact, whereas a value below 300 has been obtained without polymer impregnation. The voltage thresholds of EL are identical whether polymer is incorporated or not. However, the EL intensity as a function of voltage rises in a much steeper way when polymer is used. The EPE is improved by a factor of 3. Halliday *et al* [132] have chemically deposited polyaniline into and on to PS made from n-type Si. The polymer acts as a hole injector. EL is obtained at 0.5 A/cm^2. Bsiesy *et al* [131] have deposited polyaniline (PANI) into n^+ PS by chemical oxidation of aniline by persulphate ions. Optimal results are obtained with deposition of two layers of PANI. More layers of PANI reduce the EL intensity. The test device without polymer shows a voltage threshold of EL of about 9 V. When two PANI layers are deposited in PS, the voltage threshold of EL becomes 3 V and EL intensity increases with a much higher slope than when no polymer is present in PS. However, when more PANI layers are incorporated, the performance of the diode drops, even though it is still better than when no polymer exists in PS. EL intensity is found to be six times lower than that from the liquid junction cell. The current–voltage characteristics and the voltage and current dependence of EL are significantly improved in comparison with a control sample including a gold top contact. Li *et al* [133] have also deposited polyaniline and found that the EL intensity is increased compared with a control device.

It is worth noticing that good impregnation of a material much more conductive than PS could short-circuit the PS layer, resulting in low carrier injection into Si nanocrystals. Such a situation would be similar to regime 1 described in section 10.5.3 for liquid contact [189–191], in the sense that the contact would prevent carrier injection from the Si substrate into PS. Solid-state emulation of regime 2 or even better regime 3 of liquid contacted systems would be a promising approach for enhancing the efficiency.

Filling of the pores with materials that are inert from the electrical and optical point of view has been proposed to enhance the mechanical stability of the PS skeleton [136]. Various polymers have been tried, such as PVC, polyamide, PMMA, polystyrene, polypropylene. The hardness of the nanocomposite increases with the density of the base polymers. Maximum increase in hardness achieved is 50%.

10.5.8 Influence of the top contact

Most devices in this section are listed in table 10.7. Several top contact configurations have been investigated. An ideal contact should be transparent to visible light in order to guarantee maximum light extraction. The first devices usually included a thin semi-transparent gold top contact, as in the case of the first injection mode device from Koshida and Koyama [3].

Later on, ITO (indium tin oxide) has been preferred to gold. It is more transparent than gold. A thicker film can be used, improving the mechanical strength of the contact. In addition, the efficiency is higher when using ITO. Simons *et al* [156] have compared ITO and gold as top contact. One device included a thermally evaporated 40 nm thick gold contact. The gold film showed 40% visible transparency. The efficiency was in the range 0.01–0.02%. Another device included a radio frequency sputtered 100 nm thick ITO showing 95% visible transparency. The efficiency was in the range 0.1–0.2%. It is also shown that LEDs with a semi-transparent Au contact are much less stable in air than LEDs with an ITO contact. It is attributed to Au being more permeable to air than ITO. The Au-contacted device gets oxidized and degraded much faster than the ITO-contacted one.

Some devices include Al top contacts. In the case of Lazarouk *et al* [157, 162], aluminium contact is deposited on to PS. Then, parts of the Al are electrochemically oxidized into Al_2O_3 to create transparent windows through which EL is observed. The stability is reported to be more than a month. However, there are two drawbacks in their device. First, the efficiency is limited by the fact that a significant part of the light (originating from under Al) cannot be extracted. Second, the device emits white colour, and the EL is likely to originate from some oxide-related centres rather than from the PS red band.

If the top contact is directly deposited on to the rough and uneven surface of a highly porous and optically active PS layer, conduction peculiarities may arise and reproducibility can be bad [159]. To solve these problems, some authors [123, 134–136, 138, 159] have included a superficial compact porous layer between the optically active porous layer and the top contact. This superficial layer smoothes the mechanical and electrical transition between the top contact and the active porous layer. It should provide a better electrical contact and greater mechanical stability to the optically active PS layer. In the case of Gelloz and Koshida [138, 159], the superficial layer consists of a PS layer whose surface conserves the mirror property of the silicon substrate and which is more compact than the underlying active porous layer. This layer is formed in the dark, which usually leads to lower porosity than when illuminated. Figure 10.24 shows a schematic cross-section of the device. In

Table 10.7. Some characteristics of most devices discussed in section 10.5.8, about porous Si-based LEDs in which a particular top contact has been used.

Contact	Structure	Post-treatment	EL threshold (V mA/cm^2)	Stability of EL intensity	Emission peak (nm)	Highest efficiency EQE-EPE (%)	Year	Ref.
Au	p(D)		4–50		680	10^{-3}–	1992	3
ITO/n-type SiC	p(D)	KOH dip					1992	203
Al/Al_2O_3	n- or n^+-poly		5–400	>1 month		–0.01	1996	157, 162
Au	n-poly-Si (L)				860–930	0.04–	1996	128
Al/poly-Si	P^+p(D)	Anneal in N_2	1.5–2	1 month	620–770	0.1–	1996	135, 136
		Or in 10% O_2 in N_2					1997	
Au	p^+n(L)			Minutes	700	0.02–	1997	156
ITO	p^+n(L)			Hours	700	0.2–	1997	156
Au/Al/pin (a-Si:H)	np(D)				3 peaks: 455, 590, 670	0.13–	1997	160
Au/Al/npin (a-Si:H)	p(D)		6–				1997	202
Au/Al/npnin (a-Si:H)	p(D)		3.6–				1997	202
ITO	n^+(L)	ECO	$3.5–4 \times 10^{-4}$		640	0.51–0.05	1998	123
ITO	p^+n^+(L)	ECO	$5–1.5 \times 10^{-4}$		650	1.1–0.08	1998	123
ITO	n^+(D)n^+(L)	ECO	$2–1.8 \times 10^{-3}$	Hours	680	0.5–0.2	1999	159
ITO	n^+(D)n^+(L)	ECO	$2.2–7 \times 10^{-4}$	Days, EQE is stable for months	680	1.07–0.37	2000	138
ITO	n^+(D)n^+(L)	ECO	0.5–	Minutes	Voltage-tunable		2003	204

D and L mean that anodization has been conducted in the dark and under illumination, respectively. ECO refers to electrochemical oxidation, also referred to as anodic oxidation. EQE and EPE mean external quantum efficiency and external power efficiency, respectively. Stability data usually means that EL has dropped more than an order of magnitude after the given time period. When an exact value cannot be given, what is stated in the report is given.

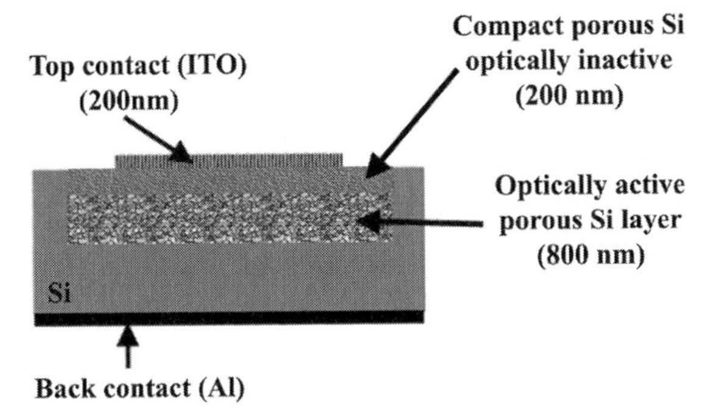

Figure 10.24. Cross-sectionnal view of a device which structure is ITO/superficial compact porous Si/optically active porous Si. Not to scale. Substrate is n^+-type Si. This device can be operated under both forward and reverse bias. Such a device has been shown to exhibit an ext. quantum efficiency of more than 1% and a record of 0.37% of ext. power efficiency below 6 V [138].

some other cases, a p^+ PS superficial layer [123, 134–136] is used. PS formed from p^+ Si shows much better mechanical stability than PS made from other Si substrates. Therefore such a layer is a good choice as a buffer layer between the top contact and the active PS layer.

EL colours from red to blue have been obtained by variation of the contact metal [119, 126–128]. In/Au, Al/Au, Ga/Au, Sn/Au and Sb/Au contacts lead to EL emission at 455, 455, 520, 555 and 700 nm, respectively [119]. The metals were also impregnated into PS. These devices are listed in table 10.6. However, the oxide in the PS layers is believed to be responsible for the short wavelength EL emission, rather than from the PS red band.

A 300 nm thick n^+ poly-Si layer has been used between the Al top contact and PS [136]. This device has already been discussed in section 10.5.6. The incorporation of the poly-Si layer reduces the surface-state concentration at the interface between PS and the top contact [136].

More sophisticated structures have been considered with a view to enhancing the carrier injection into PS. One group has studied devices in which a p-i-n [160] or a n-p-i-n [201] or a n-i-p-n [202] or a n-i-n-p-n [202] a-Si:H multi-layer structure has been deposited on to PS. The n-i-n-p-n device has a lower threshold voltage for EL detection (3.6 V) than the n-i-p-n LED (6 V) [202], which in turn is better than PS LED (>10 V). The a-Si:H multi-layer structure is believed to enhance the carrier injection into PS. Red-orange EL can be seen by naked eye in the dark [202]. Brightness of about 30–50 cd/m^2 at 600 mA/cm^2 and an EL efficiency of 0.13% have been reported [160]. The p-i-n PS LED [160] is said to be voltage-tunable between 30 and 90 V, as a result of three peaks present in the EL spectrum. The EL peak at 675 nm dominates at 30 V. The intensity of a peak at 590 nm increases with voltage but does not become stronger than the previous peak intensity. Another EL maximum, at 455 nm, becomes the strongest emission at 50 V and increases more than the peak at 675 nm until 90 V.

Futagi *et al* [203] have fabricated an LED with microcrystalline SiC deposited on to PS. The diode structure was p-type Si/PS/n-type SiC/ITO. The EL efficiency was very low. EL was observed by naked eye in the current range from 200 mA to 619 mA (contact area equals 1 mm^2), at a voltage above 20 V.

Very recently, Gelloz and Koshida [204] studied the effect of a few nanometres thick amorphous carbon layer deposited (by sputtering) on to PS before the deposition of the ITO top contact. Visible EL in room lighting was obtained at low operating voltages (3 V) from n^+-type silicon–electrochemically oxidized thin nanocrystalline porous silicon (600 nm)–amorphous carbon–ITO junctions. The EL voltage threshold was below 1 V and about −0.5 V under forward and reverse operation, respectively. The EL excitation is believed to be impact ionization. The carbon film enhances the stability. Furthermore, the EL efficiency is improved due to a reduction in current density and an increase in EL intensity. In addition, the reproducibility from device to device is very much improved by the carbon film. The enhancement in stability is attributed to the capping of PS by the carbon film and the high chemical stability of carbon and Si–C bonds, which should prevent PS oxidation. The carbon film acts as an efficient mechanical and electrical buffer layer between PS and ITO, resulting in enhanced mechanical, electrical and chemical stability of the top contact and providing high reproducibility. There are no drawbacks in the use of the carbon film as long as it is not thick enough to absorb significantly the light emission. There is one more interesting feature of the device: the EL peak wavelength is voltage-tunable between 700 and 630 nm for a voltage ranging from 2 to 5 V (a property not related to the carbon film). This behaviour may have its origin in the field-induced EL generation mechanism.

10.5.9 Porous Si microcavities

Most devices in this section are listed in table 10.8. There are three main advantages of inserting a luminescent PS layer between two reflective media. First, a significant PL line narrowing is achievable in this way [165]. Second, high luminescence directionality can be achieved by using PS in microcavities [165, 166]. Finally, although not experimentally demonstrated yet, a reduction of the EL response time is potentially possible by using PS in a microcavity [165] since the PL lifetime may be reduced in this way.

Tuning the EL emission has been achieved by placing the luminescent PS layer between two multi-layer reflective media (a Fabry–Pérot resonator). The reflectors could be Bragg reflectors made of PS layers of alternating refractive index [165]. Figure 10.25 shows a typical example of such a

Table 10.8. Some characteristics of most devices discussed in section 10.5.9, about LEDs including an optically active porous Si layer in a porous Si microvavity.

Contact	Structure	EL threshold (V mA/cm^2)	Emission peak (nm)	FWHM (nm)	Year	Ref.
Au	p(D) between two Bragg reflectors	6–0.02			1996	165
Au	p(D) on one Bragg reflector	40–10^{-4}	690	36	1996	167
Au	p(D) between two Bragg reflectors	>30–	780 to 700	19	1999	166

D means that anodization has been conducted in the dark. FWHM refers to full width at half maximum of the EL spectrum.

structure. In this case, a luminescent PS layer is sandwiched between two Bragg reflectors (containing four periods of PS layers of alternating porosities), so that the operating central wavelength of the cavity is λ_c. The central wavelength of the Bragg reflectors is λ. Araki *et al* [167] have used the gold top contact in place of the top reflector and achieved a reduction of the FWHM by a factor of 3. The FWHM was about 100 meV for emission

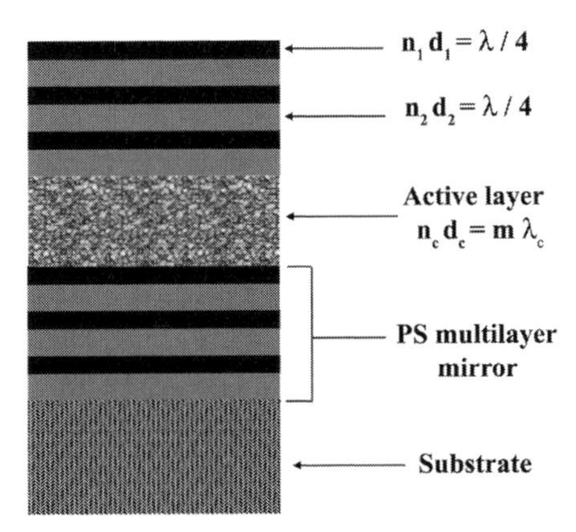

Figure 10.25. Tuning of EL emission is possible by placing the luminescent PS layer between two Bragg reflectors, thus obtaining a Fabry–Pérot resonator. A very high optical line narrowing is thus achievable. In this example, the mirrors consist in two Bragg reflectors made of three periods of PS layers of alternating refractive index. Subscripts 1 and 2 refer to the low-porosity PS layer (high refractive index), and the high-porosity PS layer (low refractive index), respectively.

energy of 1.8 eV. The same group has demonstrated the possible tuning of the PL emission from 1.5 to 2.2 eV [164]. Narrow spectra (10–40 meV in FWHM) are possible by using this approach, compared with the wide typical FWHM of PS PL (0.3 eV).

More recently, Chan and Fauchet [166] have demonstrated narrow and tunable EL, depending on the anodization parameters, using an active layer sandwiched between two Bragg reflectors. The substrate was p^+ Si. The multi-layer mirrors have alternating porosities and thicknesses of 43%; 80 nm and 62%; 160 nm. Nearly 100% reflectivity is achieved with a multi-layer mirror containing 10 periods. However, only six periods are used in the device, reaching a reflectivity of about 88% at 760 nm. The reason for the choice of fewer periods is that too many periods would result in too high a series resistance for the final device. With six periods per mirror, the devices are typically operated at a reverse bias as high as 100 V due to the thick total PS layer involved (two Bragg reflectors and the luminescent layer). The FWHMs are about 50 meV. The EL can be tuned from 1.65 to 1.85 eV (see figure 10.26) by changing the porosity of the active layer from 76% to 94%. In addition, high directionality is observed for these microcavity LEDs, concentrating within a 30° cone around the axis.

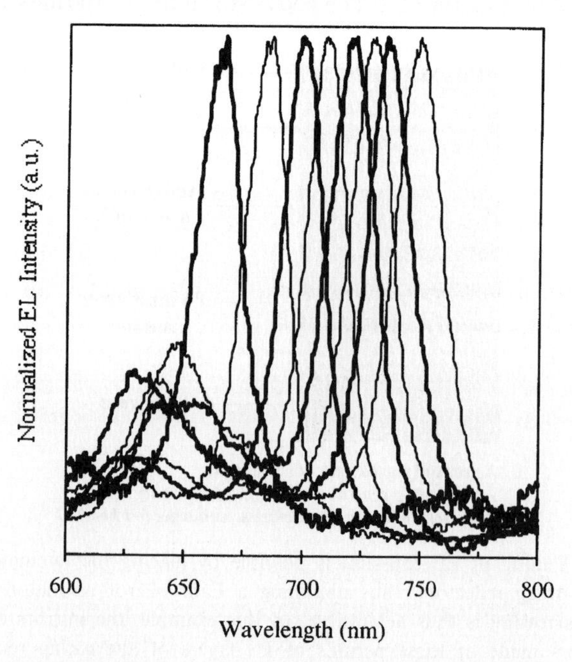

Figure 10.26. Electroluminescence from oxidized PS microcavity resonators with varying active layer porosity. Devices were reverse biased at about 100 V [166].

Microcavities are very useful for tuning the emission within the PL spectrum. However, there are some drawbacks. First, absorption in the Bragg reflectors may be a serious limitation for short wavelengths. In addition, a microcavity needs PS which can be made without depth gradients, in order to ensure good periodicity. Therefore, the fabrication of a microcavity based on n-type Si should be very difficult. n^+-Type Si may be easier since anodization in the dark is possible. Finally, the total thickness of PS included in a microcavity is quite large. As a result, high series resistance cannot be avoided and high voltages are necessary for EL operation. As an example, the device from Chan and Fauchet [166] requires about 100 V to operate.

10.5.10 Porous Si stabilization (surface chemistry and capping of porous Si)

Substitution of the hydrogen atoms terminating the PS surface by various species has been proposed in order to increase the EL stability. EL from deuterium-terminated PS exhibits better stability than hydrogen-terminated PS, but do not solve the problem for the long run [154]. Substitution of hydrogen atoms by CH_3 groups has been achieved via an electrochemical route [205]. About 80% of the hydrogen atoms are substituted. The treatment increases the PL stability by a factor of 10. There is no report of this treatment concerning EL. It does not appear to be enough for long life EL.

More recently, surface modification of PS by hydrosilylation [206] or by covalend bonding [207] has proved very effective to preserve the intensity and the emission band of the photoluminescence. Stabilization of electroluminescence from nanocrystalline porous silicon has been achieved by substituting most hydrogen atoms terminating fresh PS surface by organic groups [155, 208]. Surface modification is performed by thermal reaction of PS surface with 1-decene and *n*-caprinaldehyde above 90 °C. Devices in which PS surface has been modified show no degradation of EL efficiency and EL output intensity. The EL efficiency of an untreated device only decreases from the start. The organic groups represent a physical barrier that prevents water molecules from accessing the porous Si surface, preventing Si oxidation. The high stability is also explained by the high chemical stability of Si–C bonds (in the case of devices treated with 1-decene) resulting from the organic groups being covalently attached to the Si surface.

Layer capping and encapsulation techniques have been proposed in order to prevent the contamination of PS by all sorts of molecules present in ambient atmosphere. Especially, PS should be protected from water with a view to avoiding the ineluctable oxidation that occurs when it is left in air. The top contact itself could act as a capping layer. For example, Simons *et al* [156] have shown that LEDs with a semi-transparent Au contact are much less stable in air than LEDs with an ITO contact. This is explained by Au being more permeable to air than ITO. Then, the Au-contacted device gets oxidized and degraded much faster than the ITO-contacted one. However, the capping effect is not complete in this case since the top contact cannot cover all PS outer surface to avoid the direct contact with the substrate.

More complete capping of PS has been achieved by Lazarouk *et al* [157]. Aluminium contact is deposited on to PS. Then, part of the Al is electrochemically oxidized into Al_2O_3 to create transparent windows through which EL is observed. The stability is reported to be more than a month. However, there are two drawbacks in their device. First, the capping process decreases the EQE since EL originating from under the non-oxidized Al cannot get out of the device. Second, the device emits white colour, and the EL is likely to originate from some oxide-related centres rather than from the PS red band.

Recently, Koshida *et al* [158] have studied the capping of ECO-treated PS by ECR (electron cyclotron resonance)-sputtered SiO_2 films. The EQE is more effectively stabilized with increasing the capping layer thickness. Two deposition modes are studied: the oxide and metal modes. The metal mode leads to much better stabilization than the oxide mode, because it presents lower permeability to water molecules. The microscopic defects in SiO_2 films deposited by metal mode possibly act as anti-diffusion trapping sites for water molecules.

10.5.11 EL modulation speed

The decay and rise times of all devices are below a millisecond and do not represent any problem for display applications. However, the modulation speed of all devices is below 1 GHz, making application in optical interconnects very challenging. There are great variations among devices. The modulation speed is usually influenced by the carrier mobility in PS, the radiative recombination processes and charge trapping. The PL decay following pulsed excitation is usually in the microsecond range (see section 10.0.0). The EL modulation speed is typically found to be of the order of about tens of microseconds [87–89, 121, 128, 138, 169]. This would limit the frequency to about 100 kHz. However, one group has reported an efficient (0.1% of EQE) device based on partially oxidized PS that can be modulated at a frequency greater than 1 MHz [170]. The high response speed may be due to a recombination mechanism that does not involve the interior of the Si nanocrystals. Modulation frequency of 200 MHz has been reported [170] for a device showing lower efficiency and in which EL probably does not originate from recombination of excitons in Si nanocrystals. The device speed is said to be limited by the junction capacitance. A reduction of the EL response time is potentially possible by using PS in a microcavity [165] since the PL lifetime may be reduced in this way.

Another promising route for faster devices is the use of blue luminescence. Indeed, the PL lifetime decreases when emission energy increases. Therefore devices emitting high energies could be quite fast. As a matter of fact, Mizuno and Koshida [32] have demonstrated PL lifetimes of a few nanoseconds in the blue spectral region.

10.5.12 Integration issue

Besides luminescence devices, useful optical devices such as optical waveguides, optical cavities and non-volatile memories based on PS were fabricated on silicon substrates by simple processing (see section 10.7). Besides, optical nonlinearity in a PS Fabry–Pérot resonator has been demonstrated. This optical phenomenon leads to the availability of PS for optical switches and optical logic gates (see section 10.7). All these phenomena had only been observed separately (on different substrates) for a long time. However, progress has been made in the integration of some of these characteristics, especially the EL.

A bipolar device fully compatible with conventional Si micro-electronic processing has been demonstrated [135] and implemented [171]. The LED is based on oxidized PS using thermal oxidation in the range 800–900 °C, in N_2 with 10% O_2 [134–136]. It consists in a porosified p^+p^--type junction. It includes a poly-Si layer deposited on to the p^+ side to act as the top contact. The room temperature characteristics were the following: weeks of stable EL, highest EQE of about 0.1%, rectifying ratio of 105, EL peak from 1.7 to 2.0 eV, threshold voltage of about 2 V and a current density of about 10 mA/cm^2, maximum light output of about 1 mW/cm^2, and modulation speed by a square-wave current pulse with frequencies greater than 1 MHz. The high stability is due to replacement of the fragile hydrogen passivation by a thin Si oxide coverage. The relatively high response speed may be related to a different recombination mechanism involving the oxide rather than the interior of the Si nanocrystals. The driving transistor, connected in the common-emitter configuration, can modulate light emission by amplifying a small base input signal and controlling the current flow through the LED. The LED can be turned on and off by applying a small current pulse to the base of the bipolar transistor. Arrays of such integrated structures have also been fabricated.

EL from polycrystalline Si and its possible integration in large-area applications have also been demonstrated [172]. Porous polycrystalline Si diodes were shown to operate with efficiencies comparable with conventional PS diodes. The EL mechanism is believed to be the same in both cases. It has also been confirmed that porous polycrystalline LEDs can be driven by a poly-Si-based switching TFT [172].

In the case where the EL emitter would have to be fabricated using silicon on an insulator, it should be possible to fabricate PS without back

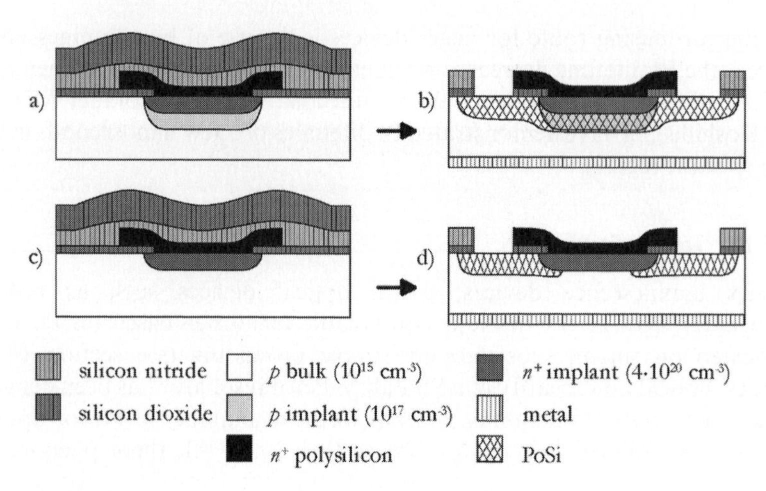

Figure 10.27. Schematic sections [(a) and (b)] of the fabrication of a polysilicon LED. Sections (c) and (d) show the same structure without higher *p* doping under the contact: with the same anodization parameters, the sample would show a crystalline stem [174].

contact. El-Bahar and Nemirovsky [173] demonstrated the formation of PS with the same characteristics as conventional PS by an a.c. electrochemical process.

Barillaro *et al* [174] fabricated a PS LED using a process compatible with an industrial bipolar + complementary MOS + diffusion MOS technology. The LED is based on a p/n^+ junction. Figure 10.27 shows the schematic cross-section of the fabrication process.

The monolithic integration of a PS-based light emitter and a PS-based waveguide, using a simple anodization and standard silicon IC process techniques, has been demonstrated [175]. The difficulty resides in the different types of doping needed in the different parts of the chip, and the necessity of a wet process step. Indeed, the most efficient LEDs require n^+-type Si [138], whereas the waveguide technology is most mature using p-type Si (core is p^+-type PS and clad is p-type PS, see section 10.7). The monolithic integration of an n^+-type PS-based light emitter and a PS-based waveguide has been achieved using several implantation steps, starting from an n-type Si substrate and a one-step anodization. Figure 10.28 shows the cross-sectional view of the chip. Refractive indexes of clad and core were 1.54 and 1.89, respectively. The EL has been observed from the edge of the device, confirming that the EL signal can be transmitted through the optical waveguide and emitted from the edge. Further work is under way for improving the EL efficiency, reducing the loss of optical waveguide and integrating other optical components into the chip. Silicon-based monolithic OEIC should be conceivable if the EL efficiency is improved and a photodetector can also integrated into the chip.

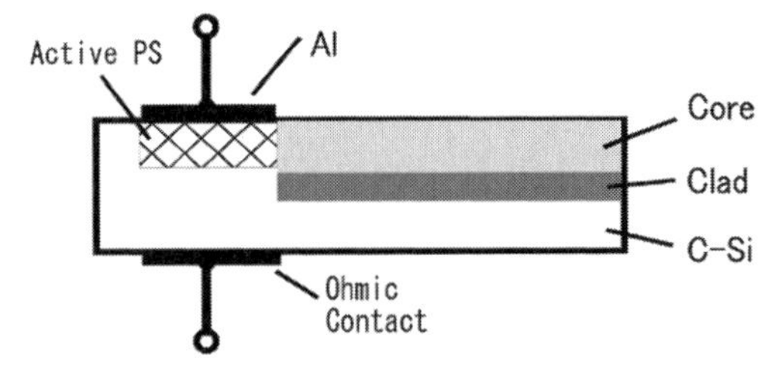

Figure 10.28. Schematic representation of a porous Si-based LED integrated with an optical waveguide. The core, clad and LED (active PS field) are formed in a single-step anodization process [175].

10.6 Ballistic EL using porous Si

10.6.1 Electron emission from porous Si and its mechanism

Cold electron emission is another useful function of PS diodes [209–217]. The first experimental PS diode consisted in a thin Au film, PS, n^+-type Si substrate, and ohmic back contact. When a positive voltage is applied to the Au electrode, with respect to the substrate, electrons as well as photons are uniformly emitted through the Au contact. Figure 10.29 illustrates the experimental setup.

The PS diode as surface-emitting cold cathode has several advantages compared with conventional cold emitters [218–222]. First, electrons are emitted perpendicularly from the diode surface. Second, the cathode operates at relatively low voltage, and third the emission current is not

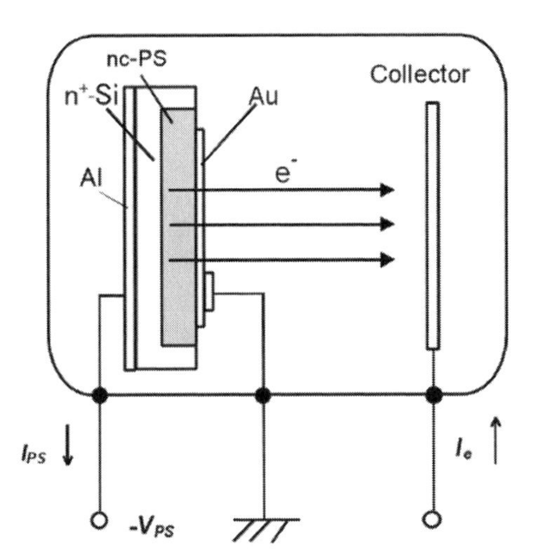

Figure 10.29. Device structure and experimental setup for ballistic electron emission from porous silicon.

System	BSD	MIM	Spindt	CNT	SCE
Basic Structure	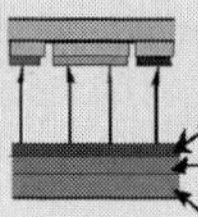 Au, PPSi, Si	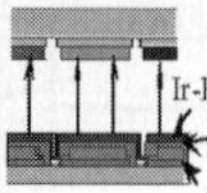 Ir-Pt-Au, Al_2O_3, Al	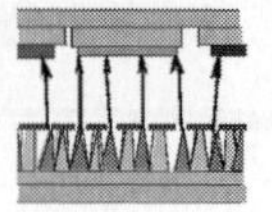	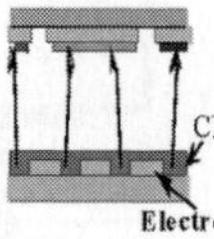 CNT, Electrode	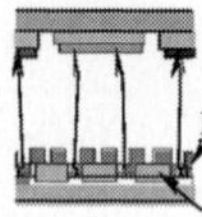 Pd O, Pt
Emission Mechanism	Ballistic Emission	Hot-electrton Tunnelling	Field Emission	Field Emission	Lateral Tunnelling
Voltage	15 ~ 30 V	~10 V	30 ~ 80 V	Several hundred to kV	10 ~ 20 V
Emission	8 mA/cm^2	80 mA/cm^2	50 A/cm^2	0.01 ~ 1 A/cm^2	2 mA/cm^2
Vacuum Level	Stable up to 1 ~ 10 Pa	0.1 Pa	< 10^{-5} Pa	10^{-5} ~ 10^{-6} Pa	< 10^{-6} Pa

Figure 10.30. Characteristics of different types of electron emitters. BSD: Ballistic Surface emitting Device (T. Komoda, *et al*: *J. Vac. Sci. & Technol.* **B17** 1076 (1999)). MIM: Metal Insulator Metal (T. Kusunoki, *et al*: *Jpn. J. Appl. Phys.*, **32**, L1695 (1993)). Spindt: C. A. Spindt, *et al*: *J. Appl. Phys.*, **47**, 5248 (1976). CNT: Carbon Nanotube (A. G. Rinzler, *et al*: *Science*, **269**, 1550 (1995), W. A. de Heer, *et al*: *Science*, **270**, 1179 (1995)). SCE: Surface Conduction Electron emitter (E. Yamaguchi, *et al*: SID Digest Tech. Papers, 52 (1997)).

sensitive to ambient pressure. These advantages are summarized in figure 10.30. Figure 10.31 shows a typical behaviour of the diode current density (I_{PS}) and the emitted current density (I_e) as a function of the voltage.

The emission mechanism has been explained as follows. PS is composed of electrically isolated or interconnected Si nanocrystals surrounded by a thin wide bandgap layer. This latter layer consists of Si oxide. Figure 10.32 illustrates the electron emission process. Under positive bias, a large potential drop is produced in the PS layer, especially near the PS surface. Electrons are thermally injected from the heavily doped n-type substrate and drifted through PS towards the top contact. In PS, electrons are accelerated especially across regions between Si nanocrystals by multi-tunnelling processes and eventually become ballistic electrons. This process is supported by Fowler–Nordheim analysis, numerical analysis (Monte Carlo) of electron drift length from time-of-flight measurements and voltage dependence of the

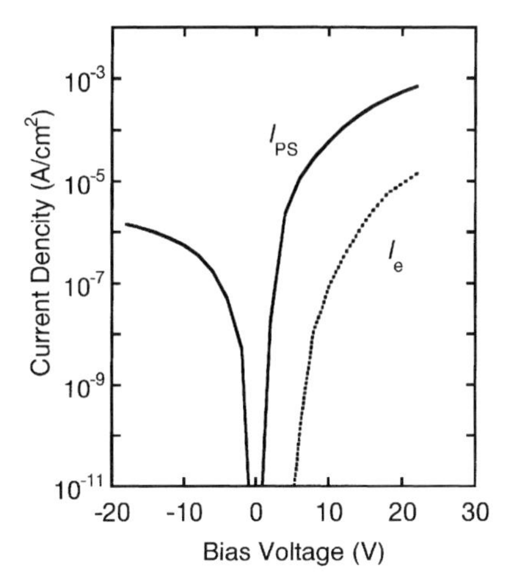

Figure 10.31. Typical bias voltage dependence of the diode current and emitted electron current for an n^+-type porous Si ballistic electron emitter.

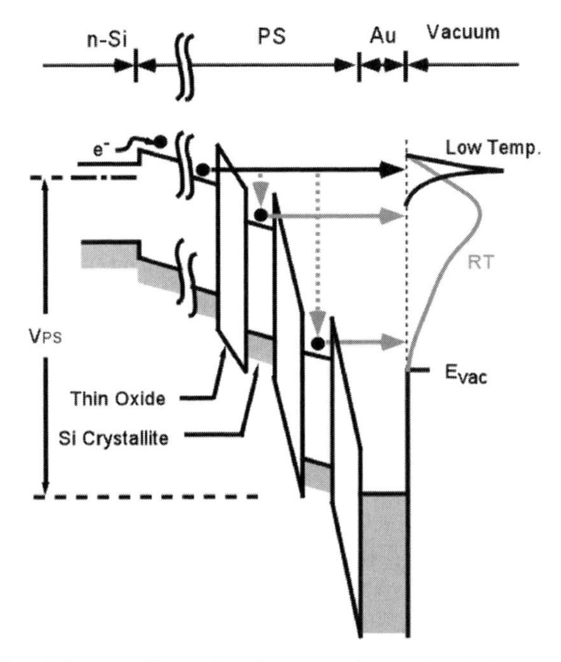

Figure 10.32. Band diagram illustrating the generation and emission processes of quasi-ballistic electrons in porous Si diodes under biased condition. The curve in the vacuum part of the diagram shows the shape of the energy-dependence of the electron emission at room temperature and at 100 K. The narrowing of the curve at low temperature is attributed to significant reduction of thermal scattering in PS [223].

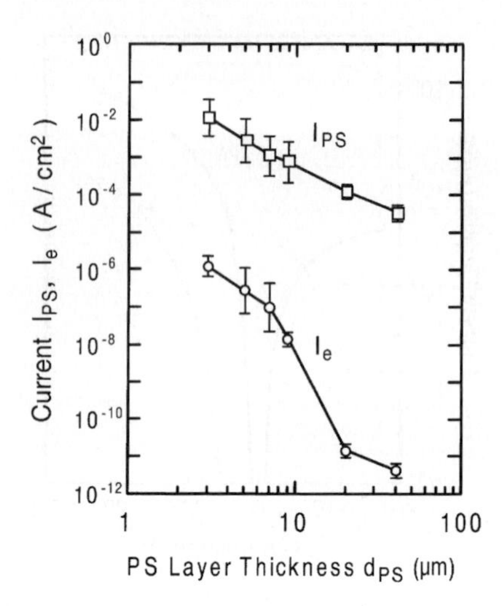

Figure 10.33. Effect of the porous Si layer thickness on the diode current and emitted electron current for an n^+-type porous Si ballistic electron emitter [211].

energy of emitted electrons [223]. Impact ionization processes occur as well. At sufficiently high bias, the situation near the Au/PS interface can be considered as a state of voltage-controlled negative electron affinity. Electrons in PS near the outer surface by the field-induced carrier generation cascade can be emitted into vacuum through a thin oxide layer and the thin Au film.

10.6.2 Further developments of electron emission from porous Si

The effect of PS thickness has been studied [211]. Figure 10.33 shows I_{PS} and I_e at a voltage of 20 V as a function of PS thickness. By reducing PS thickness from 40 to 3 μm, I_e is increased by about six orders of magnitude whereas IPS increases by less than three orders of magnitude. The result is an improvement in efficiency. In addition, the threshold voltage is greatly reduced by reducing the PS thickness, and tends to reach a constant value at about 5 V.

The efficiency of electron emission has been significantly improved by oxidizing PS using a rapid thermal oxidation (RTO) procedure [211]. Figure 10.34 shows the emission efficiency as a function of oxidation time for device operation at 20 V (PS thickness of 3 μm) and an oxidation temperature of 900 °C. The efficiency increases with oxidation time. A significant increase in quantum efficiency is achieved for an oxidation time of 70 min.

Further improvements of the surface-emitting cold cathodes based on PS diodes have been achieved using a multilayered PS structure [212, 213] and a multilayered graded PS structure [213, 214], in addition to the RTO process. Figure 10.35(a–c) show the diode band structure and the current

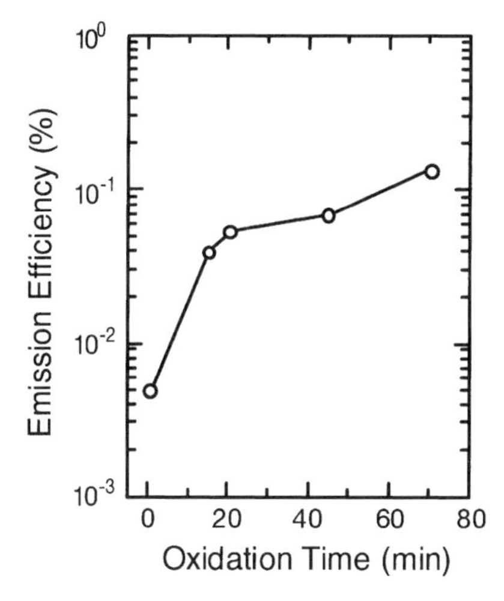

Figure 10.34. Electron emission efficiency as a function of oxidation time for a device operated at 20 V (PS thickness of 3 μm) and an oxidation temperature of 900 °C [211].

density used during PS formation for a normal, a multilayered and a graded-multilayered structure, respectively.

With the multilayered structure, the efficiency is improved due to a significant reduction of the diode current, and reaches 12% at 33 V. Also, the fluctuations of I_e are decreased compared with the normal structure. The improvement has been explained as follows. The low-porosity compact PS layer (formed under low current density) has a high electrical

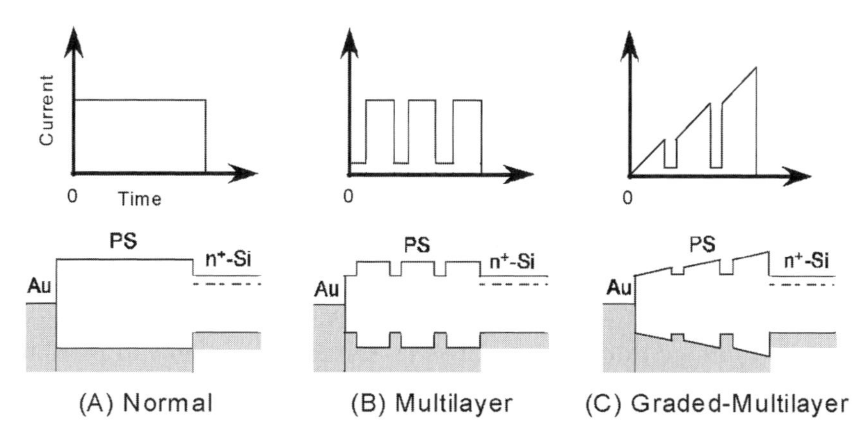

Figure 10.35. Anodization current condition and corresponding porous Si diode band structure for a normal structure (a), a multi-layer structure (b), and a graded-multi-layer structure (c).

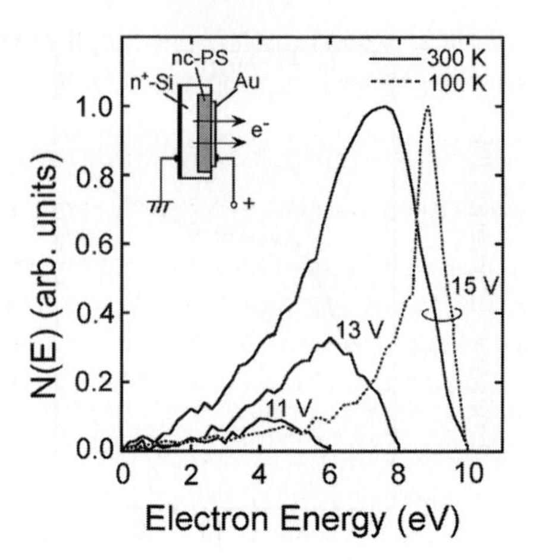

Figure 10.36. Bias voltage dependence of the energy distribution of electrons emitted into vacuum from a diode at room temperature and at 100 K [227].

conductivity and acts as an equipotential plane. It self-regulatedly reforms the electric field distribution in the PS layer, providing a better uniformity of the electric field across the PS layer. The low porosity layers also act as heat-sinks owing relatively high thermal conductivity. As a result of efficient heat sinking, the stability of the device is greatly enhanced. Spikes in I_e are significantly reduced and the thermal increase in I_{PS} is suppressed.

The graded-multilayered structure provides the same advantages as the multi-layered structure. In addition, PS layers formed under higher current densities (high-porosity PS layers) exhibit a linear porosity gradient across the structure, as illustrated by the obtained graded band structure (figure 10.35). In the normal and multilayered situations, the electric field is higher near the PS–Au interface. The graded-multilayered structure tends to make the electric field across the whole PS layer uniform. This is because the electric field in high-porosity regions tends to be higher than in low-porosity regions. As a result, the spikes in I_e are completely suppressed. Another noticeable characteristic of graded-multi-layered structures is the voltage tunability of the energy of emitted electrons, as seen in figure 10.36. This indicates that there is little serial scattering loss during drift in PS. On the contrary, when using normal or multi-layered structures, this voltage tunability of the emitted electron energy is not observed, suggesting that electrons are thermalized after serial scattering losses due to possible potential fluctuations. Figure 10.36 also includes a spectrum measured at low temperature. This spectrum is much narrower than its room temperature counterpart. This is attributed to significant reduction of thermal scattering in PS. The quasiballistic emission process is illustrated in figure 10.32.

The dynamics of electron emission is limited by the capacitance of PS. It depends on the surface of the emitter. The response time of a device with an 0.2 cm^2 surface is in the microsecond range.

10.6.3 Electron emission and flat panel display based on porous poly-Si

Electron emission has been obtained with porous poly-Si (PPS), using the graded-multilayered structure [215] (see previous section). 1.5 μm thick poly-Si film was deposited on to an n$^+$-type Si wafer by low pressure chemical vapour deposition. Efficiency of 1% was reached above 15 V. Over 200 μA/cm^2 was obtained at 30 V. The stability of I_e and I_{PS} were very good.

4×4 pixels prototype display panel has been fabricated [216, 217]. Four stripes of n$^+$ diffusion patterns were formed on a Si wafer as negative electrodes and poly-Si layer was formed on top of them. After anodization of poly-Si, a thin gold layer was deposited on to PPS. Orthogonal to negative electrodes, a thin Au electrode was patterned with four lines as positive electrodes by an ion milling technique. Each Au line was 4 mm wide. An anode electrode was coated with phosphors (ZnO:Zn;P-15) on ITO glass plate. This anode plate was set at a distance of 5 mm from the cathode. The panel was pumped at a pressure of 10^{-4} Pa and sealed. Uniform brightness was obtained on the phosphor screen for each pixel. In addition, the shape of each pixel was identical and equal to the pixel emitter size (4 mm), showing that electrons are emitted perpendicularly to the cathode without deflection, without using any focusing electrode.

Koshida's group, in collaboration with Japanese companies, have developed a ballistic electron surface-emitting display (BSD) on a glass substrate using a low temperature process [218]. The flat panel display was 168 (RGB) $\times$ 126 pixels, 2.6 inches diagonal full-colour. Sub-pixel size was 320×107 μm. First, 126 stripes of tungsten metal patterns were formed on to the glass substrate as X-address electrodes. Then, non-doped poly-Si was grown on the substrate by plasma-enhanced chemical vapour deposition below 500 °C. Poly-Si was then made porous by anodization in HF. Since RTO requires too high temperature, another solution has been found. In order to develop a low-temperature process (<500 °C), the ECO technique, which was introduced by Gelloz in the Koshida laboratory to enhance the PS EL characteristics [137, 138], has been selected as a natural choice by the authors. ECO is then conducted at room temperature. Then, the Au layers were deposited on the surface as electrodes and, orthogonal to X-address electrodes, metal layers were patterned with 168×3 lines as Y-address electrodes by an ion milling technique. A line by line addressing method with 30 Hz frequency operations was used for driving the BSD matrix panel, with a driving voltage below 30 V. P22 phosphors were coated on an ITO-deposited glass substrate as an anode screen placed at 3 mm from the BSD cathode surface. The anode voltage for excitation of the phosphor screen was 6 kV. The time response for moving pictures was sufficient for display panel. The device could operate at relatively low vacuum level (10 Pa) and without any focusing electrodes even the distance between pixels is only 40 μm.

10.6.4 Solid-state planar luminescent devices

A principle of planar-type visible light emission has been reported using ballistic electrons as excitation source [225–227]. Both poly-Si [225] and c-Si [226, 227] have been used. For the PS oxidation process, both RTO and ECO were used. The device is composed of a surface-emitting cold cathode (discussed in the previous sections) and a luminescent material directly deposited on to the electron emitter, as illustrated in the inset of Figure 10.37 in the case of c-Si [227]. In this case tris(8-hydroxyquinoline) aluminium (Alq_3) is used as fluorescent film, though any type of luminescent material could be used. The point is that there is no need for a vacuum spacing between the cold cathode and the fluorescent material. Electrons injected into the nc-PS layer are accelerated via multiple-tunnelling through interconnected silicon nanocrystallites, and reach the outer surface as energetic hot or quasiballistic electrons. They directly excite the fluorescent film, and then induce uniform visible luminescence. This process is illustrated in figure 10.38. This solid-state light-emitting device is regarded as a 'vacuum-less cathode-ray tube' and has many technological advantages over the conventional luminescent devices.

Figure 10.37 shows the current density and the EL intensity as a function of the voltage [227]. No EL could be recorded under reverse operation

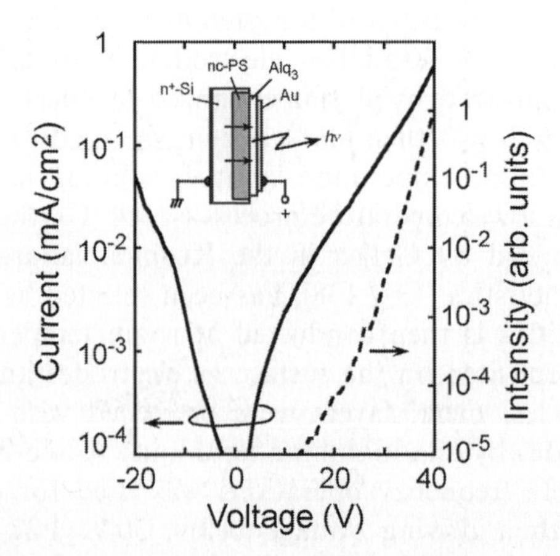

Figure 10.37. Diode current and electroluminescence intensity as a function of voltage for the device shown in the inset [227].

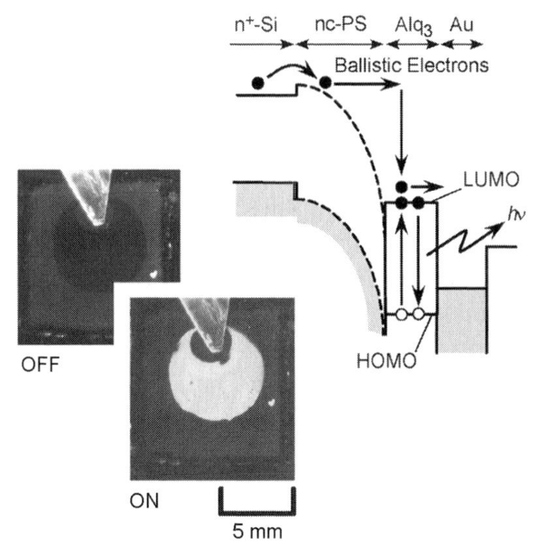

Figure 10.38. Schematic band diagram of the ballistic-electron-induced light emission process and images of a top view of the device under off and on states [227].

since no electrons could be emitted under this condition. Figure 10.39 shows the PL and EL spectra at different voltages as well as the EL intensity versus current density relationship in the inset [227]. The spectra show that the EL comes from Alq_3 and no signs of luminescence from PS could be detected.

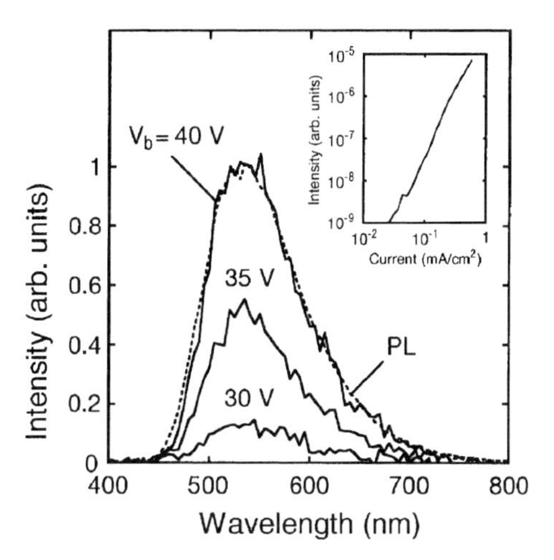

Figure 10.39. Photoluminescence of Alq_3 (dashed curve) and EL spectra at various bias voltages for the device shown in the inset of figure 10.37. The inset shows the emitted light intensity as a function of the diode current [227].

The superlinear behaviour of the EL intensity versus current density relates to the voltage-controlled characteristic property of this device. Indeed, as the driving current is increased, the relative number of energetic emitted electrons is significantly enhanced (as suggested from figure 10.36).

10.7 Related optical components based on porous Si

This section briefly introduces some properties of PS related to EL. First, passive optical components, such as filters and waveguides are presented. Then, valuable active devices, such as photodiodes, memories and optical switches are introduced.

10.7.1 Passive optical components: filters and waveguides

Interference filters like Bragg reflectors or Fabry–Pérot filters consist of alternating layers with different refractive index. The refractive index of PS can be varied by varying the porosity. A simple way to vary the porosity is to vary the anodization current during PS formation. Thus, mirrors and interference filters using PS multi-layers have been fabricated for the infrared, visible and ultraviolet [228–231]. In contrast to conventional dielectric materials (e.g. $SiO_2/Si_3N_4/TiO_2$), the refractive index of PS can be varied continuously over a wide range. The refractive index can be varied in depth following almost any function (e.g. sine wave, square wave). These filters are well-documented in some reviews [231, 232]. In section 10.5.9, PS-based LEDs including Fabry–Pérot microcavities are discussed.

Optical waveguides have been demonstrated using PS multi-layers [233]. The first instance of waveguide was functional in the infrared (1.28 μm). Extension in the visible region (632.8 nm) was also achieved by subsequent PS oxidation [233]. Two-dimensional optical confinement was also realized in reactive ion-etched strip-loaded waveguides. Compared with other waveguide technologies (e.g. doped silica-on-silicon, silicon-on-insulator), propagation losses in the multiplayer waveguides are relatively high. The major loss mechanism is scattering due to macro-roughness at both the waveguide/lower cladding interface and the rib walls.

Another waveguide has been fabricated using selectively formed and oxidized PS in order to obtain a channel SiO_2 optical waveguide [234]. Light guiding was observed in the whole visible range.

The group of Koshida [235–237] has fabricated and characterized a three-dimensionally buried PS waveguide for the visible, which fabrication process flow can be seen in figure 10.40 [237]. A high and low refractive index PS layer is used as the core and the clad, respectively. The difference in porosity and structure between p-type PS and p^+-type PS is used to automatically generate the index difference in a single anodization step.

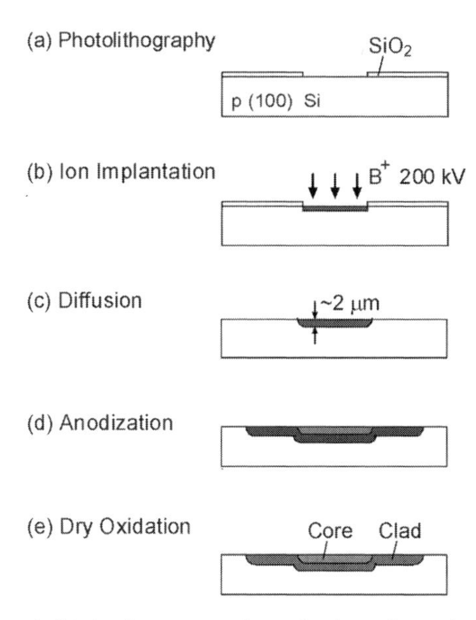

Figure 10.40. Schematic fabrication process flow of a three-dimensionally buried porous Si optical waveguide [237].

The attenuation losses in the waveguide are found due to self-absorption of light by residual silicon, structural and optical inhomogeneities in the core region, and roughness at interfaces between the core and the cladding layers. The light travelling in the waveguide is TE-polarized due to attenuation of the electric field component preferentially at the upper and lower interfaces between the core and the cladding layers. It has also been demonstrated that a buried bent PS waveguide with a curvature as small as 250 μm could be fabricated using the same simple technique and could be functional. Due to the high refractive indices contrast between the core and the cladding layers, the bending losses are found to be considerably lower than that in conventional optical fibre scheme [237].

Ferrand and Romestain [238, 239] have fabricated a PS optical planar waveguide with a sub-micronic periodic modulation of the optical index along one direction of plane using a holographic process. First, a two-layer configuration waveguide was fabricated. Light could propagate in the top PS layer thanks to total internal reflection on the underlying cladding layer and air. Then index modulation in one direction of plane was achieved by photo-assisted chemical etching using two-beam interference illumination. Near-infrared continuous transmission spectra of guided light shows several stop bands, with a decrease of intensity by two orders of magnitude. Maximum index modulation of 0.5 near the top interface has been estimated through calculation.

Photonic crystals are dielectric structures with a forbidden gap for electromagnetic waves. Si-based photonic crystals are briefly introduced here, the reader may refer to more specialized literature for in-depth information [240–242]. Photonic crystals have been considered with a view to guiding light, and for different other purposes (e.g. super-prism effect [243]). Two-dimensional waveguides have been fabricated using Si pillars [244] or macro-porous Si [245–250]. In the macro-porous configuration, periodic cylindrical holes have been etched into Si. Progress in this field has led to a waveguide operating in the optical communication wavelength [250], and in the visible (using Si_3N_4) [251]. These waveguides allows very sharp bends. For example, recently, Talneau *et al* [252] reported experimental data for bends of 60° (in GaInAsP). Along the pore axis, a short-pass filter characteristic has been observed for ultraviolet and visible light [253]. In-depth modulation of the cylinder diameter provides enhancement in the bandgap features [254].

We briefly introduce three-dimensional photonic crystals. Some structures potentially exhibiting a full bandgap (in all three directions) have been suggested. These include the so-called yablonovite structure [255], the woodpile structure [256–259], and the inverse opal structure [260]. Sub-micrometre yablonovite structure has been fabricated in a photoresist [261], which was then used as a template to fabricate dielectric and metallic yablonovite structures [262]. Almost any kind of photoresist template can be fabricated using holographic lithography [263]. The woodpile structure has been fabricated in Si using rather complicated and time-consuming techniques [256, 257, 259]. Very sharp bend three-dimensional waveguides can be obtained from these structures as demonstrated by calculations [258].

10.7.2 Active optical components: photodetectors, memories, switching

Photodetectors are introduced first. Zheng *et al* [264] have fabricated a highly sensitive photodetector made from a metal–PS junction. Close to unity quantum efficiency could be obtained in the wavelength range of 630–900 nm without any anti-reflective coating. The detector reponse time was about 2 ns, with a 9 V reverse bias. Tsai *et al* [265] have used a metal–semiconductor–metal photoconductor and a pn photodiode based on RTO-treated PS. The photoconductor exhibited 2.8 higher responsivity at 350 nm than a ultraviolet-enhanced Si photodiode. The photodiode showed an EQE of 75% at 740 nm. Both devices exhibited better reponsivity when PS is RTO-treated than without the RTO treatment. A colour-sensitive photodetector has been fabricated by Krueger *et al* [266]. The colour-sensitivity was achieved by using a PS Fabry–Pérot filter on top of a p^+n Si photodiode. The Fabry–Pérot filter was made in the p^+ part of the initial p^+n junction.

Forward-biased PS LEDs formed from p-type and n-type Si exhibit a reversible negative resistance effect [267–269]. At about 19 V, the reversible

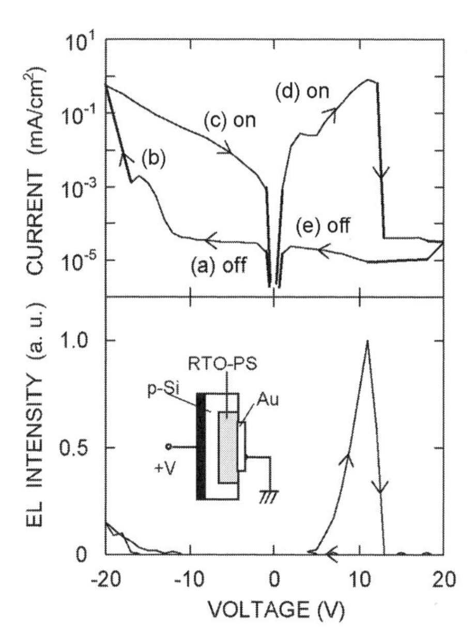

Figure 10.41. Current density and EL intensity as a function of voltage for the memory device shown in the inset [269].

negative resistance effect occurs with a peak-to-valley current ratio larger than 2 at room temperature [269]. The onset of the negative resistance effect activates the EL emission. This phenomenon has been explained in the following way. Considering p-type Si, under forward bias, holes are injected from the substrate into PS, and drift towards the top Au contact. Beyond a critical voltage, the local electric field become so strong that holes are injected into the Si nanocrystals by tunnelling. These injected holes then act as fixed charges in PS. As a result, the initial potential distribution (almost uniform across PS) becomes much stronger close to the Au contact and almost flat in PS close to the substrate. Under this new condition, hole injection from the substrate should be much less efficient, and the current density decreases. The high field region near the PS–Au electrode promotes EL emission.

When PS is oxidized by RTO, the negative resistance effects are increased and a hysteresis of the dark current sets in, as shown in figure 10.41 [269]. Figure 10.42 schematically shows the memory mechanism at different stages. The letters from a to e in figures 10.41 and 10.42 are coupled. As the bias is swept from 0 towards the reverse direction, the current at first remains relatively low (a: off-mode). At a threshold voltage of about −15 V, however, the diode operation changes from the off-mode to the on-mode (b), and much higher current flows in the diode (c). The

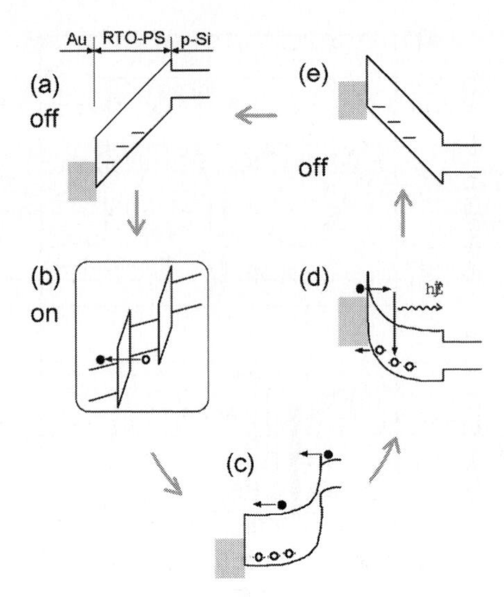

Figure 10.42. Schematic model of the memory effect. The notations (a)–(e) correspond to those in figure 10.41 [269].

on-mode is non-volatile, and is kept for at least one week until the erasing voltage is applied. The on-mode (d) can be turned to the off-mode when a sufficient forward bias voltage is applied. At this point, the high resistivity state appears again (e). The RTO-treated PS diode operates as a non-volatile memory device, and the reverse and forward bias voltages correspond to the writing and erasing voltages, respectively. In the on-state, the EL is also observed. Figure 10.43 illustrates the memory function.

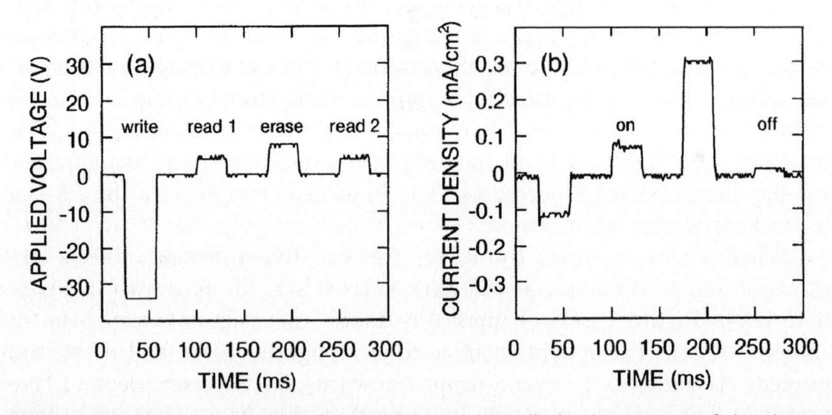

Figure 10.43. Pulsed operation of an RTO-treated PS diode. A sequence of signal access is indicated in terms of the applied voltage and the corresponding diode current density [269].

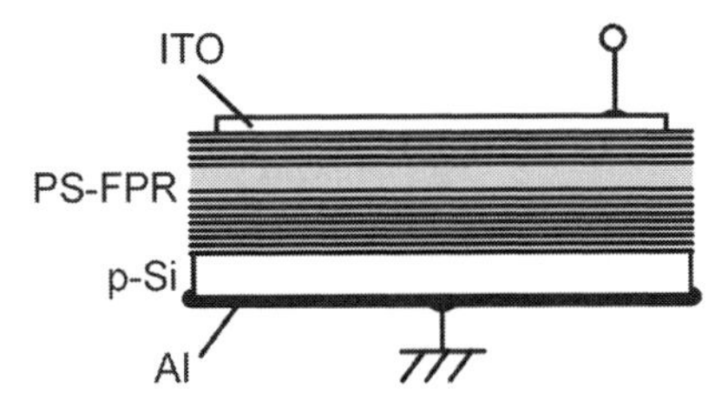

Figure 10.44. Schematic representation of a device including a PS Fabry–Pérot resonator.

After the demonstration of the photo-induced nonlinear index change in PS Fabry–Pérot resonator [270, 271], electrically-induced optical switching has been investigated using nonlinear refractive index change in PS induced by carrier injection [272]. The device is schematically shown in figure 10.44. The active PS layer is the micro-cavity in a Fabry–Pérot resonator. The reflectance spectra can be shifted in the PS micro-cavity by increasing the diode current. The refractive index of PS is increased by carrier injection and subsequent accumulation possibly in localized states. At 250 K, under 40 mA/cm^2, the refractive index of PS was increased by about 0.02%. Operated under pulsed operation, the device operates as a current-induced nonlinear optical switching device. There are fast ($<$10 ms) and slow ($>$1 s) components in the refractive index change.

Finally, Weiss and Fauchet [273] have demonstrated electrically-tunable PS active mirrors using liquid crystals impregnated into a PS Bragg mirror. It is shown theoretically that the PS active mirror can be switched on (high reflectance) and off (low reflectance) by applying a voltage. The device uses the electro-optic properties of liquid crystals and the high sensitivity of PS micro-cavity resonance position to small changes in optical thickness. The voltage applying to the diode rotates the molecules of the liquid crystal. Experimental change of 40% in reflectance and a reversible 12 nm shift of the reflectance spectrum have been demonstrated so far.

10.8 Conclusion

In the past decade, EL from Si has been extensively studied, especially since the discovery of luminescent PS [1]. LEDs based on PS have been made efficient up to 1.1% in EQE and 0.37% in EPE [138], as fast as 200 MHz [170], and stable during months of operation. However, no one LED exhibits all these record characteristics. Much work remains to be done to fabricate a practical LED based on PS. The speed of PS based LEDs is enough for displays but is far too low for interconnects. Thus, the most likely application of PS-based LED would be in the display area. The stability and EPE are still to be enhanced even for display purposes.

The ballistic electron emission from PS diodes is very promising for display applications. Prototypes have already been proven functional [216]. A solid-state planar luminescent device not requiring any vacuum intermediate region between the electron emitter and the luminescent phosphor has also been demonstrated [227].

The field of nanocrystalline Si has been doped by the recent demonstration of optical gain [4]. A race for an Si-based laser is under way. LEDs based on nc-Si in various matrices and in superlattices are also promising even though efficiencies are still very low.

Recently, band-to-band recombination in bulk-Si has been made as efficient as 1% using several improvements in the diode conception [9]. This breakthrough revives the hope of Si optoelectronics using conventional techniques.

LEDs based on Er combined with nc-Si are the most ready for possible applications, since high efficiency has been obtained with high stability [71]. Light amplification is possible with these systems so that Er-based laser is a possiblity in the future.

Solutions have been found for the integration of PS-based LEDs in VLSI technology. MOS-based LEDS have also been made compatible with MOS technology. For integrated interconnects, some possibilities for optical waveguiding, switching, and even memories have been successfully proposed. However, much work remains to be done.

References

[1] Canham L T 1990 *Appl. Phys. Lett.* **57** 1046

[2] Halimaoui A, Bomchil G, Oules C, Bsiesy A, Gaspard F, Herino R, Ligeon M and Muller F 1991 *Appl. Phys. Lett.* **59** 304

[3] Koshida N and Koyama H 1992 *Appl. Phys. Lett.* **60** 347

[4] Pavesi L, Negro L D, Mazzoleni C, Franzo G and Priolo F 2000 *Nature* **408** 440

[5] Nayfeh M H, Barry N, Therrien J, Aksakir O, Gratton E and Belomoin G 2001 *Appl. Phys. Lett.* **78** 1131

[6] Khriachtchev L, Rasanen M, Novikov S and Sinkkonen J 2001 *Appl. Phys. Lett.* **79** 1249

[7] Ruan J, Fauchet P M, Negro L D, Cazzanelli M and Pavesi L 2003 *Appl. Phys. Lett.* **83**(26) 5479

[8] Valenta J, Pelant I and Linnros J 2002 *Appl. Phys. Lett.* **81**(8) 1396

[9] Green M A, Zhao, J, Wang A, Reece P J and Gal M 2001 *Nature* **412** 805

[10] Nag W L, Lourenco M A, Gwilliam R M, Ledain S, Shao G and Homewood K P 2001 *Nature* **410** 192

[11] Gaburro Z and Pavesi L 2003 *Handbook of Luminescence, Display Materials and Devices* ch. C ed. H S Nalwa and L S Rohwer ISBN 1-58883-010-1

[12] Gelloz B and Koshida N 2003 *Handbook of Luminescence, Display Materials and Devices* ch. V ed. H S Nalwa and L S Rohwer ISBN 1-58883-010-1

[13] Smith R L and Collins S D 1992 *J. Appl. Phys.* **71** R1 7586

[14] Canham L T, Cox T I, Loni A and Simons A J 1996 *Appl. Surf. Sci.* **102** 436
[15] Cullis A G, Canham L T and Calcott P D J 1997 *J. Appl. Phys.* **82**(3) 909
[16] Canham L T (ed.) 1997 *Properties of Porous Silicon*, EMIS Datareviews Series No. 18 INSPEC, The Institution of Electrical Engineers, London
[17] Bisi O, Campisano S U, Pavesi L and Priolo F 2000 *Surf. Sci. Rev.* **38** 1
[18] Halimaoui A 1997 *Properties of Porous Silicon*, EMIS Datareviews Series No. 18 ed. L T Canham, INSPEC, The Institution of Electrical Engineers, London p 12
[19] Setzu S, Lerondel G and Romestain R 1998 *J. Appl. Phys.* **84**(6) 3129
[20] Canham L T, Cullis A G, Pickering C, Dosser O D, Cox T I and Lynch T P 1994 *Nature* **368** 133
[21] Belmont O, Bellet D and Brechet Y 1996 *J. Appl. Phys.* **79**(10) 7586
[22] Bellet D 1997 *Properties of Porous Silicon* EMIS Datareviews Series No. 18 ed. L T Canham, INSPEC, The Institution of Electrical Engineers, London p. 127
[23] Gruning U and Yelon A 1995 *Thin Solid Films* **255** 135
[24] Amato G, Bullara V, Brunetto N and Boarino L 1996 *Thin Solid Films* **276** 204
[25] Amato G and Brunetto N 1996 *Mater. Lett.* **26** 295
[26] St. Frohnhoff R, Arens-Fischer T, Heinrich T, Fricke J, Arntzen M and Theiss W 1995 *Thin Solid Films* **255** 115
[27] Von Berhen J, Fauchet P M, Cimowitz E H and Lira C T 1997 *Mater. Res. Soc. Symp. Proc.* **452** 565
[28] Canham L T 1997 *Properties of Porous Silicon* EMIS Datareviews Series No. 18 ed. L T Canham, INSPEC, The Institution of Electrical Engineers, London p. 83
[29] Wolkin M V, Jorne J, Fauchet P M, Allan G and Delerue C 1999 *Phys. Rev. Lett.* **82** 1 197
[30] Mizuno H, Koyama H and Koshida N 1996 *Appl. Phys. Lett.* **69**(25) 3779
[31] Mizuno H, Koyama H and Koshida N 1997 *Thin Solid Films* **297** 61
[32] Mizuno H and Koshida N 1999 *Mater. Res. Soc. Symp. Proc.* **536** 179
[33] Chelikowski J R and Cohen M L 1974 *Phys. Rev. B* **10** 5095
[34] Macfarlane M, McLean T P and Quarrington J E 1958 *Phys. Rev.* **111** 1245
[35] Iyer S S and Xie Y-H 1993 *Science* **260** 40
[36] Kimerling L C, Kolenbrander K D, Michel J and Palm J 1997 *Solid State Phys.* **50** 333
[37] Pankove J I 1971 *Optical Processes in Semiconductors* (New York: Dover)
[38] Landsberg P T 1970 *Phys. Status Solidi* **41** 457
[39] Dziewior J and Schmid W 1977 *Appl. Phys. Lett.* **31** 346
[40] Sinton R A and Swanson R M 1987 *IEEE Trans. Electron. Devel.* **34** 1380
[41] Altermatt P P, Schmidt J, Heiser G and Aberle A G 1997 *J. Appl. Phys.* **82** 4938
[42] Shockley W and Read W T 1952 *Phys. Rev.* **87** 835
[43] Hall R N 1952 *Phys. Rev.* **87** 387
[44] Shockley W 1950 *Electrons and Holes in Semiconductors* (New York: Van Nostrand)
[45] Haynes J R and Briggs H B 1952 *Phys. Rev.* **86** 647
[46] Haynes J R and Westphal W C 1956 *Phys. Rev.* **101** 1676
[47] Schlangelotto H, Maeder H and Gerlach W 1974 *Phys. Status Solidi A* **21** 357
[48] Alex V, Finkbeiner S and Weber J 1996 *J. Appl. Phys.* **79** 6943
[49] Dean P J, Haynes J R and Flood W F 1967 *Phys. Rev.* **161** 711
[50] Kramer J, Seitz P, Steigmeier E F, Auderset H, Delley B and Baltes H 1993 *Sensors Actuators A* **37–38** 527
[51] Liu C W, Lee M H, Chen M-J, Lin I C and Lin C F 2000 *Appl. Phys. Lett.* **76**(12) 1516

[52] Lin C F, Chen M-J, Liang E-Z, Liu W T and Liu C W 2001 *Appl. Phys. Lett.* **78**(3) 261
[53] Dittrich Th, Timoshenko V Y, Rappich J and Tsybescov L 2001 *J. Appl. Phys.* **90**(5) 2310
[54] Timoshenko V Y, Rappich J and Dittrich Th 1998 *Appl. Surf. Sci.* **123/124** 111
[55] Priolo F, Franzo G, Coffa S and Carnera A 1998 *Phys. Rev. B* **57** 4443
[56] Reittinger A, Stimmer J and Abstreiter G 1997 *Appl. Phys. Lett.* **70** 2431
[57] Zheng B, Michel J, Ren F Y G, Kimerling L C, Jacobson D C and Poate J M 1994 *Appl. Phys. Lett.* **64**(21) 2842
[58] Franzo G, Priolo F, Coffa S, Polman A and Carnera A 1994 *Appl. Phys. Lett.* **64**(17) 2235
[59] Coffa S, Franzo G and Priolo F 1996 *Appl. Phys. Lett.* **69**(14) 2077
[60] Coffa S, Franzo G, Priolo F, Pacelli A and Lacaita A 1998 *Appl. Phys. Lett.* **73**(1) 93
[61] Coffa S, Franzo G, Priolo F, Polman A and Serna R 1994 *Phys. Rev. B* **49** 16313
[62] Michel J, Benton L J, Ferrante R F, Jacobson D C, Eaglesham D J, Fitzgerald E A, Xie Y-H, Poate J M and Kimerling L C 1991 *J. Appl. Phys.* **70** 2672
[63] Markmann M, Neufeld E, Sticht A, Brunner K and Abstreiter G 1999 *Appl. Phys. Lett.* **75** 2584
[64] Du C-X, Duteil F, Hansson G V and Ni W-X 2001 *Appl. Phys. Lett.* **78**(12) 1697
[65] Markmann M, Sticht A, Bobe F, Zandler G, Brunner K, Abstreiter G and Muller E 2002 *J. Appl. Phys.* **91**(12) 9764
[66] Wang W, Eckau A, Neufeld E, Carius R and Buchal Ch 1997 *Appl. Phys. Lett.* **71**(19) 2824
[67] Franzo G, Pacifici D, Vinciguerra V, Priolo F and Iacona F 2000 *Appl. Phys. Lett.* **76**(16) 2167
[68] Zacharias M, Heitmann M S J and Streitenberger P 2001 *Physica E* **11** 245
[69] Tsybescov L, Duttagupta S P, Hirschman K D, Fauchet P M, Moore K L and Hall D G 1997 *Appl. Phys. Lett.* **70**(14) 1790
[70] Lopez H A and Fauchet P M 1999 *Appl. Phys. Lett.* **75**(25) 2167
[71] Iacona F, Pacifici D, Irrera A, Miritello M, Franzo G, Priolo F, Sanfilippo D, Di Stefano G and Fallica P G 2002 *Appl. Phys. Lett.* **81**(17) 3242
[72] Gusev O B, Kuznetsov A N, Terukov E I, Bresler M S, Kudoyarova V Kh, Yassievich I N, Zakharchenya B P and Fuhs W 1997 *Appl. Phys. Lett.* **70**(2) 240
[73] Curry R J, Gillin W P, Knights A P and Gwilliam R 2000 *Appl. Phys. Lett.* **77**(15) 2271
[74] DiMaria D J, Kirtley J R, Pakulis E J, Dong D W, Kuan T S, Pesavento F L, Theis T N, Cutro J A and Brorson S D 1984 *J. Appl. Phys.* **56**(2) 401
[75] Belomoin G, Therrien J, Smith A, Rao S, Twesten R, Chaieb S, Nayfeh M H, Wagner L and Mitas L 2002 *Appl. Phys. Lett.* **80**(5) 841
[76] Iacona F, Franzo G and Spinella C 2000 *J. Appl. Phys.* **87**(3) 5682
[77] Lu Z H, Lockwood D and Baribeau J-M 1996 *Nature* **378** 258
[78] Nassiopoulou A G, Grigoropoulos S and Papadimitriou D 1996 *Appl. Phys. Lett.* **69**(15) 2267
[79] Nassiopoulou A G, Grigoropoulos S and Papadimitriou D 1997 *Thin Solid Films* **297** 176
[80] Toyama T, Kotani Y, Okamoto H and Kida H 1998 *Appl. Phys. Lett.* **72**(12) 1489
[81] Yoshida T, Yamada Y and Orii T 1998 *J. Appl. Phys.* **83**(10) 5427
[82] Tong S, Liu X N, Wang L-C, Yan F and Bao X M 1996 *Appl. Phys. Lett.* **69**(5) 596

[83] Park N-M, Kim T-S and Park S-J 2001 *Appl. Phys. Lett.* **78**(17) 2575
[84] Song H-Z, Bao X M, Li N-S and Zhang J-Y 1997 *J. Appl. Phys.* **82**(8) 4028
[85] Luterova K, Pelant I, Valenta J, Rehspringer J-L, Muller D, Grob J J, Dian J and Honerlage B 2000 *Appl. Phys. Lett.* **77**(19) 2952
[86] Lalic N and Linnros J 1999 *J. Lumin.* **80** 263
[87] Linnros J and Lalic N 1995 *Appl. Phys. Lett.* **66**(22) 3048
[88] Lalic N and Linnros J 1996 *J. Appl. Phys.* **80**(10) 5971
[89] Lalic N and Linnros J 1996 *Thin Solid Films* **276** 155
[90] Qin G G, Li A P, Zhang B R and Li B-C 1995 *J. Appl. Phys.* **78**(3) 2006
[91] Irrera A, Pacifici D, Miritello M, Franzo G, Priolo F, Iacona F, Sanfilippo D, Di Stefano G and Fallica P G 2002 *Appl. Phys. Lett.* **81**(10) 1866
[92] Qin G G, Wang Y Q, Ma Z C and Zong W H 1999 *Appl. Phys. Lett.* **74**(15) 2182
[93] Heng C L, Sun Y K, Wang S T, Chen Y, Qiao Y P, Zhang B R, Ma Z C, Zong W H and Qin G G 2000 *Appl. Phys. Lett.* **77**(10) 1416
[94] Photopoulos P and Nassiopoulou A G 2000 *Appl. Phys. Lett.* **77**(12) 1816
[95] Osaka Y, Kohno K, Mizuno H and Koshida N 2002 *Jpn. J. Appl. Phys. Part 1 No. 12* **41** 7481
[96] Tsu R, Zhang Q and Filios A 1997 *Proc. SPIE* **3290** 246
[97] Tsu R, Filios A, Lofgren C, Dovidenko K and Wang C G 1998 *Electrochem. Solid State Lett.* **1** 80
[98] Heikkila L, Kuusela T and Hedman H-P 1999 *Superlatt. Microstruct.* **26** 157
[99] Heikkila L, Kuusela T and Hedman H-P 2001 *J. Appl. Phys.* **89**(4) 2179
[100] Photopoulos P, Nassiopoulou A G, Kouvatsos D N and Travlos A 2000 *Mat. Sci. Eng. B* **69–70** 345
[101] Gaburro Z, Pavesi L, Pucker G and Bellutti P 2001 *Mater. Res. Soc. Symp. Proc.* **638** F18.5.1
[102] Heng C L, Chen Y, Ma Z C, Zong W H and Qin G G 2001 *J. Appl. Phys.* **89**(10) 5682
[103] Wu W, Huang X F, Chen K J, Xu J B, Gao X, Xu J and Li W 1999 *J. Vac. Sci. Tecnol. A* **17**(1) 159
[104] Arnaud d'Avitaya F, Vervoort L, Bassani F, Ossicini S, Fasolino A and Bernardini F 1995 *Europhys. Lett.* **31** 25
[105] Vervoort L, Bassani F, Mihalcescu I, Vial J C and Arnaud d'Avitaya F 1995 *Phys. Status Solidi B* **190** 123
[106] Filonov A B, Kholod A N, Novikov V A, Borisenko V E, Vervoort L, Bassani F, Saul A and Arnaud d'Avitaya F 1997 *Appl. Phys. Lett.* **70** 744
[107] Ioannou-Sougleridis V, Tsakiri V, Nassiopoulou A G, Photopoulos P, Bassani F and Arnaud d'Avitaya F 1997 *Phys. Status Solidi A* **165** 97
[108] Kholod A N, Danilyuk A L, Borisenko V E, Bassani F, Menard S and Arnaud d'Avitaya F 1999 *J. Appl. Phys.* **85**(10) 7219
[109] Maruyama T, Nakamura N and Watanabe M 1999 *Jpn. J. Appl. Phys. Part 2 No. 8B* **38** 1996
[110] Maruyama T, Nakamura N and Watanabe M 2000 *Jpn. J. Appl. Phys. Part 1 No. 4B* **39** 1996
[111] Ioannou-Sougleridis V, Ouisse T, Nassiopoulou A G, Bassani F and Arnaud d'Avitaya F 2001 *J. Appl. Phys.* **89**(1) 610
[112] Ioannou-Sougleridis V, Nassiopoulou A G, Ouisse T, Bassani F and Arnaud d'Avitaya F 2001 *Appl. Phys. Lett.* **79**(13) 2076
[113] Namavar F, Muraska H P and Kalkhoran N M 1992 *Appl. Phys. Lett.* **60** 2514

[114] Muraska H P, Namavar F and Kalkhoran N M 1992 *Appl. Phys. Lett.* **61** 1338

[115] Kalkhoran N M, Namavar F and Maruska H P 1992 *Mater. Res. Soc. Symp. Proc.* **256** 89

[116] Kozlowski F, Sauter M, Steiner P, Richter A, Sandmaier H and Lang W 1992 *Thin Solid Films* **222** 196

[117] Steiner P, Kozlowski F and Lang W 1993 *Appl. Phys. Lett.* **62** 2700

[118] Loni A, Simons A J, Cox T I, Calcott P D J and Canham L T 1995 *Electronics Lett.* **31** 1288

[119] Lang W, Kozlowski F, Steiner P, Knoll B, Wiedenhofer A, Kollewe D and Bachmann T 1997 *Thin Solid Films* **297** 268

[120] Nishimura K, Nagao Y and Ikeda N 1998 *Jpn. J. Appl. Phys.* **37** L303

[121] Peng C and Fauchet P M 1995 *Appl. Phys. Lett.* **67**(17) 2515

[122] Chen Z, Bosman G and Ochoa R 1993 *Appl. Phys. Lett.* **62**(7) 708

[123] Gelloz B, Nakagawa T and Koshida N 1998 *Mater. Res. Soc. Symp. Proc.* **536** 15

[124] Lang W, Kozlowski F, Steiner P, Knoll B, Wiedenhofer A, Kollewe D and Bachmann T 1997 *Thin Solid Films* **297** 268

[125] Nishimura K, Nagao Y and Ikeda N 1998 *Jpn. J. Appl. Phys.* **37** L303

[126] Steiner P, Kozlowski F, Wielunski M and Lang W 1994 *Jpn. J. Appl. Phys.* Part 1 No. 11 **33** 6075

[127] Steiner P, Wiedenhofer A, Kozlowski F and Lang W 1996 *Thin Solid Films* **276** 159

[128] Kozlowski F, Sailer C, Steiner P, Knoll B and Lang W 1996 *Thin Solid Films* **276** 164

[129] Koshida N, Koyama H, Yamamoto Y and Collins G J 1993 *Appl. Phys. Lett.* **63**(19) 2655

[130] Koshida N, Mizuno H, Koyama H and Collins G J 1995 *Jpn. J. Appl. Phys.* **34**(1) 92

[131] Bsiesy A, Nicolau Y F, Ermolieff A and Muller F 1995 *Thin Solid Films* **255**(1/2) 43

[132] Halliday D P, Holland E R, Eggleston J M, Adams P N, Cox S E and Monkman A P 1996 *Thin Solid Films* **276** 299

[133] Li K, Diaz D C, He Y and Campbell J C 1994 *Appl. Phys. Lett.* **64**(18) 2394

[134] Tsybeskov L, Duttagupta S P and Fauchet P M 1995 *Sol. State Commun.* **95**(7) 429

[135] Tsybescov L, Duttagupta S P, Hirschman K D and Fauchet P M 1996 *Appl. Phys. Lett.* **68** 2058

[136] Fauchet P M, Tsybescov L, Duttagupta S P and Hirschman K D 1997 *Thin Solid Films* **297** 254

[137] Gelloz B, Nakagawa T and Koshida N 1998 *Appl. Phys. Lett.* **73**(14) 2021

[138] Gelloz B and Koshida N 2000 *J. Appl. Phys.* **88**(7) 4319

[139] Simons A J, Cox T I, Loni A, Canham L T, Uren M J, Reeves C, Cullis A G, Calcott P D J, Houlton M R and Newey J P 1997 *Electrochem. Soc. Proc.* **PV 95**(25) 73

[140] Beckmann K H 1965 *Surf. Sci.* **3** 314

[141] Canham L T, Houlton M R, Leong W Y, Pickering C and Keen J M 1991 *J. Appl. Phys.* **70** 422

[142] Canham L T and Blackmore G 1992 *Mater. Res. Soc. Symp. Proc.* **256** 31

[143] Xie Y H, Wilson W L, Ross F M, Mucha J A, Fitzgerald E A, Macauley J M and Harris T D 1992 *J. Appl. Phys.* **71** 2403

[144] Zhu W X, Gao Y X, Zhang L Z, Mao J C, Zhang B R, Duan J Q and Qin G G 1992 *Superlattices Microstruct.* **12** 409

[145] Borghesi A, Guizetti G, Sassella A, Bisi O and Pavesi L 1994 *Solid State Commun.* **89** 615

[146] Loni A, Simons A J, Canham L T, Phillips H J and Earwaker L G 1994 *J. Appl. Phys.* **76** 2825
[147] Feng Z C, Wee A T S and Tan K L 1994 *J. Phys. D* **27** 1968
[148] Hadj Zoubir N, Vergnat M, Delatour T, Burneau A, de Donato Ph and Barres O 1995 *Thin Solid Films* **255** 228
[149] Theiss W, Arntzen M, Hilbrich S, Wernke M, Arens-Fischer R and Berger M G 1995 *Phys. Status Solidi B* **190** 15
[150] Yamada M and Kondo K 1992 *Jpn. J. Appl. Phys.* **31** L993
[151] Gelloz B 1997 *Appl. Surf. Sci.* **108** 449
[152] Kozlowski F, Wagenseil W, Steiner P and Lang W 1995 *Mater. Res. Soc. Symp. Proc.* **358** 677
[153] Pavesi L, Chierchia R, Bellutti P, Lui A, Fuso F, Labardi M, Pardi L, Sbrana F, Allegrini M and Trusso S 1999 *J. Appl. Phys.* **86**(11) 6474
[154] Matsumoto T, Masumoto Y, Nakagawa T, Hashimoto M, Ueno K and Koshida N 1997 *Jpn. J. Appl. Phys. Part 2 No 8B* **36** L1089
[155] Gelloz B, Sano H, Boukherroub R, Wayer D D M, Lockwood D J and Koshida N 2003 *Appl. Phys. Lett.* **83**(12) 2342
[156] Simons A J, Cox T I, Loni A, Canham L T and Blacker R 1997 *Thin Solid Films* **297** 281
[157] Lazarouk S, Jaguiro P, Katsouba S, Masini G, La Monica S, Maiello G and Ferrari A 1996 *Appl. Phys. Lett.* **68**(15) 2108
[158] Koshida N, Kadokura J, Takahashi M and Imai K 2001 *Mater. Res. Soc. Symp. Proc.* **638** F18.3.1
[159] Gelloz B and Koshida N 1999 *Electrochem. Soc. Proc.* **PV 99-22** 27
[160] Chen Y A, Chen B F, Tsay W C, Laih L H, Chang M N, Chyi J I, Hong J W and Chang C Y 1997 *Solid State Electron.* **41**(5) 757
[161] Mimura H, Matsumoto T and Kanemitsu Y 1996 *J. Non-cryst. Sol.* **198–200** 961
[162] Lazarouk S, Jaguiro P, Katsouba S, La Monica S, Maiello G, Masini G and Ferrari A 1996 *Thin Solid Films* **276** 168
[163] Kuznetsov V A, Andrienko I and Haneman D 1998 *Appl. Phys. Lett.* **72**(25) 3323
[164] Araki M, Koyama H and Koshida N 1996 *J. Appl. Phys.* **80**(9) 4841
[165] Pavesi L, Guardini R and Mazzoleni C 1996 *Solid State Commun.* **97**(12) 1051
[166] Chan S and Fauchet P M 1999 *Appl. Phys. Lett.* **75**(2) 274
[167] Araki M, Koyama H and Koshida N 1996 *Appl. Phys. Lett.* **69**(20) 2956
[168] Kozlowski F, Sauter M, Steiner P, Richter A, Sandmaier H and Lang W 1992 *Thin Solid Films* **222** 196
[169] Cox T I, Simons A J, Loni A, Calcott P D J, Canham L T, Uren M J and Nash K J 1999 *J. Appl. Phys.* **86**(5) 2764
[170] Balucani M, La Monica S and Ferrari A 1998 *Appl. Phys. Lett.* **72**(6) 639
[171] Hirschman K D, Tsybeskov L, Duttagupta S P and Fauchet P M 1996 *Nature* **6607** 338
[172] Koshida N, Takizawa E, Mizuno H, Arai S, Koyama H and Sameshima T 1998 *Mater. Res. Soc. Symp. Proc.* **486** 151
[173] El-Bahar A and Nemirovsky Y 2000 *Appl. Phys. Lett.* **77**(2) 208
[174] Barillaro G, Diligenti A, Pieri F, Fuso F and Allegrini M 2001 *Appl. Phys. Lett.* **78**(26) 4154
[175] Li L, Gelloz B and Koshida N 2003 *Electrochem. Soc. Symp. Proc.* To be published
[176] Chazalviel J N and Ozanam F 1992 *Mater. Res. Soc. Symp. Proc.* **283** 359

[177] Bsiesy A, Vial J C, Gaspard F, Herino R, Ligeon M, Muller F, Romestain R, Wasiela A, Halimaoui A and Bomchil G 1991 *Surf. Sci.* **254** 195
[178] Billat S, Bsiesy A, Gaspard F, Herino R, Ligeon M, Muller F, Romestain R and Vial J C 1991 *Mater. Res. Soc. Symp. Proc.* **256** 215
[179] Hory M A, Herino R, Ligeon M, Muller F, Gaspard F, Mihalcescu I and Vial J C 1995 *Thin Solid Films* **255**(1–2) 200
[180] Cantin J L, Schoisswohl M, Grosman A, Lebib S, Ortega C, Von Bardeleben H J, Vazsonyi E, Jalsovszky G and Erostyak J 1996 *Thin Solid Films* **276**(1–2) 76
[181] Billat S 1996 *J. Electrochem. Soc.* **143**(3) 1055
[182] Bressers P M M C, Knappen J W J, Meulenkamp E A and Kelly J J 1992 *Appl. Phys. Lett.* **61**(1) 108
[183] Canham L T, Leong W Y, Beale M I J, Cox T I and Taylor L 1992 *Appl. Phys. Lett.* **61**(21) 2563
[184] Memming R 1969 *J. Electrochem. Soc.* **116**(6) 785
[185] Bsiesy A, Muller F, Ligeon M, Gaspard F, Herino R, Romestain R and Vial J C 1993 *Phys. Rev. Lett.* **71** 637
[186] Romestain R, Vial J C, Mihalcescu I, Bsiesy A 1995 *Physica Status Solidi B* **190**(1) 77
[187] Kooij E S, Despo R W and Kelly J J 1995 *Appl. Phys. Lett.* **66** 2552
[188] Green W H, Lee E J, Lauerhaas J M, Bitner T W and Sailor M J 1995 *Appl. Phys. Lett.* **67** 1468
[189] Gelloz B, Bsiesy A, Gaspard F and Muller F 1996 *Thin Solid Films* **276** 175
[190] Bsiesy A, Gelloz B, Gaspard F and Muller F 1996 *J. Appl. Phys.* **79**(5) 2513
[191] Bsiesy A, Gelloz B, Hory M A, Gaspard F, Herino R, Ligeon M, Muller F, Romestain R and Vial J C 1996 *Electrochem. Soc. Proc.* **95**(25) 1
[192] Gelloz B and Bsiesy A 1997 *Electrochem. Soc. Proc.* **97**(7) 92
[193] Gelloz B and Bsiesy A 1998 *Appl. Surf. Sci.* **135** 15
[194] Gelloz B, Bsiesy A and Herino R 1999 *J. Lumin.* **82**(3) 205
[195] Ben-Chorin M 1997 *Properties of Porous Silicon*, EMIS Datareviews Series No. 18 ed. L T Canham, INSPEC, The Institution of Electrical Engineers, London p. 165
[196] Oguro T, Koyama H, Ozaki T and Koshida N 1997 *J. Appl. Phys.* **81**(3) 1407
[197] Grosman A and Ortega C 1997 *Properties of Porous Silicon*, EMIS Datareviews Series No. 18 ed. L T Canham, INSPEC, The Institution of Electrical Engineers, London p. 328
[198] Baratto C, Faglia G, Sberveglieri G, Boarino L, Rossi A M and Amato G 2001 *Thin Solid Films* **391**(2) 261
[199] Geobaldo F, Onida B, Rivolo P, Borini S, Boarino L, Rossi A, Amato G and Garrone E 2001 *Chem. Commun.* **21** 2196
[200] Richter A, Steiner P, Kozlowski F and Lang W 1991 *IEEE Electron. Device Lett.* **12** 691
[201] Chen Y-A, Liang N-Y, Laih L-H, Tsay W-C, Chang M-N and Hong J-W 1997 *Jpn. J. Appl. Phys. Part 136 No 3B* **457** 1574
[202] Chen Y-A, Liang N-Y, Laih L-H, Tsay W-C, Chang M-N and Hong J-W 1997 *Elec. Lett. IEE33* **17** 1489
[203] Futagi T, Matsumoto T, Katsuno M, Ohta Y, Mimura H and Kitamura K 1992 *Jpn. J. Appl. Phys. Part 2 No 5B* **31** L616
[204] Gelloz B and Koshida N 2004 *Jpn. J. Appl. Phys.* In press
[205] Dubois T, Ozanam F and Chazalviel J-N 1997 *Electrochem. Soc. Proc.* **97**(7) 296
[206] Buriak J M and Allen M J 1998 *J. Amer. Chem. Soc.* **120**(6) 1339

[207] Boukherroub R, Morin S, Wayner D D M and Lockwood D J 2000 *Physica Status Solidi A* **182**(1) 117

[208] Gelloz B, Kadokura J, Boukherroub R, Wayner D D M, Lockwood D J and Koshida N 2002 *Electrochem. Soc. Symp. Proc.* **9** 195

[209] Koshida N, Ozaki T, Sheng X and Koyama H 1995 *Jpn. J. Appl. Phys.* **34**(2) L705

[210] Sheng X, Koyama H, Koshida N, Yoshikawa T, Yamaguchi M and Ogasawara K 1997 *Thin Solid Films* **297** 314

[211] Sheng X, Koyama H, Koshida N, Iwasaki S, Negishi N, Chuman T, Yoshikawa T and Ogasawara K 1997 *J. Vac. Sci. Technol. B* **15** 1661

[212] Sheng X, Koyama H and Koshida N 1998 *J. Vac. Sci. Technol. B* **16**(2) 793

[213] Sheng X and Koshida N 1998 *Mat. Res. Soc. Symp. Proc.* **509** 193

[214] Koshida N, Sheng X and Komoda T 1999 *Appl. Surf. Sci.* **146** 371

[215] Komoda T, Sheng X and Koshida N 1999 *J. Vac. Sci. Technol. B* **17**(3) 1076

[216] Komoda T, Honda Y, Hatai T, Watanabe Y, Ichihara T, Aizawa K, Kondo Y, Sheng X, Kojima A and Koshida N 1999 *IDW'99* 939

[217] Komoda T, Ichihara T, Honda Y, Aizawa K and Koshida N 2001 *Mater. Res. Soc. Symp. Proc.* **638** F4.1.1

[218] Komoda T, Honda Y, Ichihara T, Hatai T, Takegawa Y, Watanabe Y, Aizawa K, Vezin V and Koshida N 2002 *SID '02 Digest* **33**(2) 1128

[219] Spindt C A 1968 *J. Appl. Phys.* **39** 3504

[220] Spindt C A, Holland C E, Rosengreen A and Brodie I 1993 *J. Vac. Sci. Technol. B* **11** 468

[221] Yokoo K, Tanaka H, Sato S, Murota J and Ono S 1993 *J. Vac. Sci. Technol. B* **11** 429

[222] Suzuki M and Kusunoki T 1996 *IDW'96* 529

[223] Geis M W, Twishell J C and Lyszczarz T M 1996 *J. Vac. Sci. Technol. B* **14** 2060

[224] Kojima A, Sheng X and Koshida N 2001 *Mater. Res. Soc. Symp. Proc.* **638** F3.31

[225] Nakajima Y, Kojima A and Koshida N 2002 *Jpn. J. Appl. Phys.* **41** 2707

[226] Nakajima Y, Kojima A and Koshida N 2001 *Mater. Res. Soc. Symp. Proc.* **638** F4.2.1

[227] Nakajima Y, Kojima A and Koshida N 2002 *Appl. Phys. Lett.* **81**(13) 2472

[228] Berger M G, Thoenissen M, Arens-Fischer R and Muender H 1995 *Thin Solid Films* **255** 313

[229] Thonissen M and Berger M G 1997 *Properties of Porous Silicon*, EMIS Datareviews Series No. 18 ed. L T Canham, INSPEC, The Institution of Electrical Engineers, London p. 30

[230] Berger M G, Arens-Fischer R, Thoenissen M, Krueger M, Billat S, Lueth H, Hilbrich S, Theiss W and Grosse P 1997 *Thin Solid Films* **297** 237

[231] Cunin F, Schmedake T A, Link J R, Li Y, Koh J, Bhatia S and Sailor M 2002 *Nature Materials* **1** 39

[232] Thonissen M, Kruger M, Lerondel G and Romestain R 1997 *Properties of Porous Silicon*, EMIS Datareviews Series No. 18 ed. L T Canham, INSPEC, The Institution of Electrical Engineers, London p. 349

[233] Loni A, Canham L T, Berger M G, Arens-Fisher R, Munder H, Luth H, Arrand H F and Benson T M 1996 *Thin Solid Films* **276** 143

[234] Maiello G, La Monica S, Ferrari A, Masini G, Bondarenko V P, Dorofeev A M and Kazuchits N M 1997 *Thin Solid Films* **297** 311

[235] Araki M, Takahashi M, Koyama H and Koshida N 1998 *Mat. Res. Soc. Symp. Proc.* **486** 107
[236] Takahashi M, Araki M and Koshida N 1998 *Jpn. J. Appl. Phys.* Part 2 No. 9A/B **37** L1017
[237] Takahashi M and Koshida N 1999 *J. Appl. Phys.* **86**(9) 5274
[238] Ferrand P and Romestain R 2001 *Mat. Res. Soc. Symp. Proc.* **638** F4.4
[239] Ferrand P and Romestain R 2000 *Appl. Phys. Lett.* **77** 3535
[240] Yablonovitch E 1994 *J. Modern Optics* **41**(2) 173
[241] Joannopoulos J D, Meade R D and Winn J N 1995 *Photonic Crystals: Molding the Flow of Light* Princeton University Press ISBN 0-691-03744-2
[242] Yablonovitch E 2000 *Science* **5479** 557
[243] Kosaka H, Kawashima T, Tomita A, Notomi M, Tamamura T, Sato T and Kawakami S 1998 *Phys. Rev. B* **58**(160) R10097
[244] Poborchii V V, Tada T and Kanayama T 1999 *Appl. Phys. Lett.* **75** 3276
[245] Lehmann V 1993 *J. Electrochem. Soc.* **140**(10) 2836
[246] Gruening U, Lehmann V, Ottow S and Busch K 1996 *Appl. Phys. Lett.* **68**(6) 747
[247] Gruening U and Lehmann V 1996 *Thin Solid Films* **276**(1/2) 151
[248] Burner A, Gruening U, Ottow S, Schneider A, Mueller F, Lehmann V, Foell H and Goesele U 1998 *Physica Status Solidi A* **165**(1) 111
[249] S Leonard W, Van Driel H M, Busch K, John S, Birner A, Li A-P, Mueller F, Goesele U and Lehmann V 1999 *Appl. Phys. Lett.* **75**(20) 3063
[250] Rowson S, Chelnokov A and Lourtioz J-M 1999 *IEE Electronics Lett.* **35**(9) 753
[251] Netti M C, Charlton M D B, Parker G J and Baumberg J J 2000 *Appl. Phys. Lett.* **76**(8) 991
[252] Talneau A, Gouezigou L Le, Soukoulis C M and Agio M 2002 *Appl. Phys. Lett.* **80**(4) 547
[253] Lehmann V 2001 *Mat. Res. Soc. Symp. Proc.* **638** F8.1.1
[254] Schilling J, Muller F, Matthias S, Wehrspohn R B, Gosele U and Busch K 2001 *Appl. Phys. Lett.* **78**(9) 1180
[255] Yablonovitch E, Gmitter T J and Leung K M 1991 *Phys. Rev. Lett.* **67**(17) 2295
[256] Lin S Y, Fleming J G, Hetherington D L, Smith B K, Biswas R, Ho K M, Sigalas M M, Zubrzycki W, Kurtz S R and Bur J 1998 *Nature* **6690** 251
[257] Fleming J G and Lin S-Y 1999 *Opt. Lett.* **24**(1) 49
[258] Chutinan A and Noda S 1999 *Appl. Phys. Lett.* **75**(24) 3739
[259] Noda S, Tomoda K, Yamamoto N and Chutinan A 2000 *Science* **5479** 604
[260] Johnson N P, McComb D W, Richel A, Treble B M and De La Rue R M 2001 *Synthetic Metals* **116**(1–3) 469
[261] Cuisin C, Chelnokov A, Lourtioz J-M, Decanini D and Chen Y 2000 *Appl. Phys. Lett.* **77**(6) 770
[262] Cuisin C, Chelnokov A, Decanini D, Peyrade D, Chen Y and Lourtioz J M 2002 *Optical Quantum Electronics* **34**(1/3) 13
[263] Campbell M, Sharp D N, Harrison M T, Denning R G and Turberfield A J 2000 *Nature* **6773** 53
[264] Zheng J P, Jiao K L, Shen W P, Anderson W A and Kwok H S 1992 *Appl. Phys. Lett.* **61**(4) 459
[265] Tsai C, Li K-H, Campbell J C and Tasch A 1993 *Appl. Phys. Lett.* **62**(22) 2818
[266] Krueger M, Marso M, Berger M G, Thoenissen M, Billat S, Loo R, Reetz W, Lueth H, Hilbrich S and Arens-Fischer R 1997 *Thin Solid Films* **297**(1/2) 241

[267] Ueno K, Ozaki T, Koyama H and Koshida N 1997 *Mat. Res. Soc. Symp. Proc.* **452** 699

[268] Ueno K and Koshida N 1998 *Jpn. J. Appl. Phys. Part 1* **37** 1096

[269] Ueno K and Koshida N 1999 *Appl. Phys. Lett.* **74**(1) 93

[270] Takahashi M, Toriumi Y, Matsumoto T, Masumoto Y and Koshida N 1999 *Electrochem. Soc. Symp. Proc.* **99**(22) 35

[271] Takahashi M, Toriumi Y and Koshida N 2000 *Appl. Phys. Lett.* **76**(15) 1

[272] Toriumi Y, Takahashi M and Koshida N 2001 *Mat. Res. Soc. Symp. Proc.* **638** F8.3.1

[273] Weiss S M and Fauchet P M 2002 *SPIE—The International Society for Optical Engineering* **4654** 36

Chapter 11

Silicon/germanium superlattices

Hartmut Presting
DaimlerChrysler Research (REM/C), Ulm, Germany

11.1 Introduction

Modern silicon technology is best characterized by the development of the dynamic random access memory chips (DRAM). The 256 Mbit DRAM with its structural size of 0.2 μm contains already more than 100 million components on an area of about $0.5\,cm^2$. The development of 1 or even 4 Gbit DRAMs seems feasible with conventional concepts before the year 2005.

Beyond these predictable developments, however, it appears that further advances of microelectronics require new principles which, for example, could combine storage and amplification of electrical and optical functions. These completely new and tailorable properties are expected from a class of Si devices which are based on silicon/germanium (Si/Ge) superlattice structures. The classical Si_mGe_n superlattice (SL) is formed by depositing alternate m atomic monolayers of Si and n monolayers (m and n are integers >1) of Ge on top of each other which is repeated N times where the number of periods N is large compared with n and m (ideally infinite). The Si_mGe_n SL structure is usually deposited directly on a Si substrate or on a $Si_{1-x}Ge_x$ buffer layer which itself rests on a Si substrate.

New electronic and optical properties are expected from this artificial semiconductor if the period of the superlattice is small [2 to 50 atomic monolayers (ML); 1 ML ≈ 1.4 Å], i.e. if the SL period is in the range of the natural lattice constant of the constituting elements Si and Ge. The electrical and optical properties of these superlattices are controllable by geometrical dimensions, the chemical composition and the strain of the SL structure. Alternating layers of the lattice mismatched materials Si ($a_0^{Si} = 0.5431$ nm) and Ge ($a_0^{Ge} = 0.565$ nm) can be grown with high perfection by modern epitaxy techniques such as molecular beam epitaxy (MBE) or chemical vapour deposition (CVD) if the individual layer thickness is small. In the past few years important advances have been achieved in Si-based hetero-epitaxy using group IV elements like C, Ge and α-Sn. Especially, the Si/Ge heterostructures have matured during this time. Devices such as the Si/Ge based hetero-bipolar transistor (HBT) are now in the market for more than five years, and almost every Si manufacturer has several SiGe devices in his product list in order to remain competitive. Examples can be seen on the manufacturers' home pages (see e.g. www.atmel.com, www.intel.com, www.ibm.com, www.motorola.com).

The Si/Ge material system can be considered as a model system for the study of strain effects because of the complete chemical miscibility and similar chemistry of its constituents Si and Ge, which reduce overlaying chemical effects. A typical example for that is the similar electron affinity which leads to a small conduction band offset in SiGe heterostructures if strain effects are neglected.

11.2 Theory

11.2.1 Bandstructure and Brillouin zone folding of Si_mGe_n superlattices

The ultimate goal for synthesizing a Si_mGe_n superlattice is to convert the indirect band-gap of the Si and Ge band structure to a direct band-gap of a SiGe superlattice via Brillouin zone folding in the reciprocal space [1]. This would create the possibility of fabricating Si-based active and passive optical devices (e.g. LEDs and photodiodes) on a Si substrate with a much higher efficiency/sensitivity compared with conventional Si optical devices. The Si_mGe_n SL optical devices could then be integrated on a complex Si integrated circuit (IC) chip together with their electronic driver circuits on a common substrate as schematically depicted in figure 11.1.

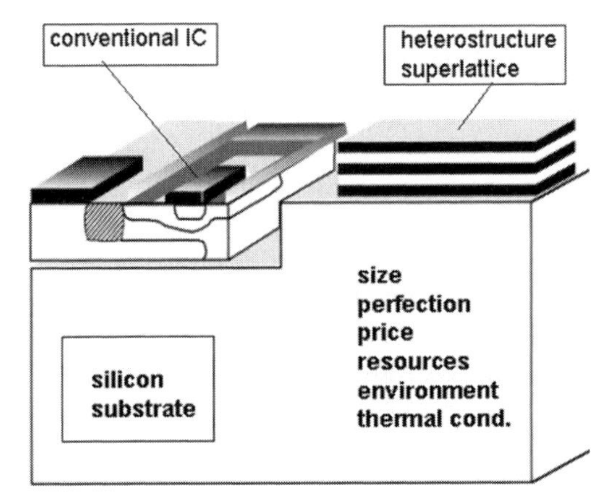

Figure 11.1. New Si-based superlattice concept for future microelectronic integrated circuits (ICs) [12].

The envisaged applications range from optical interconnections within a Si-based electronic circuit chip to optical chip-to-chip coupling within a complex Si IC board. This would tremendously enhance the variety of possible Si devices and offer completely new solutions to the bottleneck of the steadily increasing miniaturization and complexity of the topology for electronic interconnections, and would add a new degree of freedom in the design of Si IC chips. The performance of future microelectronics would be strongly enhanced by the monolithic integration of superlattice devices with conventional ICs on top of a Si substrate.

$Si_{1-x}Ge_x$ alloys grown on Si are an ideal material system, because due to their 100% complete miscibility ($0 \leq x \leq 1$) the SiGe band-gap can be continuously tuned from the Si ($E_g = 1.1\,eV$) down to the Ge band-gap ($E_g = 0.66\,eV$). Both, Si and Ge crystallize in the diamond lattice having an fcc (face centred cubic) crystallographic unit cell with a cubic point group [Oh, (or '$4/m\,3\,2/m$')] [2]. Since Ge has a 4.17% larger lattice constant the lattice mismatch between Si and Ge leads to strained hetero-epitaxy which means that, for a given Ge-content x, a $Si_{1-x}Ge_x$ alloy can only be deposited on Si up to a critical thickness t_c without the formation of dislocations [3]. The strain also leads to a substantial change in the SiGe band structure (especially, shift of the band-gap) because it lifts the six-fold degeneracy of the conduction band minimum in the $\langle 100 \rangle$ direction (Δ-direction) of the Brillouin zone into a two-fold degenerate set of electronic states extending in the $\langle 001 \rangle$ growth direction and a set of four-fold degenerate states which extend in the x–y plane perpendicular to growth direction. The band-gap curve of a (strained and unstrained) $Si_{1-x}Ge_x$ alloy as a function of x is illustrated in figure 11.2. One can see the influence of the strain on the band-gap when one compares the unstrained curve (dashed) with the strained curve (solid) which assumes that the SiGe is pseudomorphically grown (e.g. fully strained) on Si [4]. The dotted lines of the electronic states in figure 11.2 include the effect of size quantization due to the small critical thickness in the order of several atomic monolayers. A pronounced quantization effect occurs for higher Ge concentrations, which pushes the fundamental band-gap upwards. This shows that all energy gaps relevant for optical fibre communication can be achieved in this material system.

The sharp drop of the band-gap for $x > 0.85$ for the unstrained alloy indicates the transition point to the Ge-like band-structure for this high Ge content where the conduction band minimum changes from the Δ-direction ($k_{min} \approx \pi/a_0\ (0, 0, 0.8)$) to the L-point ($\pi/a_0\ (1, 1, 1)$) of the Brillouin zone.

As shortly outlined above, short-period Si_mGe_n strained layer superlattice (SLS) consists of an alternating sequence of m monolayers (ML) of Si and n ML of Ge stacked on top of each other with this sequence repeated N times ($N \geq 100$). By introducing a new super-period p_{SLS} ($p_{SLS} = n + m$) in real space, which is usually larger than the unit cell of the constituting bulk materials (a_0^{Si} equivalent to thickness of 4 ML in the $\langle 001 \rangle$ direction), a

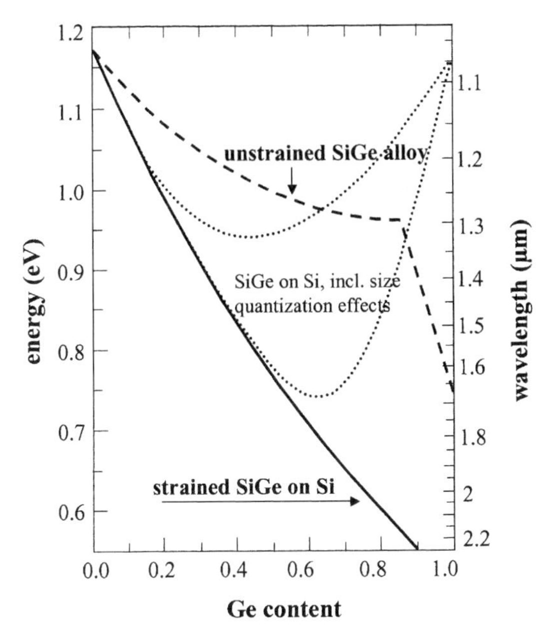

Figure 11.2. Fundamental band-gap of SiGe alloys as a function of Ge content. The dashed line corresponds to unstrained bulk SiGe samples, the solid line corresponds to fully strained bulk samples. For the dotted lines the effect of critical layer thickness is included (see text) [4].

correspondingly smaller Brillouin zone (BZ) size in reciprocal or k-space occurs which leads to a back-folding of the Si band-structure from the first BZ of the diamond body centred cubic (bcc) type BZ [the face centred cubic (fcc) diamond lattice in real space converts to bcc BZ in reciprocal space!] to the smaller SL BZ according to the calculations of Gnutzmann and Klausecker [1] (see figure 11.3). Under certain periods in real space, and corresponding SL BZ sizes, it occurs that the conduction band minimum in the Δ-direction of the Si band-structure at $k = \pi/a_0$ $(0, 0, 0.8)$ in k-space is folded back to the Γ-point [i.e. centre of the BZ $k = (0, 0, 0)$] creating a direct band-gap semiconductor in one dimension. For appropriate values of the SL period p_{SLS} ($p_{\mathrm{SLS}} = s \times 10$, $s = 1, 2, 3, \ldots$) a direct band-gap semiconductor in one dimension (parallel to the growth direction) occurs by the zone folding process which is schematically shown in figure 11.3.

Additionally, according to more recent pseudopotential calculations of Turton and Jaros [5] a direct band-gap in a $\mathrm{Si}_m\mathrm{Ge}_n$ SLS occurs when the period of the SLS fulfils $p = m + n = 10$, or integer multiples of it. The individual layer thickness being in the range $3 \leq m \leq 7$, the substrate can vary between unstrained Si and Ge. The principal results of this calculation concerning the transitions across the SL band gap are summarized in figure 11.4. According to these calculations the oscillator strength (proportion to

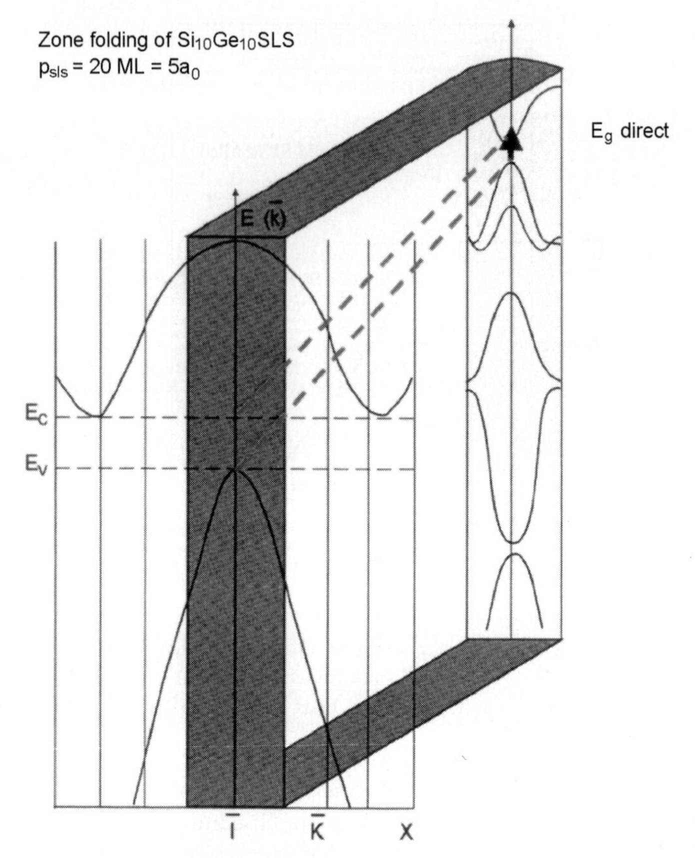

Figure 11.3. Brillouin zone folding of a Si-like band structure introduced with a period of $L = 5a_0$ ($a_0 = 0.5431$ nm, lattice constant of Si) or 20 ML. The band structure is folded into the first minizone (background) which has the extension of $2K_{SL} = 2\pi/5a_0$ being one-fifth of the original zone. By this process the original band structure with its indirect band-gap transforms into a band structure with a direct band-gap at $k = 0$ [11].

the square of the interband matrix element) for the direct band-gap transition of a 10 ML SLS is highest when $m = n = 5$ (Si_5Ge_5) and varies considerably for other combinations.

The principal results [6] concerning the transitions across the SLS band-gap with $m + n = 10$ can be summarized as follows:

(i) The minimum of the conduction miniband ground state is found to be very close to Γ in all cases, e.g. Si_3Ge_7 to Si_7Ge_3 for all substrates. Consequently, all of these systems are truly direct with the appropriate choice of the substrate.
(ii) The oscillator strength across the band-gap for a given well/barrier combination is approximately constant as a function of substrate.

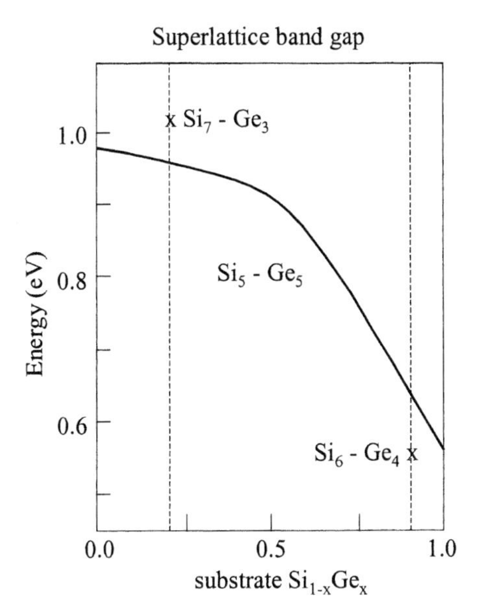

Figure 11.4. Transition energy $v1$–$c1$ for a Si_5Ge_5 superlattice on a $Si_{1-x}Ge_x$ substrate. Dashed lines indicate the region in which the system is direct. The crosses indicate extrema in the energy range when well/barrier widths are also altered [42].

This is somewhat surprising in view of the fact that the polarization dependence of the band-gap transition changes with the substrate, from ⟨110⟩ [parallel (‖) to the interface (TE)] on a Si substrate to ⟨001⟩ (perpendicular (⊥) to the interface and ‖ to growth direction) on a Ge substrate. It can, therefore, be concluded that the ground state conduction miniband contains k_x, k_y and k_z components. This observation implies that in the region of the heavy-hole to light-hole crossover the photoluminescence spectra around the band-gap energy will not exhibit a strong polarization dependence. In all of the systems considered here the crossover is achieved close to the condition for symmetric strain.

(iii) The oscillator strength across the band-gap is strongest for the Si_5Ge_5 and Si_4Ge_6 structures and weakest for the Si_7Ge_3 structure. This can be understood in terms of a decrease in the spatial overlap between the electron states in the Si layers and the hole states in the Ge layers. In particular, this overlap is due mainly to the tunnelling of the lighter mass hole states into the barriers. Thus, reducing the width of the Ge layers produces the most noticeable decrease in the oscillator strength, i.e. the transition strengths in Si_6Ge_4 and Si_7Ge_3 tend to be smaller than those in Si_4Ge_6 and Si_3Ge_7, respectively. Generally the oscillator strength decreases with increasing superlattice period. However, the strength also depends on the degree to which the conduction band

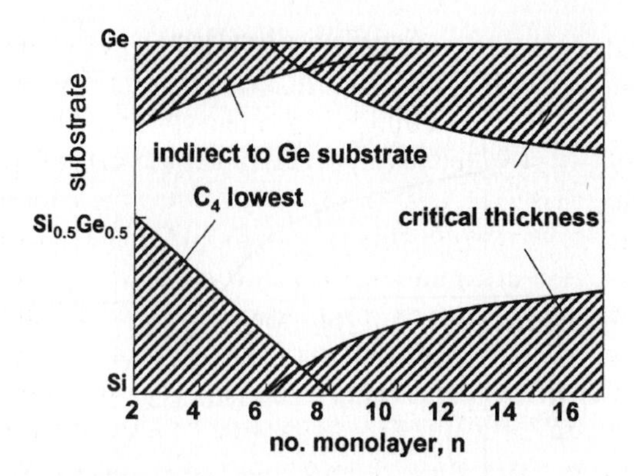

Figure 11.5. Sketch indicating the choice of Si_nGe_n SLS systems as a functon of period, n, and substrate for which the superlattice valence maximum and zone folded conduction minima are extrema within the system [42].

minimum is located at the centre of the BZ. Among the short-period SLS this is most conspicuous when $n = 5$ (10 ML period). Accordingly, there is a peak in the oscillator strength when $n = m = 5$. On the other hand the conduction band (CB) minimum moves far away from $k = 0$ in structures with $n = 4$ or $n = 8$.

(iv) In the perfect and infinite superlattice the band-gap transitions are to the second conduction band minimum in the Si_7Ge_3 and Si_4Ge_6 structures. Consequently, luminescence should not be observable in these systems since the matrix elements to the higher conduction band states are very weak compared with the matrix elements across the fundamental band-gap. However, any imperfections at the interface, or growth of only a finite number of periods, will alter the symmetry, and may therefore alter the selection rules.

(v) The transition energy across the band-gap can be varied considerably by changing the substrate's Ge composition (figure 11.4). Imposing the limitations required for a direct superlattice (indicated as hatched area in figure 11.5) shows that the band-gap energy of a Si_5Ge_5 SLS can be altered from ~950 to ~650 meV as the Ge content of the substrate is increased. The range can be extended slightly by varying the well and the barrier widths. Confining the choice of substrate to that required to produce a symmetrically strained SLS restricts the energy variation considerably.

11.2.2 Strain adjustment by a $Si_{1-y_b}Ge_{y_b}$ buffer layer

The lattice mismatch between Si and Ge of 4.17% can be accommodated either by strain or by formation of misfit dislocations. According to the equilibrium theory [7, 8], Ge deposited on Si is fully compressively strained up to the critical thickness t_c. Above t_c misfit dislocations form which partly relax the strain. However, the experimentally observed critical thickness is higher than predicted by equilibrium theory, especially, at low growth temperatures [9] which defines a metastable regime of pseudomorphic growth. A straightforward way would be to place the superlattice directly on the Si substrate. Although this direct placing of the superlattice on the Si substrate has some advantages, such as high crystal perfection and ease of processing, it cannot be used for a general approach [10] because of the limited thickness of the superlattice and non-adjustable strain distribution (SiGe layers compressed, Si layers unstrained [10]). The strain can be adjusted by a virtual substrate [11] consisting of a Si substrate and a SiGe buffer layer on top. With this intermediate buffer layer a virtual substrate for the SLS can be synthesized by adjusting the strain in the SLS which is deposited on top of the buffer layer. In the following section, thick and fully relaxed buffer layers as well as thin, homogeneous and partly relaxed buffer layers will be discussed and design rules for the growth of these buffer layers [12] will be given.

11.2.3 Effective Ge content and the virtual substrate

The thin, homogeneous SiGe buffer layer with thickness t_b and Ge content y_b is compressively strained (negative value of ε_b) by the substrate which leads to a tetragonal distortion of the cubic lattice cell. The in-plane lattice constant $a_\parallel$ of the buffer layer is somewhere between the natural lattice constant of Si and $Si_{1-y_b}Ge_{y_b}$ buffer layer. For the subsequent superlattice the virtual substrate offers the same in-plane lattice constant $a_\parallel$ as a SiGe substrate with an effective Ge content y^*. Applying Vegard's law for the natural lattice constants of the SiGe alloys results in

$$a_\parallel = a_0(1 + 0.042y + \varepsilon_b) \tag{11.1}$$

(a_0 being the Si lattice constant, 0.5431 nm). The effective Ge constant y^* of the virtual substrate is given by

$$y^* = y + 24\varepsilon_b. \tag{11.2}$$

For a buffer design chart of virtual substrates the strain has to be measured as a function of growth temperature T_g, Ge content y_b and thickness t_b of the buffer layer. Such a design chart has been constructed [12] with Bean's experimental data for MBE growth at $T_g = 550\,^\circ\mathrm{C}$ [13]. According to the design chart shown in figure 11.6(b) a virtual SiGe substrate with 25% effective Ge content ($y^* = 0.25$) requires a 0.25 µm thick buffer layer with 37.5% Ge or a 50 nm thick buffer layer with 42% Ge or a 100 nm thick buffer with 66% Ge alternatively. As this example shows, a wide range of thin, homogeneous buffer layers can be used for a virtual substrate with a definite in-plane lattice constant $a_\parallel$.

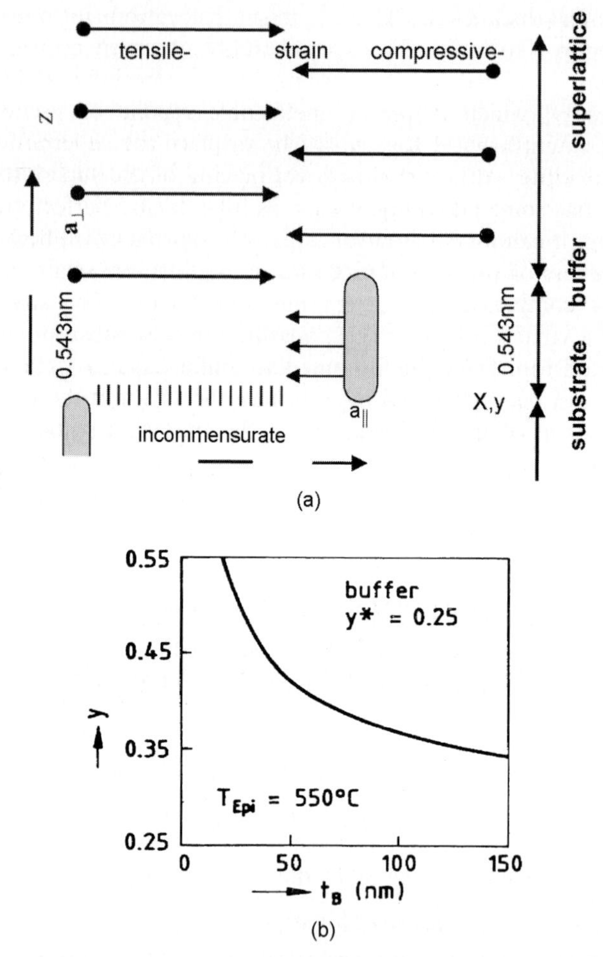

Figure 11.6. (a) Strain symmetrization by a virtual substrate (Si + $Si_{1-y}Ge_y$ buffer) in a Si_mGe_n SLS. Dots depict the natural lattice spacing, arrows the lattice spacing distortion by strain. In the strain symmetrized state the average strain of the SLS over one period is zero [11]. (b) Design chart of the corresponding buffer layer with effective Ge content of 25% ($y^* = 0.25$). Shown is the actual Ge content y versus thickness t_b of the buffer. Experimental strain values of $Si_{1-y}Ge_y$ grown layers at 550 °C were used for calculation of this design chart [11, 43].

To achieve a meaningful thickness for device applications of the strained layer superlattice (>200 nm) and to provide sufficient stability of the superlattice layer as a whole, one has to adopt the strain symmetrization condition of the SLS which requires that the average strain within one period adds up to zero. As discussed above, the strain in the SLS is adjusted by the choice of the virtual substrate with its in-plane lattice constant $a_\parallel$ and its magnitude of the individual layers is given as

$$\varepsilon_i^{\mathrm{SLS}} = \frac{a_{\parallel} - a_i}{a_i} \tag{11.3}$$

with a_i being the lattice constant of the ith layer. Strain symmetrization of a Si/SiGe superlattice on a Si substrate is obtained when the in-plane lattice constant of the buffer layer is chosen half between the lattice constants of the Si and the SiGe or the Ge layers. Under the approximation of equal elastic constants of Si and Ge [14], equation (11.3) leads to

$$\varepsilon_{\mathrm{Si}}^{\mathrm{SLS}} + \varepsilon_{\mathrm{Ge}}^{\mathrm{SLS}} = 0 \tag{11.4}$$

with $\varepsilon_{\mathrm{Si}}^{\mathrm{SLS}} = ft_{\mathrm{Ge}}/T$, $\varepsilon_{\mathrm{Ge}}^{\mathrm{SLS}} = -ft_{\mathrm{Si}}/T$, and $f = 2(a_{\mathrm{Ge}} - a_{\mathrm{Si}})/(a_{\mathrm{Si}} + a_{\mathrm{Ge}})$ being the lattice mismatch between Si and Ge ($f = 0.0417$), $t_{\mathrm{Si}} = ma_{\perp}^{\mathrm{Si}}/4$, $t_{\mathrm{Ge}} = na_{\perp}^{\mathrm{Ge}}/4$ and $T = t_{\mathrm{Si}} + t_{\mathrm{Ge}}$ being the individual and total layer thickness of one period of the SLS [$a_{\perp}$ is the lattice constant in growth direction (z), $a_{\parallel}$ is the in-plane lattice constant in the interface plane (x–y)]. $\varepsilon_{\mathrm{Si}}^{\mathrm{SLS}}$ and $\varepsilon_{\mathrm{Ge}}^{\mathrm{SLS}}$ are the magnitude of the strain of the Si and Ge layers of the SLS. For the most important case of equally thick layers ($t_{\mathrm{Si}} = t_{\mathrm{Ge}} = T/2$) equation (11.4) reduces to

$$\varepsilon_{\mathrm{Si}}^{\mathrm{SLS}} = -\varepsilon_{\mathrm{Ge}}^{\mathrm{SLS}} \equiv f/2 \quad (= 0.021) \tag{11.5}$$

The symmetrical strain is of special importance because in this strain situation the superlattice has its lowest energy content (the SLS produces zero average strain) and can be grown stable up to infinite thickness. However, to fulfil this condition for arbitrary superlattices one needs to grow an intermediate relaxed $Si_{1-y_b}Ge_{y_b}$ alloy buffer layer between substrate and superlattice [11] which provides the average in-plane lattice constant of the subsequent superlattice ($a_{\parallel}^{\mathrm{B}} \approx a_{\parallel}^{\mathrm{SLS}}$). Figure 11.6(a) depicts schematically the strain relation (compressive or tensile) of the single layers in a Si_mGe_n SLS grown on a strain adjusting buffer layer.

If, as already outlined in the previous section, the sum of the Si and Ge ML in an SLS is a multiple of 10 [$(m + n) = s \times 10, s = 1, 2, \ldots$] strong interband transitions have been predicted by theory [5] and found experimentally [15]. Further, on electroluminescence (EL) up to room temperature has been observed from mesa diodes, fabricated as shown in detail in section 11.4, containing a strain symmetrized Si_6Ge_4 SLS [16].

11.3 Growth and characterization of Si_mGe_n superlattices

11.3.1 MBE growth of buffer and Si_mGe_n SLS layers

All samples discussed in this chapter were grown by molecular beam epitaxy (MBE) on 4-inch (100 mm) diameter, $\langle 001 \rangle$ oriented, float zone Si substrates. Details of the MBE apparatus and the growth process can be found in the

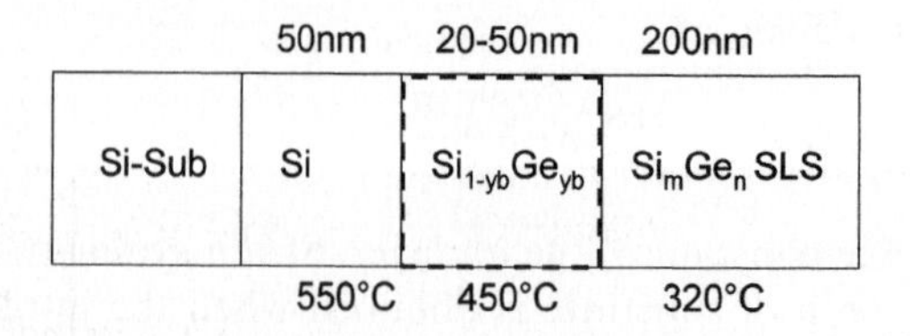

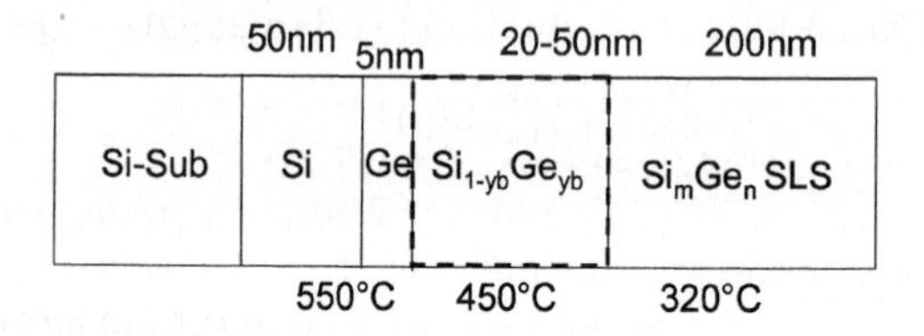

Figure 11.7. Layer structure of partially relaxed SiGe alloy buffer layer for strain adjustment of subsequent SL: (a) (simple, type 's'), (b) alloy buffer layer on Ge (type 'Ge'), (c) layer structure with pn junction of a Si_6Ge_4 SLS grown on two different buffer layers: (c1) on thin, partially relaxed 's'-type buffer, (c2) on Ge-graded, fully relaxed SiGe buffer ('grd'-type).

literature [17, 18]. Most of the SLSs are grown on a thin intermediate alloy buffer layer in which the Ge content is adjusted to provide the symmetrical strain situation of the subsequent SLS (see section 11.2.3). SLSs, usually up to a thickness of 200 nm at temperatures between 320 and 500 °C, have been grown by the author's group.

As outlined in the previous section, the relaxation in the $Si_{1-y_b}Ge_{y_b}$ alloy buffer layers with a fixed Ge content y_b takes place at the Si/SiGe interface which are designed to provide a thermally stable virtual substrate with the mean in-plane lattice constant of the subsequent SLS. All these buffer layers start with a 50–100 nm thick Si layer on a $\langle 001 \rangle$ Si substrate deposited at a temperature of around $T_g = 600$ °C to provide a well defined and clean epitaxial surface. They can be classified according to their composition, strain distribution and relaxation process. Figures 7(a) and (b) show two different types of a thin and partly relaxed SiGe alloy buffer layer with constant composition. The most frequently used buffer construction (designated as 'simple' or type 's', figure 11.7(a)) consists of a partly relaxed $Si_{1-y_b}Ge_{y_b}$ alloy layer with fixed Ge content y_b which is adjusted to the strain symmetrized point according to the design chart [3, 11] and using the experimental data of People and Bean [9] and others [19]. The SiGe layer is grown at $T_g = 450$ °C with a thickness of around 50 nm and provides an effective in-plane lattice constant $a_{\|_b}$ at the buffer/SLS interface which matches $a_{\|_{SLS}}$ of the SLS. Design rules for this buffer layer can be found in

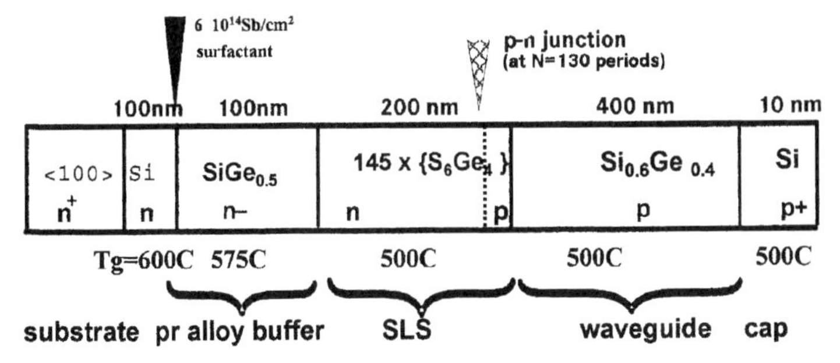

c1) Strain symmetrized Si_6Ge_4 SLS pn diode on thin, partly relaxed („s") SiGe buffer layer (B2804)

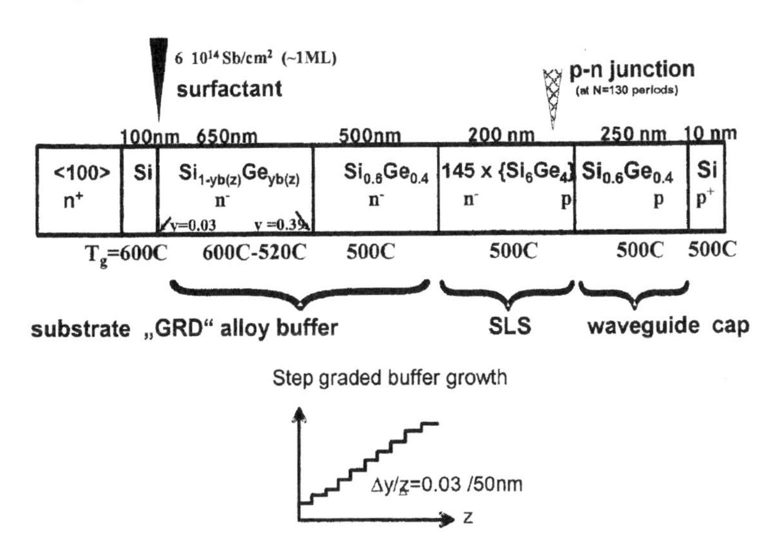

c2) Strain symmetrized Si_6Ge_4 SLS pn diode on thick, compos. graded SiGe buffer („grd") layer (B2805)

Figure 11.7. *Continued.*

[11, 12]. For the buffer type 'Ge' (figure 11.7(b)), a thin (5 nm) Ge layer is deposited on the Si start layer at $T_g = 300\,^\circ C$ being above the critical thickness [9]. On the partly relaxed Ge layer, the SiGe layer is deposited at $T_g = 300\,^\circ C$ which in this case is tensilely strained ($\varepsilon_b > 0!$). In addition, there is evidence that SiGe growth on a Ge substrate leads to a more two-dimensional surface and an enhanced growth quality [20].

Lately, a more frequently used buffer construction is a fully relaxed SiGe alloy buffer and consists of two parts, namely, a compositionally graded, and a fully relaxed constant composition part. The first one has a $Si_{1-z}Ge_{y(z)}$ alloy layer with a graded Ge composition which slowly increases from the

Si substrate towards the SLS, the second part keeps the highest Ge content of the graded part fixed as shown in figure 11.7(c2). Both parts are grown at rather high temperatures of $520 \leq T_g \leq 750\,°C$. A thickness of around 500 nm for the second constant composition (cc) part has been chosen. Due to construction peculiarities of the MBE machine, we used for the graded part a step-wise increase of the Ge flux with a ramp of $\Delta y/z = 0.03/50$ nm as depicted on the bottom of figure 11.7(c2) [18]. The growth temperature is lowered as the Ge content increases from the substrate/buffer to the buffer/SLS interface from 650 °C down to 520 °C. This temperature is held constant during the cc-buffer growth. Other authors use a higher growth temperature and a lower Ge increase for the graded part [21, 22]. Due to the slow Ge increase the strain level in the buffer layer can never reach such a high value where the high temperature enhances relaxation. At the same time the formation or nucleation of existing threading dislocations is suppressed. This combination of a compositionally graded (cg) and a cc SiGe alloy buffer has been proven to result in a 3 to 4 orders lower threading dislocation density compared to the thin and partly relaxed buffer layers discussed above. A Si_6Ge_4 SLS on both types of buffer layers mentioned above has been grown, namely on a thin ($\leq$50 nm), 's'-type alloy buffer (figure 11.7(c1)) and on a 'grd' (figure 11.7(c2)) SiGe alloy buffer layer. An n^+-doped Si substrate (n^+ 5×10^{19} Sb/cm^3) has been used and grown an n-doped buffer and superlattice layer with a rather low n-doping level caused by the spontaneous antimony (Sb) incorporation ($\sim 2 \times 10^{17}$ cm^{-3}; see figure 11.7(c)). To be able to measure electroluminescence from these samples, a pn junction has been created by depositing a p-doped $SiGe_{0.4}$ alloy cap layer with a thickness of 0.25 μm and 10 nm Si cap layer, p^+-doped on top of the n-doped SLS (see figure 11.7(c)). The depletion region of the pn junction extends mainly across the low n-doped SLS region which is important for the fabrication of Si_mGe_n SLS detectors and LEDs.

The structural parameters of the buffer as well as the superlattice layers were characterized by transmission electron microscopy (TEM), X-ray diffraction, secondary ion mass spectroscopy (SIMS) and Rutherford back-scattering (RBS) and is treated in the next section. A good overview of structural characterization of SLS samples grown with various buffer types can be found in the literature [23]. To enhance the growth quality and the SiGe interface sharpness, about one monolayer of Sb ($\sim 6 \times 10^{14}$ Sb/cm^2) was deposited prior to SiGe epitaxy and the alloy layer was grown at rather high temperatures (500–600 °C).

11.3.2 Characterization of Si_mGe_n SLS by XRD, TEM and Raman spectroscopy

Figure 11.8(a) shows for illustration a cross-sectional bright field transmission electron microscope (TEM) picture of a Si_5Ge_5 SLS grown on a

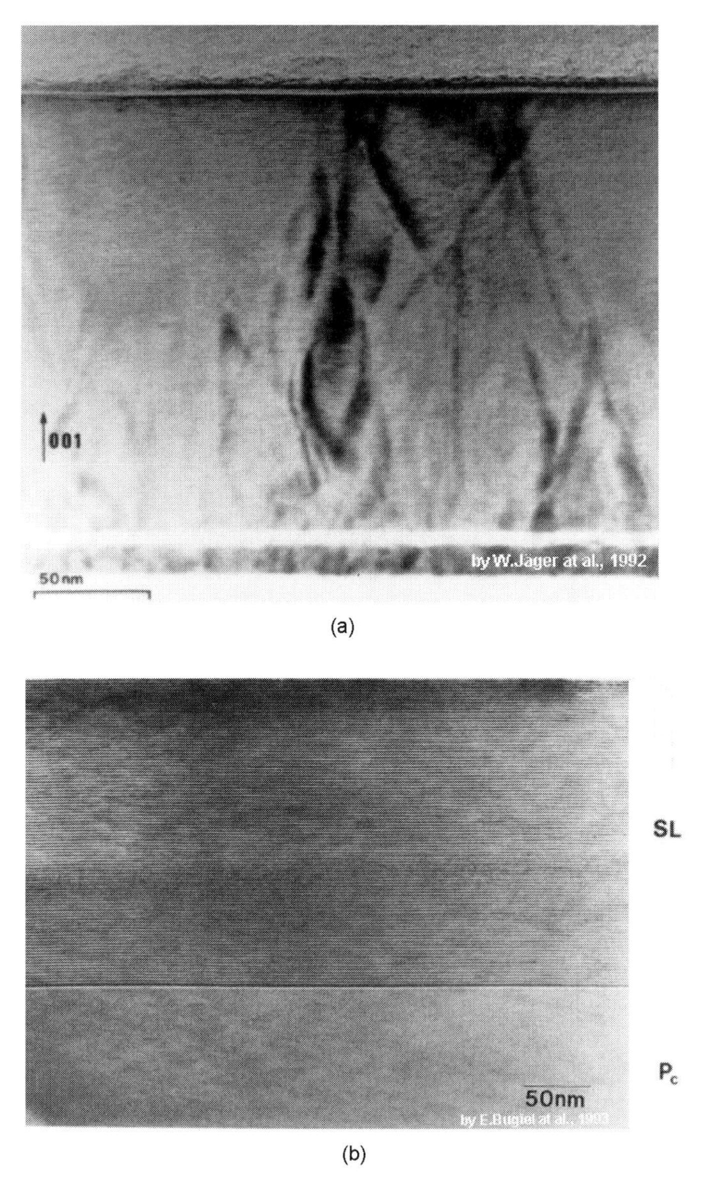

Figure 11.8. (a) Bright field cross-sectional TEM picture of a Si_5Ge_5 SLS grown on a partially relaxed buffer with intermediate thin Si layer. From the bottom the Si substrate, the 20 nm thick $SiGe_{0.75}$ alloy, the 1.6 nm thin Si sandwich layer in between and the Si_5Ge_5 SLS on top can be seen [23]. (b) Si_6Ge_4 SLS grown on a compositionally graded and constant composition, fully relaxed $Si_{1-y}Ge_y$ alloy buffer layer. Very good Si/Ge interface sharpness and layer quality can be seen of the MBE grown SLS. Recorded by E Bugiel and Chr. Quick, Institute of Semiconductor Physics, Frankfurt (Oder), Germany, 1993.

partly relaxed, thin $Si_{1-y_b}Ge_{y_b}$ buffer. On top of the Si substrate and the Si start layer (white bottom part), the 15 nm thick $Si_{0.25}Ge_{0.75}$ alloy layer (dark), the thin Si cap layer and finally the Si_5Ge_5 SLS can be seen. Figure 11.8(b) shows a cross-sectional transmission electron micrograph of a Si_6Ge_4 SLS sample also deposited on a fully relaxed, compositionally graded, $Si_{1-y(z)}Ge_{y(z)}$ Ge-type buffer layer. In the latter picture (figure 11.8(b)), a high resolution TEM micrograph with a very good interface sharpness and growth quality can be seen in the SLS region with no dislocations in contrast to the SLS shown in figure 11.8(a).

The growth and the characterization of many Si_mGe_n superlattice samples are discussed by a number of authors [23, 24]. The strain distribution, interface sharpness, period length, and composition in a Si_mGe_n SLS can be studied by x-ray analysis [25, 26] and Raman scattering [15]. The growth quality, type and density of dislocations and also the composition and period can be investigated by TEM [27, 28] and RBS [29]. The optical and electronic band-structure properties can be probed by photoluminescence (PL), photo- and electro-reflectance [30,31] and even ellipsometry measurements [32].

11.3.2.1 X-ray diffraction

X-ray diffraction is a versatile and very powerful tool for analysing the superlattice material because it is a non-destructive method which reveals the strain distribution, the composition and the periodicity of the SLS at the same time. The tension (dilatation or compression) between the superlattice and the alloy buffer layer and between the buffer and the substrate leads to a bending of the Si substrate which can be measured by x-ray curvature measurements. Here the wafer is adjusted in angle to the Bragg position and scanned stepwise through an x-ray beam. The measured curvature K results in a strain value $\varepsilon_{\|_{\rm tot}}$ of the total epitaxial thickness which in turn is related to the strain distribution of the individual layers according to

$$K = 1/r = \left(\frac{6}{t_{\rm sub}^2}\right)\varepsilon_{\|_{\rm tot}} t_{\rm tot} = \left(\frac{6}{t_{\rm sub}^2}\right)\varepsilon_{\|_i} t_i \qquad (11.6)$$

where r is the curvature radius, $\varepsilon_{\|_i}$ and t_i are the in-plane strain values and the thickness of the ith layer and $t_{\rm sub}$ is the substrate thickness. Thus, by measuring the curvature K one can easily and non-destructively check if the strain symmetrized state of the SLS has been achieved. By recording the sign of the curvature one can also determine to which side the Ge content of the buffer has to be corrected. Figure 11.9 shows an x-ray diffraction (XRD) plot (diffraction intensity versus diffraction angle Θ) of a $Si_{10}Ge_{10}$ SLS recorded by a double-crystal diffractometer in the Si (400) reflection geometry. The main peak at $\Theta = \Theta_B = 34.56°$ is caused by the (400) reflection of the Si substrate and serves as the reference peak. The rather broad peak

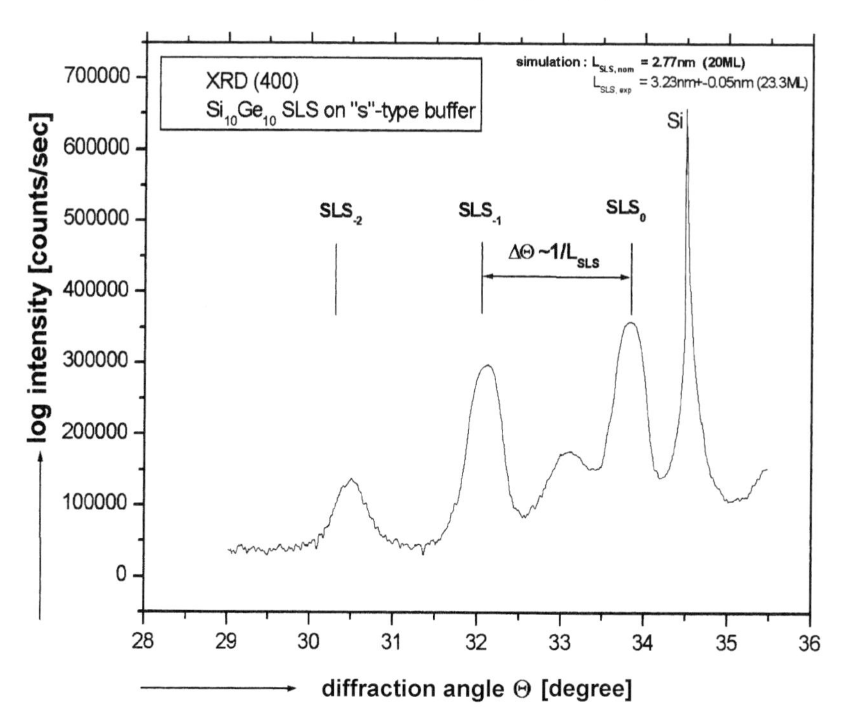

Figure 11.9. (004) x-ray diffraction plot from a $Si_{10}Ge_{10}$ SLS (B2188; log of diffracted intensity versus diffraction angle Θ). Besides the substrate reference peak at $\Theta_{sub} = 34.56°$ and the rather weak buffer peak, the zeroth order superlattice peak (marked SLS_0) and satellites (first order SLS_{-1}, second order SLS_{-2}) can be seen.

with very low intensity around 33° is caused by the 50 nm thin and partly relaxed buffer layer. The superlattice itself produces a zero-order reflection (marked SLS_0) and satellites (the first marked SLS_1 and respective SLS_{-1}) to both sides of the zeroth peak. The angular distance $\Delta\Theta$ between the satellites is determined by the SLS period which is defined as

$$\Delta\Theta = \lambda \sin\Theta / L \sin(2\Theta_B) \tag{11.7}$$

where Θ_B is the Bragg angle, Θ is the diffraction angle, and λ is the x-ray wavelength used (Cu K_α radiation, $\lambda = 0.154$ nm). Hence the period length L of the SLS is related to the angular spread of the XRD satellite peaks by a simple relation. However, it may be pointed out that equation (11.7) is an approximation, which becomes inaccurate for large angular distances $\Delta\Theta$ (small period length L). The angular distance of the SLS_0 peak from the Si substrate reference peak in the XRD spectrum is given by the parallel strain $\varepsilon_{\|_{tot}}$ of the SLS in total. To determine the individual layer thickness in the SLS as well as the strain, for an arbitrary SiGe substrate, one has to employ two independent reflection geometries in the XRD experiment where t_{Ge} (or t_{Si}) for example is determined by an independent measurement of L (equation (11.7)) and $\varepsilon_{\|Si}$ and $\varepsilon_{\|Ge}$ are related via equation (11.3). In the pseudomorphic case, i.e. when the SLS layer directly placed on the Si substrate is full elastically strained, the individual layer thickness t_{Si} and t_{Ge} can be uniquely determined by one (400) reflection measurement (pseudomorphic $\varepsilon_{\|Si} = 0$, $\varepsilon_{\|Ge} = -f_{Ge} = -4.17 \times 10^{-2}$). For a quantitative interpretation of a diffraction profile a comparison of the experimental curve with a numerical simulation is essential. The number of satellites occurring, their intensities, their lineshape and their linewidth reflect the structural quality of the SLS which can be seen to be rather good for the sample investigated.

11.3.2.2 Raman and photoluminescence spectroscopy

Optical characterization methods, which have been successfully performed for superlattices so far, are Raman scattering [15, 33], photoluminescence [34], ellipsometry [32] and modulation spectroscopy measurements [31]. Raman scattering reveals information on strain, periodicity, material quality and composition of the superlattice; photoluminescence and modulation spectroscopy (e.g. electro-reflectance spectroscopy), however, probe the electronic band structure of the superlattice.

Raman scattering by phonons is a local probe of the lattice dynamics and thus leads to information on the local crystalline structure, the orientation, composition, built-in strain etc. of the investigated sample. The probed sample volume is determined by the diameter of the laser focus, which can be as small as 1 μm^2, and the penetration depth of the laser light into the semiconductor, which can be easily varied from a few μm down to a few tens of nm. It is, therefore, an ideal tool to investigate thin epitaxial layers grown on various substrates. Figure 11.10 shows a typical Raman spectrum of a Si_6Ge_4 superlattice (B1859, upper curve) in comparison with a $Si_{0.6}Ge_{0.4}$ alloy sample (B1856), both grown on the same type of strain-relieving buffer layer. The epitaxial layers have a total thickness of about 200 nm. The Raman spectrum of sample B1856 is typical for an alloy with higher Si than Ge concentration, superimposed on the spectrum of the silicon substrate. The substrate gives rise to a small optical phonon mode of bulk Si at about 520 cm^{-1}, the other three main features at about 300, 400 and 500 cm^{-1} of the alloy spectrum correspond to Ge–Ge, Si–Ge and Si–Si vibrations in the alloy. The Ge content y and the built-in strain of the alloy can be extracted from the exact positions and intensity ratios of the Raman peaks [15]. The spectrum of the Si_6Ge_4 SLS sample B1859 is qualitatively different due to the ordered layer structure. In the wavenumber region near 120 cm^{-1} an additional strong phonon peak occurs due to the so-called folded acoustic longitudinal modes. The exact position of this mode is strongly dependent on the superlattice

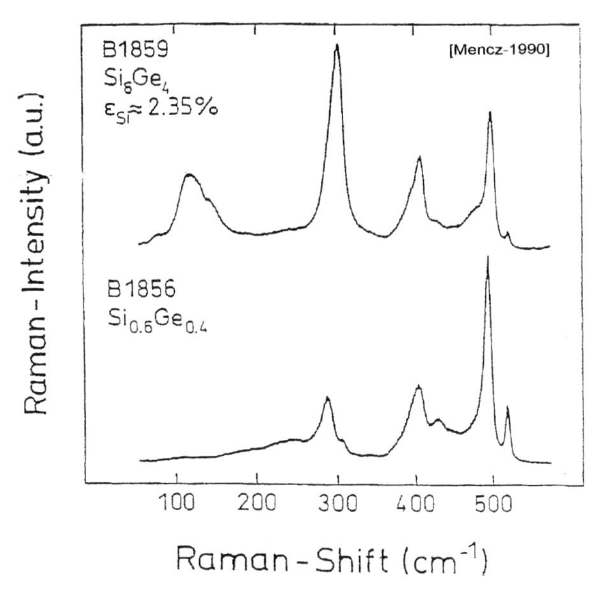

Figure 11.10. Raman spectra of a Si_6Ge_4 SLS (B1859) and the corresponding alloy (B1856) For the SLS the in-plane strain of the Si layers in the superlattice is given [44].

period and can be seen as experimental proof of zone folding. The width of the folded mode is sensitive to interface roughness and thickness fluctuations. The optical phonon peaks of the SLs are also drastically different from those of the alloy spectrum. The Ge mode is increased in intensity compared with the Si–Ge mode. The energetic positions of all three modes are sensitive to the built-in strain. The Si and Ge modes also depend on the individual layer thickness due to confinement of the optical vibrations. In addition, higher order confined modes are observed on the low energy side especially of the Si mode. The intensity of the Si–Ge mode at about 400 cm^{-1} can be seen as a probe of the interface sharpness of the SLS.

Photoluminescence spectroscopy is a technique well suited for studying interband transitions in the expected energy range of the quasi-direct fundamental energy gap of the SLS ($0.5 < h\omega < 1.0$ eV). Early PL measurements on strain-symmetrized 10 ML period superlattices (Si_6Ge_4 or Si_4Ge_6) [34] grown by the author's group in Ulm gave indications of direct band-gap transitions at $h\omega \approx 0.8$ eV. Figure 11.11(a) compares the PL signal of a Si_6Ge_4 SLS with the corresponding $Si_{0.6}Ge_{0.4}$ alloy sample with the same integral Ge content. Both samples were grown on a 20 nm thin and strain adjusting 's-type' $Si_{0.4}Ge_{0.6}$ alloy buffer layer prepared under identical MBE process conditions (growth temperature of SLS $T_g = 450\,^{\circ}C$, $t_{SLS} = 200$ nm). The clearly dominating signal of the Si_6Ge_4 SLS compared with the $SiGe_{0.4}$ alloy at $h\omega = 0.86$ eV is believed to stem from a quasi-direct transition of the SLS as a consequence of the zone folding. Figure

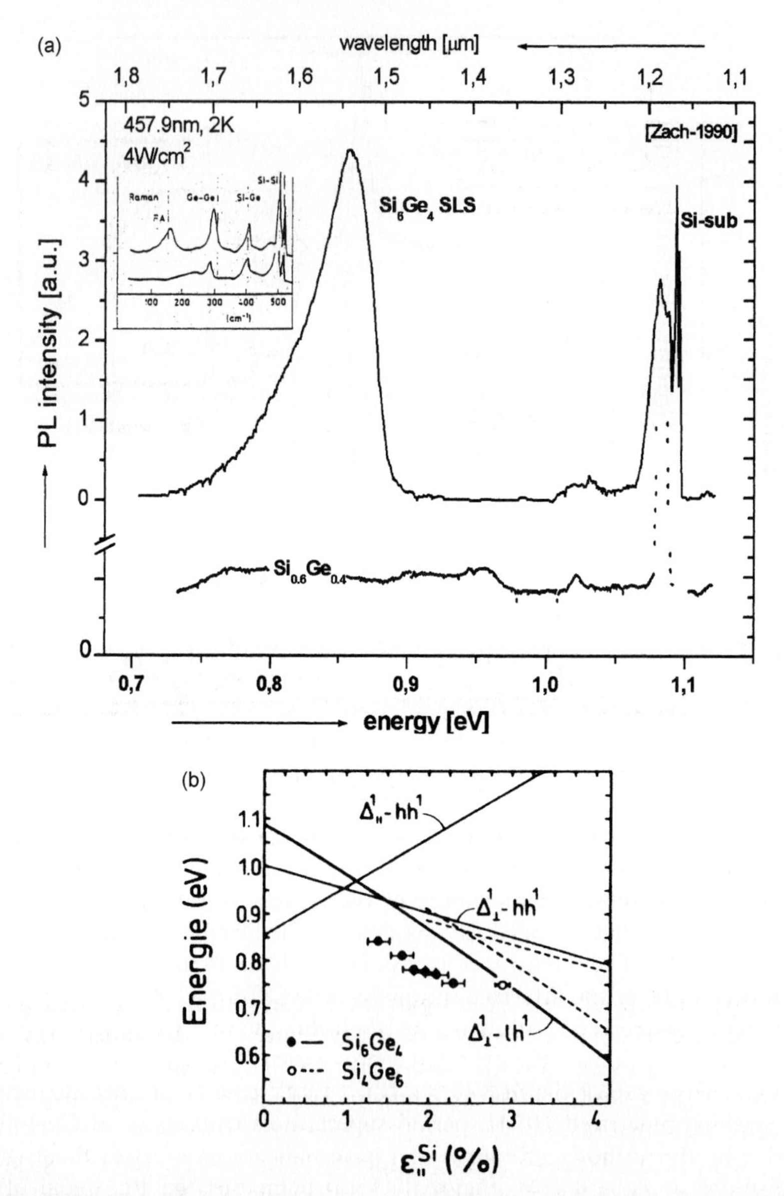

Figure 11.11. (a) Comparison of the PL signal of a Si_6Ge_4 SLS with a $Si_{0.6}Ge_{0.4}$ alloy sample The photoluminescence was excited by the blue line (457.9 nm) of an Ar^+ laser with an intensity of roughly 100 mW cm^2 and measured at $T = 5$ K. The inset shows the corresponding Raman spectrum where the occurrence of the folded acoustic mode proves the existence of the superlattice periodicity. (b) PL peak energies of Si_6Ge_4 (full circles) and Si_4Ge_6 (open circles) SLS as function of biaxial strain in the Si layers. Also shown are the relevant band-gaps of the Si_6Ge_4 (full line) and the Si_4Ge_6 (broken line) as calculated from a Kronig–Penney model [34].

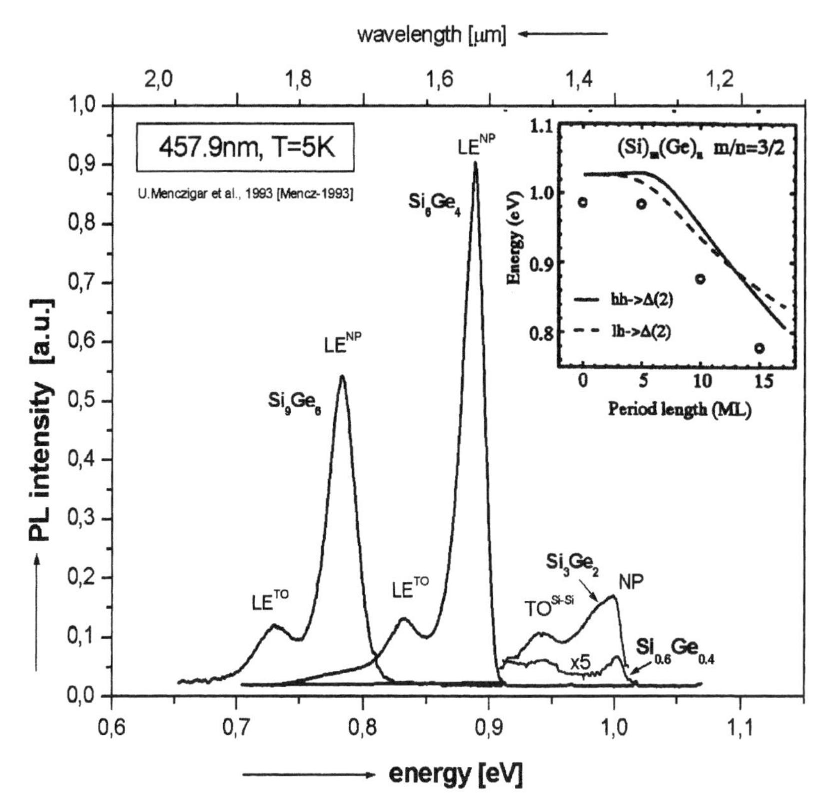

Figure 11.12. PL spectra for the Si_9Ge_6, Si_6Ge_4, Si_3Ge_2 SLS and the corresponding $SiGe_{0.4}$ alloy. The inset shows the calculated energies of the interband transitions as a function of period length [41].

11.11(b) plots basically the PL energies of a Si_6Ge_4 (full circles) and a Si_4Ge_6 SLS (open circles) as a function of biaxial strain in the superlattice as calculated by a simple Kronig–Penney band-structure calculation. As can be seen, the PL peak shifts to lower energies with increasing strain as expected from theory.

Figure 11.12 compares the PL spectra of a Si_3Ge_2 (B2516), Si_6Ge_4 (B2416), Si_9Ge_6 (B2512) SLS, with the PL spectrum of a $Si_{0.6}Ge_{0.4}$ reference sample (B2414) which are plotted on the same scale in this figure. All superlattices including the reference alloy have been grown on the same step-graded buffer layer where only the period length of these SLS samples is different, but the relative Ge content is fixed at 40%. The luminescence intensities are normalized to the thickness of the superlattice and the thickness of the alloy layer with constant Ge content for the alloy reference sample, respectively. The PL spectra for the Si_3Ge_2 SLS and the $Si_{0.6}Ge_{0.4}$

alloy are almost identical which gives clear evidence that the Si_3Ge_2 superlattice is more an alloy due to the intermixing of the Si and Ge layers at growth temperatures of 500 °C. The PL intensity for the Si_3Ge_2 SLS, however, is already enhanced by about a factor of 10 compared with the alloy reference sample. The PL signals for the Si_3Ge_2 SLS and the $Si_{0.6}Ge_{0.4}$ alloy can be attributed to a no-phonon (NP) transition and an associated phonon replica involving a transverse optical Si–Si mode ($TO^{Si\text{-}Si}$), respectively [35]. For the random alloy sample the energy position for the NP line is in excellent agreement with the calculated band-gap energy of the alloy taking into account the built-in lateral strain. For the Si_6Ge_4 and the Si_9Ge_6 SLSs much stronger PL signals are observed at 0.877 and 0.778 eV, respectively. As shown below, these PL signals can be attributed to an NP transition of localized excitons (LE^{NP}). Associated phonon replica (LE^{TO}) are also observed at 0.825 and 0.726 eV, respectively. The normalized PL intensity is strongest for the Si_6Ge_4 SLS.

For both Si_6Ge_4 and Si_9Ge_6 the $\Delta(2)$ minimum is folded close to the Γ point which results in an enhanced oscillator strength. The reduced PL intensity and band-gap energy for Si_9Ge_6 compared with Si_6Ge_4 can be explained by the staggered band line-up of the CB and VB edges in the Si and Ge layers. The holes (electrons) are more confined in the Ge (Si) layers with increasing period length. The separation of electrons and holes in real space with increasing period length results in a decreasing oscillator strength and a reduced band-gap. The energy of the interband transitions, calculated by a simple envelope function approximation [36], between the light hole (lh) and heavy hole (hh) states of the VB and the $\Delta(2)$ states of the CB for strain symmetrized $(Si)_m(Ge)_m$ SLSs with $m/n = 3/2$ are shown in the inset of figure 11.12, and compared with the energy positions of the LE^{NP} lines. In this calculation, the Si concentration (x_{Si}) in the Si layers and the Ge concentration in the Ge layers (x_{Ge}) of the SLSs are assumed to be reduced with decreasing period length due to the intermixing of the Si and Ge layers during growth at $T = 500$ °C. The actual concentration profile in the SLS is not known. The x_{Si} is assumed to depend on the Si layer thickness (d_{Si}) as

$$x_{Si} = 1 - \frac{0.4}{1 + \exp[(d_{Si} - 4.0\,\text{Å})/1.0\,\text{Å}]}. \tag{8.11}$$

The x_{Ge} can be deduced from equation (11.7) (making use of $x_{Si} + x_{Ge} = 1$). For the intermixed Ge and Si layers, the effective masses, deformation potentials, and band offsets used for the calculation were linearly interpolated between those of Si and Ge. The energy positions of the NP lines for the alloy and the Si_3Ge_2 SLS are almost identical, which indicates that the Si_3Ge_2 SLS is more like an alloy due to the intermixing. The systematic shift of the NP lines for Si_6Ge_4 and Si_9Ge_6 superlattices to lower energies with increasing period length is well described by the calculation. Deviations for the SLSs from the nominal structural data result in slightly different calculated band-gap energies. These deviations, however, are of the same order of magnitude as the theoretical errors arising from the uncertainties in deformation potentials given in the literature.

11.4 Electroluminescence and related properties

For the observation of electroluminescence (EL) from Si_mGe_n SLS samples one needs to grow pn-doped samples such that the superlattice region is situated inside the depletion zone of the pn junction. The author's group have fabricated circular mesa diodes as well as waveguide mesas from SLS p-i-n-doped samples using standard semiconductor processing techniques. The diodes have been processed in a three-mask process using conventional contact photolithography. The individual processing steps are shown in figures 11.13(a) and (b).

In the first step the mesa structures are defined photolithographically and etched by a CF_4/SF_6 plasma etch up to a height of roughly 0.5 μm using photoresist as etch mask. The mesa diameters range from 100 to 1000 μm. The mesa surfaces were passivated by a 200 nm PECVD oxide deposited at roughly 300 °C. With the second lithography mask, a ring with a width ranging from 10 to 100 μm was defined on top of the mesa and subsequently etched into the oxide. The contact metal is evaporated (30 nm Ti, 300 nm Au) and the whole structure is electroplated by a 3–5 μm thick Au layer. Then the metal is removed by a lift-off process and the mesa structure as shown in figure 11.14(b) (step 6) remains. In the third lithography step the oxide inside the ring contact is opened and wet chemically etched away by buffered HF. In the final step the full substrate area is metallized as rear-side contact (step 8, figure 11.14(b)) using again Ti/Au for n- and p-type contacts. After the wafer processing chips are sawed with a size of 3 mm × 4 mm and mounted on a TO_5 housing. The chip is glued to a ceramic pedestal on the housing with a conductive adhesive. Mesa and substrate contact are bonded by a 25 μm thick aluminium wire by thermo-compression bonding. Figure 11.14 shows a micrograph of a SiGe SLS diode chip mounted on a TO_5 housing.

Another possibility to fabricate an electroluminescent device is an integrated waveguide/detector or light emitting diode (LED) combination with light impinging or being emitted from the side facet. Figure 11.15 shows a schematic sketch of this device deposited on a SIMOX wafer. The technological processing of the waveguide/photodetector combination is done by a four layer process, the size of the processed pieces are 18 mm × 18 mm. Its steps are schematically shown in figure 11.16. In the first step the ridge of the detector/LED region is photolithographically defined by the first mask layer. The waveguide mesa of the detector is

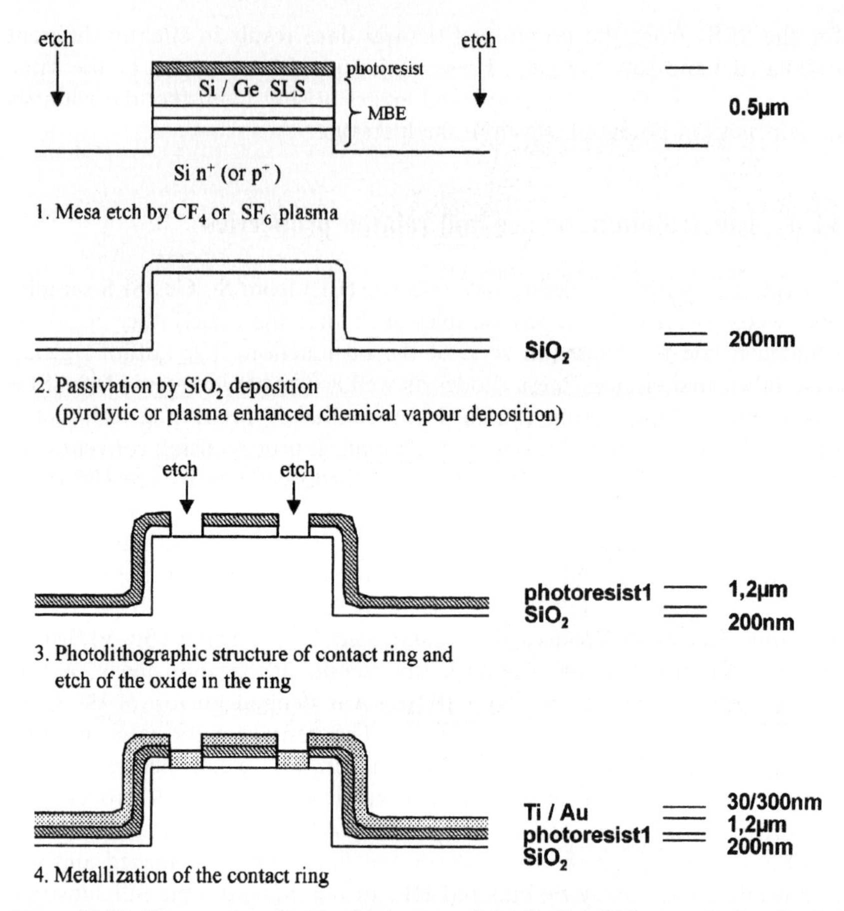

Figure 11.13. Process steps for the fabrication of the Si_mGe_n SLS mesa diodes. Typical dimensions of the mesa are shown in step 8.

etched by a SF_6/O_2 plasma in a parallel plate plasma reactor to the bottom contact layer (~700 nm etch depth).

Then 1 µm of SiO_2 is deposited by plasma-enhanced chemical vapour deposition (step 'a' in figure 11.16). In the second mask step the waveguide ridges are photolithographically defined. The oxide is wet etched by buffered hydrofluoric acid (NH_4F/HF, 87.5% : 12.5%) and after removal of the photoresist the SiO_2 serves as etch mask for the following waveguide etch. The waveguide is etched by potassium hydroxide (KOH) to a depth of 6 µm in roughly 1 h (step 'b' in figure 11.16). Due to the KOH etch the ridge waveguide is formed with bevelled edges of 56° (relative to the substrate interface) of the $\langle 111 \rangle$ planes because of their high etch resistance. This means that the waveguide width increases from the top to the bottom

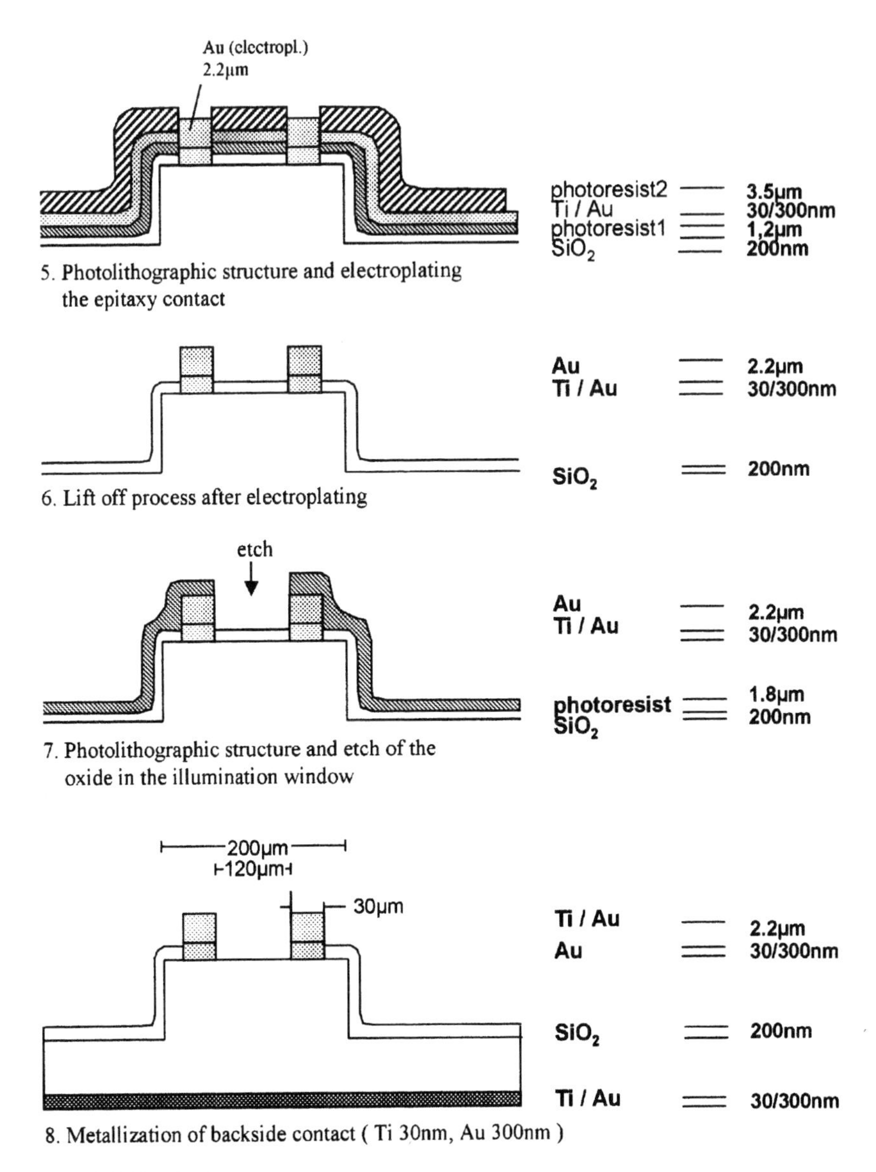

Figure 11.13. *Continued.*

(substrate) such that the width at half height is roughly one and a half times the photolithographically defined width. After the waveguide etch the photolithography step of mask layer 3 is done defining the contact windows for the top and the bottom contacts. A subsequent oxide wet etch (buffered HF) is done which is succeeded by the contact metallization of the top

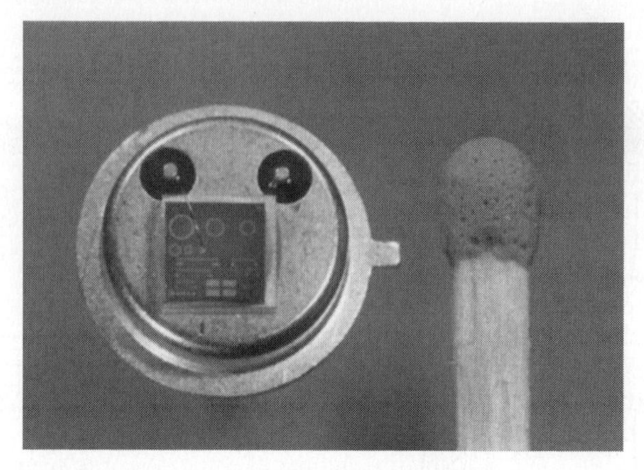

Figure 11.14. Top view of processed Si_mGe_n SLS mesa diode.

and the bottom contacts with aluminium (500 nm thickness, step 4 in figure 11.17). Aluminium has been chosen as contact metal for the n- and p-contact because it provides good ohmic contact for the low p-doped bottom contact layer (5×10^{-17} B/cm^3). In addition the bottom contact area has been made

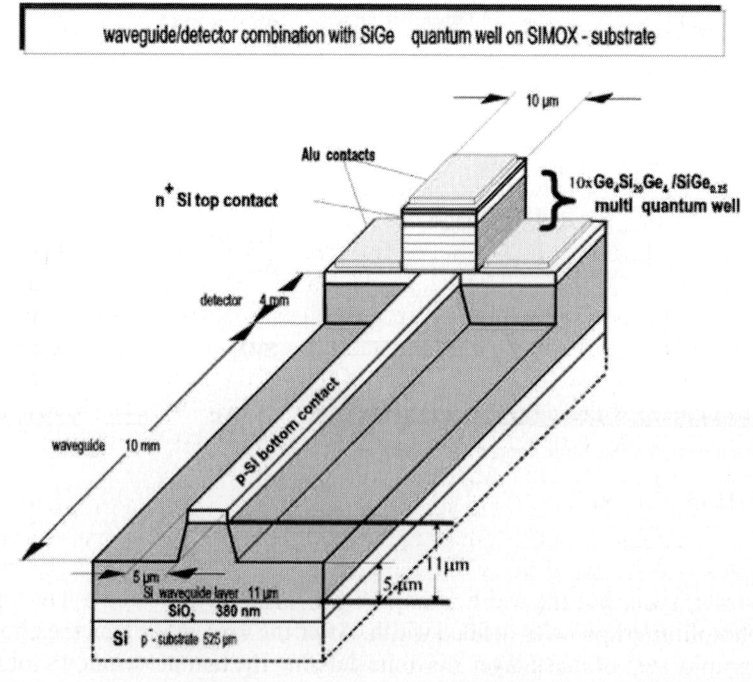

Figure 11.15. Schematical sketch of ridge waveguide/detector (LED) combination device.

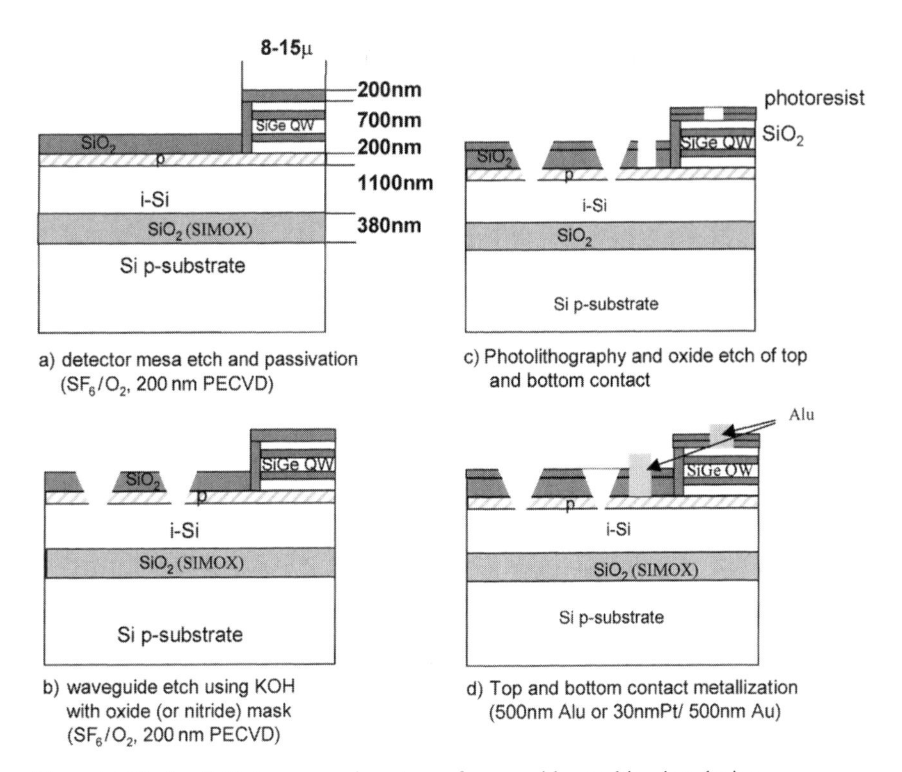

Figure 11.16. (a–d) Major processing steps of waveguide combination device.

large to decrease the contact series resistance. After metallization, the 18 mm × 18 mm pieces are mechanically polished with a diamond paste (grain size ≤0.5 μm) and are sawed into chips (8 mm × 16 mm). They are mounted on a ceramic carrier with gold-plated contact pads. Bottom and top contact pads on the chip are finally ball bonded by a 25 μm diameter aluminium wire to the gold-plated contact pins of the carrier. Figure 11.16(d) shows schematically the cross-section of the processed device.

Figure 11.17 plots the EL spectrum from a Si_6Ge_4 SLS strain adjusted superlattice with a p^+n junction in the region 0.7–1.1 eV (1.3–1.7 μm) which has been extensively discussed [16].

The EL spectra of a fabricated SLS mesa diode, schematically depicted in the inset of figure 11.17, is shown at three different heat sink temperatures, namely at 159 K, 258 K and room temperature (RT ≈ 300 K) with the same injection current density (3 mA/mm^2). The EL signal consists, in principle, of two broad features with peak energies of 0.77 and 0.88 eV respectively, whereas the relative intensity of the two peaks are temperature dependent. To determine the nature of the emission processes the intensity was measured as a function of the injection current (L-I characteristics) which is plotted in

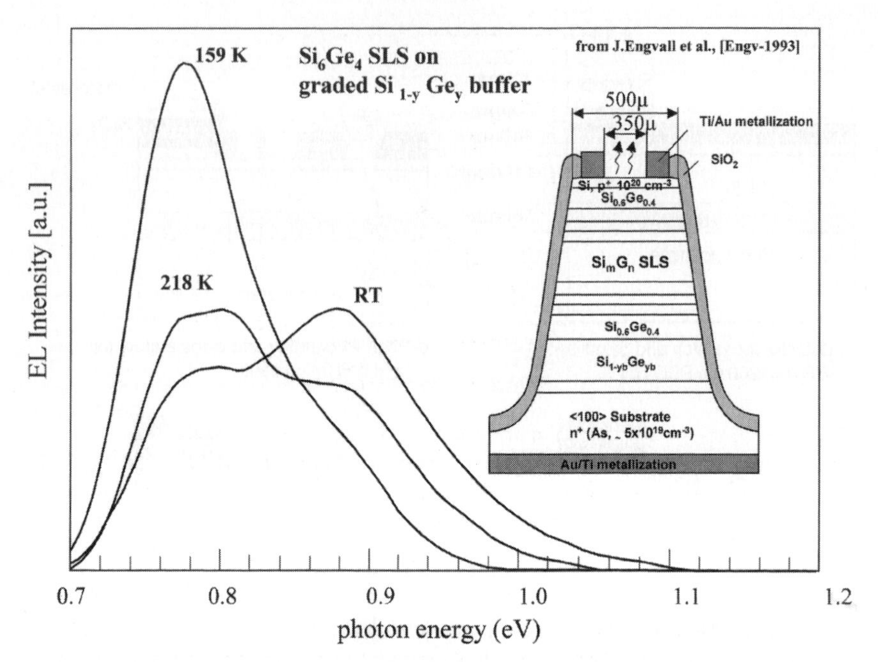

Figure 11.17. EL spectra of a strain-adjusted Si_6Ge_4 SLS taken with an injecting current density of $3\,mA/mm^2$ at three different heat sink temperatures [16]. The inset shows a sketch of the processed mesa diode.

figure 11.18 at a temperature of 156 K. To minimize the heating of the sample, the mean injected power was reduced by using a lower duty cycle at high intensities. Simultaneously, the voltage over the sample was measured. In this diagram the intensity of the 0.77 and 0.88 eV peaks are marked with full and open circles, respectively. The intensity of both peaks in the EL spectrum shows a superlinear dependence ($s > 1$) in a certain range of current densities, the onset of superlinearity occurs at similar current densities for both peaks. The observed superlinearity is characteristic for a multi-state model of recombination channels in addition to the observed EL transitions.

11.4.1 Optical and electrical characterization of 10 ML Si_mGe_n SLS

In the following are presented the different electrical and optical characterization results of a series of three p–n diodes from ten monolayer, strain adjusted Si_nGe_m superlattices (Si_6Ge_4, Si_5Ge_5, Si_4Ge_6) with respect to current–voltage, capacitance–voltage, electroluminescence, short-circuit photocurrent (I_{sc}) and photo-capacitance measurements. The SLS interband transitions were measured by short-circuit current spectroscopy, and

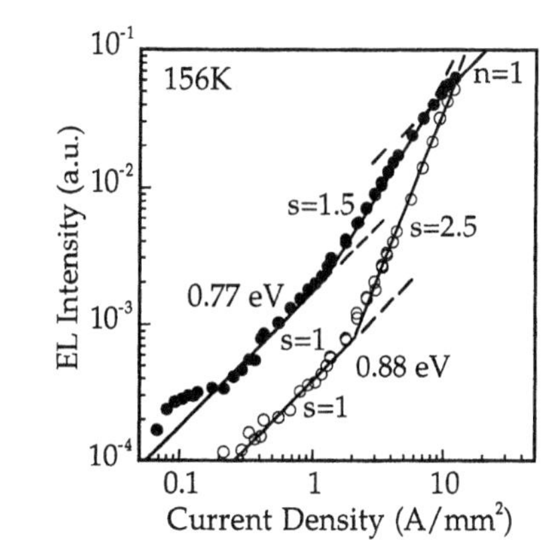

Figure 11.18. EL intensity at 156 K from the previous figure plotted as a function of injection current density according to data of ref. [16]. The two peaks are marked with full (at 0.77 eV) and open circle (0.88 eV peak). The superlinearity is characteristic for a multi-state model of recombination channels across the band-gap of the semiconductor.

band-gap energies were determined by a fitting procedure. The temperature dependence of the junction capacitance, and of the short-circuit current, indicates a potential barrier for electrons at the superlattice–buffer interface that impedes electron transport from the p–n junction at lower temperatures. Spectrally resolved EL for the Si_6Ge_4 sample was measured up to room temperature, originating probably from intraband transitions.

The SLS samples in this series have been grown on n^+ Si substrates ($n^+ \approx 10^{-19}\,cm^{-3}$). On the substrate, a 50 nm thick Si layer, doped with Sb by secondary implantation (DSI) to a concentration of more than $1 \times 10^{17}\,cm^{-3}$ was grown. A 100 nm $Si_{1-y_b}Ge_{y_b}$ buffer layer was deposited on the top at 575 °C, followed by a 145 period Si_nGe_m superlattice (200 nm) grown at 500 °C. Sb has been predeposited (prior to buffer and prior to SLS growth) as surfactant [37, 38]. On top of the superlattice a p-layer was grown at 500 °C as a $Si_{1-x}Ge_x$ alloy with the same Ge concentration as the SLS. p-Doping was done by boron co-evaporation with a level in excess of $1 \times 10^{19}\,cm^{-3}$. The alloy was capped with 10 nm of p^+ Si. Circular mesa diodes with different sizes (100–1000 μm diameter) were fabricated using the above described technology. The layer composition is listed in table 11.1. The p-layer was contacted with Au/Ti contact ring with an illumination window inside to allow the optical measurements.

11.4.2 Capacitance–voltage

To determine the doping profile and the depletion layer width the diodes have been characterized by capacitance–voltage measurements. In figure 11.19 the doping profile versus depletion layer width of the Si_5Ge_5 and

Table 11.1. Layer structure of Si_nGe_m superlattice series grown on 's-type' buffer layer.

Sample	Buffer	SLS	p Layer
B2460	$Si_{0.50}Ge_{0.50}$	Si_6Ge_4	$Si_{0.60}Ge_{0.40}$
B2517	$Si_{0.40}Ge_{0.60}$	Si_5Ge_5	$Si_{0.50}Ge_{0.50}$
B2518	$Si_{0.25}Ge_{0.75}$	Si_4Ge_6	$Si_{0.40}Ge_{0.60}$

Si_6Ge_4 SLS from the above mentioned series is plotted as obtained from the capacitance–voltage plot ($N_D \approx d(1/C^2)/dU$) [39]. As can be seen from this figure, the depletion width can be tuned from 90 nm for zero external bias up to roughly 300 nm for $U_{bias} \approx 4$ V which more or less extends over the whole region of the superlattice. This means that the measured photocurrent only stems from the SLS region itself.

11.4.3 Short-circuit current and electroluminescence

Short-circuit photocurrent (I_{sc}) originates only from the space charge region of the superlattice p–n diode. The I_{sc} is related to the absorption inside the

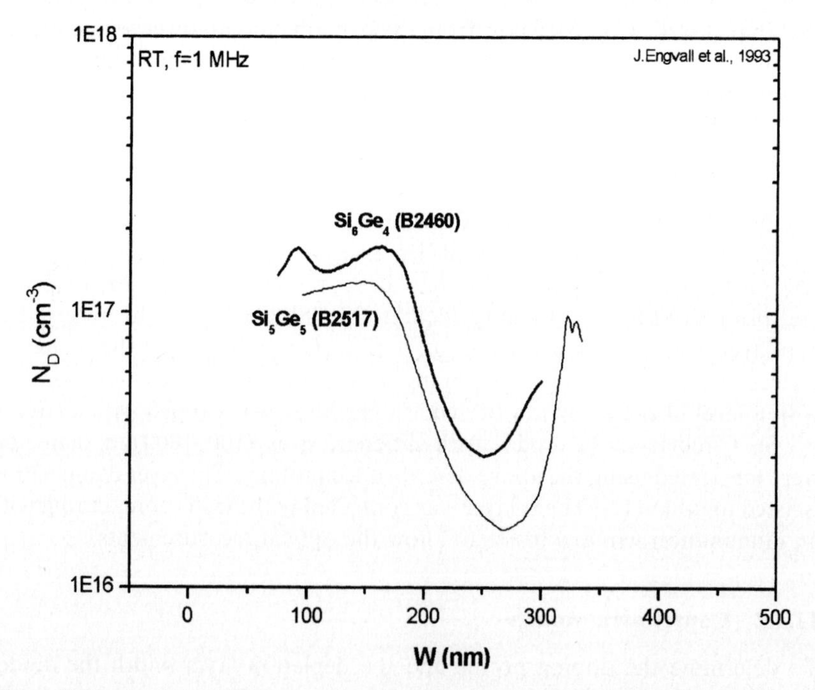

Figure 11.19. Doping profile as measured from the capacitance–voltage experiment for the Si_6Ge_4 and the Si_5Ge_5 SLS [45].

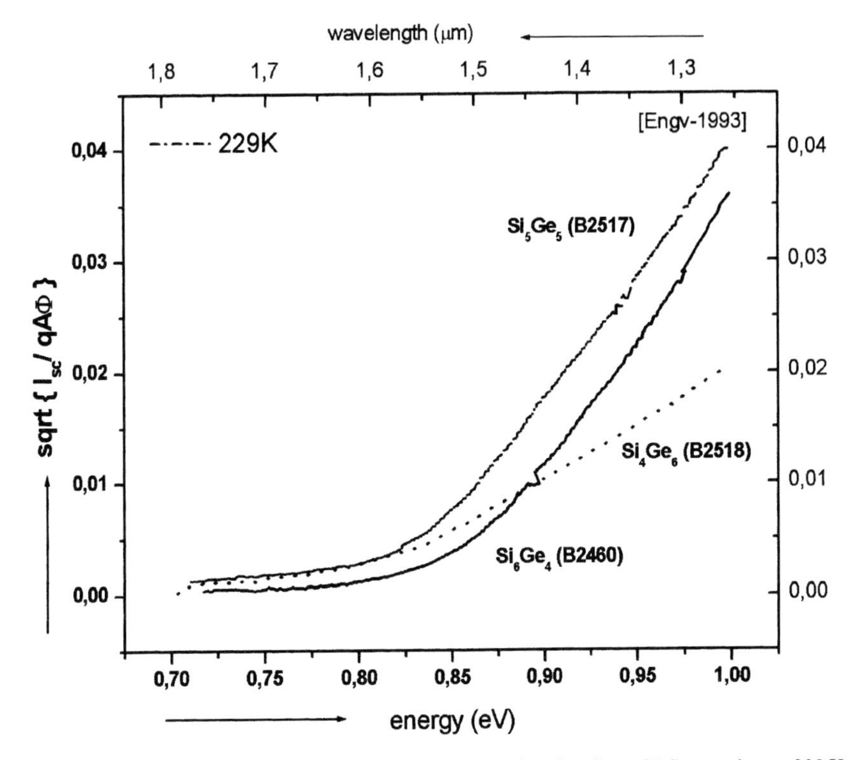

Figure 11.20. Square root of short-circuit current I_{sc} for the three SLS samples at 229 K over a large energy range. The dots are the square root of the measured current, the solid lines are the square root of the fitted expression [45].

SLS. One can use I_{sc} to determine the fundamental energy gap of the superlattices. In figure 11.20, the square root of the I_{sc} is plotted versus photon energy for all three samples studied at a temperature of 229 K. The linear relationship in this plot for a region of the spectra above the threshold current indicates a quadratic dependence on photon energy in agreement with the reported findings [40].

In figure 11.21, the energy positions of the PL and EL peaks, as well as the onset of absorbance, are compared. If the assignment of the low-temperature PL peak is correct, these data suggest that the EL peak at higher temperatures has an energy larger than the band-gap. At a first glimpse, this is surprising.

However, considering that the PL vanished already at about 40 K and that the EL is still observed at room temperature, it is not unreasonable to assume that different recombination processes are involved. The PL at low temperatures was studied by Menczigar *et al* [41], who found the PL properties to be consistent with a model in which the luminescence originated from

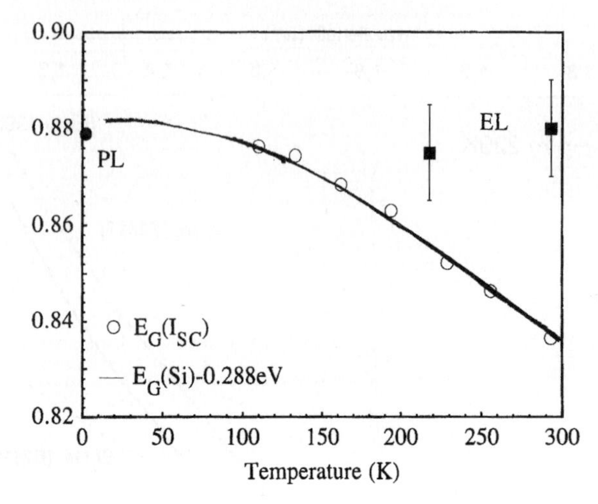

Figure 11.21. EL and PL peak positions from the strain adjusted Si_6Ge_4 SLS discussed above compared to the band-gap determined from the I_{sc} measurements [45].

localized excitons with a typical binding energy of 10 meV. The PL signal was observed up to about 40 K where the phonon assisted, non-radiative recombination processes begin to dominate. The localized exciton is not expected to be seen at higher temperatures. The fact that the EL at room temperature is observed above the band-gap is not unusual in forward biased p–n junctions. There are a number of processes that result in an energy shift as observed in the sample. In the EL experiment, a much higher injected power is applied (typically 3 W/mm^2), whereas only about 5–10 mW/mm^2 is used in the PL. This may cause a so-called band-filling effect with an attributed blue shift. The elevated temperature (150–300 K) could also account for the shift of the observed EL signal to somewhat higher energies. An estimate of the carrier distribution at 300 K gives a luminescence peak position at about 50 meV above the band-edge, as seen in figure 11.22. The diagram shows the room temperature EL compared with calculated line-shapes using different power laws for the absorption coefficient which are given in the inset of this figure. However, a detailed calculation should involve the joint density of states for these SLS structures, which at present is not available.

Finally, the EL spectrum has also been measured for differently fabricated mesa diodes of a Si_6Ge_4 SLS sample. Figure 11.23 shows room temperature EL spectra of different fabricated mesa diodes of Si_6Ge_4 SLS: waveguide facet emitter, mesa emitter, and also different buffer layer designs. For the mesa diodes, at normal incidence it is seen that a considerable increase in the band-gap emission at about 0.9 eV is obtained from samples grown on step-graded buffer for a constant current. This is

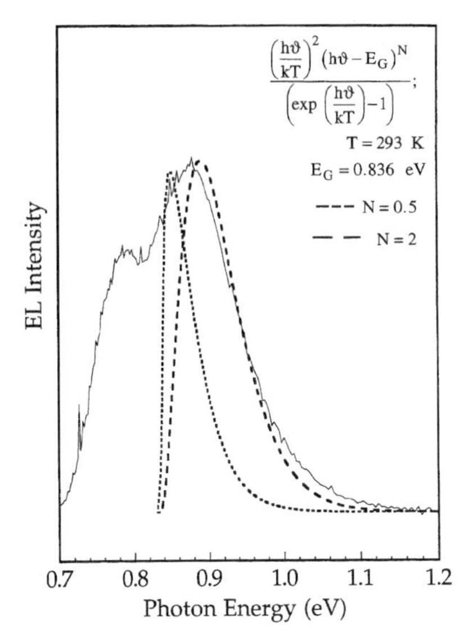

Figure 11.22. EL spectrum of the Si_6Ge_4 SLS at room temperature (RT) compared with the expected lineshape for two different functional dependences of the absorption. The different dependences are given in the inset of this figure [45].

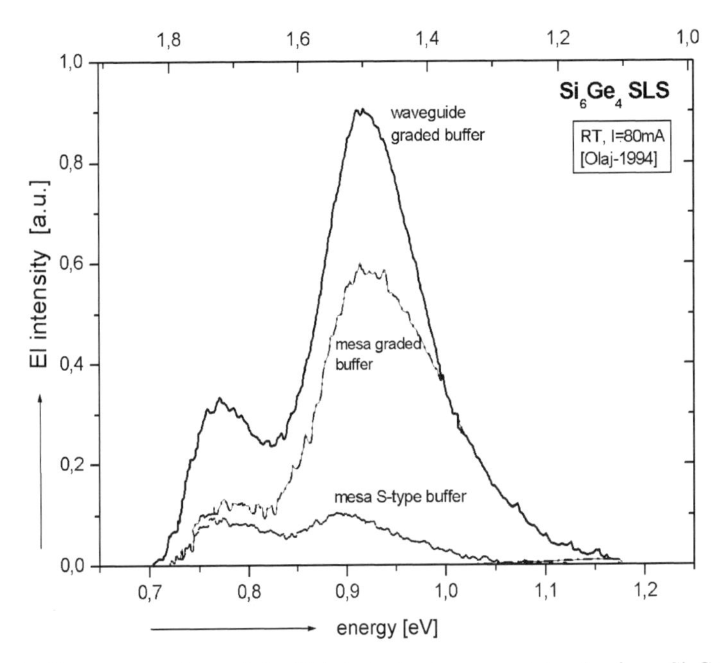

Figure 11.23. A comparison of the RT electroluminescence intensity from Si_6Ge_4 SLS samples grown under different conditions and fabricated with different technologies [40].

interpreted as a lower concentration of non-radiative recombination channels due to defects for these samples. Further increase in the light output is obtained if the superlattice is overgrown with a thick $Si_{0.6}Ge_{0.4}$ alloy waveguide layer and fabricated as edge emitter. The band-gap EL intensity is temperature dependent and decreases exponentially at higher temperatures with an activation energy of about 29 meV. The absolute efficiency of both types of EL emissions has been estimated to be around 10^{-4}–10^{-5}. However, this figure drops at about one order of magnitude at RT.

11.5 Outlook

Ultrathin Si_mGe_n superlattices based on Si substrates are good candidates for optical devices such as light-emitting diodes (LEDs) and photodetectors since their emission and detection wavelength (i.e. their band-gap) can be tuned in the long haul optical communication spectral band. Therefore, a great research effort has been undertaken worldwide to investigate and exploit the optical properties of the superlattices. However, the vision of an all-silicon world, with Si optical devices monolithically integrated on a Si IC with their electronic driver units, will be still very far reaching and may never be accomplished because of the strong competition and good performance of the direct band-gap III–V hetero-devices. Due to the nowadays widespread availability and cost reduction of $InGa_{1-x}As_x$ and $In_{1-y}Ga_{1-x}As_xP_y$ emitting devices it could make sense to fabricate an integrated Si detector chip and a separate III–V laser which are integrated into the same package. This trend will depend on future developments and research in this field as well as on the cost share of packaged emitter chips. However, for very small local area optical networks (e.g. within a mainframe computer) the need for integration and cost reduction of the whole optical transmitter chip will be enormous and SiGe might there be a serious candidate.

Acknowledgments

The author would like to acknowledge the work of many of his colleagues; especially for reading the manuscript we are indebted to Johannes Konle. The author also acknowledges the sample growth by Horst Kibbel and the XRD measurements by H-J Herzog. Several of the EL and PL spectra have been taken by Dr U Menczigar and Dr J Engvall, at that time working at the University of Munich, Germany, and Lund, Sweden. A major part of this research has been funded by the EU Commission under grant ESPRIT Basic Research, Project No. 7128.

References

[1] Gnutzmann U and Clausecker K 1974 *Appl. Phys.* **9** 3

[2] See any textbook on Solid State Physics, e.g. Ashcroft N W and Mermin N D (eds) 1976 *Solid State Physics* (New York: Holt, Rinehart and Winston)

[3] Kasper E 1987 *Advances in Solid State Physics* (*Festkörperprobleme*) **27** 265

[4] Abstreiter G 1993 *Physica Scripta* **48**

[5] Turton R J and Jaros M 1989 *Appl. Phys. Lett.* **54** 1986

[6] Turton R J and Jaros M 1990 *Appl. Phys. Lett.* **56** 767

[7] van der Merwe 1972 *Surf. Science* **31** 198

[8] Matthews J W and Blakeslee A E 1974 *J. Crystal Growth* **27** 118

[9] People R and Bean J C 1985 *Appl. Phys. Lett.* **47** Erratum 322; 1986 *Appl. Phys. Lett.* **49** Erratum 229

[10] E Kasper 1986 *Surface Science* **174** 630

[11] Kasper E, Herzog H-J, Jorke H and Abstreiter G 1987 *Superlattices and Microstructures* **3** 141

[12] Kasper E 1987 *Physics and Applications of Quantum Wells and Superlattices* ed. E E Mendez and K von Klitzing, NATO ASI series B, vol. 170 (New York: Plenum Press)

[13] Bean J C 1986 *Si Based Heterostructures in Si MBE* ed. E Kasper and J C Bean (Boca Raton, FL: CRC Press)

[14] Brantley W A 1973 *J. Appl. Phys.* **44** 534

[15] Abstreiter G, Eberl K, Friess K, Wegscheider W and R Zachai 1989 *J. Crystal Growth* **95** 432

[16] Engvall J, Olajos J, Grimmeiss H, Presting H, Kibbel H and Kasper E 1993 *Appl. Phys. Lett.* **63** 491

[17] Kasper E and Wörner K 1985 *J. Electrochem. Soc.* **132** 2481

[18] Kibbel H and Kasper E 1990 *Vacuum* **41** 929

[19] Kasper E and Herzog H-J 1977 *Thin Solid Films* **44** 357

[20] Fujita L, Fukatsu S, Yaguchi H, Shiraki Y and Ito R 1991 *Appl. Phys. Lett.* **59** 367

[21] Le Goues F K, Meyerson B S and Morat J F 1991 *Phys. Rev. Lett.* **66** 2903

[22] Fitzgerald E A, Xie Y-H, Green M L, Brasen D, Kortan A R, Michel J, Mii Y-J and Weir B E 1991 *Appl. Phys. Lett.* **59** 811

[23] Jäger W, Stenkamp D, Erhart P, Sybertz W, Leifer K, Kibbel H, Presting H and Kasper E 1992 *Thin Solid Films* **222** 221

[24] Presting H, Kibbel H, Jaros M, Turton R M, Menczigar U, Abstreiter G and Grimmmeiss H G 1992 *Semic. Sci. Technology* **7** 1127

[25] Baribeau J-M 1988 *Appl. Phys. Lett.* **52** 105

[26] Tanner B K 1989 *J. Electrochem. Soc.* **136** 3438

[27] Houghton D C, Lockwood D C, Dharma-Wardana D C, Fenton M C, Baribeau J-M and Denhoff M W 1987 *J. Crystal Growth.* **81** 434

[28] Hull R, Beran J C, Eaglesham D J, Bonar J M and Buescher C 1989 *Thin Solid Films* **183** 117

[29] Mantl S, Kasper E and Jorke H *Proc. Mat. Research* 1987 *Soc.* **91** 305

[30] Menczigar U, Eberl K, and Abstreiter G 1991 *Proc. Mat. Research Soc.* **220** 361

[31] Pearsall T P, Vandenberg J M, Hull R and Bonar J M 1989 *Phys. Rev. Lett.* **63** 2104

[32] Schmid U 1992 *Phys. Rev. B* **45** 6793

[33] Cardona M 1990 *Superlatt. Microstructures* **7** 183

[34] Zachai R, Eberl K, Abstreiter G, Kasper E and Kibbel H 1990 *Phys. Rev. Lett.* **64** 1055

[35] Weber J and Alonso M I 1989 *Phys. Rev. B* **40** 5683

[36] Bastard G 1988 *Wave Mechanics Applied to Semiconducting Heterostructures* (Paris: Les Editions de Physique)

[37] Thornton J M C, Williams A A, McDonmald J E, van Silfhout R G, van der Veen J F, Finney M and Norris C 1991 *J. Vac. Sci. Technology B* **92** 146

[38] Copel M, Reuter M C, Horn van Hoegen M and Tromp R M 1990 *Phys. Rev. B* **42** 11682

[39] Engvall J, Olajos J, Grimmeiss H, Presting H, Kibbel H and Kasper E 1993 *Appl. Phys. Lett.* **63** 126

[40] Olajos J, Engvall J, Grimmeiss H G, Menczigar U, Gail M, Abstreiter G, Kibbel H, Kasper E and Presting H 1994 *Semic. Sci. Technology* **9** 2011

[41] Menczigar U, Abstreiter G, Olajos J, Grimmeiss H, Kibbel H and Presting H 1993 *Phys. Rev. B* **47** 4099

[42] Turton R J and Jaros M 1989 *Proceedings of workshop I* Esprit Basic Research action 3174, Ulm

[43] Kasper E, Kibbel H and Presting H 1990 *Proceedings of workshop II* Esprit Basic Research action 3174, Ulm

[44] Menczigar U, Schorer R and Abstreiter G 1990 *Periodic Progress Report I* ESPRIT Basic Research action 3174

[45] Engvall J, Olajos J and Grimmeiss H G 1993 *Periodic Progress report I* ESPRIT Basic Research action 7128

Chapter 12

Diamond films

Dean M Aslam
Electrical and Computer Engineering, Michigan State University, East Lansing, USA

12.1 Introduction

In a typical luminescence process, when the electrons are excited to higher energies by light (photoluminescence), electron beam (cathodoluminescence) or electronic current (electroluminescence), a subsequent transition to lower energies leads to emission of light. While semiconductors, such as GaAs, with direct band gap (electronic transitions involve no momentum changes) are used for light-emitting devices, the indirect band gap semiconductors, such as diamond and Si, typically cannot be used for light-emitting devices such as lasers. It is interesting to note that specially prepared material and device structures made from Si or diamond are capable of light emission. Defects in such materials are believed to be responsible for their unexpected luminescence properties. Electroluminescence in both single crystalline and polycrystalline diamond, reported in a number of studies, is of interest due to its possible applications in opto-electronic devices capable of functioning in infrared, visible and ultraviolet regions of the electromagnetic spectrum.

This chapter includes discussion on diamond film growth techniques, IC-compatible polycrystalline diamond (poly-C) processing, electroluminescence in diamond and field emission electroluminescence (FEEL) in poly-C. The FEEL in poly-C and carbon nanotubes is reported in a book chapter for the first time.

In general, light emission results by electron transition from higher to lower energy states. In contrast to electrons in isolated atoms, the electron transitions in crystal lattices, such as those found in Si and GaAs, are much more complex. For example, an electron transition from conduction band to valence band results in light emission in GaAs but not in Si. Energy versus momentum (E–k) band diagrams shown in figure 12.1 [1]

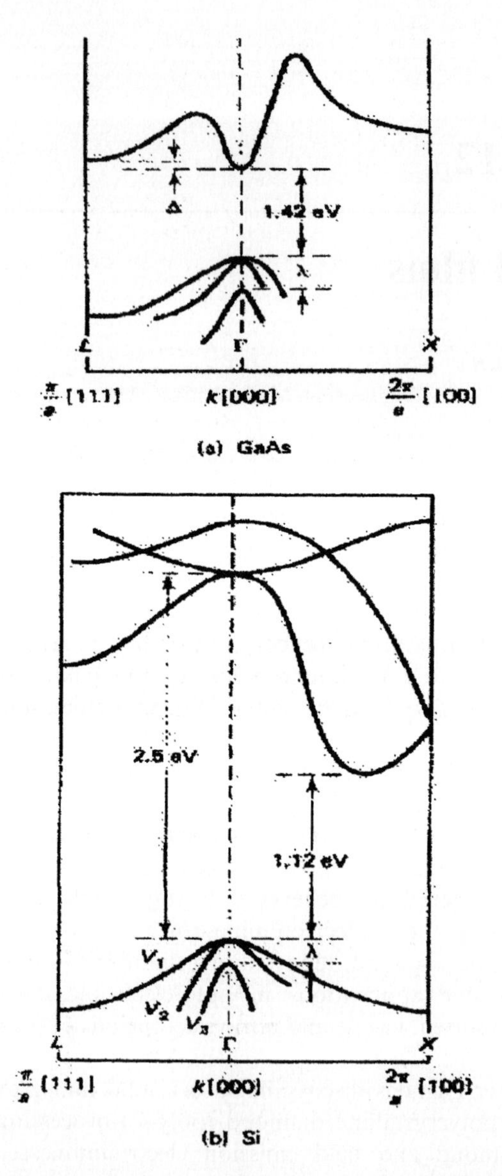

Figure 12.1. Energy versus momentum band diagrams (E–k) of (a) GaAs and (b) Si.

help understand this difference between two distinct classes of semiconductors; direct band gap type semiconductors (only energy changes in an electronic transition) and indirect band gap type semiconductors (both energy and momentum change in an electronic transition). For example, an electronic transition from conduction to valence band in GaAs involves a change (decrease) in energy only (no change in momentum). Consequently, the released energy appears as light. However, for such a transition in the case of Si, both energy and momentum of electrons change and, as a photon of light cannot carry the momentum change, there is no appreciable light emission. The release of extra energy and momentum in Si is, therefore, taken away by another type of quantum particle called a phonon (a quantum of thermal energy). A multi-phonon process takes place in accordance with the laws of conservation of energy and momentum.

12.1.1 Luminescence in indirect gap semiconductors

In a luminescence process, the electron must first be excited to higher energy, which can be accomplished by light (photoluminescence), electron beam (cathodoluminescence) or electrical energy (electroluminescence). A subsequent electron transition to lower energies leads to emission of light. A common example of electroluminescence is a forward biased p–n junction made out of a direct band-gap type semiconductor such as GaAlAs. The light-emitting diodes (LEDs) and LASERS are based on electroluminescence. Red, green and blue LEDs are made from GaAlAs, GaP and SiC/GaN, respectively. For reasons discussed above, semiconductors such as Si and diamond typically do not exhibit electroluminescence. The reader might be surprised why the subject of this chapter is electroluminescence in diamond.

The fact that electroluminescence has experimentally been observed in crystalline diamond and poly-C seems to be, at first sight, a contradiction of its indirect band gap (figure 12.2). In fact, defects in semiconductors with indirect band gaps (diamond and Si) are believed to be responsible

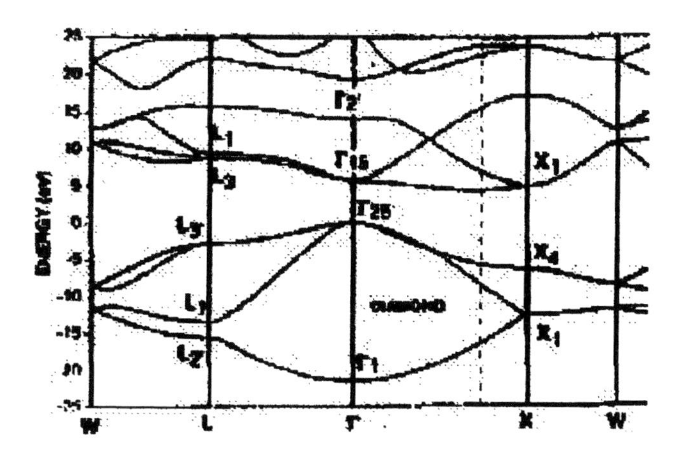

Figure 12.2. *E–k* band diagram of diamond; the lowest electron energy in the conduction band is not at $k = 0$ (Γ on the x-axis).

Table 12.1. Comparison of semiconductor properties.

Property	Diamond		Si	GaAs	β-SiC
	Poly-C	Crystalline			
Figure of merit					
Johnson, 10^{23} Ω s^{-2}		73 856	9	62.5	10 240
Keyes, 10^2 W cm^{-1} s^{-2} C		444	13.8	6.3	90.3
Sat. electron velocity, 10^7 cm s^{-1}		2.7	1	1	2.5
Carr. mobilities, cm^2 V^{-1} s^{-1}					
Holes	1–165	1600	600	400	50
Electrons		2200	1500	8500	400
Electrical resistivity, Ω cm	10^{-2}–10^8	10^{-3}–10^{13}	10^{-4}–10^8	?–10^8	150
Breakdown field, MV cm^{-1}	0.1–1	10	0.3	6	4
Band gap, eV		5.45	1.12	1.42	3
Dielectric constant	6.7	5.5	11.7	12.5	9.7
Thermal expansion coefficient, 10^{-6} C^{-1}	2.6	1.1	2.6	5.9	4.7
Thermal cond., W cm^{-1} K^{-1}	20	4–22	1.5	0.5	5
Lattice constant, Å		3.57	5.43	5.65	4.36
Density, g cm^{-3}		3.52	2.32	5.31	3.215
Melting point, °C		4000[a]	1412	1240	2540
Hardness, kg mm^{-2}		10 000	1000	600	3500
Poisson ratio	0.11	0.15	0.23	0.315	
Young's modulus, 10^{12} Pa	0.8–1.2	1.1–1.2	0.155	0.085	0.7

[a] Diamond may convert to graphite well below 4000 °C depending upon the ambient and it oxidizes at 650 °C to form CO_2/CO in O_2.

for their luminescence properties. Even for Si, specially prepared Si samples, such as porous Si and nanocrystalline Si, exhibit luminescence properties. All types of diamond are found to have a number of defects related to vacancies and impurities. Due to a unique combination of properties of diamond (table 12.1), its electroluminescence properties offer a very unique application potential in optical microelectromechanical systems (MEMS), displays and other opto-electronic devices in infrared, visible and ultraviolet regions. The unique properties of diamond stem from the so-called sp^3 bonding properties of carbon atoms when they form a crystal lattice.

12.1.2 Hybrid bonding structure of carbon

Perhaps due to its unique hybrid bonding structure (*sp*, sp^2 and sp^3 C–C bonds), carbon has become a crucial part of all known living systems. The *sp* bonds in carbon, though frequently used, are not always well understood. Carbon has four electrons in its outermost shell (the valence shell), which is

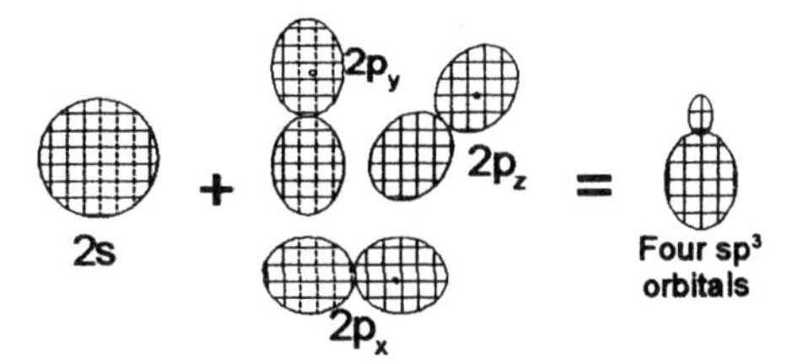

Figure 12.3. Hybridization of s and p orbitals in carbon to form sp^3 orbitals, which form the so-called sigma bonds responsible for the strong diamond lattice.

the second shell and it has one $2s$ and three $2p$ orbitals. Here is a very important question related to the formation of diamond lattice: Are the electrons in $2s$ and $2p$ orbitals responsible for the four bonds that each carbon atom makes with its four neighbours in the diamond lattice? One possible answer is that each of the $2s$, $2p_x$, $2p_y$ and $2p_z$ orbitals forms a bond with the neighbours. There are two problems with this answer. First, these bonds will have different strengths, which is in contradiction to the experimental fact indicating equal strength for all the four bonds. Second, the strength of any of such bonds will be lower than that found experimentally. Then, how can the important question raised earlier be answered?

The answer to this difficult question was provided by Linus Pauling (the winner of two Nobel prizes) in 1931, who showed mathematically how an s orbital and three p orbitals can combine, or hybridize, to form four equivalent atomic orbitals with tetrahedral orientation. The resulting hybrid orbital is called the sp^3 orbital, shown in figure 12.3. The sp^2 or sp^1 orbitals result if two or one p orbital combine with the s orbital, respectively [2]. While sp^3 C–C bonds lead to diamond lattice, the sp^2 C–C bonds lead to graphite and a material which contains both sp^2 and sp^3 C–C bonds is called diamond-like carbon (DLC). If a sheet consisting of a mono-layer of graphite, which consists of sp^2 bonds, is wrapped to form a pipe; the resulting structure is called a single wall carbon nanotube. A multi-wall carbon nanotube has many such layers.

12.2 Diamond film growth technologies

The first evidence of diamond growth by chemical vapour deposition (CVD) by Eversole in 1952–53 led to the use of H_2 and CH_4 in the hot filament CVD (HFCVD) of diamond by Angus in 1971 on diamond substrates [3]. The inexpensive CVD polycrystalline diamond (poly-C) was grown on non-diamond substrates by Deryagin in 1976, Spitsyn in 1981, and by Matsumoto *et al* in 1983. Today, it is believed that, during the CVD of diamond, the CH_3 is responsible for deposition of C as diamond and non-diamond phases. The atomic hydrogen removes the non-diamond phases leaving behind the

Table 12.2. Different diamond deposition methods; HFCVD stand for hot filament CVD.

Methods	HFCVD	MPCVD	DC-arc jet CVD	Combustion synthesis	Multiple pulsed laser
Deposition rate (μm/h)	0.1–10	0.1–10	30–150	4–40	3600
Substrate temperature (°C)	300–1000	300–1200	800–1100	600–1400	50
Deposition area (cm^2)	5–900	5–100	<2	<3	N/A
Advantages	Simple, large area	Quality, stability	High rate, good quality	Simple, high rate	Ultra-high rate
Disadvantages	Contamination, fragile filament	Rate	Contamination, small area	Small area	Expensive

diamond phase. Near the end of 1990s, the basic science of CVD diamond was well understood, and today diverse plasma and thermal techniques have been developed to produce poly-C films several mm thick and over 12 inches in diameter. Optically smooth 300 μm thick undoped poly-C wafers are available in the market. Although there are some reports of n-type poly-C and crystalline diamond growth, the well established techniques exist only for *in-situ* doping of p-type poly-C and crystalline diamond [4].

A number of diamond growth techniques are currently available for the growth of diamond and carbon nanotubes. Microwave plasma CVD (MPCVD) can be used for the growth of both diamond and carbon nanotubes. As shown in table 12.2 [5], the deposition rate of diamond shows a large variation (0.1–3600 μm) depending on the growth technique. For applications of diamond in opto-electronic devices and systems it is very important to develop IC-compatible diamond film technology.

12.2.1 Diamond film microfabrication technologies

The application of crystalline diamond in active electronic devices has so far been limited due to lack of large-area hetero-epitaxial films, a reliable n-type doping and IC-compatible p-type doping technology. Consequently, most diamond-related studies have focused on doping and film fabrication techniques [3, 4]. However, to produce inexpensive diamond films for electronic devices including opto-electronic devices, it is crucial to develop a film fabrication technology that is compatible with the current micro-fabrication and micro-machining techniques. Inexpensive poly-C films with typical seeding density [6], thickness, surface roughness, deposition temperature and resistivity

Figure 12.4. Poly-C films with surface roughnesses of 500 and 16 nm.

in the ranges of 10^8–10^{11} cm^{-2}, 0.1–200 μm, 15–2000 nm, 400–900 °C and 0.001–1000 Ω cm^{-1}, respectively, have been successfully produced. Figure 12.4 shows atomic force microscope (AFM) micrographs of films with surface roughnesses of 500 and 16 nm achieved using initial nucleation densities of 10^8 and 10^{11} cm^{-2}, respectively.

12.2.2 Diamond sensor and microsystems technologies

Recent studies related to possible applications have focused on technologies for sensors, MEMS and wireless integrated microsystems (WIMS) [7-14]. Figure 12.5 shows a surface micromachined poly-C bridge structure, fabricated using 2 μm thick SiO_2 as the sacrificial layer [9]. The patterning of this bridge structure was accomplished through an IC-compatible patterning technique [9, 15]. The use of poly-C in sensors for temperature, pressure, acceleration, radiation and gases has also been investigated [16–19].

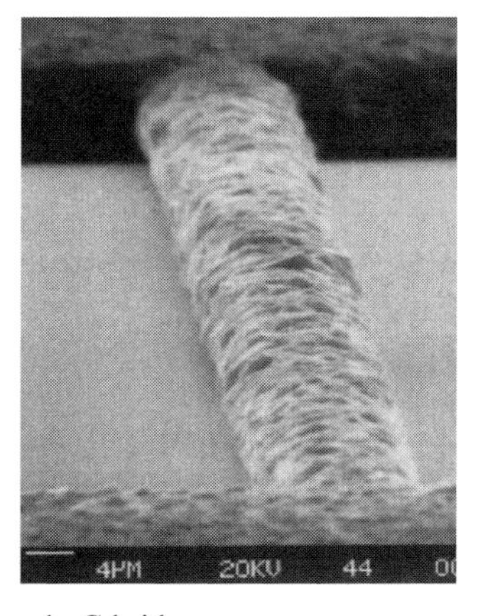

Figure 12.5. An 8 μm wide poly-C bridge.

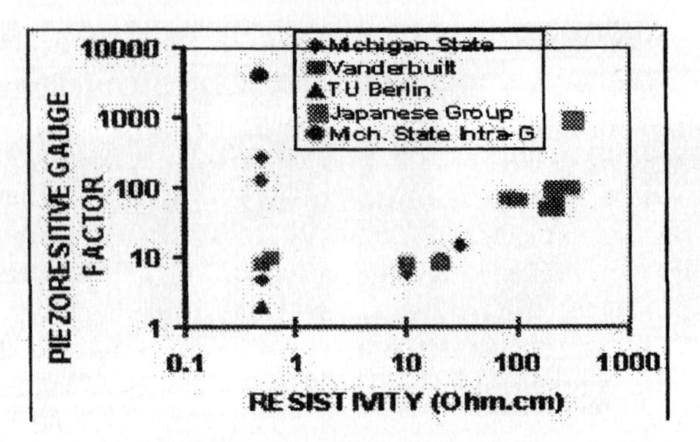

Figure 12.6. Reported piezoresistive gauge factor values for poly-C.

Piezoresistivity was reported for the first time in vapour-deposited diamond in 1992 by researchers from Michigan State University [9, 10]. Although these findings triggered a huge interest for poly-C piezoresistive sensors, one problem was soon realized. The high gauge factor of up to 4000 results only if very expensive single-crystal diamond is used. However, in a breakthrough work in 1996 [16] [this work was supported by NSF MRSEC during 1994–97], MSU researchers reported an intra-grain gauge factor of over 4000 for a sensor device fabricated within a grain of a large-grain (5–150 μm) optically-smooth free-standing diamond wafer. Figure 12.6 summarizes the inter- and intra-grain gauge factor values reported for poly-C.

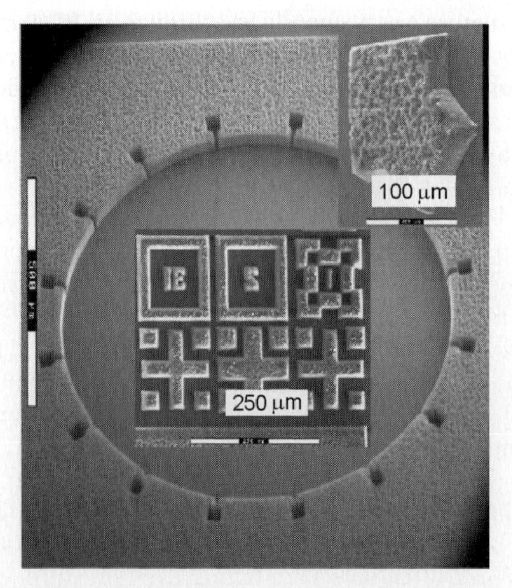

Figure 12.7. Free-standing poly-C structures with thicknesses in the range 30–80 μm.

The structural reliability and lifetime of MEMS are critical for their use in opto-electronic devices/systems, biomedical instruments, automobiles, aircraft and satellites. Due to limitations of conventional MEMS materials, new MEMS materials have been the subject of a number of studies [14, 20–23]. Poly-C, with very low friction, extreme hardness, excellent chemical immunity and highest thermal conductivity, has the potential to be the best material for MEMS. It can be used as an electrical insulator (undoped) and as a semiconductor (p-type) in a single structure. The reported diamond structures are only a few micrometres tall. Researchers at Michigan State University developed an IC-compatible technology to fabricate poly-C gears using metal as a mould [14] and 50–100 μm tall structures (figure 12.7) using dry-etched Si as a mould [5].

12.3 Electroluminescence in carbon-based materials

The major motivation for the study of luminescence properties of diamond has been the understanding of defects in diamond. The light emission properties are due to defects in diamond which in pure form does not emit light because it is an indirect-type semiconductor. Cathodoluminescence (CL), produced by an electron beam incident on the diamond surface, has been used to study defects near the diamond surface. Typical electron beam acceleration voltages are in the range 10–50 keV and the corresponding emission depths are in the range 0.5–17.7 μm. The CL spectrum typically consists of sharp peaks and broad bands. The sharp peaks are indicative of discrete transitions between energy levels whereas broad bands indicate variable transition energies [24–26]. Two broad peaks are often observed and are referred to as green and blue band A because they occur in the green and blue wavelengths, respectively [27]. A comparison of the CL spectra and x-ray topographic images revealed a strong correlation of the 2.85 eV band to the dislocation density in diamond.

The 2.2 eV band was thought to be attributed to a boron-related centre [28]. The sharp peaks were often associated with impurity atoms or aggregates. Nitrogen-containing defects generated numerous CL peaks. The most common nitrogen peaks occurred at 2.156, 2.33, 2.807 and 3.188 eV. The 2.156 eV defect was thought to consist of nitrogen and a vacancy [29]. A doublet line appearing near 2.33 eV might also be related to nitrogen [28]. Separate lines at 2.807 and 3.188 eV were thought to be caused by a single nitrogen atom and a carbon interstitial [30]. Another defect that was believed to be related to a carbon interstitial occurred at 4.582 eV [31]. The 1.681 eV defect was attributed to silicon [32]. The emission intensity of one defect may vary with the concentration of other defects and the efficiency of individual defects differs. Therefore, CL does not provide a quantitative determination of the defect density.

Table 12.3. Luminescence spectra in diamond.

Peak position		Description
eV	nm	
5.15, 4.97, 4.65	240.7, 249.4, 266.6	Boron related acceptors [37]
4.582	270	Carbon interstitial [31]
3.188	388.8	Carbon interstitial, single nitrogen atom [30]
2.85	435	Blue band A, related to dislocations [28]
2.8	442	Growth induced dislocations [38]
2.33	532	Doublet line, nitrogen related [28]
2.36, 2.3	526.3, 543	Nitrogen–carbon interstitial complex [39]
2.2	563	Green band A, possibly related to boron [28]
2.156	574.9	Single N and vacancies [29]
1.945	638	Vacancy [40]
1.9	652.4	Related to phosphorus [41]
1.682	737	Silicon [32]
1.673	741	Neutral vacancy called GR1 [32]

Photoluminescence (PL) is typically produced by laser light [33–35]. As the PL occurs from defects with lower energy than the incident photons, it allows selective excitation of a defect by controlling the incident laser energy. Higher beam intensities can excite electrons even for lower concentrations of defects. The PL spectra are similar to the CL spectra for most defects. A summary of luminescence studies in diamond is provided in table 12.3.

Electroluminescence (EL) is produced by excitation of electrons by electronic current. Yellow-green EL was observed in a Schottky barrier point contact device measured in the temperature range 300–750 °C [36]. EL was also observed from a Schottky diode made of boron-doped poly-C [27]. Figure 12.8 shows EL spectra of the lightly doped and heavily doped diamond films, respectively. EL was observable when the forward bias across the diode exceeded 25 V or the reverse bias exceeded 20 V. The EL spectra were similar in forward and reverse biases. The EL and CL luminescent properties seem to be similar.

Blue-green EL has been observed from free-standing poly-C [37]. The In_2O_3 transparent conductor and conducting epoxy were used for the front and back contacts on the 15 μm thick free-standing poly-C. EL was observed from the individual 5–10 μm grains of the poly-C film. The peak position of EL occurred at 485 nm. It was proposed that the recombination of donor–acceptor (nitrogen–boron) pairs was responsible for the emission of light. As the samples were not intentionally doped with nitrogen or boron, possible sources of nitrogen contamination were possibly from the wall of the reactor

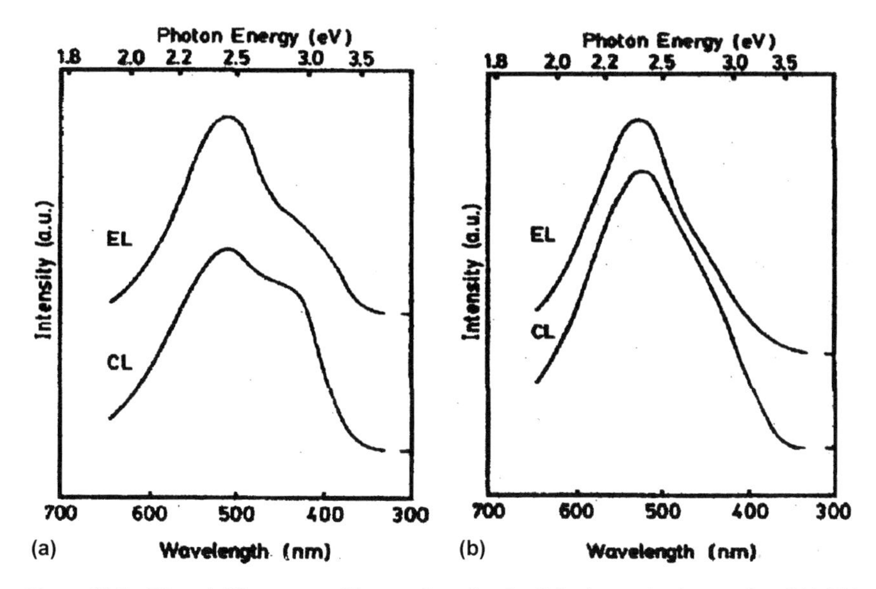

Figure 12.8. EL and CL spectra of boron-doped poly-C for boron/carbon ratio of (a) 200 and (b) 1000 ppm [37].

or from the relatively poor vacuum during deposition. The trace of boron was likely to have been carried over from one deposition to the other.

Blue-violet EL from a 400 μm thick free-standing poly-C film has been reported [38]. The electrical contacts were provided by Ti/Pt/Au (labelled as 'contact' in the inset of figure 12.9). The EL was observed at voltages above 800 V corresponding to applied fields of 2 V/μm. Figure 12.9 shows EL spectra at different applied voltages. The main luminescence band was centred at 3.0 eV. Additional peaks at about 3.65 and 3.85 eV appeared at

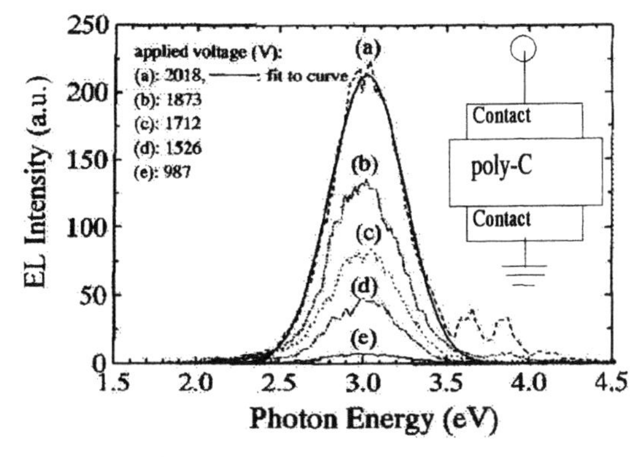

Figure 12.9. EL spectra of a free-standing poly-C film [38] subjected to different voltages applied to the upper contact of the schematic shown in the inset.

voltages above 1850 V. The subgap photocurrent spectrum in the photon energy range from 1.7 to 5.0 eV revealed that there was a 3.0 eV transition of electrons or holes between the conduction or valence band and the energy states in the band gap.

EL has been observed in other carbon-based materials such as carbon nanotubes [39–41], diamond like carbon (DLC) [42], amorphous carbon [43] and poly-C [44]. One type of EL, observed in poly-C [44] and carbon nanotubes [40, 41] and is called field emission EL (FEEL), is the subject of next section. FEEL, which is observed only during the field emission process, has potential applications in flat panel displays not requiring phosphor but requiring vacuum.

12.3.1 Field emission electroluminescence

Recent studies have revealed a very interesting type of EL in poly-C and carbon nanotubes. In a recent paper [41] it has been abbreviated as FEEL. An effect similar to FEEL was observed by Bonard *et al* [40] during electron field emission from single- and multi-wall carbon nanotubes (SWNT and MWNT). The intensity of light emitted from the vicinity of the apex of the tip increased with the emission current (figure 12.10). Luminescence was not homogeneous on the emitting surface as the light intensity variations were detected. For SWNT films, light was emitted at higher energies as compared with MWNTs, but their other luminescence characteristics were comparable with those of MWNTs. Based on theoretical calculations of the local density of states at the tip it was suggested that the luminescence was due to electronic transitions between energy levels at the emitting surface of the tip.

Kim and Aslam [41] studied FEEL in poly-C and MWNT using a setup shown in figure 12.11. The measurement system, capable of field emission mapping of current over the sample surface, was placed in a vacuum chamber held at a pressure of 10^{-7} torr. The luminescence spectra collected through an optical fibre (made of 1000 μm thick fused silica supplied by Ocean Optics) were analysed by a spectrometer (PC2000 Ocean Optics). The spectrometer, with a resolution of 5 nm full width at half maximum (FWHM) and a spectral response in the range of 350–1000 nm, was connected to the optical fibre using an SMA 905 connector. The optical fibre was positioned 2 mm above the emitting surface. The field emission current measurements were carried out at room temperature in a dark chamber.

In the case of FEEL observed in doped and undoped polycrystalline diamond (poly-C) films, not all field emission experiments were accompanied by FEEL. Figure 12.12 shows two field emission curves measured at two different points on a poly-C film. FEEL was observed only for sites which showed a rapid increase of emission current. For sites that exhibited FEEL, an emission current of 1 μA was required. Why some areas of the

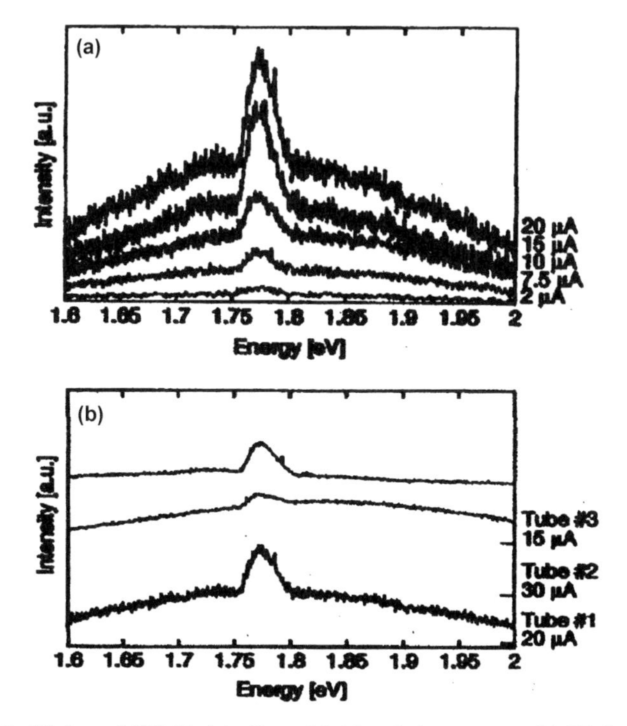

Figure 12.10. EL from MWNT: (a) effect of field emission current, (b) EL from different samples.

sample show FEEL while others do not is currently not well understood. However, this behaviour which is reproducible in poly-C samples used in the study [44], might indicate that only certain type of defect in poly-C, with a number of different type of defects, might be responsible for FEEL.

It was found that boron-doped, undoped and free-standing (B-doped) poly-C films exhibited FEEL. Figure 12.13 shows FEEL for an undoped

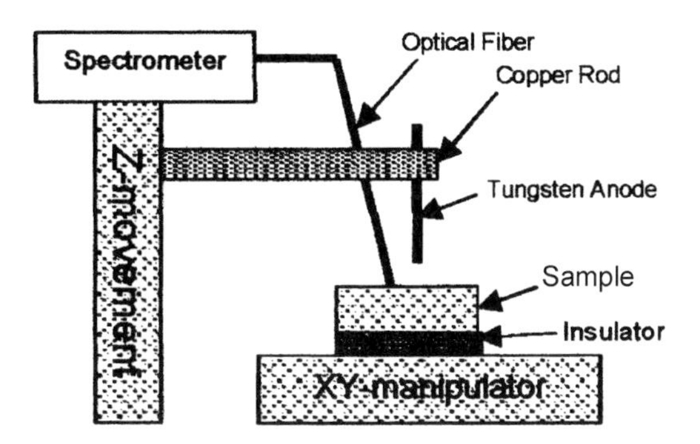

Figure 12.11. FEEL characterization setup.

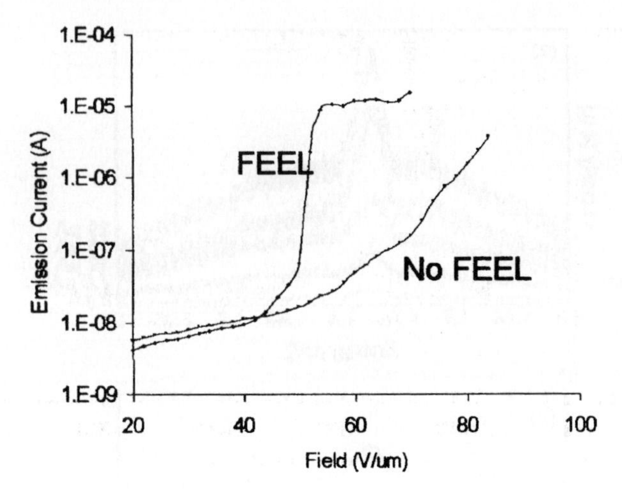

Figure 12.12. FEEL observed only for certain spots on the sample.

poly-C film. The EL intensity increases with the emission current but the peak position remains unaffected by the emission current. The FEEL from undoped poly-C was similar to that from boron-doped free-standing poly-C (figure 12.14) except that luminescence was observed at higher fields than B-doped free-standing poly-C film. The free-standing poly-C, which has grain sizes in the range of 5–100 μm, exhibits a peak at 482 nm and a small peak appears at 638 nm. A peak at 482 nm was also observed from

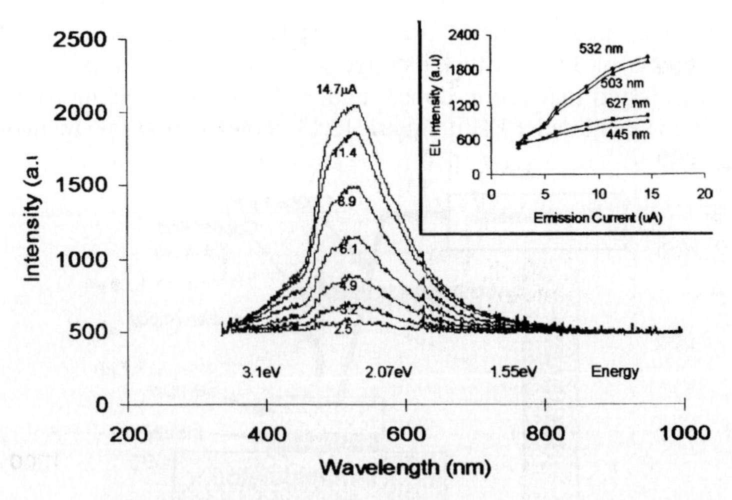

Figure 12.13. FEEL spectra in poly-C as function of wavelength and emission current (inset).

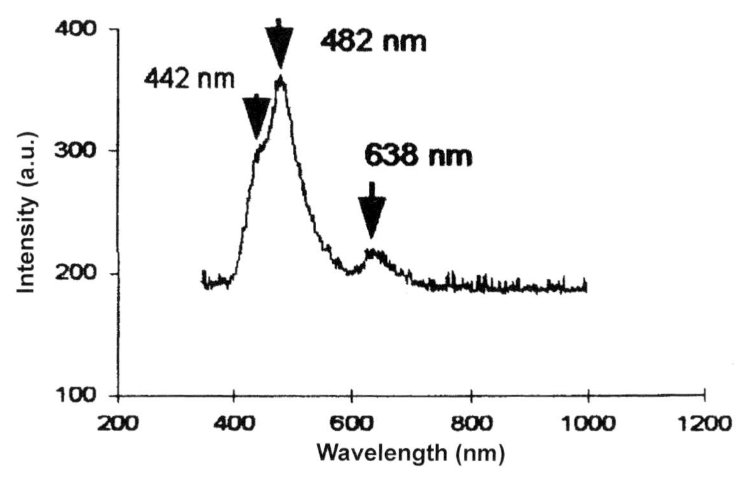

Figure 12.14. FEEL spectra of free-standing poly-C film; the field emission current is 17.2 μA.

the electro-luminescence study of diamond films [38]. The peaks at 442 and 638 nm could be associated with growth-induced dislocation and vacancy, respectively [45, 46].

Kim and Aslam [41] also observed FEEL in multi-wall carbon nanotubes as shown in figure 12.15. Highly oriented carbon nanotubes (figure 12.16), deposited uniformly on the Si substrate, had density and diameter in the range 10^8–10^9 cm^{-2} and 20–100 nm, respectively. The sharp and broad peaks are in the visible range (430–690 nm). The different intensities

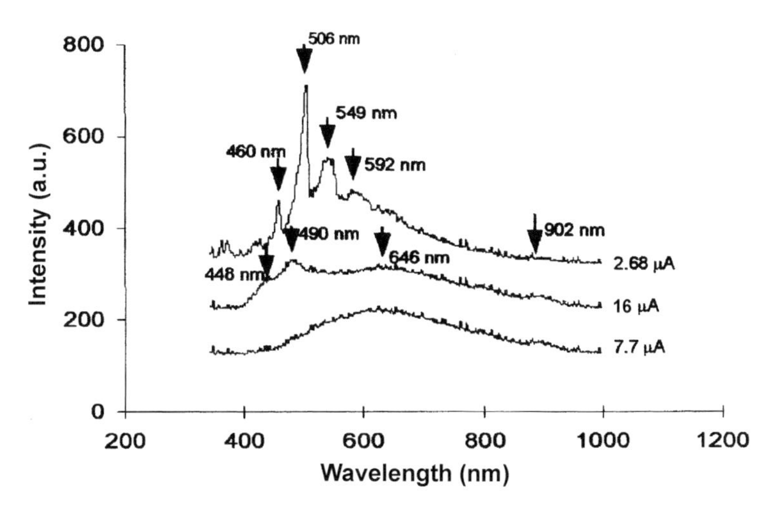

Figure 12.15. FEEL spectra in carbon nanotubes.

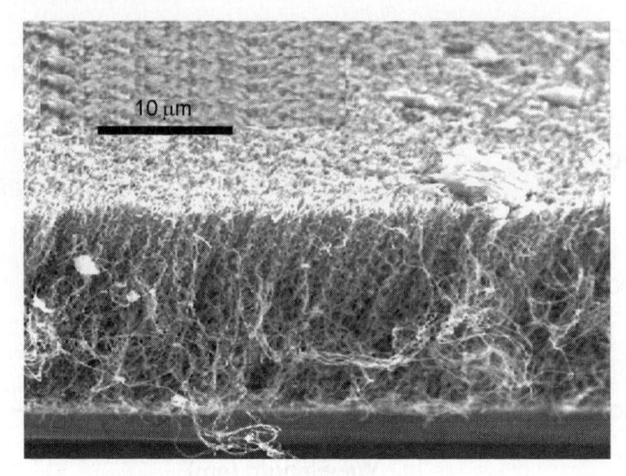

Figure 12.16. Scanning electron microscope picture of high density carbon nanotubes.

of peaks in the curves shown in figure 12.15 may suggest different defect densities near the emitting surfaces of the multi-wall nanotubes.

12.3.1.1 FEEL model

The FEEL from poly-C is currently not well understood. Kim has proposed the following qualitative model [44]. The model is based on explanations of the field emission mechanisms suggested by a number of studies [47–49].

The electron emission from p-type diamond, which typically has no electrons available in its conduction band, has been very difficult for theoreticians to explain. As the field emission process is due to quantum mechanical tunnelling of electrons from diamond to vacuum, the emission of valence band electrons into vacuum requires an applied field of over 70 MV cm^{-1} (the corresponding work function is in the range of 4–5 eV) at the emitting surface if the electron affinity of diamond is assumed to be zero. Experimental studies reveal that typical emission fields for p-type diamond (poly-C or single crystal) are in the range of 0.01–0.1 MV cm^{-1} which correspond to work function values below 1 eV.

It was speculated that the grain boundary defects might form a conduction channel [47, 48] in the band gap of diamond. However, electron emission at low field was also achieved on mono-crystalline diamond. As impurity conduction may exist in synthetic diamond [49] and the impurity level in B-doped diamond is less than 0.4 eV above the valence band edge, one might think of impurities being involved in the field emission process. However, these impurity levels are located several eV below the vacuum level and would require very high fields for electron emission.

As the vacancy-related defects in diamond films can be substantially high [50, 51], carbon dangling bonds and other defects can lead to electron

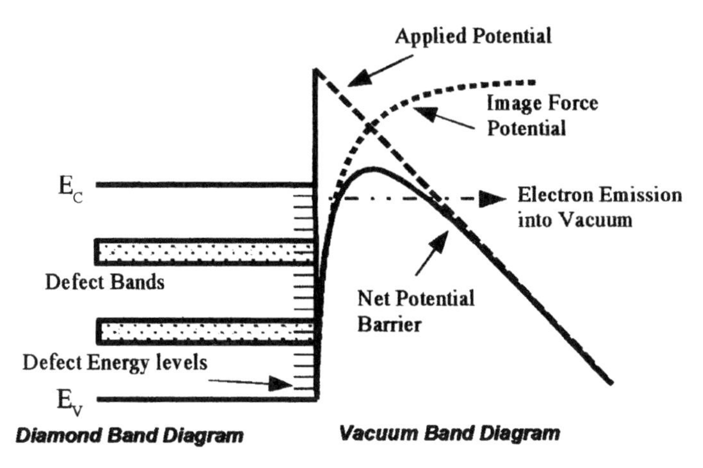

Figure 12.17. Energy band diagram of a diamond–vacuum system showing electron emission into vacuum.

states in the band gap of diamond. If the densities of such defects are high enough the formation of defect bands in the band gap of diamond may raise the Fermi level into the upper part of the band gap, reducing the energy barrier that the electrons must tunnel through (figure 12.17). The electrons can transport in defect bands to the surface states found near the surface of diamond. If the surface states are closely spaced, electrons can be hopping to the higher surface states, during a field emission process, getting closer to vacuum level. This combined with the image force lowering of the energy barrier can contribute to the field emission current.

According to the model proposed by Kim, a fraction of electrons found in the surface states may drop to lower energy levels before they are emitted. This fraction may be found in direct-type defects that are related to light emission. The direct energy transitions between the surface states can lead to FEEL. This would suggest the dependence of FEEL intensity on the field emission current supporting the results shown in figure 12.13. Another interesting feature of this model is its ability to qualitatively explain the unusual results shown in figure 12.12. As the defects in diamond (found in poly-C and crystalline diamond) could be divided into direct (high photon emission probabilities) and indirect (high phonon emission probabilities) types, one can understand why all points on a sample do not exhibit FEEL.

Further work is needed to study the nature of FEEL-related defects in diamond films. One interesting experiment could be to prepare and characterize the samples with enhanced densities of direct-type defects. Films with high densities of direct-type defects would be very interesting for FEEL-based flat panel displays.

12.4 Future trends

It is expected that electroluminescence, in particular FEEL, will find applications in optical microsystems, RFMEMS (radio frequency micro-electromechanical systems) and wireless integrated microsystems (WIMS) [52–55]. An understanding of both the physics and the fabrication technologies of poly-C would be very important for its applications in MEMS, RFMEMS and WIMS.

Acknowledgments

The NSF-funded WIMS centre is involved in microsystems technologies including poly-C technolgies; see the centre website for more information, www.wimserc.org

References

[1] Wang S 1989 *Fundamental Semiconductor Theory and Device Physics* (Englewood Cliffs, NJ: Prentice-Hall)
[2] McMurry J 1984 *Organic Chemistry* 4th edition (Brookes/Cole)
[3] L S Plano 1944 *Diamond: Electronic Properties and Applications* ed. L S Pan and D R Pania (Amsterdam: Kluwer Academic) p. 62
[4] Okano K 1944 *Diamond: Electronic Properties and Applications* ed. L S Pan and D R Pania (Amsterdam: Kluwer Academic) p. 139
[5] Zhu X, Guillaudeu S, Aslam D M, Kim U, Stark B and Najafi K 2003 *Proceedings of the 16th IEEE International Conference on Micro Electro Mechanical Systems (MEMS 2003)* p. 658, Kyoto, Japan, January 2003; 2003 Zhu X, Tang Y, Aslam D M, Stark B and Najafi K 'All diamond packaging for wireless integrated microsystems using ultra-fast diamond growth' *J. MEMS* submitted in April 2004
[6] Yang G S and Aslam M 1995 *Appl. Phys. Lett.* **66**(3) 311
[7] Dreifus D L 1944 *Diamond: Electronic Properties and Applications* ed. L S Pan and D R Pania (Amsterdam: Kluwer Academic) p. 371
[8] Fiegl B, Kuhnert R and Schartzbauer H 1994 *Diamond Rel. Mater.* **3** 658
[9] Taher I, Aslam M and Tamor M 1994 *Sensors and Actuators A* **45**(1) 35
[10] Aslam M, Taher I, Masood A, Tamor M A and Potter T J 1992 *Appl. Phys. Lett.* **60** 2923
[11] Sahli S and Aslam D M 1998 *Sensors and Actuators A* **69**(1) 27
[12] Wur D R *et al* 1993 *Tranducers '93*, Yokohama (Japan), p. 722
[13] Dorsch *et al* 1992 *Diamond'92 (ICNDST-3)*, Heidelberg, Germany, p. 20.2
[14] Aslam M and Schulz D 1995 *Technical Digest: 8th International Conference on Solid-State Sensors and* Actuators, Stockholm, Sweden, vol. 2, p. 222
[15] Masood M, Aslam M, Tamor T and Potter T J 1992 *J. Electrochem. Soc.* **138** L67
[16] Sahli S and Aslam D M 1998 *Sensors and Actuators A* **71**(3) 193
[17] Kania D R, Landstrass M I, Plano M A, Pan L S and Han S 1993 *Diamond Rel. Mat.* **2** 1012

[18] Gurbuz Y, Kang W O, Davidson J L, Kerns D V and Henderson B 1997 *Technical Digest, Transducers'97* p. 979

[19] Yang G and Aslam D M 1996 *IEEE Electron Dev. Lett.* **17**(5) 250

[20] Ramesham R 1999 *Thin Solid Films* **340**(1–2) 26

[21] Ramesham R, Ellis C D, Olivas J D and Bolin S 1998 *Thin Solid Films* **330**(2) 62

[22] Bjorkman H, Rangsten P, Simu U, Karlsson J, Hollman P and Hjort K 1998 *Proc. MEMS '98*, Heidelberg, Germany, p. 34

[23] Aslam M and Tamor M A 1995 US patent 5,413,668 (Ford Motor Company)

[24] Lang A R, Makepeace A P W and Butler J E 1998 *Diamond Relat. Mater.* **7** 1698

[25] Faggio G, Marinelli M, Messina G, Milani E, Paoletti A, Santangelo S, Tucciarone A and Rinati G V 1999 *Diamond Relat. Mater.* **8** 640

[26] Hayashi K, Yamanaka S, Watanabe H, Sekiguchi T, Okushi H and Kajimura K 1997 *J. Appl. Phys.* **81** 744

[27] Kawarada H, Yokota Y, Mori Y, Nishimura K and Hiraki A 1990 *J. Appl. Phys.* **67** 983

[28] Robins L H and Black D R 1994 *J. Mater. Res.* **9** 1298

[29] Collins A T and Lawson S C 1989 *J. Phys. Condens. Mater.* **1** 6929

[30] Collins A T and Woods G S 1987 *J. Phys. C* **20** L797

[31] Collins A T, Davies G, Kanda H and Woods G S 1988 *J. Phys. C* **21** 1363

[32] Collins A T, Kamo M and Y Sato 1990 *J. Mater. Res.* **5** 2507

[33] Feng T and Schwartz B D 1993 *J. Appl. Phys.* **73** 1415

[34] Glinka Y D, Lin K W, Chang H C and Lin S H 1999 *J. Phys. Chem. B* **103** 4251

[35] Diederich L, Kuttel O M, Aebi P and Schlapbach L 1998 *Surf. Sci.* **418** 219

[36] Tatarinov V S, Mukhachev Y S and Parifianovich W A 1979 *Sov. Phys. Semicond.* **13** 956

[37] Zhang B, Shen S, Wang J, He J, Shanks H R, Leksono M W and Girvan R 1994 *Chin. Phys. Lett.* **11** 235

[38] Manfredotti C, Wang F, Polesello P, Vittone E, Fizzotti F and Scacco A 1995 *Appl. Phys. Lett.* **67** 3376

[39] Rinzler A G, Hafner J H, Nikolaev P, Lou L, Kim S G, Tomanek D, Nordlander P, Colbert D T and Smalley R E 1995 *Science* **269** 1550

[40] Bonard J M, Stockli T, Maier F, de Heer W A and Chatelain A 1998 *Phys. Rev. Lett.* **81** 1441

[41] Kim U and Aslam D M 2003 *J. Vac. Sci. Technol.*; 2003 *Virtual J. Nanocsale Sci. Technol. B* **21**(4) 1291

[42] Kim S B and Wager J F 1988 *Appl. Phys. Lett.* **53**(19) 1880

[43] Liu S, Gangopadhyay S, Sreenivas G, Ang S S and Naseem H A 1997 *J. Appl. Phys.* **82**(9) 4508

[44] Kim U 2002 PhD Dissertation, Michigan State University

[45] Yang X, Barnes A V, Albert M M, Albridge R G, McKinley J T, Tolk N H and Davidson J L 1995 *J. Appl. Phys.* **77** 1758

[46] Steeds J W, Charles S, Davis T J, Gilmore A, Hayes J, Pickard D and Butler J E 1999 *Diamond Relat. Mater.* **8** 94

[47] Muto Y, Sugino T and Shirafuji J H 1991 *Appl. Phys. Lett.* **59** 843

[48] Karabutov A V, Frolov V D, Pimenov S M and Konov V I 1999 *Diamond Relat. Mater.* **8** 763

[49] Williams A W S, Lightowlers E C and Collins A T 1970 *J. Phys. C* **3** 1727

[50] Fanciulli M and Moustakas T D 1993 *Phys. Rev. B* **48** 14982

[51] Allers L and Mainwood A 1998 *Diamond Relat. Mater*. **7** 261
[52] Wise K D, Najafi K, Aslam D M, Brown R B, Giachino J M, McAfee L C, Nguyen C T-C, Warrington R O and Zellers E T 2001 Session Invited Keynote *Digest Sensors Expo*, Chicago, p. 175
[53] Aslam D M, Najafi K, Wise K D and Zellers T 2002 *Proc. COMS 2002*, Ypsilanti, MI
[54] Tang Y, Sahli S, Aslam D M, Merriam D and Wise K D 2002 *Proc. AICHE Annual Conference on Sensors*, Indianapolis
[55] Nguyen C T-C 2001 *Digest of Papers*, Topical Meeting on Silicon Monolithic Integrated Circuits in RF Systems, 12–14 Sept, p. 23

PART 4

OTHER MATERIALS

Chapter 13

Polymeric semiconductors

***Jie Liu*[1], *Yijian Shi*[2], *Tzung-Fang Guo*[3] and *Yang Yang*[4]**
[1]General Electric Global Research, New York, USA
[2]Opsys Corporation, San Jose, California, USA
[3]Nitto-Denko Corporation, San Diego, USA
[4]Department of Materials Science and Engineering, University of California, Los Angeles, USA

13.1 Introduction

It has been known for more than two decades that certain plastic materials possess the properties of semiconductors. However, they did not attract tremendous research interest until 1990 when the electroluminescence (EL) from conjugated polymers was first reported [1]. Usually, opto-electronic devices using the same polymeric semiconductor show much different performances because, among other things, they are processed differently. Prospects of currently emerging applications in display [2] and other potential ones, such as solid-state lighting [3], have resulted in worldwide research competitions in perfecting the polymer light-emitting diode (PLED) technology. The advance of polymer semiconductors has resulted in a Nobel prize for chemistry in 2000 to Alan J Heeger, Alan G Mac Diarmid and Hideki Shirakawa for their discovery of semiconductor behaviour of polymeric materials.

In the past 10 years and more, efficiency of those polymer light-emitting devices has been increased by several orders in magnitude. Major contributions to the efficiency enhancement come from development and optimization of materials to meet specific functions, such as emitting, charge-injecting and charge-transferring, and equally important, the detailed device design (most adopted principles from related matured inorganic LED technology, such as hetero-junction structures). Furthermore, another substantial efficiency boost is expected with the more recent realization of that by emission from triplet states [4], 100% internal efficiency is achievable.

Conjugated polymers are a class of materials with unique properties. From the physics point of view, they are semiconductors with the optical and electrical properties similar to the traditional inorganic semiconductors. From the chemistry point of view, they are macromolecules, which can be designed and synthesized to achieve the desired chemical and physical properties. From the materials engineering point of view, since they are amorphous, unlike their inorganic counterparts, i.e. well-ordered crystals, simple, low-cost, high-throughput processing and manufacturing is achievable. The combination of these unique characteristics makes conjugated polymers a charming and yet very useful material. One of the greatest benefits of polymer electronics lies in polymers' solution processability. The polymer materials can be dissolved in ordinary organic solvents and deposited on to substrates as thin films through simple coating techniques such as spin-coating, ink-jet printing, and screen printing. Thus, complex high vacuum/high temperature processes and expensive photolithography processes critical to conventional inorganic semiconductor production are not required for polymer electronic devices. In theory, the whole production process of polymer integrated circuits can be a continuous web processing, which makes polymer electronic devices significantly more cost-effective than traditional inorganic semiconductor ones.

Although the advantages of being amorphous can never be overstated, among all aspects studied, efforts to understand their amorphous nature, processing dependence and effects on performance had for a long time been under-investigated, if not ignored. However, in the earlier stage, several experimental observations have suggest that many physical properties of those polymer-based devices are subject to change as varying processing conditions. For example, Yang *et al* [5] observed that the threshold for gain narrowing of polymer films prepared with one solvent, tetrahydrofuran (THF), is apparently lower than that of the films prepared using others, such as chlorobenzene (CB) and *p*-xylene. The difference in threshold values was attributed to the differences in orientations of polymer chains present in the solid-state films.

Among others, two research groups of University of California at Los Angeles, led by Professor Yang Yang and Professor B J Schwartz, respectively, have conducted pioneered work on this subject. Their independent work, aiming at different aspects, brought up the same message, i.e. polymers' semiconducting characteristics are highly related to how polymer chains arrange themselves (also-called polymer morphology) in their solid-state thin films.

The goal of this article is to provide a comprehensive summary of previous morphological studies, including the origin of morphology difference, correlations between the film morphology and processing conditions, and correlations between the film morphology and physical properties of polymer thin films. The detailed discussion of photophysics is not the objective of this article, but can be found elsewhere [6, 7].

In the body of this chapter, we will first discuss in section 13.2 some fundamental aspects regarding how the processing conditions could change the polymer morphology; in section 13.3, we present a general discussion on how the film morphology would affect the electrical and optical properties of the polymer thin films; and in section 13.4, the use of dilution effects to improve the device performance is discussed. The devices discussed in this chapter are mainly polymer-based light emitting devices fabricated by spin-coating. It is expected that the fundamental principles obtained for these studies can be also applied to other polymer-based electronic devices fabricated using other solution processing techniques such as ink-jet printing and screen printing technologies.

13.2 The control of polymer morphology

13.2.1 Effects of concentration/spin-speed

It is well known that polymer molecules in solutions tend to aggregate when the concentration reaches a critical point. This critical concentration depends on the nature of the polymer molecules, such as the molecular weight and the chemical structure, as well as the environment, mainly the physical and chemical properties of the solvent and the temperature. The origin of aggregation is the inter-molecular forces (mainly van der Waals' forces) among polymer chains. Since these forces are short-range attraction forces, such inter-molecular attraction forces among the individual polymer molecules can be significantly reduced in highly diluted solutions, where the polymer chains are isolated from each other by a vast amount of solvent molecules (as few spaghetti in a 10-quart boiling pot). Therefore, the probability of the individual polymer chains to entangling with each other in dilute solutions is small. As the concentration increases and the effective distance between the individual polymer chains becomes smaller, such interchain interactions become more and more significant. As a result, aggregation of polymer chains becomes more feasible. Further increases of the concentration will lead to a higher extent of aggregation and eventually lead to polymer gelling, a result of heavy entanglement of the polymer chains (as cooked spaghetti served in a bowl).

Simha and Ultrachi [8] suggested that these concentration regions could be characterized using the product of concentration (c) and intrinsic viscosity (η_{in}) of a polymer solution. The greater the product, the more severe the aggregation of polymer chains, the farther the behaviour of solution deviates from an ideal solution. A schematic diagram showing this aggregation process as a function of concentration is shown in figure 13.1.

More recently, Shi *et al* [9] proposed a new method for characterizing such concentration regimes based on the reduced viscosity. Figure 13.2 shows reduced viscosities η/η^*, where η and η^* are the viscosities of

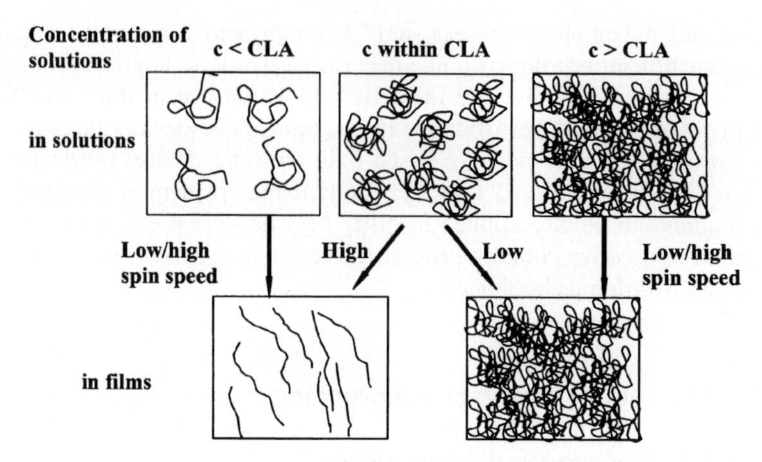

Figure 13.1. A schematic demonstration of the correlations between the concentration of the polymer solution, the aggregation in solution, and the spin speed dependence of the film morphology.

the polymer solution and the solvent, respectively, of MEH-PPV solutions versus their concentrations. It is found that MEH-PPV solutions have three distinct regimes, a linear region at low concentrations, a curved region at middle concentrations, and another linear region at

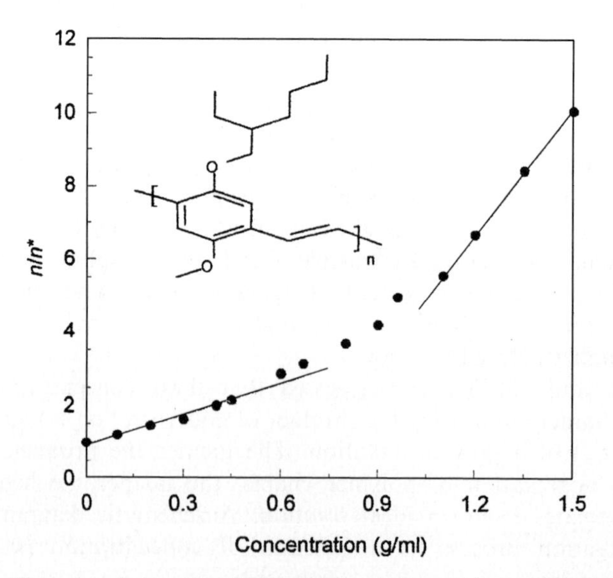

Figure 13.2. The reduced viscosity of MEH-PPV solutions (solvent: cyclohexanone) as a function of the concentration of the polymer solution. The curved region represents the concentrations for loose aggregation (CLA). The chemical structure of MEH-PPV is shown as the inset [9].

high concentrations. It is suggested that the polymer chains are not aggregated at the linear region of low concentrations ($c < 0.4\%$) and heavily aggregated at the linear region of high concentrations ($c > 1\%$). The middle region is referred to as the concentrations for loose aggregation (CLA).

This concentration dependence of aggregation is also observable using ultraviolet–visible absorption spectroscopy measurement. For example, the ultraviolet–visible absorption λ_{max} of a highly diluted MEH-PPV solution is ~510 nm, significantly larger than that of a more concentrated solution (~495 nm, depending on the concentration) indicating that the polymer chains have better conjugation in dilute solutions than in higher concentrations. A similar phenomenon is also observable in the spin-coated polymer thin film. Figure 13.3 shows normalized absorption spectra of two films spun at the same speed (8000 rpm), but using different concentrations: a thinner film (180 Å) spun from a 0.3 wt% MEH-PPV solution, and a thicker film (900 Å) spun from a 1 wt% MEH-PPV solution in cyclohexanone (CHO). The difference in the λ_{max} of the two spectra is obvious: $\lambda_{max} = 510$ nm for the 0.3 wt% and $\lambda_{max} = 496$ nm for the 1 wt%. The spectral shift suggests that the polymer chains in the film spun from the more dilute solution are more extended and the π-electrons in the polymer backbone are more conjugated. As will be discussed in more detail below,

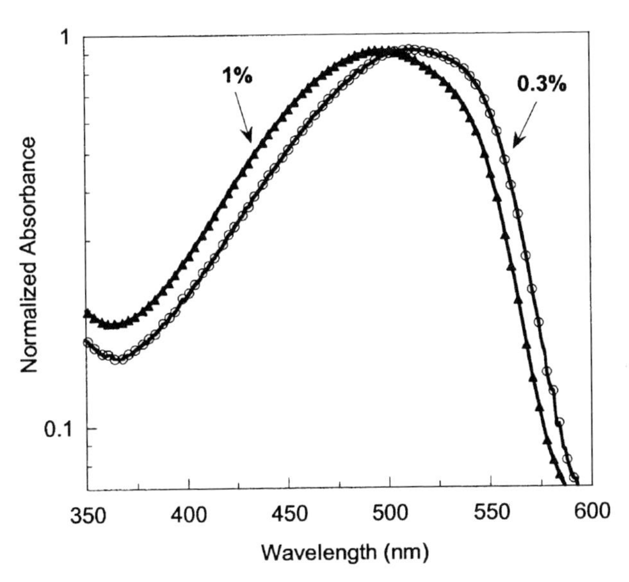

Figure 13.3. Normalized absorption spectra for MEH-PPV films spin-coated on glass plates using 0.3 wt% and 1 wt% MEH-PPV solutions (solvent = CHO; spin speed = 8000 rpm) [9].

in highly dilute solutions the solvent effects is minimal due to the absence of significant van der Waals' forces between the polymer chains.

Similar phenomena were also observed by varying the spin-speed upon spin-coating, one of the most widely-used fabrication techniques. Concentration controls the strength of aggregation of polymer chains, while spin-speed, or more specifically centrifugal force, reflects the tendency to break the aggregations.

During the spin-coating process, the centrifugal force and the radial flow of solvent have a tendency of stretching the polymer chains radically against the cohesive forces of the solution. As depicted in figure 13.1, if the centrifugal force is larger than the cohesive forces of the solution, one would expect that spin-coating should result in a more extended/stretched conformation of the polymer molecules. In contrast, if the cohesive force is stronger than the centrifugal force, one will expect that there is less conformational change due to the spinning. The effectiveness of these processes is also affected in certain extent by the spin time and the solvent evaporation rate during the spin-coating process. Therefore, depending on the processing conditions of the spin-coating, the resulted film morphology could be significantly different from that of the original solution state.

A spectral red-shift similar to that previously shown in figure 13.3 can also be demonstrated by varying the spin speed if a polymer concentration within the CLA region is used. For example, a 0.7 wt% MEH-PPV solution in CHO spun at 2000 rpm resulted in a film with an absorption peak at $\lambda_{max} = 499$ nm (film thickness = 700 Å), whereas the same solution spun at 8000 rpm resulted in a film with $\lambda_{max} = 509$ nm (film thickness = 300 Å) (figure 13.4). At higher (≥ 1 wt%) or lower (<0.4 wt%) concentrations, however, the ultraviolet–visible spectra were not observed to shift with spin speed.

Based on the above discussion, the fact that the spin speed dependence is not observable for films spun at concentrations >1 wt% can be explained by the formation of strong aggregates. These aggregates are so strong that spinning the solution at up to 8000 rpm (the upper limit of the spinner used) is insufficient to break them apart. Since high concentrations and the heavy entanglement of the polymer chains also result in reduced effective conjugation lengths of the polymer backbone, it is not surprising that the films resulting from high concentrations (≥ 1 wt%) have smaller absorption λ_{max} values (496 nm). As the concentration decreases, the cohesive force of the polymer solutions decreases, thus the aggregation becomes 'looser'. It is therefore expected that such 'loose aggregates' can be more easily torn apart by the centrifugal force. This explains the observation that within the CLA regime, the absorption spectrum (λ_{max}) of the polymer film is strongly affected by the spin speed. When the spin speed is lower than a 'threshold', the cohesive force dominates and a smaller absorption λ_{max} value is expected. In contrast, when the spin speed is high enough to

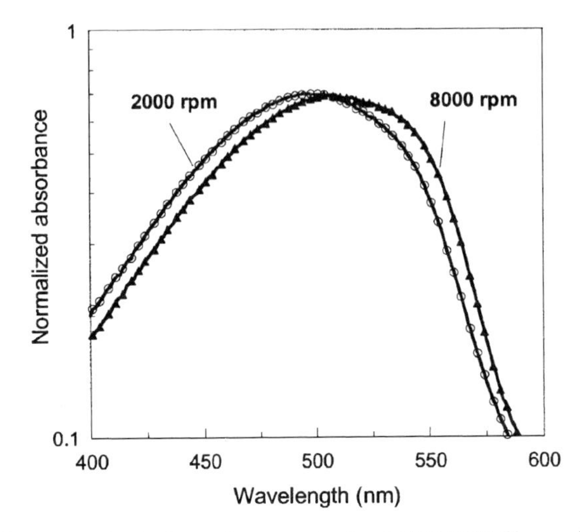

Figure 13.4. The normalized absorption spectra for MEH-PPV films spin-cast on glass plates using different spin speeds (2000 and 8000 rpm). The MEH-PPV solution used for spin-coating was 0.7 wt% in cyclohexanone [9].

overcome the cohesive force, a spectral red-shift (larger λ_{max}) is expected (figure 13.4).

At concentrations below the CLA region (e.g. <0.4 wt% in figure 13.2), the λ_{max} of the resulting films is usually close to 510 nm and is nearly independent of the spin speed (within 1000–8000 rpm). This suggests that the polymer chains are easily stretched to the more extended conformations and/or the polymer chains are already in the most extended conformations. In addition, polymer films spun at these lower concentrations are usually so thin that they dry almost instantaneously during the spin-coating process, and therefore this more extended (and thus more conjugated) metastable conformation is 'locked-in' upon the vaporization of the solvent. As can be seen from the next section, the λ_{max} of these films is reduced after thermal annealing due to the partial recovery from these metastable states (see below) to the thermodynamically more stable states. On the other hand, the 510 nm λ_{max} value seems to reflect the maximum conjugation one could achieve in this polymer.

13.2.2 Effects of solvent

13.2.2.1 The thermodynamics of solvation effect

The rule-of-thumb regularity for discussing the solvent–solute interactions, or the solvent effects, is the principle of 'like dissolves like'. The fundamental basics of this principle are the second law of thermodynamics: the driving force for the mixing of two species is the loss of the Gibbs free energy. When a polymer is dissolved in an organic solvent, the polymer chains should achieve the conformations that can minimize the free energy. Generally, the change in entropy is always positive for such a mixing process. The change in enthalpy, however, can be either positive or negative depending on chemical species present in the polymer and solvent. When a polymer molecule has multiple functional groups, it is expected that these functional groups will behave differently in regards to interaction energies with the solvent molecules. Consequently, some of the functional groups are preferentially solvated more heavily than the others. For instance, the chemical structure of MEH-PPV molecules consists of an aromatic polymer backbone and many ethyl-hexyloxy side chains. It is thus expected that the aromatic solvents can solvate the polymer backbone better than the alkyl side chains. In contrast, the 'staying together' or the aggregation of the alkyl side chains in aromatic solvents may lower the ΔH_M. It is therefore expected that strands of MEH-PPV tend to aggregate lengthwise in the form of a spiral cylinder; the aromatic backbones of the long molecules form the shell of the cylinder due to greater solvation. The alkyl side chains of the molecules point radially inwards inside the cylinder (figure 13.5). We call this the Ar-type aggregation style.

Results from molecular dynamic calculations suggest that for a MEH-PPV strand, a twisted conformation shown in figure 13.6, with the side chains pointing out radially (with respect to the polymer backbone) to all directions, is the most stable conformation. Such a conformation may also benefit from gaining more configurational entropy since the side chains have higher freedom of rotation in comparison with those shown in figure 13.5. However, the twisting of the polymer backbone will interfere with the conjugation along the phenyl-vinyl main chain and thus lead to an extra internal energy increase. Therefore, the final conformation of the polymer chains should reflect a state, which could balance all these factors for the system to reach the minimum free energy. Although accurate thermodynamic data for MEH-PPV in many common organic solvents are not known at this time, a qualitative rationalization can be made based on the heats of mixing for similar, but smaller molecular systems. For example, mixing of the non-aromatic THF or $CHCl_3$ with many aromatic compounds yields relatively large negative heats of mixing (hundreds to nearly one thousand J/mol) [10], suggesting especially strong solvent–solute interactions between these species. In contrast, the heats of mixing between two aromatic compounds are usually zero or only slightly positive. Therefore, the twisted conformation (defined as non-Ar-type conformation) shown in figure 13.6 is more likely to be attained in non-aromatic solvents, such as THF and $CHCl_3$.

According to the above discussion, it is expected that the MEH-PPV molecules should have a more planar (longer conjugation length)

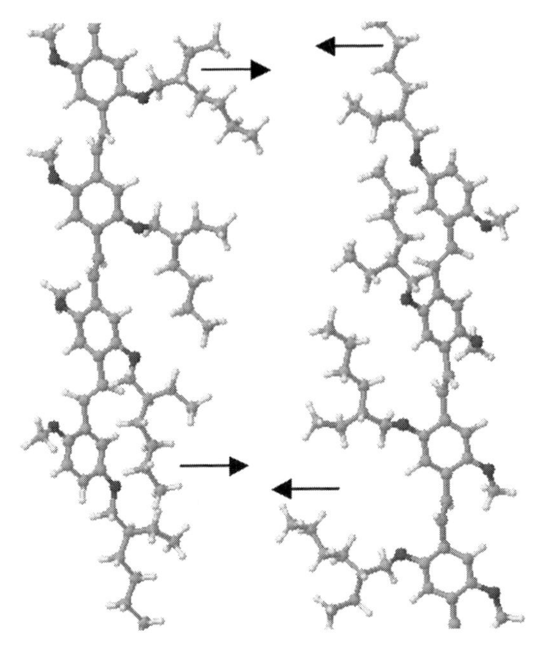

Figure 13.5. The Ar-type aggregation style of MEH-PPV molecules in an aromatic solvent: the polymer backbones are solvated by the solvent molecules while the side chains entangle with each other, resulting in an aggregate with the conducting backbones arranged outside and the insulating side chains pointing inwards towards each other [9].

conformation in aromatic solvents, and attain a more twisted conformation in non-aromatic solvents such as THF and $CHCl_3$. Experimentally, it is observed that the absorption λ_{max} of a MEH-PPV solution in THF is significantly smaller than that observed in aromatic solvents [11], which is consistent with better conjugation in aromatic solvents. As can be seen from the following sections, many other physical properties of the polymer are also solvent dependent.

In addition to the differences observed in the λ_{max} of ultraviolet–visible spectra, the solvation effects were also evidenced by measuring surface energy of spin-coated polymer films. Shi *et al* [9] have observed that the contact angles (86–87 °) between H_2O and the polymer films spun from THF and $CHCl_3$ are significantly smaller than those (average ~95°) spun from aromatic solvents. The different contact angle values indicate that these films have different surface energies. On the other hand, a MEH-PPV film spun from a solution using cyclohexanone (CHO) as solvent has a contact angle of 94° with water, which is close to that of the aromatic solvents. This is consistent with its aromatic-like behaviour observed in the spin-speed dependent electroluminescence spectrum experiments [9].

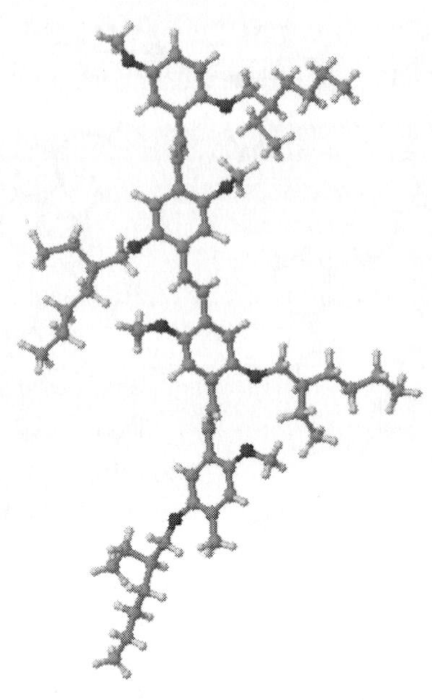

Figure 13.6. The non-Ar type of aggregation style: non-aromatic solvents (THF and $CHCl_3$) result in a twisted conformation of the MEH-PPV molecules with the side chains arranged around the polymer backbone, which hinders the interchain interactions [9].

Although not technically an aromatic solvent, it is obvious that the six-member ring structure of this molecule leads to an aromatic-like solvation behaviour. The fact that films spun from THF and $CHCl_3$ have smaller contact angles with water indicates that these films are less hydrophobic (or more hydrophilic).

It should be noted that the solvation effects are also concentration dependent mainly due to the concentration dependence of the entropy of mixing. These effects become more significant at higher concentrations and less pronounced in more dilute solutions. It is expected that the polymer molecules in dilute solutions should attain the more extended/open conformations in order to reach the maximal entropy. This explains the fact that the absorption λ_{max} of MEH-PPV solutions reaches a maximum value ($\sim$510 nm) in highly dilute solutions and this value is essentially independent of the solvent at low concentrations. The above solvent dependence of the contact angle should become less significant when more dilute solutions are used. It is observed that the contact angle between water and an MEH-PPV film spun from a 0.4% solution in THF has the same value (95 °) as that spun from *p*-xylene under the same concentration (0.4%) and the same spin speed (4800 rpm).

13.2.2.2 Orientation of polymer chains (FT-IR study)

These morphological differences observed in the above MEH-PPV films are further supported by the reflection absorption Fourier transform infrared (FT-IR) spectroscopy measurements [12]. In the reflection absorption mode FT-IR measurement, a vertically polarized source beam is used. Thus, the absorption from a vibration mode of the sample is expected to reach maximum when its transition dipole is normal to the sample surface, and reach minimum when the transition dipole is parallel to the sample surface. An example demonstrating the spectral differences due to the morphological changes in the MEH-PPV films spun from different solvents is shown in figure 13.7. At the high frequency regime of figure 13.7, the absorption peaks at 2958, 2930, 2872 and 2857 cm^{-1} correspond to the $-CH_3$ asymmetric stretching, the $-CH_2-$ asymmetric stretching, the $-CH_3$ symmetric stretching, and the $-CH_2-$ symmetric stretching vibrations, respectively. It is obvious that the relative intensities of these absorption peaks are markedly changed as the processing solvent is changed from dichlorobenzene (DCB) to tetrahydrofuran (THF).

More insights related to the molecular orientation can be found from the fingerprint region of the spectra. The absorptions at 969 and 859 cm^{-1} (group 1) have been assigned to the out-of-plane *trans*-vinyl C–H twisting and the out-of-plane wagging of phenyl C–H, respectively; the absorptions at 1042 and 1028 cm^{-1} (group 2) are assigned to the symmetrical and asymmetrical C(aromatic)–O–C stretching vibration modes, respectively. Since the group 1 transitions (969 and 859 cm^{-1}) have dipoles normal to

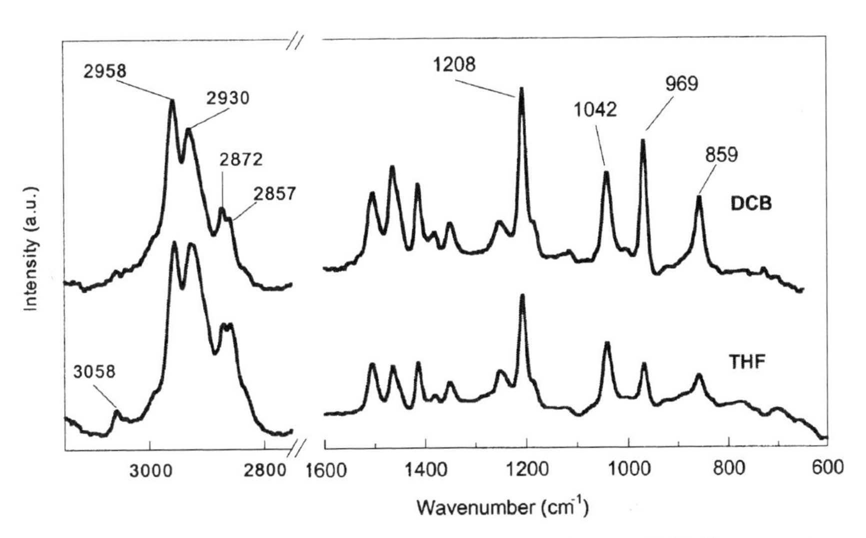

Figure 13.7. The reflection absorption FT-IR spectra of MEH-PPV films spun from different solvents [12].

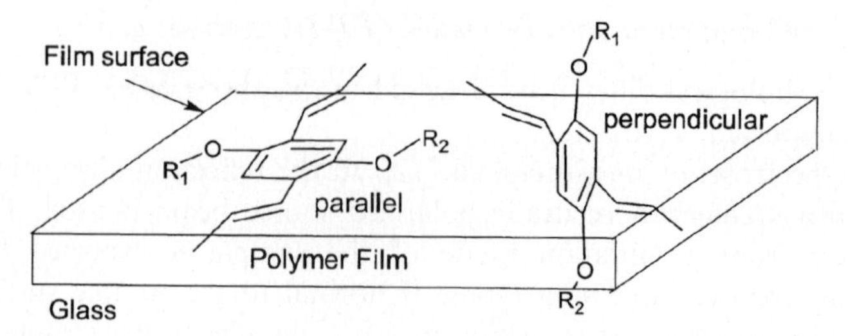

Figure 13.8. Two possible orientations of the aromatic ring on the substrate surface [9].

the phenyl–vinyl plane, the intensities of such transitions are expected to be higher when the phenyl–vinyl planes are aligned parallel to the sample surface and smaller when aligned perpendicular to the substrate surface. In contrast, the transition dipoles of group 2 (1042 and 1028 cm^{-1}) are 'in plane'. Thus, the intensities of the group 2 transitions are expected to be more intense when the phenyl planes are normal to the substrate and less intense when the phenyl rings are parallel to the substrate surface. It is observed experimentally that the relative intensities of group 1 (969 and 859 cm^{-1}) to group 2 (1042 and 1028 cm^{-1}) are significantly different for polymer samples processed with different solvents. By comparing the transitions at 969 and 1042 cm^{-1} (figure 13.7), for example, it can be seen that the relative intensity decreases in the order of $969 > 1042\,cm^{-1}$ when the film is spun from DCB, and $1042 > 969\,cm^{-1}$ when the film is processed with THF. This suggests that in the film processed with DCB, there are higher factions of phenyl and vinyl groups aligned parallel to the substrate surface if compared with the film processed with THF (refer to figure 13.8). Other polarized optical measurements have also shown that chains in MEH-PPV films fabricated using similar aromatic solvents primarily lie in the plane of the film [13], thus showing a high degree of uniaxial anisotropy [14].

It should be pointed out that the twisted conformation shown in figure 13.6 is a metastable state, which should only exist in the presence of a proper solvent. Once the solvent is removed, the polymer chains should spontaneously recover the more conjugated conformations, although such recovery is perhaps limited by the restricted motion of the polymer molecules in the solid state. This prediction is supported by the experimental observations that although the absorption λ_{max} of the polymer solutions in THF is significantly smaller than the λ_{max} in aromatic solvents [11], the films spun with aromatic and non-aromatic solvents have essentially the same λ_{max} value.

13.2.3 Effects of thermal annealing

For spin-coated polymer thin films, temperature could change the polymer morphology dramatically. As discussed earlier, the polymer chains are stretched radially and laid 'flat' during the spin-coating process. Upon evaporation of the solvent, such conformations are 'locked-in', leaving some internal stress in the polymer film. Thus, such conformations are thermodynamically not the most stable states. Upon heating to elevated temperatures, the polymer chains are subject to relaxation to the more thermodynamically preferred conformations. It can be expected that such relaxations of polymer films should be relatively minor at temperature far below the glass transition temperature (T_g) of the polymer. This effect becomes much more significant at temperature near to or greater than the T_g since the heat generated during the normal operation of a device may also induce such morphological changes and thus alters the properties of the device, in practical applications the spin-coated polymer films are usually pre-heated (at temperature $<T_g$) before use, namely thermal annealing, to eliminate such potential effects that could possibly develop during the device's normal operation. It should be noticed, however, that the observed T_g of a polymer thin film is usually lower than that of the bulk material [15, 16]. Thus, it is possible that the morphological changes could happen below the T_g of the bulk material.

These temperature-induced conformational changes can be monitored by a number of analytical tools, such as an ultraviolet–visible spectrophotometer and FT-IR [17]. An example of monitoring such a relaxation process using a reflection/absorption mode FT-IR spectrophotometer is shown in figure 13.9, in which the spectra of MEH-PPV film dried at room temperature, thermally annealed at 70 °C, and at 140 °C are shown. Since the T_g of MEH-PPV is approximately 75 °C, it is expected that significant morphological changes would take place near 75 °C and above. The actual spectral changes with different annealing temperature can be easily seen from figure 13.9, i.e. the absorptions at 969 and 859 cm^{-1} increase and those at 1042 and 1208 cm^{-1} decrease as the annealing temperature becomes higher. For example, the absorption intensity ratio of the peak at 969 cm^{-1} to that at 1042 cm^{-1} is 0.91 when the film is dried at room temperature under vacuum. It is 1.04 after annealing at 70 °C and 1.31 after annealing at 140 °C. The fact that the intensities at 969 and 859 cm^{-1} increase with annealing temperature indicates that transition dipoles for both the out-of-plane *trans*-vinyl C–H twisting (969 cm^{-1}) and the out-of-plane wagging of phenyl C–H (859 cm^{-1}) become more oriented normal to the reflection plane, or the substrate surface, since the phenyl C–H and vinyl C–H bonds are co-planar due to the conjugation between the phenyl and vinyl groups. The observed absorption enhancement for such vibration modes reveals that both the double bonds and the phenyl rings become more parallel to the substrate plane, upon annealing at high temperatures. In contrast, the observed reductions in absorption intensity at 1042 and 1028 cm^{-1}, assigned to the symmetrical and asymmetrical C(aromatic)–O–C stretching vibration modes, respectively, is also consistent with that the phenyl rings becomes

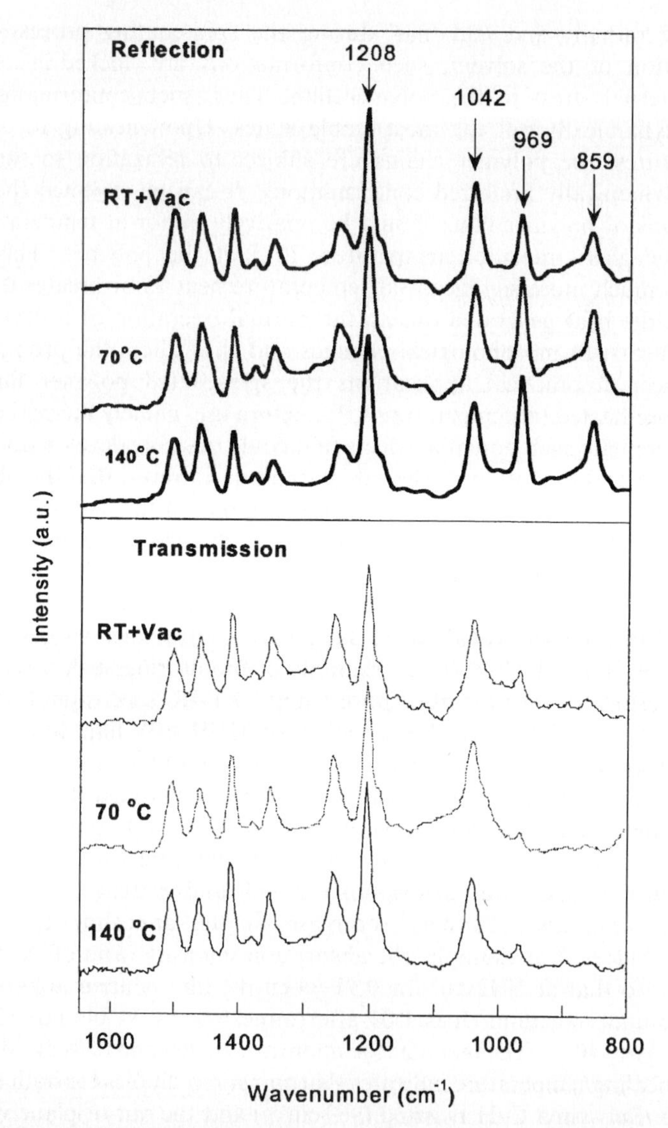

Figure 13.9. Reflection and transmission mode FT-IR spectra of MEH-PPV films spun from THF and annealed under different temperatures where RT + Vac refers to at room temperature and in vacuum [17].

more parallel to the substrate plane upon thermal annealing. The lack of significant changes in transmission spectra before and after thermal annealing further confirms that the changes in reflection spectra are unlikely due to any thermal induced changes of the chemical composition of the polymer film.

13.3 The morphological dependence of device performance

From the above discussions, we have already learned that the solvent, the polymer concentration, the spin speed, and the baking time and temperature all affect the final film morphology. In this section, we will discuss how these morphological changes would alter the electrical and optical properties.

13.3.1 The film conductivity

It is known that in the spin-coated films most polymer chains lie in the plane of the substrate surface [13]. In a typical polymer thin-film device structure 'anode/polymer/cathode', the charge carriers travel across the polymer film, which is perpendicular to the plane of the film. Therefore, the conductivity of the film is to a large extent dependent on the rate for interchain hopping of the carriers, or the rate for the interchain electron transfer, which is perpendicular to the film. If a polymer film is dominated by the Ar-type aggregation style shown in figure 13.5, where the bulky side chains (the insulators) are trapped inside the aggregate and the conducting polymer backbones are exposed, it is expected that the interchain π–π interaction, and thus the interchain electron transfer, should be favourable (smaller energy barrier for electron hopping). In contrast, if the bulky side chains are arranged around the conducting polymer backbones (figure 13.6), the interchain electron transfer will be hindered by the side chains since the conducting polymer backbones are separated farther apart by the side chains. Therefore, a higher energy barrier for the interchain electrons hopping is expected.

Thus, it can be predicted according to the above discussion that the MEH-PPV films spun from aromatic solvents should have better conductivity than those spun from non-aromatic solvents such as THF and $CHCl_3$. Greater current densities are expected for devices spun from aromatic solvents than those spun from non-aromatic solvents under the same electrical field. An example is shown in figure 13.10, where the current-voltage (I–V) curves of a device spun from DCB and a device spun from THF are plotted in the same chart for easy comparison [18]. For example, the current of the device spun from DCB reaches 13 mA at 4 V, while at the same applied voltage the device spun from THF only reaches 2 mA, a 6.5-fold difference! Although this difference may also involve a contribution from the different PEDOT/MEH-PPV interface similar to that observed in the 'polymer on metal' contacts, as will be discussed in the following sections, it is expected that this contribution is relatively small, especially at higher current densities or higher applied voltages (see below). Thus, the fact that these devices have different current densities at the same applied bias should be mainly due to the differences in the conductivity of the MEH-PPV films.

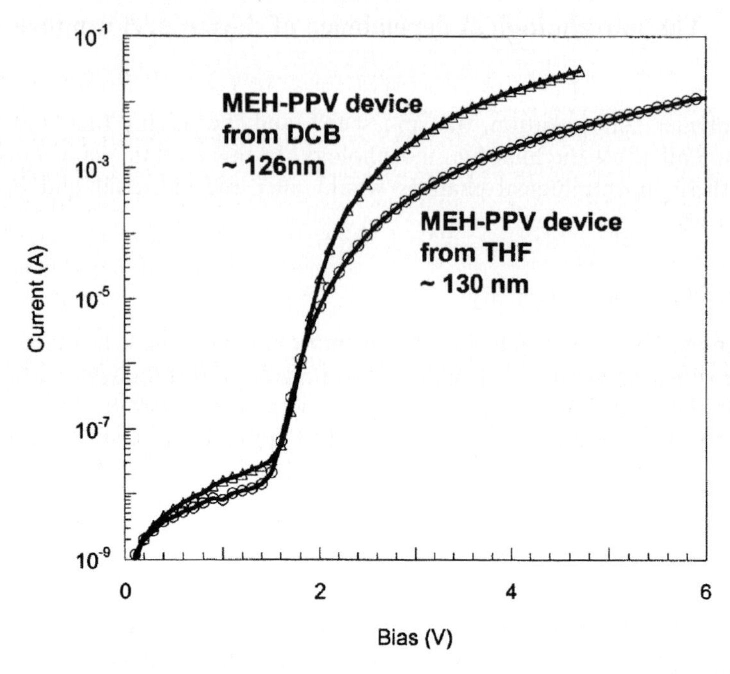

Figure 13.10. *I–V* curves for devices spun from DCB and THF. Although the two MEH-PPV films have essentially the same thickness, the film spun from DCB has significantly higher current injection than that spun with THF.

13.3.2 Metal/polymer interfaces

The current density of the device may also be affected to some extent by the charge injection properties of the polymer/electrode interfaces or, more specifically, the energy barriers for the injection of the charge carriers into the polymer thin film.

Typically for PLEDs, an electrode with high work function is used as the anode for hole injection, and the polymer thin film is spin-coated on top of the anode. The low work function cathode, however, is deposited on the polymer film via thermal evaporation under high vacuum. Hole-only devices were fabricated and tested to check processing conditions dependence of the metal/polymer interfaces. In those devices, both electrodes consist of the same metal element (high work function metals). Therefore, there are no logical cathodes and anodes in these devices. However, for the sake of consistency and ease of discussion, the term 'cathode' is still used to define the electrode formed by deposition of metal on the polymer film (metal on polymer), and the term 'anode' to define the electrode on to which the polymer film is spin-coated (polymer on metal).

13.3.2.1 The 'metal on polymer' contacts

Conjugated polymers rely on the conjugated π-electrons to conduct electrical currents. Thus, feasible electron or hole injection from a metal to the polymer is expected if there is a good physical contact between the metal and the π-electrons of the polymer backbone. In an ideal circumstance where the polymer metal interface is a perfect ohmic contact, the barrier for charge injection depends only on the energy gap between the work function of the metal and the HOMO level (for hole injection) or the LUMO level (for electron injection) of the polymer. This 'ideal' barrier has been defined as the intrinsic energy barrier (ϕ_i) for the charge injection. It is thus expected that such an ohmic contact is only possible when there exists a direct contact of the electrode metal with the π-electrons of the conjugated polymer backbone. This condition can be satisfied or nearly satisfied in a metal-on-polymer (MOP) type contact, in which the metal electrode is evaporated and the metal atoms are condensed on to the polymer film surface. It is found that the evaporated metal atoms can diffuse into the polymer film up to several nanometres in depth [19, 20] during the deposition process. Thus formation of a direct metal/π-electrons contact is expected in a MOP contact. In addition, the fact that the metal atoms can diffuse into the polymer film producing an 'inter-penetrated' regime which will physically increase the polymer–metal contact area and thus should also help to lower the barrier for charge injection. It has been observed in the MEH-PPV based PLED devices that the electron injection from a Ca cathode into the MEH-PPV film was almost barrierless [21]. Since the LUMO level of MEH-PPV and the work function of Ca is nearly identical (2.9–3.0 eV), the observed zero barrier for electron injection suggests that this Ca/polymer contact has characteristics of an ohmic contact. It is thus expected that the MOP type of contacts should belongs to an ohmic (or nearly ohmic) contact type. Since the HOMO/LUMO energy levels of a polymer material and the work function of a metal element are generally considered to be the intrinsic properties of the materials, the energy barrier is expected to be independent of the processing conditions, such as solvent and spin speed, of the polymer film. This is found to be nearly true in room temperature (see discussion in the following sections).

13.3.2.2 The 'polymer on metal' contacts

When the polymer/metal contact is formed by deposition of the polymer solution on to a metal surface, the polymer-on-metal (POM) contact, the interfacial properties are greatly dependent on the processing conditions. Since the polymer is deposited on a smooth and dense metal surface, it is now impossible to form an 'inter-penetrated' area as in the case of the MOP contact. Furthermore, the evaporation of the solvent molecules creates a large amount of empty spaces inside the polymer film as well as in the metal/polymer interfacial area. Therefore, a poorer polymer/metal contact is expected. On the other hand, these empty spaces could also give room for the polymer molecules to relax during device operation, which might break down the existing contact and thus prevent the efficient charge injection. Therefore, the actual energy barrier for a POM contact is expected to be higher than the intrinsic energy barrier and to be processing conditions dependent. The effective energy barrier ϕ will be the sum of the intrinsic barrier ϕ_i and a contact-dependent component $\Delta\phi$, as follows:

$$\phi = \phi_i + \Delta\phi.$$

According to this definition, the intrinsic energy barrier ϕ_i represents the minimum energy required for the charge injection from the metal into the polymer molecule, which is a constant for a given polymer/metal pair. On the other hand, this contact-dependent component $\Delta\phi$ depends on the quality of the metal/polymer interface, which is processing or morphology dependent.

The direct evidence for the existence of $\Delta\phi$ and its morphological dependence comes from the observation of the unsymmetrical *I–V* curves for a series of hole-only devices consisting of a polymer thin film sandwiched in between two metal electrodes. In these devices, both electrodes used the same high work function metal. According to the above definitions, each device has a POM-type anode and a MOP-type cathode. If $\Delta\phi = 0$ or it is independent on processing conditions, both POM and MOP interfaces should have the same ϕ values. Thus the *I–V* characteristics of these devices should be identical under forward and reversed bias. However, this is in contradiction to experimental observations. For example, it is observed that the forward bias *I–V* characteristics of such a hole-only device consisting of Au (anode: POM contact)/polymer/Au (cathode: MOP contact) is significantly different from that under reversed bias [22]. This phenomenon is also observable using other high work function metals such as Cu and Ag. All these devices have non-zero built-in potentials. The ϕ value is generally in the order of tens to hundreds of millivolts [22]. More interestingly, a hybrid hole-only device consisting of a Cu POM anode and an Al MOP cathode, Cu (anode: POM contact)/MEH-PPV/Al (cathode: MOP contact), was found to have almost identical forward and reversed bias [23]. Since the HOMO level of MEH-PPV is $\sim$5.1 eV, the barrier for hole injection from Cu (work function = 4.5 eV) is expected to be lower than that from Al (work function = 4.3 eV) if both electrodes are in ohmic contact with the polymer film. However, the nearly identical *I–V* characteristics observed for forward and reversed bias suggested that both electrodes have practically the same effective energy barrier values for hole injection. These results strongly suggest that ϕ could not be simply treated as the energy difference between the work function of the electrode metal and the

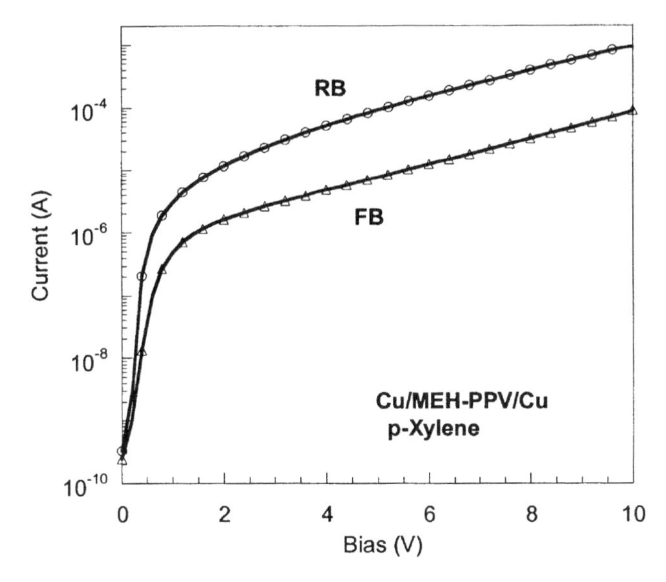

Figure 13.11. The *I–V* curves under forward and reversed bias for a hole-only device using Cu electrodes. The MEH-PPV film was spun from *p*-xylene [12].

HOMO energy level of the polymer. This phenomenon can only be explained by introducing the contact-dependent component $\Delta\phi$. According to the previous discussion, the $\Delta\phi$ for a MOP cathode is either very small or zero, while that of the POM anode is larger and is morphology dependent. Thus, the energy barrier for hole injection under forward bias (hole injection from the POM anode) is expected to be higher than under reverse bias (hole injection from the MOP cathode). Therefore it is not a surprise that the current is higher at reversed bias for such devices (figure 13.11). The fact that the above hybrid device, the Cu (POM contact)/polymer/Al (MOP contact) device, has almost symmetrical *I–V* curves is a coincidence that the $\Delta\phi$ value for the POM anode is just large enough to compensate the work function difference between Al and Cu, resulting in both electrodes having the same effective energy barrier values. Since the work function of Cu is $\sim$0.2 eV higher than Al, therefore $\Delta\phi$ for the Cu anode can be estimated to be approximately 0.2 eV for the above device (assuming $\Delta\phi$ of the MOP contact is zero).

From the above discussion, one would expect that in the case of POM contacts, the Ar-type of aggregation style (figure 13.5), which has the conducting polymer backbones exposed, should readily form a better electrical contact with the metal electrode (figure 13.12(a)). The non-Ar-type of aggregation style (figure 13.6), however, is expected to form poorer contact with the metal surface since the metal is now separated from the π-electrons by the insulating ethyl-hexyloxy side-chains (figure 13.12(b)). In

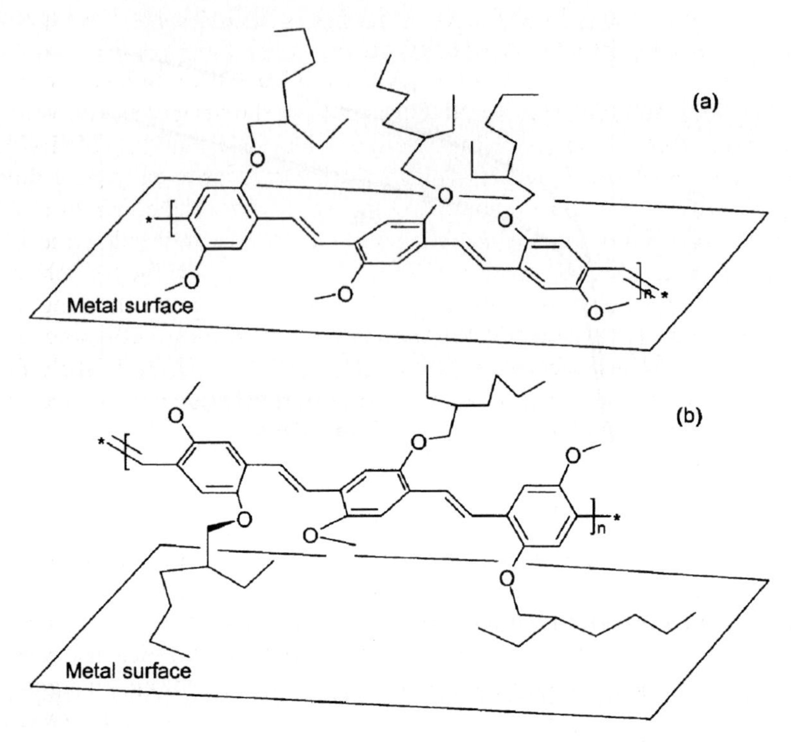

Figure 13.12. A graphical demonstration of the morphologically dependent POM contact: (a) polymer processed with aromatic solvents; (b) polymer processed with non-aromatic solvents [12].

other words, the contact shown in figure 13.12(b) should have higher $\Delta\phi$ value than that shown in figure 13.12(a). This is consistent with experimental observations. It is found that the built-in potential for devices processed using THF is significantly larger than those processed using an aromatic solvent.

For a regular PLED device, a thin layer of PEDOT is used in between the ITO and the MEH-PPV film to help improve hole injection. Thus, similar morphological effects are expected in the ITO/PEDOT and the PEDOT/MEH-PPV interfaces. Due to the ionic nature of PEDOT (doped with poly(4-styrenesulphonate)), it is expected that there should be a relatively strong dipole interaction between ITO and the PEDOT layer. Additionally, the non-conductive ethylene groups of the molecule are much smaller than the side chains of a MEH-PPV molecule, therefore the $\Delta\phi$ of the ITO/PEDOT interface should be small. Since the PEDOT layer is also much more conductive than the MEH-PPV film, it is expected that the major energy barrier for hole injection, if there is any, will be at the PEDOT/MEH-PPV interface. If the solvent used in the MEH-PPV solution does not dissolve the PEDOT layer, which is true in most cases, it is expected that the resulting PEDOT/MEH-PPV interface will have the POM contact like characteristics and morphological dependence. This rationale is consistent with the observed spin-speed dependence of the device turn-on voltage (see discussion in the following sections). The fact that the MEH-PPV films generally have poor adhesion to the PEDOT layer may be a direct cause of this strong morphological dependence. Due to the lack of strong binding forces between the PEDOT and MEH-PPV molecules, the hole injection from PEDOT to the MEH-PPV relies on the 'loose' physical contact of the polymer molecules. Therefore, it is not surprising that the interfacial properties are likely to vary significantly with processing conditions. In this regard, it is expected that the PEDOT/MEH-PPV interface can be improved if a mixed region can be created between the bulk MEH-PPV and the bulk PEDOT layers.

13.3.3 Turn-on voltages

It is commonly observed in PLED devices that the threshold voltage for current injection is different from that required for the device to emit photons. Generally, the threshold voltage for current injection, or the current-on voltage ($V_{I\text{-ON}}$), is defined as the voltage at which the current 'switches on' in a semi-log plot. Similarly, the light-emitting (or light-on) voltage ($V_{L\text{-ON}}$) is defined as the onset voltage at which the light 'switches on' in a semi-log plot. Based on numerical simulations, Malliaras and Scott [24]. suggested that the carrier injection efficiency in a PLED device is primarily dominated by the carrier injection rate. The carrier mobility only matters if the injection capabilities are similar. Thus the light-emitting voltage $V_{L\text{-ON}}$ is more related to the injection of minority carrier. For a classical MEH-PPV device using ITO/PEDOT as the anode and Ca as cathode, it is believed that the hole is the minority carrier. Therefore, $V_{I\text{-ON}}$ should reflect the voltage for the electron injection and $V_{L\text{-ON}}$ for hole injection. Thus, the PLED device is a single-carrier (electron-only) device when operated between $V_{I\text{-ON}}$ and $V_{L\text{-ON}}$, and the voltage difference $\Delta V = V_{L\text{-ON}} - V_{I\text{-ON}}$ reflects the energy barrier for hole injection.

13.3.3.1 The current-on voltage

It is well accepted that the $V_{I\text{-ON}}$ is related to the built-in potential V_{bi}, which is the difference in the work function of the cathode and the anode in addition to a correction term primarily due to interfacial effects,

$$V_{I\text{-ON}} = V_{\text{bi}} = \Delta\Phi + \varphi$$

where $\Delta\Phi$ is the work function difference between the anode and the cathode, and φ is a correction term primarily determined by the quality of the

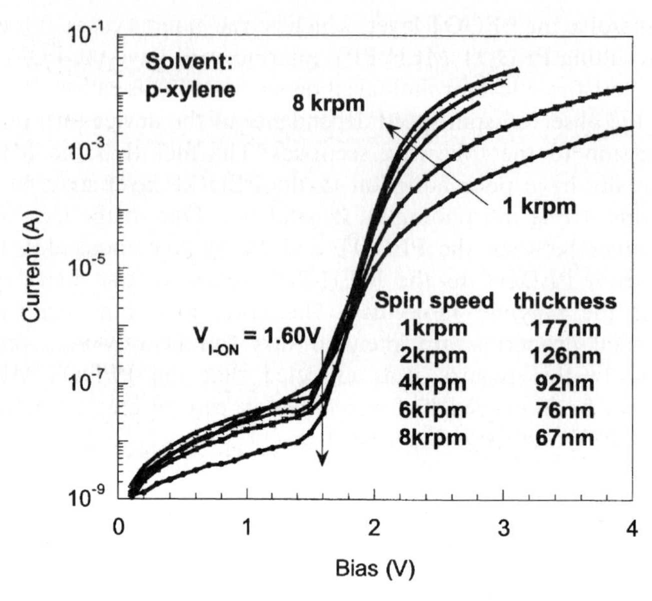

Figure 13.13. *I–V* curves for a series of MEH-PPV based PLED devices spun at different spin speeds using p-xylene as the solvent.

interfaces. φ is expected to be temperature dependent. When there is an ideal ohmic contact in the interface, $\varphi = 0$, the above equation can be rewritten as $V_{I\text{-ON}} = \Delta\Phi = V_{bi}$. When operated at $V_{I\text{-ON}} < V < V_{L\text{-ON}}$, the MEH-PPV PLED device is an electron-only device. Therefore, φ has the same meaning as the energy barrier for electron injection ($\Delta\phi$) from the cathode. As discussed above, $\Delta\phi$ for a MOP contact is expected to be very small and be independent of the polymer morphology. It is indeed observed that the $V_{I\text{-ON}}$ of the MEH-PPV LED devices is essentially independent of the spin speed and the solvent used for the spin-coating [25] (figures 13.13 and 13.14). Additionally, it can be seen from figure 13.15 that $V_{I\text{-ON}}$ at room temperature is independent of the thickness of the MEH-PPV film and its value is approximately equal to $\Delta\Phi$ (~1.6 V). These observations strongly suggest that the electron injection energy barrier is very small at room temperature, which is consistent with the observations of others [25, 26].

13.3.3.2 The light-on voltage

In contrast to $V_{I\text{-ON}}$, which is essentially independent of the spin-coating conditions at room temperature, the light-on voltage $V_{L\text{-ON}}$ is strongly dependent on the spin-coating conditions. Since $V_{L\text{-ON}}$ is the threshold

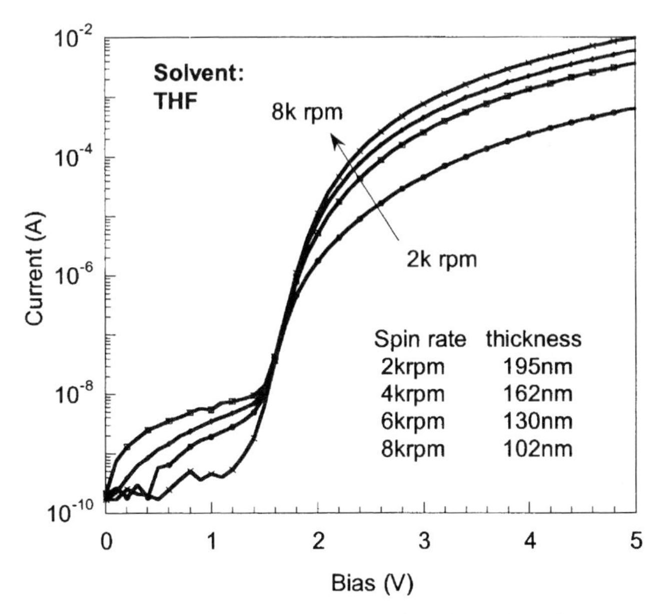

Figure 13.14. *I–V* curves for a series of MEH-PPV based PLED devices spun at different spin speeds using TFH as the solvent.

voltage for hole injection as discussed previously, the difference between $V_{L\text{-ON}}$ and $V_{I\text{-ON}}$ reflects the energy barrier for hole injection from the anode into the MEH-PPV film. As discussed earlier, this will be mainly determined by the PEDOT/MEH-PPV interface. Since this is a POM type of contact, it is not surprising that the barrier ($\Delta\phi$) for hole injection and thus the $V_{L\text{-ON}}$ of the device is more susceptible to morphology or process related changes. Therefore, according to the previous discussion the $V_{L\text{-ON}}$ should be higher if the polymer is processed with non-aromatic solvents (i.e. THF and $CHCl_3$) and lower if an aromatic solvent is used. This is exactly what has been observed experimentally. For example, the $V_{L\text{-ON}}$ for a device processed with DCB is 1.75 V, while that for a device processed with THF is 1.94 V (figure 13.16).

On the other hand, a poor anode/polymer contact such as that shown in figure 13.12(b) can be improved to some extent by intensively stretching the polymer molecules, i.e. spin-coating the polymer solution at very high spin speeds. At high spin speeds, the polymer coils are stretched open, allowing the conducting polymer backbone to settle closer to the PEDOT molecules on the surface. This results in a better contact and thus lowers the hole injection barrier and thus $V_{L\text{-ON}}$. This has been demonstrated experimentally in figure 13.16. For a device processed with aromatic solvents, however, the $V_{L\text{-ON}}$ value is much less sensitive to the spin

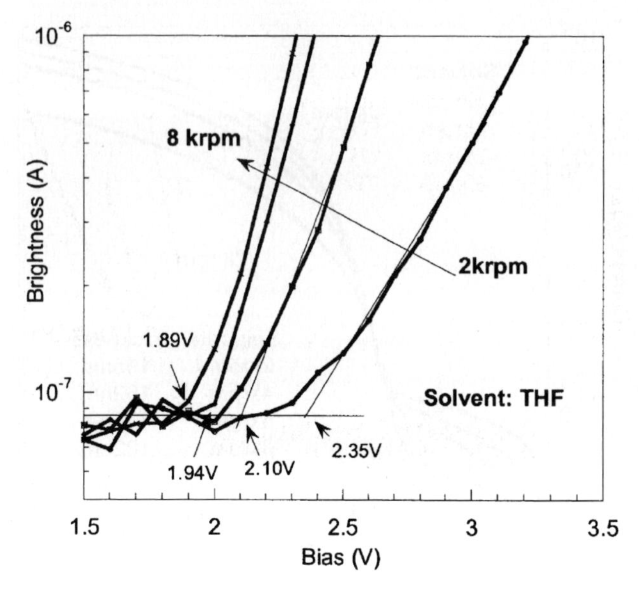

Figure 13.15. The spin-speed dependence of the light-emitting voltage for MEH-PPV based PLED devices processed with non-aromatic solvents. Large spin-speed dependence is observed.

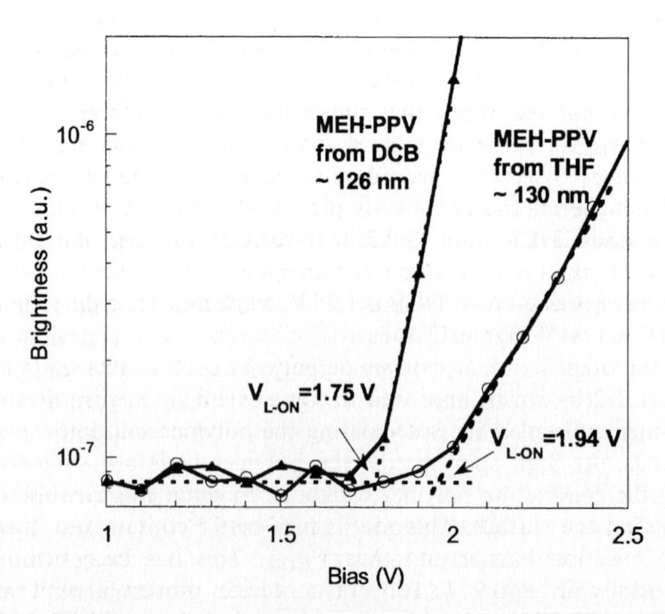

Figure 13.16. Brightness–voltage curves for devices fabricated with DCB and THF.

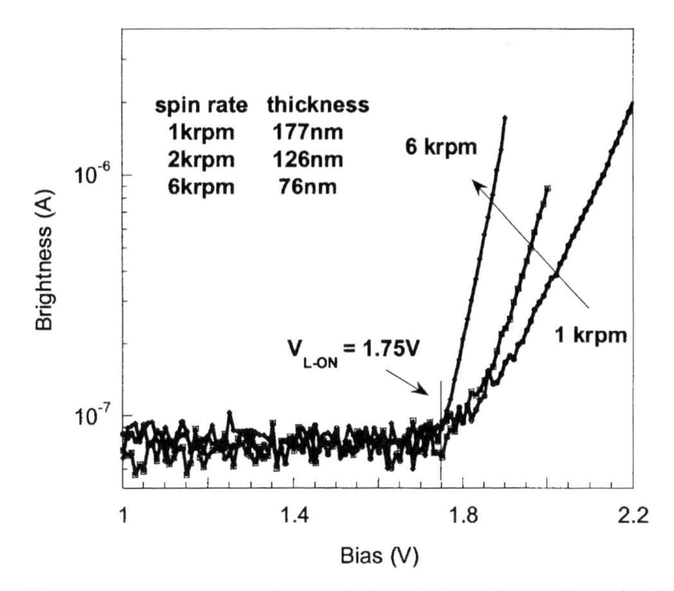

Figure 13.17. The spin-speed dependence of the light-emitting voltage for MEH-PPV based PLED devices processed with aromatic solvents. Little spin-speed dependence is observed.

speed. This can be seen from figure 13.17, which is a set of brightness–voltage (B–V) curves, i.e. brightness (represented by the photocurrent of the detector) versus applied voltage, for a series of devices processed with DCB solvent but the MEH-PPV films were spun at different speed (1000–6000 rpm). It can be easily seen that these devices have almost identical $V_{L\text{-ON}}$ values (~1.75 V), corresponding to a $\Delta\phi$ value of approximately 0.15 eV for hole injection.

As discussed in section 13.2.3, it is expected that the film morphology can also be changed by thermal annealing. Since the $V_{L\text{-ON}}$ value is morphology dependent, it is expected that thermal annealing should also vary the $V_{L\text{-ON}}$ values. The B–V curves of a series of PLED devices annealed at different temperatures are shown in figure 13.18. It can be seen from these plots that it is generally true that higher annealing temperatures result in lower $V_{L\text{-ON}}$ values. This is true for devices processed with all solvents studied.

As discussed earlier, the $V_{L\text{-ON}}$ is a measure of the capability of hole injection from ITO/PDOT anode into MEH-PPV. The fact that higher annealing temperatures result in lower $V_{L\text{-ON}}$ values indicates that a better contact between PEDOT and PEH-PPV is obtained upon thermal annealing. As a result, devices annealed at higher temperatures show a greater current

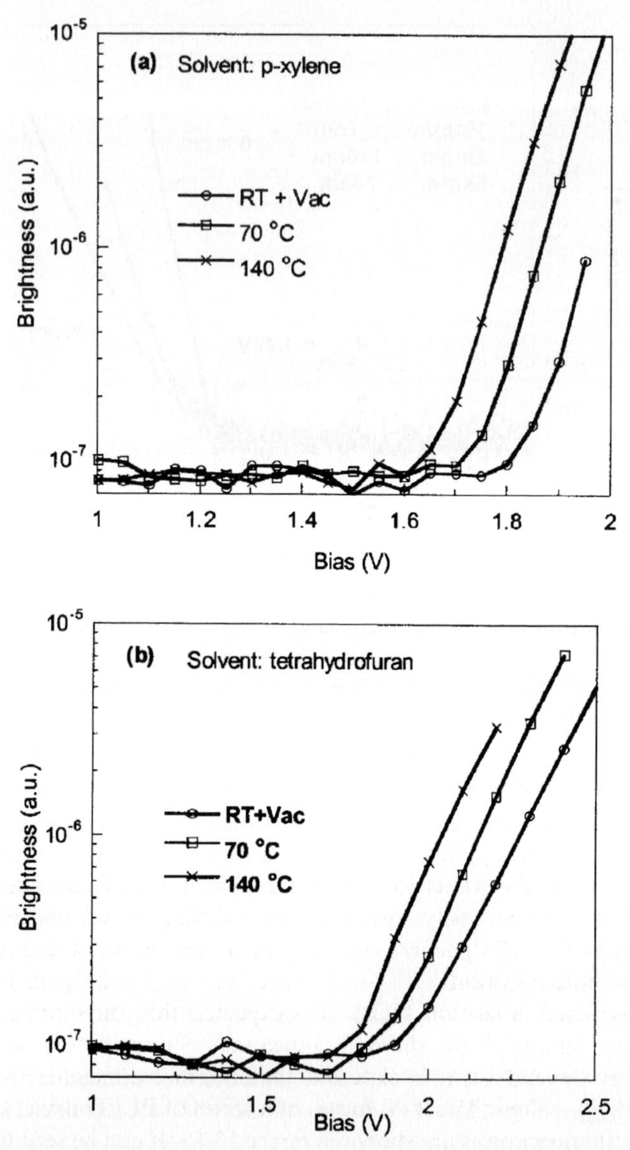

Figure 13.18. Plots of brightness versus applied voltage (B–V curves) for MEH-PPV based PLED devices annealed under different conditions and processed with different solvents (a) *p*-xylene and (b) tetrahydrofuran [17].

under the same applied electrical field (figure 13.19). This effect is more pronounced at the low applied field region since at this region the current is dependent more on the carrier injection efficiency (or the energy barrier for carrier injection) and less on the film resistivity.

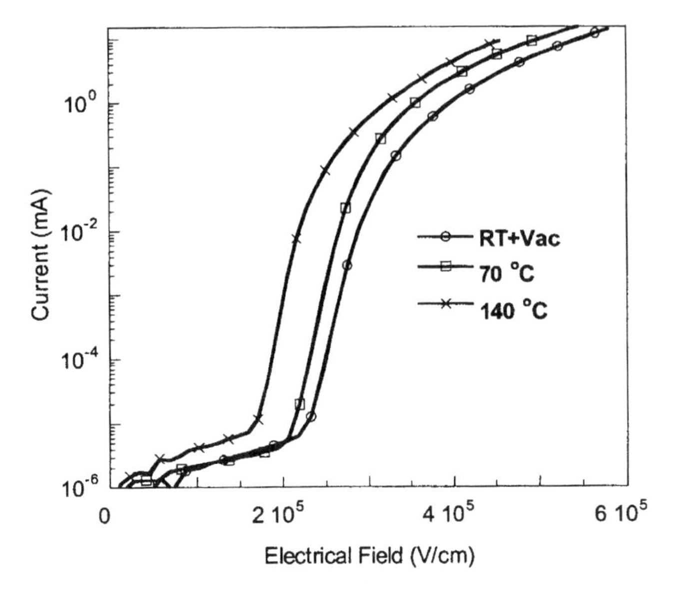

Figure 13.19. The current–electrical field curves of devices annealed at different conditions [17].

13.3.4 The emission spectrum

13.3.4.1 The solvent and spin speed dependence

13.3.4.1.1 MEH-PPV processed with aromatic solvents. As mentioned earlier, the EL and PL spectra of spin-cast films are also morphologically dependent. It is found that within the CLA regime, the EL and PL spectra of spin-cast MEH-PPV films are strongly dependent on spin speed. Examples of spin-speed and concentration dependence are shown in figure 13.20. It is consistently observable that when the polymer solution is coated at high speeds (e.g. 4000–8000 rpm, 0.7 wt%), the resulting devices have a strong yellow emission peak ($\lambda_{max} \approx 575$ nm) and a weak red shoulder (~630 nm) (figure 13.20(a). The spectrum of this yellow emission is similar to the PL spectrum of a highly dilute MEH-PPV solution, corresponding to the un-aggregated single chain exciton emission. As discussed earlier, it is expected that films resulting from high spin speeds should consist mostly of the more extended and less coiled polymer chains. Therefore, this yellow EL emission (~575 nm) is also assigned to the single chain exciton from the more extended polymer chains of the film. At lower spin speeds, the spectrum red-shifts, and the intensity of the red emission peak (630 nm) increases. This effect is observable in all aromatic solvents studied, such as chlorobenzene, 1,2-dichlorobezene, toluene, and *p*-xylene, and is independent of the applied

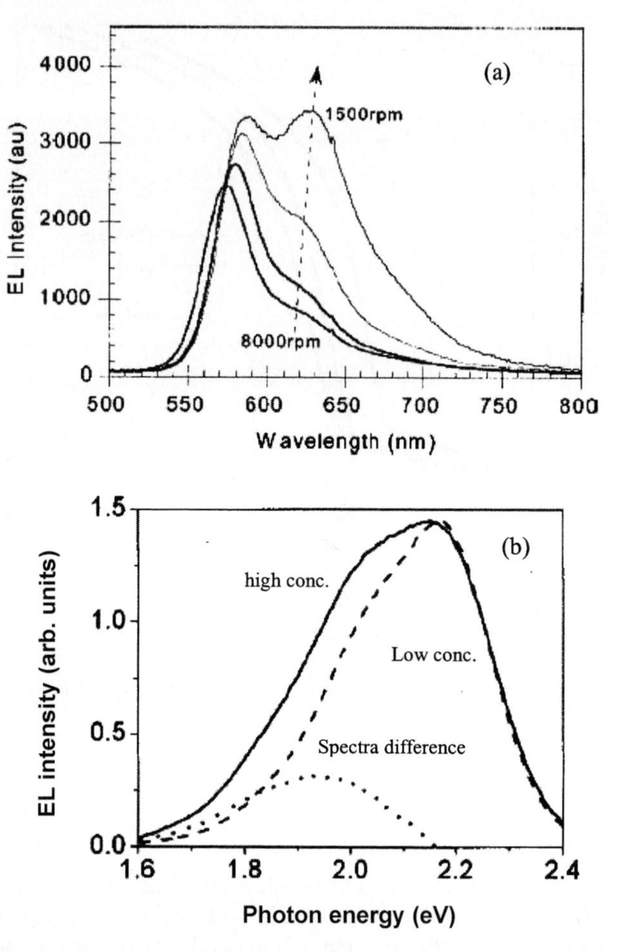

Figure 13.20. Spin-speed (a) [9] and concentration (b) ([27], figure 7) dependence of MEH-PPV EL spectra.

electric field. A similar trend is also observed in the PL spectra of the corresponding thin films. However, these changes in the emission spectra are limited to those devices spun within the CLA. At concentrations above the CLA, the red emission (630 nm) always dominates; at concentration below the CLA, the yellow emission (575 nm) dominates. Shown in figure 13.20(b) is an example of concentration dependence of MEH-PPV EL spectra ([27], figure 7).

The 630 nm peak in the emission spectrum of the MEH-PPV film has traditionally been considered as the intrinsic vibronic structure of the spectrum, and the above spectrum changes were previously correlated to the microcavity effect [28–30]. This traditional belief was questioned by Shi *et al* [9], when they observed that the above effect is independent of the thickness of the polymer film, which rules out the possibility of it being an optical effect. They suggested that the 630 nm emission was a result of the formation of an interchain species (named as Ex-I in the following discussion). The assignment of this species to an interchain species is also supported by the observation that the intensity of this peak is significantly reduced upon diluting the MEH-PPV with polyfluorene (PF), as shown in the following section. Huser and Yan [31] also suggested that the yellow and the red peaks are due to different emitting species based on a study of the PL spectrum of MEH-PPV films using the microscopic fluorescence technique. As discussed earlier, the film morphology varies with spin speed when the polymer concentration is within the CLA. At lower spin speeds, more coiled aggregates survive the spin-coating process. At higher spin speeds, these aggregates are torn apart by the centrifugal force. It is thus clear that the Ar-type of aggregation style favours the formation of the Ex-I species while breaking apart such aggregates results in more emission from the single chain excitons.

It is often observed for many polymers that the aggregation of the polymer chains leads to a spectral red-shift in their absorption spectra due to the formation of ground state complexes. Quantum mechanics calculations also suggest that a π–π stacking of the polymer backbones can red-shift the absorption spectrum [32]. If this also applies to MEH-PPV films, a spectral red-shift in the absorption spectrum is expected when there are more aggregations in the MEH-PPV film. However, the experimental observations are the opposite. As discussed earlier, the MEH-PPV films spun at lower spin speeds, where more aggregates have survived the spin-coating, have smaller absorption λ_{max} values; the films spun at higher spin-speed, where fewer aggregates are expected, have larger absorption λ_{max} values. This indicates that such interchain complexes are not formed in the ground state in these MEH-PPV films. In addition, the observed spectral shifts resulting from the spin speed are also too small ($\sim$5 nm) to be associated with the formation of complexes. It is therefore concluded that the π–π stacking of the polymer backbones in MEH-PPV is hindered in the ground state, probably due to the bulky 2-ethyl-hexyloxy side chains. Such interactions only become more pronounced in the excited state.

The PL decay dynamics of MEH-PPV in the film as well as in the solution have been studied extensively. However, results obtained by different research groups are often inconsistent. For example, Samuel *et al* [33] have shown that the kinetics of the MEH-PPV films at 600 nm is dominated by exponential decay with a time constant of 580 ps, while Rothberg and co-workers [34, 35] have reported that the PL decay dynamics of the MEH-PPV film at room temperature is non-exponential and consists of a fast component ($t \approx 300$ ps) and a slower component. These authors suggested that the non-exponential dynamics were probably due to the inhomogeneity of emission rates and to the dynamics of excited-state diffusion to the quenching defects. Although it is possible that this discrepancy is due to the intrinsic difference between the polymer samples used by the two groups, the fact that they also have markedly different PL spectra should not be ignored. The PL spectrum reported by Samuel *et al* [33] closely resembles the EL spectrum of the orange-red devices, while that reported by Rothberg and co-workers is similar to that of a yellow device. According to the above discussion, the most logical explanation for this discrepancy is that the spectrum of Samuel *et al* was dominated by the Ex-I species, while that reported by Rothberg and co-workers contained more emission from the single chain exciton. Therefore it is not surprising that the Ex-I dominated spectrum decays significantly slower than the typical single chain exciton decay (300 ps) observed in dilute solutions and in films [36, 37].

13.4.1.2 MEH-PPV processed using non-aromatic solvents. In non-aromatic solvents such as THF and $CHCl_3$, the spin speed dependence of the emission spectrum is more complicated. At the lower end of the CLA, the observed spin speed effect is similar to that observed for aromatic solvents and cyclohexanone. That is, higher spin speeds result in stronger yellow emission, and lower speeds result in stronger red emission. This effect is demonstrated in figure 13.21 using a 0.4% polymer solution in THF (CLA $\approx$ 0.3–0.7%, in THF). Interestingly, the spin speed effect is reversed at the higher end of the

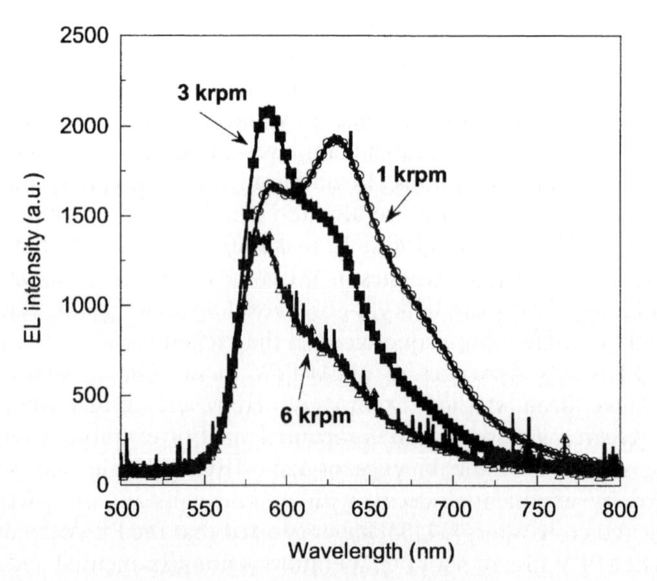

Figure 13.21. The spin-speed dependent EL spectra (normalized) of devices spun from a 0.4% MEH-PPV solution (the lower end of CLA) in THF [9].

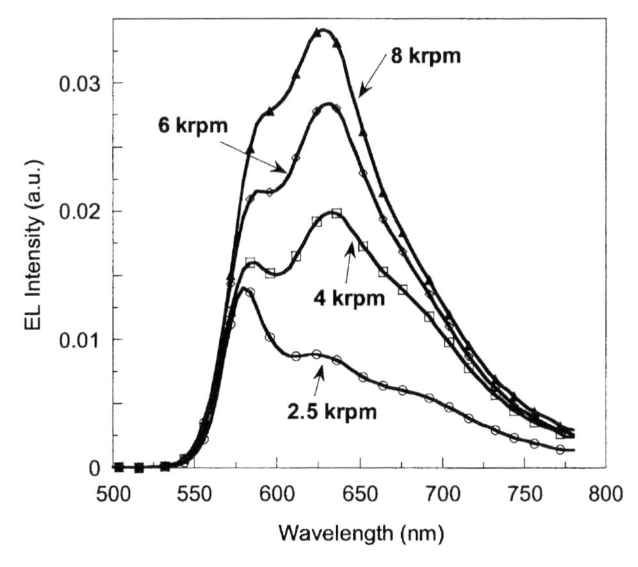

Figure 13.22. The spin-speed dependent EL spectra (normalized) of MEH-PPV films observed at the higher end of CLA (0.7 wt%) in THF. The film thicknesses are 1400 Å (2500 rpm), 1200 Å (4000 rpm), 1000 Å (6000 rpm) and 800 Å (8000 rpm) [9].

CLA. For example, spin-coating a 0.7 wt% MEH-PPV solution (THF) at high spin speeds (e.g. 6000–8000 rpm) results in orange-red devices while at lower speeds (e.g. ≤2500 rpm) results in yellow dominated devices (figure 13.22). The absorption λ_{max} of the film is almost unaffected by the spin-speed at this concentration (0.7 wt%, THF). A similar effect is also observed when the polymer is spun from $CHCl_3$. It has been noticed by Heeger and co-workers [38] that the polymer films spun from THF and *p*-xylene have different morphology. This 'strange' behaviour observed in THF and $CHCl_3$ has added more interesting aspects to this solvent dependence of polymer morphology.

It is thus obvious that the Ar-type aggregation favours the formation of the Ex-I species while the non-Ar-type (at the high end of CLA) inhibits the Ex-I formation. This phenomenon can be rationalized using the structure shown in figure 13.6. With all the bulky side chains dangling around the main chain in the non-Ar-type aggregation, the formation of the Ex-I interchain species is inhibited. At higher concentrations (e.g. 0.7–0.8 wt%, THF or $CHCl_3$) and lower spin speeds (e.g. ≤2500 rpm), this non-Ar-type conformation (figure 13.5) is 'memorized' by the polymer film. Since the π–π stacking of the polymer backbones is hindered, the single chain exciton emission dominates the spectrum (figure 13.22). When higher spin speeds are used the polymer coils/aggregates are forced to open to a certain extent, maybe just enough to allow the cross insertion of other polymer chains, thereby resulting in more feasible Ex-I formation. The direct evidence indicating that the polymer coils are not completely open is from the ultraviolet–visible absorption spectra measurement. Although a substantial change is observed in the emission spectrum, there is no noticeable spectral change observed in the absorption spectra, nor in the contact angle measurement. At the lower end of the CLA, however, the MEH-PPV coils are completely (or nearly completely) torn apart by the high spin speeds to give mostly the open chain polymer molecules. Since the Ex-I formation is now impossible, the emission spectrum is again dominated by the single chain exciton.

It should be noted that due to the complexity of the polymer system, the emission spectrum of a device might include many species with similar emission wavelengths. Thus, the actual emission spectrum observed is the overlap of many different species. In fact, Rothberg and co-workers also observed an interchain species at longer wavelength ($\lambda_{max} \approx 700$ nm). They have suggested that the formation of this excimer species quenches the single chain exciton fluorescence. In the following discussion, we address this species as the Ex-II excimer species. Therefore, strictly speaking the Ex-I (as well as the Ex-II) species should be considered as a series of closely resembled excimer species rather than a single species. Since the Ex-II has much longer emission wavelength and longer fluorescence lifetime (820 ps) [33] than that of the Ex-I species, one could logically suggest that the Ex-I is an excimer species with only limited π–π overlapping (i.e. cross overlapping of the main polymer chains) while the Ex-II is probably receiving a better π–π overlapping.

13.3.4.2 The effects of thermal annealing

The effect of thermal annealing on the film morphology has previously been studied using FT-IR and ultraviolet–visible spectroscopy. It is thus expected that the effect of thermal annealing should also be reflected in the emission spectra. Figure 13.23 demonstrated the spectral changes causing by thermal annealing. Shown in figure 13.23(a) are the PL spectra (normalized) of a series of MEH-PPV films spin-coated under identical conditions but annealed at different temperatures; in figure 13.23(b), however, the spectra are from the same MEH-PPV film after being annealed at different temperature (from room temperature to 140 °C). It can be easily seen that both experiments show the same effect, i.e. higher annealing temperature resulting in red-shift of the emission spectrum. According to the previous assignment of the emission peaks, it can be concluded that thermal annealing at elevated temperature results in pronounced morphological changes, which are favourable for excimer formation. When the MEH-PPV was heated from ambient temperature to 140 °C, the 575 nm yellowish peak was progressively reduced in intensity and at the same time two shoulders at ~630 nm and ~670 nm, corresponding to the Ex-I and Ex-II species, respectively, become more

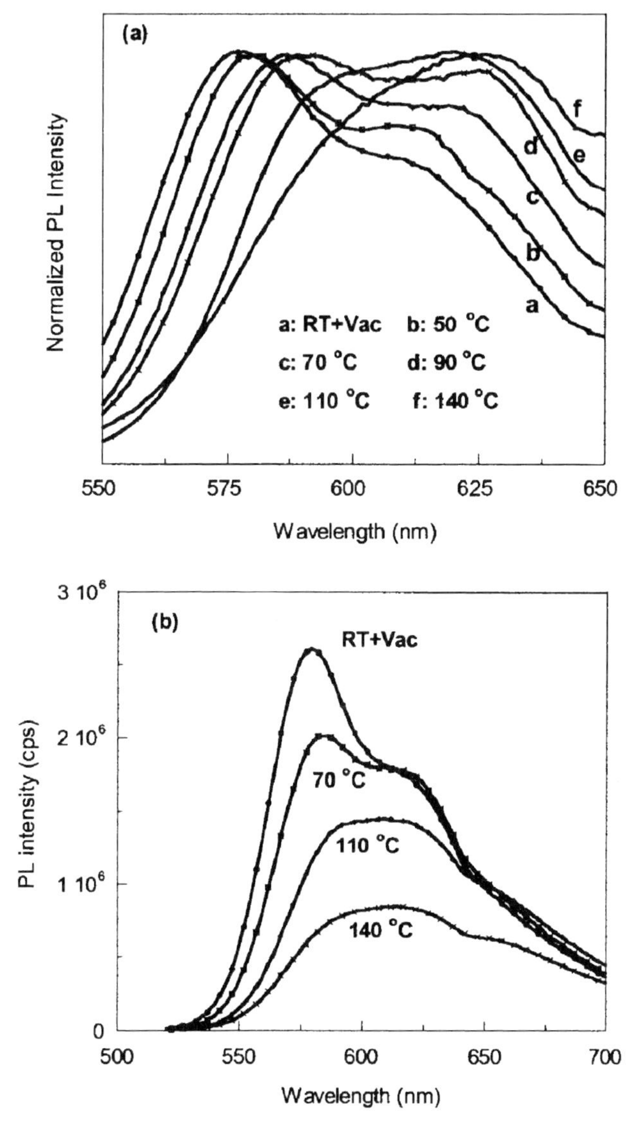

Figure 13.23. Annealing temperature dependence of PL spectra of MEH-PPV films: (a) normalized PL spectra for MEH-PPV films annealed under different temperatures; (b) the PL spectra of a MEH-PPV film recorded after being annealed at different temperatures. The time for annealing is 2 h at each temperature [17].

pronounced (figure 13.23(b)). The PL quantum efficiency also decreases with the annealing temperature. This is consistent with the formation of the Ex-II species, which has significantly quenched the single chain exciton emissions [34].

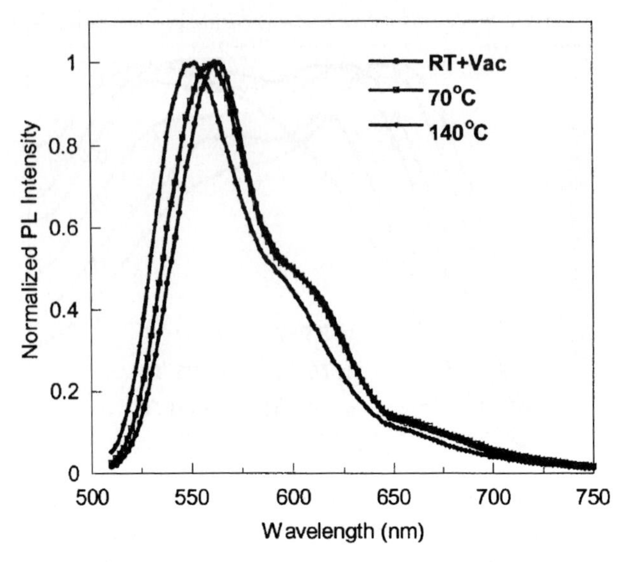

Figure 13.24. Normalized PL spectra of BCHA-PPV films annealed under different temperatures [17].

As discussed earlier, the morphological change in a spin-coated polymer film is expected to be smaller when annealed below T_g and becomes more pronounced when annealed at or above T_g. The spectra in figure 13.23(a) have revealed exactly such a trend. It can be seen that annealing at 50 °C only results in a minor change in the PL spectrum and much more pronounced changes are observed at 70 °C or higher.

If the above spectral changes are indeed due to the formation of excimer, it is expected that these spectral changes should be suppressed when the interaction are hindered. It is known that poly(2,5-bis(cholestanoxy)-1,4-phenylene vinylene (BCHA-PPV), an analogue of MEH-PPV with two bulkier side chains attached to the phenyl groups, can hardly form excimer species. Thus, the PL spectrum of the BCHA-PPV film is expected to be less sensitive to annealing temperature. As can be seen in figure 13.24, annealing up to 140 °C results in only slight blue shift of the spectrum without decreasing the main peak intensity nor growing in of new peaks at longer wavelengths.

13.3.5 The quantum efficiency

13.3.5.1. Solvent and spin speed

There have been extensive discussions in recent years regarding the role of aggregation and excimer formation in conjugated polymers. It is generally believed that aggregation quenches the excited state [34, 39]. The mechanism

Table 13.1. Solvent and spin-speed-dependent quantum efficiency and emission colour [9].

Solvent	Speed (rpm)	Thickness (Å)	QE (%)	CIE		λ_{max} (nm)
				X	*Y*	
DCB	2500	860	3.0	0.62	0.38	632
	4000	720	2.4	0.59	0.40	588
	6000	630	2.1	0.57	0.43	584
	8000	590	2.0	0.56	0.43	580
THF	2500	1150	2.1	0.59	0.41	580
	4000	970	2.2	0.61	0.39	630
	6000	850	2.8	0.61	0.39	632
	8000	780	3.8	0.62	0.38	632

for this aggregation quenching has been attributed to interchain interactions [34, 40–42]. As a result, synthetic chemists have greatly dedicated their intelligence to chemical engineering of the polymers at the molecular level to suppress the interchain interactions and to achieve better quantum efficiencies [43–48]. However, this traditional belief has recently been questioned by Shi *et al* [9]. They suggested that in certain circumstance proper aggregation of the polymer chains could actually enhance the device quantum efficiency (QE). They also indicated that the QE of a device has a strong correlation with the emission spectrum and the polymer morphology rather than the thickness of the polymer film. They observed that the orange-red devices always give higher QE (30–80% higher) than the yellow devices (table 13.1). This observation is reproducible in all the solvents that have been studied, and it is essentially independent of the thickness (within a range of 500–1500 Å) of the MEH-PPV films. Though the exact QE values for the single chain exciton, the Ex-I excimer, and the Ex-II excimer are not known at this time, these data clearly indicated that the formation of the Ex-I species actually enhances the quantum efficiency of the device. Based on the above discussion, one of the possible mechanisms for this QE enhancement via the formation of the Ex-I species is that it reduces the amount of the Ex-II species formed.

The formation of the excimer species has traditionally been associated with aggregation [39] while the possibility for the un-aggregated chains to form excimers has rarely been addressed. In dilute solutions, excimer formation in the un-aggregated polymer chains is unlikely, since the small diffusion rates of these macromolecules could not compete with the decay of the excited state. In the solid state, however, the polymer chains are closely packed next to each other. Thus, the excimer species can in theory be formed easily. Rothberg and co-workers [34, 39] have observed a weakly emissive excimer species at $\lambda_{max} \approx 700\,\text{nm}$, corresponding to the Ex-II species, in the PL spectrum of MEH-PPV films. This emission is also observed in the EL spectra of MEH-PPV films (figure 13.22). It is consistently observed that the Ex-II emission is always accompanied by the appearance of a reasonably strong yellow emission ($\lambda_{max} \approx 575$ nm, refer to figure 13.22). This observation may imply that there is a connection between the Ex-II species and the yellow emissive species. Rothberg and co-workers [34, 39] have shown by time resolved fluorescence experiments that the exciton fluorescence is quenched by the rapid formation of the much less emissive Ex-II species. If this is also the case in the EL emission, the correlation between the QE and the colour of the device can be rationalized by the relative rates for the formation of the Ex-I, the single chain exciton, and the Ex-II species.

13.3.5.2 Thermal annealing

It has been discussed in sections 13.3.3 and 13.3.4 that annealing MEH-PPV films at an elevated temperature could result in two effects: lower device turn-on voltage ($V_{L\text{-ON}}$) due to a better POM interface and a change in QE due to formation of the Ex-I and the Ex-II species. As seen previously in figure 13.18, heat treatment at elevated temperature, e.g. 140 °C, can lower the $V_{L\text{-ON}}$. However, at this temperature the QE of the device is significantly quenched due to formation of the Ex-II species (figure 13.23). Thus, the optimal annealing temperature should be adjusted to balance the requirements for low $V_{L\text{-ON}}$ and high QE. Shown in figure 13.25 are two examples of thermal-treatment dependence of devices' performance. For example, thermal-annealing treatments prior to and post cathode deposition significantly enhance the external quantum efficiency of MEH-PPV based devices compared with the untreated control (figure 13.25(a) and ([49], figure 3)). In addition, it has shown that for a polyfluorene-based green polymer, proper thermal annealing treatments prior to the cathode deposition dramatically change devices' initial performance (figure 13.25(b) and (c) and ([50], figures 2(c) and 3)) as well as operation stability [50].

Although annealing below or close to the T_g gives the best efficiency (figure 13.26(a)), a higher annealing temperature is required to achieve the minimum $V_{L\text{-ON}}$ value. It is obvious that these two requirements could not be satisfied by using only one annealing temperature. This problem is solved by introducing the 'double layer' approach. In this approach, there are logically two layers of MEH-PPV film in the device: one annealed at a high temperature to achieve the lowest $V_{L\text{-ON}}$ possible and the other annealed at a lower temperature to achieve a better efficiency. This is achieved by first spin-coating a very thin layer (hole injection layer) of MEH-PPV on top of the PEDOT/ITO substrate and annealing at a high temperature, such as 140 °C. Then a second layer of MEH-PPV (the emitting layer) was spin-coated on top of the first layer, which is annealed at a lower temperature, i.e. ~70 °C. Since the bottom thinner layer of MEH-PPV has a better

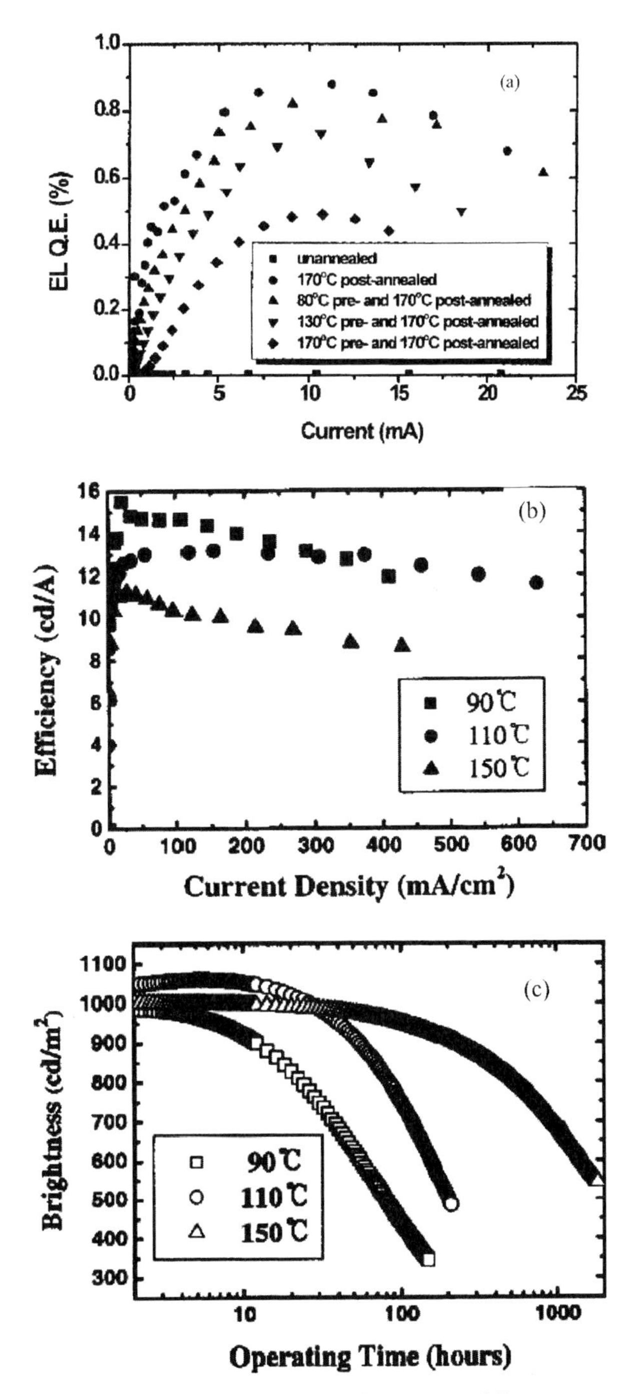

Figure 13.25. (a) ([49], figure 3); (b and c) ([50], figures 2c and 3).

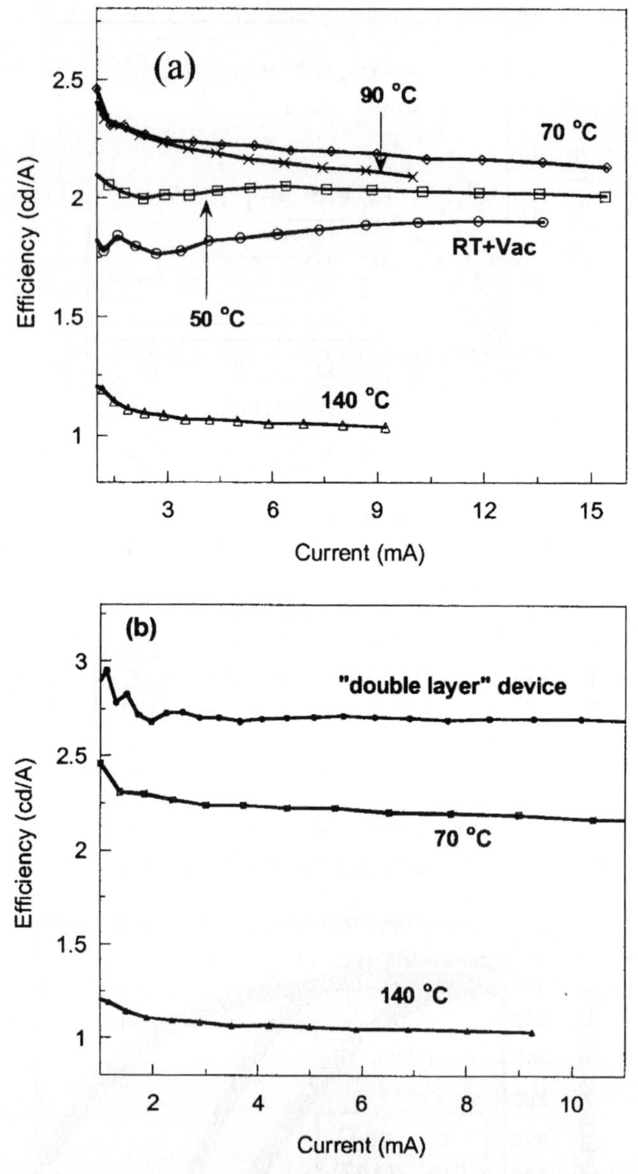

Figure 13.26. Efficiency versus current curves for (a) single layer devices and (b) a 'double layer' device annealed at different temperatures [17].

POM contact, it improves the hole-injection from the ITO/PEDOT anode. In this 'double layer' device configuration, since the first layer is much thinner than the emissive layer, the recombination zone is mostly located within the second MEH-PPV layer. Thus, the device could achieve the lowest $V_{L\text{-ON}}$ and the highest efficiency at the same time. Figure 13.26(b) showed that there is a ~20% efficiency improvement in 'double layer' device if compared with a regular device with a single layer of MEH-PPV annealed at only one temperature (70 °C).

13.4 Reduction of the interchain species using solid solutions

The above discussion has already shown that the formation of interchain species can broaden the emission spectrum and sometimes reduce the quantum efficiency of the device. Thus, the formation of interchain species in general should be avoided. This can be achieved by either chemically engineering the molecular structure, such by as introducing bulky substituents to the polymer main chain, or physically separating the polymer chains from each other to prevent the formation of the interchain species. For example, the BCHA-PPV is much less subjected to the formation of interchain species when compared with its analogues bearing smaller side chain groups such as MEH-PPV and BEH-PPV [43]. However, these bulky side groups also lead to poor charge injection and transport capability, thus offsetting the overall EL efficiency. The typical efficiency of BCHA based PLEDs reported is only 0.5 cd/A [51]. The nano-composite technology [52, 53] and the dilution effect using a solid-state solution [54–56] are also used to prevent the formation of the interchain species by physically isolating the light-emitting polymer (LEP) molecules. Since the molecular engineering of the polymer molecules and the use of nano-composite technology are beyond the scope of this chapter, only the dilution effect using polymer blends will be addressed in the following sections. In this approach, a physically 'inert' (does not form an exciplex with the LEP molecules) host polymer is used as the 'solvent' of the solid-state film, and the LEP molecules are uniformly distributed into the host matrix. Since the LEP molecules are physically isolated from each other by the host polymer molecules, the excimer formation is reduced. Therefore, the successful use of this approach relies on the selection of a proper host polymer, which will allow the uniform distribution of the LEP molecules (the guest molecules) into its matrix without forming exciplex species with the LEP molecules.

13.4.1 Inert spacer—polystyrene as the host

Since the formation exciplex species depends on the electronic structures, the steric hindrance, as well as the geometric conformations (or the morphology) of the two species, it is usually difficult to predict whether a given pair of conjugated polymers could form an exciplex or not. However, if the basic energy requirement for the exciplex formation, i.e. the two species have similar HOMO and LUMO energy levels, is not met, it is usually true that

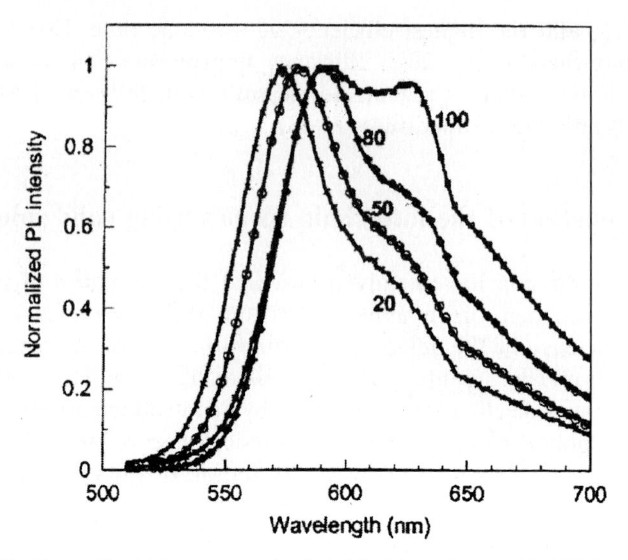

Figure 13.27. Normalized PL spectra of solid thin films of pure MEH-PPV and MEH-PPV/PS solid solutions of different MEH-PPV concentrations. The numbers shown on the graph are the wt% of the MEH-PPV contents in the film [55].

the two species are not likely to create exciplex species. For example, it will be hard to imagine that a conjugated polymer such as MEH-PPV could form an exciplex with a non-conjugated polymer such as polystyrene (PS) since their HOMO and LUMO energy levels are so different. Yan *et al* [54] studied the photophysics of some solid-state films consisting of dilute MEH-PPV dispersed in the PS matrix. It is shown that the photophysics of these dilute MEH-PPV films are very similar to those observed in dilute MEH-PPV solutions in chlorobenzene. They observed simple exponential decay dynamics for these polymer blend films. This suggests that the interchain interactions are indeed reduced by the PS host molecules.

Yang and co-workers [55] have shown by steady-state fluorescence spectroscopy and the EL spectra of PLED devices that the suppressed interchain interactions are not only observable in dilute MEH-PPV films but also observable in relatively high MEH-PPV concentrations. Shown in figure 13.27 are the PL spectra of a series of MEH-PPV/PS films of different concentrations. It can be seen from these spectra that as little as 20% of PS (80% of MEH-PPV) is sufficient to reduce the longer wavelength emissions, i.e. the Ex-I and Ex-II species, significantly. At the meantime, the full width at half maximum (FWHM) of the emission spectra changed from 110 nm for a 100% MEH-PPV film to ~75 nm for 80/20 MEH-PPV/PS film. Further increases of PS concentration resulted in only small changes in the shape and the FWHM of the spectrum (figure 13.27). As a result of reduced interchain species, using the PS host molecules increases the efficiency of the device. It

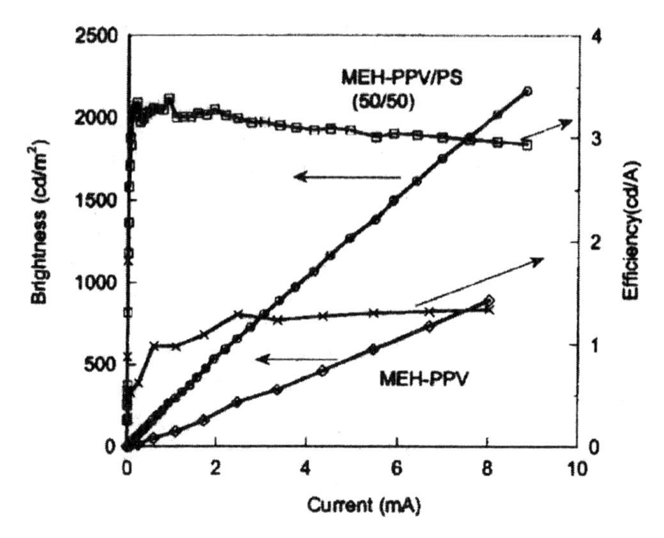

Figure 13.28. Brightness–current–efficiency plots of PLED devices fabricated from a 100% MEH-PPV film and a 50/50 MEH-PPV/PS film [55].

can be seen in figure 13.28 that the efficiency of a PLED device made from a 50/50 MEH-PPV/PS film is almost 2.5 folds higher if compared with a device using 100% MEH-PPV. The increased thermal stability of the polymer blend is another benefit for this approach. Shown in figure 13.29 is a comparison of

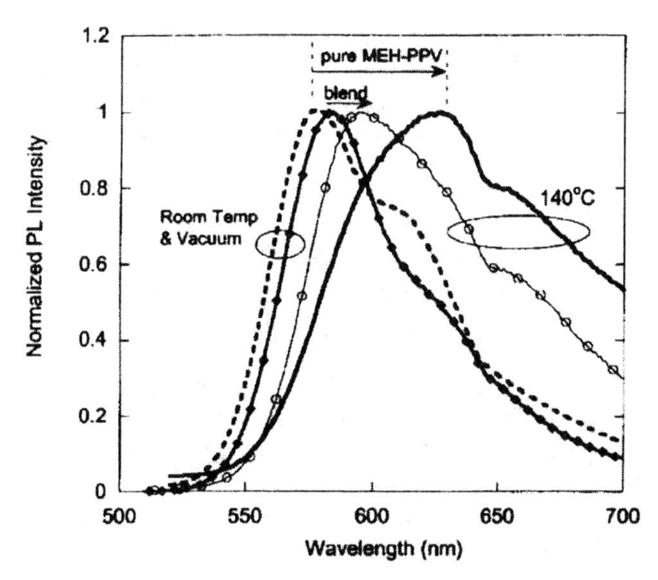

Figure 13.29. Comparison of PL spectra of 100% MEH-PPV film versus 50/50 MEH-PPV/PS blended film before and after thermally annealed at 140 °C [55].

the PL spectral changes of a pure MEH-PPV film versus a 50/50 MEH-PPV/PS blended film before and after thermally annealed at 140 °C. Although the annealing temperature is much higher than the T_g of both materials, the spectral shift for the blended film is much smaller than the pure MEH-PPV film, consisting with less morphological changes in the blended film.

13.4.2 Energy/charge transfer poly(9,9-dioctyfluorene) as the host

Although improved performance has been achieved in MEH-PPV/PS, using higher PS concentration is not practical due to its nature of insulator. In addition to the requirements mentioned previously, an ideal host material should also be able to conduct an electrical current. Such materials are generally conjugated polymers. So far most blend systems studied mainly consist of polymer hosts with low-concentration small molecules as the guests. Only a small number of polymer–polymer blend systems have been studied [57–60].

When a conjugated polymer is used as the host, it is generally required that the HOMO/LUMO band gap of the host should be larger than that of the emitter. In addition, the host should also facilitate efficient energy transfer or charge transfer from the host molecules to the emitter.

13.4.2.1 The MEH-PPV/PF system

It is shown by Yang and co-workers [61] that poly(9,9-dioctylfluorene) (PF), a blue emitter, is an ideal host polymer for MEH-PPV. Figure 13.30 shows PL and EL spectra of the pure PF film, pure MEH-PPV film, and the MEH-PPV/PF blend films of different concentrations. The emission spectra of the MEH-PPV/PF blends show strong concentration dependence. The pure MEH-PPV film has a reddish emission with a main peak at 630 nm, which has been previously assigned to an interchain species (Ex-I). As the MEH-PPV concentration decreases, the yellow emission peak (~575 nm) is progressively growing in with the accompanying reduction of the 630 nm reddish emission peak. The FWHM of the emission spectrum also reduces as the MEH-PPV concentrations in the film decreases (figure 13.30). Maximum single chain emission (575 nm) is obtained at the device containing 4% MEH-PPV. The FWHM value of the device also reaches its lowest value (FWHM = 40 nm). This is approximately one-third of that (FWHM = 110 nm) of the device containing 100% MEH-PPV. Furthermore, the PL spectra of the MEH-PPV components in the blends are independent of excitation wavelength. Identical emission spectra are obtained when the films are excited at 380 nm (where both PF and MEH-PPV are excited) and 480 nm (where only the MEH-PPV component is excited), which suggest that the energy transfer from the PF molecules to the MEH-PPV molecules is very efficient. The PL spectrum of this 4%MEH-PPV/96%PF blend closely

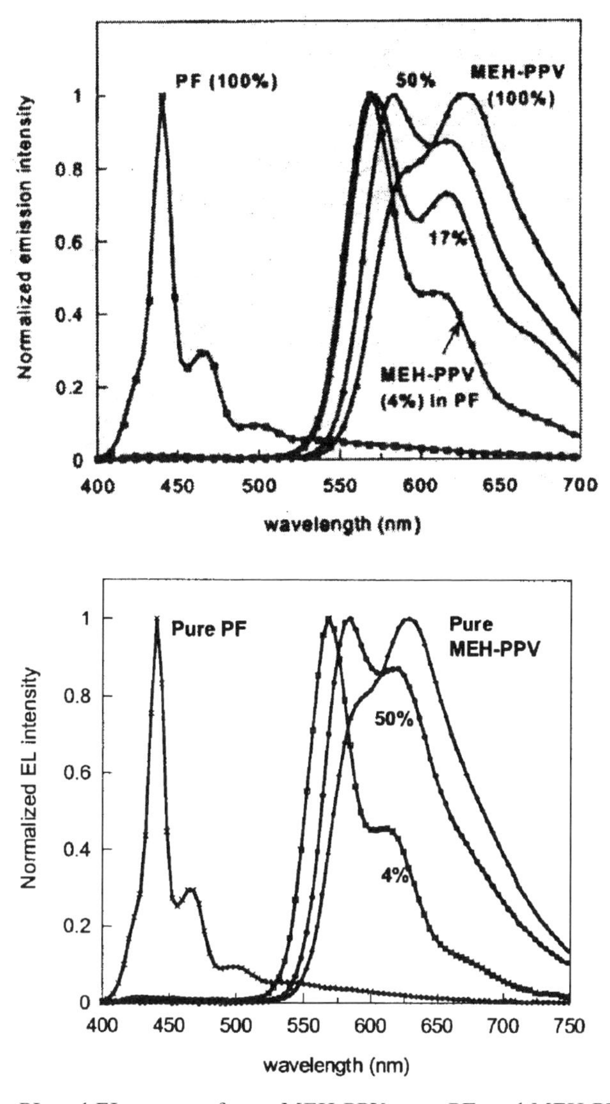

Figure 13.30. PL and EL spectra of pure MEH-PPV, pure PF, and MEH-PPV/PF solid solutions of different MEH-PPV concentration. The percentages shown on the graphs correspond to the weight concentrations of MEH-PPV in a MEH-PPV/PF blend [61].

resembles the photoluminescence spectrum of a diluted MEH-PPV solution in a regular organic solvent. The fact that no new emission peaks can be found in the PL and EL spectra of the blended films suggests that either no exciplex species is formed or the exciplex species is non-emissive. The possibility of forming ground state complexes between MEH-PPV and PF molecules are also ruled out, since the ultraviolet–visible absorption spectra of the blended

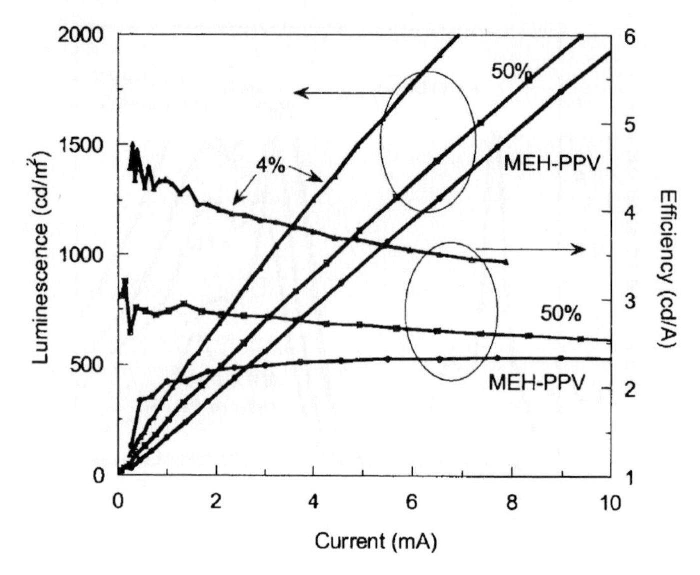

Figure 13.31. The luminescence–current–efficiency curves of devices based on the pure MEH-PPV, and 4% and 50% MEH-PPV/PF solid solutions [61].

films can be simply reproduced by overlapping the weighted spectrum of the pure MEH-PPV and that of the pure PF.

When the MEH-PPV content of the film is below 4%, the emission spectrum consists of a shorter wavelength portion due to the PF fluorescence and a longer wavelength portion due to the MEH-PPV emission. The shape of the MEH-PPV portion of the spectrum is not affected by the MEH-PPV concentrations. When the MEH-PPV content of the film is $\geq$4%, no emission from PF can be detected. This suggests that 4% is the minimum MEH-PPV concentration required to guarantee nearly 100% energy transfer from the host molecules to the MEH-PPV molecules.

The EL spectra of PLED devices consist of pure MEH-PPV, pure PF, and their blends (figure 13.30) also show similar trends as seen in their PL spectra. This indicates that the same light-emitting species are involved in both the photo-excitation and electro-excitation of the MEH-PPV molecules. It can be easily seen from the luminescence–current–efficiency curves shown in figure 13.31 that the device efficiency becomes higher at lower MEH-PPV contents in the film. The highest device efficiency is achieved at 4%MEH-PPV/96%PF blend, which reaches 3.9 Cd/A at a current of 3 mA (device area = 12 mm^2), which is ~70% higher than that of the pure MEH-PPV device (2.2 Cd/A).

The CIE coordinates also change with the MEH-PPV content of the film (figure 13.32). Figure 13.32 shows the route in which the device colour drifting with the MEH-PPV percentage of the film. As can be seen from

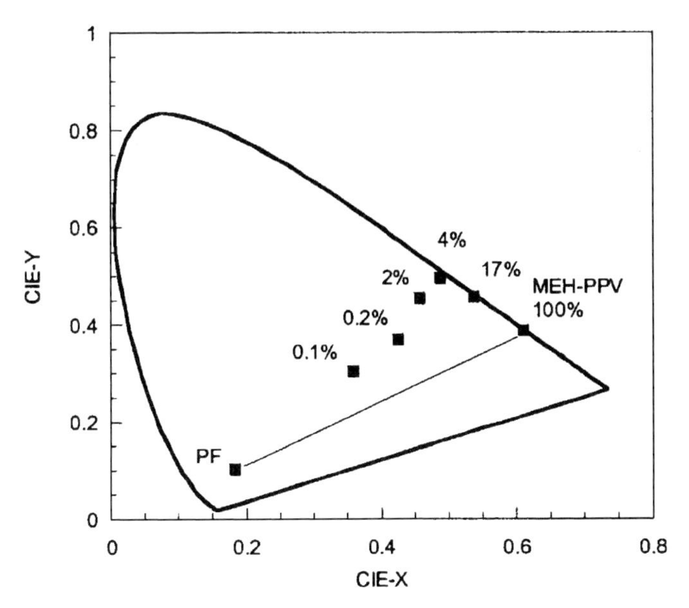

Figure 13.32. CIE coordinators of the EL spectra vary with the MEH-PPV contents in PLED devices made from pure PF, MEH-PPV/PF blends, and pure MEH-PPV.

this graph, the CIE coordinates of the emission spectra of the MEH-PPV/PF blends do not follow the straight connecting-line between the pure PF (blue) and the pure MEH-PPV (red). Instead, they first detoured upwards into the white region (i.e. $\mathrm{CIE}_x = 0.3578$, $\mathrm{CIE}_y = 0.3045$ at 0.2% MEH-PPV), and end at the yellow region (4% MEH-PPV content). When the MEH-PPV content is $>4\%$, they go downwards towards the pure MEH-PPV with increasing MEH-PPV content. This behaviour can be easily interpreted by the effect due to reduction of the interchain species.

As shown in figure 13.30, the PL spectrum of the 17%MEH-PPV film showed strong emission from the Ex-I species (~630 nm). A study by AFM on this film also indicates that aggregation of the MEH-PPV molecules does occur at this concentration. The AFM phase images of polymer blends reflect differences in the properties of their constituents, thereby allowing surface compositional mapping by AFM in polymer blends [62]. Shown in figure 13.33 are the AFM phase images of the MEH-PPV/PF blend films containing 17% and 4% MEH-PPV. The contrast covers phase angle variation in the 50° range. The phase separation can be easily seen in the film having high (17%) MEH-PPV content. The minor phase, i.e. the MEH-PPV component, is dispersed within the PF matrix as aggregates with a feature size about 30 nm in width. In contrast, a uniformly homogenous phase image is observed for film containing 4% MEH-PPV, which

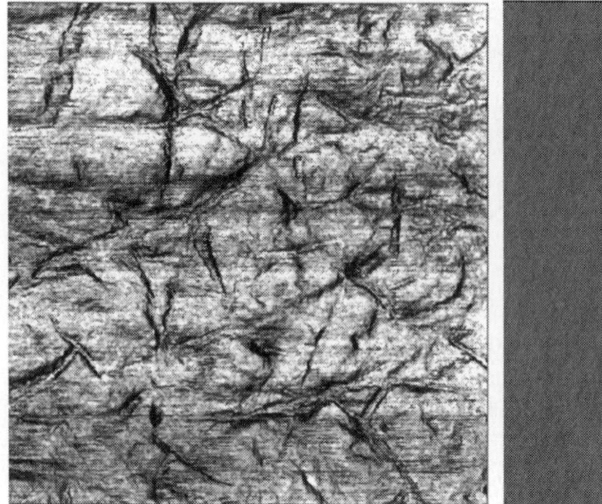

Figure 13.33. AFM phase images of a MEH-PPV/PF film containing 17% (left) and 4% (right) of MEH-PPV. Both scans are 3 μm × 3 μm. The contrast covers phase angle variation in the 50° range [61].

suggest that the MEH-PPV molecules are well 'dissolved' by the PF host in this case.

13.4.2.2 The BCHA-PPV/PF system

Although PF can facilitate efficient energy transfer to the MEH-PPV molecules, the energy transfer become much less efficient in the BCHA-PPV/PF system [56]. Figure 13.34 shows PL and EL spectra of thin films consist of pure BCHA-PPV, pure PF and their blends of different compositions. As can be seen from these spectra when the films are photo-excited at 380 nm, emission from both PF and BCHA-PPV components are observable. The EL spectra also showed emission from both the PF and the BCHA-PPV species, which indicates less effective energy transfer from PF to BCHA-PPV if compared with MEH-PPV. Since the energy levels of the BCHA-PPV ($E_{\mathrm{HOMO}} = 5.6$ eV and $E_{\mathrm{LUMO}} = 3.1$ eV) are within that of the PF ($E_{\mathrm{HOMO}} = 5.9$ eV and $E_{\mathrm{LUMO}} = 2.3$ eV) [63], the insufficient energy transfer is most likely due to the steric effect caused by the bulky side groups of the polymer.

Although no benefit is observable from the emission spectra, the most important improvement for using a BCDA-PPV/PF blend is perhaps in the improved hole injection. As already discussed, PLED devices made from pure BCHA-PPV have low efficiency due to poor hole injection and poorer electrical conductivity. The efficiency and the device operating voltage were significantly improved by using this polymer blend. As can be seen in figure 13.35, the device turn-on voltage (for a luminance threshold

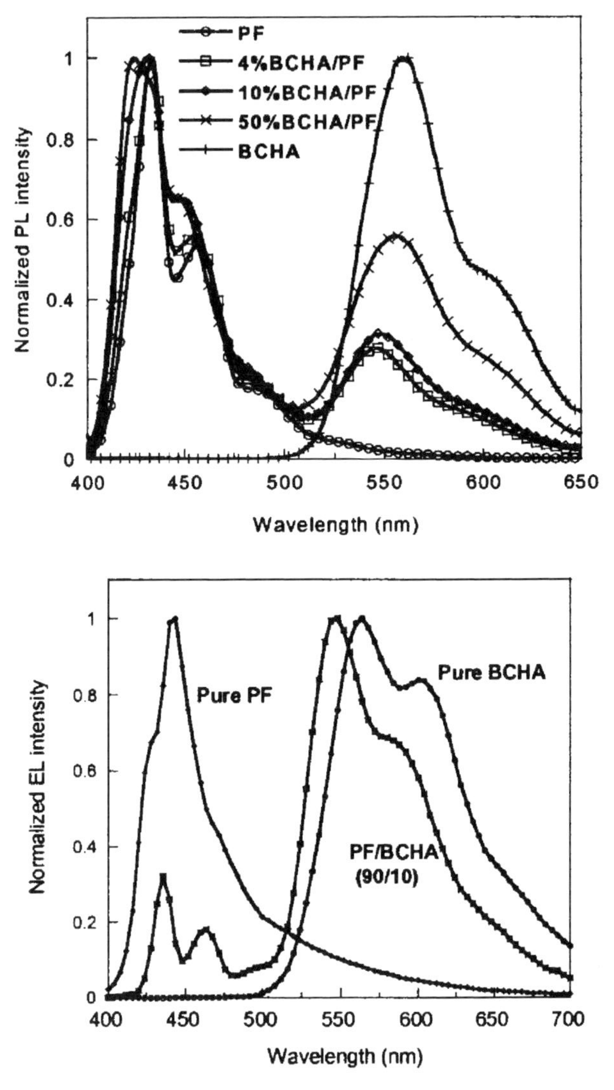

Figure 13.34. Normalized PL and EL spectra of pure BCHA-PPV, pure PF and their blends [56].

of 2×10^{-2} Cd/m^2) of the BCHA-PPV (10%)/PF device is only 3.2 V, which is much smaller than that of the devices of pure BCHA-PPV (5.0 V) and of pure PF device (4.0 V). The EL efficiency of this blended device is also much higher (3.0 Cd/A) than that of a pure BCHA-PPV device (0.5 Cd/A).

In summary, this chapter presents systematic studies on the polymer morphology formation and its influence on the electronic and photonic properties of conjugated polymer. One of the major advantages of conjugated

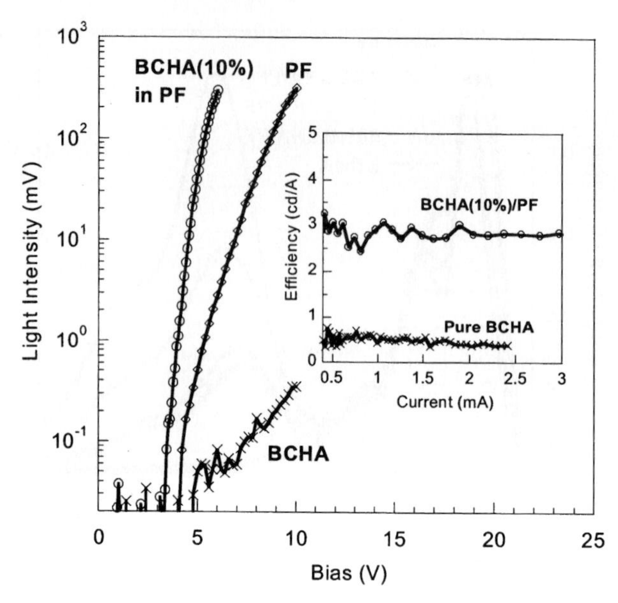

Figure 13.35. Device performance of PLED devices consist of pure BCHA-PPV, pure PF and BCHA (10%)/PF blend devices. Main figure shows light intensity–voltage curves. The inset shows efficiency–current curves of pure BCHA-PPV and BCHA (10%)/PF devices [56].

polymers over their organic small molecular as well as inorganic counterparts is their solution (wet) processing capability, which makes low-cost manufacturing possible. However, the morphology of polymers, and consequently device performance, is subject to change of detailed process parameters, as well as device operating conditions. Understanding this subject as well as developing suitable controls of film morphology are of both academic and industrial interest.

References

[1] Burroughes J H, Bradley D D C, Brown A R, Marks R N, Mackay K, Friend R H, Burns P L and Holmes A B 1990 *Nature* **347** 539
[2] Forrest S R 2000 *IEEE J. Selected Topics in Quantum Electronics* **6** 1072
[3] Duggal A R, Shiang J J, Heller C M and Foust D F 2002 *Appl. Phys. Lett.* **80** 3470
[4] Adachi C, Baldo M A, Thompson M E and Forrest S R 2001 *J. Appl. Phys.* **90** 5048
[5] Yang C Y, Hide F, Diaz-Garcia M A, Heeger A J and Cao Y 1998 *Polymer* **39** 2299
[6] Nguyen T Q, Doan V and Schwartz B J 1999 *J. Chem. Phys.* **110** 4068
[7] Nguyen T Q, Martini I, Liu J and Schwartz B J 2000 *J. Phys. Chem. B* **104** 237
[8] Simha R and Ultrachi L 1967 *J. Polymer Sci. A* **2**(5) 853
[9] Shi Y, Liu J and Yang Y 2000 *J. Appl. Phys.* **87** 4254

[10] Christensen J J, Hanks R W and Izatt R M 1982 *Handbook of Heats of Mixing* (New York: Wiley)
[11] Zheng M, Bai F and Zhu D 1998 *J. Photochem. Photobiol. A* **116** 143
[12] Liu J, Guo T F, Shi Y and Yang Y 2001 *J. Appl. Phys.* **89** 3668
[13] McBranch D, Campbell I H and Smith D L 1995 *Appl. Phys. Lett.* **66**(10) 1175
[14] Tammer M and Monkman A P 2002 *Adv. Mater.* **14** 210
[15] Keddie J K, Jones R A L and Cory R A 1994 *Europhys. Lett.* **27** 59
[16] DeMaggio G B, Frieze W E, Gidley D W, Zhu M, Hristov H A and Yee A F 1997 *Phys. Rev. Lett.* **78** 1524
[17] Liu J, Guo T F and Yang Y 2002 *J. Appl. Phys.* **91** 1595
[18] Liu J, Shi Y and Yang Y 2000 *J. Appl. Phys.* **88** 605
[19] Salaneck W, Logdlund M, Birgersson J, Barta P, Lazzaroni R and Bredas J 1997 *Synth. Met.* **85** 1219
[20] Hung L S and Tang C W 1999 *Appl. Phys. Lett.* **74** 3209
[21] Bharathan J M and Yang Y 1998 *J. Appl. Phys.* **84** 3207
[22] Malliaras G G, Salem J R, Brock P J and Scott J C 1998 *Phys. Rev. B* **58** R13411
[23] Roman L S, Berggren M and Inganas O 1999 *Appl. Phys. Lett.* **75** 3557
[24] Malliaras G G and Scott J C 1998 *J. Appl. Phys.* **83** 5399
[25] Yu G, Zhang C and Heeger A J 1994 *Appl. Phys. Lett.* **64** 1540
[26] Campbell I H, Hagler T W, Smith D L and Ferraris J P 1996 *Phys. Rev. Lett.* **76** 1900
[27] Sinha S and Monkman A P 2003 *J. Appl. Phys.* **93** 5691
[28] Wittmann H F, Grüner J, Friend R H, Spencer G W C, Moratti S C and Holmes A B 1995 *Adv. Mater.* **7** 541
[29] Dodabalapur A, Rothberg L J and Miller T M 1994 *Appl. Phys. Lett.* **65**(18) 2308
[30] So S K and Choi W K 1999 *Appl. Phys. Lett.* **74** 1939
[31] Huser T and Yan M 2000 The Fourth International Topical Conference on Optical Probes of Conjugated Polymers and Photonic Crystals, abstract *P1-46*, 15–19 February 2000, Salt Lake City, Utah, USA
[32] Cornil J, dos Santos D A, Crispin X, Silbey R and Bredas J L 1998 *J. Am. Chem. Soc.* **120** 1289
[33] Samuel I D W, Rumbles G, Collison C J, Friend R H, Moratti S C and Holmes A B 1997 *Synth. Met.* **84** 497
[34] Jakubiak R, Collison C J, Wan W C, Rothberg L J and Hsieh B R 1999 *J. Phys. Chem. A* **103** 2394
[35] Jakubiak R, Rothberg L J, Wan W and Hsieh B R 1999 *Synth. Met.* **101** 230
[36] Smilowitz L, Hays A, Heeger A J, Wang G and Bowers J E 1993 *J. Chem. Phys.* **98**(8) 6504
[37] Samuel I D W, Crystall B, Rumbles G, Burn P L, Holmes A B and Friend R H 1993 *Chem. Phys. Lett.* **213** 472
[38] Diaz-Garcia M A, Hide F, Schwartz B J, Andersson M R, Pei Q and Heeger A J 1997 *Synth. Met.* **84** 455
[39] Yan M, Rothberg L J, Papadimitrakopoulos F, Galvin M E and Miller T M 1994 *Phys. Rev. Lett.* **73** 744
[40] Cao Y, Parker I D, Yu G, Zhang C and Heeger A J 1999 *Nature* **397** 414
[41] Conwell E M 1998 *Phys. Rev. B* **57** 14200
[42] Conwell E M, Perlstein J and Shaik S 1996 *Phys. Rev B* **54** R2308
[43] Gettinger C L, Heeger A J, Drake J M and Pine D J 1994 *J. Chem. Phys.* **101** 1673

[44] Hsieh B R, Yu Y, Forsythe E W, Schaaf G M and Feld W A 1998 *J. Am. Chem. Soc.* **120** 231
[45] Rothberg L J, Yan M, Son S, Galvin M E, Kwock E W, Miller T M, Katz H E, Haddon R C and Papadimitrakopoulos F 1996 *Synth. Met.* **78** 231
[46] Son S, Dodabalapur A, Lovinger A J and Galvin M E 1995 *Science* **269** 376
[47] Hu B and Karasz F E 1998 *Synth. Met.* **92** 157
[48] Sun B J, Miao Y-J, Bazan G C and Conwell E M 1996 *Chem. Phys. Lett.* **260** 186
[49] Lee T W and Park O O 2000 *Adv. Mater.* **12** 801
[50] Kim J, Lee J, Han C W, Lee N Y and Chung I J 2003 *Appl. Phys. Lett.* **82** 4238
[51] Cao Y, Yu G and Heeger, A J 1998 *Adv. Mater.* **10** 917
[52] Winkler B, Dai L and Mau A W-H 1999 *J. Mater. Sci. Lett.* **18** 1539
[53] Lee T-W, Park O, Yoon J and Kim J-J 2001 *Adv. Mater.* **13** 211
[54] Yan M, Rothberg L J, Kwock E W and Miller T M 1995 *Phys. Rev. Lett.* **75** 1992
[55] He G, Li Y, Liu J and Yang Y 2002 *Appl. Phys. Lett.* **80** 4247
[56] He G, Liu J, Li Y and Yang Y 2002 *Appl. Phys. Lett.* **80** 1891
[57] Yu G, Nishino H, Heeger A J, Chen T A and Rieke R D 1995 *Synth. Met.* **72** 249
[58] Gupta R, Stevenson M, Dogariu A, McGehee M D, Park J Y, Sradanov V, Heeger A J and Wang H 1998 *Appl. Phys. Lett.* **73** 3492
[59] List E J W, Holzer L, Tasch S, Leising G, Scherf U, Mullen K, Catellani M and Luzzati S 1999 *Solid State Commun.* **109** 455
[60] Kim Y C, Lee T W, Park O O and Cho H N 2001 *Adv. Mater.* **13** 646
[61] Liu J, Shi Y and Yang Y 2001 *Appl. Phys. Lett.* **79** 578
[62] Whangbo M-H, Magonov S N and Elings V 1997 *Surf. Sci.* **375** L385
[63] Janietz S, Bradley D D C, Grell M, Giebeler C, Inbasekaran M and Woo E P 1998 *Appl. Phys. Lett.* **73** 2453

Chapter 14

Liquid crystalline materials

***Stephen M Kelly*[1] *and Mary O'Neill*[2]**
[1]Department of Chemistry, University of Hull, Hull, UK
[2]Department of Physics, University of Hull, Hull, UK

14.1 Introduction

Several different classes of electroluminescent organic materials, broadly defined as small molecules, oligomers, polymers, elastomers and polymer networks, are being investigated for use in emissive electronic and photonic organic devices, such as organic light-emitting devices (OLEDs) and organic lasers. It is possible that the self-assembling properties and supramolecular structures of the equivalent liquid crystalline materials can be used to improve the performance or simplify the fabrication of such electro-optic devices using luminescent organic materials [1–3]. For example, nematic and smectic liquid crystals (LCs) with a calamitic, rod-like, molecular structure can exhibit high charge-carrier mobility, polarized emission and enhanced output-coupling in OLEDs; the anisotropic transport and high carrier mobilities of columnar liquid crystals with a disk-like molecular shape also make them promising candidates for charge-transport layers in OLEDs and an all-columnar multilayer OLED has been demonstrated; the photonic and emissive properties of chiral liquid crystals with a helical supramolecular structure may lead to their use as mirror-less lasers; the formation of anisotropic polymer networks by the photochemical crosslinking of polymerizable liquid crystals (reactive mesogens) gives rise to the added benefits of multilayer capability and lithographic photo-patternability. The unique aspects of light-emitting liquid crystals in diverse forms are the subject of this chapter. We begin with a general review of the material requirements for OLEDs and organic lasers. We then describe how liquid crystallinity offers the benefits of polarized light emission, enhanced light outcoupling and improved carrier transport. In the last part of this chapter the properties of liquid crystalline small molecules, oligomers, polymers and polymer networks of relevance to OLEDs and organic lasers are reviewed.

14.2 Material properties for OLEDs

Liquid crystals have been synthesized and studied for more than 115 years and over the past four decades in particular an enormous number of new LCs have been synthesized for many different types of liquid crystal displays (LCDs) used in watches, calculators, computer monitors etc. [1, 2]. In order to meet the common specifications of such LCDs these LCs are generally required to be chemically, photochemically and electrochemically stable as well as being transparent to visible light, highly insulating and easily re-oriented in an electric field. However, just as the LCD has slowly overtaken the cathode ray tube (CRT) as the predominant display of any kind, so OLEDs are beginning to challenge the continued dominance of LCDs as a flat panel display [2, 3]. OLEDs represent an emissive and potentially more efficient display technology for certain applications, such as in portable telephones and digital cameras, where brightness, wide viewing angle, low power consumption and the potential for video-rate addressing are essential [4, 5]. OLEDs require conjugated organic materials with appropriate molecular energies for electronic injection, high charge mobility and light emission. In addition amorphous glassy phases having high viscosity are required with high glass transition temperatures. The simplest type of OLED where light is generated by electrical excitation is shown schematically in figure 14.1. A thin film of an organic light-emitting material is sandwiched between a transparent anode and a metallic cathode.

The electrons and holes, which are injected into the lowest unoccupied molecular orbital (LUMO) and highest occupied molecular orbital (HOMO), respectively, drift through the organic film under the influence of the applied electric field (see figure 14.2). The coulombic attraction between an electron and hole at the same chromophore site results in the formation of an exciton (a bound electron–hole pair), which recombines radiatively to produce luminescence. More sophisticated and efficient OLEDs consist of multiple layers with the luminescent layer sandwiched between hole and electron injection/transport layers. Such efficient OLEDs require the matching of energy levels to minimize the barriers for carrier injection and to trap both electrons and holes in the luminescent region.

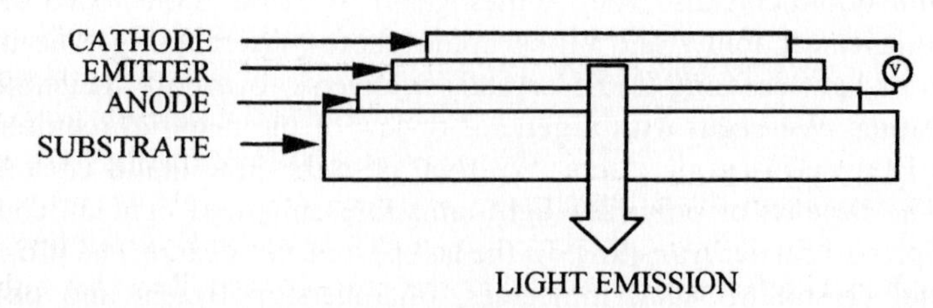

Figure 14.1. Schematic of a single-layer OLED.

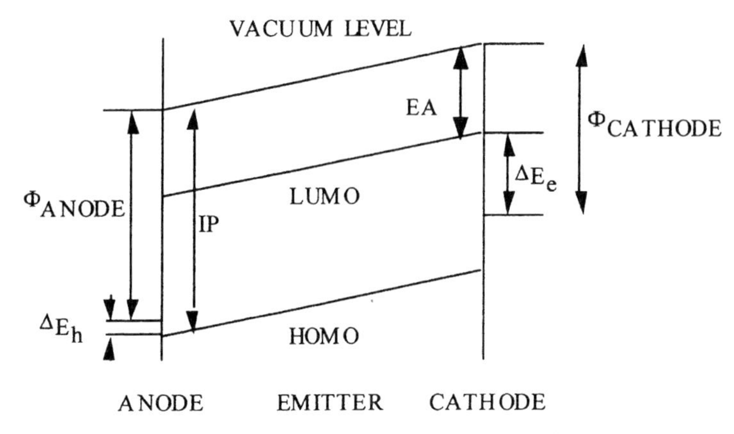

Figure 14.2. Energy level diagram of single-layer OLED under forward bias. $\Phi_{cathode}$ and Φ_{anode} are the work-functions of the cathode and anode respectively. ΔE_e and ΔE_h are the barriers for electron and hole injection respectively.

The efficiency of any type of OLED depends very strongly on the nature of the materials used [6–10]. A high carrier mobility (μ), defined as the electron or hole velocity per unit electric field, is required to avoid the build-up of space charge in the device which increases the operating voltages and power consumption. Furthermore, it is important to balance the injection of electron and holes for efficient exciton generation and recombination. The luminescent material must also exhibit a high quantum efficiency of emission. Moreover, the optical mode pattern and recombination zone must be matched for efficient out-coupling of the light. Finally the organic materials must be chemically ultra-pure and also free from ions to avoid trapping as well as also being photochemically, thermally and electrochemically stable in order to achieve long-term operation. In addition to these properties, solid-state organic lasers must sustain high carrier densities [11]. Excited state absorption must be minimized for optical pumping and polaron absorption in the PL spectral region must also be reduced before electrically pumped lasing can be obtained. The first type of OLED to be developed made use of the thermal evaporation of materials of low molecular weight (small molecules), which form uniform thin films on the substrate as thermally stable organic glasses. Discrete thin films of different small molecules are sublimed in sequence to optimize the OLED performance. This multilayer approach has resulted in the best OLED performance reported to date with an external quantum efficiency of 19% [12]. The second approach involves the use of light-emitting main-chain conjugated polymers. These allow cheaper and scaleable deposition methods, such as spin-casting and ink-jet printing, to be used to form electroluminescent and charge-transporting films also stabilized in the glassy state below the glass transition temperature. External quantum

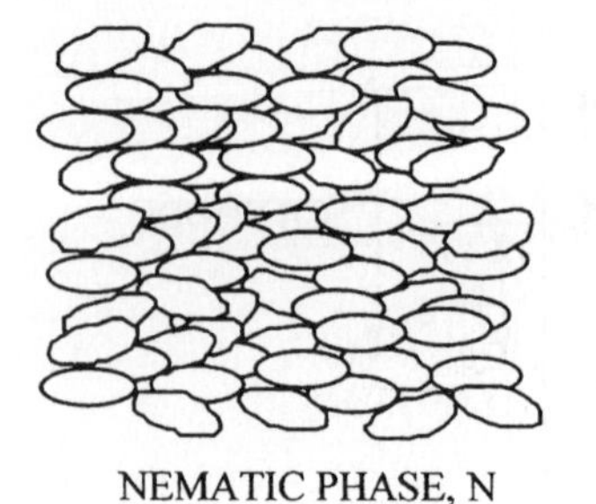

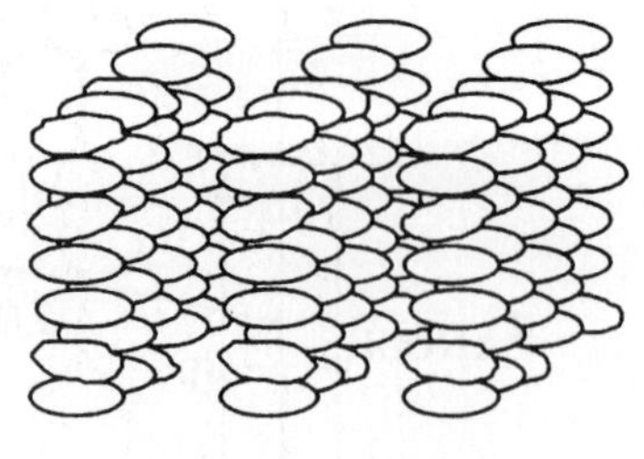

Figure 14.3. Schematic of a nematic phase with orientational order and of a smectic A phase, SmA, with positional and orientational order.

efficiencies of 6% have been obtained using this approach [13]. Oligomers have only really been used so far as models for the corresponding polymers.

Very recently liquid crystals have emerged as a third class of organic material for OLEDs. In some cases liquid crystalline properties were found in well known OLED materials, for example polyfluorene, whereas in others liquid crystallinity was deliberately designed in order to tailor the device properties. Although these coloured conjugated liquid crystals clearly have different molecular properties to those used in LCDs the same range of supramolecular, macroscopic structures (phases or mesophases) are found. To date, the nematic and smectic phases of calamitic (rod-like) LCs and the various columnar (discotic) phases produced by phasmidic or disk-shaped molecules see (figures 14.3 and 14.4), and their crosslinked polymer network equivalents have been used in OLEDs. Figure 14.5 shows an idealized representation of a polymer network formed from a reactive mesogen [14, 15]. Optically active chiral nematic small molecules in the

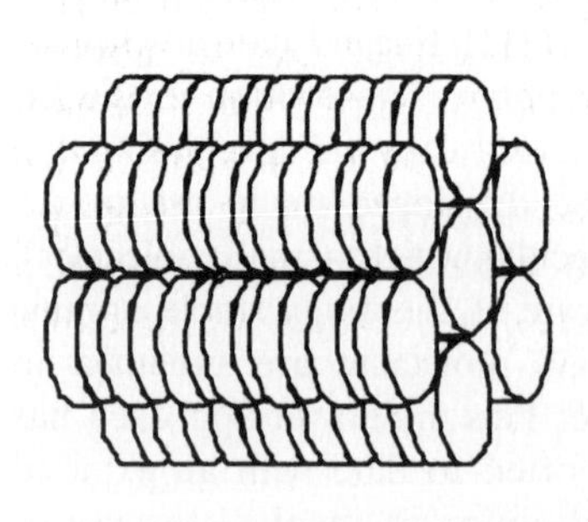

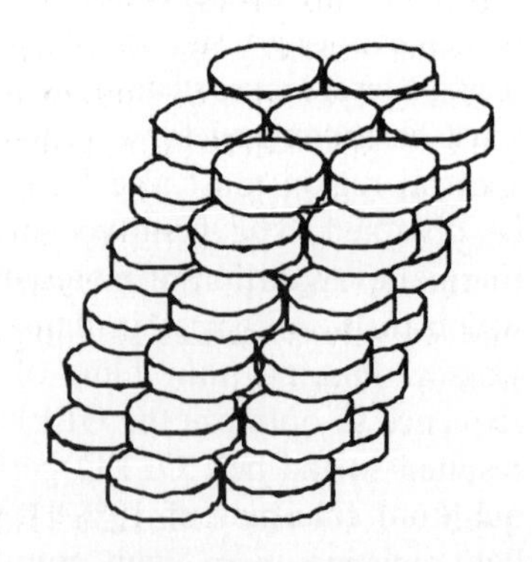

Figure 14.4. Schematic of a discotic (columnar) phase with homoeogeneous and tilted homoeotropic orientation.

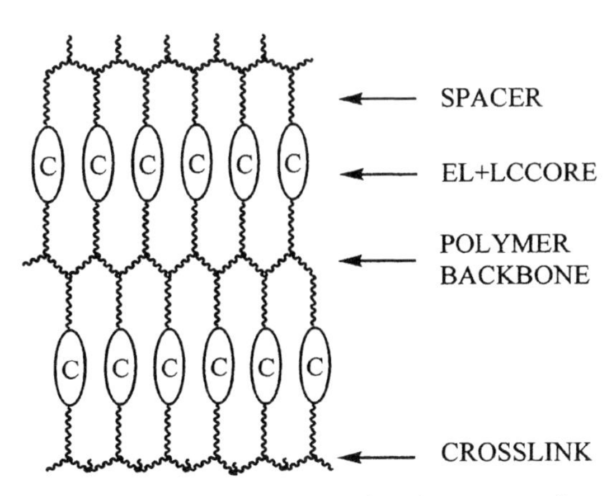

Figure 14.5. Schematic of a crosslinked anisotropic polymer network.

glassy state have been used in OLEDs with circularly polarized emission and the photonic band-structure of (optically active) chiral nematic LCs has recently been used to provide feedback for mirror-less, solid-state organic lasers [16].

This third LC approach to OLEDs combines some of the advantages of both of these commercial approaches using amorphous electroluminescent organic materials, avoids some of the drawbacks and makes use of some aspects unique to LCs. For example, the physical vapour deposition of low-molar-mass organics is expensive and not easily scaleable to large devices; the deposition of multiple polymer layers using conventional spin coating techniques is problematic, due to interlayer mixing when the solvents used for spin coating subsequent layers dissolve the underlying films, although this problem has been ameliorated to some degree by the use of polymer blends, some of which exhibit a degree of vertical segregation [17]. Indeed, thermal crosslinking of polymers has also been used to fabricate graded hole-transporting polymer layers [18]. Although the use of nematic and smectic LCs in OLEDs is a very recent phenomenon, it appears that they offer at least three main potential advantages over the isotropic materials used to date: polarized emission, high charge carrier-mobility and enhanced output-coupling. These potential advantages are described in more detail below.

14.3 LCs for polarized luminescence

Light-emitting and liquid crystalline small molecules, oligomers, polymers and polymer networks can be used to generate plane polarized light, which is promising for many applications. A typical PL spectrum from a

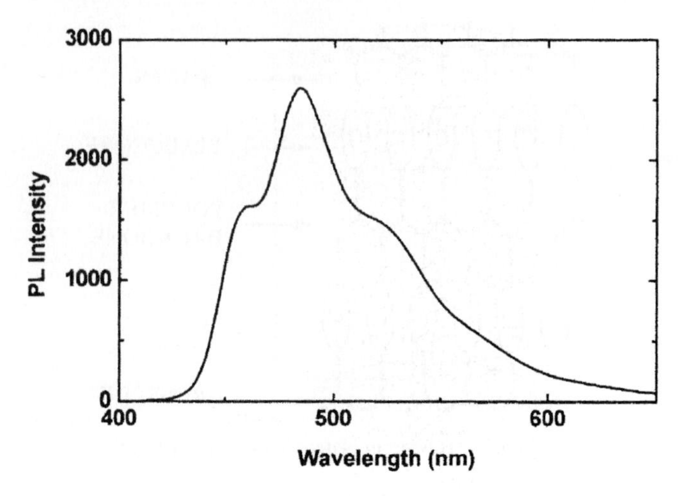

Figure 14.6. PL spectrum of light-emitting LC.

light-emitting liquid crystal is shown in figure 14.6. The emission originates from the radiative recombination of a singlet exciton and the well-resolved vibronic structure results from monodispersity and high order. The self assembling properties of calamitic LCs provides orientational order and an underlying alignment layer can be used to give uniform orientation.

The anisotropy of absorbance is defined by

$$\frac{A_{\text{para}} - A_{\text{perp}}}{A_{\text{para}} + 2A_{\text{perp}}} = \langle P_2 \rangle (1 - \tfrac{3}{2} a)$$

where A_{para} and A_{perp} are the absorbance parallel and perpendicular to the orientation direction of the liquid crystalline director. $\langle P_2 \rangle$ is the second rank orientational order parameter and a the molecular anisotropy factor. The latter depends on the alignment of the transition dipole moment and its value increases with absorbance perpendicular to the molecular long axis [19]. The orientational order parameter depends on molecular size with small molecules having a maximum value of $\sim$0.7. Oligomers and polymers, on the other hand, can exhibit very high orientational order parameters, due to the linear nature of the extensive oligomer or polymer backbone, although the molecular anisotropy also depends on the supramolecular conformation of the backbone [20–24]. For convenience the anisotropy of luminescence is often described by $(I_{\text{para}} - I_{\text{perp}})/(I_{\text{para}} + 2I_{\text{perp}})$ although the tensor describing luminescence is more complicated: it has both $\langle P_2 \rangle$ and $\langle P_4 \rangle$ components since it originates from the product of the absorption and emission second rank tensors. In general, the measured anisotropy is greater in luminescence than in absorption.

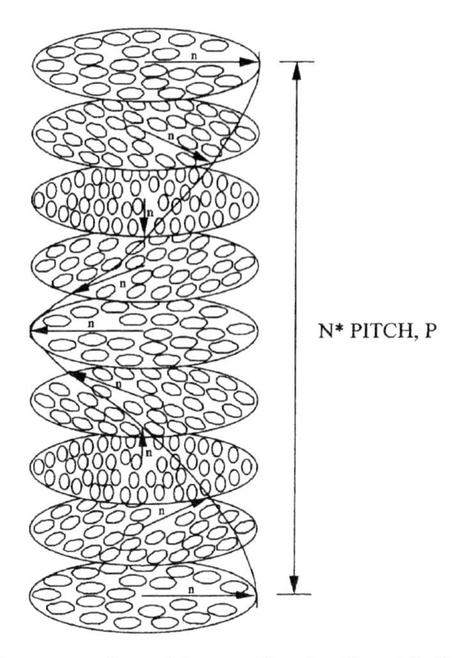

Figure 14.7. Helical structure formed by a uniformly aligned bulk chiral nematic (N^*) phase.

The chiral nematic phase is the optically active version of the nematic phase. It forms a macroscopic helical structure with a regular periodicity (see figure 14.7). Circularly polarized light, rather than plane polarized light, can be generated from macroscopically aligned bulk samples of (monodomain) luminescent chiral nematic LCs in two ways. One way exploits the photonic band structure of chiral nematic LCs and the other relies on coupling between the electric and magnetic transition dipole moments. The vectors are normally orthogonal so no coupling between them occurs. Circularly polarized emission is characterized by the circular dissymmetry factor:

$$g_e = 2(I_L - I_R)/(I_L + I_R)$$

where $I_{R(L)}$ is the right-handed (left-handed) circularly polarized luminescence. The colours observed in reflection from uniformly aligned chiral nematic LCs result from their spontaneous self-assembly to form the supramolecular helix shown in figure 14.7 in which the nematic director is continuously rotated to form a helical structure with the optic axis orthogonal to the nematic director. The birefringent nature of the chiral nematic phase gives rise to a periodic modulation of refractive index establishing a one-dimensional photonic stop-band. This prevents the propagation along the helical (optic) axis of circularly polarized light having the same handedness as the helix, over the wavelength range

$$\Delta\lambda_R \approx \Delta n \lambda_R / n,$$

where the stop-band is centred at

$$\lambda_R = np,$$

and the pitch, p, is defined as shown in figure 14.7. Δn is the birefringence and n the average refractive index given by

$$n = [(n_0^2 + n_e^2)/2]^{1/2}.$$

Highly circularly polarized PL has been obtained by making use of the photonic properties of light-emitting chiral LCs whose PL spectrum matches the stop band (wavelength of selective reflection) [25, 26]. Low threshold mirrorless lasing has been reported using this technique [16].

A quantitative description of luminescence in the stop-band spectral region has recently been given [25]. According to Fermi's Golden Rule, the rate of spontaneous emission is proportional to the density of photon states (DOS) in the medium. For a very thick chiral nematic film with a right-handed helix, the DOS of right-hand circularly polarized light is zero in the stop-band and diverges at the edges. Hence spontaneous emission in the stop-band is left-handed circularly polarized. For a thin chiral nematic film, as illustrated in figure 14.8, multiple reflections between the front and

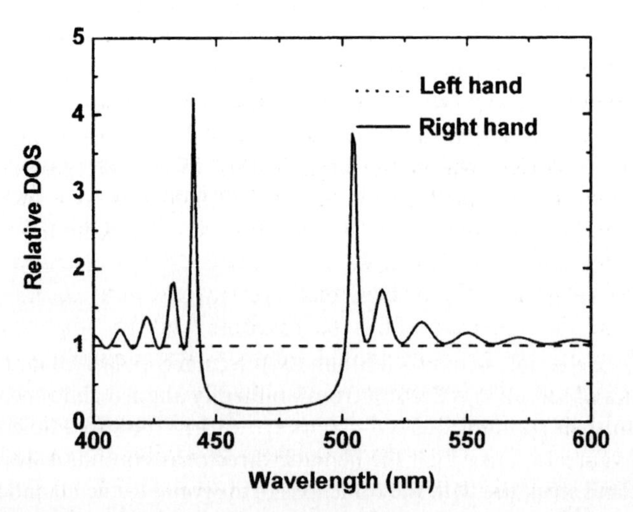

Figure 14.8. Density of photon states of a right-handed chiral nematic LC versus wavelength for a uniformly aligned helix of length 15*a*. The DOS is normalized to that of a uniform dielectric.

back interfaces of the film result in Fabry–Pérot like peaks in the DOS at resonant wavelengths near the stop-band edge. At these resonant wavelengths the intensity of right-handed circularly polarized PL is enhanced so that the sign of g_e is reversed at the stop-band edge. The group velocity is inversely proportional to the DOS so that these emitted photons have a long dwell time through the film. This increases the probability of stimulated emission so that the one-dimensional photonic bandgap structure can be used as a mirrorless laser resonator. The DOS of left-hand circularly polarized light is unaffected by the helix and is independent of wavelength. Lasing in chiral nematic LCs was first proposed in 1973 [27] and fifteen years later was demonstrated at the stop-band edge of a dye-doped chiral nematic LC [16]. The helical structure of the chiral LC generates the optical feedback and the gain originates from direct excitation of a dye-dopant in the LC. Distributed feedback lasers fabricated from layered dielectrics have been known for a considerable amount of time [28]. A chiral nematic resonator exhibits several inherent advantages as well as that of the spontaneous self-assembly of the resonator structure. The electric field of the resonator modes at the band edges has the same helicity as the chiral medium. This indicates that there is optimized coupling with the transition dipole moment of an anisotropic emitter whose orientation mimics that of the helix [29]. It is not easy to directly compare the threshold fluence for the various systems discussed above, since the excitation beam size is often not defined. However, all of these systems are optically pumped with high-energy pulsed lasers. A new mode of operation for lasing has been proposed recently in which the defect mode results from a phase discontinuity at the interface between two chiral helical structures. This discontinuity results in lasing in the centre of the band edge [30, 31]. The threshold fluence for lasing is substantially lower for this mode than for band-edge lasing [32].

The external quantum efficiency of OLEDs is also limited by total internal reflection, which traps photons, emitted at large angles to the normal of the device, within the organic thin films and substrate [6, 33]. The total intensity coupled out of the OLED increases by a factor of 1.6 when the emitter is oriented in the plane compared with that from the same emitter with an isotropic configuration. The director of low-molar-mass calamitic LCs in the nematic or smectic state can be aligned in the plane of the substrate. This macroscopic uniform alignment of the chromophores in the LC state should enhance output coupling. The orientational order of the mesophases of calamitic LCs provides another advantage in guest–host systems based on Förster energy transfer, which involves a dipole–dipole interaction between an excited donor and ground-state acceptor so that excitation is spatially transferred from the donor to the acceptor. The rate of energy transfer is proportional to the orientational factor for dipole–dipole interaction, which is six times greater for perfect parallel alignment of the donor and acceptor compared with a random spatial orientation [34]. Calamitic LCs in the nematic or smectic state adopt a parallel configuration of their long molecular axes (LC director) and should, therefore, exhibit an enhanced Förster transfer rate. Efficient energy transfer in a LC polymer guest–host system using a microcavity to provide feedback resulted in an extremely low threshold fluence of 3 nJ cm^{-2} per pulse for optically pumped lasing [35]. The orientational order of calamitic LCs may increase the tendency to aggregate and care must be taken in the design of LC chromophores to avoid this. Intermolecular interaction between rod-like molecules can split energy levels and red-shift the PL spectrum from that of the isolated molecule whilst quenching quantum efficiency. Excimer formation also produces a broadened red-shifted PL spectrum. The transition dipole moment of the aggregate is different from that the singlet exciton so that polarization properties also change. Various molecular strategies are used to maintain exciton emission in the solid state and so avoid quenching configurations [8, 36]. For example, aggregation in fluorene polymers is influenced by the solvent used in spin-casting and by the annealing condition post-deposition.

14.4 Charge transporting LCs for OLEDs

One of the primary motivations for investigation of LCs for OLEDs is that they promote anisotropic carrier transport. The mobility of amorphous organic small molecules and conjugated polymers is limited by disorder. Even with recent improvements resulting from molecular design, amorphous molecular materials are not available with hole and electron mobilities much greater than 10^{-3} and 10^{-4} $cm^2\,V^{-1}\,s^{-1}$ [37]. Rod-like smectic liquid crystals self-organize into layers, as illustrated in figure 14.1, and show mobilities up to 1×10^{-2} $cm^2\,V^{-1}\,s^{-1}$ [38]. The orientational order in the nematic state (see also figure 14.1) facilitates the formation of a two-dimensional conductor. Columnar liquid crystals self-organize into stacks as shown in figure 14.2 with large carrier mobilities (10^{-3}–10^{-1} $cm^2\,V^{-1}\,s^{-1}$) resulting from the overlap between the π-electrons of the aromatic cores. The charge transport is anisotropic in that the mobility parallel to the core of the columns of self-assembled disks is orders of magnitude higher than that measured from column to column. All of these configurations have also been shown to promote anisotropic carrier transport due to the increased overlap of aromatic cores [38–41]. Holes represent the majority charge carriers in the columnar phase of most discotic LCs due to their limited electron affinity and low ionization potential. This is a result of the presence of many delocalized π-electrons in the aromatic core of columnar LCs.

The relatively low values of mobility ($\ll 1$ $cm^2\,V^{-1}\,s^{-1}$) in most organic materials mean that incoherent hopping transport rather than coherent (band-like) models are mainly used. This is true for LCs as well as for amorphous materials. In the Gaussian disorder model proposed by Bässler and co-workers [42, 43], static disorder results in random fluctuations of the energy of the carrier which give rise to variations in the intermolecular hopping rate. Carriers tend to relax to the bottom of the Gaussian density of state (DOS) distribution of energies so that the log of the mobility is inversely dependent on the temperature squared. A number of different approaches have been used to investigate the mechanism of hole transport in discotic LCs [44–48]. Particular attention has been paid to explaining the low temperature dependence of mobility and confirming the one-dimensional nature of the transport [45]. Monte-Carlo calculations of thermally activated hopping predict a diffusivity, D, proportional to temperature [47]. This gives a temperature-independent mobility since $\mu \propto D/T$. The incoherent Haken–Strobl–Reineker model suggests that the D depends on dynamic fluctuations, which are caused by lattice movements, in the matrix elements that govern intermolecular hopping along the stack. Temperature-independent mobilities can be obtained when the amplitudes of the fluctuations are small [48]. A polaronic hopping model also gives a similar result by balancing the temperature-enhancing and temperature-inhibiting terms [47]. The photocurrent transients of unsubstituted triphenylenes have been simulated on the basis that shallow trapping further reduces the carrier mobility [45]. The simulations show a field-dependent hole diffusivity at early times so confirming the one-dimensional nature of the transport. A one-dimensional static disorder model is considered where carriers hop between shallow traps provided by random fluctuations in the site energies. Both uncorrelated and correlated disorder is considered in systems of finite length [45]. Quantum mechanics has also been used to study hole transport in discotics [49]. The hopping rate between two molecular cores on the stack is proportional to the square of the intermolecular transfer energy, which is estimated as half the splitting energy of the HOMO on formation of a dimer. This approach overestimates the actual mobility but shows why rotational disorder of the molecules in the stacks can have a large impact on mobility—the splitting energy varies enormously with rotation—and how this can be reduced by molecular design.

Temperature-independent hole mobilities were also found in the individual smectic phases of phenylnaphthalene LCs. Polaron hopping models are applied to fit the data and suggest that the charge carrier interaction between the aromatic cores as well as the binding energy of the polaron changes with phase [50]. Monte-Carlo simulations based on a static disorder model suggest that the mobility is field-independent at low fields because of the low variation in site energy resulting from smectic ordering [51]. Electronic rather than ionic transport in low-mass nematics was first observed in glassy LC phases at room temperature [41]. Hole mobilities up to $4 \times 10^{-4}\,\mathrm{cm^2\,V^{-1}\,s^{-1}}$ were obtained. The mobility has a large temperature and a small field variation.

This is explained using a static disorder model that includes spatial correlations in the carrier energies.

14.5 Electroluminescent semiconductor LCs

The tens of thousands of liquid crystals synthesized specifically for use in LCDs are not generally suited to application in OLEDs: for example, the vast majority of LCDs require nematic liquid crystals, which retain a fluid nematic phase with a low viscosity over a broad temperature range (e.g. from −35 °C to 90 °C). OLEDs, such as those shown in figure 14.3, use substantially thinner active layers, <100 nm, of a solid organic material rather than those of LCDs using a liquid and thicker (e.g. 5 μm) layer. OLEDs require solid layers of organic materials for long-term device stability and robustness, and to avoid liquid flow, inter-mixing of layers and in-diffusion of the overlying electrode. Furthermore, LCs used in LCDs are designed to be insulators and to be transparent to the visible light modulated by the LCD, i.e. they should not be fluorescent. The presence of traps formed at crystal grain boundaries also makes polycrystalline films of LCs in the solid state unsuitable for OLEDs. Therefore, new light-emitting and charge-transporting LCs have had to be designed and synthesized specifically for use in OLEDs. Furthermore, the macroscopic order present in the fluid liquid crystalline state of a small molecule, oligomer or polymer has somehow to be rendered immobile in an OLED. This stabilization of conformation and configuration is achieved by either quenching the LC state into the glassy state or chemically crosslinking polymerizable LCs to form a rigid polymer network. The three principal types of LC synthesized so far for use in OLEDs are small molecules, oligomers and main-chain, conjugated polymers. Each can be converted by suitable chemical modification into polymerizable analogues, which can be polymerized to form polymer networks. The advantages and disadvantages of using one or the other of these three types of LC are described in the following part of this chapter.

14.6 LC polymers

The carrier mobility of main-chain conjugated polymers is intrinsically low, e.g. polyphenylene vinylene (PPV) (**1**), the standard polymer for OLEDs, has a mobility of 10^{-5} $cm^2\,V^{-1}\,s^{-1}$ (see table 14.1) [52].

It appears that the mobility in the nematic state is higher than that in the corresponding amorphous solid, e.g. an exceptionally large hole mobility of 9×10^{-3} $cm^2\,V^{-1}\,s^{-1}$ was obtained for 9,9-dioctyl polyfluorene (PFO) (**2**; Cr ~170 °C, N ~270–280 °C, I) as a uniformly aligned nematic glass at room

Table 14.1. Chemical structure of main-chain polymers with charge-transport and/or luminescence properties.

1	n
2	C_8H_{17} C_8H_{17} n
3	$OC_{16}H_{33}$ $C_{16}H_{33}O$ n
4	O O $C_{10}H_{20}$ O O n

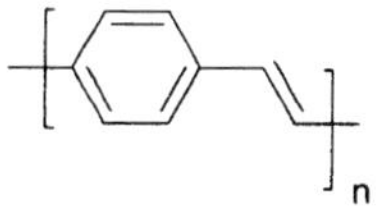

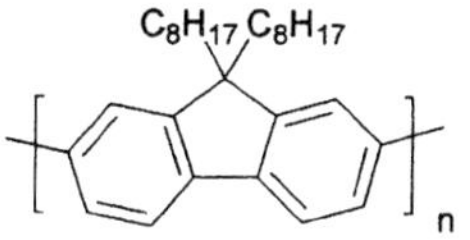

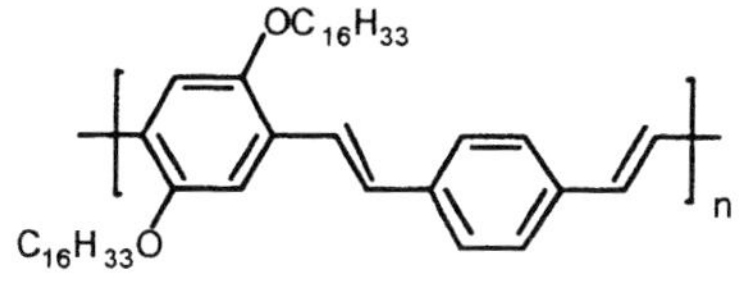

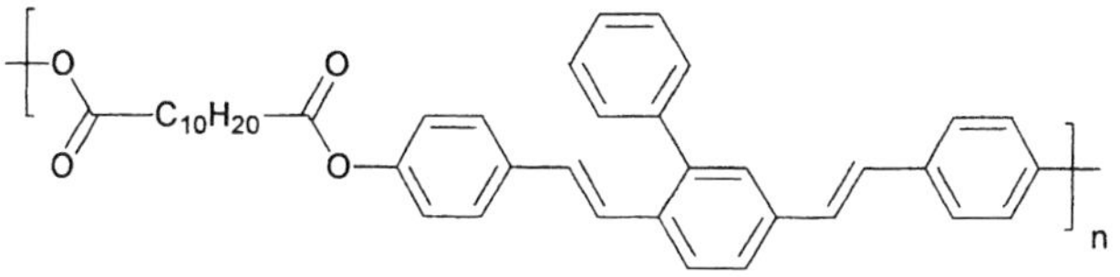

temperature [53]. This compares with a hole mobility of $4 \times 10^{-4}\,\mathrm{cm^2\,V^{-1}\,s^{-1}}$ when the same polymer (**2**) is deposited as an isotropic film [54]. The self-annealing and self-assembling properties of the LC state are presumed to reduce the number of traps. The aliphatic side-chains of polyfluorenes, many of which are liquid crystalline, reduce the intermolecular forces of attraction due to steric effects, which can result in high solid-state PL quantum efficiencies of up to 60% [36, 55–57]. It is probable that many light-emitting polymers are liquid crystalline [36]. Unfortunately the high melting points and glass transition temperatures typical for such long conjugated and rigid molecules means that the LC character of these polymers is often not observed and many of these main-chain conjugated polymers are intractable solids. However, the presence of long aliphatic substituents in a lateral position increases the intermolecular distance and can result in a significantly lower melting point as shown for the PPV analogue (**3**; Cr 103 °C, N 133 °C, I, t_g 66 °C) [58, 59].

Many different approaches to the generation of polarized light using LCs have been described. [3, 55, 57]. Plane polarized electroluminescence from the LC state was first demonstrated using conjugated main-chain polymers with isolated phenylenevinylene segments in the main-chain [36, 57, 60–62]. Blue-green polarized light with a high PL anisotropy of 0.64 was obtained for EL using the segmented PPV polymer (**4**) oriented in a smectic phase on a rubbed polyimide alignment layer in an OLED [62]. A polarization ratio ($I_{para}:I_{perp}$) of 15:1 was obtained for EL using a LC polyfluorene aligned on a rubbed polyimide alignment layer doped with a hole-transporting (triarylamine) small molecule [63]. Doping with hole transporting materials is necessary as most organic polymers are insulating in nature. A higher ratio of 25:1 and a brightness of 250 cd m^{-2} was obtained using a similar polyfluorene oriented on a thin insoluble substrate of PPV, which was rubbed during conversion from its soluble precursor [64]. Polarized photoluminescence with a high polarization ratio of 22:1 was achieved using mechanically stretched polymer blends consisting of a small amount of a luminescent LC poly(phenylene-ethynylene) contained within a non-emitting ultrahigh molecular weight polyethylene host [57, 65–68]. A sensitizer was also added to increase the efficiency of the device [65]. This system could be used to fabricate LC polarizers [65, 67]. A microcavity containing an aligned PFO sample has been used to spectrally separate the cavity modes for PL polarized parallel and perpendicular to the alignment direction in order to increase the polarization ratio [69]. A polarization ratio of >300:1 was obtained for the most intense PL mode. However, approaches using PL to generate polarized emission are less attractive than EL as PL still requires an additional, external light source. Azobenzene polymers, which undergo anisotropic *cis–trans* isomerization on irradiation with polarized light, have also been used as (photo)alignment layers instead of rubbed polyimide, which is not particularly attractive for OLEDs (see below) [70]. The LC alignment induced by the azo-polymers is not permanently fixed and the orientation direction can be reversed by subsequent irradiation of the photo-alignment polymer with orthogonally polarized light. However, the original alignment direction of the nematic phase of an electroluminescent LC polyfluorene is retained, albeit with a lower dichroic ratio, on reorientation of the underlying azobenzene alignment layer [70]. This fact suggests that azobenzene polymers, although intrinsically conformationally labile, could provide stable alignment for LC polymers with high glass transition temperatures, e.g. $I_{para}:I_{perp}$ of 14:1 was observed from a uniformly oriented polyfluorene on an azobenzene photo-alignment layer. A luminescence efficiency of 0.66 cd A^{-1} was achieved at 8 V using an ultrathin photo-alignment layer, although with a lower polarization ratio [71]. An unavoidable disadvantage of using electroluminescent LC main-chain polymers for polarized EL is the necessity for annealing at high temperature (~200 °C) for a number of hours in order to align the LC director and the transition moment. The alignment is then fixed in position by quenching to form the LC glassy state and also to prevent subsequent crystallization. OLEDs have also been demonstrated using main-chain polymers incorporating discotic chromophores in the main-chain and a triarylamine dopant [72].

Circularly-polarized emission has also been generated from light-emitting LC polymers. Polyfluorene adopts a helical configuration and intra-chain coupling between twisted monomer units rotates the magnetic transition dipole moment. This gives rise to partially circularly-polarized emission [73]. The g_e factor of 0.28 is low but still substantially greater than that obtained by inter-chain coupling of PPV with chiral side-chains [74]. A maximum g_e value of 1.3 was obtained from a luminescent chiral nematic poly(*para*-phenylene) [75]. In the latter case, the PL spectrum overlaps the one-dimensional photonic stop-band.

14.7 LC oligomers

Oligomers may be defined as short polymers depending on the number of monomer units making up the oligomer [21, 22]. Light-emitting LC oligomers are potentially interesting for application in OLEDs due to their ability to combine the high-order parameter of LC polymers and their tendency to form the glassy state with the solubility and processability of small molecules. The large number of repeat units in monodisperse oligomers, i.e. those where all the oligomers possess the same number of monomer units, such as the oligomers (**4**–**8**) shown in table 14.2, represents a chemically more challenging approach than that using either small molecules or polymers [23–26]. Short oligomers, such as the pentameter (**5**; Cr 109 °C, N 116 °C, I) [23] and heptameter (**6**; Cr 183 °C, N 204 °C, I) [24], which could also be reasonably classified as large molecules, generally exhibit a high melting point. Therefore the blue PL reported the pentameter (**5**) and heptameter (**6**) is from the solid state. Oligomers with a longer molecular core are required to form the glassy state typical of polymers, e.g. the monodisperse fluorene pentamer (**7**; t_g 100 °C, LC >175 °C, I).

Polydisperse oligomers consist of a collection of molecules made up of similar repeat units, but the number of repeat units differs, for example the polydisperse fluorene oligomer (**8**; t_g 123 °C, LC >375 °C, I). Polydisperse oligomers exhibit a distribution of molecular length and weight. Using such either monodisperse or polydisperse LC oligomers, high-orientation-order parameters responsible for polarized emission are possible without the high-temperature liquid crystalline phases or the coiled conformations of polymers, e.g. the oligomer (**8**) shows a PL anisotropy up to 18:1 [76]. The highest polarization ratio for aligned LC of any kind reported so far of 25:1 has been recently described using a dodecamer on a rubbed PEDPT/PSS alignment layer [76, 77]. Other physical properties of OLEDs containing these oligomers are reported to be superior to those incorporating the corresponding polyfluorenes. However, relatively long annealing times (30–60 min) at high temperatures just above the glass transition temperatures are still required. The polarization ratio increases with decreasing thickness of the oligomer layer due to influence of the alignment layer across the thinner emission layer.

Table 14.2. Chemical structure of oligomeric LCs with charge-transport and/or luminescence properties.

5	OC_8H_{17} ; $C_8H_{17}O$
6	$C_{14}H_{29}O$; OC_4H_9 ; C_4H_9O ; $OC_{14}H_{29}$
7	5
8	4 ; 2 ; 4

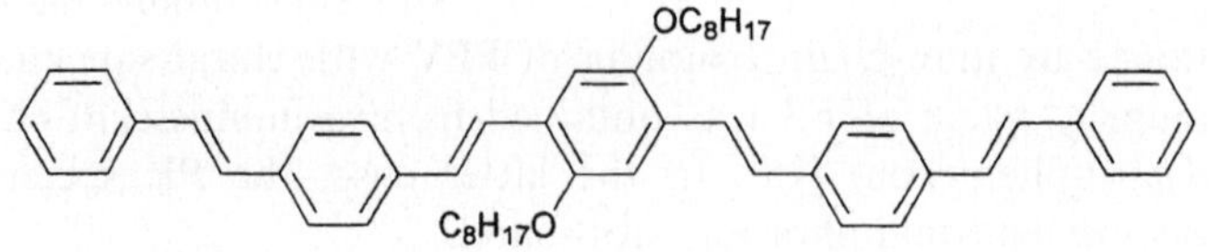

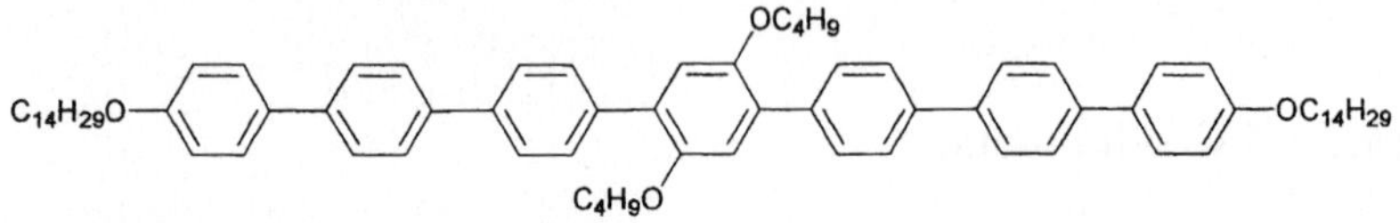

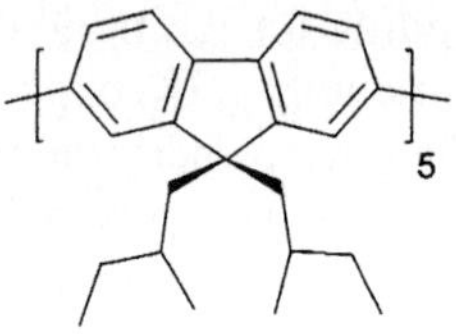

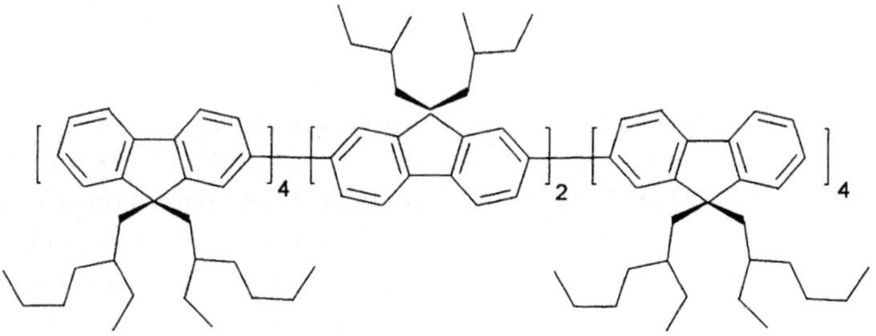

14.8 Small molecule LCs

The nematic and smectic phases of calamitic LCs with a rod-like structure (see figure 14.1), and the columnar phases of discotic LCs with a columnar structure composed of stacks of molecular discs (see figure 14.2), have all been made use of in prototype OLEDs (see figure 14.3). The one-dimensional columns of discotic LCs and the two-dimensional layers of smectic LCs promote anisotropic carrier transport due to the increased overlap of aromatic cores [34–36]. The high charge carrier mobility of the LC state was first observed in the columnar phases of hexa-alkoxytriphenylene (HAT) discotic LCs. A range of asymmetrical triphenylene materials such as those (**9**–**12**) shown in table 14.3 with glassy or plastic columnar phases had been developed during the 1990s to overcome the problems of the high melting points and the tendency for columnar samples to recrystallize above room temperature of normal symmetrical triphenylene LCs [78, 79]. The transport properties of triphenylenes were improved substantially by adding a stoichiometric amount of a substituted phenyltriphenylene (PTP)

Table 14.3. Chemical structure of columnar LCs with charge-transport and/or luminescence properties.

	Core	R
9	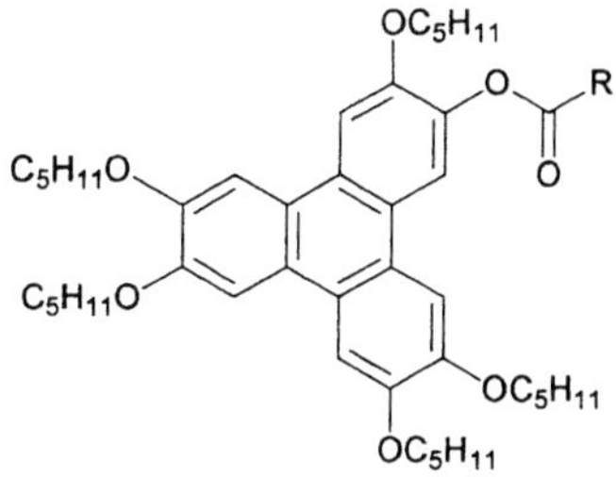 OC_5H_{11} / $C_5H_{11}O$ substituted triphenylene ester (O–C(=O)–R)	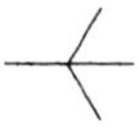
10		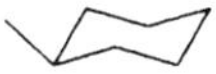
11		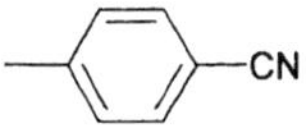 —CN
12		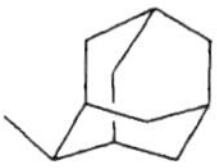
13	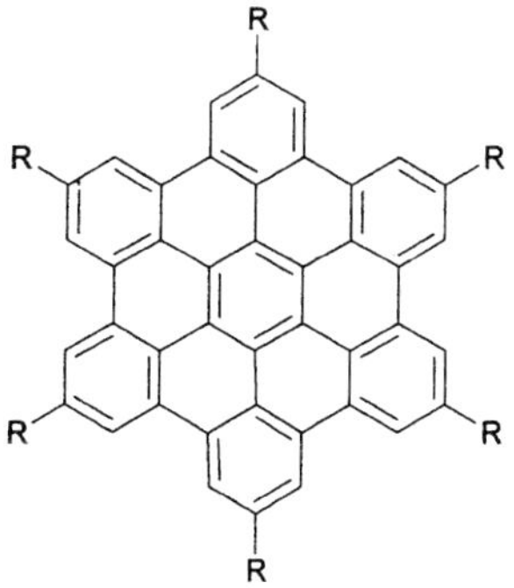 R-substituted hexabenzocoronene	$-C_6H_4C_{12}H_{25}$
14		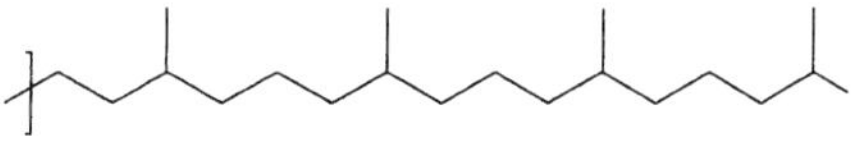
15	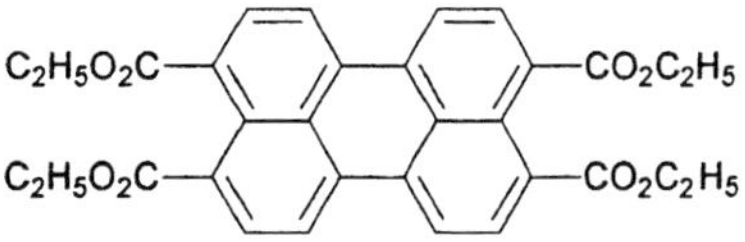 $C_2H_5O_2C$– / $-CO_2C_2H_5$ substituted perylene	

to them to form alternating interleaved columnar stacks, thereby increasing the order and stability of the resultant columnar mesophases. A columnar glass, which retains the macroscopic ordering of the columnar phase and its high hole mobility of $1.6 \times 10^{-2}\,cm^2\,V^{-1}\,s^{-1}$, is formed on cooling to room temperature [80].

Discotics of higher molecular mass usually have LC phases at elevated temperatures and a recent important advance is the development of hexacoronene materials (**13** and **14**) with room temperature columnar phases. The low melting point was induced in two ways: either a phenylene group is used as the coupling agent between the core and the aliphatic side-chains to form **13** (Col_1 18 °C, Col_2 83 °C, Col_3) [81] or six multi-branched aliphatic side chains are attached to a hexa-peri-hexabenzocoronene to form (**14**; Cr −36 °C, Col_{ho} 231 °C, I) [82]. The high degree of chain branching present in compound (**14**) results in the formation of a low-melting hexagonal columnar phase with an extraordinary broad phase range. These developments are particularly interesting for commercial applications because of the large size of the hexacoronene core and the empirical observation that the maximum mobility obtainable for discotic LCs increases with the size of the aromatic core [83]. Indeed a room temperature mobility of $0.22\,cm^2\,V^{-1}\,s^{-1}$ was obtained in the LC phase of the hexacoronene (**13**). However, chemical and ionic purity may be an issue as columnar LCs can be difficult to purify. There are a few electron-transporting discotic materials. A perylene (**15**) having high-temperature columnar and smectic phases (Cr 244 °C, Col 313 °C, I) exhibits a mobility $>0.1\,cm^2\,V^{-1}\,s^{-1}$ in the LC phase [84]. However, perylenes generally exhibit a high melting point, in this case >180 °C, and often low solubility in organic solvents. An electron-deficient columnar plastic crystal at room temperature has also been synthesized recently [85].

The generation of uniform orientation of the columnar phases of discotic LCs in a simple reproducible fashion is problematic. Homeotropic alignment is required for OLEDs with the columnar discs parallel to the substrate so that the molecular stacks conduct between the top and bottom electrodes in a direction orthogonal to the device substrate. Solution processing of discotic LCs often gives rise to homogenous or tilted homeotropic alignment with tilt disclinations between the domains (see figure 14.2). Annealing can improve the molecular ordering and enhance non-tilted homeotropic orientation. Multilayer OLEDs with low luminance but promising lifetimes have been fabricated by Langmuir–Blodgett deposition techniques of discotic LCs with columnar phases [86]. Monolayer OLEDs were reported using the the triphenylene (**9**) and also guest–host mixtures of the triphenylene (**9**) and a triaryl amine [87]. Light emission from the triphenylene was observed in both cases [87]. An all-columnar bilayer light-emitting diode was produced by sequential thermal evaporation of HAT and an electron-deficient perylene [88]. Both materials exhibit columnar phases at elevated temperatures and the columnar packing is retained in the crystalline phase at room temperature. The wavelength of emission can be tuned by the synthesis of asymmetrically substituted HAT derivatives [89].

Charge transport in the nematic phase of calamitic LCs with a rod-like, rather than a disc-shaped, molecular structure was studied several decades ago and found at that time to be primarily of ions. The transport properties in the more ordered lamellar smectic phases of calamitic LCs was studied decades later (see table 14.4).

The transport in the smectic phases of these LCs was found to be non-ionic and surprisingly high, e.g. for terthiophene (**16**; SmG 72 °C, SmF 88 °C, SmC 91 °C, I) the hole mobility, μ_h, and electron mobility, μ_e, $>1 \times 10^{-2}\,cm^2\,V^{-1}\,s^{-1}$ in the smectic G phase [38]. The magnitude of the

Table 14.4. chemical structure of low-molar-mass LCs with charge-transport and/or luminescence properties.

16 C_8H_{17} S S S C_8H_{17}

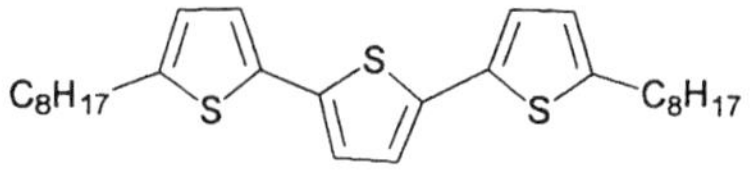

17 C_8H_{17} $OC_{12}H_{25}$

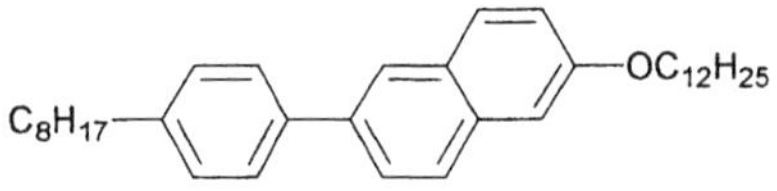

18 $C_7H_{15}O$ S N $SC_{12}H_{25}$

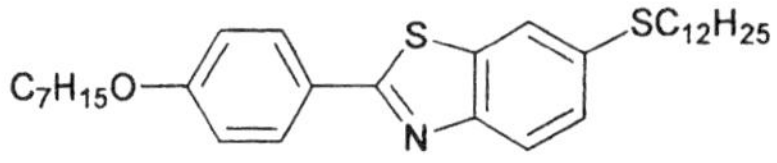

19 $C_5H_{11}CH{=}CHC_3H_6O_2C$ S S $CO_2C_3H_6CH{=}CHC_5H_{11}$

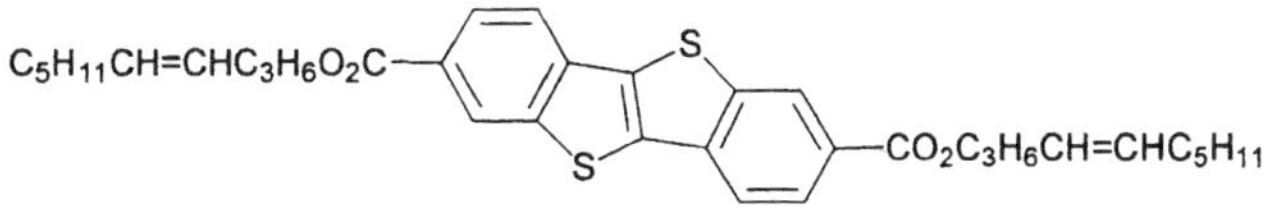

20 O O $C_5H_{10}O$ N N OC_5H_{10} O O

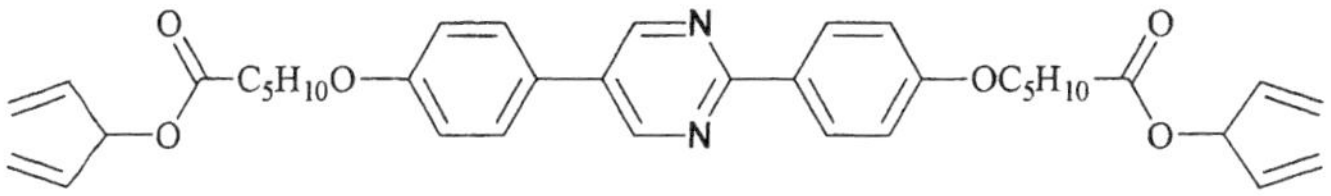

21 O O $C_5H_{10}O$ H_7C_3 C_3H_7 OC_5H_{10} O O

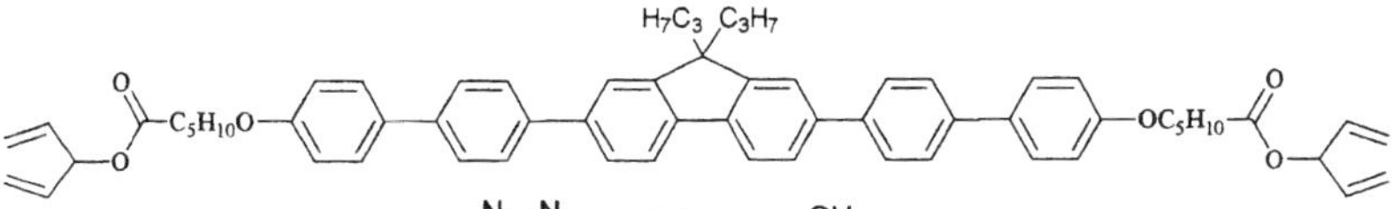

22 C_7H_{17} N—N O N CH_3 CH_3

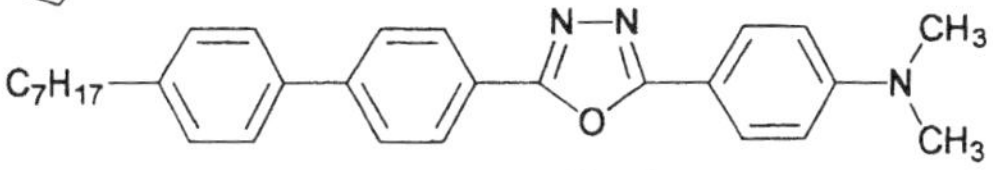

23 $C_6H_{13}O$ N—N O OC_6H_{13}

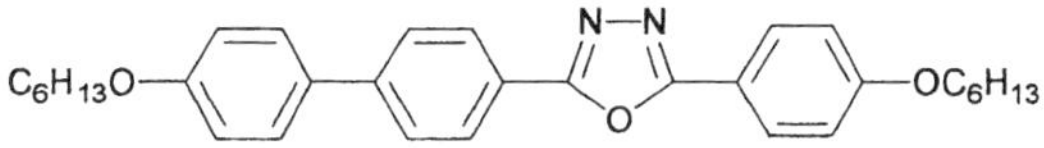

carrier mobility increased with increasing degree of order within the three smectic phases investigated, i.e. $\mu_{SmG} > \mu_{SmF} \times \mu_{SmC}$. These smectic phases are tilted, i.e. the director makes a non-zero angle with the layer plane. Several phenylnaphthalenes such as (**17**; Cr 79 °C, SmB 101 °C, SmA 121 °C, I) possess orthogonal smectic phases. The hole mobilities are temperature-independent within the individual phases [90]. It is unclear whether the molecular tilt influences the charge transport through smectic layers in any way. The benzothiazole (**18**; Cr 90 °C, SmA 100 °C, I) [91] and the sanidic (board-like) (**19**; Cr 90 °C, SmA 100 °C, I) [92] are further examples of a low-mass calamitic (rod-like) LCs with a high charge mobility for potential use in OLEDs. Unfortunately, these LCs are all solids at room temperature and, as a direct consequence, are not suitable for use in commercial OLEDs due to the presence of crystal grain boundaries, which act as traps. The reactive mesogen (**20**) was the first compound to exhibit a smectic phase at room temperature (Cr 25 °C, SmC 124 °C, I) and a high electron mobility of $1.5 \times 10^{-5}\, cm^2\, V^{-1}\, s^{-1}$ [93]. However, this material is a viscous fluid at room temperature and was designed for the formation of solid polymer networks (see below).

A number of low-molar-mass liquid crystals have now been found to be luminescent [38, 94–97]. Unfortunately, there has been very little effort to investigate their luminescent behaviour and to quantify their PL quantum efficiency. Low-molar-mass LCs stabilized in the glassy state are potentially an attractive alternative to LC polymers and oligomers as charge-transport layers and to generate polarized electroluminescence due to the lower processing temperatures and spontaneous alignment of a lower-viscosity smectic phases and especially the nematic phase. The orientational order inherent in the nematic state facilitates the formation of a two-dimensional semiconductor. In contrast to the results of earlier studies of LCs for LCD applications, electronic rather than ionic transport was found in the glassy nematic phase of appropriate low-molar-mass nematic LCs at room temperature [41]. High hole mobilities up to $4 \times 10^{-4}\, cm^2\, V^{-1}\, s^{-1}$ were reported. Improved molecular design and synthesis has led to (**21**) with a hole mobility $>1 \times 10^{-3}\, cm^2\, V^{-1}\, s^{-1}$ in the glassy nematic state [98]. The oxadiazole (**22**; Cr 143 °C, N 138 °C, I) also exhibits a nematic phase, which exhibits blue PL [99]. Surprisingly there are still very few reports of polarized EL from the nematic or smectic phases of low-molar-mass calamitic LCs. In one such report compound (**23**) was aligned in an LCD cell and polarized EL with a low polarization ratio observed at 70 °C in the smectic SmX phase [100]. The extremely large cell gap, $d = 3\, \mu m$, led to a very high operating voltage, $V = 280$ V, and a low brightness of $0.8\, cd\, m^{-2}$. A similar LCD cell configuration was used to generate plane polarized light from a guest–host mixture of a coumarin dye dissolved in the smectic B phase of the phenylnaphthalene (**17**). The homogeneous alignment was achieved by using rubbed polyimide. Again a low polarization ratio, very high operating voltage, $V = 210$ V, and a low brightness of $0.7\, cd\, m^{-2}$ were observed at 90 °C [101]. Diene-reactive mesogens have been aligned on a photo-alignment layer [102] and polarized emission with a reasonable ratio of 12:1 was obtained from the glassy nematic state (see below).

Circularly polarized PL with a g_e value approaching 1.8 over a limited range of the emission spectrum was achieved from a dye-doped chiral nematic glass (**24**; t_g 77 °C, N* 147 °C, I) (see table 14.5) [103]. Subsequently commercially available chiral nematic mixtures were used to obtain circularly polarized PL in a similar guest–host mode [104]. In all the above cases the PL linewidth is greater than the spectral width of the stop-band. The low average value of g_e results from the reversal in the sign of g_e at the edges of the stop-band [25]. Different methods have been developed to improve the wavelength dependence of g_e. A chiral nematic film with a graded pitch was prepared to give a broad stop-band from about 370 to 580 nm. A maximum g_e value ~1.4. was obtained [104]. Alternatively PL from a dye with a narrow spectral

Table 14.5. Chemical structure of chiral LCs with for circularly polarized luminescence.

24	$R_1 = CO_2C_2H_4O$–C$_6$H$_4$–COO–C$_6$H$_4$–C$_6$H$_4$–OCH$_3$; $R_2 = CO_2C_2H_4O$–C$_6$H$_4$–COO–C$_6$H$_4$–CO–NH–C*H(CH$_3$)–C$_6$H$_5$
25	$O_2CC_8H_{16}$–C≡C–C≡C–$C_8H_{16}CO_2$
26	H_7C_3 C_3H_7

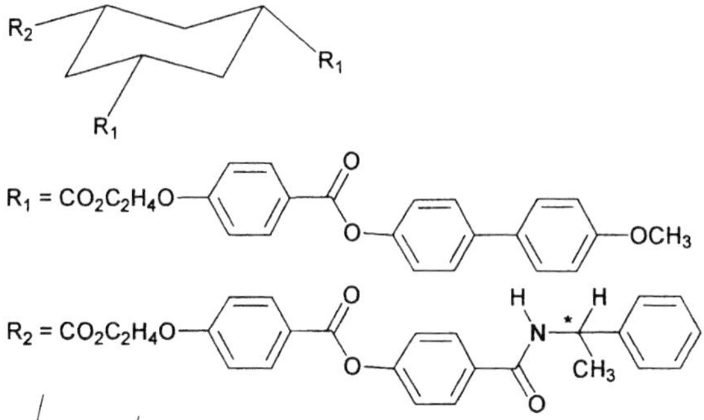

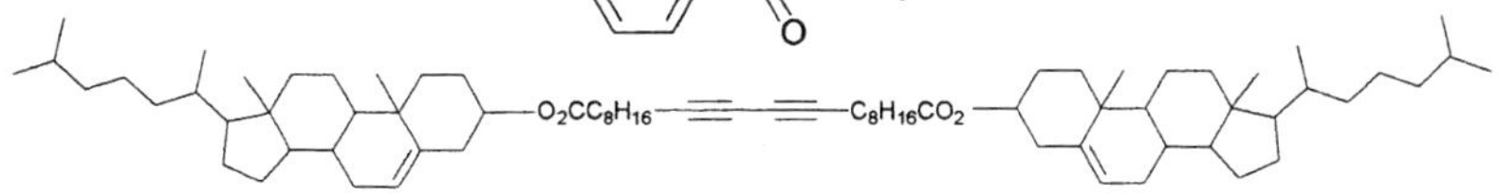

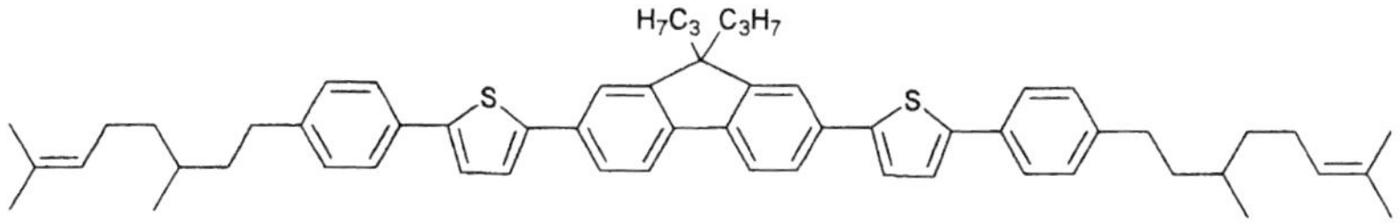

width gives an almost wavelength-independent g_e. A maximum value of 1.27 was achieved using this method [105]. The dicholesteryl ester of a diacetylene-carboxylic acid (**25**) also exhibits a glassy chiral nematic glass that could be used for circularly polarized PL [106, 107]. The fluorene (**26**) represents a new class of calamitic chiral nematic glassy LC with a unique spectrum of properties giving rise to circularly polarized light emission with an unusually high polarization ratio (maximum circular extinction ratio of 16 : 1) across a surprisingly broad bandwidth (>110 nm) with the highest reported average ratio (>8 : 1) reported so far [108]. The PL spectrum also overlaps the exceptionally broad one-dimensional photonic stop-band. This highly circularly polarized light is generated from a sample of (**26**) in the helical Grandjean planar texture, which is only 3 μm thick. This is an order of magnitude thinner than state-of-the-art materials. Guest–host cells using the fluorene (**26**) and a dichroic dye, which are indirectly excited by non-radiative energy transfer from the luminescent chiral nematic host to the guest dye, exhibit an even higher ratio of circular polarization (maximum ratio 19 : 1) with quenching of emission from the host. These results are for PL. However, the hole mobility of this new class of materials is also unusually high ($2 \times 10^{-4}\,cm^2\,V^{-1}\,s^{-1}$ at room temperature) [41] and should facilitate the fabrication of electrically pumped devices. Photochemically tunable circularly polarized fluorescence using a chiral nematic mixture of small molecules has recently been reported [109].

Since the first demonstration of lasing in chiral nematic systems in 1998 [16], rapid progress has been made in a wide range of systems, including lyotropic and ferroelectric LCs [110, 111]. The wavelength of laser emission was tuned by mechanical deformation of a dye-doped elastomer [112]. An applied electric field has also been used to move the stop-band edge in a chiral nematic LC and hence tune the laser wavelength [113]. The temperature stability was improved by using a photopolymerized chiral nematic polymer network [114]. The formation of the network results in a solid film that can be removed from the substrate [115]. The first report of lasing in a three-dimensional photonic crystal involved lasing in the blue phase of a chiral nematic LC [116]. All these lasers use guest–host systems whereby the guest dye in the LC host is directly excited by an optical pulse. Lasing has also been achieved using an undoped luminescent chiral nematic (N^*) LC, although the threshold is high [117]. Förster energy transfer has also been made use of to indirectly excite a dye in a two-dye doped chiral nematic LC [118, 119].

14.9 LC polymer networks

Luminescent and charge-transporting LC polymer networks formed from LC monomers (reactive mesogens) represent an elegant solution to the morphological problems associated with low-molar-mass LCs without having to resort to using oligomers or polymers. Reactive mesogens are polymerizable equivalents of small molecule LCs, but with two additional polymerizable groups one at each end of a flexible aliphatic spacer attached to the aromatic core. The ultraviolet polymerization of small liquid crystalline molecules, such as the acrylate (**27**; Cr 116 °C, N 150 °C, I) shown in table 14.6, is a well-known technique to develop thermally stable passive optical devices such as optical retarders [57, 120–124]. and is an extremely promising approach for OLEDs because it renders the thin film insoluble so that multilayer devices are easily formed and transport and recombination zones vertically defined. Figure 14.5 is an idealized illustration of an (anisotropic) LC polymer network. Reactive mesogens can be deposited on a substrate from solution, e.g. by spin coating or doctor blade techniques. The liquid crystalline state self-assembles spontaneously on the substrate surface as the solvent evaporates, although a thermal treatment is usually adopted to improve the quality of the alignment. If the substrate surface is covered by an appropriate alignment layer, then the LC state becomes macroscopically oriented, e.g. to form a monodomain or a defined series of patterned microdomains. This orientation can then be fixed in position by thermal or photochemical polymerization without the orientation being disturbed at all in some cases [16, 17]. Polymerization and crosslinking occur either by the thermal or photo-induced generation of free radicals or by ionic photo-initiation.

Polarized PL from an LC network was first demonstrated using a polymerizable mixture (Cr 116 °C, N 150 °C, I) of guest luminescent diacrylate

Table 14.6. Chemical structure of reactive mesogens with charge-transport and/or luminescence properties.

27	$H_2C{=}CHCO_2C_6H_{12}O$–, O, CH_3, O, O, O, –$OC_6H_{12}O_2CCH{=}CH_2$
28	$OC_6H_{12}OCOC(CH_3){=}CH_2$, $H_2C{=}C(CH_3)CO_2C_6H_{12}O$
29	$H_2C{=}CHCO_2C_3H_6O$–, CH_3, –$OC_3H_6O_2CCH{=}CH_2$
30	$(H_2C{=}C)_2CHO_2CC_5H_{10}O$–, H_7C_3, C_3H_7, S, S, –$OC_5H_{10}CO_2CH(CH{=}CH_2)_2$

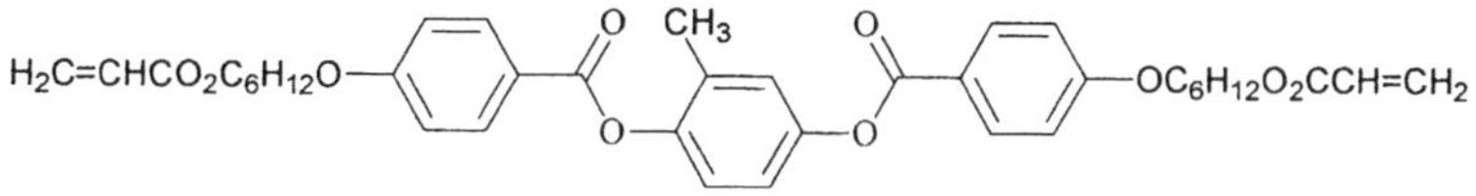

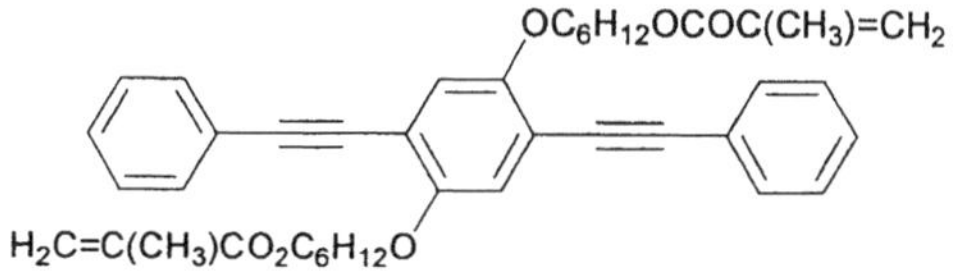

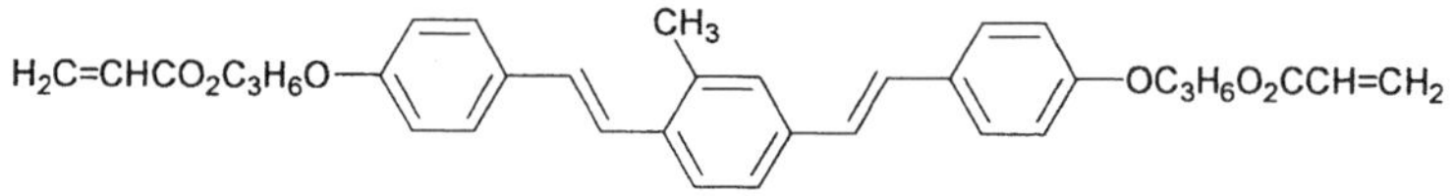

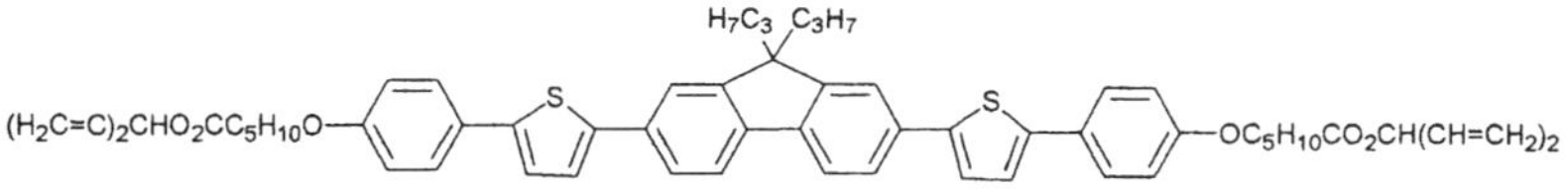

(**28**), a photo-initiator and a thermal inhibitor dissolved in the non-luminescent reactive mesogen (**27**) as a host [125]. A low polarization ratio was obtained. The acrylate (**29**; Cr 90 °C, SmA 144 °C, N 210 °C, I) was synthesized by adding polymerizable acrylate end-groups to a light-emitting phenylene-vinylene molecular fragment to form a similar reactive mesogen to the diacrylate (**27**) [57, 126]. However, the acrylate (**29**) is luminescent due to the presence of the carbon–carbon double bonds (vinylene groups, –CH=CH–) in place of the carboxy groups (COO). It was aligned in the liquid crystalline state and then polymerized by heating up to a temperature of 180 °C to form an anisotropic polymer network [57, 126]. A low PL ratio of polarized emission was determined for the polymerized network. The chromophore of this phenylenevinylene may well be too small for EL and reactive mesogens with a larger aromatic core containing more unsaturated rings have been found to be more suitable, for example the diene (**30**; Cr 92 °C, N 108 °C, I, t_g 39 °C). Suitable molecular engineering allows the synthesis of reactive mesogens with much lower melting points and a nematic phase at room temperature. This allows processing to be carried out at room temperature [14, 15].

Photopolymerization offers a further advantage of pixellation by photolithography. This approach was demonstrated with a crosslinked triphenylene which formed a hole-transporting material for an OLED [127]. The light-emitting and hole transporting calamitic reactive mesogen (**30**) was photopolymerized to form a LC polymer network used in an OLED [14, 157]. Devices consisting of a hole-transporting/luminescent polymer network with and without an overlying electron transporting polymer

Figure 14.9. Photograph of blue PL from a patterned polymer network, irradiated with ultraviolet light at 351 nm [129].

deposited on top of the insoluble network show almost identical electroluminescent spectra confirming that emission originates from the crosslinked reactive mesogen [15]. The three-layer device is more efficient perhaps as a result of an improved balance of electron and hole injection and/or of a shift of the recombination region away from the absorbing cathode. The combination of high charge mobility, low-cost multilayer fabrication, lithographic photopatternability, polarized emission and enhanced output-coupling render calamitic LC polymer networks viable organic materials for OLEDs. Furthermore, LC polymer networks maintain their physical properties up until (very high) decomposition temperatures [57, 120–124].

Figure 14.9 shows PL from a patterned polymer network formed by polymerization and crosslinking of the diene reactive mesogen (**30**). The bright regions were selectively crosslinked by exposure to ultraviolet radiation with the top pixel being only partially crosslinked. The substrate was subsequently washed in chloroform to remove the non-irradiated material. The success of the photopolymerization approach requires minimal photodegradation of the chromophores during crosslinking and no long-term damage from free radicals generated either by the initial irradiation or by subsequent breakdown of the photo-initiator over time. Initial studies of the polymer network formed by polymerization of the diene reactive mesogen (**30**) without the use of a photo-initiator indicate that the PL quantum efficiency is actually increased by the crosslinking procedure [128]. The hole mobility is once again doubled following polymerization (see above) to a value of $3 \times 10^{-5}\,cm^2\,V^{-1}\,s^{-1}$ [41]. Time-resolved PL shows that the quantum efficiency of the diene reactive mesogen (**30**) is limited not by aggregation or excimer formation but by spatial diffusion to traps [128].

The diene reactive mesogen (**20**) shown in table 13.4 was the first organic compound to exhibit a smectic C phase at room temperature (Cr 25 °C, SmC 124 °C, I) with a high electron mobility of $1.5 \times 10^{-5}\,cm^2\,V^{-1}\,s^{-1}$. It can also be photopolymerized to form an electron-transporting polymer network as a thin uniform film [93]. Polymerization of the diene reactive mesogen (**30**) shown in table 13.6, to form a polymer network doubled the hole mobility [41].

Polarized EL and PL from reactive mesogens in the LC state was first demonstrated on mechanically rubbed polyimide alignment layers. However, the process of mechanical rubbing produces dust and scratching, which can be expected to severely affect the performance of thin layer EL devices. These problems are avoided completely using non-contact photo-alignment techniques developed for LCDs and recently adapted to OLEDs [130]. Illumination of a suitable photo-active polymer with polarized ultraviolet light generates a surface anisotropy in the alignment layer and hence a preferred in-plane orientation of an overlying LC phase. An added advantage is that the polarization direction and pattern of EL can be determined photolithographically. Side-chain polymers with photocrosslinkable coumarin pendants, which undergo an anisotropic depletion by 2 + 2 cycloaddition on irradiation with polarized ultraviolet light, were first used in a photo-aligned polarized OLED [14]. Such photo-active polymers have been used to provide photochemically stable alignment for reactive mesogens with strong azimuthal orientation. The luminescent reactive mesogen (**30**) was deposited from solution by spin casting on the anisotropically crosslinked alignment layer. The LC director spontaneously aligns parallel to the polarization direction of the aligning ultraviolet light as the solvent evaporates and the nematic phase self-assembles on the substrate surface. An anisotropic polymer network formed by the photopolymerization of the reactive mesogen (**30**) on a doped photo-alignment layer gives an EL polarization ratio of 11:1, limited by the order parameter of small molecules [14].

Oligomers with a polymerizable groups at the end of two aliphatic spacers, such as the fluorene pentamer (**31**; t_g −10 °C, N 123 °C, I) shown in table 14.7, which are liquid crystalline at room temperature, were aligned on a thin rubbed PPV alignment layer and subsequently crosslinked

Table 14.7. Chemical structure of a polymerizable LC oligomer with charge-transport and/or luminescence properties.

31	$H_2C{=}CHCO_2C_6H_{12}O$–[fluorene]$_2$–fluorene(H_7C_3, C_3H_7)–[fluorene]$_2$–$OC_6H_{12}O_2CCH{=}CH_2$

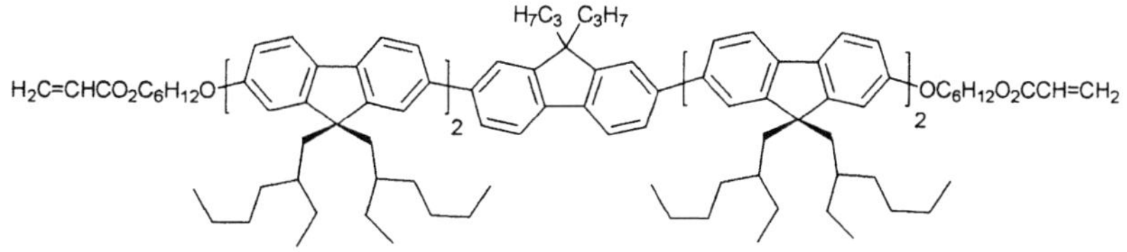

to form a light-emitting polymer network with polarized emission [20]. A *D* ratio of 22:1 was obtained in EL, but it was associated with some broad excimer or exciplex emission.

Polymers with polymerizable endgroups have also been thermally cross-linked in the LC state in order to render them insoluble and thereby facilitate the fabrication of multilayer OLEDs [55, 57]. Polarized emission has not been reported so far. A polymer-stabilized smectic gel, rather than a cross-linked polymer network, containing the phenylnaphthalene (**17**) (see table 14.3) was found to retain the high hole mobility values, $>10^{-3}\ cm^2\ V^{-1}\ s^{-1}$, of the unstabilized pure material at elevated temperatures [90]. Photopoly-merized polymer networks have also been formed from fluorene oligomers with acrylate, rather than diene, end-groups [15, 20]. Photocrosslinking does not require LC phases and a red/green/blue OLED has been reported recently based on crosslinkable oxetane polymers [131]. The hole mobility of a HAT columnar LC was also increased by gelation in a hydrogen-bonded fibre. This enhancement was attributed to an increase in the colum-nar order and to suppression of the molecular fluctuations of the columnar phase by the gel. This is consistent with the fact that the EL threshold voltage of HAT columnar glasses is reduced on annealing OLEDs containing them [37]. The high hole mobility is retained with an improved surface morphology when HAT columnar glasses are dispersed in a polymer matrix [132]. However, crosslinking of discotic LC reactive mesogens in the columnar phase leads to lower mobility values due to the disruption of the columnar stacks in order to facilitate the formation of the polymer chain [133]. Very promising vitrifying chiral nematic reactive mesogens for circularly polarized emission from anisotropic networks have also been reported [134].

14.10 Conclusions

The use of electroluminescent LCs in OLEDs and organic lasers is a relatively new area of research and its full potential will only begin to be realized with optimized LC molecular structure and appropriate macroscopic morphology. The columnar phases of discotic LCs show high mobility resulting from enhanced molecular order when the discs self-organize into one-dimensional columns. However, the materials used do not possess an optimal morphology for use in OLEDs, since they are either crystalline solids or liquid crystalline fluids at room temperature. The former leads to trapping at crystal grain boundaries and the latter to layer flow and problems with OLED fabrication. A possible approach to improved performance is to use glassy, plastic or more sophisticated cross-linked discotic columnar films. The in-plane spontaneous alignment of rod-like LCs gives higher output-coupling efficiencies as well as anisotropic transport for OLEDs. The lower mobilities of rod-like LCs compared with disc-shaped columnar LCs is not a major disadvantage for OLEDs, where the balance of electrons and hole densities is a more important parameter. LC polymer networks formed by the photopolymerization of reactive mesogens is a promising approach for efficient multilayer OLEDs and has the added advantage of lithographic photopatternability for pixel formation. The first LC-PN OLEDs have been demonstrated using nematic LCs, but more research is required to improve the performance of this new type of electro-optic device and to establish its long-term viability. The analogous smectic phases may exhibit higher mobility. Other challenges include the extension of the colour range and efficiency of electroluminescent LCs. Improved conducting alignment layers with high hole mobility are also required for more efficient polarized EL. The mirrorless lasing of one-dimensional photonic bandgap structures using chiral LCs has been demonstrated. The threshold fluence for lasing is too high, but there is a lot of scope for reduction, for example, by operating in the defect mode configuration. So far optical pumping has been exclusively used and electrically pumped lasing in chiral systems would represent a major breakthrough.

References

[1] Kelly S M and O'Neill M 2000 in *Handbook of Advanced Electronic and Photonic Materials* vol. 2 ed. H S Nalwa (San Diego: Academic Press) ch 1

[2] Kelly S M 2000 *Flat Panel Displays* in Advanced Organic Materials RSC Materials Monograph Series, ed. J A Conner (Cambridge: Royal Society of Chemistry)

[3] O'Neill M and Kelly S M 2003 *Adv. Mater.* **15** 1135

[4] Tang C W and Van Slyke S A 1987 *Appl. Phys. Lett.* **51** 913

[5] Burroughs J H, Bradley D D C, Brown A R, Marks N, Mackay K, Friend R H, Burn P L and Holmes A B 1990 *Nature* **347** 539

[6] Forrest S R, Bradley D D C and Thompson 2003 *Adv. Mater.* **15** 1043

[7] Patel N K, Cina S and Burroughes J H 2002 *IEEE J. Select. Topics Quantum Electron.* **8** 346

[8] Cornil J, Beljonne D, Calbert J-P and Bredas J L 2001 *Adv. Mater.* **13** 1053

[9] Friend R H, Gymer R W, Holmes A B, Burroughes J H, Marks R N, Taliani C, Bradley D D C, Dos Santos D A, Bredas J L, Logdlund and M Salaneck W R 1999 *Nature* **397** 121

[10] Chen C H, Shi J and Tang C W 1997 *Macromol. Symp.* **125** 1

[11] Hide F, Diaz-Garcia M A, Schwartz B J, Andersson M R, Pei Q B and Heeger A J 1996 *Science* **273** 1833

[12] Adachi C, Baldo M A, Thompson M E and Forrest S R 2001 *J. Appl. Phys.* **90** 5048

[13] Kim Ho J S, Greenham N C and Friend R H 2000 *J. Appl. Phys.* **8** 1073

[14] Contoret A E A, Farrar S R, Jackson P O, May L, O'Neill M, Nicholls J E, Richards G J and Kelly S M 2000 *Adv. Mater.* **12** 971

[15] Contoret A E A, Farrar S R, O'Neill M, Nicholls J E, Richards G J, Kelly S M and Hall A W 2002 *Chem. Mater.* **14** 1477

[16] Kopp V I, Fan B, Vithana H K M and Genack A Z 1998 *Optics Lett.* **21** 1707

[17] Corcoran N, Arias A C, Kim J S, MacKenzie J D and Friend R H 2003 *Appl. Phys. Lett.* **82** 299
[18] Müller D C, Braig T, Nothofer W G, Arnoldi M, Gross M, Scherf U, Nuygen O and Meerholz K 2000 *Chem. Phys. Chem.* **1** 207
[19] Schartel B, Wachtendorf V, Grell M, Bradley D D C and Hennecke M 1999 *Phys. Rev. B* **60** 277
[20] Jandke M, Hanft D, Strohriegl P, Whitehead K, Grell M and Bradley D D C 2001 *Proc. SPIE* **4105** 338
[21] Müllen K and Wegner G 1992 *Synthesis* **2**
[22] Scherf U and Müllen K G 1998 *Adv. Mater.* **10** 433
[23] Gill R E, Meetsma A and Hadziioannou G 1996 *Adv. Mater.* **8** 212
[24] Larios-Lopez L, Navarro-Rodriguez D, Ariaz-Marin E M, Moggio I, Reyes-Castaneda, Donnio B, LeMoigne J and Guillon D 2003 *Liq. Cryst.* **30** 423
[25] Schmidtke J and Stille W 2003 *Euro Phys. J. B* **31** 179
[26] Voigt M, Chambers M and Grell M 2001 *Chem. Phys. Lett.* **347** 173
[27] Goldberg L S and Schnur J M 1973 US Patent 3771065
[28] Kogelnik H and Shank C V 1972 *J. Opt. Soc. Amer. B* **43** 2327
[29] Kopp V I, Zhang Z Q and Genack A Z 2000 *Proc. SPIE* **3939** 39
[30] Kopp V I and Genack A Z 2002 *Phys. Rev. Lett.* **89** 033901
[31] Kopp V I, Bose R and Genack A Z 2003 *Optics Lett.* **28** 349
[32] Schmidtke J, Stille W and Finkelmann H 2003 *Phys. Rev. Lett.* **90** 083902
[33] Kim J S, Ho P K, Greenham N C and Friend R H 2000 *J. Appl. Phys.* **88** 73
[34] Rabek J F 1987 *Mechanisms of Photophysical Process and Photochemical Reactions in Polymers: Theory and Applications* (Wiley) ch 2
[35] Lee T-W, Park O O, Cho H N, Kim D Y and Kim Y C 2003 *J. Appl. Phys.* **93** 1367
[36] Neher D 2001 *Macromol. Rapid Commun.* **22** 1365
[37] Strohriegl P and Grazulevicius J V 2002 *Adv. Mater.* **14** 1439
[38] Funahashi M and Hanna J 2000 *Appl. Phys. Lett.* **76** 2574
[39] Adam D, Schuhmacher P, Simmerer J, Haussling L, Siemensmeyer K, Etzbach K H, Ringsdorf H and Haarer D 1994 *Nature* **371** 141
[40] Bushby R J and Lozman O R 2002 *Curr. Opin. Colloid Surface Sci.* **7** 343
[41] Farrar S R, Contoret A E A, O'Neill M, Nicholls J E, Richards G J and Kelly S M 2002 *Phys. Rev. B* **66** 125107
[42] Bässler H 1993 *Phys. Stat. Sol. B* **175** 15
[43] Im C, Bässler H, Rost H and Horhold H H 2000 *J. Chem. Phys.* **113** 3802
[44] Pecchia A, Lozman O R, Movaghar B, Boden N, Bushby R J, Donovan K J and Kreouzis T 2002 *Phys. Rev. B* **65** 104204
[45] Cordes H, Baranovskii S D, Kohary K, Thomas P, Yamasaki S, Hensel F and Wendorff J H 2001 *Phys. Rev. B* **63** 94201
[46] van de Craats A M, Siebbeles L D A, Bleyl I, Haarer D, Berlin Y A, Zharikov A A and Warman J M 1998 *J. Phys. Chem. B* **102** 9625
[47] Kreouzis T, Donovan K J, Boden N, Bushby R J, Lozman O R and Liu Q 2002 *J. Chem. Phys.* **114** 1797
[48] Palenberg M A, Silbey R J, Malagoli M and Bredas J L 2002 *J. Chem. Phys.* **112** 1541
[49] Cornil J, Lemaur V, Calbert J-P and Bredas J-L 2002 *Adv. Mater*. **14** 726
[50] Shiyanovskaya I, Singer K D, Twieg R J, Sukhomlinova L and Gettwert V 2002 *Phys. Rev. E* **65** 41715
[51] Ohno A and Hanna J 2003 *Appl. Phys. Lett.* **82** 751

[52] Geens W, Shaheen S E, Wessling B, Brabec C J, Poortmans J and Sariciftci N S 2002 *Org. Electronics* **3** 105
[53] Redecker M, Bradley D D C, Inbasekaran M and Woo E P 1998 *Appl. Phys. Lett.* **74** 1400
[54] Redecker M, Bradley D D C, Inbasekaran M and Woo E P 1998 *Appl. Phys. Lett.* **73** 1565
[55] Miller R D, Klaerner G, Davey M H, Chen W D and Scott J C 1998 *Adv. Mater.* **10** 993
[56] Scherf U and List E J W 2002 *Adv. Mater.* **14** 477
[57] Grell M and Bradley D D C 1999 *Adv. Mater.* **11** 895
[58] Bao Z, Chen Y, Cai R and Yu L 1993 *Macromolecules* **26** 5281
[59] Hamaguchi M and Yoshino K 1994 *Jpn. J. Appl. Phys.* **33** L1478
[60] Oberski J, Festag R, Schmitt C, Lüssem G, Wendorff J H, Greiner A, Hopmeier M and Motamedi F 1995 *Macromolecules* **28** 8676
[61] Lüssem G, Festag R, Greiner A, Schmitt C, Unterlechner C, Heitz W, Hopmeier M, Wendorff J H and Feldmann J 1995 *Adv. Mater.* **7** 923
[62] Lüssem G, Geffarth F, Greiner A, Heitz W, Hopmeier M, Oberski M, Unterlechner C and Wendorff J H 1996 *Liq. Cryst.* **21** 903
[63] Grell M, Knoll W, Meisel A, Miteva T, Neher D, Nothofer H G, Scherf U and Yasuda A 1999 *Adv. Mater.* **11** 671
[64] Whitehead K S, Grell M, Bradley D D C, Jandke M and Strohriegl P 2000 *Appl. Phys. Lett.* **76** 2946
[65] Weder C, Wrighton M S, Spreiter R, Bosshard C and Günter P 1996 *J. Phys. Chem.* **100** 18931
[66] Steiger D, Smith P and Weder C 1997 *Macromol. Rapid Commun.* **18** 643
[67] Weder C, Sarwa C, Montali A, Bastiaansen C and Smith P 1998 *Science* **279** 837
[68] Montali A, Bastiaansen C, Smith P and Weder C 1998 *Nature* **392** 261
[69] Virgili T, Lidzey D G, Grell M, Walker S, Asimakis A and Bradley D D C 2001 *Chem. Phys. Lett.* **341** 219
[70] Sainova D, Zen A, Nothofer H H, Asawapirom U, Scherf U, Hagen R, Bieringer T, Kostromine S and Neher D 2002 *Adv. Funct. Mater.* **12** 49
[71] Yang X H, Neher D, Lucht S, Nothofer H G, Guntner R, Scherf U, Hagen R and Kostromine S 2002 *Appl. Phys. Lett.* **81** 2319
[72] Christ T, Stumpflen and Wendorff J H 1997 *Macromol. Rapid Commun.* **18** 93
[73] Oda M, Nothofer H G, Scherf U, Sunjic V, Ricter D, Regenstein W and Neher D 2002 *Macromolecules* **35** 6792
[74] Peeters E, Christiaans M P T, Janssen R A J, Schoo H F M, Dekkers H P J M and Meijer E W J 1997 *J. Am. Chem. Soc.* **119** 9909
[75] Katsis D, Kim D U, Chen H P, Rothberg L J, Chen S H and Tsutsui T 2001 *Chem. Mater.* **13** 643
[76] Geng Y H, Culligan S W, Trajkovska A, Wallace J U and Chen S H 2003 *Chem. Mater.* **15** 542
[77] Culligan S W, Geng Y H, Chen S H, Klubeck K, Vaeth K M and Tang C W 2003 *Adv. Mater.* **15** 1176
[78] Lüssem G and Wendorff J H 1998 *Polym. Adv. Technol.* **9** 443
[79] van de Craats A M 1998 *J. Phys. Chem.* B **102** 9625
[80] Kreouzis T, Scott K, Donovan K J, Boden N, Bushby R J, Lozman O R and Liu Q 2000 *Chem. Phys.* **262** 489

[81] van de Craats A M, Warman J M, Fechtenkotter A, Brand J D, Harbison M A and Müllen K 1999 *Adv. Mater.* **11** 1469
[82] Liu C-Y, Fechtenkotter A, Watson M D, Müllen K and Bard A J 2003 *Chem. Mater.* **15** 124
[83] van de Craats A M and Warman J M 2001 *Adv. Mater.* **13** 130
[84] Struijk C W, Sieval A B, Dakhorst J E J, van Dijk Kimkes M, Koehorst P R B M, Donker H, Schaafsma T J, Picken S J, van de Craats A M, Warman J M, Zuilhof H and Sudholter E J R 2000 *J. Am. Chem. Soc.* **122** 11057
[85] Bock H, Babeau A, Seguy I, Jolinat P and Destruel P 2002 *Chem. Phys. Chem.* **3** 532
[86] Mizoshita N, Monobe H, Inoue H, Ukon M, Watanabe T, Shimizu Y, Hanabusa K and Kato T 2002 *JCS Chem. Commun.* **428**
[87] Christ T, Glüsen B, Greiner A, Kettner A, Sander R, Stümpflen V, Tsukruk V and Wendorff J H 1997 *Adv. Mater.* **9** 48
[88] Seguy I, Destruel P and Bock H 2000 *Synth. Met.* **111** 15
[89] Rego J A, Kumar S and Ringsdorf H 1996 *Chem. Mater.* **8** 1402
[90] Yoshimoto N and Hanna J 2002 *Adv. Mater.* **14** 988
[91] Funahashi M and Hanna J 1997 *Phys. Rev. Lett.* **78** 2184
[92] Haristoy D, Mery S, Heinrich B, Mager L, Nicoud J F and Guillon D 2000 *Liq. Cryst.* **27** 321
[93] Vlachos P, Kelly S M, Mansoor B and O'Neill M 2002 *JCS Chem. Commun.* **874**
[94] Benning S A, Hassheider T, Keuker-Baumann S, Bock H, Della Sala F, Frauenheim T and Kitzerow H S 2001 *Liq. Cryst.* **28** 1105
[95] Boardman F H, Dunmur D A, Grossel M C and Luckhurst G R 2002 *Chem. Lett.* **1** 60
[96] Sato M, Ishii R, Nakashima S, Yonetake K and Kido J 2001 *Liq. Cryst.* **28** 1211
[97] Sentmann A C and Gin D L 2001 *Adv. Mater.* **13** 1398
[98] Mansoor B, O'Neill M, Aldred M P and Kelly S M private communication
[99] Mochizuki H, Hasui T, Kawamot M, Shiono T, Ikeda T, Adachi C, Taniguchi Y and Shirota Y 2000 *JCS Chem. Commun.* **1923**
[100] Tokuhisa H, Era M and Tsutsui T 1998 *Appl. Phys. Lett.* **72** 2639
[101] Kogo K, Goda T, Funahashi M and Hanna J-I 1998 *Appl. Phys. Lett.* **72** 1595
[102] Aldred M P, Eastwood A J, Kelly S M, Vlachos P, Mansoor B, O'Neill M and Tsoi W C 2003 *Chem. Mater.* in press
[103] Chen S H, Katsis D, Schmid A W, Mastrangelo J C, Tsutsui T and Blanton T N 1999 *Nature* **397** 506
[104] Katsis D, Chen H P, Chen S H, Rothberg L J T and Tsutsui 2000 *Appl. Phys. Lett.* **77** 2982
[105] Voigt M, Chambers M and Grell M 2002 *Liq. Cryst.* **29** 653
[106] Tamaoki N, Kruk G and Matsuda H 1999 *J. Mater. Chem.* **9** 2384
[107] Tamaoki N 2001 *Adv. Mater.* **13** 1135
[108] Woon K L, O'Neill M, Richards G J, Aldred M P, Kelly S M and Fox A M 2003 *Adv. Mater.* **15** 1555
[109] Bobrovsky A Y, Boiko N I, Shibaev V P and Wendorff J H 2003 *Adv. Mater.* **15** 282
[110] Shibaev P V, Tang K, Genack A Z, Kopp V and Green M M 2002 *Macromolecules* **35** 3022
[111] Ozaki M, Kasano M, Ganzke D, Haase W and Yoshino K 2002 *Adv. Mater.* **14** 306
[112] Finkelmann H, Kim S T, Mnoz A, Palffy-Muhoray P and Taheri B 2001 *Adv. Mater.* **13** 1069

[113] Furumi S, Yokoyama S, Otomo A and Mashiko S 2003 *Appl. Phys. Lett.* **82** 16
[114] Schmidtke J, Stille W, Finkelmann H and Kim S T 2002 *Adv. Mater.* **14** 746
[115] Matsui T, Ozaki R, Funamoto K, Ozaki M and Yoshino K 2002 *Appl. Phys. Lett.* **81** 3741
[116] Cao W Y, Munoz A, Palffy-Muhoray P and Taheri B 2002 *Nature Mater.* **1** 111
[117] Taheri B, Munoz A F, Palffy-Muhoray P and Twieg R 2001 *Mol. Cryst. Liq. Cryst.* **358** 73
[118] Alvarez E, He M, Munoz A F, Palffy-Muhoray P, Serak S V, Taheri B and Twieg R 2001 *Mol. Cryst. Liq. Cryst.* **369** 75
[119] Chambers M, Fox M, Grell M and Hill J 2002 *Adv. Funct. Mater.* **12** 808
[120] Broer D J, Finkelmann H and Kondo H 1988 *Makromol. Chem.* **189** 185
[121] Broer D J, Boven J, Mol G N and Challa G 1989 *Makromol. Chem.* **190** 2255
[122] Kelly S M 1995 *J. Mater. Chem.* **5** 2047
[123] Kelly S M 1998 *Liq. Cryst.* **24** 71
[124] Schadt M 1997 *Annu. Rev. Sci.* **27** 305
[125] Davey A P, Howard R G and Blau W J 1997 *J. Mater. Chem.* **7** 417
[126] Bacher A, Bentley P G, Bradley D D C, Douglas L K, Glarvey P A, Grell M, Whitehead K S and Turner M L 1999 *J. Mater. Chem.* **9** 2985
[127] Bacher A, Erdelen C H, Paulus W, Ringsdorf H, Schmidt H-W and Schuhmacher P 1999 *Macromolecules* **32** 4551
[128] Contoret A E A, Farrar S R, Khan S M, O'Neill M, Richards G J, Aldred M P and Kelly S M 2003 *J. Appl. Phys.* **93** 1465
[129] Contoret A E A private communication
[130] O'Neill M and Kelly S M 2000 *J. Phys. D: Appl. Phys.* **33** R67
[131] Müller C D, Falcou A, Rojahns M, Wiederhirn V, Rudati P, Frohne H, Nuykens O, Becker H and Meerholz K 2003 *Nature* **421** 829
[132] Bayer A, Kopitzke J, Noll F, Seifert A and Wendorff J H 2001 *Macromolecules* **34** 3600
[133] Disch S, Finkelmann H, Ringsdorf H and Schumacher P 1995 *Macromolecules* **28** 2424
[134] Pfeuffer T and Strohriegl P 2001 *Proc. SPIE* **4463** 40

Chapter 15

High T_c oxide superconductors

D D Shivagan, B M Todkar and S H Pawar
School of Energy Studies, Department of Physics, Shivaji University, Kolhapur, India

15.1 Introduction

The discovery of high-temperature superconductivity in La-based cuprate by Bednorz and Muller in 1986 [1] stimulated unprecedented excitement and led to the discovery of different families of cuprates, which remain superconducting at appreciable temperature. Some of the important ones were Y(RE)-123, Bi-2122/2223, Tl-2122/2223 and Y-124, Hg-1223 and MgB_2. Though superconductivity is a transport phenomenon, there are several good reasons for the investigations of luminescence in these materials. One of them is that many of these high-T_c materials belong to a class of materials of oxygen-dominated lattices. The luminescence of oxygen-dominated lattices has been observed for many years in naturally occurring phosphates, silicates, carbonates and other materials. Their tremendous commercial importance has had a strong influence on the nature of the work done on oxygen-dominated materials. Another most important reason for studying luminescence is the sensitivity of this technique to defect properties of the materials like defect concentration, disorder, substitution or contamination of a special system with other phases. It is particularly this latter reason which makes luminescence studies attractive for the research work in the field of high-temperature cuprate materials. The data available on luminescence aspects, however, remain relatively scant. This is due to the fact that these materials are uniformly black and of course opaque to visible radiation. Luminescence observed is thus strictly confined to the surface regions and the comparatively less intense as found in various other systems such as sulphide and sulphate phosphors doped with rare earth impurities [2–14]. Nevertheless, the new materials have been studied for a variety of luminescence phenomena like photo-, thermo-, electro- and cathodo-luminescence. This chapter deals with the different types of luminescence found in these novel systems and in particular the electroluminescence (EL) of (123) high-T_c oxide superconductors.

15.2 Luminescence in high-T_c oxide superconductors

15.2.1 Oxygen-dominated lattices

Oxygen-dominated lattices are those in which oxygen is a major chemical constituent of host lattice. There are number of luminescent oxide compounds which can be classified into five groups as (a) simplest compounds with generic formula M_xO_z, (b) binary oxides $M_xA_yO_z$, (c) ternary oxides $M_xA_yB_{y'}O_z$, (d) quaternary oxides $M_xA_yB_{y'}C_{y''}O_z$ and (e) others containing halides [15]. The rare earth (RE)-based high-T_c oxide superconductors are of the type of ternary oxides while the bismuth based high-T_c superconductors are of quaternary type.

The basic crystal structure of RE-based oxide superconductors is either tetragonal or orthorhombic. For perfect tetragonal lattice, the base is $RE_1Ba_2Cu_3O_6$, while for perfect orthorhombic lattice the base is $RE_1Ba_2Cu_3O_7$. Though these two phases of oxides are structurally perfect, the charge neutrality requires the value of Cu to fluctuate from 1^+ to 3^+. The processing parameters, in particular the processing environment, i.e. oxygen, argon, air annealing etc., give rise to structural defects in the form of oxygen vacancies, which play an important role not only in superconductivity but also in many other properties; one, which is very sensitive, is the phenomenon of luminescence.

The oxygen-dominated lattices showing luminescence are generally wide-band gap semiconductors or insulators. They are commonly doped with RE elements (bismuth, thallium, manganese, copper etc.) as impurities, which serve as luminescence centres. Depending on the preparation methods and conditions, the high-T_c superconductors may exhibit metallic, semiconducting or even insulating behaviour and simultaneously also show luminescence. Non-optimized processing conditions can lead to non-homogeneous materials which may in addition contain unreacted and other compounds as impurity phases, which have long been known as luminescent materials. Thus, the sample processing is expected to be of overriding importance in controlling the luminescence exhibited by the samples. Because of the complexity of oxygen-dominated superconducting lattices, the investigation of luminescence mechanisms has led mainly to the phenomenological theory. The origin of luminescence in high-T_c oxygen-dominated lattices may be due to (a) the presence of unreacted luminescent grade simple compound oxides used in systems and (b) host lattice defects mainly due to oxygen vacancies, in otherwise pure cuprate materials. Further, the luminescence in high-T_c superconductors is restricted to a few layers at the surface. The light emitted by the luminescent centres within the interior of the bulk does not come out due to the high optical absorption coefficient of the sample. In the case of polycrystalline high-T_c superconductors, on the microscopic level, the luminescence yield is not uniform. It is higher at the grain boundaries than at the interior of the grain.

15.2.2 Defects

The phenomenon of luminescence is very sensitive to the defects in solids. These defects give rise to different glow peaks in thermoluminescence: origin of phosphorescence (PL) decay, shift in fluorescence spectra, changes in intensities of luminescence due to the phenomena of sensitization and quenching, EL etc. The defects in high-T_c superconducting cuprates have been reviewed by Zhu *et al* [16]. Considering the structure of these materials, there are mainly five types of defects, namely: faults in stacking sequences, twin boundaries, dislocations, grain boundaries, and oxygen/vacancy ordering defects. Out of these, oxygen/vacancy-ordering defects have played important role in T_c values. The variation of T_c values with oxygen concentration is shown in figure 15.1 for $Y_1Ba_2Cu_3O_{7-\delta}$.

The variation in oxygen concentration gives rise to the change in oxygen ordering in the basal plane, containing –O–Cu–O– chains along the *b* direction, which is also responsible for the tetragonal to the orthorhombic transition. The characteristic feature of the oxygen/vacancy ordering is that the oxygen vacancies will form long vacancy chains (–□–Cu–□–) along the *b* direction (□ represents a vacant oxygen site in the O–Cu–O chain) rather than a random distribution. Furthermore, the chains of –O–Cu–O– and of –□–Cu–□– will order to form a superstructure. de Fontain *et al* [17] have shown that a series of stable ordered structures,

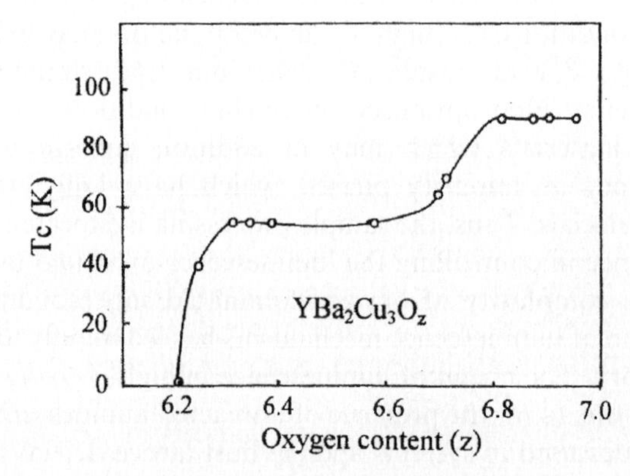

Figure 15.1. The variation of critical temperature (T_c) with oxygen content in $Y_1Ba_2Cu_3O_{7-\delta}$ [16].

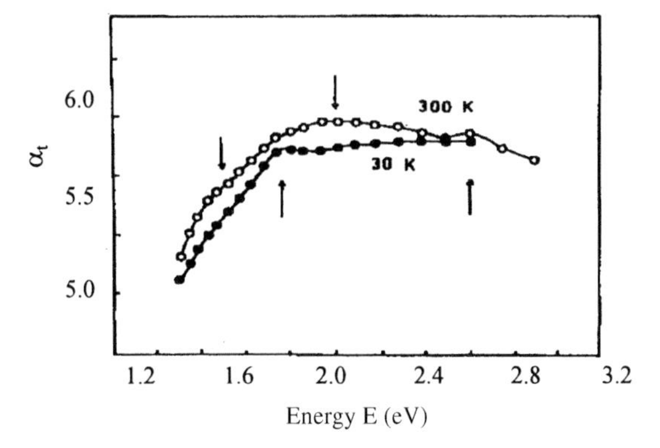

Figure 15.2. Absorption spectra: variation of α_t with the energy of the photon for Y–Ba–Cu–O superconducting films at 300 and 30 K [22].

e.g. O□, □OO and O□□, OOO□, O□□□ etc. (here O and □ represent (010) planes of filled and empty oxygen sites, respectively) can be predicted by extending the range of effective pair interaction perpendicular to the chain direction among the filled and unfilled O–Cu–O chains. The evidence for these various extended ordered structures was observed in diffraction and in dark field images [18–21].

15.2.3 Optical transitions

Luminescence is mainly associated with the optical transitions of the ions either of the host or impurity in a system. It is well known that the high-T_c materials behave in different ways in normal state and superconducting state. To some extent, this has also been observed in case of optical behaviour and in the phenomenon of luminescence [22–30]. Recently Fugol *et al* [22] have reported the optical absorption studies of $Y_1Ba_2Cu_3O_{7-\delta}$ superconducting thin films in the optical range 1.5–3.0 eV. The optical absorption spectra recorded at two different temperatures (30 and 300 K) are shown in figure 15.2.

It has been observed that the four different overlapping absorption bands peaking at around energy values of 1.7, 1.95, 2.2 and 2.8 eV behave in different ways in normal and superconducting states indicating their different origins. In order to understand the origin of bands precisely the difference absorption spectrum has been estimated as shown in figure 15.3.

It is seen that the optical absorption bands corresponding to 1.95 and 2.8 eV are least disturbed, while bands at about 1.7 and 2.2 eV are most perturbed; by changing the system from its normal to superconducting state. The analysis of the data reveals that these bands are of two different types [23]. The 1.7 and 2.2 eV bands belong to metallic type, and are associated with the transition to Fermi level inside the hybridized Cu–O valance

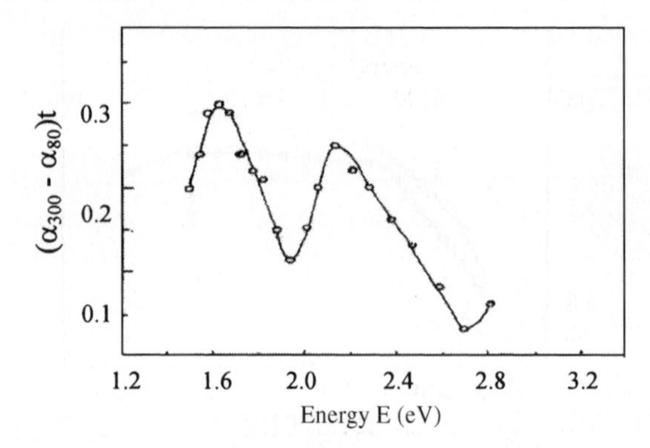

Figure 15.3. Difference absorption spectra: variation of $(\alpha_{300}-\alpha_{80})t$ for $Y_1Ba_2Cu_3O_{7-\delta}$ [22].

band. The 1.95 and 2.8 eV bands belong to dielectric type and related to the charge transfer transitions through the optical gap E_g.

15.2.4 Photoluminescence

The RE-doped [31–34] as well as RE-based host lattice materials [35] are well known for their photoluminescence (PL) properties. On the similar line, Tissue and Wright [36] have reported the first observation of sharp line RE fluorescence in the $La_{1.85-x}RE_xSr_{0.15}CuO_4$ type of superconductors for RE = Pr and Eu and $x = 0.02$, 0.18, 0.36 and 1.85. This observation was surprising since the materials are uniformly black and opaque with no identification of small amounts of green or insulating phases. However, the fluorescence was not intense in comparison with other doped RE systems but it was still large enough to perform interesting spectroscopy. The opaque nature of the sample restricted the fluorescence to the material near surface. The fluorescence spectra at 13 K of $La_{1.67}Pr_{0.18}Sr_{0.15}CuO_4$ after exciting with 474.6 nm and of $La_{1.67}Eu_{0.18}Sr_{0.15}CuO_4$ after exciting with 526.2 nm are shown in figures 15.4(a) and (b), respectively.

The most significant aspect of this work is the realization that RE-excited fluorescence can be used as a local probe of the environment in the same manner that has proven so useful in insulating materials.

Tissue and Wright [37] have further studied the phenomenon of photoluminescence in doped superconducting cuprates to identify the impurity and insulating phases. It is revealed that the impurity phases are residual hexagonal La_2O_3 starting material and a La-silicate phase, which forms during sintering from contamination with boat or furnace material. The phenomenon of photoluminescence has been successfully used to monitor the solid-state reaction rate of the starting materials and to observe heterogeneity in the distribution of impurity phases in the final samples.

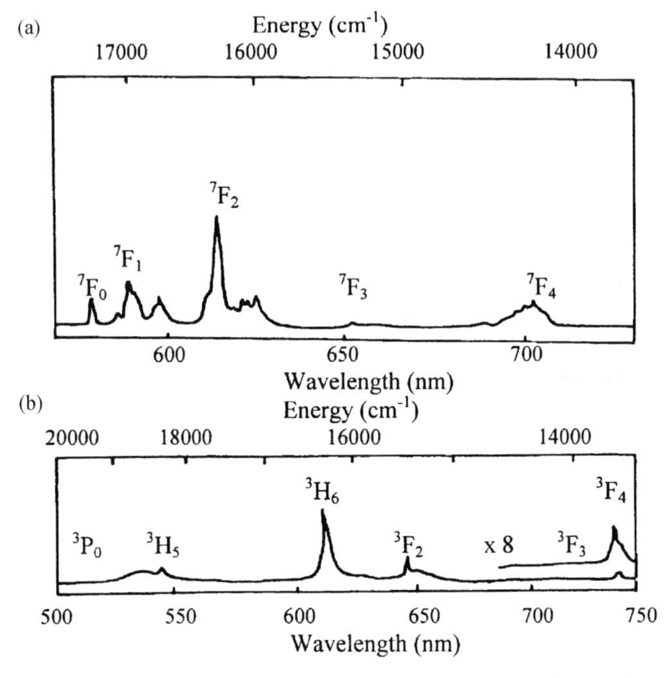

Figure 15.4. Fluorescence spectra at 13 K of: (a) $La_{1.67}Pr_{0.18}Sr_{0.15}CuO_4$ after excitation with 474.6 nm, (b) $La_{1.67}Eu_{0.18}Sr_{0.15}CuO_4$ after excitation with 526.2 nm [36].

The solid-state reaction rate of the starting materials in synthesizing the high-T_c superconductors was studied by monitoring the disappearance of photoluminescence in $La_2O_3:Eu^{3+}$. The photoluminescence intensity of $^5D_0 \rightarrow {}^7F_0$ transition at 581.0 nm in figure 15.4(a) was measured for different heat-treated samples to determine the amount of unreacted La_2O_3 in the samples. The variation of photoluminescence intensity with heating time is shown in figure 15.5.

From the figure, it is seen that the solid-state reaction is almost completed in 12–18 h. Thus, Tissue and Wright have successfully demonstrated that the PL has proved to be very useful as an analytical tool to monitor the reaction rate of the starting materials and to detect small amounts of impurity phases in high-T_c cuprate superconductors.

Thomas and Nampoori [38] studied PL spectra of supercon-ducting and non-superconducting $YBa_2Cu_3O_7$ at room and liquid nitrogen temperatures by using 420 nm radiations for excitation. The excitation energy of 3 eV (420 nm radiations) is the same for both superconducting and non-superconducting samples, and it is attributed to the transitions of an oxygen-derived valance band corresponding to $3d^9 \rightarrow 3d^{10}\underline{L}$ ($\underline{L}$ is a hole in an oxygen-derived valance band). The PL spectra both in superconducting and non-superconducting states are shown in figure 15.6.

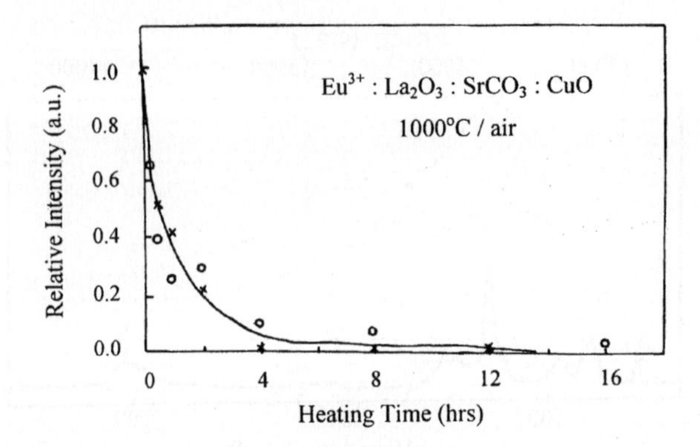

Figure 15.5. Kinetics of the solid state reaction of $La_{1.85}Sr_{0.15}CuO_4$ monitoring the disappearance of $La_2O_3:Eu^{3+}$ [37].

With the same excitation energy, significant changes are observed in the spectra of superconducting and non-superconducting samples. At room temperature, the non-superconducting sample shows the emission peak at about 640 nm corresponding to radiative de-excitation from the lower edge of $3d^{10}\underline{L}$ to $3d^9$ state. For superconducting sample, in addition to $3d^9$ and $3d^{10}\underline{L}$ states, a new configuration involving $3d^9\underline{L}$ is formed [39].

The PL peaks observed at about 580 and 540 nm in super-conducting sample are attributed respectively to the transitions $3d^{10}\underline{L} \rightarrow 3d^9\underline{L}$ and $3d^{10}\underline{L} \rightarrow 3d^9$. From figure 15.6, it is further seen that there is a quenching in PL of both superconducting and non-superconducting samples at liquid nitrogen temperature. The authors, however, could not give an explanation for these observations. Similar quenching of PL has also been observed by

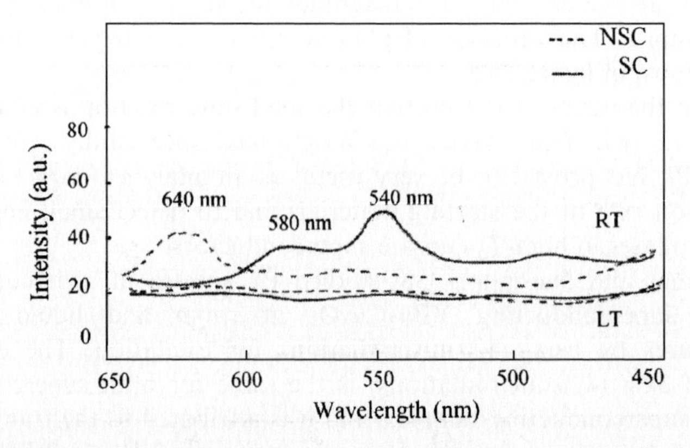

Figure 15.6. Photoluminescence spectra at room temperature and liquid nitrogen temperature for superconducting and non-superconducting $YBa_2Cu_3O_{7-\delta}$ samples with 420 nm excitation [38].

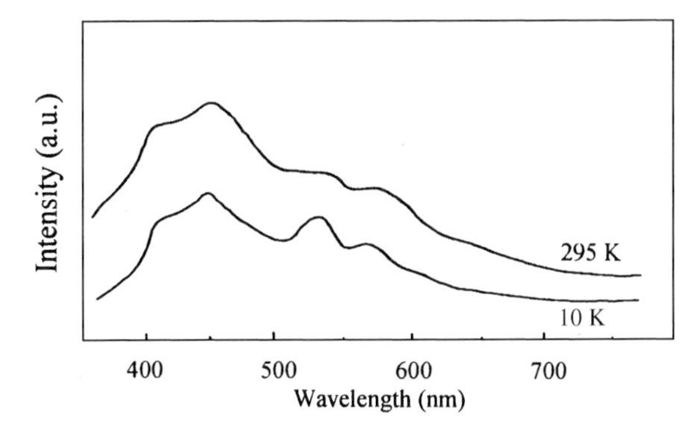

Figure 15.7 Photoluminescence spectra of $YBa_2Cu_3O_{7-\delta}$ at 10 and 295 K. Excitation 308 nm [40].

Yongan *et al* [40]. PL spectra of $YBa_2Cu_3O_{7-x}$ recorded at 10 and 295 K are shown in figure 15.7.

The quenching in PL can be attributed to the optical absorption behaviour with temperature. As shown in figure 15.2, the optical absorption (α_t) is smaller at low temperature. The systematic variation of differential absorption with temperature is shown in figure 15.8.

From figure 15.8 it can be observed that as the temperature of $YBa_2Cu_3O_{7-x}$ decreases, the optical absorption decreases, and hence the number of photons utilized for excitation decreases, giving rise to quenching in PL.

An attempt has been made to understand the origin of PL in $Y_1Ba_2Cu_3O_{7-x}$ by Yongan *et al* [40]. From figure 15.7, it is seen that there are four emission bands peaking at about 408, 445, 525 and 565 nm. Further precise experiments have revealed that the 525 nm band excited by

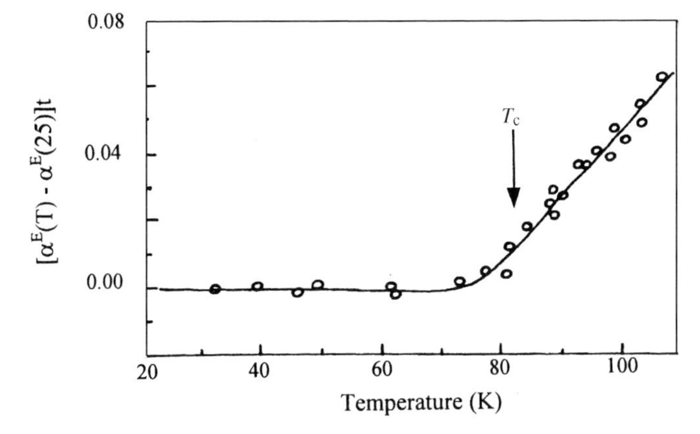

Figure 15.8 Temperature behaviour of difference absorption spectra [22].

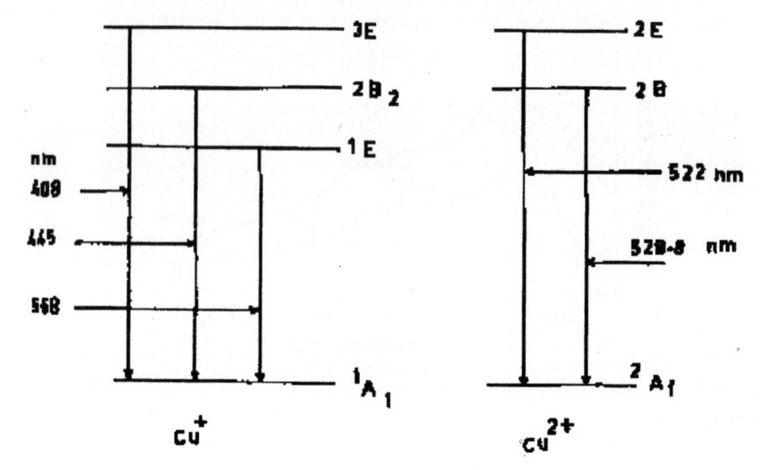

Figure 15.9. The energy level diagram for Cu^+ and Cu^{2+} ions indicating the optical transition for the five emission bands observed in the photoluminescence of $YBa_2Cu_3O_{7-\delta}$ [40].

the 308 nm laser line can be apparently deconvoluted into two bands, i.e. the 523 and 528.8 nm bands. Thus there are five bands totally belonging to emissions of Cu^+ and Cu^{2+} ions. Having considered the energy levels of Cu ions in the crystal field, and compared the emissions with the emissions of Cu_2O and CuO, the optical transitions for five emission bands are shown in figure 15.9.

The studies on temperature dependence of emission intensity showed that the 408, 445 and 568 nm bands of Cu^+ become intense compared with the 523 and 528.8 nm bands of Cu^{2+}, when the temperature rises from 10 to 300 K.

It is well known that the doping in $Y_1Ba_2Cu_3O_{7-x}$ alters the superconducting properties depending on the occupation of impurity in the host lattice site. The substitution of Mn in place of Cu alters the superconducting properties. Similarly the substitution of Mn has altered the PL behaviour. Besides the five Cu bands, four emission bands appear in the PL spectrum. Among the five emission bands, the 568.4 nm band is greatly reduced, indicating that Mn ions mainly replace the Cu ions on the Cu–O planes. The temperature dependence of emission intensity of Mn band peaking at around 618.8 nm is shown in figure 15.10.

It can be noticed that the intensity reduces sharply when the temperature is dropped to 90 K, when the sample goes from its normal state to the superconducting state.

15.2.5 Thermoluminescence

The phenomenon of thermoluminescence (TL) was reported for the first time in high-T_c superconductors by Cooke *et al* [41] and later by others [42–47].

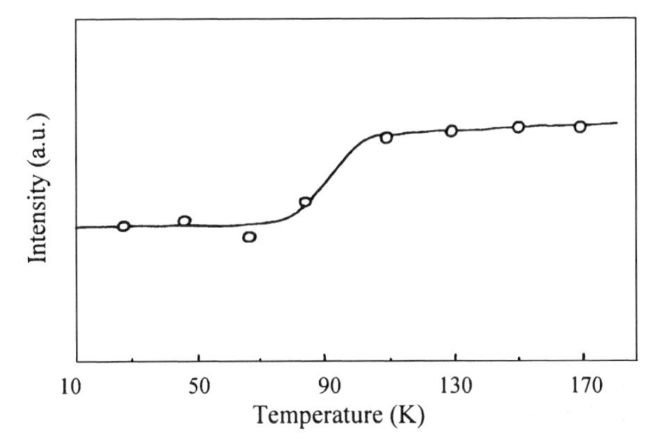

Figure 15.10. Temperature dependence of Mn^{2+} bands peaking at about 618.8 nm [40].

Cooke *et al* [41] have reported TL in single-phase $GdBa_2Cu_3O_7$ and two-phase $RE_{1.5}Ba_{1.5}Cu_2O_x$ (RE=Ho, Eu) compounds. They have observed TL glow peaks between 80 and 67.5 K after excitation with x-rays at low temperature. They have proposed tentative explanations for the defect mechanism involved, e.g. recombination of electrons trapped as F-type colour centres with thermally released V-type holes, or de-excitation of RE ions.

Roth *et al* [43] have reported TL in 2*N* (low purity) and 5*N* (high purity) grade samples of $Y_1Ba_2Cu_3O_7$ with x-irradiation at low temperatures (20 K). High-purity samples exhibit two intense glow peaks at 90 and 195 K (figure 15.11) corresponding to their activation energies of 0.16 and 0.42 eV respectively.

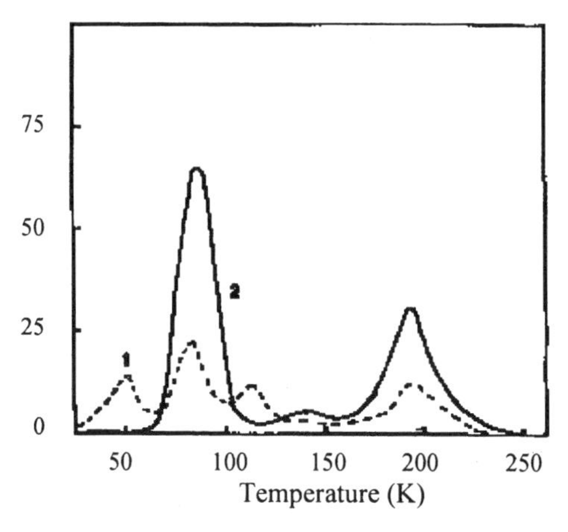

Figure 15.11. Thermoluminescence glow curves of 2*N* grade (curve 1) and high purity 5*N* grade (curve 2) $Y_1Ba_2Cu_3O_7$ [43].

The low purity grade sample shows four glow peaks at about 51, 85, 112 and 195 K. The two glow peaks at about 90 and 195 K are commonly observed for all the samples and related to the intrinsic lattice defects possibly related to oxygen vacancies.

15.2.6 Cathodoluminescence

The excitation by electron bombardment and getting luminescence is known as cathodoluminescence (CL). The CL in scanning electron microscope (SEM) is a well-established tool concerning semiconducting materials. Its application for high-T_c superconductor is a new subject of interest. Recently Miller *et al* [48] have performed CL measurements in SEM on Y–Ba–Cu–O ceramic pellets. They have reported that the high-T_c superconducting phase is less luminescent than its related semiconducting and insulating phase. Similar observations have been made in other types of luminescence. However, the CL technique gives direct physical evidence and hence is more useful in distinguishing the different impurity phases in high-T_c superconductors.

Barkay *et al* [49] have performed CL measurements on Bi–Sr–Ca–Cu–O and Y–Ba–Cu–O high–T_c superconductors in thin film forms. The depth and the lateral locations of different phases have been revealed. In thin films, unlike the bulk superconductors, the role of substrate plays an important role in CL, when the energy of the electron beam is higher. The generation depth of CL increases in proportion to the beam electron range (R) in the sample, which depends on the electron energy (E) as $R \approx E^{1.5-1.7}$. The CL measurements were made at different temperatures down to liquid nitrogen temperature and it was found that CL intensity increases with decrease in temperature. This observation is in contradiction to that of the behaviour made in photoluminescence. The increase in CL intensity with decrease in temperature can be understood as follows: the CL phenomenon is not only restricted to the surface of the sample but also to the bulk of the sample as the electron beam penetrates more in depth than that of the photo-excitations. From the optical behaviour of high-T_c superconductors, it is known that the optical absorption decreases with decrease in temperature (figures 15.2 and 15.3) and become relatively more transparent to the visible light at low temperature. This gives rise to the enhancement in CL as there is contribution not only from the surface but also from the interior of the sample. Thus the CL signal can be used for mapping the three-dimensional location of phases in the thin film of high-T_c superconductors.

15.2.7 Electroluminescence

The nature of EL depends on the physical, optical and electrical properties of the substance. Thus, by studying the EL behaviour, one can throw some light on these properties of the substance. The authors have studied the EL behaviour of oxide high-T_c superconductors such as Y–Ba–Cu–Zr–O, Bi–Sr–Ca–Cu–O, Y–Ba–Cu–O, Y–Gd–Ba–Cu–O, Y–Ba–Cu–Mn–O etc., and predicted that there is some correlation between EL and high-T_c superconductivity.

15.3 $Y_1Ba_2Cu_3O_{7-\delta}$ superconductors

The voltage and frequency dependence of EL brightness in a Y–Ba–Cu–O superconductor system may give a clue to help one understand the mechanism of superconductivity in the high-T_c superconductors. The EL emission is influenced by various factors such as applied voltage and its frequency, temperature, magnetic field, design of the cell and its time of operation, dielectric media etc. The study of voltage dependence of EL brightness provides valuable information regarding the mode of excitation of charge carriers in the phenomena of EL. When the a.c. voltage is applied across Y–Ba–Cu–O sample, it is found to emit light. The intensity of the light emitted due to the application of the electric field is the function of the amplitude and frequency of the applied voltage. In order to investigate the dependence of brightness on voltage and frequency, the EL measurements were carried out in the voltage range (from zero to the above threshold voltage V_{th} at which the EL emission starts) and at ambient frequencies, and the graphs of brightness (in arbitrary units) were plotted against the applied voltage. From the drawn brightness–voltage curves for $Y_1Ba_2Cu_3O_{6.94}$, it has been observed that the EL intensity is an increasing function of the applied voltage and this increase of light output is more rapid at higher frequencies. It can also be noted from these plots that some minimum voltage is required, above which EL starts, called the threshold voltage, V_{th}. Above V_{th} the brightness increases with an increase in exciting voltage. This increase in EL brightness can be understood on the basis that initially the number of particles in which EL takes place is small, but on increasing the voltage, more and more active regions are exposed to voltage gradient above the threshold level, giving rise to increase in brightness.

It is found [50–56] that there are several equations to explain the behaviour of brightness with voltage at a fixed frequency. The nature of the possible relationship existing between brightness B and the applied voltage V is investigated by plotting the following graphs.

1. The logarithm of B/V as a function of $1/V$.
2. The logarithm of B/V as function of $1/\sqrt{V}$.
3. The logarithm of B as function of $1/V$.
4. The logarithm of B as function of V.
5. The logarithm of B as function of $1/\sqrt{V}$.

Oxygen content also plays an important role in the EL of high-T_c Y–Ba–Cu–O superconductor. An EL study has been made in the authors' laboratory on $Y_1Ba_2Cu_3O_{6.94}$, $Y_1Ba_2Cu_3O_{6.61}$ and $Y_1Ba_2Cu_3O_{6.29}$ samples in the voltage range 0–1500 V_{rms} and frequency range of 50–2000 Hz. In the following sections, an attempt has been made to discuss the dependence of EL brightness of $Y_1Ba_2Cu_3O_{6.94}$ on applied voltage and its frequency, and later the effect of oxygen content on EL brightness have been discussed.

Further, the EL emission is also found to be the frequency-dependent in the case of an applied a.c. field. The variation of brightness with frequency for a high-T_c sample at fixed voltage exhibits brightness which increases with increase in frequency of the applied voltage. The increase in the brightness with increase in frequency can be explained as follows.

The applied electric field to the sample obeys the relation

$$E = E_0 \sin(ft)$$

where E_0 is the electric field at $t = 0$ and f is the frequency of applied field.

In one cycle of the applied field, the sample gets exposed twice to the peak value of E, emitting two quanta by the luminescence centre per cycle. As the frequency f increases, the number of emitting quantas also increases. This should yield linearity between f and B. However, non-linearity has been reported in many phosphor systems [57–63] at high frequencies, wherein the saturation effects set up.

15.3.1 Role of oxygen content

The oxygen content in Y–Ba–Cu–O superconductors has been found to affect both the T_c values and the EL output [64]. The role of host lattice defects in the form of oxygen vacancies towards EL behaviour of these samples has been studied. The variation of EL output with oxygen content is shown in figure 15.12. It is seen from the figure that EL intensity increases with decrease in the oxygen content of the samples. This may imply that the host lattice defect (oxygen vacancies) act as luminescence centres.

The EL behaviour of different samples as a function of the a.c. excitation voltage has also been investigated. Typical variation of brightness with applied a.c. voltage at a frequency of 500 Hz is shown in figure 15.13. At a fixed frequency, the brightness B/A was found to be an increasing function of voltage. The linearity of $\log(B/A)$ versus $\log(V/t)$ plots (figure 15.14) indicates the power law relation of the form $B = AV^n$, where A and n are constants.

The oxygen content in Y–Ba–Cu–O has also been found to influence the threshold voltage (V_{th}) and is found to decrease gradually from 255 to 150 V/mm with the decrease in oxygen content from $O_{6.94}$ to $O_{6.29}$. This is in accordance with the fact that the excitation potential of the luminescence centre is dependent on its surrounding in the host lattice.

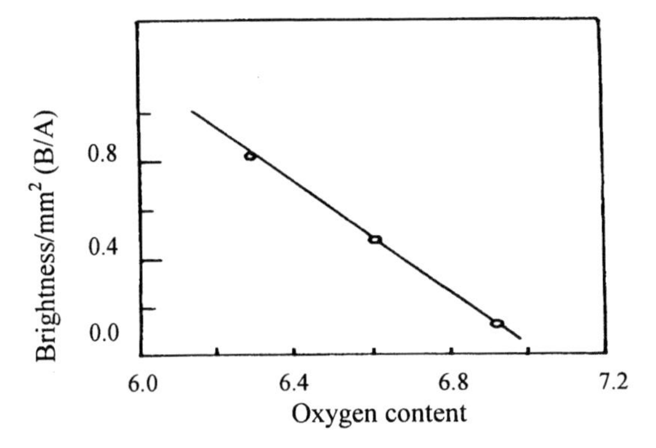

Figure 15.12. Variation of EL output with oxygen content [64].

15.3.2 Effect of Gd concentration

An understanding of the superconducting and structural properties of the doped material is important, not only to elucidate the nature of possible mechanism, but also because potentially deleterious phases are commonly formed during synthesis of composites. In the case of the complete substitution of Y^{3+} by the magnetic and isoelectronic RE ions [65, 66], that is $RBa_2Cu_3O_7$ (where R=La, Nd, Sm, Gd, Ho, Er and Lu), there is no substantial change in the superconducting transition temperature T_c. Moreover, the magnetic moment of these lanthanides obviously does not seem to interfere with the occurrence of superconductivity in these systems. Because of the presence of oxygen vacancies and dopant activator elements in this system, EL studies have been thought to be quite interesting. EL in

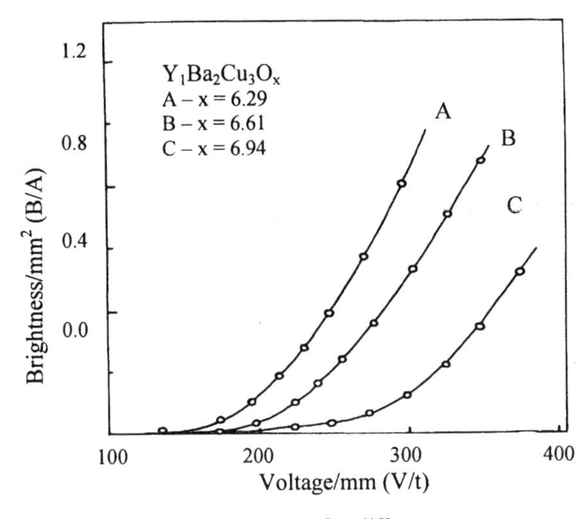

Figure 15.13. Brightness versus voltage curves for different oxygen content samples [64].

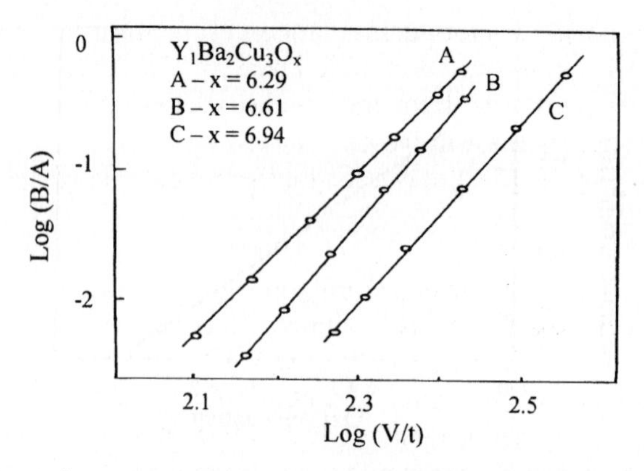

Figure 15.14 Log(B/A) versus log(V/t) plots for different oxygen content samples [64].

oxygen-dominated lattices has been studied earlier at length [15]. Oxygen plays an important role in EL and superconducting properties of high-T_c Y–Ba–Cu–O superconductors. Both superconducting as well as luminescence properties were found to be dependent on host lattice defects (oxygen vacancies). The EL output is seen to increase with oxygen deficiency.

Different Y–Ba–Cu–O samples containing Gd up to 100 at% and having nearly the same oxygen content of 6.94 were studied for their resistance and EL behaviours. Not much change has been found in the T_c ($R = 0$) values with Gd addition. However, their EL behaviour is seen to systematically vary. The EL behaviour of the samples undoped and doped with Gd has been studied as a function of a.c. voltage and frequency.

EL studies made on Gd-substituted Y–Ba–Cu–O superconductors, in the voltage range 0–1500 V_{rms} and frequency range 50–2000 Hz show systematic change in EL behaviour. However, there is no much change in their T_c ($R = 0$) values with Gd addition.

It is seen that the EL intensity is an increasing function of the applied voltage as well as dopant concentration. At lower dopant concentration, the threshold voltage V_{th} has been found to decrease with the increase in dopant concentration. On the other hand, at higher dopant concentration, the value of threshold voltage has been observed to increase with the increase in Gd concentration. It has been found that at 0.25 dopant concentration, the value of the threshold voltage is minimum.

At a fixed frequency, beyond a certain threshold voltage, the brightness was found to be the increasing function of applied voltage. Curves showing the voltage dependence of brightness were essentially similar for different samples though the magnitude varies. The nature of the possible relationship existing in this system between brightness B and the applied voltage V has been investigated to be nonlinear. However, $\log B$ versus $\log(V/t)$ plots exhibit linear behaviour. This nature of the plots suggests that the excitation of charge carriers is by field ionization of impurity ion electrons.

Log B versus log F plots drawn for all the samples of $Y_{1-x}Gd_xBa_2$-$Cu_3O_{6.94}$ have been shown to exhibit linearity. As explained earlier, this increase in brightness with frequency is due to an increase in the number of quanta with the increase in frequency.

The nature of the plots obeys the mathematical relation

$$B = Af^n$$

where A and n are constants depending on the concentration of Gd.

The EL brightness has been found to decrease with increase in Gd content up to a certain value (0.25 at.wt%). Beyond this concentration, the EL brightness has been increased continuously, showing quenching of EL brightness at higher dopant concentration, known as concentration quenching. The threshold voltage V_{th} is also found to be the function of Gd concentration. At lower values of Gd, the threshold voltage has been observed to decrease with increase in Gd content. However, for higher Gd content, V_{th} has been seen to increase with increase in Gd content.

The EL intensity first increases with increase in Gd concentration up to 0.25 at.wt% and beyond this value a drop in intensity was observed. The increase in EL brightness with Gd concentration can be understood as follows. In the undoped Y–Ba-Cu–O samples the EL is attributed to the host lattice defects corresponding to the oxygen vacancies. A broad emission band has been observed in the blue region. The intensity of this band is found to increase with the addition of Gd^{3+} doping. This is attributed to the energy transfer phenomena from Gd^{3+} ions to the host lattice luminescence centres. The Gd^{3+} emission has a peak at about 311.9 nm corresponding to the transition from state $^6P_{7/2}$ to the state $^8S_{7/2}$. Thus, with increase in Gd content the EL brightness of the samples with constant oxygen stoichiometry is expected to increase which is in conformity with the above observations. At higher concentration the decrease in EL brightness with Gd concentration might be due to the killing effect.

15.3.3 Effect of Mn substitution

Numerous investigations regarding the effects of the substitution in place of copper in $Y_1Ba_2Cu_3O_{7-\delta}$ by a great variety of metallic elements have appeared in the literature [67–74]. The situation of substitution at copper is complicated by the presence of two chemically inequivalent sites for Cu, Cu(1) and Cu(2), and the dopant ion occupies either one of the two sites or both, depending on its preferential coordinate number. Since the ionic radii of all the transition elements are close to that of Cu, they can be substituted for the latter. The question of the occupation by dopant is central in the interpretation of primary features such as the degree of degradation of superconducting properties, the apparent orthorhombic to tetragonal phase transformation, oxygen site occupancy, substitutional solubility and annealing effects [75, 76].

Substitutes of 3*d* transition elements for Cu in $Y_1Ba_2Cu_3O_{7-\delta}$ superconductor decrease the value of T_c considerably and the magnitude of the effect depends on the concentration and the characteristic of the dopant ion [77–79]. To investigate this phenomenon, with the goal of obtaining more insight into the unique role of Cu, Mn has been substituted for Cu in $Y_1Ba_2Cu_3O_{7-\delta}$ superconductor. Attempts have been made to study the effect of Mn substitution on the EL and superconductivity of high-T_c Y–Ba–Cu–O superconductor. It has been found that both the EL and superconducting properties are sensitive to the doping concentration of Mn.

The brightness–voltage curves for $YBa_2Cu_{3-x}Mn_xO_{7-\delta}$ superconducting samples at fixed frequency show that, above V_{th}, the light output is an increasing function of the applied voltage. The voltage dependence of brightness was essentially similar for different samples though the magnitude may vary.

The nature of the possible relationship existing, in the present investigation, between brightness B and the applied voltage V has been investigated by plotting the possible graphs. All other curves, except $\log B$ versus $\log V$, have been found to deviate from the linearity. However, with $\log B$ plotted as a function of $\log V$, linearity has been observed (figure 15.15).

The linearity of $\log B$ versus $\log V$ plots observed in figure 15.15 obeys the 'power law' relation of the form

$$B = AV^n$$

where A is a constant and n corresponds to the slope of the $\log B$ versus $\log V$ plot. The existence of power law relationship between brightness and voltage indicates the possibility of a bi-molecular process in the recombination of electrons with luminescence centres. The linearity of $\log B$ versus $\log V$ plots shows that the excitation of charge carriers is due to the field ionization of impurity ions.

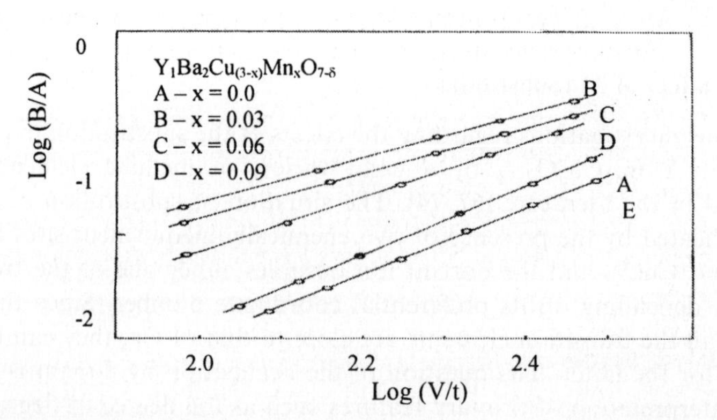

Figure 15.15. $\log(B/A)$ versus $\log(V/t)$ plots for $Y_1Ba_2Cu_{3-x}Mn_xO_{7-\delta}$ samples [77].

Table 15.1. Effect of Mn concentration on the T_c values, threshold voltage and EL brightness of $YBa_2Cu_{3-x}Mn_xO_{7-\delta}$ superconductor.

Mn concentration at% of Cu	Transition temperature T_c $(R = 0)$ (K)	EL intensity (B/A) at 400 V/mm and at 100 Hz (a.u.)	Threshold voltage (V_{th}) (V/mm)
0	95	0.15	110
1	92	0.38	40
2	87	0.295	55
3	82	0.21	95
4	65	0.08	140

An increase in the brightness with the frequency, provided that voltage is kept constant, has been noticed. This increase in the brightness is due to the application of an a.c. electric field to the sample. In one cycle of the applied field, the sample gets exposed twice to the peak value of field, emitting two quanta by the luminescence centre per cycle. As the frequency F increases, the number of emitting quanta also increases.

In order to investigate the nature of B–F curves, various relations reported in the literature have been tried. For all the $YBa_2Cu_{3-x}Mn_xO_{7-\delta}$ samples, it was found that $\log B$ versus $\log F$ plots shows linearity. The nature of the plots obeys the mathematical relation of the form

$$B = Af^n$$

where A and n are constants depending on the concentration of Mn.

The Mn concentration has been found to have a significant effect on the threshold voltage (V_{th}) (table 15.1).

At the lower doping concentration, the threshold voltage is found to decrease with increase in the Mn concentration. The decrease in the threshold voltage at lower concentration can be explained on the basis that the increased activator concentration offers increased numbers of activator atoms, presenting a number of capturing sites, suitably located to absorb, by collision, a significant fraction of the energy extracted from the field by the charge carriers. Thus, increase in activator concentration increases the capture probability of the charge carriers. Moreover, increase in activator concentration also helps to add the donor levels increasing the availability of charge carriers. This may help to start EL even at lower applied voltage.

Acknowledgments

Authors wish to thank Dr A V Narlikar and his group at National Physical Laboratory, New Delhi, for making available the samples in pellet form for

EL studies. The financial assistance from UGC DRS (SAP) is gratefully acknowledged. DDS wishes to thank CSIR, New Delhi for the award of SRF.

References

[1] Bednorz J G and Muller K A 1986 *Z. Phys. B* **64** 189
[2] Pawar S H, Lawangar R D, Shalgaonkar C S and Narlika A V 1971 *Phil. Mag.* **24** 727
[3] Lawangar R D, Shalgaonkar C S, Pawar S H and Narlikar A V 1972 *Solid State Commun.* **10** 1241
[4] Pawar S H and Narlikar A V 1973 *Indian J. Pure and Appl. Phys.* **10** 892
[5] Pawar S H and Narlikar A V 1974 *J. Luminescence* **9** 52
[6] Pawar S H and Narlikar A V 1976 *Mat. Res. Bull.* **11** 821
[7] Lawangar R D and Pawar S H 1977 *Indian J. Phys.* **51A** 219
[8] Lawangar R D, Pawar S H and Narlikar A V 1971 *Mat. Res. Bull.* **12** 341
[9] Shalgaonkar C S, Lawangar R D, Pawar S H and Narlikar A V 1977 *Mat. Res. Bull.* **12** 523
[10] Sabnis S G and Pawar S H 1977 *Indian J. Pure and Appl. Phys.* **15** 817
[11] Mulla M R and Pawar S H 1977 *Mat. Res. Bull.* **12** 929
[12] Pawar S H and Narlikar A V 1978 *J. Luminescence* **16** 429
[13] Mulla M R and Pawar S H 1979 *Pramana* **12** 593
[14] Sabnis S G and Pawar S H 1980 *Indian J. Pure and Appl. Phys.* **18** 562
[15] Jonson P D 1966 *Luminescence of Inorganic Solids* ed. Paul Goldberg (New York: Academic Press) p. 287
[16] Zhu Y, Taffo J and Suenaga M 1991 *MRS Bull.* November p. 54
[17] de Fontain D, Ceder G and Asta M 1980 *Nature* **43** 544
[18] Reyes-Gasega I, Crekele T, Van Tendeloo G 1989 *Physica C* **159** 831
[19] Beyers R, Ahn B T and Gorman G 1989 *Nature* **340** 619
[20] Chen C H, Weber D L and Schneemeyer L F 1988 *Phys. Rev. B* **38** 2888
[21] Tetenbaum M, Tani B, Czech B and Blander M 1989 *Physica C* **158** 371
[22] Fugol I, Saemann-Ischenko G and Samovarov V 1991 *Solid State Commun.* **80** 201
[23] Litovchenko V G, Frolov S I, Gavrilenko V I and Guangcheng X 1992 *Thin Solid Films* **207** 270
[24] Wang X, Nanba T and Ikezawar M 1988 *Jap. J. Appl. Phys.* **27** 1913
[25] Kelly M K, Barboux P and Tarascon J M 1988 *Phys. Rev. B* **3891** 870
[26] Bozovic I, Char K and Yoo S J B 1988 *Phys. Rev. B* **387** 5077
[27] Herr S L, Kamaras K and Porter C D 1987 *Phys. Rev. B* **36** 733
[28] Frolov S I, Litovchenko V G and Tanatar M A 1991 *Solid State Commun.* **79**(1) 39
[29] Hirochi K, Mizuno K and Matsushima T 1990 *Jap. J. Appl. Phys.* **29**(7) 1104
[30] Dagys R, Babonas G J, Pukinskas G and Leonyak L 1991 *Solid State. Commun.* **79**(11) 955
[31] Mulla M R and Pawar S H 1980 *Indian J. Pure and Appl. Phys.* **18** 936
[32] Sabnis S G and Pawar S H 1988 *Pramana* **14** 143
[33] Mulla M R and Pawar S H 1980 *Indian J. Pure and Appl. Phys.* **19** 24
[34] Mulla M R and Pawar S H 1980 *Indian J. Pure and Appl. Phys.* **55A** 249
[35] Pawar S H and Bargale B B 1976 *Indian J. Pure and Appl. Phys.* **14** 960

[36] Tissue B M and Wright J C 1987 *J. Luminescence*, **37** 117
[37] Tissue B M and Wright J C 1988 *J. Luminescence*, **42** 173
[38] Thomas R and Nampoori V P N 1988 *Proc. Int. Conf. on High T_c Superconductivity*, Jaipur, India, p. 241
[39] Blanconi *et al* 1987 *Int. J. Mod. Phys. B* **1** 835
[40] Yongan W, Xinyl Z, Yingxue Z and Jingbai W 1990 *J. Luminescence* **45** 165
[41] Cooke D W, Rempp H, Fish Z and Smith J J 1987 *Phys. Rev. B* **36** 2287
[42] Cooke D W *et al* 1991 *J. Luminescence* **48** 819
[43] Roth M, Halprin A and Katz S 1988 *Solid State Commun.* **67** 105
[44] Jahan M S *et al* 1991 *J. Luminescence* **48** 823
[45] Jahan M S *et al* 1990 *J. Luminescence* **47** 655
[46] Reddy K N 1990 *Phys. Stat. Solidi (a)* **119** 655
[47] Popova M N, Puyats A V, Springls M E and Khybov E P 1988 *JEPT Lett.* **48** 667
[48] Miller J H, Hynn J D and Holder S L 1990 *Appl. Phys. Lett.* **56** 89
[49] Barkay Z, Axoulay J and Leheah Y 1990 *Appl. Phys. Lett.* **57** 1808
[50] Waymouth J F and Bitter F, 1954 *Phys. Rev.* **95** 941
[51] Destriau G 1947 *Phil. Mag.* **38** 700
[52] Lehmann W 1957 *J. Electrochem. Soc.* **104** 45
[53] Zalm P, Diamer G and Klasens N 1955 *Philips Res. Reports* **10** 205
[54] Alfrey A F and Taylor J B 1955 *Proc. Phys. Soc.* **368** 775
[55] Nagy E 1956 *J. Phys. Radium* **17** 1498
[56] Harman G G and Raybold R L 1956 *Phys. Rev.* **104** 1498
[57] Piper W W and William F E 1955 *Phys. Rev.* **98** 1809
[58] Zalm P 1956 *Phil. Res. Reports* **11** 353
[59] Halsted R E and Koller L R 1954 *Phys. Rev.* **93** 349
[60] Curie D 1952 *J. Phys. Radium* **13** 317
[61] Thronton W A 1956 *Phys. Rev.* **102** 38
[62] Taylor J B and Alfrey F G 1955 *Brit. J. Appl. Phys. Suppl.* **4** 44
[63] Ballentine D W G and Ray B 1963 *Brit. J. Appl. Phys.* **14** 157
[64] Pawar S H, Todkar B M, Awana V P S, Agarwal S K and Narlikar A V 1992 *Indian J. Pure and Appl. Phys.* **30** 332
[65] Rosen H J, Macfariane R M, Engler E M, Lee V Y and Jacowitz R D 1988 *Phys. Rev. B* **38** 2460
[66] Hor P H, Meng R L, Wang Y Q, Gao L, Huang Z J, Bechtold J, Forster K and Chu C W 1987 *Phys. Rev. Lett.* **58** 1891
[67] Tarascon J M, Barboux P, Miceli P F, Greene L H and Hull G W 1988 *Phys. Rev. B* **37** 7458
[68] Maeno Y, Tomita T, Kyogoku M, Awaji S, Aoki Y, Hosino K, Minami A and Fujita T 1987 *Nature* **328** 512
[69] Jayaram B, Agarwal S K, Rao C V N and Narlikar A V 1988 *Phys. Rev. B* **38** 2903
[70] Tang H, Qiv Z Q, Du Y W, Xiao U, Chien C V L and Wallace J C 1987 *Phys. Rev. B* **36** 4018
[71] Mehbad M, Wydier P, Deltour R, Duvigneaud P H and Naessens G 1987 *Phys. Rev. B* **36** 8819
[72] Zolliker P, Cox D E, Tranquada J M and Shiraue G 1988 *Phys. Rev. B* **38** 6575
[73] Horn S, Reilly K, Fisk Z, Kwok R S, Thompson J D, Borges H A, Chang C L and den Boer M L 1988 *Phys. Rev. B* **38** 2930

[74] Blue C, Elgaid K, Zitkousky I, Boolchand P, McDaniel D, Joiner W C H, Dosten J and Huff W 1988 *Phys. Rev. B* **37** 5905

[75] Jorgensen J D, Beno M A, Hinks D G, Soderholm L, Volin K J, Hitterman R L, Grace J D, Schuller I K, Segre C U, Zang K and Kleefish M S 1987 *Phys. Rev. B* **36** 3680

[76] Market J T, Dunlap B D and Maple M B 1989 *Mater. Res. Bull.* **14** 37

[77] Xiao G, Strelitz F H, Gavrin A, Du Y W and Chein C L 1987 *Phys. Rev. B* **35** 8782

[78] Takayama-Muromachi E, Uchida Y and Koto K 1987 *Jap. J. Appl. Phys.* **26** 2087

[79] Tarascon J M, Barboux P, Miceli P F, Greene L H, Hu G W, Eibschutz M and Sunshine S A 1988 *Phys. Rev. B* **37** 7510

Index

Q